CALCULUS AND LINEAR ALGEBRA

CALCULUS AND LINEAR ALGEBRA

VOLUME II

Vector Spaces, Many-Variable Calculus, and Differential Equations

Wilfred Kaplan

Donald J. Lewis

Department of Mathematics
University of Michigan

Scholarly Publishing Office, University of Michigan University Library Ann Arbor

Published by The Scholarly Publishing Office
The University of Michigan University Library

1st paperback edition, 2007 The Scholarly Publishing Office
First published by Wiley & Sons, Inc.

Scholarly Publishing Office
300 Hatcher North
Ann Arbor, MI 48109
lib.spo@umich.edu

ISBN: 978-1-4255-8934-9 (softcover)
978-1-4255-8935-6 (electronic)

The University of Michigan University Library, through its Scholarly Publishing Office, is committed to providing academic publishing services that are responsive to the needs of both producers and users, foster a sustainable economic model, and support robust author rights.

The Scholarly Publishing Office seeks to disseminate high-quality scholarly content through both print and electronic publishing. As part of this mission, the Library is pleased to re-issue Calculus and Linear Algebra by Wilfred Kaplan and Donald Lewis. This text was an invaluable part of the education of a past generation of math students, and through this republication it will become a vital part of teaching a new generation.

Paul N. Courant
University Librarian and Dean of Libraries
Harold T. Shapiro Collegiate Professor of Public Policy
Professor of Economics and of Information
The University of Michigan

Wilfred Kaplan
Emeritus Professor of Mathematics
University of Michigan

Photo courtesy of the Mathematics Department, University of Michigan

Wilfred Kaplan, born in 1915, is a native of Boston. He graduated from Harvard College in 1936, then remained at Harvard for a Ph.D. in mathematics in 1939. From 1940 until his retirement in 1986, he taught mathematics at the University of Michigan. His research has dealt with differential equations, complex function theory, and applied mathematics. He is the author of many influential textbooks, including Advanced Calculus (1952 to 2003), Ordinary Differential Equations (1958), Operational Methods for Linear Systems (1962), Advanced Mathematics for Engineers (1981), and Maxima and Minima with Applications (1999). Over a long period of time, Professor Kaplan has been active in the affairs of the American Association of University Professors (AAUP), a national organization dedicated to defending faculty rights, primarily academic freedom and tenure. An accomplished musician, he has enjoyed playing piano or violin in chamber music groups, and has done much to promote the arts, especially music and drama, in the Ann Arbor area.

Donald J. Lewis
Emeritus Professor of Mathematics
University of Michigan

Photo courtesy of the Bentley Historical Library, University of Michigan (Box 6, T. H. Hildebrandt Papers)

Donald J. Lewis was born in Adrian, Minnesota in 1926. He majored in mathematics at the College of St. Thomas (St. Paul, Minnesota), then earned a Ph.D. in mathematics from the University of Michigan in 1950. He taught at Ohio State University and the University of Notre Dame, and held several research appointments before joining the faculty of the University of Michigan in 1961. He served as Chair of the Mathematics Department for ten years, followed by four years as Director of the Division of Mathematical Sciences of the National Science Foundation in Washington, D.C. Professor Lewis has done extensive research in algebraic number theory and diophantine equations, and has directed 27 Ph.D. theses at Michigan. He has played an active role in mathematical educational innovations and is author of the textbook Introduction to Algebra (1965). Outside of mathematics, he has a particular interest in gardening and has enjoyed maintaining an elaborate garden at his home in Ann Arbor.

Biographical notes written by Peter Duren, Professor of Mathematics, University of Michigan

REISSUANCE OF CALCULUS TEXT BY KAPLAN AND LEWIS

By Wilfred Kaplan

The power of the Internet in meeting scholarly demands is illustrated by the Google Book Project, which aims to make books of all kinds available online. A particular need arose at the University of Michigan, when Associate Professor Stephen DeBacker proposed to use a long out-of-print work by his colleagues Wilfred Kaplan and Donald J. Lewis (now both emeritus professors) for a course he was to teach. The work he selected was Calculus and Linear Algebra, originally published by John Wiley & Sons; volume I appeared in 1970 and volume II in 1971. The first volume covers vectors in the plane and one-variable calculus; the second treats vector space, many-variable calculus and differential equations. The two volumes provide material for a freshman-sophomore course in calculus in which linear algebra is gradually introduced and blended with the calculus. The work introduces many novel ideas and proofs.

In order to make the work available for instruction, assistance was needed. The Scholarly Publishing Office of the University of Michigan Library has agreed to partner with the University's Mathematics Department to reissue it as originally printed. Thus two volumes will be published, of the quality needed for classroom use, but with costs held down as far as possible. The authors have agreed to this and are receiving no royalties for the publication. Both proclaim their happiness that the work, a labor of love created 40 years ago, should again be used by students.

As a byproduct of the production of the books, an online version of the texts is being created, to be listed in Google Books. Thus the work becomes freely available to the world in its entirety.

PREFACE

In Volume I we developed one-variable calculus, along with vectors in the plane and a few fundamental ideas about general vector spaces. In this second volume, we develop linear algebra much further and then apply it to geometry, many-variable calculus, and differential equations. These topics are so closely related that the subject matter is revealed here as one well-defined, tightly knit body of mathematics. Linear algebra is concerned with relations whose graphs are *linear:* lines, planes, and their counterparts in higher dimensions. In geometry we view these graphs as structures in Euclidean space R^3, or more generally in R^n. The calculus is concerned, in part, with relations whose graphs are *curved* objects: paths and surfaces. Differential calculus is essentially a tool for "linearizing" these relations (through the differential) and their graphs (through tangent lines and planes). Once they have been linearized, the relations and graphs can be treated by linear algebra and by geometry. The calculus also is concerned with classes of functions: for example, the class of all functions continuous on an interval or in a region; or the class of all functions having a continuous *n*th derivative on an interval; or the set of all polynomials; or the set of all rational functions; or the set of all functions representable by power series on an interval (analytic functions); or the set of solutions of a homogeneous linear differential equation on an interval. Each of these classes constitutes a vector space, and the ideas of linear algebra again find application. Perhaps, the most beautiful example of this application is shown in Figure 13-49, showing the kernel and range of the four linear operators ∇^2, ∇, curl, and div.

We find that the central ideas are most vividly revealed in geometry, and hope that the users of this book can devote adequate time to Chapter 11, in which Euclidean geometry is studied in detail.

The following is a brief chapter-by-chapter summary with a few comments.

Chapter 9. Vector Spaces. Vector spaces are defined axiomatically, but it is made clear that in almost all applications verification of the axioms requires merely checking closure under the basic operations. The essential machinery of linear algebra is developed: subspaces, addition of sets, linear varieties (cosets), linear independence, basis, dimension, linear mappings, kernel, range, rank, nullity, linear transformations (mappings of a space into itself), vector spaces of linear mappings, algebras of linear transformations, inverse mapping.

In many cases, geometric illustrations are given (in anticipation of Chapter 11), and many examples are chosen from the calculus.

Chapter 10. Matrices and Determinants. Matrices are introduced as linear mappings of V_n (the vector space of real n-tuples) into V_m. Hence, the general results for such mappings apply and one can, for example, at once speak of the rank of a matrix. The simple ideas of linear algebra permit one to obtain all the main results concerning simultaneous linear equations (with the exception of a few results depending on determinants). These equations are studied carefully, and are related to geometry. The operations on matrices are then developed fully—in all cases, as an application of the previously developed theory of linear mappings. Special attention is given to square matrices (linear transformations) and their inverses. Determinants are developed systematically, and their significance in geometry is pointed out. The concluding optional sections concern matrices of functions, the technique of elimination, eigenvalues, and similarity.

Chapter 11. Linear Euclidean Geometry. Here the emphasis is on 3-dimensional space, but brief optional sections indicate the generalization to n dimensions. The inner product is introduced, its properties are deduced, and R^3 is defined in terms of the corresponding distance function. The cross product is also developed and is shown to be a powerful tool for studying lines and planes. Many problems of geometry are treated by linear algebra, but the emphasis is on demonstrating how effective this method is, rather than giving a complete theory. The treatment of area and volume is related to the calculus. Linear mappings of R^3 into R^3 are considered; the idea of Jacobian matrix and that of its determinant appear, and their geometric meaning is stressed. (Here, in fact, differential and integral calculus, linear algebra, and geometry all come together in one of the central ideas of mathematics.) Surfaces in space and cylindrical and spherical coordinates are discussed. Change of coordinates is treated rather briefly.

Chapter 12. Differential Calculus of Functions of Several Variables. Partial derivatives and differentials are developed and shown to be part of a theory emphasizing vector functions and vector operations: in particular, the gradient and the Jacobian matrix. The various chain rules are shown to be cases of one very simple rule for vector functions. Implicit and inverse functions are treated, with emphasis on the linear approximation provided by the calculus; there are corresponding applications to tangents and normals. Maxima and minima are discussed, including the case of side conditions (Lagrange multipliers); here again linear algebra is important.

Chapter 13. Integral Calculus of Functions of Several Variables. The double and triple integral are studied, with emphasis on the essential properties needed for applications. Integration in curvilinear coordinates (especially cylinder and spherical) is considered, with reference to the Jacobian formula. Numerous applications are discussed. Line integrals are studied systematically, with emphasis on Green's theorem and independence of path. The operations of divergence and curl are introduced and, through Green's theorem, shown to have geometric and physical meaning. The extensions of these ideas to space is considered briefly.

Chapter 14. Ordinary Differential Equations. This is, in effect, a short course in the subject, with emphasis on linear equations and matrix methods. The Existence Theorem is discussed (but not proved here). Practical applications are discussed, and the idea of stability is given much weight. Phase plane analysis, series methods, and numerical techniques are considered rather briefly.

Suggested Minimal Course. The following outline provides a complete course, but one treating only the most essential features of each topic: Sections 9-1 to 9-9, 9-11 to 9-14, 9-16 to 9-21, 10-1 to 10-13, 11-1, 11-2, 11-4, 11-6 to 11-8, 11-10, 11-12, 11-14, 11-15, 11-17, 11-19, 11-20, 12-1 to 12-14, 12-17 to 12-19, 13-1, 13-2, 13-4 to 13-9, 14-1 to 14-9, 14-11 to 14-13. One may even wish to omit Chapter 14 completely; for many purposes the sections on differential equations in Chapter 7 are adequate. By postponing a thorough treatment of differential equations to the junior year, one gains time for a more thorough treatment of Chapters 9 to 13.

Daggers and Double Daggers. As in Volume I, a dagger (†) denotes an optional section, and a double dagger (‡) indicates a section that is both optional and exceptionally difficult. A few proofs and problems are also marked with a double dagger as an indication of difficulty.

Acknowledgments. We express our appreciation to the publisher for the fine support and encouragement given throughout the preparation of this volume; we especially thank John B. Hoey for his untiring efforts on behalf of the project. To Helen M. Ferguson and Anna Church we express our appreciation for their fine work in typing the manuscript.

Wilfred Kaplan

Ann Arbor, 1970 *Donald J. Lewis*

CONTENTS

CHAPTER 9
VECTOR SPACES

CHAPTER 10
MATRICES AND DETERMINANTS

CHAPTER 11
LINEAR EUCLIDEAN GEOMETRY

CHAPTER 12
DIFFERENTIAL CALCULUS OF FUNCTIONS OF SEVERAL VARIABLES

CHAPTER 13
INTEGRAL CALCULUS OF FUNCTIONS OF SEVERAL VARIABLES

CHAPTER 14
ORDINARY DIFFERENTIAL EQUATIONS

CALCULUS AND LINEAR ALGEBRA

9
VECTOR SPACES

9-1 THE CONCEPT OF A VECTOR SPACE

In Chapter 1, we studied vectors in the plane, learned how to add them, and to multiply each vector by a scalar (real number). We learned that these vectors obeyed the rules:

For all vectors **u**, **v**, **w** and all scalars a, b:

1. $\mathbf{u} + \mathbf{v} = \mathbf{v} + \mathbf{u}$.
2. $(\mathbf{u} + \mathbf{v}) + \mathbf{w} = \mathbf{u} + (\mathbf{v} + \mathbf{w})$.
3. $\mathbf{u} + \mathbf{0} = \mathbf{u}$, and $\mathbf{u} + \mathbf{z} = \mathbf{u}$ implies $\mathbf{z} = \mathbf{0}$.
4. For each $\mathbf{u}$, $\mathbf{u} + \mathbf{z} = \mathbf{0}$ has a unique solution for $\mathbf{z}$, denoted by $-\mathbf{u}$.
5. $a(b\mathbf{u}) = (ab)\mathbf{u}$
6. $(a + b)\mathbf{u} = a\mathbf{u} + b\mathbf{u}$.
7. $a(\mathbf{u} + \mathbf{v}) = a\mathbf{u} + a\mathbf{v}$.
8. $1\mathbf{u} = \mathbf{u}$.

Also each vector **u** in the plane can be expressed uniquely in terms of the orthogonal basis **i**, **j** by an equation $\mathbf{u} = a\mathbf{i} + b\mathbf{j}$ (see Figure 9-1).

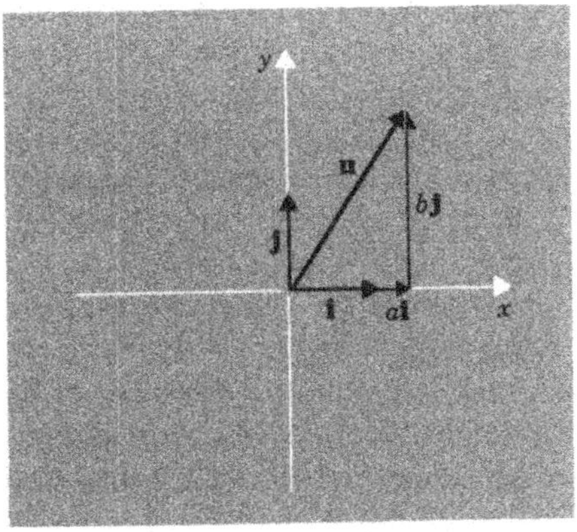

Figure 9-1 Vectors in the plane.

Now vectors in the plane are not the only system obeying the rules (9-10). For example, the vectors in 3-dimensional space can be treated in just the same way as the vectors in the plane, and all the rules (9-10) can be verified in similar fashion. For vectors in 3-dimensional space, each vector **u** can be represented uniquely in terms of an orthogonal basis consisting of *three* unit vectors **i**, **j**, **k**, as in Figure 9-2. We shall develop the theory of vectors in 3-dimensional space in Chapter 11. Here we shall rely on intuition to discuss these vectors and shall frequently use them to illustrate the theory.

Because we now have two systems obeying the same rules, it is convenient to give a name to such systems: we call them *vector spaces*.

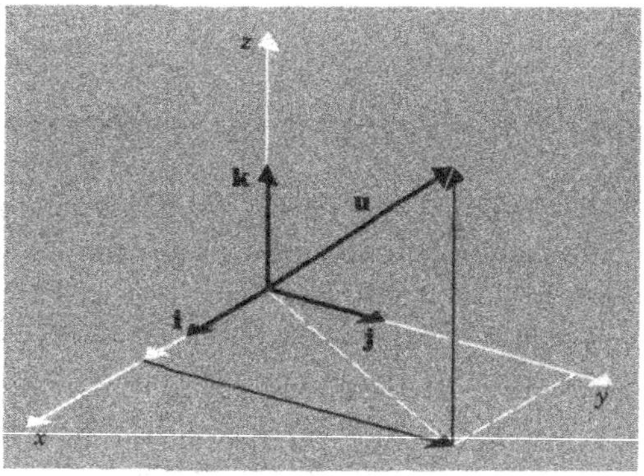

Figure 9-2 Vectors in 3-dimensional space.

Definition. A vector space V is a set of elements such that *addition is defined in* V: each two elements **u**, **v** in V have a sum $\mathbf{u} + \mathbf{v}$ in V; *elements of* V *can be multiplied by scalars* (real numbers): if c is a scalar and **u** is in V, then $c\mathbf{u}$ is in V; V contains a unique zero element **0** and all the rules (9-10) hold true.

Elements of a vector space are frequently called *vectors*.

We have given two examples of vector spaces: the vectors in the plane and the vectors in space. But there are many more. We saw in Section 2-9 that there are sets of functions that satisfy all the rules (9-10) and, anticipating the general definition, we called them *vector spaces of functions*. For example, all the continuous functions $f, g, \ldots$ on an interval form a vector space of functions. Here the "zero element" is the zero function 0, equal to 0 throughout the interval. Throughout the rest of this book we shall denote this vector space by $\mathcal{C}$. Where it is necessary to specify the interval, we shall write, for example, $\mathcal{C}[a, b]$ or $\mathcal{C}(a, b)$ (continuous functions on the closed interval $[a, b]$ and continuous functions on the open interval (a, b), respectively). Similarly, we write $\mathcal{C}^{(n)}$ (or $\mathcal{C}^{(n)}[a, b]$ and so on) for the vector space formed of all functions having continuous derivatives through order n on the chosen interval. We also write $\mathcal{C}^{(\infty)}$ (or $\mathcal{C}^{(\infty)}[a, b]$ and so on) for the vector space formed of all functions having continuous derivatives of every order on the chosen interval. That these functions form a vector space, in each case, follows at once from the rules for the derivative of $f + g$ and cf.

The Vector Space V_n. For each positive integer n, we can form a vector space whose elements are the ordered n-tuples of real numbers. For example, V_2 consists of all ordered pairs of real numbers: (2, 3), (7, −2), $(\sqrt{2}, \sqrt{3})$, (0, 0), ... The fact that the pairs are ordered means that we distinguish between (2, 3) and (3, 2). The elements of V_2 are usually called vectors. We define addition and multiplication by scalars in V_2 by the rules

$$(p, q) + (r, s) = (p + r, q + s)$$
$$c(p, q) = (cp, cq)$$

The vector (0, 0) serves as the zero vector in V_2. One can now verify all the rules (9-10). For example,

$$(p, q) + (r, s) = (p + r, q + s) = (r + p, s + q)$$
$$= (r, s) + (p, q)$$

so that the first rule (commutative law of addition) is proved. The other rules are proved in the same way (Problem 10 below). Also, we can associate with each vector (a, b) in V_2 the vector $a\mathbf{i} + b\mathbf{j}$, and the definitions we have just given for addition and multiplication by scalars are in agreement with the usual rules for vectors in the plane. Hence, *V_2 can be represented geometrically as the system of vectors in the plane.* We shall often take advantage of this representation.

The vector space V_3 of ordered triples (p, q, r) can be discussed in exactly the same way. Addition and multiplication by scalars are defined by the equations

$$(p, q, r) + (s, t, u) = (p + s, q + t, r + u)$$
$$c(p, q, r) = (cp, cq, cr)$$

The zero vector is the vector (0, 0, 0). By analogy with the results obtained in Chapter 1 for V_2, we can show that *the vectors of V_3 can be represented as the vectors in 3-dimensional space.* Without having proved this (a proof is given in Chapter 11), we shall take advantage of this representation for illustrations of the theory.

All of this carries over to V_n: the set of all ordered n-tuples of real numbers $(v_1, \ldots, v_n)$. We call $v_1, \ldots, v_n$ the *components* or *coordinates* of the vector $\mathbf{v} = (v_1, \ldots, v_n)$ (and refer to v_1 as the first component, v_2 as the second, and so on). Two vectors of V_n are added by adding corresponding components and a vector is multiplied by the scalar c by multiplying each component by c:

$$(u_1, \ldots, u_n) + (v_1, \ldots, v_n) = (u_1 + v_1, \ldots, u_n + v_n)$$
$$c(v_1, \ldots, v_n) = (cv_1, \ldots, cv_n)$$

The zero vector is (0, . . . , 0). Verification of the rules (9-10) is exactly the same as for V_2. The vectors of V_n can be represented as the vectors in n-dimensional Euclidean space R^n. This is shown in Chapter 11; occasionally, we shall anticipate the result in order to suggest the geometric meaning of an algebraic statement.

The vector space V_1 consists of all single real numbers, with the usual addition and multiplication. Hence, V_1 is the same as the real number system itself (except that we shall never speak of division of vectors). We can represent V_1 as the vectors on a line; for example, as all vectors in the xy-plane which happen to lie along the x-axis.

Vector Spaces of Polynomials. The set of all polynomials can be considered as a vector space of functions, all defined on a given interval. We shall denote this vector space by $\mathcal{P}$, and normally think of the interval as $(-\infty, \infty)$. Similarly, we denote by $\mathcal{P}_m$ the vector space of all polynomials of degree at most m.

Vector Spaces in Daily Life. Vectors occur in many ways in our lives. In preceding chapters we gave many examples of the vectors occurring in physics: force, velocity, acceleration, and so on. We often work with the vector space V_2. For example, a husband and a wife may each have a checking account, and we can denote by (h, w) the pair of balances [for example, $(-\$10.03, \$27.15)$—husband overdrawn!]; we also use the pair notation for deposits and withdrawals. Then each month's activities consist in adding a vector (h, w) to the previous balance (h_0, w_0). If both husband and wife double their previous month's achievements, the amount added is twice the previous vector (h, w) and so on. Other more sophisticated examples can be given. The many dials in the control panel of a modern engineering system—such as a jet plane—can be regarded as giving the components of a vector in V_n, where n may be as large as 50. In modern engineering design, vectors are indeed used in precisely this way.

Throughout this chapter we shall consider vector spaces in general. We begin with our first theorem:

THEOREM 1. *Let V be a vector space. Then for all elements* $\mathbf{u}, \mathbf{v}$ *in V and all scalars a we have*

$$\text{(i) } 0\mathbf{u} = \mathbf{0} \qquad \text{(ii) } a\mathbf{0} = \mathbf{0}$$

$$\text{(iii) } (-1)\mathbf{u} = -\mathbf{u} \qquad \text{(iv) } -(\mathbf{u} + \mathbf{v}) = (-\mathbf{u}) + (-\mathbf{v})$$

PROOF. By rules 6 and 8 in (9-10) we have

$$\mathbf{u} = 1\mathbf{u} = (1 + 0)\mathbf{u} = 1\mathbf{u} + 0\mathbf{u} = \mathbf{u} + 0\mathbf{u}$$

By rule 3 in (9-10) $\mathbf{u} + \mathbf{z} = \mathbf{u}$ implies that $\mathbf{z} = \mathbf{0}$; therefore, $0\mathbf{u}$ must be $\mathbf{0}$ and (i) is proved. Property (ii) follows from (i) and rule 5 in (9-10); since we have

$$\mathbf{0} = 0\mathbf{u} = (a \cdot 0)\mathbf{u} = a(0\mathbf{u}) = a\mathbf{0}$$

Next we observe that, by rule 8 in (9-10),

$$\mathbf{0} = 0\mathbf{u} = [1 + (-1)]\mathbf{u} = 1\mathbf{u} + (-1)\mathbf{u} = \mathbf{u} + (-1)\mathbf{u}$$

and, hence, by the fourth rule, $(-1)\mathbf{u} = -\mathbf{u}$. Property (iv) follows from (iii) and rule 7 in (9-10); its proof is left as an exercise.

V_n as a Vector Space of Functions. We can consider each vector of V_n as a sequence of n elements, hence, as a real function whose domain is the set $1, \ldots, n$ (see Section 2-12). The definition of addition and multiplication by scalars, for vectors in V_n, is then in agreement with the way we add functions and multiply functions by scalars. Hence, V_n is in fact one more case of a vector space of functions. More generally, the set of all real functions whose domain is any given set E on the x-axis is a vector space; in fact, the set E can be any set whatsoever—what is crucial is that the functions considered have real values and, hence, the functions can be added and multiplied by scalars (real numbers).

Complex Vector Spaces. For some applications we must change our definition of vector space to allow the scalars to be *complex numbers*. One then refers to a *complex vector space*. To a remarkable extent, the theory of "real vector spaces," as defined above, is the same as the theory of "complex vector spaces." Hence, one can generally replace the words vector space by "complex vector space" throughout the chapter. There are a few exceptions to this statement, which we shall point out as they occur. Examples of complex vector spaces are the vector spaces V_n^c of n-tuples of *complex numbers;* V_1^c is the complex number system itself. One can also consider the vector space $\mathcal{C}^c$ of all continuous complex functions of t on an interval; that is, all functions $f(t) + ig(t)$, where f and g are continuous real functions on the interval.

PROBLEMS

1. Let $\mathbf{u} = (3, 2)$, $\mathbf{v} = (4, 1)$ be vectors in V_2. Evaluate:
   (a) $\mathbf{u} + \mathbf{v}$ (b) $\mathbf{u} - \mathbf{v}$ (c) $3\mathbf{u}$ (d) $2\mathbf{v}$ (e) $2\mathbf{u} + 5\mathbf{v}$
   (f) $6\mathbf{u} - 2\mathbf{v}$ (g) $0\mathbf{u}$ (h) $\mathbf{u} + \mathbf{0}$
2. Let $\mathbf{u} = (2, 1, 2)$, $\mathbf{v} = (3, 2, 1)$ and $\mathbf{w} = (1, 2, -5)$ be vectors in V_3.
   (a) Represent $\mathbf{u}$, $\mathbf{v}$, and $\mathbf{w}$ graphically as vectors in 3-dimensional space.
   (b) Calculate $\mathbf{u} + \mathbf{v}$ and show $\mathbf{u}$, $\mathbf{v}$ and $\mathbf{u} + \mathbf{v}$ graphically in 3-dimensional space.
   (c) Evaluate $3\mathbf{u}$, $2\mathbf{v}$ and $-\mathbf{w}$.
   (d) Evaluate $\mathbf{u} + \mathbf{v} + 5\mathbf{w}$ and $3\mathbf{u} - \mathbf{v} - 2\mathbf{w}$.
   (e) If possible, determine scalars c_1, c_2, c_3, not all 0, such that $c_1\mathbf{u} + c_2\mathbf{v} + c_3\mathbf{w} = \mathbf{0}$.
3. Let $\mathbf{u} = (5, 1, 7, 2)$, $\mathbf{v} = (0, 1, 0, -4)$ and $\mathbf{w} = (0, 0, 0, 0)$ be vectors in V_4. Evaluate: (a) $2\mathbf{u} - 3\mathbf{v} + \mathbf{w}$, (b) $\mathbf{u} + 2\mathbf{v} + \mathbf{w}$, (c) $0\mathbf{u} + 0\mathbf{v} + 1765\mathbf{w}$, (d) $(\mathbf{u} + \mathbf{w}) - (\mathbf{u} + 3\mathbf{w})$.
4. Determine whether each of the following sets of functions is a vector space. [*Hint.* As pointed out in Section 2-9, one need only verify that, for every scalar c and every pair f, g in the set, cf and $f + g$ are in the set. The algebraic rules (9-10) hold true automatically.]
   (a) On the interval $(-\infty, \infty)$, the polynomials of degree 3.
   (b) On $(-\infty, \infty)$, the polynomials of degree at least 2.
   (c) On $(-\infty, \infty)$, the polynomials having a zero at $x = 5$.

(d) On $(-\infty, \infty)$, the polynomials of form $a_0 + a_1x^3 + \cdots + a_nx^{3n}$.

(e) All f in $\mathcal{C}^{(2)}[0, 1]$ such that $f''(x) = x^2f(x)$.

(f) All f in $\mathcal{C}^{(1)}[0, 1]$ such that $f'(x) = [f(x)]^2$ (for example, $f(x) = (2 - x)^{-1}$).

(g) All f in $\mathcal{C}[-1, 1]$, such that f is monotone strictly increasing.

(h) All f in $\mathcal{C}^{(2)}[-1, 1]$ such that $f''(x) > 0$.

5. Let $\mathbf{u} = (1 - i, 2i)$, $\mathbf{v} = (1 + i, -2)$, $\mathbf{w} = (2, -2 + 2i)$ be vectors in the complex vector space V_2^c. (a) Evaluate $(3 + i)\mathbf{u}$. (b) Evaluate $(1 + i)\mathbf{v}$. (c) Determine, if possible, a complex scalar c such that $\mathbf{v} = c\mathbf{u}$. (d) Determine, if possible, a complex scalar c such that $\mathbf{w} = c\mathbf{u}$.

6. Consider infinite sequences $\{s_n\}, \{t_n\}, \ldots, n = 1, 2, \ldots,$ with addition and multiplication by scalars defined as follows: $\{s_n\} + \{t_n\} = \{s_n + t_n\}$, $c\{s_n\} = \{cs_n\}$.

   (a) Show that the convergent real sequences form a vector space and interpret this vector space as a space of functions.

   (b) Show that the set of all complex sequences forms a complex vector space. Can this vector space be interpreted as a vector space of functions?

7. Prove part (iv) of Theorem 1.

‡8. (a) Let V be a vector space and let W be the set of all functions mapping a given set S into V. If F, G are in W, define $F + G$ to be the function that assigns the vector $F(x) + G(x)$ to the point x in S, and define cF to be the function which assigns $cF(x)$ to the point x in S. Show that W is a vector space.

   (b) Let V be a vector space. For a fixed positive integer k, let W be the set of all ordered k-tuples of vectors of V. Thus each object of W is a k-tuple $(\mathbf{v}_1, \ldots, \mathbf{v}_k)$, where each of $\mathbf{v}_1, \ldots, \mathbf{v}_k$ is in V. Define addition and multiplication by scalars in W as for V_n and show that the resulting system is a vector space.

‡9. (a) Let V and W be vector spaces. Let U be the set of all ordered pairs $(\mathbf{v}, \mathbf{w})$, where $\mathbf{v}$ is in V and $\mathbf{w}$ is in W. Define addition and multiplication by scalars in U by analogy with the definition for V_2. Show that the resulting system is a vector space. We usually denote U by the symbol $V \times W$.

   (b) Show that $V_2 = V_1 \times V_1$.

   (c) Show that, if V, W, Z are vector spaces, then $(V \times W) \times Z = V \times (W \times Z)$ and, hence, we can drop parentheses and denote this vector space by $V \times W \times Z$.

   (d) Show that $V_4 = V_2 \times V_1 \times V_1$.

10. (a) Complete the proof that V_2 satisfies all the rules (9-10).

    (b) Verify that the rules (9-10) are satisfied by the set of all real functions defined on the same set S.

11. Show that the set consisting of the real number 0 alone is a vector space, under the usual rules for addition and multiplication of numbers.

9-2 SUBSPACES

It often occurs that one vector space is contained in a second one, and that addition and multiplication by scalars in the first vector space are carried out in exactly the same way as in the second one. When this happens, we say that the first vector space is a *subspace* of the second.

EXAMPLE 1. If V is a vector space, let V_0 be the subset of V consisting of the zero element, $\mathbf{0}$, of V alone. Then V_0 is a subspace of V.

PROOF. In V we have $\mathbf{0} + \mathbf{0} = \mathbf{0}$ and $c\mathbf{0} = \mathbf{0}$ for all scalars c. Thus addition and multiplication by scalars is defined for V_0 and is the same as for V. The eight rules (9-10) hold true for V_0, since they hold true in V. Hence, V_0 is a vector space, and it is a subspace of V.

We call the subspace V_0 the *zero-space* (of V).

EXAMPLE 2. The vector space $\mathcal{C}^{(1)}[a, b]$ is a subspace of $\mathcal{C}[a, b]$. For every function having a continuous derivative on $[a, b]$ must be continuous on $[a, b]$.

EXAMPLE 3. Let V be a vector space and let $\mathbf{w}$ be a fixed vector in V. Let W be the set of *all* scalar multiples of $\mathbf{w}$; that is, W consists of all vectors of V of form $c\mathbf{w}$. Then W, with the addition and multiplication by scalars as in V, is a vector space and therefore, is a subspace of V.

PROOF. To prove the assertion, we note first that, if c is a scalar and if $a\mathbf{w}$ and $b\mathbf{w}$ are in W, then $(a\mathbf{w}) + (b\mathbf{w})$ and $c(a\mathbf{w})$ are also in W. For, by (9-10),

$$(a\mathbf{w}) + (b\mathbf{w}) = (a + b)\mathbf{w}, \qquad c(a\mathbf{w}) = (ca)\mathbf{w}$$

Also $0\mathbf{w} = \mathbf{0}$ is in W. The rules (9-10) must hold true in W, since W is contained in V and the rules hold true in V. Therefore, W is a vector space, a subspace of V.

If $\mathbf{w} = \mathbf{0}$, then W reduces to V_0.

In Figure 9-3 we illustrate Example 3 for the case $V = V_2$ and $\mathbf{w}$ a nonzero

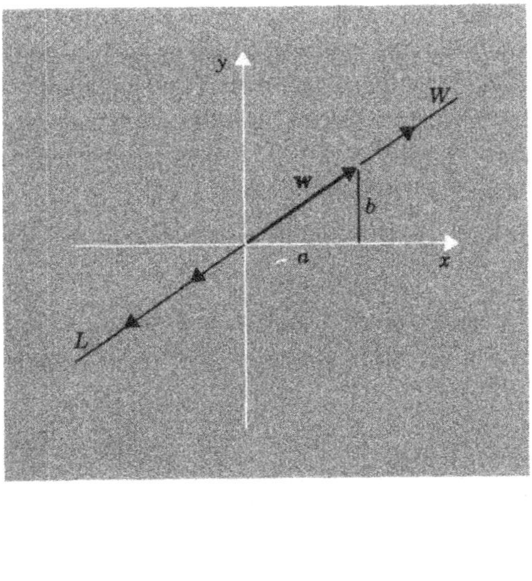

Figure 9-3 Subspace W of V_2.

vector of V_2. The vectors of the subspace W are simply the vectors along a line L through the origin. If $\mathbf{w} = (a, b)$, the line L has slope b/a.

In general, a subspace U of a vector space V must contain all scalar multiples of each vector $\mathbf{u}$ in U. Hence, if $\mathbf{u}$ is in U, then so is $\mathbf{0} = 0\mathbf{u}$ (the zero vector) and so is $-\mathbf{u} = (-1)\mathbf{u}$ (the negative of $\mathbf{u}$ in V). Thus the zero vector of V is the zero vector for each subspace of V, and the negative of a vector in a subspace is the same as the negative of that vector in V.

THEOREM 2. *Let V be a vector space. Let W be a subset of V. Then W is a subspace of V if, and only if, the following three conditions hold true:*
(i) *W is not empty—that is, W contains at least one vector;*
(ii) *the sum of each two vectors in W is also in W;*
(iii) *each scalar multiple of each vector of W is in W.*

PROOF. (a) Let W be a subspace of V. Then W is a vector space and hence (ii) and (iii) hold true. Also W cannot be empty, since W must have a zero vector. Therefore, (i) holds true also.

(b) Conversely, let W satisfy (i), (ii), and (iii). Then by (i) W contains a vector, say $\mathbf{v}$, and, hence, by (iii) W contains $\mathbf{0} = 0\mathbf{v}$ and $-\mathbf{v} = (-1)\mathbf{v}$. The eight rules (9-10) now follow immediately for W, since they hold true in V. For example, if $\mathbf{u}$ is in W, then by rule 4 for V, $\mathbf{u} + \mathbf{v} = \mathbf{0}$ has a unique solution for $\mathbf{v}$, namely $-\mathbf{u}$; but we just saw that $-\mathbf{u}$ is also in W. Therefore, rule 4 holds true in W.

When a subset X of a vector space V has the property that the sum of each pair of vectors in X is also in X, we say that X is *closed under addition.* Similarly, when X has the property that all the scalar multiples of each vector in X are in X, we say that X is *closed under scalar multiplication.* Thus Theorem 2 can be stated: *A nonempty subset X of a vector space V is a subspace exactly when X is closed both under addition and under scalar multiplication.* For example, the set of scalar multiples of a single vector is closed under addition and scalar multiplication and so is a subspace (Example 3 above).

Function Spaces. The set of *all* real functions on a given interval (or set) is a vector space V, as one verifies at once. Each particular subset of this set of functions, with the usual addition and multiplication by scalars, may or may not be a vector space. To decide whether it is, we must find out whether it is a subspace of V—that is, as above, we must find out whether it is nonempty and closed under addition and scalar multiplication. This principle was used in Section 2-9. (See also Problem 4 following Section 9-1.)

EXAMPLE 4. The set X of differentiable functions f on $[0, 1]$ such that $f' = 2f$ is a vector space.

PROOF. Clearly X contains the zero function and so is a nonempty set of functions on $[0, 1]$. If f and g are in X, then $f + g$ and cf are differentiable on $[0, 1]$ and

$$(f+g)' = f' + g' = 2f + 2g = 2(f+g), \qquad (cf)' = cf' = c(2f) = 2(cf)$$

Hence, X is closed under addition and scalar multiplication and is, therefore, a vector space.

Given a vector space V, we can always regard V as a subspace of itself. Thus each vector space V always contains the subspaces V_0 and V; these subspaces are commonly referred to as the *trivial subspaces* of V. A subspace of V which is not one of the trivial subspaces of V is called a *nontrivial subspace* of V.

Nontrivial Subspaces of V_2. We represent the vectors of V_2 as vectors in the plane, as in Figure 9-1. Let W be a nontrivial subspace of V_2. Then W contains a nonzero vector $\mathbf{u}$, and W contains all scalar multiples of $\mathbf{u}$. If W were to contain a vector $\mathbf{v}$, which is not a scalar multiple of $\mathbf{u}$, then $\mathbf{u}$, $\mathbf{v}$ would be linearly independent, and W must contain all vectors in V_2 of the form $a\mathbf{u} + b\mathbf{v}$, where a, b are arbitrary scalars. But every vector in the plane can be so represented and, hence, W would have to coincide with V_2. Therefore, there can be no such vector $\mathbf{v}$, and W consists of the scalar multiples of $\mathbf{u}$, as in Figure 9-3. Thus each nontrivial subspace W of V_2 corresponds to a line through the origin in the plane.

Nontrivial Subspaces of V_3. We represent V_3 geometrically as the set of all vectors $\overrightarrow{OP}$ from the origin in 3-dimensional space and reason intuitively (full proofs appear later in the chapter). Let W be a nontrivial subspace of V_3. Then W contains a nonzero vector $\mathbf{u}_1 = \overrightarrow{OP_1}$, hence contains all the scalar multiples $t\mathbf{u}_1$—that is, all vectors $\overrightarrow{OP}$, for P on the line L_1 through O and P_1 (see Figure 9-4). This may be all of W. If not, then W contains a vector $\mathbf{u}_2 = \overrightarrow{OP_2}$ where P_2 is not on L_1. Hence, W also contains all vectors $\mathbf{u} = \overrightarrow{OP}$ of the form $t_1\overrightarrow{OP_1} + t_2\overrightarrow{OP_2}$. As t_1 and t_2 take on all real values, P varies over a *plane* H through O, as in Figure 9-4. This may be all of W. If not, then W contains a vector $\mathbf{u}_3 = \overrightarrow{OP_3}$, where P_3 is not in H. Hence, W contains all vectors $\mathbf{u} = \overrightarrow{OP}$ of the form $t_1\overrightarrow{OP_1} + t_2\overrightarrow{OP_2} + t_3\overrightarrow{OP_3}$. But since P_3 is not in H, the points P sweep out all of 3-dimensional space, and W is all of V_3. But W would then no longer be a nontrivial subspace of V_3. Therefore, there are only two types of nontrivial subspaces of V_3: those corresponding to lines through O and those corresponding to planes through O. It is also clear that every line through O and every plane through O does correspond to a nontrivial subspace W of V_3.

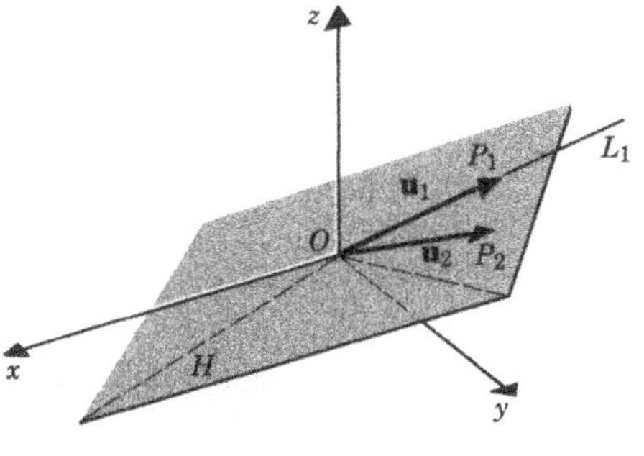

Figure 9-4 Subspaces of V_3.

EXAMPLE 5. Let W be the set of all (x, y, z) in V_3 such that $3x + 2y - 5z = 0$. Show that W is a nontrivial subspace of V_3 and that W corresponds to a plane through O in 3-dimensional space.

Solution. Let c be a scalar and let $\mathbf{u}_1 = (x_1, y_1, z_1)$ and $\mathbf{u}_2 = (x_2, y_2, z_2)$ be in W. Then $3x_1 + 2y_1 - 5z_1 = 0$ and $3x_2 + 2y_2 - 5z_2 = 0$. Hence

$$3cx_1 + 2cy_1 - 5cz_1 = 0$$

and

$$3(x_1 + x_2) + 2(y_1 + y_2) - 5(z_1 + z_2) = 0$$

Therefore, $c\mathbf{u}_1$ and $\mathbf{u}_1 + \mathbf{u}_2$ are in W. We note that W contains $\mathbf{u} = (2, -3, 0)$ and $\mathbf{v} = (0, 5, 2)$ and W does not contain $(1, 0, 0)$. Therefore, W is a subspace of V_3 and must be a nontrivial subspace. Since $\mathbf{v}$ is not a scalar multiple of $\mathbf{u}$, W cannot correspond to a line L through O. Accordingly, W corresponds to a plane H through O.

Remark. By similar reasoning, we show in general: if a_1, a_2, a_3 are scalars, not all 0, and W is the set of all vectors (x_1, x_2, x_3) in V_3 such that $a_1x_1 + a_2x_2 + a_3x_3 = 0$, then W is a nontrivial subspace of V_3 and W corresponds to a plane H through O in 3-dimensional space (see Problem 11 below).

PROBLEMS

1. Show that, under the usual rules for addition and multiplication by scalars, the following sets of functions are vector spaces.
   (a) The set of differentiable functions on $[a, b]$.
   (b) The set of all functions having a second derivative on $[0, 1]$.
   (c) The set of functions defined on $[0, 2]$ which have zeros at 0, 1 and 2.
   (d) The set of all polynomials without constant term.
   (e) The set of real polynomials having $\pm i$ as zeros.
   (f) The set of real polynomials divisible by $x^2 + x + 1$.
   (g) The set of all functions on $[0, 10]$ which are zero on $[2, 3]$.
   (h) The set of all functions f on $[0, 1]$, having a third derivative on this interval, and such that $f''' - xf' + (\sin x)f = 0$.
   (i) The set of all functions having a second derivative for all real values and such that $f''(x) \equiv 0$.
   (j) The set of all rational functions whose denominator is $x^2 + x + 1$.
   (k) The set of all polynomials such that $p(0) = p(1)$.
   (l) The set of all step-functions on $[0, 3]$ (see Section 4-14).
   (m) The set of all piecewise continuous functions on $[0, 3]$.
   (n) The set of all functions representable as sum of a convergent power series $\Sigma a_n x^n$ on $(-1, 1)$.
2. Show that the following sets of functions are *not* vector spaces:
   (a) The set of all differentiable functions on $[0, 1]$ whose derivative is $3x^2$.
   (b) The set of all differentiable functions f on $[0, 1]$ such that $f' = f - 1$.

(c) The set of all functions f on $[0, 2]$ having the property that $x \leq |f(x)|$ for $0 \leq x \leq 2$.

(d) The set of all functions f in $[0, 2]$ such that $f(1) = 1$.

3. We represent the vectors in V_2 as vectors $\overrightarrow{OP}$ in the plane, as in Figure 9-1. Indicate graphically the following subsets of V_2 and state whether each is a subspace of V_2.

(a) All vectors $t\mathbf{i} + 2t\mathbf{j}$, where $t \geq 0$.

(b) All vectors $3t\mathbf{i} - t\mathbf{j}$, where $-\infty < t < \infty$.

(c) All vectors $(1 - t)\mathbf{i} + (2 - 2t)\mathbf{j}$, where $-\infty < t < \infty$.

(d) All vectors $t\mathbf{i} + (3 - 2t)\mathbf{j}$, where $-\infty < t < \infty$.

(e) All vectors $\ln t\,\mathbf{i} + \ln t^2\mathbf{j}$, where $0 < t < \infty$.

(f) All vectors $\sin n\pi\,\mathbf{i} + \cos(n\pi/2)\mathbf{j}$, where $n = 0, \pm 1, \pm 2, \ldots$.

4. We represent the vectors in V_3 as vectors $\overrightarrow{OP}$ in space, as in Figure 9-2. Indicate graphically the following subsets of V_3 and state whether each is a subspace of V_3.

(a) All vectors $t\mathbf{i} + t\mathbf{j} + t\mathbf{k}$, $-\infty < t < \infty$.

(b) All vectors $t\mathbf{i} + (t + u)\mathbf{j} + u\mathbf{k}$, $-\infty < t < \infty$, $-\infty < u < \infty$.

(c) All vectors $(2 + t)\mathbf{i} + t\mathbf{j} + t\mathbf{k}$, $-\infty < t < \infty$.

(d) All vectors $\mathbf{i} + t\mathbf{j} + u\mathbf{k}$, $-\infty < t < \infty$, $-\infty < u < \infty$.

(e) All vectors $\sin 2t\,\mathbf{i} + \sin t \cos t\,\mathbf{j} + 3 \sin 2t\,\mathbf{k}$, $-\infty < t < \infty$.

(f) All vectors $t^3\mathbf{i} + \tan u\,\mathbf{j} + \ln s\,\mathbf{k}$, $-\infty < t < \infty$, $-\pi/2 < u < \pi/2$, $0 < s < \infty$.

5. (a) Show that $\mathcal{C}^{(2)}(-\infty, \infty)$ is a subspace of $\mathcal{C}^{(1)}(-\infty, \infty)$. Is $\mathcal{C}^{(2)}(-\infty, \infty)$ a nontrivial subspace of $\mathcal{C}^{(1)}(-\infty, \infty)$?

(b) Show that $\mathcal{C}^{(h+1)}(-\infty, \infty)$ is a subspace of $\mathcal{C}^{(h)}(-\infty, \infty)$. Is it a nontrivial subspace?

6. Let $\mathcal{I}$ be the set of all infinite sequences of real numbers. $\mathcal{I}$ can be considered as the set of all real functions on the set $1, 2, \ldots, n, \ldots$ and, hence, $\mathcal{I}$ forms a vector space (see problem 6 following Section 9-1). Let A be the set of all convergent sequences, B the set of all sequences that converge to 0, N the set of sequences $\{a_n\}$ such that $\{|a_n|\}$ converges. Prove: (a) A is a nontrivial subspace of $\mathcal{I}$. (b) B is a nontrivial subspace of A. (c) A is a nonempty subset of N. (d) N is not a subspace of $\mathcal{I}$.

7. V_3 is not a subspace of V_4, since one is a set of 3-tuples while the other is a set of 4-tuples. Let W be the set of 4-tuples with the last coordinate 0. Show that W is a subspace of V_4 and that there is a one-to-one correspondence between V_3 and W, say $\mathbf{u} \leftrightarrow \mathbf{u}^*$ such that if $\mathbf{u} \leftrightarrow \mathbf{u}^*$, $\mathbf{v} \leftrightarrow \mathbf{v}^*$, then $\mathbf{u} + \mathbf{v} \leftrightarrow \mathbf{u}^* + \mathbf{v}^*$ and $c\mathbf{u} \leftrightarrow c\mathbf{u}^*$. Are there other subspaces of V_4 which are also in such a one-to-one correspondence with V_3?

8. For each of the following subsets of V_4, determine whether the subset is a subspace:

(a) W: all $\mathbf{x} = (x_1, x_2, x_3, x_4)$ such that $x_1 = x_2$.

(b) U: all $\mathbf{x}$ such that $x_1 = x_2$ and $x_1 + x_2 + x_3 + x_4 = 0$.

(c) J: all $\mathbf{x}$ such that x_1 is rational.

(d) K: all $\mathbf{x}$ such that $x_1 + x_2 + x_3 + x_4 \leq 0$.

(e) L: all $\mathbf{x}$ such that $x_1 = x_2{}^2$.

(f) M: all $\mathbf{x}$ such that either $x_1 = x_2$ or $x_3 = x_4$.

(g) N: all $\mathbf{x}$ such that $|x_1| + |x_2| + |x_3| + |x_4| \neq 0$.

9. (a) Let U and W be subspaces of a vector space V. Show that if U is a subset of W, then U is a subspace of W.

   (b) Let W be a subspace of V and let U be a subspace of W. Show that U is a subspace of V.

10. Let U be the vector space of all real functions f on $[-1, 1]$. Determine whether each of the following is a subspace of U.

   (a) U_1: the set of all f such that $f(0) = 0$.

   (b) U_2: the set of all f such that $f(x) = 0$ for $-1 \leq x < \frac{1}{2}$.

   (c) U_3: the set of all f such that f is continuous at $x = \frac{1}{2}$.

   (d) U_4: the set of all f such that $f(x) = f(-x)$ for $-1 \leq x \leq 1$.

   (e) U_5: the set of all f such that f is monotone strictly increasing on $[-1, 1]$.

11. (a) Prove the assertion in the Remark at the close of Section 9-2. [*Hint.* If, for example, $a_1 \neq 0$, then show that $(1, 0, 0)$ is not in W and find two nonzero vectors in W such that neither is a scalar multiple of the other.]

   (b) Prove: If $a_1, \ldots, a_k$ are scalars, not all 0 and W is the set of all $(x_1, \ldots, x_k)$ such that $a_1x_1 + \cdots + a_kx_k = 0$, then W is a subspace of V_k and W is a nontrivial subspace if $k \geq 2$.

9-3 INTERSECTION OF SUBSPACES

Let U and W be subsets of a vector space V and let X be the set of all vectors in V that are in both U and W (see Figure 9-5). The set X is called the *intersection of U and W*. We usually denote intersection by $\cap$ and then write

$$X = U \cap W$$

THEOREM 3. *If U and W are subspaces of a vector space V then their intersection $U \cap W$ is a subspace of V.*

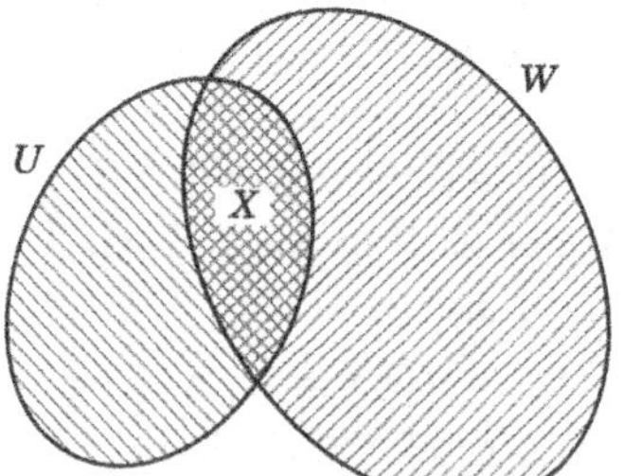

Figure 9-5 Intersection of subsets.

PROOF. Let $X = U \cap W$. If $\mathbf{x}, \mathbf{y}$ are in X, then $\mathbf{x}, \mathbf{y}$ are in U and $\mathbf{x}, \mathbf{y}$ are in W. Since U and W are subspaces of V, it follows that $\mathbf{x} + \mathbf{y}$ and $a\mathbf{x}$ (for any scalar a) are in U and in W and, therefore, are in X. We have thus shown that X is closed under addition and under scalar multiplication. Also X is nonempty, since U and W contain $\mathbf{0}$. It follows that $X = U \cap W$ is also a subspace of V.

Clearly $U \cap W$ is the *largest* subspace of V common to both U and to W, since it is the largest common subset and it also happens to be a subspace.

EXAMPLE 1. The set U of all triples in V_3 having the first component 0 is a subspace of V_3, as is the set W of all triples (x, y, z) with $x = y$. By Theorem 3, the set $X = U \cap W$ is a subspace of V_3; X consists of the triples $(0, 0, z)$ where z is arbitrary (see Figure 9-6). The reader can verify, independently, that X is indeed a subspace of V_3.

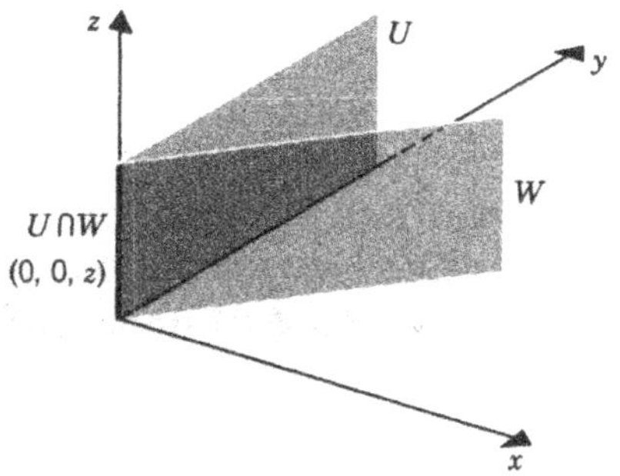

Figure 9-6 Intersection of two subspaces (planes) in V_3.

EXAMPLE 2. Let V_2 be interpreted as the set of vectors in the plane. Then the intersection of two different nontrivial subspaces of V_2 is always the zero-space V_0. This is easily seen by recalling that the nontrivial subspaces of V_2 correspond to lines through the origin (Figure 9-3), and two distinct lines through the origin necessarily intersect only in the origin. Thus the space of the intersection consists of the zero vector alone; that is, the intersection is V_0.

When two subspaces of a vector space V intersect in the zero-space, we say that they *intersect only trivially*.

If U_1, U_2, U_3 are three subsets of V, one can form the multiple intersection $U_1 \cap (U_2 \cap U_3)$. This consists of the elements of U_1 which are also in both U_2 and U_3. Hence, it is simply the set of all elements common to U_1, U_2 and U_3. For that reason, we drop the parentheses and denote the set by $U_1 \cap U_2 \cap U_3$. On applying Theorem 3 first to $U_2 \cap U_3$ and then to $U_1 \cap (U_2 \cap U_3)$, we see that if U_1, U_2, U_3 are subspaces of a vector space V, then so also is $U_1 \cap U_2 \cap U_3$. In general, if $U_1, \ldots, U_h$ are subspaces of V, then the set $U_1 \cap U_2 \cap \cdots \cap U_h$ consists of those vectors that are common to all the subspaces $U_1, \ldots, U_h$, and $U_1 \cap \cdots \cap U_h$ is a subspace of V.

EXAMPLE 3. The set W of all vectors $(x_1, \ldots, x_5)$ in V_5 such that

$$\begin{aligned} x_1 + x_2 + x_3 - x_4 &= 0, \\ 2x_1 - 3x_3 + x_4 + x_5 &= 0, \\ x_1 - x_2 + x_3 - 3x_5 &= 0 \end{aligned}$$

is a subspace of V_5.

PROOF. Let U_1 be the subset of V_5 consisting of all vectors in V_5 satisfying the first equation: $x_1 + x_2 + x_3 - x_4 = 0$. As in Example 5 in Section 9-2 (see also Problem 11, following Section 9-2), we verify that U_1 is a subspace of V_5. Similarly, U_2, the set of all vectors in V_5 satisfying the second equation, and U_3, the set of all vectors satisfying the third equation, are subspaces of V_5. Now W consists of all vectors satisfying all three equations. Hence, $W = U_1 \cap U_2 \cap U_3$, and therefore, W is a subspace of V_5.

Example 3 is a special case of the following general rule: *The common solutions $(x_1, \ldots, x_n)$ of the homogeneous linear equations*

$$a_{11}x_1 + \cdots + a_{1n}x_n = 0, \quad \ldots, \quad a_{h1}x_1 + \cdots + a_{hn}x_n = 0$$

form a subspace of V_n. (See Problem 3**a** below.)

PROBLEMS

1. Describe the intersections of the following subsets U_1, U_2 of $\mathcal{C}(-\infty, \infty)$ and determine whether the intersection is a subspace.
   (a) U_1: all f such that $f(0) = 0$; U_2: all f such that $f(1) = 0$.
   (b) U_1: all polynomials; U_2: all even functions.
   (c) U_1: all polynomials; U_2: all bounded functions.
   (d) U_1: all f having period 3π; U_2: all f having period 2π.
   (e) U_1: all f having limit 0 as $x \to \infty$; U_2: all f having limit 1 as $x \to \infty$.
   (f) U_1: all f such that $\int_0^\infty f(x)\,dx$ exists; U_2: all f such that $\int_{-\infty}^0 f(x)\,dx$ exists.
2. Discuss geometrically the possible type of intersection of two nontrivial subspaces U_1, U_2 of V_3 for each of the following cases:
   (a) U_1 and U_2 correspond to lines through O.
   (b) U_1 corresponds to a line through O, U_2 to a plane through O.
   (c) U_1 and U_2 correspond to planes through O.
3. (a) Prove: if $h \geq 1$, then the set W of common solutions of the h homogeneous linear equations:

   $$a_{11}x_1 + \cdots + a_{1n}x_n = 0, \quad \ldots \quad , a_{h1}x_1 + \cdots + a_{hn}x_n = 0$$

   is a subspace of V_n. Furthermore, if at least one a_{ij} is nonzero, then W is not V_n itself. [Hint: Consider Problem 11-b following Section 9-2.]
   (b) Give an example, with $h = 3$, of h linear equations in x_1, x_2, x_3 whose only common solution is $(0, 0, 0)$.
   (c) How small can h be in part (b)? How large can it be?

(d) Is it true that (0, 0, 0) is the only possible common solution of four homogeneous linear equations in x_1, x_2, x_3? Explain.

4. We saw (Example 2) that any two nontrivial subspaces of V_2 intersect in V_0. Show by example that it is possible for two subsets of V_2 which are not subspaces to intersect in a nontrivial subspace of V_2.

9-4 ADDITION OF SUBSETS

Let X, Y be subsets, not necessarily subspaces, of a vector space V. We denote by $\{X + Y\}$ the set of all vectors $\mathbf{v}$ in V which can be expressed as the sum of a vector from X and a vector from Y. Thus $\mathbf{v}$ is in $\{X + Y\}$ exactly when there exist a vector $\mathbf{x}$ in X and a vector $\mathbf{y}$ in Y such that $\mathbf{v} = \mathbf{x} + \mathbf{y}$. The set $\{X + Y\}$ is called the *sum of the sets X and Y*. Its meaning will be made clear by several examples.

EXAMPLE 1. In V_2, represented as usual as the set of vectors in the plane, let X consist of the two vectors $3\mathbf{i}$ and $5\mathbf{j}$, let Y consist of the two vectors $4\mathbf{i} + 8\mathbf{j}$, $10\mathbf{i} + 20\mathbf{j}$. Then $\{X + Y\}$ consists of the four vectors

$$7\mathbf{i} + 8\mathbf{j}, \qquad 13\mathbf{i} + 20\mathbf{j}, \qquad 4\mathbf{i} + 13\mathbf{j}, \qquad 10\mathbf{i} + 25\mathbf{j}$$

EXAMPLE 2. In V_2, let X be the vector $\mathbf{u} = \overrightarrow{OQ}$ and let Y be the set of all vectors $\overrightarrow{OP}$ from the origin O to point P lying on the line segment AB. Then $\{X + Y\}$ consists of all vectors $\overrightarrow{OS} = \overrightarrow{OQ} + \overrightarrow{OP}$, where Q is fixed and P varies on AB. Hence, $\{X + Y\}$ corresponds to a line segment $A'B'$, obtained from AB by translating each point by the vector $\mathbf{u}$, as in Figure 9-7; in particular, $\overrightarrow{AA'} = \mathbf{u}$, $\overrightarrow{BB'} = \mathbf{u}$.

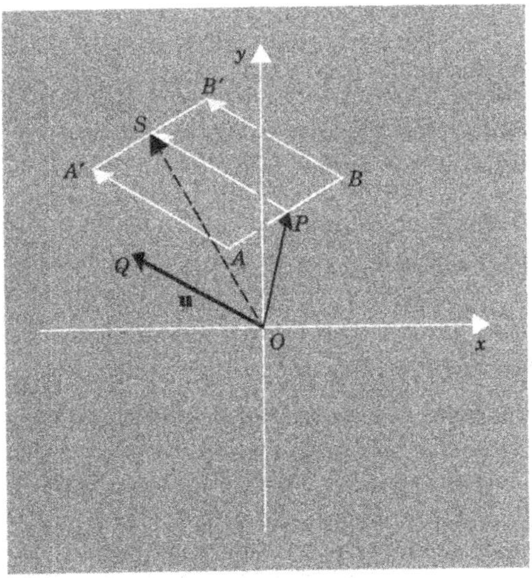

Figure 9-7 Sum of sets of vectors corresponding to a point and a line segment.

EXAMPLE 3. In V_2, let X consist of $\mathbf{u} = \overrightarrow{OQ}$, as in Example 2, but let Y consist of all vectors from O to a point on a line L through O. As in Exam-

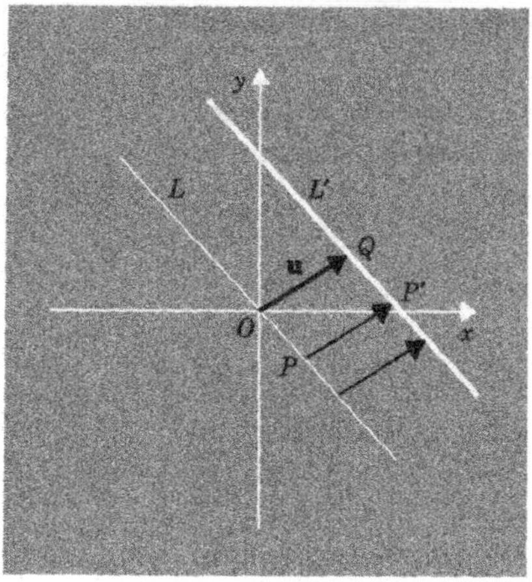

Figure 9-8 Sum of sets of vectors corresponding to a point and a line.

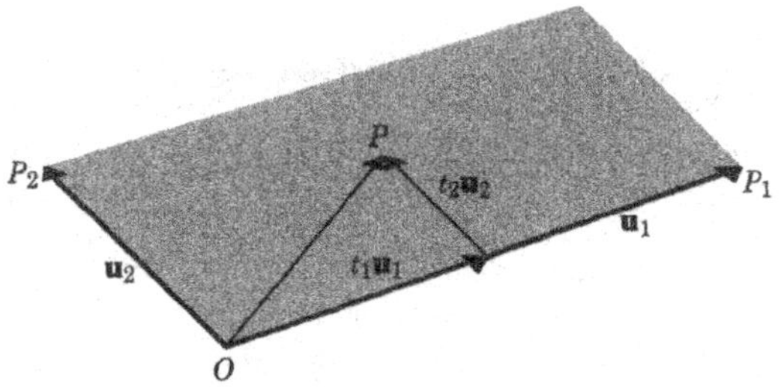

Figure 9-9 Sum of sets of vectors corresponding to two line segments.

ple 2, the set corresponding to $\{X + Y\}$ is L translated by the vector $\mathbf{u}$, and we obtain a line L', as in Figure 9-8.

Notations. When a set X consists of a single vector $\mathbf{u}$, we shall write $\{\mathbf{u} + Y\}$ for the sum $\{X + Y\}$. As Examples 2 and 3 show, in V_2, $\{\mathbf{u} + Y\}$ corresponds to a set obtained from the set corresponding to Y by translation through the vector $\mathbf{u}$. When a set Y is a subspace consisting of all multiples $t\mathbf{v}$ of one vector $\mathbf{v}$, we also denote Y by the symbol $\{t\mathbf{v}\}$. Thus in Example 3, $\{X + Y\}$ can be denoted by $\{\mathbf{u} + \{t\mathbf{v}\}\}$. In this example, $\{X + Y\}$ corresponds to the line L' and our writing $\{X + Y\}$ as $\{\mathbf{u} + \{t\mathbf{v}\}\}$ is simply another way of saying that the line L' has the vector equation $\overrightarrow{OP} = \mathbf{u} + t\mathbf{v}$, $-\infty < t < \infty$, as in Chapter 1.

EXAMPLE 4. In V_2 let X consist of all vectors $\overrightarrow{OP}$, where P varies on a segment OP_1, and let Y consist of all vectors $\overrightarrow{OP}$, where P varies on a segment OP_2. Let $\mathbf{u}_1 = \overrightarrow{OP_1}$, $\mathbf{u}_2 = \overrightarrow{OP_2}$ be linearly independent. Then $\{X + Y\}$ consists of all vectors

$$\mathbf{u} = \overrightarrow{OP} = t_1\mathbf{u}_1 + t_2\mathbf{u}_2, \qquad \text{where } 0 \le t_1 \le 1, 0 \le t_2 \le 1$$

Here the points P corresponding to $\{X + Y\}$ fill the parallelogram two of whose edges are OP_1 and OP_2 (see Figure 9-9).

EXAMPLE 5. In V_3, let $\mathbf{u}_1$, $\mathbf{u}_2$ be nonzero vectors, neither of which is a scalar multiple of the other. Let X consist of all scalar multiples of $\mathbf{u}_1$, Y of all scalar multiples of $\mathbf{u}_2$. Then $\{X + Y\}$ consists of all vectors $t_1\mathbf{u}_1 + t_2\mathbf{u}_2$. Here X corresponds to a line L_1 through O, Y to a line L_2 through O, $\{X + Y\}$ to a plane H through O and containing L_1 and L_2 (see Figure 9-10).

EXAMPLE 6. In V_3, let X consist of one vector $\mathbf{w}$, Y of all linear combinations $t_1\mathbf{u}_1 + t_2\mathbf{u}_2$ of two nonzero vectors $\mathbf{u}_1$, $\mathbf{u}_2$, neither of which is a scalar multiple of the other. Here Y corresponds to a plane H through O and, as in Examples

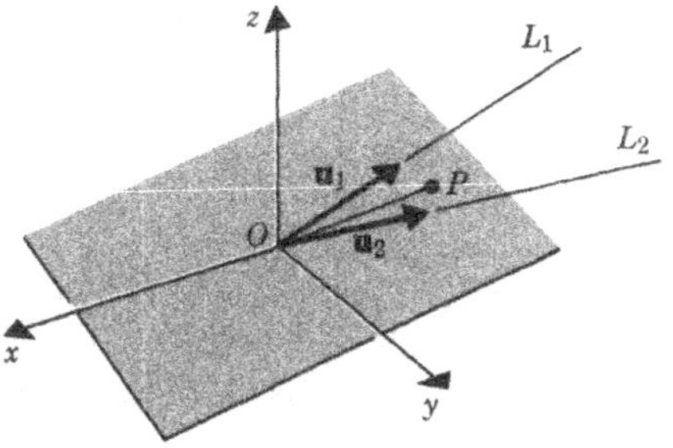

Figure 9-10 Sum of sets of vectors corresponding to two lines.

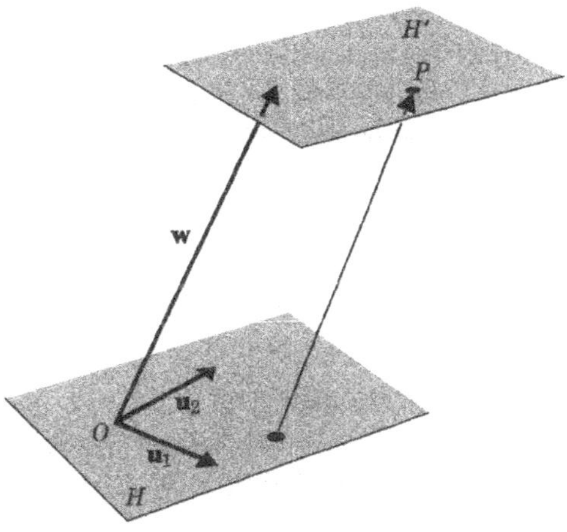

Figure 9-11 Sum of sets of vectors corresponding to a point and a plane.

2 and 3, $\{X + Y\}$ corresponds to a plane H' obtained from H by translating through the vector $\mathbf{w}$ (see Figure 9-11).

EXAMPLE 7. In V_3, let X consist of one vector $\mathbf{u}$ and let Y be of the form $\{t\mathbf{v}\}$, where $\mathbf{v} \neq \mathbf{0}$. Then, as in Example 3, Y corresponds to a line L through O and $\{X + Y\} = \{\mathbf{u} + \{t\mathbf{v}\}\}$ to a line L' obtained from L by translation through the vector $\mathbf{u}$ (see Figure 9-12).

Remark. From Examples 3, 6, and 7 we can state in general that, if Y is a nontrivial subspace of V_2 or V_3, then $\{\mathbf{u} + Y\}$ corresponds to a line or a plane; it corresponds to a line when Y is of form $\{t\mathbf{v}\}$, $\mathbf{v} \neq \mathbf{0}$; it corresponds to a plane in space when Y is a nontrivial subspace of V_3 formed of all vectors $\{t_1\mathbf{u}_1 + t_2\mathbf{u}_2\}$ as in Example 6. It is clear that all lines and planes are obtainable in this way. When $\mathbf{u} = \mathbf{0}$, $\{\mathbf{u} + Y\}$ reduces to Y and the corresponding set becomes a line or plane through the origin.

EXAMPLE 8. The set X of 4-tuples (x_1, x_2, x_3, x_4) of real numbers such that $x_1 - x_2 + x_3 + 5x_4 = 1$ can be expressed as $\{\mathbf{e}_1 + W\}$, where W is the subspace of V_4 such that $x_1 - x_2 + x_3 + 5x_4 = 0$, and $\mathbf{e}_1 = (1, 0, 0, 0)$.
To show this we note that if (x_1, x_2, x_3, x_4) is in W, then $(x_1 + 1) - x_2 +$

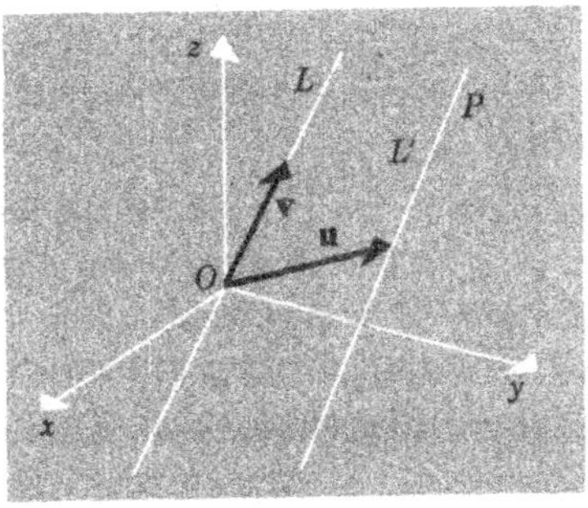

Figure 9-12 Sum of sets of vectors corresponding to a point and a line in space.

$x_3 + 5x_4 = 1$ and therefore $\{\mathbf{e}_1 + W\}$ is contained in X. Conversely if (x_1, x_2, x_3, x_4) is in X, then $x_1 - x_2 + x_3 + 5x_4 = 1$ and $(x_1 - 1) - x_2 + x_3 + 5x_4 = 0$ and $(x_1, x_2, x_3, x_4) - \mathbf{e}_1$ is in W. Thus X is contained in $\{\mathbf{e}_1 + W\}$ and hence $X = \{\mathbf{e}_1 + W\}$.

EXAMPLE 9. Let Z be the subset of $\mathcal{C}(-\infty, \infty)$, consisting of all functions f such that $f'(x) = 2x$. Then Z consists of all functions $x^2 + C$, where C is an "arbitrary constant." We can interpret Z as $\{X + Y\}$, where X consists of the one function x^2 and Y consists of all constant functions—that is, of all scalar multiples of the function 1.

THEOREM 4. *If U and W are subspaces of a vector space V, then $\{U + W\}$ is a subspace of V. The subspace $\{U + W\}$ contains both U and W. Furthermore, if Z is a subspace of V which contains both U and W then Z contains $\{U + W\}$. Thus $\{U + W\}$ is the smallest subspace of V which contains both U and W.*

PROOF. Suppose that $\mathbf{v}_1$ and $\mathbf{v}_2$ are in $\{U + W\}$. Then there exist $\mathbf{u}_1, \mathbf{u}_2$ in U and $\mathbf{w}_1, \mathbf{w}_2$ in W such that

$$\mathbf{v}_1 = \mathbf{u}_1 + \mathbf{w}_1, \qquad \mathbf{v}_2 = \mathbf{u}_2 + \mathbf{w}_2$$

It follows from the properties of addition in vector spaces that

$$\mathbf{v}_1 + \mathbf{v}_2 = (\mathbf{u}_1 + \mathbf{w}_1) + (\mathbf{u}_2 + \mathbf{w}_2) = (\mathbf{u}_1 + \mathbf{u}_2) + (\mathbf{w}_1 + \mathbf{w}_2)$$

Since U is a subspace, $\mathbf{u}_1 + \mathbf{u}_2$ is in U. Similarly, $\mathbf{w}_1 + \mathbf{w}_2$ is in W. Hence, $\mathbf{v}_1 + \mathbf{v}_2$ is in $\{U + W\}$; that is, $\{U + W\}$ is closed under addition. Also, if a is a scalar, then

$$a\mathbf{v}_1 = a(\mathbf{u}_1 + \mathbf{w}_1) = a\mathbf{u}_1 + a\mathbf{w}_1$$

since $\mathbf{u}_1, \mathbf{w}_1$ are in the vector space V. But U is a subspace and, hence, $a\mathbf{u}_1$ is in U. Similarly, $a\mathbf{w}_1$ is in W. Hence, $a\mathbf{v}_1$ is in $\{U + W\}$; that is, $\{U + W\}$ is closed under multiplication by scalars. It then follows that $\{U + W\}$ is a subspace of V.

Since U is a subspace, $\mathbf{0}$ is in U; hence, $W = \{\mathbf{0} + W\}$ is in $\{U + W\}$ and must be a subspace of $\{U + W\}$. Similarly, U is a subspace of $\{U + W\}$.

Let Z be a subspace of V containing both U and W. Since Z is a subspace it must be closed under addition and, hence, it must contain all sums of the form $\mathbf{u} + \mathbf{w}$, where $\mathbf{u}$ is in U and $\mathbf{w}$ is in W. But then Z contains the set $\{U + W\}$. This completes the proof of the theorem.

Theorem 4 is illustrated by Example 5 above. Here X and Y correspond to subspaces represented by lines through O, $\{X + Y\}$ to a subspace represented by a plane through O, as in Figure 9-10.

Now let X and Y be subsets of the vector space V and let Z be the set consisting of all vectors $\mathbf{v}$ in V which are in X or Y or both. The set Z is called the *union* of X and Y. We denote union by $\cup$ and write $Z = X \cup Y$. Theorem 4 asserts that $\{U + W\}$ contains $U \cup W$, and that it is the smallest subspace containing $U \cup W$. In general $\{U + W\}$ is a much larger set than

$U \cup W$. For example, in V_2, if U is the set of scalar multiples of $\mathbf{i}$ and W is the set of scalar multiples of $\mathbf{j}$, then $\{U + W\} = V_2$. But $\mathbf{v} = a\mathbf{i} + b\mathbf{j}$ is in $U \cup W$ precisely when, at least, one of a, b is 0. Thus $U \cup W$ consists of the vectors corresponding to points on the coordinate axes, while $\{U + W\}$ is the set corresponding to all points in the plane. This example also shows that in contrast to intersection, the union of two subspaces need not be a subspace.

If $X_1, \ldots, X_h$ is a finite collection of subsets of a vector space V, we define their sum $\{X_1 + \cdots + X_h\}$ to be all vectors in V of the form $\mathbf{x}_1 + \cdots + \mathbf{x}_h$, where $\mathbf{x}_1$ is in $X_1, \ldots,$ and $\mathbf{x}_h$ is in X_h. For example, the set $\{X + Y\}$ of Example 6 above can also be written as $\{\mathbf{w} + \{t_1\mathbf{u}_1\} + \{t_2\mathbf{u}_2\}\}$. If $U_1, \ldots, U_h$ are subspaces of V, then $\{U_1 + \cdots + U_h\}$ is a subspace of V; in fact, it is the smallest subspace of V containing all of the subspaces $U_1, \ldots, U_h$ (Problem 11 below).

PROBLEMS

1. In V_2 describe $\{X + Y\}$ and graph for each of the following choices of the corresponding sets in the xy-plane:
   (a) X: the points $(0, 0)$ and $(1, 1)$; Y: the points $(2, 2)$ and $(3, 2)$.
   (b) X: the points $(0, 0)$ and $(2, 1)$; Y: the points $(1, 2)$ and $(2, -1)$.
   (c) X: the point $(1, 1)$; Y: the segment joining $(0, 0)$ and $(1, 0)$.
   (d) X: the point $(2, 3)$; Y: the segment joining $(1, 2)$ and $(3, 4)$.
   (e) X: the point $(1, 1)$; Y: the line $y = 1$.
   (f) X: the point $(3, 2)$; Y: the line $x + 2y = 3$.
   (g) X: the point $(2, 1)$; Y: the line $x = 3t$, $y = 5t$.
   (h) X: the point $(3, 1)$; Y: the line $x = t$, $y = 4t$.
   (i) X: the point $(2, 3)$; Y: the circle $x^2 + y^2 = 1$.
   (j) X: the point $(5, 2)$; Y: the circle $(x + 5)^2 + (y + 2)^2 = 1$.
   (k) X: the segment joining $(0, 0)$ and $(1, 0)$; Y: the segment joining $(0, 0)$ and $(0, 1)$.
   (l) X: the segment joining $(0, 0)$ and $(1, 1)$; Y: the segment joining $(0, 0)$ and $(-1, 1)$.
   (m) X: the line $y = x$; Y: the line $y = 2x$.
   (n) X: the line $x = t$, $y = 3t$; Y: the line $x = -t$, $y = t$.
2. In V_3 describe $\{X + Y\}$ and graph for each of the following choices for the corresponding sets in xyz-space:
   (a) X: the point $(0, 0, 1)$; Y: all points $(x, y, 0)$ (the xy-plane).
   (b) X: the point $(1, 1, 1)$; Y: all points $(0, y, z)$ (the yz-plane).
   (c) X: the point $(2, 2, 2)$; Y: the line $\overrightarrow{OP} = t(1, 1, 0)$.
   (d) X: the point $(1, 3, 1)$; Y: the line $\overrightarrow{OP} = t(0, 1, 1)$.
   (e) X: the point $(0, 0, 1)$; Y: the plane $\overrightarrow{OP} = t_1(1, 0, 0) + t_2(0, 1, 1)$.
   (f) X: the point $(1, 0, 0)$; Y: the plane $\overrightarrow{OP} = t_1(1, 1, 0) + t_2(0, 1, 1)$.

(g) X: the line $\overrightarrow{OP} = t(1, 1, 2)$, Y: the line $\overrightarrow{OP} = t(1, 0, 0)$.

(h) X: the line $\overrightarrow{OP} = t(1, 0, 1)$: Y: the line $\overrightarrow{OP} = t(0, 1, 1)$.

(i) X: the x-axis; Y: the yz-plane.

(j) X: the y-axis; Y: the xz-plane.

3. Let X, Y be subsets of V_2, the vectors in the plane, and let each correspond to a line. Prove each of the following:

(a) If X, Y correspond to parallel lines, then there is a vector $\mathbf{a}$ such that $X = \{\mathbf{a} + Y\}$.

(b) If X, Y correspond to parallel lines then $\{X + Y\}$ corresponds to a line parallel to that for X.

(c) If X, Y correspond to nonparallel lines then $\{X + Y\} = V_2$.

4. In $\mathcal{C}(-\infty, \infty)$ represent each of the following sets in the form $\{X + Y\}$, where X consists of one function and Y is a subspace:

(a) All f such that $f'(x) = \cos x$.

(b) All f such that $f'(x) = 5x^4$.

(c) All f such that $f''(x) = 12$.

(d) All f such that $f''(x) = 18e^{3x}$.

(e) All f such that $f'''(x) = 1$.

(f) All f such that $f'''(x) = \sin 2x$.

(g) All f such that $f(1) = 1$.

(h) All f such that $f(n) = n$ for all integers n.

5. If (a_1, a_2, a_3, a_4) is such that $a_1 - a_2 + a_3 + 5a_4 = 1$, show that the set X of Example 8 can be expressed as $X = \{(a_1, a_2, a_3, a_4) + W\}$.

6. Let W be the set of all vectors of the form $(s, 2s - t, t + s, t)$ in V_4, with s, t arbitrary. Let U be the set of all vectors of the form $(2a + 2b, b, -b, 3a + 2b)$, with a, b arbitrary.

(a) Show that $\{W + U\}$ is the set of all vectors of the form

$$(u + 2w, 2u - v, u + v, v + 3w)$$

(b) Show that U, W, and $\{U + W\}$ are subspaces of V_4.

(c) Find $U \cap W$.

7. Prove: if X, Y, Z are subsets of a vector space V and X is contained in Y, then $\{X + Z\}$ is contained in $\{Y + Z\}$.

8. Let U and W be subspaces of V. Prove:

(a) If U is contained in W then $\{U + W\} = W$.

(b) If $\{U + W\} = W$ then U is contained in W.

(c) If U is contained in W then $U \cap W = U$.

(d) If $U \cap W = U$ then U is contained in W.

9. In V_2, let X be those pairs having first coordinate 1 and let Y be those pairs having second coordinate 2. Show that $\{X + Y\} = V_2$. *Remark.* This problem shows that $\{X + Y\}$ can be a subspace, even though X and Y are not both subspaces.

10. (a) Prove: if W is a subspace of V and $\mathbf{w}$ is an element of W then $\{\mathbf{w} + W\} = W$.

(b) Prove: if W is a subspace of V and $\mathbf{x}$ is an element of V such that $\{\mathbf{x} + W\} = W$ then $\mathbf{x}$ is in W.

(c) Show that (a) and (b) no longer hold true if W is an arbitrary subset of V.

11. Prove: If $U_1, \ldots, U_h$ are subspaces of a vector space V, then $\{U_1 + \cdots + U_h\}$ is a subspace of V which contains each of the subspaces $U_1, U_2, \ldots, U_h$. Furthermore, each subspace Z of V which contains $U_1, U_2, \ldots, U_h$ must contain $\{U_1 + \cdots + U_h\}$. Thus $\{U_1 + \cdots + U_h\}$ is the smallest subspace of V containing the subspaces $U_1, \ldots, U_h$.

9-5 LINEAR VARIETIES

A subset L of a vector space V is said to be a *linear variety* of V provided $L = \{\mathbf{u} + W\}$ for some vector $\mathbf{u}$ in V and some subspace W of V. The vector $\mathbf{u}$ is called a *leader* of L and the subspace W is called a *base space* of L. We encountered a number of such sets in the preceding section; see Examples 3, 6, 7 and 8. As the vector equation: $\overrightarrow{OP} = \mathbf{a} + t\mathbf{v}$, $-\infty < t < \infty$, shows, a line in the plane corresponds to a linear variety $\{\mathbf{a} + \{t\mathbf{v}\}\}$ in V_2. Conversely, each linear variety $\{\mathbf{a} + \{t\mathbf{v}\}\}$ in V_2 corresponds to a line (Example 3 in Section 9-4). Similarly, lines and planes in 3-dimensional space correspond to linear varieties of V_3. Thus linear varieties are generalizations of lines and planes. The linear variety $L = \{\mathbf{u} + W\}$ can be considered to be obtained from the subspace W by translating it by the vector $\mathbf{u}$. We say that L is a *translate* of W.

The set consisting of a single vector $\mathbf{u}$ alone is a linear variety, since $\mathbf{u} = \{\mathbf{u} + V_0\}$, where V_0 is the zero-space. Hence, in the plane and in space, a single point also corresponds to a linear variety.

Each subspace W of V is a linear variety, since $W = \{\mathbf{0} + W\}$. In particular, V is itself a linear variety of V.

EXAMPLE 1. Let L be the set of all real polynomials whose constant term is 2: $2 + 3x + x^2, 2 + 5x + x^2 - x^4, \ldots$ Then L is a linear variety of the vector space $\mathcal{P}$ of all polynomials, since $L = \{2 + W\}$, where W is the subspace consisting of all polynomials with 0 as a zero; that is, W is the subspace of the polynomials whose constant term is 0.

EXAMPLE 2. If $a_1, \ldots, a_n$ (not all 0) and b are real numbers, then the set L of vectors $(x_1, \ldots, x_n)$ in V_n such that

$$a_1x_1 + \cdots + a_nx_n = b$$

is a linear variety of V_n.

PROOF. First we note that L contains at least one vector $\mathbf{u} = (u_1, \ldots, u_n)$. For example, if $a_1 \neq 0$, then $\mathbf{u} = (b/a_1, 0, \ldots, 0)$ is in L, since

$$a_1 \cdot \frac{b}{a_1} + 0 + \cdots + 0 = b$$

We reason similarly if $a_i \neq 0$ for some i other than 1.

Now let $\mathbf{u} = (u_1, \ldots, u_n)$ be one definite vector in L, so that $a_1u_1 + \cdots$

$+ a_n u_n = b$. Let $\mathbf{v} = (v_1, \ldots, v_n)$ be an arbitrary vector of V_n. If $\mathbf{v}$ is also in L, then we have

$$a_1 u_1 + \cdots + a_n u_n = b$$
$$a_1 v_1 + \cdots + a_n v_n = b$$

and, hence, by subtraction,

$$a_1(v_1 - u_1) + \cdots + a_n(v_n - u_n) = 0$$

Therefore, $\mathbf{v} - \mathbf{u} = \mathbf{w} = (w_1, \ldots, w_n)$ satisfies the homogeneous linear equation

$$a_1 w_1 + \cdots + a_n w_n = 0 \tag{9-50}$$

Conversely, if $\mathbf{w}$ satisfies Equation (9-50), then the equations

$$a_1 u_1 + \cdots + a_n u_n = b, \qquad a_1 w_1 + \cdots + a_n w_n = 0$$

imply, by addition, that

$$a_1(u_1 + w_1) + \cdots + a_n(u_n + w_n) = b$$

so that $\mathbf{v} = \mathbf{u} + \mathbf{w}$ is in L. Therefore, $\mathbf{v}$ is in L if, and only if, $\mathbf{v} = \mathbf{u} + \mathbf{w}$ where $\mathbf{w} = (w_1, \ldots, w_n)$ satisfies equation (9-50). But we saw (Example 3 in Section 9-3) that the set of all solutions of (9-50) forms a subspace W of V_n. Hence, L is the linear variety $\{\mathbf{u} + W\}$, where W is a subspace of V_n.

THEOREM 5. *Let $L = \{\mathbf{v} + W\}$, be a linear variety of a vector space V. If $\mathbf{y}$ is in L then $L = \{\mathbf{y} + W\}$; that is, any vector in a linear variety can be chosen as the leader of the linear variety. A linear variety of V is a subspace of V exactly when $\mathbf{0}$ is in the linear variety.*

PROOF. If $\mathbf{y}$ is in L, then $\mathbf{y} = \mathbf{v} + \mathbf{w}$, where $\mathbf{w}$ is some vector in W. We have $\{\mathbf{y} + W\} = \{(\mathbf{v} + \mathbf{w}) + W\} = \{\mathbf{v} + \{\mathbf{w} + W\}\}$. But W is a subspace and, hence, is closed under addition; therefore $\{\mathbf{w} + W\} = W$ (see Problem 10 following Section 9-4). Hence, $\{\mathbf{y} + W\} = \{\mathbf{v} + W\} = L$.

If a linear variety is a subspace it must contain $\mathbf{0}$ (see Section 9-2). Also, if $\mathbf{0}$ is in a linear variety $\{\mathbf{v} + W\}$, then by the first part of the theorem we have $\{\mathbf{v} + W\} = \{\mathbf{0} + W\} = W$; that is, the linear variety is a subspace.

THEOREM 6. *A linear variety of a vector space has a unique base space.*

PROOF. Let L be a linear variety and suppose $L = \{\mathbf{v} + W\} = \{\mathbf{y} + W'\}$. Since any vector in L can be chosen as the leader, we have $L = \{\mathbf{v} + W\} = \{\mathbf{v} + W'\}$. If $\mathbf{w}$ is in W, then $\mathbf{v} + \mathbf{w}$ is in L, whence $\mathbf{v} + \mathbf{w} = \mathbf{v} + \mathbf{w}'$ for some $\mathbf{w}'$ in W'. It follows that each $\mathbf{w}$ in W is in W'; that is, W is contained in W'. Similarly, W' is contained in W. Therefore, $W = W'$ and L has but one base space.

COROLLARY. *Two linear varieties of a vector space having the same base space are equal if and only if they have an element in common.*

The proof is left as an exercise (Problem 6 below).

Remark. It follows from Theorem 6 that a linear variety is a translate of a unique subspace W. Thus in V_2 and V_3 every linear variety is parallel to a unique line or plane through the origin. From the Corollary it follows that, in V_2 or V_3, two parallel lines having a common point are coincident.

THEOREM 7. *Let $L = \{\mathbf{x} + W\}$, $M = \{\mathbf{y} + U\}$ be two linear varieties of a vector space V. Then either $L \cap M$ is empty or $L \cap M$ is a linear variety with $W \cap U$ as base space.*

PROOF. If $L \cap M$ is not empty, let $\mathbf{z}$ be a common element. Then, by Theorem 5, $L = \{\mathbf{z} + W\}$ and $M = \{\mathbf{z} + U\}$. Clearly, each of these sets contains the set $\{\mathbf{z} + (U \cap W)\}$; hence, $L \cap M$ contains $\{\mathbf{z} + (U \cap W)\}$. On the other hand, let $\mathbf{w}$ be a vector in $L \cap M$. Then $\mathbf{w}$ is in L, whence $\mathbf{w} - \mathbf{z}$ is in W. Similarly, $\mathbf{w} - \mathbf{z}$ is in U and, therefore, $\mathbf{w} - \mathbf{z}$ is in $U \cap W$, or equivalently $\mathbf{w}$ is in $\{\mathbf{z} + (U \cap W)\}$. Since $\mathbf{w}$ was any vector in $L \cap M$, we conclude that $L \cap M$ is contained in $\{\mathbf{z} + (U \cap W)\}$. Consequently, $L \cap M = \{\mathbf{z} + (U \cap W)\}$.

In the plane, our theorem is merely asserting that either 2 lines are disjoint or they meet in a linear variety, namely, the common point of intersection.

EXAMPLE 3. The set of vectors (x_1, x_2, x_3, x_4) such that

$$x_1 - x_2 + 2x_3 - x_4 = 2 \qquad \text{and} \qquad 2x_1 + x_2 - x_3 + 2x_4 = 7$$

is a linear variety of V_4.

PROOF. By Example 2 above, the set L of 4-tuples satisfying the first equation is a linear variety of V_4. Similarly, the set M of 4-tuples satisfying the second equation is a linear variety. The 4-tuples satisfying *both* equations are the vectors in $L \cap M$, which by Theorem 7 is either empty or a linear variety. Since $(3, 1, 0, 0)$ satisfies both equations, $L \cap M$ is a linear variety. By solving the two simultaneous equations for x_1 and x_2, we conclude that $L \cap M$ consists of those vectors $(x_1, \ldots, x_4)$ such that

$$x_1 = 3 - \frac{1}{3}(x_3 + x_4), \qquad x_2 = 1 - \frac{1}{3}(4x_4 - 5x_3)$$

If we write $x_3 = 3s$, $x_4 = 3t$, then

$$\begin{aligned}(x_1, x_2, x_3, x_4) &= (3 - s - t, 1 - 4t + 5s, 3s, 3t)\\ &= (3, 1, 0, 0) + s(-1, 5, 3, 0) + t(-1, -4, 0, 3)\end{aligned}$$

so that $(3, 1, 0, 0)$ is a leader and

$$L \cap M = \{(3, 1, 0, 0) + \{s(-1, 5, 3, 0) + t(-1, -4, 0, 3)\}\}$$

By induction one can generalize Theorem 7 and prove: if $L_1, \ldots, L_k$ are linear varieties of a vector space V, then either $L_1 \cap \cdots \cap L_k$ is empty or else it is a linear variety whose base space is the intersection of the base spaces for $L_1, \ldots, L_k$. As an immediate consequence, we obtain the theorem:

THEOREM 8. *If $a_{11}, \ldots, a_{1k}, a_{21}, \ldots, a_{2k}, \ldots, a_{m1}, \ldots, a_{mk}, b_1, \ldots, b_m$ are real numbers, then the set X of vectors $(x_1, \ldots, x_k)$ in V_k such that*

$$a_{11}x_1 + \cdots + a_{1k}x_k = b_1, \quad \ldots, \quad a_{m1}x_1 + \cdots + a_{mk}x_k = b_m$$

is either empty or is a linear variety of V_k.

Remark. We must allow for the possibility that X is empty since we could have all the $a_{ij} = 0$ and some $b_j \neq 0$; then no $(x_1, \ldots, x_k)$ would satisfy the equations.

The proof is left as an exercise (Problem 5).

PROBLEMS

1. Represent each of the following as a linear variety, specifying a leader, and describing the base space. Throughout, $-\infty < t < \infty$, $-\infty < s < \infty$.
   (a) In V_2: all (x, y) such that $x = 1 + 3t$, $y = 2 - 5t$.
   (b) In V_2: all (x, y) such that $x = 3 - 5t$, $y = 7 - 2t$.
   (c) In V_2: all (x, y) such that $3x - 2y = 6$.
   (d) In V_2: all (x, y) such that $2x + y = 5$.
   (e) In V_3: all (x, y, z) such that $x = 2 - t$, $y = 1 + t$, $z = 3 + t$.
   (f) In V_3: all (x, y, z) such that $x = 7 + t$, $y = 2 - 2t$, $z = 5 + 4t$.
   (g) In V_3: all (x, y, z) such that $x + y + z = 3$.
   (h) In V_3: all (x, y, z) such that $2x + 3y + 4z = 12$.
   (i) In V_3: all (x, y, z) such that $x = 2 + 5t + 2s$, $y = 3 - 2t + 4s$, $z = 4 + t - s$.
   (j) In V_3: all (x, y, z) such that $x = 1 + 7t$, $y = 2 - 5s$, $z = 3 + t + s$.
   (k) In V_4: all (x_1, x_2, x_3, x_4) such that $x_1 = 1 + t$, $x_2 = 2 - t$, $x_3 = 5 + t$, $x_4 = 2 + 2t$.
   (l) In V_4: all (x_1, x_2, x_3, x_4) such that $x_1 = 3 - t$, $x_2 = 4 + t$, $x_3 = 5 + t$, $x_4 = 2 + t$.
   (m) In V_2: all (x, y) such that $x = 1 + t + s$, $y = 2 - 3t + 2s$.
   (n) In V_3: all (x, y, z) such that $x = 1$, $y = 2$, $z = 3$.
   (o) In V_3: all (x, y, z) such that $x = 2 + 2s$, $y = 5 + 4t$, $z = 5$.
   (p) In V_4: all (x_1, x_2, x_3, x_4) such that $x_1 = 2 + 4s - 5t$, $x_2 = 3 + t - s$, $x_3 = s - t$, $x_4 = s$.
   (q) In V_4: all (x_1, x_2, x_3, x_4) such that $x_1 = 1 + 3s$, $x_2 = 1 + 5t$, $x_3 = s$, $x_4 = t$.
   (r) In V_4: all (x_1, x_2, x_3, x_4) such that $x_1 + x_2 + x_3 - x_4 = 0$.
   (s) In V_4: all (x_1, x_2, x_3, x_4) such that $2x_1 + x_2 + x_3 + x_4 = 4$.
2. Decide which of the following are linear varieties of $\mathcal{P}$.
   (a) The set X of polynomials $A(x)$ such that $A(0) = 1$, $A'(0) = 0$.
   (b) The set Y of polynomials $B(x)$ such that $B(1) = B(2) = 3$.
   (c) The set Z of polynomials with nonzero constant term.
3. Decide which of the following are linear varieties of the space $\mathcal{C}$ of real functions continuous for all x.
   (a) The set F of functions f such that $f' = 3x^2 - 1$.
   (b) The set G of functions g such that $g(1) = g(-1) = 2$.

(c) The set H of functions h such that $h(0) = h(1) = h(2)$.

(d) The set K of functions k such that $k(n) = n$, for all integers n.

(e) The set L of functions p which are 0 outside an interval $|x| < B_p$, where B_p varies with the function p.

4. Decide which of the following are linear varieties of $\mathcal{I}$, the space of infinite sequences of real numbers.

(a) The set S of infinite sequences which converge to -1.

(b) The set T of infinite sequences which have $+\infty$ as a limit.

(c) The set R of infinite sequences which converge.

(d) The set J of infinite sequences having all but finitely many terms equal to 1.

5. (a) Prove the generalization of Theorem 2 stated in the paragraph preceding the statement of Theorem 8.

(b) Prove Theorem 8.

(c) Give an example with $m = k = 2$, where the a_{ij} are not 0 and the set X of Theorem 8 is empty.

6. Prove the Corollary to Theorem 6.

9-6 SPAN OF A SET

Let V be a vector space. A vector $\mathbf{v}$ in V is said to be a *linear combination* of a finite set of vectors $\mathbf{u}_1, \ldots, \mathbf{u}_k$ in V, if there exist scalars $a_1, \ldots, a_k$ (possibly all 0) such that

$$\mathbf{v} = a_1\mathbf{u}_1 + \cdots + a_k\mathbf{u}_k \tag{9-60}$$

For example, each polynomial of degree 5 is a linear combination of $1, x, x^2, x^3, x^4, x^5$. Each vector (a, b) in V_2 is a linear combination of $\mathbf{i}, \mathbf{j}$; we can write: $(a, b) = a\mathbf{i} + b\mathbf{j}$. We call the scalars $a_1, \ldots, a_k$ in a linear combination (9-60) the *coefficients of the linear combination*.

The set of *all* linear combinations of $\mathbf{u}_1, \ldots, \mathbf{u}_k$ is called the *span* of $\mathbf{u}_1, \ldots, \mathbf{u}_k$ and this set is denoted by Span $(\mathbf{u}_1, \ldots, \mathbf{u}_k)$. For each $\mathbf{u}$, Span $(\mathbf{u}) = \{t\mathbf{u}\}$ and Span $(\mathbf{0}) = V_0$. Also Span $(\mathbf{i}, \mathbf{j}) = V_2$, Span $(\mathbf{i} - \mathbf{j}, \mathbf{i} + \mathbf{j}) = V_2$ and, in general, if $\mathbf{u}, \mathbf{v}$ are linearly independent vectors in the plane, then Span $(\mathbf{u}, \mathbf{v}) = V_2$.

If A denotes the set of vectors $\{\mathbf{u}_1, \ldots, \mathbf{u}_k\}$, we shall also write Span A for Span $(\mathbf{u}_1, \ldots, \mathbf{u}_k)$.

The span of a set $\mathbf{u}_1, \ldots, \mathbf{u}_k$ is always a subspace of V. For the sum of two linear combinations is a linear combination:

$$(a_1\mathbf{u}_1 + \cdots + a_k\mathbf{u}_k) + (b_1\mathbf{u}_1 + \cdots + b_k\mathbf{u}_k) = (a_1 + b_1)\mathbf{u}_1 + \cdots + (a_k + b_k)\mathbf{u}_k$$

and each scalar multiple of a linear combination is a linear combination:

$$c(a_1\mathbf{u}_1 + \cdots + a_k\mathbf{u}_k) = (ca_1)\mathbf{u}_1 + \cdots + (ca_k)\mathbf{u}_k$$

Furthermore, if W is a subspace of V containing $\mathbf{u}_1, \ldots, \mathbf{u}_k$, then W contains all linear combinations of these vectors; that is, W contains Span $(\mathbf{u}_1, \ldots, \mathbf{u}_k)$. Thus we can say: Span $(\mathbf{u}_1, \ldots, \mathbf{u}_k)$ is the smallest subspace of V containing $\mathbf{u}_1, \ldots, \mathbf{u}_k$.

If W is a subspace of V and A is a subset of V such that Span $A = W$, we say that A *spans* W. Different subsets may span the same subspace W. Thus in V_2 each of the sets $\{(1, 0), (0, 1)\}$, $\{(2, 3), (1, -1)\}$, $\{(2, 4), (1, 1), (-1, 2)\}$ spans V_2.

Remark. We can also speak of the span of an infinite subset A of V. In that case, Span A is the set of *all* linear combinations of *all* finite subsets of A. Just as for finite sets, we can show that Span A is the smallest subspace of V containing the set A. The span of the set $\{x, x^2, \ldots, x^m, \ldots\}$ in the vector space $\mathcal{P}$ of all polynomials is the subspace consisting of all polynomials without constant term.

9-7 BASES, LINEAR INDEPENDENCE

Inclusion of sets. If a set B is contained in a set A, we say that B is a *subset* of A and write $B \subset A$. It may happen that B is A itself. When this is not true—that is, when B is contained in A, but B is not all of A, we shall say that B is a *proper subset* of A. For example, $\{1, 2\}$ is a proper subset of $\{1, 2, 3\}$; the rational numbers are a proper subset of the real numbers. A subspace of a vector space V which is a proper subset of V is called a *proper subspace* of V.

A *basis* for a vector space V is a set A such that Span $A = V$ and such that no proper subset of A spans V; that is, if B is contained in A but is not all of A, then Span $B \neq V$.

The sets $\{(1, 0), (0, 1)\}$ and $\{(2, 3), (1, 1)\}$ are each a basis for V_2; for they clearly span V_2 and no proper subset can span V_2, since otherwise V_2 would consist of the scalar multiples of one vector. The set $\{(2, 4), (1, 1), (-1, 2)\}$ also spans V_2, but it is not a basis for V_2, since the subset $\{(2, 4), (1, 1)\}$ spans V_2.

In V_n let $\mathbf{e}_1 = (1, 0, \ldots, 0)$, $\mathbf{e}_2 = (0, 1, 0, \ldots, 0), \ldots, \mathbf{e}_n = (0, \ldots, 0, 1)$, so that $\mathbf{e}_i$ is the ordered n-tuple with 1 in the ith coordinate and 0 elsewhere. Then $\mathbf{e}_1, \ldots, \mathbf{e}_n$ is a basis for V_n. For clearly $(a_1, \ldots, a_n) = a_1\mathbf{e}_1 + \cdots + a_n\mathbf{e}_n$, so that this set does span V_n. Also, we cannot express any one of $\mathbf{e}_1, \ldots, \mathbf{e}_n$ as a linear combination of the others and, therefore, no proper subset of $\mathbf{e}_1, \ldots, \mathbf{e}_n$ spans V_n. We call $\mathbf{e}_1, \ldots, \mathbf{e}_n$ *the standard basis of* V_n.

The infinite set $\{1, x, x^2, \ldots, x^m, \ldots\}$ spans the vector space $\mathcal{P}$ of all polynomials, and no proper subset can. For, if some proper subset B spanned $\mathcal{P}$, there would be some integer n such that x^n is not in B; but x^n is in $\mathcal{P}$ and, hence, for some m

$$x^n = a_0 1 + a_1 x + \cdots + a_{n-1}x^{n-1} + a_{n+1}x^{n+1} + \cdots + a_m x^m$$

This implies that the polynomial

$$a_0 + a_1x + \cdots + a_{n-1}x^{n-1} - x^n + a_{n+1}x^{n+1} + \cdots + a_mx^m$$

is the zero polynomial, so that all coefficients must be 0; but one coefficient is -1, and we have a contradiction.

In Section 4-11 we saw that the set J of proper rational functions with given denominator $g(x)$ was a vector space. If $g(x) = (x - 1)^2(x - 2)^3$, we saw that

$$\{1/g, x/g, x^2/g, x^3/g, x^4/g\}$$

and

$$\{1/(x - 1), 1/(x - 1)^2, 1/(x - 2), 1/(x - 2)^2, 1/(x - 2)^3\}$$

were each a basis for J.

A finite set $\{\mathbf{u}_1, \ldots, \mathbf{u}_k\}$ of vectors of the vector space V is said to be *linearly dependent*, if there exist scalars $a_1, \ldots, a_k$, not all 0, such that

$$\mathbf{0} = a_1\mathbf{u}_1 + \cdots + a_k\mathbf{u}_k$$

If one of the $\mathbf{u}_i$, say $\mathbf{u}_1$, is the zero vector $\mathbf{0}$, the set $\mathbf{u}_1, \ldots, \mathbf{u}_k$ must be linearly dependent, since $\mathbf{0} = 1 \cdot \mathbf{0} + 0 \cdot \mathbf{u}_2 + \cdots + 0 \cdot \mathbf{u}_k$. If one of the $\mathbf{u}_i$ is a linear combination of the others, then the set $\{\mathbf{u}_1, \ldots, \mathbf{u}_k\}$ is linearly dependent. For if, for example,

$$\mathbf{u}_1 = a_2\mathbf{u}_2 + \cdots + a_k\mathbf{u}_k$$

then $\mathbf{0} = (-1)\mathbf{u}_1 + a_2\mathbf{u}_2 + \cdots + a_k\mathbf{u}_k$, and here the coefficient of $\mathbf{u}_1$ is not 0. The converse is also true (Problem 7 below). Thus we can state: *the set* $\{\mathbf{u}_1, \ldots, \mathbf{u}_k\}$ *is linearly dependent if, and only if, one of the vectors in the set is a linear combination of the others.*

We can also speak of linearly dependent infinite sets. In general, a set A of a vector space is linearly dependent exactly when one vector in the set is a linear combination of a finite number of the other vectors.

A subset of a vector space is said to be *linearly independent* if it is not linearly dependent. Thus a set $\{\mathbf{u}_1, \ldots, \mathbf{u}_k\}$ is linearly independent if, whenever $\mathbf{0}$ is expressed as a linear combination of $\mathbf{u}_1, \ldots, \mathbf{u}_k$, *all* the coefficients are 0. We saw in Chapter 1 that $\{\mathbf{i}, \mathbf{j}\}$ is a linearly independent set in the space of vectors in the plane, whereas any set of three vectors from this space is linearly dependent. The infinite set $\{1, x, \ldots, x^n, \ldots\}$ is a linearly independent set in the vector space $\mathcal{P}$ of all polynomials, since $a_0 1 + a_1 x + \cdots + a_m x^m$ is the zero polynomial exactly when all a_i are 0.

It should be noted that the linear dependence or linear independence of a subset depends on the scalars used. For example: when the set C of complex numbers is considered to be a complex vector space, any two elements of C are linearly dependent (for complex scalars). However, C can be considered to be a real vector space (Problem 13 below), and in this space the set $\{1, i\}$ is a linearly independent set (for real scalars). Usually, it is clear from the context whether we are allowing the scalars to be complex numbers or only real numbers. However, if confusion is possible, we shall speak of "linear independence over the reals" and of "linear independence over the complex numbers."

THEOREM 9. *Let A be a subset of vector space V and let A contain two or more elements. Then A is linear dependent if, and only if, there is a proper subset B of A such that* Span A = Span B.

PROOF. Suppose such a subset B exists. Then there must be a vector $\mathbf{v}$ in A but not in B. Now $\mathbf{v}$ is in Span A and, since Span A = Span B, $\mathbf{v}$ is also in Span B. Thus

$$\mathbf{v} = a_1\mathbf{u}_1 + \cdots + a_k\mathbf{u}_k$$

where $\mathbf{u}_1, \ldots, \mathbf{u}_k$ are in B. Consequently, the $\mathbf{u}_i$'s are in A and are different from $\mathbf{v}$. But then

$$a_1\mathbf{u}_1 + \cdots + a_k\mathbf{u}_k + (-1)\mathbf{v} = \mathbf{0}$$

and we conclude that A is linearly dependent.

Next let A be a linearly dependent set. Then there exist vectors $\mathbf{v}_1, \ldots, \mathbf{v}_s$ in A such that

$$a_1\mathbf{v}_1 + \cdots + a_s\mathbf{v}_s = \mathbf{0}$$

with not all coefficients 0, say $a_1 \neq 0$. We can then express $\mathbf{v}_1$ as a linear combination of $\mathbf{v}_2, \ldots, \mathbf{v}_s$. Therefore, every linear combination of elements in A can be expressed as such a linear combination without using $\mathbf{v}_1$. Accordingly, if we let B be A with $\mathbf{v}_1$ removed, then Span B = Span A, and B is a proper subset of A. Thus the theorem is proved.

COROLLARY. *A basis for a vector space W is a linearly independent subset of W which spans W.*

The proof is left to the reader.

We saw in Chapter 1 that any set of two linearly independent vectors in the plane was a basis for the space of vectors in the plane. The following example extends this to vectors in space.

EXAMPLE 1. Each linearly independent set of three vectors from V_3 is a basis for V_3.

An algebraic proof is given in the next section. Here we remark that, if $\mathbf{u} = \overrightarrow{OP}$, $\mathbf{v} = \overrightarrow{OQ}$, $\mathbf{w} = \overrightarrow{OR}$ are linearly independent, then O, P, Q, R do not lie in a plane and, hence, each vector $\mathbf{z} = \overrightarrow{OS}$ can be expressed as a linear combination of $\mathbf{u}, \mathbf{v}, \mathbf{w}$, as suggested in Figure 9-13.

EXAMPLE 2. The set $\{1, e^x, e^{2x}\}$ is a linearly independent set in $\mathcal{C}^{(2)}(-\infty, \infty)$.

PROOF. Suppose that the set $\{1, e^x, e^{2x}\}$ is linearly dependent, then

$$a + be^x + ce^{2x} = 0$$

where one, at least, of a, b, c is nonzero. On differentiating twice in succession, we find that

$$be^x + 2ce^{2x} = 0 \qquad \text{and} \qquad be^x + 4ce^{2x} = 0$$

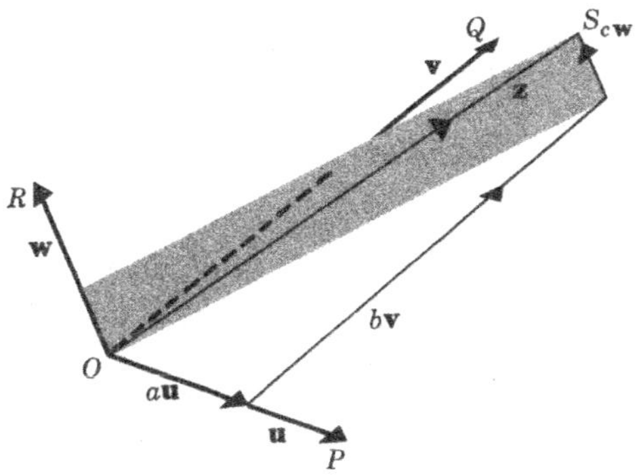

Figure 9-13 Basis for V_3.

On setting $x = 0$, in the last three equations, we obtain the following equations:

$$a + b + c = 0, \qquad b + 2c = 0, \qquad b + 4c = 0$$

Clearly, the last two equations imply $c = 0$ and $b = 0$, and then the first would imply $a = 0$, contrary to the assumption that one, at least, of a, b, c is nonzero.

One could also show that $\{1, e^x, e^{2x}\}$ is a linearly independent set, by observing that if $a + be^x + ce^{2x}$ is the zero function, then on taking $x = 0$, 1, 2 we obtain:

$$a + b + c = 0, \qquad a + be + ce^2 = 0, \qquad a + be^2 + ce^4 = 0$$

But this system of linear equations has $a = b = c = 0$ as its only solution.

EXAMPLE 3. The set $\{(2, 1, 3), (-1, 4, 1), (8, -5, 7)\}$ is a linearly dependent subset of V_3.

PROOF. We need to find a, b, c (not all 0) so that

$$a(2, 1, 3) + b(-1, 4, 1) + c(8, -5, 7) = (0, 0, 0)$$

This is equivalent to solving the system:

$$2a - b + 8c = 0, \qquad a + 4b - 5c = 0, \qquad 3a + b + 7c = 0$$

We solve by eliminating a from the first two equations to obtain: $-9b + 18c = 0$. Hence, $b = 2c$, $a = -3c$ is a solution. We take $c = 1$ and have

$$-3(2, 1, 3) + 2(-1, 4, 1) + (8, -5, 7) = (0, 0, 0)$$

Remark. Observe that the problem of determining linear dependence or linear independence in practice reduces to solving systems of linear equations.

EXAMPLE 4. Let W be the subset of functions g in $\mathcal{C}^{(2)}(-\infty, \infty)$ such that $g'' = 0$. Then W is a subspace of $\mathcal{C}^{(2)}(-\infty, \infty)$ and $\{1, x\}$ is a basis for W.

We leave for the reader to verify that W is a subspace and that $\{1, x\}$ is a linearly independent subset of $\mathcal{C}^{(2)}(-\infty, \infty)$. By integrating twice, we find that g in W is necessarily of the form $cx + k$. Hence, $\{1, x\}$ is a linearly independent set spanning W and, therefore, $\{1, x\}$ is a basis for W.

THEOREM 10. *Let $\{\mathbf{u}_1, \ldots, \mathbf{u}_n\}$ be a basis for a vector space V. Then each vector in V is expressible uniquely as a linear combination of $\mathbf{u}_1, \ldots, \mathbf{u}_n$. Conversely, if $\mathbf{u}_1, \ldots, \mathbf{u}_n$ are vectors in V and each vector in V is expressible uniquely as a linear combination of $\mathbf{u}_1, \ldots, \mathbf{u}_n$ then $\{\mathbf{u}_1, \ldots, \mathbf{u}_n\}$ is a basis for V.*

PROOF. If $\{\mathbf{u}_1, \ldots, \mathbf{u}_n\}$ is a basis for V and $\mathbf{u}$ is in V, then $\mathbf{u} = a_1\mathbf{u}_1 + \cdots + a_n\mathbf{u}_n$. If also, $\mathbf{u} = b_1\mathbf{u}_1 + \cdots + b_n\mathbf{u}_n$, then $(a_1 - b_1)\mathbf{u}_1 + \cdots + (a_n - b_n)\mathbf{u}_n = \mathbf{0}$. Since $\{\mathbf{u}_1, \ldots, \mathbf{u}_n\}$ is a basis, it is a linearly independent set and, therefore, $a_1 - b_1 = 0, \ldots, a_n - b_n = 0$, so that $a_1 = b_1, \ldots, a_n = b_n$. Thus each $\mathbf{u}$ in V is expressible *uniquely* as a linear combination of the $\mathbf{u}_i$.

Conversely, if each $\mathbf{u}$ in V is expressible uniquely as a linear combination of $\mathbf{u}_1, \ldots, \mathbf{u}_n$, then Span $\{\mathbf{u}_1, \ldots, \mathbf{u}_n\} = V$. Furthermore, $\{\mathbf{u}_1, \ldots, \mathbf{u}_n\}$ is a linearly independent set, since otherwise some $\mathbf{u}_i$ is expressible both as $\mathbf{u}_i$ itself and also as a linear combination of the other $\mathbf{u}_j$, contrary to the uniqueness assumption. Hence, $\{\mathbf{u}_1, \ldots, \mathbf{u}_n\}$ is a linearly independent spanning set for V and so is a basis for V.

Theorem 10 implies that two linear combinations of vectors in a basis give the same vector if, and only if, the coefficient of each basis vector is the same in both expressions; that is, if $\{\mathbf{u}_1, \ldots, \mathbf{u}_n\}$ is a basis for V and

$$a_1\mathbf{u}_1 + \cdots + a_n\mathbf{u}_n = b_1\mathbf{u}_1 + \cdots + b_n\mathbf{u}_n$$

then $a_1 = b_1, a_2 = b_2, \ldots, a_n = b_n$. This is the much used method of *comparing coefficients.*

PROBLEMS

1. Show that the following are linearly independent subsets of the vector space $\mathcal{P}$ of all polynomials.
   (a) $\{1, x - 1, x^2 - x, x^3 - x^2\}$
   (b) $\{1, 1 + x, 1 + x + x^2, 1 + x + x^2 + x^3\}$
   (c) $\{x, x^2 - x, x^3 - x\}$
2. Decide whether the following subsets of $\mathcal{C}(0, \infty)$ are linearly independent:
   (a) $\{\sin x, \cos x, 1\}$ (b) $\{\sin x^2, \sin x, x\}$
   (c) $\{\sin x, \sin 2x, \sin 3x\}$ (d) $\{\sin x, \sin (x + 1), \cos x\}$
   (e) $\{e^x, xe^x, x^2e^x\}$ (f) $\{\cos x, \cos^2 x, \cos^3 x\}$
   (g) $\{\ln x, \ln x^2, \ln x^3\}$ (h) $\{e^x, \ln x, x\}$
3. Which of the following subsets of V_3 are linearly independent?
   (a) $\{(1, 1, 0), (1, 0, 1), (0, 1, 1)\}$
   (b) $\{(1, 2, 3), (3, 4, 5), (5, 6, 7)\}$
   (c) $\{(1, 2, 3), (3, 2, 1), (1, 1, 1), (2, 3, 1)\}$

(d) $\{(1, 2, 3), (3, 2, 1), (-7, -2, 1)\}$
(e) $\{(1, 1, 2), (2, 3, -1), (-1, -6, 9)\}$

4. (a) . . . (h) Find a basis for each of the spaces spanned by the sets in Problem 2.

5. (a) . . . (e) Proceed as in Problem 4 for the vector spaces spanned by the sets in Problem 3.

6. (a) Prove: if $a \neq 0$, then Span $(\mathbf{u}_1, \ldots, \mathbf{u}_k) =$ Span $(a\mathbf{u}_1, \mathbf{u}_2, \ldots, \mathbf{u}_k)$.
(b) Prove: if $a \neq 0$, then Span $(\mathbf{u}_1, \ldots, \mathbf{u}_k) =$ Span $(\mathbf{u}_1 + a\mathbf{u}_2, \mathbf{u}_2, \ldots, \mathbf{u}_k)$.

7. Prove: if $\{\mathbf{u}_1, \ldots, \mathbf{u}_k\}$ is a linearly dependent set in vector space V, then one of the $\mathbf{u}_i$ is expressible as a linear combination of the others.

8. Let A, B be subsets of a vector space V. Prove:

(a) If A is a linearly dependent set and A is contained in B, then B is a linearly dependent set.

(b) If B is a linearly independent set and A is contained in B, then A is a linearly independent set.

9. Which of the following subsets of V_3^c are linearly independent over the complex numbers?

(a) $\{(i, 1, 0), (1 + i, 2, 0), (3, 1, 0)\}$
(b) $\{(i, 1, 0), (0, 1, i), (0, i, i)\}$
(c) $\{(i, 1, 0), (2 + i, 3i, 5 - i), (2, 4 + 4i, 4 - 6i)\}$
(d) $\{(1, 0, 0), (i, 0, 0), (0, 1, 0)\}$
(e) $\{(i, 1, 3), (6 + i, 7 + 4i, 1 - i), (-6 + i, -5 - 4i, 7 + i)\}$

10. Justify on the basis of Theorem 10:

(a) If $ax^3 + bx^2 + cx + d \equiv 5x^3 - 2x^2 + 3x + 5$, where a, b, c, d are real numbers, then $a = 5$, $b = -2$, $c = 3$, $d = 5$.

(b) If $(2a - 3b) \cos x + (5a + b) \sin x \equiv \cos x + 11 \sin x$, then $a = 2$, $b = 1$.

(c) If $a(x - 1)^{-1} + b(x + 2)^{-1} \equiv (x + 8)(x^2 + x - 2)^{-1}$, then $a = 3$, $b = -2$.

11. Let $\mathcal{C}^{(2),c}$ be the space of complex-valued functions f defined on $(-\infty, \infty)$ such that f'' is continuous on $(-\infty, \infty)$ [see Section 5-8].

(a) Show that $\mathcal{C}^{(2),c}$ is a complex vector space.

(b) Show that the functions f in $\mathcal{C}^{(2),c}$ such that $f'' + f = 0$ form a subspace W.

12. (a) Prove: if W is a subspace of V and A is a linearly independent subset of W, then A is a linearly independent subset of V.

(b) Prove: if W is a subspace of V and A is a basis for W, then A is a linearly independent subset of V.

(c) Prove: if W is a subspace of V and A is a subset of W which is a linearly independent subset of V, then A is a linearly independent subset of W.

13. Let C be the set of all complex numbers, with the usual operations. Show that C can be regarded as a *real* vector space (only real scalars allowed) and that $\{2 + 3i, 4 + 5i\}$ are linearly independent. Show that if, instead, C is regarded as a complex vector space (complex scalars allowed), then $\{2 + 3i, 4 + 5i\}$ is a linearly dependent set. *Remark.* As a real vector space, C is essentially the same as V_2.

9-8 DIMENSION

We have seen that a vector space may have more than one basis. It is therefore natural to ask what property the different bases have in common. Analysis of the examples discussed in the previous section suggests that about the only possible common property is that of having the same number of elements. This is, in fact, always the case, as we show in Theorem 15 in Section 9-10: *all bases of a vector space V have the same number of elements.* We also note that the zero-space V_0 has *no basis,* for V_0 contains only the one vector **0**, and so contains no linearly independent subset. It can be shown that all other vector spaces do have bases, although the bases may be infinite sets. These facts enable us to assign a number called the *dimension* to each vector space. We shall denote the dimension of vector space V by dim V. The definition of dimension is as follows:

$$\dim V = \begin{cases} 0, & \text{if } V = V_0 \\ n, & \text{if } V \text{ has a finite basis of } n \text{ vectors} \\ \infty, & \text{otherwise} \end{cases}$$

Accordingly, if $\dim V = 5$, then each basis of V contains exactly 5 vectors; if $\dim V = \infty$, then no finite subset of V can be a basis for V. If $V \neq V_0$, then dim V is a positive integer or ∞.

EXAMPLE 1.

(a) $\dim V_2 = 2$, since $\mathbf{i}, \mathbf{j}$ is a basis for V_2.

(b) $\dim V_n = n$, since $\{\mathbf{e}_1, \ldots, \mathbf{e}_n\}$ is a basis for V_n.

(c) If $\mathcal{P}_5$ is the vector space of polynomials of degree at most 5, then $1, x, \ldots, x^5$ is a basis and $\dim \mathcal{P}_5 = 6$.

(d) If J is the vector space of proper rational functions with denominator $g(x) = (x-1)^2(x-2)^3$, then $\dim J = 5$, since $\{1/g(x), x/g(x), x^2/g(x), x^3/g(x), x^4/g(x)\}$ is a basis for J.

(e) If $\mathcal{P}$ is the vector space of all polynomials, then $\dim \mathcal{P} = \infty$, since $\{1, x, \ldots, x^m, \ldots\}$ is a basis for $\mathcal{P}$.

We saw in Sections 4-11 and 7-17 how the dimension of a vector space is a useful tool for the study of the calculus. Chapters 12 and 14 provide many more illustrations of this.

We point out that the dimension of a space depends on the number system used as scalars. The dimension of C, the set of complex numbers, as a complex vector space is 1. But C can also be considered as a real vector space (Problem 13, following Section 9-7) and then has dimension 2. We shall point out the few cases in which this question arises.

EXAMPLE 2. If a_1, a_2, a_3 are real numbers, if $a_3 \neq 0$, and if W is the set of all vectors (x_1, x_2, x_3) of V_3, such that $a_1x_1 + a_2x_2 + a_3x_3 = 0$, then W is a subspace of V_3 and $\dim W = 2$.

PROOF. As in Example 5 in Section 9-2, we see that W is a subspace of V_3. Let $b_1 = -a_1/a_3$, $b_2 = -a_2/a_3$. Then W can be described as the set of all (x_1, x_2, x_3) such that $b_1x_1 + b_2x_2 - x_3 = 0$. We now observe that $\eta_1 =$

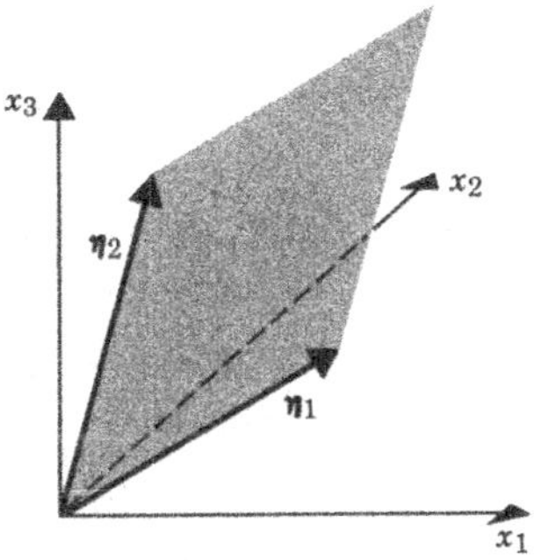

Figure 9-14 Two-dimensional vector space of Example 2.

$(1, 0, b_1)$ and $\eta_2 = (0, 1, b_2)$ are in W. Also η_1, η_2 are linearly independent, since $c_1\eta_1 + c_2\eta_2 = \mathbf{0}$ implies $(c_1, c_2, b_1c_1 + b_2c_2) = \mathbf{0}$ and, hence, implies $c_1 = 0$ and $c_2 = 0$. If $\mathbf{x} = (x_1, x_2, x_3)$ is in W, then $x_3 = b_1x_1 + b_2x_2$, so that

$$\mathbf{x} = (x_1, x_2, b_1x_1 + b_2x_2) = x_1\eta_1 + x_2\eta_2$$

Hence, Span $(\eta_1, \eta_2) = W$. Since η_1, η_2 is a linearly independent set, we conclude that it forms a basis for W.

In this example, W can be interpreted as a plane through the origin, as in Figure 9-14.

Remark 1. In the same way we show that if $a_n \neq 0$ (or if any one $a_i \neq 0$), then $a_1x_1 + \cdots + a_nx_n = 0$ describes an $(n-1)$-dimensional subspace W of V_n. For $a_n \neq 0$, a basis for W is given by $\eta_1 = (1, 0, \ldots, 0, b_1)$, $\eta_2 = (0, 1, 0, \ldots, 0, b_2), \ldots, \eta_{n-1} = (0, \ldots, 0, 1, b_{n-1})$, where $b_1 = -a_1/a_n, \ldots, b_{n-1} = -a_{n-1}/a_n$.

The meaning of dimension of a vector space is shown in the next theorem, which asserts that two vector spaces having the same finite dimension are algebraically alike.

THEOREM 11. *Let $\mathbf{u}_1, \ldots, \mathbf{u}_n$ be a basis for the vector space V, so that each vector $\mathbf{v}$ of V can be represented uniquely as $\mathbf{v} = x_1\mathbf{u}_1 + \cdots + x_n\mathbf{u}_n$. To each vector $\mathbf{v}$ assign the corresponding n-tuple $(x_1, \ldots, x_n)$ and regard $(x_1, \ldots, x_n)$ as a vector of V_n. This assignment is a one-to-one correspondence between V and V_n. If $\mathbf{v}$ corresponds to $(x_1, \ldots, x_n)$ and $\mathbf{w}$ to $(y_1, \ldots, y_n)$, then $\mathbf{v} + \mathbf{w}$ corresponds to $(x_1 + y_1, \ldots, x_n + y_n)$, $c\mathbf{v}$ to $(cx_1, \ldots, cx_n)$.*

PROOF. The correspondence is one-to-one, since each vector $\mathbf{v}$ of V is represented uniquely as a linear combination of the basis vectors. For $\mathbf{v}$ and $\mathbf{w}$ as given,

$$\mathbf{v} = x_1\mathbf{u}_1 + \cdots + x_n\mathbf{u}_n, \qquad \mathbf{w} = y_1\mathbf{u}_1 + \cdots + y_n\mathbf{u}_n$$

so that

$$\mathbf{v} + \mathbf{w} = (x_1 + y_1)\mathbf{u}_1 + \cdots + (x_n + y_n)\mathbf{u}_n, \qquad c\mathbf{v} = (cx_1)\mathbf{u}_1 + \cdots + (cx_n)\mathbf{u}_n$$

and the conclusion follows.

Remark 2. Theorem 11 shows that every (real) n-dimensional vector space V is essentially the same as V_n. However, for each choice of a basis for V, we obtain a different correspondence between V and V_n.

Remark 3. There is an analogous theorem for complex vector spaces and V_n^c.

EXAMPLE 3. Let $\mathcal{P}_n$ be the vector space of all polynomials of degree at most n. Then $1, x, \ldots, x^n$ form a basis for $\mathcal{P}_n$ and the correspondence

$$c_0 + c_1 x + \cdots + c_n x^n \leftrightarrow (c_0, c_1, \ldots, c_n)$$

is a one-to-one correspondence between $\mathcal{P}_n$ and V_{n+1} (note the dimension!), as in Theorem 11.

9-9 DIMENSION OF SUBSPACES AND OF LINEAR VARIETIES

In this section we discuss the dimension of subspaces and linear varieties of a vector space V. We shall suppose V to have finite dimension n. We would naturally expect the dimension of a proper subspace of V to be smaller than dim V. In order to prove this, we need the following theorem, whose proof is postponed to Section 9-10:

THEOREM 12. *Let V be a vector space of dimension n. Let $\{\mathbf{v}_1, \ldots, \mathbf{v}_m\}$ be a linearly independent subset of V. Then $m \leq n$. If $m = n$, then $\{\mathbf{v}_1, \ldots, \mathbf{v}_m\}$ is a basis for V. If $m < n$, there exist vectors $\mathbf{v}_{m+1}, \ldots, \mathbf{v}_n$ in V such that $\{\mathbf{v}_1, \ldots, \mathbf{v}_m, \mathbf{v}_{m+1}, \ldots, \mathbf{v}_n\}$ is a basis for V. Thus every linearly independent subset of V is part of some basis for V.*

Theorem 12 asserts that for a finite dimensional vector space V, any given set of linearly independent vectors of V is part of some basis for V. Actually, there are many such bases (see the next example). The basis $\{\mathbf{v}_1, \ldots, \mathbf{v}_m, \ldots, \mathbf{v}_n\}$ for V of Theorem 12 is said to be an *extension* of the linearly independent set $\{\mathbf{v}_1, \ldots, \mathbf{v}_m\}$ to a basis for V. A single nonzero vector $\mathbf{v}$ constitutes a linear independent subset of V and, hence, Theorem 12 asserts that each nonzero vector $\mathbf{v}$ appears in some basis for V. Among other things, Theorem 12 shows that there are many different bases for V, and that we have a certain amount of freedom in choosing a basis for V.

Theorem 12 also tells us that each set of k linearly independent vectors is a basis for V_k (as in Figure 9-13 for V_3).

EXAMPLE 1. Let W be the set of vectors (x_1, x_2, x_3, x_4) in V_4 such that $x_1 + x_2 + x_3 + x_4 = 0$. Since $(3, -1, -1, -1) = \boldsymbol{\eta}$ is in W, there is a basis for W containing $\boldsymbol{\eta}$. For example $\{\boldsymbol{\eta}, \mathbf{e}_1 - \mathbf{e}_2, \mathbf{e}_1 - \mathbf{e}_3\}$ is such a basis. The reader should verify that the set is linearly independent. It would then follow from Theorem 12 that this set is a basis for W. Also, since $\mathbf{e}_1 - \mathbf{e}_2$ and $\mathbf{e}_3 - \mathbf{e}_4$ are linearly independent vectors in W, there is a third vector $\mathbf{w}$ such that $\{\mathbf{e}_1 - \mathbf{e}_2, \mathbf{e}_3 - \mathbf{e}_4, \mathbf{w}\}$ is a basis for W. Among the possible choices for $\mathbf{w}$ are the following vectors: $\boldsymbol{\eta}, \mathbf{e}_1 - \mathbf{e}_3, \mathbf{e}_2 - \mathbf{e}_3, 2\mathbf{e}_1 - \mathbf{e}_2 - \mathbf{e}_3, \ldots$.

THEOREM 13. *Let V be a vector space of finite dimension n. If U is a subspace of V, then* dim $U \leq$ dim V *and equality holds true only when* $U = V$.

PROOF. Since U is contained in V, each linearly independent subset of U is a linearly independent subset of V. By Theorem 12, V cannot contain a linearly independent set with more than n elements. Hence, each basis for U, being a linearly independent subset of U and, hence, of V, can contain at most n elements. Thus dim $U \leq$ dim V.

If dim $U =$ dim $V = n$, a basis for U would be a linearly independent subset of V containing n elements and, by Theorem 12, it must be a basis for V. Thus U and V would be spanned by the same subset and so would be the same vector space.

As a special case of Theorem 12 we can state: *if* $\{\mathbf{w}_1, \ldots, \mathbf{w}_m\}$ *is a basis for a subspace W of a finite-dimensional vector space V, then there is a basis for V which includes* $\mathbf{w}_1, \ldots, \mathbf{w}_m$. *Thus each basis of a subspace can be extended to a basis for the entire vector space.*

Remark. A vector space of dimension n has infinitely many subspaces of dimension m, if $0 < m < n$.

THEOREM 14. *If U and W are finite dimensional subspaces of a vector space V, then* dim $\{U + W\}$ + dim $(U \cap W)$ = dim U + dim W.

The proof of this theorem is given in Section 9-10.

EXAMPLE 2. Suppose V has dimension n and that each of U and W is a subspace of V of dimension $n - 1$. Suppose further that $U \neq W$. Then dim $(U \cap W) = n - 2$.

PROOF. Since $U \neq W$, one of the spaces contains a vector not in the other. Hence, $\{U + W\}$ contains one of U, W (or perhaps both) as a proper subspace. Accordingly (see Theorem 13), $n \geq$ dim $\{U + W\} > n - 1 =$ dim $U =$ dim W; whence, dim $\{U + W\} = n$. It then follows from Theorem 14 that

$$\begin{aligned} \dim (U \cap W) &= \dim U + \dim W - \dim \{U + W\} \\ &= (n - 1) + (n - 1) - n = n - 2 \end{aligned}$$

EXAMPLE 3. Let U be the set of 4-tuples $(x_1, \ldots, x_4)$ such that $x_1 - x_2 + x_3 = 0$. Let W be the set of 4-tuples $(x_1, \ldots, x_4)$ such that $x_2 + x_3 + x_4 = 0$. Then $U \cap W = Z$ is the set of 4-tuples satisfying both $x_1 - x_2 + x_3 = 0$ and $x_2 + x_3 + x_4 = 0$, and dim $Z = 2$.

PROOF. As in Remark 1 in Section 9-8, U, W are subspaces of V_4, and dim U = dim $W = 3$. Clearly, $(1, 1, 0, 0)$ is in U and not in W, so that $U \neq W$. It now follows from Example 2 that dim $Z = 2$.

EXAMPLE 4. Show that in 3-dimensional space the intersection of 2 different planes through the origin is a line containing the origin.

Solution: A plane in V_3 containing the origin is a subspace of dimension 2. Hence, by Example 2, the intersection of two different planes through the origin is a subspace of dimension 1; that is, a line containing the origin.

The *dimension of a linear variety* is defined to be the dimension of the base space of the linear variety. Thus in V_3, the dimension of a point is 0, the dimension of a line is 1, and the dimension of a plane is 2. If we have a linear variety of dimension d, then we can expect to be able to find an element of the variety satisfying d auxiliary conditions. For example, we can ask for the point on the line (1-dimensional variety) of minimal distance from the origin (1 auxiliary condition). The set of solutions of a linear differential equation of order k is a linear variety of dimension k (see Chapter 14); hence, we can usually find a solution satisfying k conditions on the value of the function and its derivatives at certain points. For this reason the dimension of a linear variety is also called the number of *degrees of freedom* of the linear variety.

Remark. When defining properties and when proving theorems about vector spaces, it is best, and usually easiest, to do so without using a particular basis. When a property is defined using a basis, it is necessary to determine whether the property is intrinsic to the vector space itself or whether it depends explicitly on a particular basis. Thus, in defining dimension of a vector space, we made very certain it did not depend on any particular basis, but was a property possessed by each basis and, hence, a property intrinsic to the vector space. However, when it comes to computing, one does find that it simplifies matters to use a particular basis. If the quantity to be computed (for example, dimension) is known to be independent of a basis, then we can be sure that, regardless of our choice of basis, we will obtain the same result. Usually, the computation can be made very easy if an appropriate basis is chosen. Properties that are intrinsic to a vector space play a role analogous to geometric invariants.

‡9-10 PROOFS OF THEOREMS ON DIMENSION

In this section we give proofs for the theorems on dimension which were stated and discussed in the two preceding sections. We begin with several lemmas.

LEMMA 1. *If* $\{\mathbf{u}_1, \mathbf{u}_2, \ldots, \mathbf{u}_m\}$ *is a finite linearly independent set of vectors in a vector space* V *and* $a_1, \ldots, a_{m-1}$ *are scalars, then*

$$\{\mathbf{u}_1 - a_1\mathbf{u}_m, \mathbf{u}_2 - a_2\mathbf{u}_m, \ldots, \mathbf{u}_{m-1} - a_{m-1}\mathbf{u}_m, \mathbf{u}_m\}$$

is a linearly independent set and the two sets span the same subspace of V.

PROOF. Suppose that

$$c_1(\mathbf{u}_1 - a_1\mathbf{u}_m) + \cdots + c_{m-1}(\mathbf{u}_{m-1} - a_{m-1}\mathbf{u}_m) + c_m\mathbf{u}_m = \mathbf{0} \quad (9\text{-}100)$$

Then

$$c_1\mathbf{u}_1 + \cdots + c_{m-1}\mathbf{u}_{m-1} + (c_m - c_1a_1 - \cdots - c_{m-1}a_{m-1})\mathbf{u}_m = \mathbf{0} \quad (9\text{-}101)$$

Since $\{\mathbf{u}_1, \ldots, \mathbf{u}_m\}$ is a linearly independent set, (9-101) implies:

$$c_1 = c_2 = \cdots = c_{m-1} = c_m - (a_1c_1 + \cdots + a_{m-1}c_{m-1}) = 0$$

and, hence, that $c_1 = \cdots = c_{m-1} = c_m = 0$. It, therefore, follows that the set $\{\mathbf{u}_1 - a_1\mathbf{u}_m, \ldots, \mathbf{u}_{m-1} - a_{m-1}\mathbf{u}_m, \mathbf{u}_m\}$ is linearly independent.

Clearly, Span $(\mathbf{u}_1, \ldots, \mathbf{u}_m)$ contains $\{\mathbf{u}_1 - a_1\mathbf{u}_m, \ldots, \mathbf{u}_{m-1} - a_{m-1}\mathbf{u}_m, \mathbf{u}_m\}$. On the other hand, for $i = 1, \ldots, m-1$, $\mathbf{u}_i = (\mathbf{u}_i - a_i\mathbf{u}_m) + a_i\mathbf{u}_m$ and, so, $\{\mathbf{u}_1, \ldots, \mathbf{u}_m\}$ is included in Span $(\mathbf{u}_1 - a_1\mathbf{u}_m, \ldots, \mathbf{u}_{m-1} - a_{m-1}\mathbf{u}_{m-1}, \mathbf{u}_m)$. It then follows that

$$\text{Span } (\mathbf{u}_1, \ldots, \mathbf{u}_m) = \text{Span } (\mathbf{u}_1 - a_1\mathbf{u}_m, \ldots, \mathbf{u}_{m-1} - a_{m-1}\mathbf{u}_m, \mathbf{u}_m)$$

LEMMA 2. *Let $\{\mathbf{v}_1, \ldots, \mathbf{v}_n\}$ be a basis for a vector space V. If $\{\mathbf{u}_1, \ldots, \mathbf{u}_m\}$ is a linearly independent set from V, then $m \leq n$.*

PROOF. The proof is by induction on n. If $n = 1$, then each vector in V is a scalar multiple of $\mathbf{v}_1$ and any two vectors in V are linearly dependent. Thus $n = 1$ implies $m \leq 1$ and, hence, that $m \leq n$.

Now suppose that the theorem is true for vector spaces having a basis with fewer than n elements. Let V, $\{\mathbf{v}_1, \ldots, \mathbf{v}_n\}$ and $\{\mathbf{u}_1, \ldots, \mathbf{u}_m\}$ be given as in the lemma. Since the $\mathbf{u}_i$ are in V, and $V = \text{Span }\{\mathbf{v}_1, \ldots, \mathbf{v}_n\}$, we have

$$\mathbf{u}_i = a_{i1}\mathbf{v}_1 + \cdots + a_{in}\mathbf{v}_n, \qquad (i = 1, 2, \ldots, m)$$

If $a_{1n} = a_{2n} = \cdots = a_{mn} = 0$, then all the $\mathbf{u}_i$ would lie in the space $W = \text{Span }\{\mathbf{v}_1, \ldots, \mathbf{v}_{n-1}\}$. As W has a basis of at most $n - 1$ elements, our induction hypothesis implies $m \leq n - 1 < n$. If some $a_{in} \neq 0$, we may, by relabeling the vectors $\mathbf{u}_i$, suppose $a_{mn} \neq 0$. Consider the vectors

$$\begin{aligned}\mathbf{w}_i &= \mathbf{u}_i - a_{in}a_{mn}^{-1}\mathbf{u}_m \\ &= (a_{i1} - a_{in}a_{mn}^{-1}a_{m1})\mathbf{v}_1 + \cdots + (a_{i,n-1} - a_{in}a_{mn}^{-1}a_{m,n-1})\mathbf{v}_{n-1}\end{aligned}$$

By Lemma 1, $\{\mathbf{w}_1, \ldots, \mathbf{w}_{m-1}, \mathbf{u}_m\}$ is a linearly independent set in V and, hence, $\{\mathbf{w}_1, \ldots, \mathbf{w}_{m-1}\}$ is a linearly independent set. Clearly, $\mathbf{w}_1, \ldots, \mathbf{w}_{m-1}$ lie in Span $\{\mathbf{v}_1, \ldots, \mathbf{v}_{n-1}\}$ and, hence, again by the induction hypotheses, we have $m - 1 \leq n - 1$; that is, $m \leq n$. This proves our lemma.

THEOREM 15. *If $\{\mathbf{u}_1, \ldots, \mathbf{u}_m\}$ and $\{\mathbf{v}_1, \ldots, \mathbf{v}_n\}$ are finite sets and each is a basis for a vector space V, then $m = n$.*

PROOF. A basis of V is necessarily a linearly independent subset of V. It follows from Lemma 2, on taking $\{\mathbf{v}_1, \ldots, \mathbf{v}_n\}$ as the basis for V and $\{\mathbf{u}_1, \ldots, \mathbf{u}_m\}$ as the linearly independent subset from V, that $m \leq n$. Similarly, on reversing the roles of $\{\mathbf{u}_1, \ldots, \mathbf{u}_m\}$ and $\{\mathbf{v}_1, \ldots, \mathbf{v}_n\}$, we find $n \leq m$. The theorem now follows.

It should be observed that Lemma 2 assures us that if a vector space V has one basis that is finite, then all bases of V are finite. Theorem 15 assures us that if a vector space V has one basis that is finite, then all bases for V contain the same number of elements. It is this last fact that we need to define dimension of a vector space.

LEMMA 3. *Let* $\{\mathbf{v}_1, \ldots, \mathbf{v}_m\}$ *be a linearly independent set of vectors from a vector space V which do not span the space V. Then there is a vector* $\mathbf{w}$ *in V such that* $\{\mathbf{w}, \mathbf{v}_1, \ldots, \mathbf{v}_m\}$ *is a linearly independent subset of V.*

PROOF. Since Span $(\mathbf{v}_1, \ldots, \mathbf{v}_m)$ is contained in, but is not equal to V, there must be a vector $\mathbf{w}$ in V and not in Span $(\mathbf{v}_1, \ldots, \mathbf{v}_m)$. Suppose that

$$a\mathbf{w} + a_1\mathbf{v}_1 + \cdots + a_m\mathbf{v}_m = \mathbf{0} \tag{9-102}$$

for some $a, a_1, \ldots, a_m$. If $a \neq 0$, it would follow that $\mathbf{w}$ is in Span $(\mathbf{v}_1, \ldots, \mathbf{v}_m)$. Hence, $a = 0$, but then (9-102) gives $\mathbf{0}$ as a linear combination of vectors from a linearly independent set. This can only happen if $a_1 = \cdots = a_m = 0$. It follows that $\{\mathbf{w}, \mathbf{v}_1, \ldots, \mathbf{v}_m\}$ is a linearly independent subset of V.

Note. The vector $\mathbf{w}$ chosen in Lemma 3 can be any vector of V not in Span $(\mathbf{v}_1, \ldots, \mathbf{v}_m)$. Hence, if $\{\mathbf{u}_1, \ldots, \mathbf{u}_n\}$ is a basis for V, then at least one of the $\mathbf{u}_i$ is not in Span $(\mathbf{v}_1, \ldots, \mathbf{v}_m)$ (for otherwise we would clearly have $V \subset$ Span $(\mathbf{v}_1, \ldots, \mathbf{v}_m)$) and thus we can choose $\mathbf{w}$ to be one of the $\mathbf{u}_i$.

PROOF OF THEOREM 12. We are given a linearly independent subset $\{\mathbf{v}_1, \ldots, \mathbf{v}_m\}$ of a vector space V of dimension n. We must show that $m \leq n$; that for $m = n$, the subset is a basis for V; that for $m < n$, there is a basis $\{\mathbf{v}_1, \ldots, \mathbf{v}_m, \mathbf{v}_{m+1}, \ldots, \mathbf{v}_n\}$ for V. It follows from Lemma 2 that $m \leq n$. If $m = n$ and $\{\mathbf{v}_1, \ldots, \mathbf{v}_m\}$ were not a basis for V, then Span $(\mathbf{v}_1, \ldots, \mathbf{v}_m)$ $\neq V$ and, by Lemma 3, V would contain a linearly independent set of $m + 1 = n + 1 > n$ vectors, contrary to Lemma 2. Hence, $m = n$ implies that $\{\mathbf{v}_1, \ldots, \mathbf{v}_m\}$ is a basis for V.

If $m < n$, then necessarily Span $(\mathbf{v}_1, \ldots, \mathbf{v}_m) \neq V$. By Lemma 3 we can then find $\mathbf{w}_1$ so that $\{\mathbf{v}_1, \ldots, \mathbf{v}_m, \mathbf{w}_1\}$ is a linearly independent set. If Span $\{\mathbf{v}_1, \ldots, \mathbf{v}_m, \mathbf{w}_1\} \neq V$, then we can find $\mathbf{w}_2$ so that $\{\mathbf{v}_1, \ldots, \mathbf{v}_m, \mathbf{w}_1, \mathbf{w}_2\}$ is a linearly independent set. By Lemma 2 and Lemma 3 we can continue this process until we have a linearly independent set of n vectors $\{\mathbf{v}_1, \ldots, \mathbf{v}_m, \mathbf{w}_1, \ldots, \mathbf{w}_{n-m}\}$ spanning V, and hence a basis for V. This completes the proof.

PROOF OF THEOREM 14. We must prove that if U and W are finite-dimensional subspaces of a vector space V, then dim $\{U + W\}$ + dim $(U \cap W)$ = dim U + dim W. Since $U \cap W$ is a subspace of the finite dimensional space U, by Theorem 13, $U \cap W$ is finite dimensional. Since $\{U + W\}$ is spanned by the union of a basis for U and a basis for W, $\{U + W\}$ is also finite dimensional.

Let $\{\mathbf{u}_1, \ldots, \mathbf{u}_s\}$ be a basis for $U \cap W$. By Theorem 12, there are vectors $\mathbf{v}_1, \ldots, \mathbf{v}_t$ in U such that $\{\mathbf{u}_1, \ldots, \mathbf{u}_s, \mathbf{v}_1, \ldots, \mathbf{v}_t\}$ is a basis for U. Similarly, there are vectors $\mathbf{w}_1, \ldots, \mathbf{w}_r$ in W such that $\{\mathbf{u}_1, \ldots, \mathbf{u}_s, \mathbf{w}_1, \ldots, \mathbf{w}_r\}$ is a basis for W. Clearly,

$$\text{Span}\ (\mathbf{u}_1, \ldots, \mathbf{u}_s, \mathbf{v}_1, \ldots, \mathbf{v}_t, \mathbf{w}_1, \ldots, \mathbf{w}_r) = \{U + W\}$$

If

$$a_1\mathbf{u}_1 + \cdots + a_s\mathbf{u}_s + b_1\mathbf{v}_1 + \cdots + b_t\mathbf{v}_t + c_1\mathbf{w}_1 + \cdots + c_r\mathbf{w}_r = \mathbf{0} \tag{9-103}$$

then

$$\begin{aligned} \mathbf{x} &= a_1\mathbf{u}_1 + \cdots + a_s\mathbf{u}_s + b_1\mathbf{v}_1 + \cdots + b_t\mathbf{v}_t \\ &= -c_1\mathbf{w}_1 - \cdots - c_r\mathbf{w}_r \end{aligned} \tag{9-104}$$

is in $U \cap W$. Hence, there exist scalars $d_1, \ldots, d_r$ so that $\mathbf{x} = d_1\mathbf{u}_1 + \cdots + d_s\mathbf{u}_s$, whence, from (9-104), we have

$$d_1\mathbf{u}_1 + \cdots + d_s\mathbf{u}_s + c_1\mathbf{w}_1 + \cdots + c_r\mathbf{w}_r = \mathbf{0}$$

But $\{\mathbf{u}_1, \ldots, \mathbf{u}_s, \mathbf{w}_1, \ldots, \mathbf{w}_r\}$ is a linearly independent set in V and, consequently, $c_1 = \cdots = c_r = 0$. But then, from (9-103), we have

$$a_1\mathbf{u}_1 + \cdots + a_s\mathbf{u}_s + b_1\mathbf{v}_1 + \cdots + b_t\mathbf{v}_t = \mathbf{0}$$

and, since $\{\mathbf{u}_1, \ldots, \mathbf{u}_s, \mathbf{v}_1, \ldots, \mathbf{v}_t\}$ is a basis for U and, therefore, a linearly independent set, we have $a_1 = \cdots = a_s = b_1 = \cdots = b_t = 0$. We have thus shown $\{\mathbf{u}_1, \ldots, \mathbf{u}_s, \mathbf{v}_1, \ldots, \mathbf{v}_t, \mathbf{w}_1, \ldots, \mathbf{w}_r\}$ to be a linearly independent set which spans $\{U + W\}$. Thus $\dim \{U + W\} = r + s + t = (r + s) + (t + s) - s = \dim U + \dim W - \dim (U \cap W)$.

Infinite-Dimensional Spaces. An infinite-dimensional vector space can have subspaces that are finite-dimensional and subspaces that are infinite-dimensional. For example, the vector space $\mathcal{P}$ of all polynomials contains the 6-dimensional subspace $\mathcal{P}_5$ and the infinite-dimensional subspace E, consisting of polynomials which are even functions: that is, polynomials such as $1 + 3x^2 - 5x^4$ which contain no term of odd degree.

PROBLEMS

1. Which of the following subsets of V_3 are bases for V_3?
   (a) (1, 1, 0), (1, 0, 1), (0, 1, 1)
   (b) (1, 1, 1), (2, 1, 3), (1, 0, 2)
   (c) (6, 7, 8), (4, 3, 2), (1, 1, 1)
   (d) (4, 3, 6), (−1, 2, 0)
   (e) (7, −1, 3), (8, 6, 1), (4, 3, 2), (0, 2, 0)
   (f) (1, 2, 3), (3, 2, 1), (1, 1, 1)
   (g) (−1, 1, −1), (−1, −1, 1), (1, −1, −1)
2. In each case you are given a vector space and a subset X of vectors. Decide if the subset X is a linearly independent set and, when possible, find a basis for the space which contains X.
   (a) V_4: $X = \{(1, 1, 1, 0), (1, 2, 3, 4)\}$.
   (b) V_4: $X = \{(1, 1, 1, 1), (1, 2, 3, 4)\}$.
   (c) $\mathcal{P}_6$: $X = \{x + x^3, x^2 + x^6, 1 + x - x^3\}$.
   (d) V = proper rational functions with denominator $D = x^3 + 2x^2 - 5x - 6$; $X = \{(x^2 + 1)/D, 1/(x + 3) = (x^2 - x - 2)/D\}$.
   (e) V as in (d), $X = \{1/(x + 3), x/D\}$.

3. Determine the dimension of the following spaces:

(a) Span ((1, 2, 3), (3, 2, 1), (7, 7, 7)) in V_3.

(b) Span ($x^2 - x + 5, x + 17, x^3$) in $\mathcal{P}$.

(c) Span ($\sin x, \sin(x + \pi), \sin \pi x$) in $\mathcal{C}^{(\infty)}(-\infty, \infty)$.

(d) Span ($x \sin x, (x + 1)\sin x, \sin(x + 1), \cos x$) in $\mathcal{C}^{(\infty)}(-\infty, \infty)$.

4. (a) Let V be the set of real polynomials of degree at most 5 having 1 as a zero. Determine a basis for V. Find dim V. Find a basis for $\mathcal{P}_5$ which extends your basis for V.

(b) Let T be the set of real polynomials of degree at most 5 whose third derivative is zero at 0. Show that T is a subspace of $\mathcal{P}_5$. Determine a basis for T and extend to a basis for $\mathcal{P}_5$.

(c) Let W be the set of real polynomials $p(x)$ of degree at most 5 such that $p'(3) = 0$. Show that W is a subspace of $\mathcal{P}_5$. Determine a basis for W and extend that basis to one for $\mathcal{P}_5$.

(d) With V, T, W as in parts (a), (b), (c), determine the following: dim $V \cap T$, dim $V \cap W$, dim $W \cap T$, dim $V \cap T \cap W$, dim $\{V + W\}$, dim $\{V + T\}$.

5. (a) Prove: the set $\{\mathbf{v}_1, \ldots, \mathbf{v}_k\}$ is a basis for V_k if, and only if, $\mathbf{e}_1, \ldots, \mathbf{e}_k$ are each expressible as a linear combination of $\mathbf{v}_1, \ldots, \mathbf{v}_k$.

(b) Prove: if $\{\mathbf{v}_1, \ldots, \mathbf{v}_n\}$ is a basis for a vector space V, then a set of n elements $\{\mathbf{u}_1, \ldots, \mathbf{u}_n\}$ from V is a basis for V if, and only if, each $\mathbf{v}_i$ is expressible as a linear combination of $\mathbf{u}_1, \ldots, \mathbf{u}_n$.

6. (a) Let $a_1, \ldots, a_n$ be real numbers and suppose $a_1 \neq 0$. Let $\boldsymbol{\eta}_i = a_i\mathbf{e}_1 - a_1\mathbf{e}_i$ ($i = 2, 3, \ldots, n$). Show that $\{\boldsymbol{\eta}_2, \ldots, \boldsymbol{\eta}_n\}$ is a basis for the space W of n-tuples $(x_1, \ldots, x_n)$ such that $a_1x_1 + \cdots + a_nx_n = 0$.

(b) Let $a_1, \ldots, a_n, b_1, \ldots, b_n$ be real numbers and suppose $a_1b_2 - a_2b_1 \neq 0$. Let

$$\xi_j = (a_jb_2 - a_2b_j)\mathbf{e}_1 + (a_1b_j - b_1a_j)\mathbf{e}_2 - (a_1b_2 - a_2b_1)\mathbf{e}_j$$

($j = 3, 4, \ldots, n$). Show that $\{\xi_3, \ldots, \xi_n\}$ is a basis for the space U of n-tuples such that

$$a_1x_1 + \cdots + a_nx_n = 0 \quad \text{and} \quad b_1x_1 + \cdots + b_nx_n = 0$$

(c) U is a subspace of W. Find a basis for W which extends the basis $\{\xi_3, \ldots, \xi_n\}$.

(d) W is a subspace of V_n. Find a basis for V_n which extends the basis you obtain in part (c).

7. (a) Let W be those vectors $(x_1, \ldots, x_5)$ in V_5 such that $x_1 - x_2 + x_3 = 0$ and $x_2 + x_4 + x_5 = 0$. Determine a basis for W. Find a basis for V_5 which extends your basis for W.

(b) Let U be those vectors $(x_1, \ldots, x_5)$ of V_5 such that $2x_1 - x_2 + x_3 - x_5 = 0$ and $x_1 + x_2 - x_4 + 6x_5 = 0$. Determine a basis for U. Find a basis for V_5 which extends your basis for U.

(c) Show that dim $U \cap W \geq 1$.

(d) Find a basis for $U \cap W$.

8. (a) Show that $\{e^x, e^{2x}, e^{3x}, \ldots\}$ is an infinite linearly independent subset of $\mathcal{C}^{(\infty)}(-\infty, \infty)$. What is dim $\mathcal{C}^{(\infty)}(-\infty, \infty)$? dim $\mathcal{C}^{(h)}(-\infty, \infty)$? dim $\mathcal{C}(-\infty, \infty)$?

(b) Determine the dimension of the space Z of continuous functions on $0 \leq x \leq \pi$ which have 0 and π as zeros.

9. Let $a_{11}, \ldots, a_{1n}, \ldots, a_{mn}$ be real numbers. Let W be those vectors $(x_1, \ldots, x_n)$ of V_n such that $a_{11}x_1 + \cdots + a_{1n}x_n = 0, \ldots, a_{m1}x_1 + \cdots + a_{mn}x_n = 0$.
   (a) Show that W is a subspace of V_n.
   (b) Show that $\dim W \geq n - m$.
   (c) Give an example where $m = 2$ and $\dim W = n - 1 > n - m$.
10. (a) Let W be the set of vectors from V_4 having first coordinate 0. Find a basis for V_4 which contains no vector from W.
   (b) If U is a nontrivial subspace of a vector space V, show that there is a basis for V containing no vector from U. Contrast this result with Theorem 12.
11. Show that the following are linear varieties of V_4 and determine their dimension:
   (a) All (x_1, x_2, x_3, x_4) such that $x_1 + x_2 + x_3 + x_4 = 1$.
   (b) All (x_1, x_2, x_3, x_4) such that $x_1 - x_2 + x_3 - x_4 = 2$, $x_1 + 3x_2 - x_3 + x_4 = 6$.
   (c) All (x_1, x_2, x_3, x_4) satisfying the conditions in both (a) and (b).
12. Prove: if U and W are subspaces of a vector space V, then $\dim (U \cap W) \leq \min (\dim U, \dim W)$, with equality holding true only if the subspace of smaller dimension is a subspace of the other.
13. Let L and M be linear varieties of a finite dimensional vector space V. Show that if $L \cap M$ is not empty and neither contains the other, then

$$\dim (L \cap M) < \min (\dim L, \dim M)$$

14. Let V be an n-dimensional space with basis $\{\mathbf{u}_1, \ldots, \mathbf{u}_n\}$. Let $\mathbf{v} \to \mathbf{v}^*$ be the correspondence specified in Theorem 11. Prove:
   (a) $\{\mathbf{u}_1^*, \ldots, \mathbf{u}_n^*\}$ is a basis for V_n.
   (b) If $\{\mathbf{v}_1, \ldots, \mathbf{v}_h\}$ is a linearly independent set of V, then $\{\mathbf{v}_1^*, \ldots, \mathbf{v}_h^*\}$ is a linearly independent subset of V_n and, conversely, if $\{\mathbf{v}_1^*, \ldots, \mathbf{v}_h^*\}$ is a linearly independent subset of V_n, then $\{\mathbf{v}_1, \ldots, \mathbf{v}_h\}$ is a linearly independent subset of V.
   (c) A set $\{\mathbf{w}_1, \ldots, \mathbf{w}_n\}$ is a basis for V if, and only if, the set $\{\mathbf{w}_1^*, \ldots, \mathbf{w}_n^*\}$ is a basis for V_n.
   (d) If W is a subspace of V and W^* consists of all elements $\mathbf{w}^*$ for which the corresponding vector $\mathbf{w}$ is in W, then W^* is a subspace of V_n.

9-11 LINEAR MAPPINGS

If a function f maps a set X into a set Y, then we call X the *domain* of the function, or mapping, f. The set of all values of f forms the *range* of the mapping. The range of f is a subset of Y; it may coincide with Y. We shall call Y itself the *target space* of the mapping (Figure 9-15).

In the calculus we studied real functions $f(x)$, defined on an interval; here the domain is the interval (which we can consider as a subset of V_1), the target space is V_1, and the range is a subset of V_1. We also considered vector

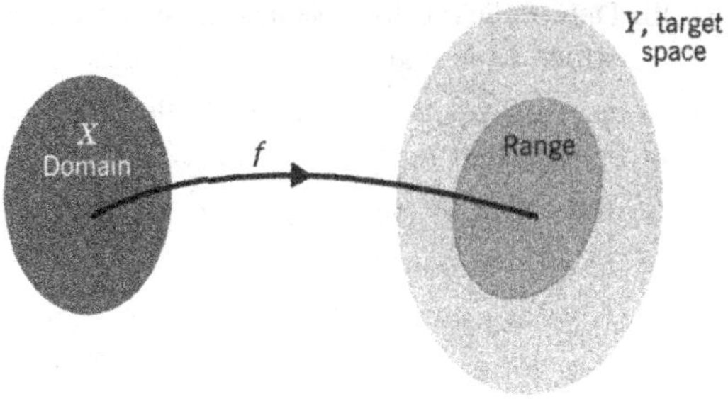

Figure 9-15 Domain, range, and target space.

functions of a real variable: $\mathbf{r} = \mathbf{f}(t)$. Here the domain was again an interval, but the target space was V_2. We have also encountered real functions of two variables: for example, $z = x^2 + y^2$. Here the domain is a set of pairs (x, y), hence, can be considered to be a subset of V_2; the target space is again V_1.

These examples suggest that it would be valuable to study the general case of a function f whose domain is a subset E of a vector space U and whose target space is another vector space V. We begin such a study here by examining the simplest such mapping: a *linear mapping* of a vector space U into a vector space V. Here the domain is U, and the target space is V. The more general mappings will be studied in Chapter 12. It is important that we understand the properties of linear mappings, for, as we shall see, they play a role in the study of the general mapping which is analogous to the role of the tangent and differential in the study of real functions of one variable.

Let U and V be vector spaces with the same set of scalars. A function or mapping T from U into V is said to be a *linear mapping* or a *linear function* if, for all vectors $\mathbf{u}_1$, $\mathbf{u}_2$ in U and all scalars c, we have

$$T(\mathbf{u}_1 + \mathbf{u}_2) = T(\mathbf{u}_1) + T(\mathbf{u}_2) \qquad \text{and} \qquad T(c\mathbf{u}_1) = cT(\mathbf{u}_1) \qquad (9\text{-}110)$$

If T is such a mapping, then by induction we find from (9-110) that,

$$T(\mathbf{u}_1 + \mathbf{u}_2 + \cdots + \mathbf{u}_k) = T(\mathbf{u}_1) + T(\mathbf{u}_2) + \cdots + T(\mathbf{u}_k)$$

and, more generally,

$$T(a_1\mathbf{u}_1 + \cdots + a_k\mathbf{u}_k) = a_1T(\mathbf{u}_1) + \cdots + a_kT(\mathbf{u}_k)$$

Perhaps, the best known linear mappings are the derivative and the integral. The derivative D is a linear mapping of $\mathcal{C}^{(1)}[a, b]$ into the space $\mathcal{C}[a, b]$. For, if f and g are in $\mathcal{C}^{(1)}[a, b]$, then both functions have continuous derivatives on $[a, b]$, so that $f' = Df$ and $g' = Dg$ are in $\mathcal{C}[a, b]$ and, by the basic properties of the derivative,

$$D(f + g) = D(f) + D(g), \qquad D(cf) = cD(f)$$

(We normally drop the parentheses and write Df for the derivative of f). The definite integral is a linear mapping of $\mathcal{C}[a, b]$ into $V_1 = R$ (the real numbers) since, if $J(f) = \int_a^b f(t)\,dt$, then $J(f)$ is a real number and

$$J(f + g) = J(f) + J(g), \qquad J(cf) = cJ(f)$$

by the basic rules of integration. Also the indefinite integral $\int_a^x f(t)\,dt$ is a linear mapping of $\mathcal{C}[a, b]$ into $\mathcal{C}^{(1)}[a, b]$.

It is customary to use the terminology "linear mapping," rather than "linear function," since in so many examples the elements in the domain of the linear mapping are themselves functions. In the literature one also finds linear mappings referred to as linear transformations, linear operators, homomorphisms, and just morphisms. We shall use the terminology "linear transformation" when the domain and target space are the same; that is, $U = V$. The terminology "linear operators" is more frequently used when the domain is a space of functions. The terminology "morphism" is commonly used in the study of abstract algebra. We sometimes abbreviate "mapping" by the word "map."

EXAMPLE 1. If a_1, a_2 are real numbers, then $T((x_1, x_2)) = a_1x_1 + a_2x_2$ is a linear mapping from V_2 to $V_1 = R$. For we have

$$\begin{aligned} T((x_1, x_2) + (y_1, y_2)) &= T((x_1 + y_1, x_2 + y_2)) \\ &= a_1(x_1 + y_1) + a_2(x_2 + y_2) \\ &= (a_1x_1 + a_2x_2) + (a_1y_1 + a_2y_2) \\ &= T((x_1, x_2)) + T((y_1, y_2)) \end{aligned}$$

$$T((cx_1, cx_2)) = a_1cx_1 + a_2cx_2 = c(a_1x_1 + a_2x_2) = cT((x_1, x_2))$$

We shall eventually show (Example 6) that each linear mapping from V_2 into $V_1 = R$ is of this form.

Suppose that one of a_1, a_2 is not 0. Then the nonzero vector $\xi = (a_2, -a_1)$ is mapped into 0 by the linear mapping T. More generally, T maps each vector in the subspace Span (ξ) into 0. Furthermore, if $T((c, d)) = 0 = a_1c + a_2d$, then $(c, d) = s(a_2, -a_1)$, for some real scalar s; hence, (c, d) is in Span (ξ). Thus, Span (ξ) is exactly the set of vectors mapped into 0 by T. If $\mathbf{u}$ is any vector in V_2, then $T(\mathbf{u}) = T(\mathbf{u} + c\xi)$, for each scalar c. Hence, T maps each vector of the line $\{\mathbf{u} + \text{Span}(\xi)\}$ into the same real number $T(\mathbf{u})$. No other vectors in V_2 are mapped onto $T(\mathbf{u})$, for if T has the same value at two vectors, then their difference is in Span (ξ) (see Figure 9-16). If we graph the points $(x_1, x_2, T((x_1, x_2)))$ in 3-dimensional space, the graph is a plane (see Figure 9-17).

Remark. We shall drop parentheses for a mapping such as T here, and write simply $T(x_1, x_2)$. In general, when T has domain V_n and $\mathbf{x} = (x_1, \ldots, x_n)$, we shall write $T(\mathbf{x})$ as $T(x_1, \ldots, x_n)$.

EXAMPLE 2. The mapping $T(x_1, x_2, x_3) = (x_1, x_2)$ is a linear mapping of V_3 into V_2, since

$$\begin{aligned} T(x_1 + y_1, x_2 + y_2, x_3 + y_3) &= (x_1 + y_1, x_2 + y_2) \\ &= T(x_1, x_2, x_3) + T(y_1, y_2, y_3) \end{aligned}$$

$$T(cx_1, cx_2, cx_3) = (cx_1, cx_2) = cT(x_1, x_2, x_3)$$

EXAMPLE 3. The mapping of U into V which maps each vector of U onto $\mathbf{0}$, is called the *zero-mapping*. The zero-mapping is a linear mapping [Problem 3(a)]. We shall usually denote the zero-mapping by O.

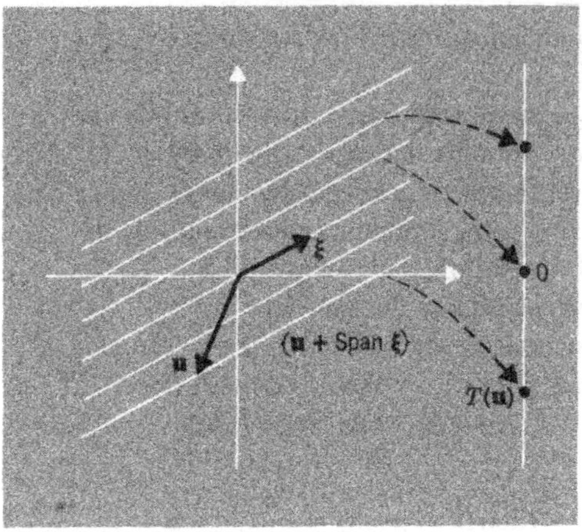

Figure 9-16 Linear mapping of V_2 into R.

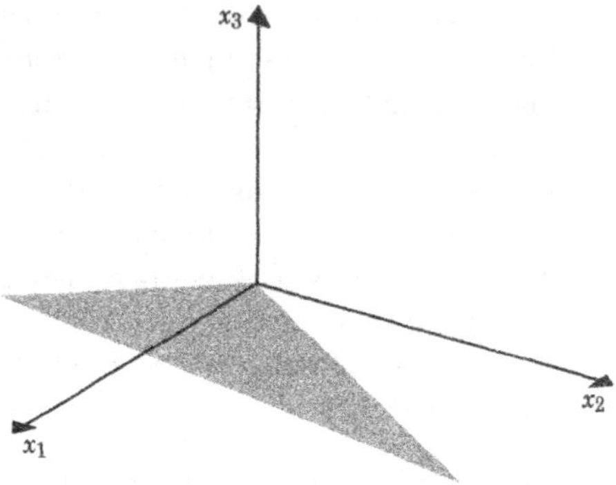

Figure 9-17 Graph of $x_3 = T(x_1, x_2)$.

EXAMPLE 4. The mapping $T(\mathbf{x}) = |\mathbf{x}|$ of V_2 into V_1 is not a linear mapping, since $T(x_1, x_2) = \sqrt{x_1^2 + x_2^2} = T(-x_1, -x_2)$ and, hence, generally, $T(c\mathbf{x}) \neq cT(\mathbf{x})$ for $c = -1$.

EXAMPLE 5. If $\mathbf{a}$ is a fixed nonzero vector in a vector space V, then $T(\mathbf{x}) = \mathbf{x} + \mathbf{a}$ is a mapping of V into V. This is *not* a linear mapping, since $T(2\mathbf{x}) = 2\mathbf{x} + \mathbf{a} \neq 2\mathbf{x} + 2\mathbf{a} = 2T(\mathbf{x})$.

THEOREM 16. *Let T be a linear mapping of U into V. Then $T(\mathbf{0}) = \mathbf{0}$ and $T(-\mathbf{u}) = -T(\mathbf{u})$, [that is, T takes the zero vector of U into the zero vector of V and T takes the negative of $\mathbf{u}$ into the negative of $T(\mathbf{u})$].*

PROOF. $$T(\mathbf{0}) = T(0\mathbf{u}) = 0T(\mathbf{u}) = \mathbf{0}$$

and $$T(-\mathbf{u}) = T((-1)\mathbf{u}) = (-1)T(\mathbf{u}) = -T(\mathbf{u})$$

We here point out a slight inconsistency in terminology. A real function $y = mx + b = f(x)$ is called a *linear function*, but it defines a linear mapping of V_1 into V_1 *only if* $b = 0$; for $f(0) = b$ and, by Theorem 16, for a linear mapping from V_1 to V_1, $f(0) = 0$. The functions $y = mx$ do define linear mappings, as one verifies at once. A similar inconsistency appears in Example 5; the functions in that example are also sometimes called linear.

Let T be a linear mapping of U into V. Let $\{\mathbf{u}_1, \ldots, \mathbf{u}_n\}$ be a basis for U. Then each vector $\mathbf{u}$ in U is expressible uniquely in the form

$$\mathbf{u} = a_1\mathbf{u}_1 + \cdots + a_n\mathbf{u}_n, \qquad \text{where the } a_i \text{ are scalars.}$$

But then

$$T(\mathbf{u}) = a_1T(\mathbf{u}_1) + \cdots + a_nT(\mathbf{u}_n)$$

Consequently, we can find the value of a linear mapping T at any vector $\mathbf{u}$ from the values of T at the vectors of a basis for U. If T and T_1 are linear mappings of U into V such that $T(\mathbf{u}_i) = T_1(\mathbf{u}_i)$ (for $i = 1, \ldots, n$), then T and

T_1 are the same mapping. Thus, *in describing a linear mapping, it is only necessary to prescribe its values at a basis for the domain space.*

EXAMPLE 6. *Let $a_1, \ldots, a_n$ be real numbers, then the mapping*

$$T(x_1, \ldots, x_n) = a_1x_1 + \cdots + a_nx_n$$

is a linear mapping of V_n into $V_1 = R$. Conversely, every linear mapping of V_n into V_1 is of this form.

PROOF. It is an easy matter to verify that T is indeed a linear mapping of V_n into V_1, and we leave it to the reader to do so.

Let S be a linear mapping of V_n into $V_1 = R$. Suppose that $S(\mathbf{e}_1) = b_1, \ldots, S(\mathbf{e}_n) = b_n$. Then

$$\begin{aligned} S(x_1, \ldots, x_n) &= S(x_1\mathbf{e}_1 + \cdots + x_n\mathbf{e}_n) \\ &= x_1S(\mathbf{e}_1) + \cdots + x_nS(\mathbf{e}_n) = b_1x_1 + \cdots + b_nx_n \end{aligned}$$

Thus the mapping S has the form asserted.

In Example 6, T is the zero mapping if, and only if, all $a_i = 0$. If some $a_i \neq 0$, then T has range R, since $T(ca_i^{-1}\mathbf{e}_i) = c$, for each real c.

EXAMPLE 7. *Let $a_1, a_2, a_3, b_1, b_2, b_3$ be real numbers. Then the mapping*

$$T(x_1, x_2, x_3) = (a_1x_1 + a_2x_2 + a_3x_3, b_1x_1 + b_2x_2 + b_3x_3)$$

is a linear mapping of V_3 into V_2. Conversely, every linear mapping of V_3 into V_2 is of this form.

PROOF. The reader should verify that T is a linear mapping and that the mapping T is the zero mapping exactly when $a_1 = a_2 = a_3 = b_1 = b_2 = b_3 = 0$. Let S be a linear mapping of V_3 into V_2. Suppose that

$$S(1, 0, 0) = (c_1, d_1), \qquad S(0, 1, 0) = (c_2, d_2), \qquad S(0, 0, 1) = (c_3, d_3)$$

Then S has the asserted form, since

$$\begin{aligned} S(x_1, x_2, x_3) &= x_1S(1, 0, 0) + x_2S(0, 1, 0) + x_3S(0, 0, 1) \\ &= x_1(c_1, d_1) + x_2(c_2, d_2) + x_3(c_3, d_3) \\ &= (c_1x_1 + c_2x_2 + c_3x_3, d_1x_1 + d_2x_2 + d_3x_3) \end{aligned}$$

EXAMPLE 8. Let T be a linear mapping of the space of real polynomials $\mathcal{P}$ into itself, where $T(1) = 0$, $T(x^m) = x^{m-1}$ for $m \geq 1$. If $\alpha(x) = a_0 + a_1x + \cdots + a_nx^n$, then

$$\begin{aligned} T(\alpha(x)) &= T(a_0) + T(a_1x) + \cdots + T(a_nx^n) = a_0T(1) + \cdots + a_nT(x^n) \\ &= a_1 + a_2x + \cdots + a_nx^{n-1} = \frac{\alpha(x) - \alpha(0)}{x} \end{aligned}$$

The range of T is $\mathcal{P}$ since $T(x\alpha(x)) = \alpha(x)$. The mapping T is not one-to-one, since $T(a + \alpha(x)) = T(\alpha(x))$ for all real numbers a. The mapping T is different from the derivative, since $T(x^2) = x \neq 2x = D(x^2)$.

Notation. For a general mapping T, we denote the set of all values of $T(x)$ for x in E by $T(E)$.

PROBLEMS

1. Show that each of the following is a linear mapping of U into V:
   (a) $U = V_3,\ V = V_3,\ T(x_1, x_2, x_3) = (x_2, x_3, x_1)$
   (b) $U = V_3,\ V = V_4,\ T(x_1, x_2, x_3) = (x_1 + x_2, x_2, x_3 + x_1, x_1)$
   (c) $U = \mathcal{C}[0, 1],\ V = V_1,\ T(f) = f(0)$
   (d) $U = \mathcal{C}[0, 1],\ V = V_1,\ T(f) = f(0) + f(1)$
   (e) $U = \mathcal{C}[0, 1],\ V = V_2,\ T(f) = (f(0), f(1))$
   (f) $U = V_2,\ V = \mathcal{C}[a, b],\ T(x_1, x_2) = x_1e^x + x_2e^{2x}$
   (g) $U = \mathcal{C}[0, 1],\ V = \mathcal{C}[0, 1],\ T(f) = f(x)\cos x$
   (h) $U = \mathcal{C}^{(1)}[a, b],\ V = \mathcal{C}[a, b],\ T(f) = f'(x)\sin x$
   (i) $U = \mathcal{C}^{(2)}[a, b],\ V = \mathcal{C}[a, b],\ T(f) = xf'' - f' + e^x f$
   (j) $U = \mathcal{C}^{(3)}[a, b],\ V = \mathcal{C}[a, b],\ T(f) = f''' + f'' + f' + f$
   (k) $U = \mathcal{C}[a, b],\ V = \mathcal{C}[a, b],\ T(f) = \int_a^x e^t f(t)\,dt$
   (l) $U = \mathcal{C}^{(1)}[a, b],\ V = \mathcal{C}[a, b],\ T(f) = \int_a^x f(t)\,dt + 3f'(x)$

2. We denote a typical polynomial by $\alpha(x)$, its degree by deg $\alpha(x)$, and define a mapping T by giving the value of T assigned to each $\alpha(x)$. Which of the following mappings T are linear mappings of $\mathcal{P}$ into $\mathcal{P}$?
   (a) $T(\alpha(x)) = \alpha(0)$; (b) $T(\alpha(x)) = \alpha(5)$;
   (c) $T(\alpha(x)) =$ constant term of $\alpha(x)$;
   (d) $T(\alpha(x)) = T(a_0 + a_1x + \cdots + a_nx^n) = a_0 + a_2x^3 + \cdots + a_nx^{n+1}$
   $= x\alpha(x) + (1 - x)\alpha(0) - x^2\alpha'(0)$;
   (e) $T(\alpha(x)) = \alpha(2) + \alpha(3)x + \alpha(4)x^2$; (f) $T(\alpha(x)) = \alpha''(x)$;
   (g) $T(\alpha(x)) = (2 + \deg \alpha(x))^{-1}\alpha(x)$; (h) $T(\alpha(x)) = (x^2 + x - 1)\alpha'(x)$;
   (i) $T(\alpha(x)) = 5 + 6x^3 - 17x^4 + x^3\alpha(x)$;
   (j) $T(\alpha(x)) = \alpha(0) + \alpha'(0)x + \frac{1}{2}\alpha''(0)x^2$.

3. Let U, V be vector spaces with the same set of scalars. Prove:
   (a) The zero function from U to V is a linear mapping of U into V.
   (b) If c is a scalar then the function $T(\mathbf{u}) = c\mathbf{u}$ is a linear mapping of U into U.
   (c) If T is a linear mapping of U into V and c is a scalar then the function T_1, where $T_1(\mathbf{u}) = cT(\mathbf{u})$, is a linear mapping of U into V.

4. Show that the mappings in Examples 6 and 7 are linear mappings.

5. For each set of conditions given, determine whether there exists a linear mapping T of U into V satisfying the conditions.
   (a) $U = V_2,\ V = V_2,\ T(1, 1) = (1, 2),\ T(1, -1) = (0, 3)$
   (b) $U = V_2,\ V = V_2,\ T(1, 1) = (1, 0),\ T(1, -1) = (3, 0),\ T(2, 3) = (1, 0)$
   (c) $U = V_2,\ V = V_2,\ T(1, 2) = (1, 3),\ T(2, 1) = (2, 0),\ T(1, 1) = (1, 1)$
   (d) $U = \mathcal{P},\ V = \mathcal{P},\ T(1) = 0,\ T(x^n) = x^{n+1}$ for $n \geq 1$

(e) $U = \mathcal{P}$, $V = \mathcal{P}$, $T(1) = x$, $T(x + 1) = x^2$, $T(x^2 - 1) = x^3$

(f) $U = \mathcal{P}$, $V = \mathcal{P}$, $T(1) = x^2$, $T(x - 1) = x$, $T(x^2 + x) = x$, $T(x^2) = x^2$

9-12 RANGE OF A LINEAR MAPPING

We recall that the range of a function f is the set of all y in the target space of f such that $f(x) = y$ for some x in the domain of f. Now let T be a linear mapping of vector space U into vector space V. Then the mapping T has a range which is a subset of V. We denote this subset by Range T. Thus Range T is the set of all elements $\mathbf{v}$ in V such that $T(\mathbf{u}) = \mathbf{v}$ for some $\mathbf{u}$ in U. Since $T(\mathbf{0}) = \mathbf{0}$, Range T always includes $\mathbf{0}$, and is not empty.

Other terms are sometimes used for Range T: the *image set for* T (abbreviated Im T) and the *cokernel of* T (abbreviated Coker T).

EXAMPLE 1

(a) Range $O = V_0$. Conversely, if Range $T = V_0$, then $T(\mathbf{u}) = \mathbf{0}$ for all $\mathbf{u}$ and, hence, $T = O$.

(b) If T is the linear mapping in Example 1 of Section 9-11 and, at least, one of a_1, a_2 is not 0, then Range $T = R = V_1$, since for any real number b we can always find numbers x_1, x_2 satisfying the equation $a_1x_1 + a_2x_2 = b$.

(c) Let J be the linear mapping: $J(f) = \int_0^1 f(x)\,dx$. Then Range $J = R$, since $J(1) = 1 \neq 0$ and, therefore, $J(c) = c$, for each c in R. [Here $J(1)$ and $J(c)$ refer to the value of J at the constant functions $f \equiv 1$ and $f \equiv c$.]

(d) If T is a nonzero linear mapping of a real vector space U into $V_1 = R$, then Range $T = R$. For $T \neq O$ implies that there is a $\mathbf{u}$ in U such that $T(\mathbf{u}) = a \neq 0$ and, hence, $T[(c/a)\mathbf{u}] = c$ for all c in R.

(e) If T is a nonzero mapping of R into V_3, then Range T is a line through the origin: Range T = Span $(T(1))$. Notice: if $T(1) = \mathbf{0}$, then $T(c) = \mathbf{0}$ for all c in R and, hence, $T = O$.

As these examples suggest, we have the following theorem:

THEOREM 17. *If T is a linear mapping of U into V, then* Range T *is a subspace of V.*

PROOF. We know that Range T is not empty. We need to show that if $\mathbf{v}_1, \mathbf{v}_2$ are in Range T and c is a scalar, then $\mathbf{v}_1 + \mathbf{v}_2$ and $c\mathbf{v}_1$ are in Range T. Since $\mathbf{v}_1, \mathbf{v}_2$ are in Range T, there must exist vectors $\mathbf{u}_1, \mathbf{u}_2$ in U such that

$$T(\mathbf{u}_1) = \mathbf{v}_1, \qquad T(\mathbf{u}_2) = \mathbf{v}_2$$

Then, since T is a linear mapping, by (9-110) we have

$$T(\mathbf{u}_1 + \mathbf{u}_2) = T(\mathbf{u}_1) + T(\mathbf{u}_2) = \mathbf{v}_1 + \mathbf{v}_2, \qquad T(c\mathbf{u}_1) = cT(\mathbf{u}_1) = c\mathbf{v}_1$$

and, hence, $\mathbf{v}_1 + \mathbf{v}_2$ and $c\mathbf{v}_1$ are, indeed, in Range T.

THEOREM 18. *If $\{\mathbf{u}_1, \ldots, \mathbf{u}_n\}$ is a basis for U and T is a linear mapping of U into V, then* Range T = Span $(T(\mathbf{u}_1), \ldots, T(\mathbf{u}_n))$.

PROOF. If $\mathbf{v}$ is in Range T, then there is a $\mathbf{u}$ in U such that $T(\mathbf{u}) = \mathbf{v}$. But $\mathbf{u} = a_1\mathbf{u}_1 + \cdots + a_n\mathbf{u}_n$ and, hence, $\mathbf{v}$ is a linear combination of $T(\mathbf{u}_1), \ldots, T(\mathbf{u}_n)$. Hence, Range T is included in Span $(T(\mathbf{u}_1), \ldots, T(\mathbf{u}_n))$. On the other hand, $\Sigma c_i T(\mathbf{u}_i) = T(\Sigma c_i \mathbf{u}_i)$ is in Range T; hence, Span $(T(\mathbf{u}_1), \ldots, T(\mathbf{u}_n))$ is included in Range T. This proves the theorem.

We say that a linear mapping T, of U into V, maps U *onto* V if Range T coincides with V. We then say that T is an *onto mapping*. From Theorem 18 it follows that if there is a linear mapping of U onto V, then dim $V \leq$ dim U.

9-13 KERNEL OF A LINEAR MAPPING

The set of all the zeros of a linear mapping T is called the *kernel of* T and is denoted by Kernel T. Thus the kernel of a linear mapping T of U into V is the set of all $\mathbf{u}$ in U such that $T(\mathbf{u}) = \mathbf{0}$. By Theorem 16, $T(\mathbf{0}) = \mathbf{0}$. Hence, Kernel T always contains $\mathbf{0}$ and is a nonempty subset of U.

EXAMPLE 1

(a) Let T be a nonzero linear mapping of $V_1 = R$ into a real vector space V. Since T is not the zero mapping, $T(1) \neq \mathbf{0}$, and hence $T(c) = cT(1) \neq \mathbf{0}$ for all $c \neq 0$. Thus Kernel $T = V_0$.

(b) If T is the linear mapping of Example 1, Section 9-11, then Kernel T = Span $((a_2, -a_1))$, provided one of a_1 and a_2 is not 0.

(c) The derivative D is a linear mapping from the space $\mathfrak{D}$ of differentiable functions on [0, 1] into the space $\mathfrak{F}$ of all real functions on [0, 1]. In this case Kernel D is the set of constant function, since $Df = 0$ exactly when f is a constant function. If f is in $\mathfrak{C}$, the space of continuous functions on [0, 1], then $D\left(\int_0^x f(t)\,dt\right) = f(x)$, and therefore $\mathfrak{C}$ is contained in Range D. Is it true that $\mathfrak{C}$ = Range D?

(d) If T maps U into V, then Kernel $T = U$ if, and only if, T is the zero mapping O. For if Kernel $T = U$, then $T(\mathbf{u}) = \mathbf{0}$ for all $\mathbf{u}$; and if $T(\mathbf{u}) = \mathbf{0}$ for all $\mathbf{u}$, then U = Kernel T.

THEOREM 19. *If T is a linear map from a vector space U into a vector space V, then* Kernel T *is a subspace of U.*

PROOF. Let $\mathbf{u}_1, \mathbf{u}_2$ be vectors in Kernel T and let c be a scalar of U, then $T(\mathbf{u}_1 + \mathbf{u}_2) = T(\mathbf{u}_1) + T(\mathbf{u}_2) = \mathbf{0} + \mathbf{0} = \mathbf{0}$, and $T(c\mathbf{u}_1) = cT(\mathbf{u}_1) = c\mathbf{0} = \mathbf{0}$. Thus Kernel T is closed under addition and multiplication by scalars and so is a subspace.

The kernel and range very nearly characterize a linear mapping. The significance of the kernel is demonstrated in the next theorem:

THEOREM 20. *Let T be a linear mapping of U into V. Then $T(\mathbf{u}_1) = T(\mathbf{u}_2)$ if, and only if, $\mathbf{u}_1$ and $\mathbf{u}_2$ belong to the same linear variety of U with base space K =* Kernel T. *Thus if $\mathbf{v}$ is in Range T, then the set of all vectors $\mathbf{u}$ in U for which $T(\mathbf{u}) = \mathbf{v}$ is the linear variety $\{\mathbf{u}_0 + K\}$, where $T(\mathbf{u}_0) = \mathbf{v}$.*

PROOF. If $T(\mathbf{u}_1) = T(\mathbf{u}_2)$, then $T(\mathbf{u}_1 - \mathbf{u}_2) = T(\mathbf{u}_1) - T(\mathbf{u}_2) = \mathbf{0}$ and, hence, $\mathbf{u}_1 - \mathbf{u}_2$ is in K. But then $\mathbf{u}_1$ and $\mathbf{u}_2$ are both in the linear variety $\{\mathbf{u}_2 + K\}$. Conversely, if $\mathbf{u}_1$ and $\mathbf{u}_2$ are in the linear variety $\{\mathbf{u}_0 + K\}$, then $\mathbf{u}_1 = \mathbf{u}_0 + \mathbf{k}_1$, $\mathbf{u}_2 = \mathbf{u}_0 + \mathbf{k}_2$ for some $\mathbf{k}_1, \mathbf{k}_2$ in K. Consequently, $T(\mathbf{u}_1) = T(\mathbf{u}_0) + T(\mathbf{k}_1) = T(\mathbf{u}_0)$ and, similarly, $T(\mathbf{u}_2) = T(\mathbf{u}_0)$, so that $T(\mathbf{u}_1) = T(\mathbf{u}_2)$. The last statement in the theorem is just a rephrasing of the earlier part.

Figure 9-18 is a diagrammatic scheme for Range T and Kernel T. See also Figure 9-19 in Section 9-17.

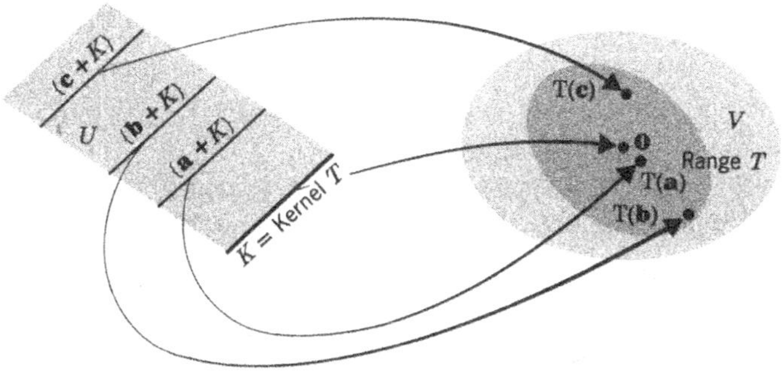

Figure 9-18 Kernel and range of a linear mapping.

Pre-image. If T is a linear mapping of U into V and $\mathbf{v}$ is in Range T, then the set of all vectors $\mathbf{u}$ in U such that $T(\mathbf{u}) = \mathbf{v}$ is called the T-pre-image of $\mathbf{v}$ (or the pre-image of $\mathbf{v}$ for the mapping T). As we saw in Theorem 20, the T-pre-image of each vector in Range T is a linear variety in U with base space $K =$ Kernel T; in particular, the T-pre-image of $\mathbf{v}$ is $\{\mathbf{u}_0 + K\}$ if $T(\mathbf{u}_0) = \mathbf{v}$. By the T-pre-image of a set E we mean the set of all $\mathbf{u}$ for which $T(\mathbf{u})$ is in E.

EXAMPLE 2

(a) Let D be the derivative. Then D is a linear mapping of $\mathcal{C}^{(1)}[a, b]$ into $\mathcal{C}[a, b]$. The kernel of D is the set of constant functions, and the range of D is $\mathcal{C}[a, b]$; for $Df = 0$ (the zero function) if and only f is a constant, and $D\int_a^x f(t)\,dt = f(x)$ for every continuous function f. Thus D is an onto mapping which is not one-to-one. The D-pre-image of f is the linear variety $\{F(x) +$ Kernel $D\} = \{F + C\}$, where $F' = f$. Thus the D-pre-image of f is just the indefinite integral of f, and the arbitrary constant in the indefinite integral is just Kernel D.

(b) In Example 8 of Section 9-11, Kernel T is W, the set of constant polynomials and the T-pre-image of $\alpha(x)$ is the linear variety $\{x\alpha(x) + W\}$.

(c) As noted in Example 7, Section 9-11, the mapping

$$T(x_1, x_2, x_3) = (x_1 + x_2 - x_3, x_1 - x_2 + 17x_3)$$

is a linear mapping of V_3 into V_2. The Kernel of T is given by all (x_1, x_2, x_3) such that $x_1 + x_2 - x_3 = 0$ and $x_1 - x_2 + 17x_3 = 0$; these equations are satisfied if, and only if, $x_1 = -8x_3$ and $x_2 = 9x_3$, where x_3 may have any real value. Hence, Kernel $T = \{t(-8, 9, 1)\}$ [Here t ranges over R]. The T-pre-image of $(1, 1)$ is then given by $\{(1, 0, 0) + \{t(-8, 9, 1)\}\}$.

(d) In Example 6, Section 9-11, suppose that $a_1 \neq 0$. Then

$$\text{Kernel } T = \text{Span }((a_2, -a_1, 0, \ldots, 0), \ldots, (a_n, 0, \ldots, -a_1))$$

For each of the vectors $\mathbf{u}$ listed has the property $T(\mathbf{u}) = \mathbf{0}$; hence, their span is included in the subspace Kernel T. Conversely, if $T(x_1, \ldots, x_n) = a_1x_1 + \cdots + a_nx_n = 0$, then

$$x_1 = -\frac{a_2}{a_1}x_2 - \cdots - \frac{a_n}{a_1}x_n$$

whence

$$(x_1, \ldots, x_n) = (-\frac{a_2}{a_1}x_2 - \cdots - \frac{a_n}{a_1}x_n, x_2, \ldots, x_n)$$

$$= -a_1^{-1}x_2(a_2, -a_1, 0, \ldots, 0) - \cdots - a_1^{-1}x_n(a_n, 0, \ldots, 0, -a_1)$$

Thus $(x_1, \ldots, x_n)$ is contained in Span $((a_2, -a_1, 0, \ldots, 0), \ldots)$. Accordingly, Kernel T has the form stated. We can now conclude that the T-pre-image of an arbitrary c is $\{ca_1^{-1}\mathbf{e}_1 + \text{Kernel } T\}$.

The T-pre-image of a point $\mathbf{b}$ may very well be the empty set. In fact it is the empty set exactly when $\mathbf{b}$ is not in Range T.

As an immediate consequence of Theorem 20 we see that *T is a one-to-one mapping of U into V, if and only if,* Kernel $T = \{\mathbf{0}\}$, where $\mathbf{0}$ is the zero vector of U. Furthermore, if Kernel $T = \{\mathbf{0}\}$, so that T is one-to-one, and $\{\mathbf{u}_1, \ldots, \mathbf{u}_k\}$ is a linearly independent set in U, then $\{T(\mathbf{u}_1), \ldots, T(\mathbf{u}_k)\}$ is a linearly independent set in V. For if $a_1T(\mathbf{u}_1) + \cdots + a_kT(\mathbf{u}_k) = \mathbf{0}$, with some $a_i \neq 0$, then $T(a_1\mathbf{u}_1 + \cdots + a_k\mathbf{u}_k) = \mathbf{0}$, and $a_1\mathbf{u}_1 + \cdots + a_k\mathbf{u}_k$ would be in Kernel T and, thus, equal to $\mathbf{0}$, contrary to the assumption of linear independence of $\{\mathbf{u}_1, \ldots, \mathbf{u}_k\}$. Thus *one-to-one linear mappings preserve linear independence.* We also see that T is a one-to-one mapping of U onto V if, and only if, both Kernel $T = \{\mathbf{0}\}$ *and* Range $T = V$ (kernel as small as possible, range as large as possible).

PROBLEMS

1. (a) . . . (h) Determine the range and kernel of the linear mappings of parts (a) . . . (h) of Problem 1, following Section 9-11.
2. (a) . . . (j) Determine the range of the mappings of parts (a) . . . (j) of Problem 2 following Section 9-11.
3. (a) . . . (f) Determine the kernel of the mappings of parts (a) . . . (f) of Problem 2, following Section 9-11.
4. (a) . . . (c) Determine the range and the kernel for each of the mappings of parts (a) . . . (c) of Problem 3, following Section 9-11.
5. (a) Show that the derivative is a linear mapping of $\mathcal{P}$ into itself which is not one-to-one but is onto.

(b) Show that the derivative is a linear mapping of $\mathcal{P}_m$ into itself which is neither one-to-one nor onto.

6. For each of the following linear mappings of a V_n into V_2, determine the range, the kernel and the pre-image of $(1, 0)$; also determine whether the mapping is one-to-one and whether it is onto.

(a) $T(x_1, x_2, x_3) = (x_1 - x_2, x_2 + x_3)$

(b) $T(x_1, x_2, x_3) = (x_1, x_2 - x_3)$

(c) $T(x_1, x_2) = (x_1 - x_2, x_1 + x_2)$

(d) $T(x_1, x_2, x_3, x_4) = (x_1 + x_4, x_2 + x_3)$

7. Prove: if T is a one-to-one linear mapping of U into V and $\{\mathbf{u}_1, \ldots, \mathbf{u}_n\}$ is a basis for U, then $\{T(\mathbf{u}_1), \ldots, T(\mathbf{u}_n)\}$ is a basis for Range T.

8. Prove: if U or V is the zero space, then O is the only linear mapping of U into V.

9. In the space $\mathcal{C}[0, 1]$, define $T(f)$ to be M_f, where M_f is the function of x defined as follows: $M_f(x)$ = maximum of f in $[0, x]$, $0 \leq x \leq 1$. This is illustrated by a thermometer which records the maximum temperature. It can be shown that $M_f(x)$ is a continuous function on $[0, 1]$ if f is.

(a) Find $T(f)$ for $f(x) = x - x^2$, $f(x) = e^{-x}$, $f(x) = \sin 3x$, $f(x) = x^2 - x$.

(b) Is T a linear mapping?

(c) Describe those functions f for which $T(f)$ is the zero function.

(d) Describe those functions f for which $T(f) = f$.

9-14 RANK AND NULLITY OF A LINEAR MAPPING

Let T be a linear mapping of a finite dimensional vector space U into a vector space V. We introduce the following terminology:

$$\text{rank } T = \dim \text{Range } T \tag{9-140}$$

and

$$\text{nullity of } T = \text{null } T = \dim \text{Kernel } T \tag{9-141}$$

Thus the derivative D is a linear mapping of $\mathcal{P}_m$ into $\mathcal{P}$ with rank $D = m$ and null $D = 1$. A linear mapping T is one-to-one precisely when null $T = 0$; it is onto precisely when rank $T = \dim V$, and it is the zero map precisely when null $T = \dim U$.

THEOREM 21. *If T is a linear mapping of a finite dimensional vector space U into a vector space V then one has the relations:*

$$\text{null } T \leq \dim U \tag{9-142}$$

and

$$\text{rank } T \leq \min\{\dim U, \dim V\} \tag{9-143}$$

PROOF. Since Kernel T is a subspace of U and null T = dim Kernel T, we must have (9-142) (see Theorem 13).

Since $T(U)$ is a subspace of V, we must have rank $T \leq \dim V$. If $\{\mathbf{u}_1, \ldots, \mathbf{u}_n\}$ is a basis for U, then (see Theorem 18) Range T = Span $\{T(\mathbf{u}_1), \ldots, T(\mathbf{u}_n)\}$ and, hence, dim Range $T \leq n = \dim U$. It follows that (9-143) must hold true.

The linear mapping $T_1(x_1, x_2, x_3) = (x_1, x_2)$ is a case in which rank T_1

$= \dim V < \dim U$. The linear mapping $T_2(x_1, x_2) = (x_1, x_2, x_1 + x_2)$ is a case where rank $T_2 = \dim U < \dim V$, since Range $T_2 = \text{Span}((1, 0, 1), (0, 1, 1))$. The linear mapping $T_3(x_1, x_2) = (x_1, 2x_1, -x_1)$ is a case in which rank $T_3 < \dim U$ and rank $T_3 < \dim V$.

EXAMPLE 1. Let T be the linear mapping of V_2 into V_3 such that $T(1, 1) = (0, 2, 2)$, $T(1, -1) = (4, 0, 0)$. Determine $T(x, y)$, Kernel T, Range T, rank T, null T, the T-pre-image of $(1, 0, 0)$.

Solution. Clearly, $\{(1, 1), (1, -1)\}$ is a linearly independent set and, hence, a basis for V_2. If we write $(x, y) = a(1, 1) + b(1, -1)$, then we find $x = a + b$, $y = a - b$, so that $a = (x + y)/2$, $b = (x - y)/2$ and, hence,

$$(x, y) = \frac{x + y}{2}(1, 1) + \frac{x - y}{2}(1, -1)$$

Accordingly, we can find $T(x, y)$:

$$\begin{aligned} T(x, y) &= \frac{x + y}{2}T(1, 1) + \frac{x - y}{2}T(1, -1) \\ &= \frac{x + y}{2}(0, 2, 2) + \frac{x - y}{2}(4, 0, 0) \\ &= (2x - 2y, x + y, x + y) \end{aligned}$$

Kernel T consists of all (x, y) such that $2x - 2y = 0$ and $x + y = 0$; hence, Kernel T consists of $(0, 0)$ alone. Thus T is one-to-one and null $T = 0$. Since $\{(1, 1), (1, -1)\}$ is a basis for V_2,

$$\begin{aligned} \text{Range } T &= \text{Span } (T(1, 1), T(1, -1)) = \text{Span } ((0, 2, 2), (4, 0, 0)) \\ &= \text{Span } ((0, 1, 1), (1, 0, 0)) \end{aligned}$$

Thus Range T consists of all vectors of form (a, b, b) in V_3. Clearly, rank $T = 2$. The equation $T(x, y) = (1, 0, 0)$ is found to be satisfied for $x = \frac{1}{4}$, $y = -\frac{1}{4}$. Since T is one-to-one, $(\frac{1}{4}, \frac{-1}{4})$ is the T-pre-image of $(1, 0, 0)$.

THEOREM 22. *Let T be a linear mapping of a finite-dimensional space U into V, then*

$$\text{null } T + \text{rank } T = \dim U \tag{9-144}$$

PROOF. If null $T = 0$, then Kernel $T = V_0$ and T is a one-to-one mapping. Hence, T carries a basis of U into a basis for Range T (see Problem 7, following Section 9-13). Hence, null $T = 0$ implies rank $T = \dim U$.

Next suppose that null $T = m > 0$. Let $\{\mathbf{u}_1, \ldots, \mathbf{u}_m)$ be a basis for Kernel T. By Theorem 12 in Section 9-9, we can extend it to a basis $\{\mathbf{u}_1, \ldots, \mathbf{u}_m, \ldots, \mathbf{u}_n\}$ for U. Clearly

$$T(a_1\mathbf{u}_1 + \cdots + a_n\mathbf{u}_n) = a_{m+1}T(\mathbf{u}_{m+1}) + \cdots + a_nT(\mathbf{u}_n)$$

Hence, Range $T = \text{Span } (T(\mathbf{u}_{m+1}), \ldots, T(\mathbf{u}_n))$. Suppose further, that $b_{m+1}T(\mathbf{u}_{m+1}) + \cdots + b_nT(\mathbf{u}_n) = \mathbf{0}$. Then

$$T(b_{m+1}\mathbf{u}_{m+1} + \cdots + b_n\mathbf{u}_n) = \mathbf{0}$$

and, hence, $b_{m+1}\mathbf{u}_{m+1} + \cdots + b_n\mathbf{u}_n$ is in Kernel T. But $\{\mathbf{u}_1, \ldots, \mathbf{u}_m\}$ is a basis for Kernel T and, hence, there exist scalars $c_1, \ldots, c_m$ such that

$$b_{m+1}\mathbf{u}_{m+1} + \cdots + b_n\mathbf{u}_n = c_1\mathbf{u}_1 + \cdots + c_m\mathbf{u}_m$$

Since $\{\mathbf{u}_1, \ldots, \mathbf{u}_n\}$ is a basis for U, we must have

$$b_{m+1}' = \cdots = b_n = -c_1 = \cdots = -c_m = 0$$

It follows that $\{T(\mathbf{u}_{m+1}), \ldots, T(\mathbf{u}_n)\}$ is a linearly independent set spanning Range T and, hence, is a basis for Range T. Thus we have shown that rank T = dim Range $T = n - m$. This completes the proof of the theorem.

As an immediate consequence of Theorem 22, we can state: *if there is a one-to-one linear mapping T of U onto V, then dim U =* dim V. For such a mapping, rank T = dim V and null $T = 0$, so that dim U = dim V.

Given finite dimensional vector spaces U and V, it is natural to inquire what subspaces Z of U can be kernels of linear mappings of U into V. We know from Theorems 21 and 22, that if Z is a kernel of a linear mapping T, then

$$\dim U = \dim Z + \dim \text{Range } T \leq \dim Z + \dim V$$

It is perhaps a pleasant surprise to find, as we show below, that the inequality

$$\dim U - \dim V \leq \dim Z$$

is the only condition a subspace Z needs satisfy in order to be the kernel of *some* linear mapping of U into V. Thus any subspace not too small, namely of dimension at least dim U − dim V, is the kernel of some linear mapping of U into V. There is a similar situation regarding subspaces of V which can be the range of a linear mapping of U into V; any subspace of V, not too large (namely of dimension at most dim U), is the range of some linear mapping of U into V. These results are summarized in the following theorem:

THEOREM 23. *Let U, V be finite dimensional vector spaces. A subspace Z of U is the kernel of a linear mapping of U into V, if and only if,* dim $Z \geq$ dim U − dim V. *A subspace W of V is the range of a linear mapping of U into V if, and only if,* dim $W \leq$ dim U.

The proof of this theorem is given in the next section.

We can also ask what linear varieties of U can be the pre-image of a fixed vector $\mathbf{b}$ in V. The answer is given again in terms of the dimension of the linear variety. If $\mathbf{b} = \mathbf{0}$, its pre-image must be the kernel of the linear mapping and Theorem 23 applies. We note also that a subspace of U can never be the pre-image of a nonzero vector $\mathbf{b}$; for all vectors in the pre-image of $\mathbf{b}$ are mapped onto $\mathbf{b}$, yet $\mathbf{0}$ is in each subspace and every linear mapping maps $\mathbf{0}$ onto $\mathbf{0}$. Apart from these cases, we have the following very general theorem.

THEOREM 24. *Let U, V be finite dimensional vector spaces. Let L be a linear variety of U which is not a subspace of U. Let* $\mathbf{b}$ *be a nonzero vector of V. Then there is a linear mapping T of U into V such that L is the T-pre-image of* $\mathbf{b}$ *if, and only if,* dim $L \geq$ dim U − dim V.

The proof is given in the next section.

‡9-15 PROOFS OF TWO THEOREMS

We use the following notation throughout this section: $\dim U = n$, $\dim V = m$, Z is a subspace of U of dimension k, W is a subspace of V of dimension p, and L is a linear variety of U of dimension k.

PROOF OF THEOREM 23. We need to prove that Z is the kernel of a linear mapping of U into V if, and only if, $k \geq n - m$. If Z is the kernel of a linear mapping T of rank r, then, by Theorems 22, and 21, we have $n = k + r \leq k + m$, and hence $k \geq n - m$.

Conversely, suppose that $k \geq n - m$ and that $\mathbf{z}_1, \ldots, \mathbf{z}_k$ is a basis for Z. By Theorem 12, Section 9-9, we can extend this basis for Z to a basis $\mathbf{z}_1, \ldots, \mathbf{z}_k, \ldots, \mathbf{z}_n$ for U. Let $\mathbf{v}_1, \ldots, \mathbf{v}_m$ be a basis for V. Now a linear mapping (see Section 9-11) is completely determined by its action on a basis of the domain space. Let T be a linear mapping of U into V such that

$$T(\mathbf{z}_1) = \mathbf{0}, \quad \ldots, \quad T(\mathbf{z}_k) = \mathbf{0}, \quad T(\mathbf{z}_{k+1}) = \mathbf{v}_1, \quad \ldots, \quad T(\mathbf{z}_n) = \mathbf{v}_{n-k}$$

Since $k = \dim Z \geq n - m$, we have $m \geq n - k$ and, hence, the images of $\mathbf{z}_{k+1}, \ldots, \mathbf{z}_{n-k}$ are a linearly independent set in V. Thus rank $T = n - k$. We observe that

$$T(a_1\mathbf{z}_1 + \cdots + a_n\mathbf{z}_n) = a_{k+1}\mathbf{v}_1 + \cdots + a_n\mathbf{v}_{n-k} = \mathbf{0}$$

if, and only if, $a_{k+1} = \cdots = a_n = 0$. Hence, the kernel of T is Z.

For the second part of the theorem, we need to show that W is the range of a linear mapping if, and only if, $p \leq n$. If W is the range of a linear mapping, then $p \leq n$ by (9-143). Conversely, if $p \leq n$, let $\mathbf{w}_1, \ldots, \mathbf{w}_p$ be a basis for W and let $\mathbf{u}_1, \ldots, \mathbf{u}_p, \mathbf{u}_{p+1}, \ldots, \mathbf{u}_n$ be a basis for U. Then the linear mapping T given by

$$T(\mathbf{u}_1) = \mathbf{w}_1, \quad \ldots, \quad T(\mathbf{u}_p) = \mathbf{w}_p, \quad T(\mathbf{u}_{p+1}) = \mathbf{0}, \quad \ldots, \quad T(\mathbf{u}_n) = \mathbf{0}$$

has W as its range.

This completes the proof of Theorem 23.

Remark. The linear mappings specified in the above theorem are not unique. In particular, the mappings constructed in each part of the proof depend very essentially on the choice of bases for Z, U, W and V. Any change in these bases leads to a different linear mapping, and the new mapping has the same desired properties.

PROOF OF THEOREM 24. If T is a linear mapping of U into V with null $T = k = \dim$ Kernel T and L is the T-pre-image of the nonzero vector $\mathbf{b}$, then L contains a nonzero vector $\mathbf{x}$ and $L = \{\mathbf{x} + \text{Kernel } T\}$. Hence L is not a subspace of U and $\dim L = k$.

Next, let $L = \{\mathbf{x} + Z\}$, where Z is the base space for L. Let $\mathbf{z}_1, \ldots, \mathbf{z}_k$ be a basis for Z. Since L is not a subspace of V_n, the vector $\mathbf{x}$ is not in Z and the set $\mathbf{z}_1, \ldots, \mathbf{z}_k, \mathbf{x}$ is a linearly independent set in U. We extend this set to a basis for U, say $\mathbf{z}_1, \ldots, \mathbf{z}_k, \mathbf{x}, \mathbf{z}_{k+2}, \ldots, \mathbf{z}_n$. Now choose a basis for V

which includes **b**, say **b**, $\mathbf{u}_2, \ldots, \mathbf{u}_m$. Let T be the linear mapping of U into V such that

$$T(\mathbf{z}_1) = \mathbf{0}, \quad \ldots, \quad T(\mathbf{z}_k) = \mathbf{0}, \quad T(\mathbf{x}) = \mathbf{b}, \quad T(\mathbf{z}_{k+2}) = \mathbf{u}_2, \quad \ldots, \quad T(\mathbf{z}_n) = \mathbf{u}_{n-k}$$

Then, as in the preceding theorem, Kernel $T = Z$ and, hence, the T-pre-image of **b** is L.

PROBLEMS

1. For each of the following linear mappings of V_3 into V_3, determine the kernel range, nullity, rank and pre-image of $(1, 0, 0)$.
   (a) $T(x_1, x_2, x_3) = (x_1, x_1 - x_2, x_1 - x_3)$
   (b) $T(x_1, x_2, x_3) = (x_1 - x_2, x_2 - x_3, x_1 + x_3)$
   (c) $T(x_1, x_2, x_3) = (x_1 + x_2, x_2 - x_3, 2x_1 + x_2 + x_3)$
   (d) $T(x_1, x_2, x_3) = (x_1 + x_2, x_2 + x_3, x_3 - x_1)$
2. Consider the following linear mappings of V_4 into V_3 specified by their values at a basis of V_4 in Table 9-1. For each mapping, determine the range, the kernel, the rank, the nullity, and the pre-image of $(1, 0, 0)$ and of $(1, 1, 1)$.

Table 9-1

	(1, 0, 0, 0)	(0, 1, 0, 0)	(0, 0, 1, 0)	(0, 0, 0, 1)
(a) T_1	(1, 2, 3)	(0, 0, 0)	(2, 1, 0)	(−1, 0, 1)
(b) T_2	(1, 2, 3)	(3, 2, 1)	(−1, −1, −1)	(−1, 0, 1)
(c) T_3	(1, 2, 5)	(1, 0, 0)	(0, 0, 7)	(−7, 6, 4)
(d) T_4	(1, 0, 1)	(0, 2, 1)	(−3, 6, 0)	(17, −20, 7)

3. (a) Prove: if T is a linear mapping of V_n into V_n, then T is onto if, and only if, $\{T(\mathbf{e}_1), \ldots, T(\mathbf{e}_n)\}$ is a basis for V_n.
   (b) Prove: if $k < n$, then there is no linear mapping of V_k onto V_n.
   (c) Prove: if $k > n$, then there is no one-to-one linear mapping of V_k into V_n.
4. The rotation of each vector **v** in the plane by a fixed angle θ is a function from V_2 to V_2. Is this function a linear mapping? Is it one-to-one? Is it onto? (Reason geometrically.)
5. (a) Show that the function of V_2 into V_2 given by $T_0(x_1, x_2) = (x_1 - 3x_2, x_2)$ is a linear mapping. Determine its rank and nullity. Determine the T_0-pre-image of (a, b).
   (b) Show that if a_1, b_1, a_2, b_2 are real numbers, then
   $$T(x_1, x_2) = (a_1x_1 + a_2x_2, b_1x_1 + b_2x_2)$$
   is a linear mapping of V_2 into V_2.
   (c) Show: if $a_1b_2 \neq b_1a_2$, then the mapping T is one-to-one and onto.
   (d) Prove: Kernel $T \neq V_0$ exactly when $a_1b_2 - a_2b_1 = 0$, and if also $a_1 \neq 0$, then Kernel $T =$ Span $(a_2, -a_1)$, and Range $T =$ Span (a_1, b_1).

(e) Prove: if S is a linear mapping of V_2 into V_2 and $S(1, 0) = (c, d)$, $S(0, 1) = (e, f)$ then $S(x, y) = (cx + ey, dx + fy)$.

6. Let $a_1, a_2, a_3, b_1, b_2, b_3$ be real numbers. Prove:

(a) $T(x_1, x_2, x_3) = (a_1x_1 + a_2x_2 + a_3x_3, b_1x_1 + b_2x_2 + b_3x_3)$ is a linear mapping of V_3 into V_2.

(b) If S is a linear mapping of V_3 into V_2 and $S(1, 0, 0) = (a_1, b_1)$, $S(0, 1, 0) = (a_2, b_2)$, $S(0, 0, 1) = (a_3, b_3)$ then T and S are the same mapping.

(c) The mapping T is onto if, and only if, for some i and j, $a_ib_j - a_jb_i \neq 0$.

(d) The mapping T is the zero mapping, if, and only if, $a_1 = a_2 = a_3 = b_1 = b_2 = b_3 = 0$.

(e) If $a_1b_2 - a_2b_1 = k \neq 0$, then Kernel T = Span $((a_2b_3 - a_3b_2, a_3b_1 - a_1b_3, a_1b_2 - a_2b_1))$ and the T-pre-image of (c, d) is $\{k^{-1}(cb_2 - da_2, da_1 - cb_1, 0) + \text{Kernel } T\}$.

(f) If T is not the zero transformation, then rank $T = 1$ if, and only if, there exist real numbers s, t, not both 0, such that $sa_1 - tb_1 = sa_2 - tb_2 = sa_3 - tb_3 = 0$. In this case, Range T = Span (t, s) and, if in addition $a_1 \neq 0$, then Kernel T = Span $((a_3, 0, -a_1), (a_2, -a_1, 0))$, and the T-pre-image of (t, s) is $\{(ta_1^{-1}, 0, 0) + \text{Kernel } T\}$.

(g) Determine Kernel T if rank $T = 1$ and $a_1 = 0$ (there are numerous cases).

(h) Interpret the results of parts (e) and (f) in terms of the geometry of the plane and of space.

7. (a) Prove: if $a_{11}, \ldots, a_{1n}, \ldots, a_{m1}, \ldots, a_{mn}$ are real numbers then

$$T(x_1, \ldots, x_n) = (a_{11}x_1 + \cdots + a_{1n}x_n, \ldots, a_{m1}x_1 + \cdots + a_{mn}x_n)$$

is a linear mapping of V_n into V_m, and conversely if T is a linear mapping of V_n into V_m, then there exist numbers $a_{11}, \ldots, a_{1n}, \ldots, a_{m1}, \ldots, a_{mn}$ such that the above relation holds true.

(b) If $m < n$, show that null $T > 0$.

(c) If $m \geq n$, show that rank $T = n$ exactly when the vectors $(a_{11}, \ldots, a_{m1}), \ldots, (a_{1n}, \ldots, a_{mn})$ are a linearly independent set.

9-16 ADDITION OF LINEAR MAPPINGS, SCALAR MULTIPLES OF LINEAR MAPPINGS

We have seen that two real-valued functions defined on the same domain can be added and multiplied to obtain new functions. The sum of f and g is the function $h = f + g$ such that $h(x) = f(x) + g(x)$ for all x in the domain. Similarly, the product of f and g is the function $F = fg$ such that $F(x) = f(x)g(x)$ for all x in the domain. It is natural to ask whether we can combine linear mappings in these ways to obtain new linear mappings. We find that this is easily done for addition. For multiplication there is trouble, since, in general, we cannot multiply two vectors. However, we do have scalar multiplication in a vector space, and this enables us to define a scalar times a mapping. If S and T are linear mappings of U into V, we thus define $S + T = M$ and

$aT = N$, where a is a scalar, by the equations:

$$M(\mathbf{u}) = S(\mathbf{u}) + T(\mathbf{u}) \quad \text{and} \quad N(\mathbf{u}) = aT(\mathbf{u}) \quad \text{for all } \mathbf{u} \text{ in } U \tag{9-160}$$

These definitions have meaning even if S and T are not linear. But here we assume that they are linear and can then verify that $S + T = M$ and $aT = N$ are linear. We have

$$\begin{aligned} M(\mathbf{u}_1 + \mathbf{u}_2) &= S(\mathbf{u}_1 + \mathbf{u}_2) + T(\mathbf{u}_1 + \mathbf{u}_2) = (S(\mathbf{u}_1) + S(\mathbf{u}_2)) + (T(\mathbf{u}_1) + T(\mathbf{u}_2)) \\ &= (S(\mathbf{u}_1) + T(\mathbf{u}_1)) + (S(\mathbf{u}_2) + T(\mathbf{u}_2)) = M(\mathbf{u}_1) + M(\mathbf{u}_2) \end{aligned}$$

$$M(c\mathbf{u}_1) = S(c\mathbf{u}_1) + T(c\mathbf{u}_1) = cS(\mathbf{u}_1) + cT(\mathbf{u}_1) = c[S(\mathbf{u}_1) + T(\mathbf{u}_1)] = cM(\mathbf{u}_1)$$

Thus $M = S + T$ is a linear mapping of U into V. Also

$$\begin{aligned} N(\mathbf{u}_1 + \mathbf{u}_2) &= aT(\mathbf{u}_1 + \mathbf{u}_2) = a[T(\mathbf{u}_1) + T(\mathbf{u}_2)] \\ &= aT(\mathbf{u}_1) + aT(\mathbf{u}_2) = N(\mathbf{u}_1) + N(\mathbf{u}_2) \\ N(c\mathbf{u}_1) &= aT(c\mathbf{u}_1) = acT(\mathbf{u}_1) = caT(\mathbf{u}_1) = cN(\mathbf{u}_1) \end{aligned}$$

Thus $N = aT$ is also a linear mapping of U into V.

EXAMPLE 1. The following are linear mappings of $\mathcal{C}^{(1)}$ into $\mathcal{C}$ (for a given interval): I (the identity mapping): $I(f) = f$, D (the derivative), and J, where $J(f)$ is the function $xf(x)$. Hence, the following mappings are also linear mappings of $\mathcal{C}^{(1)}$ into $\mathcal{C}$:

$$T_1 = 2D, \text{ where } T_1(f) = 2f', \qquad T_2 = 3J, \text{ where } T_2(f) = 3xf(x)$$

$$T_3 = D + 2I, \text{ where } T_3(f) = f' + 2f$$

$$T_4 = 3D + J - I, \text{ where } T_4(f) = 3f' + (x - 1)f$$

Now let T be a linear mapping of U into V. If $a \neq 0$, then $aT(a^{-1}\mathbf{x}) = T(\mathbf{x})$, and we conclude that Range T = Range aT; similarly, we have Kernel aT = Kernel T. If $a = 0$, then $aT(\mathbf{x}) = 0T(\mathbf{x}) = \mathbf{0}$ for all $\mathbf{x}$ and, hence, $0T = O$.

Now let $\mathfrak{F}$ be the set of all functions (not necessarily linear) mapping a given nonempty set X (not necessarily a vector space) into a vector space V. Thus, for each $\mathbf{f}$ in $\mathfrak{F}$ and for each x in X, $\mathbf{f}(x)$ is a vector in V. Then, as remarked above, we can define $\mathbf{f} + \mathbf{g}$ and $a\mathbf{f}$ to be the mappings with values $\mathbf{f}(x) + \mathbf{g}(x)$, $a\mathbf{f}(x)$, respectively, for all x in X. Hence, $\mathbf{f} + \mathbf{g}$ and $a\mathbf{f}$ are again functions in $\mathfrak{F}$. Thus $\mathfrak{F}$ is closed under addition and multiplication by scalars. It is now a simple matter to verify that $\mathfrak{F}$ is indeed a vector space (that is, $\mathfrak{F}$ satisfies the rules (9-10)). Now the set of all linear mappings of U into V is a nonempty subset of the set $\mathfrak{F}$ of all mappings of U into V and, as above, the set of linear mappings is closed under addition and multiplication by scalars. We conclude: *the set of all linear mappings of a vector space U into a vector space V is a vector space.* We can go further and prove:

THEOREM 25. *If U and V are finite-dimensional vector spaces (with the same scalars), then the set of all linear mappings of U into V is a finite-dimensional vector space of dimension* (dim U)(dim V).

PROOF. Let $\{\mathbf{u}_1, \ldots, \mathbf{u}_n\}$ be a basis for U and let $\{\mathbf{v}_1, \ldots, \mathbf{v}_m\}$ be a basis for V. For each pair i, j, let T_{ij} be the linear mapping of U into V such that

$$T_{ij}(\mathbf{u}_i) = \mathbf{v}_j, \qquad T_{ij}(\mathbf{u}_l) = \mathbf{0} \text{ for } l \neq i$$

There are mn mappings T_{ij}, one for each pair i, j, with $1 \leq i \leq n$, $1 \leq j \leq m$. If T is any linear mapping of U into V, then T is completely determined by its values at $\mathbf{u}_1, \ldots, \mathbf{u}_n$. Let

$$T(\mathbf{u}_s) = a_{s1}\mathbf{v}_1 + \cdots + a_{sm}\mathbf{v}_m \qquad (s = 1, \ldots, n)$$

The mapping $\sum_{i=1}^{n} \sum_{j=1}^{m} a_{ij}T_{ij}$ has the same value at each $\mathbf{u}_s$ as T and, therefore, this mapping coincides with T. Thus the T_{ij} together span the vector space of linear mappings of U into V. The set $T_{11}, T_{12}, \ldots, T_{nm}$ is linearly independent, since if $O = \Sigma\Sigma a_{ij}T_{ij}$, then $a_{ij} = 0$ for all subscripts i and j.

Since the linear mappings of U into V form a vector space, we have the usual rules for vector spaces:

$$\begin{aligned}
&S + T = T + S, \quad (S + T) + M = S + (T + M), \quad S + O = S \\
&S + T = S \text{ implies } T = O, \quad a(bS) = (ab)S, \quad (a + b)S = aS + bS \\
&1S = S, \quad (-1)S = -S, \quad a(S + T) = aS + aT \\
&S + T = O \text{ implies } T = -S
\end{aligned} \tag{9-161}$$

9-17 COMPOSITION OF LINEAR MAPPINGS

We saw in Chapter 2 that, if f was a function mapping X into Y, and g was a function mapping Y into Z, then we could form the composite function $g \circ f$, mapping X into Z: $(g \circ f)(x) = g[f(x)]$. Consequently, if S is a linear mapping of a vector space U into a vector space V and T is a linear mapping of V into W, then the composite mapping $T \circ S$ is defined and is a mapping of U into W. We now ask whether $T \circ S$ is a linear mapping of U into W. The answer is affirmative, as the following calculation shows. Let $M = T \circ S$. Then because of the linearity of S and T, we have

$$\begin{aligned}
M(\mathbf{u}_1 + \mathbf{u}_2) &= T[S(\mathbf{u}_1 + \mathbf{u}_2)] = T[S(\mathbf{u}_1) + S(\mathbf{u}_2)] \\
&= T(S(\mathbf{u}_1)) + T(S(\mathbf{u}_2)) = M(\mathbf{u}_1) + M(\mathbf{u}_2) \\
M(c\mathbf{u}_1) &= T(S(c\mathbf{u}_1)) = T(cS(\mathbf{u}_1)) = cT(S(\mathbf{u}_1)) = cM(\mathbf{u}_1)
\end{aligned}$$

Thus $M = T \circ S$ is a linear mapping of U into W.

EXAMPLE 1. If $f(x)$ has a continuous derivative on $(-\infty, \infty)$, then so does $e^x f(x)$. Hence, the mapping $T(f) = e^x f(x)$ is a mapping of $\mathcal{C}^{(1)}$ into itself. One can verify that T is a linear mapping. Since D is a linear mapping of $\mathcal{C}^{(1)}$ into $\mathcal{C}$, we conclude that $D \circ T$ is a linear mapping of $\mathcal{C}^{(1)}$ into $\mathcal{C}$. In particular,

$$(D \circ T)(f) = D(e^x f) = e^x f' + e^x f, \qquad \text{for all } f \text{ in } \mathcal{C}^{(1)}$$

EXAMPLE 2. Let S be the linear mapping of V_4 into V_3 given by

$$\begin{aligned}
S(1, 0, 0, 0) &= (0, 1, 1), \qquad & S(0, 1, 0, 0) &= (0, 1, 0) \\
S(0, 0, 1, 0) &= (2, 3, 1), \qquad & S(0, 0, 0, 1) &= (1, 2, 0)
\end{aligned}$$

Let T be the linear mapping of V_3 into V_2 given by

$$T(1, 0, 0) = (1, 2), \quad T(0, 1, 0) = (-2, 1), \quad T(0, 0, 1) = (3, 2)$$

Then if $M = T \circ S$, we have

$$\begin{aligned} M(1, 0, 0, 0) &= T(0, 1, 1) = T(0, 1, 0) + T(0, 0, 1) \\ &= (-2, 1) + (3, 2) = (1, 3) \end{aligned}$$

$$M(0, 1, 0, 0) = (-2, 1), \quad M(0, 0, 1, 0) = (-1, 9), \quad M(0, 0, 0, 1) = (-3, 4)$$

Since $\{(0, 1, 1), (0, 1, 0), (2, 3, 1)\}$ is a linearly independent set in V_3, S is an onto mapping. Similarly, since $\{(1, 2), (-2, 1)\}$ is a linearly independent subset of V_2, T is an onto mapping. But then $M = T \circ S$ is also an onto mapping. Thus rank $M = 2$ and, therefore, null $M = 4 - \text{rank } M = 2$.

If $\mathbf{u}$ is in Kernel M, then $T(S(\mathbf{u})) = \mathbf{0}$, and, therefore, $S(\mathbf{u})$ is in Kernel T = Span $(7, -4, -5)$. But then $\mathbf{u}$ is in the S-pre-image of Span $(7, -4, -5)$. Thus

$$\text{Kernel}(T \circ S) \subset S\text{-pre-image of Kernel } T = \text{Span}((9, 5, -4, 1), (-1, 2, 1, -2))$$

But null $T \circ S = \dim [\text{Kernel } (T \circ S)] = 2$, and, hence, Kernel $(T \circ S) =$ Span $((9, 5, -4, 1), (-1, 2, 1, -2))$.

Figure 9-19 gives a diagrammatic scheme for the behavior of the composition of two linear mappings. One can prove that rank $(T \circ S)$ *is less than or equal to the smaller of* rank T *and* rank S (Problem 10 below).

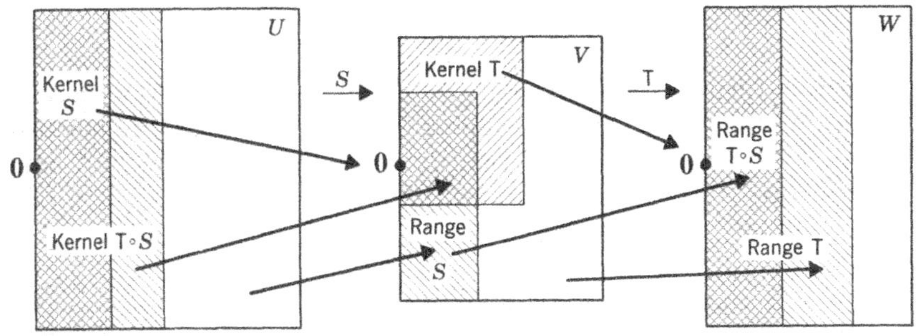

Figure 9-19 Composition of two linear mappings.

Notation. Since we do not have an ordinary product of linear mappings (see the first paragraph of Section 9-16), there can be no confusion if we denote the composite of two linear mappings by TS instead of by $T \circ S$. The former notation is more commonly used, and we shall use it from this point on. As is customary, we shall also use the phrase "the product of linear mappings" to mean "the composite of linear mappings."

Warning. We can speak of the product TS of two linear mappings only when Range S is a subset of the domain of T. In particular, TS may be defined while ST is not. In the next section we shall see that, even when ST and TS are both defined, they need not be equal.

Distributive and Associative Laws. In the next theorem we show that, if the various products and sums are defined, then the usual distributive laws hold true for products of linear mappings; also the associative law holds true.

THEOREM 26. *Let S_1, S_2 be linear mappings of U into V, T_1, T_2 linear mappings of V into W, and R a linear mapping of W into Z. Then*

$$T_1(S_1 + S_2) = (T_1S_1) + (T_1S_2) \tag{9-170}$$

$$(T_1 + T_2)S_1 = (T_1S_1) + (T_2S_1) \tag{9-171}$$

$$R(T_1S_1) = (RT_1)S_1 \tag{9-172}$$

$$(kT_1)S_1 = T_1(kS_1) = k(T_1S_1) \tag{9-173}$$

PROOF. Clearly T_1S_1 and T_1S_2 are linear mappings of U into W and, hence, their sum A is defined and is a linear mapping of U into W. Similarly, $B = T_1(S_1 + S_2)$ is a linear mapping of U into W.

If $\mathbf{u}$ is a vector in U, then

$$B(\mathbf{u}) = T_1[(S_1 + S_2)(\mathbf{u})] = T_1[S_1(\mathbf{u}) + S_2(\mathbf{u})] = T_1(S_1(\mathbf{u})) + T_1(S_2(\mathbf{u}))$$

while

$$A(\mathbf{u}) = (T_1S_1)(\mathbf{u}) + (T_1S_2)(\mathbf{u}) = T_1(S_1(\mathbf{u})) + T_1(S_2(\mathbf{u}))$$

Thus A and B agree at each $\mathbf{u}$ in U, so must be the same linear mapping. This proves (9-170). Relations (9-171), (9-172), and (9-173) are proved similarly.

9-18 INVERSE OF A LINEAR MAPPING

We saw in Chapter 2 that not every function has an inverse. More specifically, we saw that a function f mapping X into Y has an inverse function f^{-1} mapping Y into X precisely when f is a one-to-one mapping of X onto Y. In that case $f^{-1}(y) = x$ exactly when $f(x) = y$, and then

$$f^{-1}[f(x)] = x, \qquad f[f^{-1}(y)] = y$$

for all x in X and all y in Y, respectively. Thus if T is a one-to-one linear mapping of a vector space U onto a vector space V, then the inverse mapping T^{-1}, mapping V into U, exists, and

$$T^{-1}[T(\mathbf{u})] = \mathbf{u}, \qquad T[T^{-1}(\mathbf{v})] = \mathbf{v}$$

for all $\mathbf{u}$ in U and $\mathbf{v}$ in V. Furthermore, T^{-1} is a one-to-one onto mapping. Also T^{-1} is itself a linear mapping. For let $\mathbf{v}_1$, $\mathbf{v}_2$ be in V. Since T is one-to-one and onto, there exist unique vectors $\mathbf{u}_1$, $\mathbf{u}_2$ in U such that $T(\mathbf{u}_1) = \mathbf{v}_1$ and $T(\mathbf{u}_2) = \mathbf{v}_2$. Then

$$\begin{aligned} T^{-1}(\mathbf{v}_1 + \mathbf{v}_2) &= T^{-1}[T(\mathbf{u}_1) + T(\mathbf{u}_2)] = T^{-1}[T(\mathbf{u}_1 + \mathbf{u}_2)] \\ &= \mathbf{u}_1 + \mathbf{u}_2 = T^{-1}(\mathbf{v}_1) + T^{-1}(\mathbf{v}_2) \\ T^{-1}(a\mathbf{v}_1) &= T^{-1}[aT(\mathbf{u}_1)] = T^{-1}[T(a\mathbf{u}_1)] = a\mathbf{u}_1 = aT^{-1}(\mathbf{v}_1) \end{aligned}$$

This demonstrates that T^{-1} is a linear mapping.

We have shown: if T is a one-to-one linear mapping of U onto V then the inverse mapping T^{-1} is a one-to-one linear mapping of V onto U and

$$T^{-1}T = I_U, \; TT^{-1} = I_V \tag{9-180}$$

Here I_U is the identity mapping on U and I_V is the identity mapping on V. Since T^{-1} is a one-to-one, onto mapping, it also has an inverse $(T^{-1})^{-1}$. Now $(T^{-1})^{-1}(\mathbf{u}) = \mathbf{v}$ exactly when $T^{-1}(\mathbf{v}) = \mathbf{u}$, and $T^{-1}(\mathbf{v}) = \mathbf{u}$ exactly when $T(\mathbf{u}) = \mathbf{v}$. Hence T and $(T^{-1})^{-1}$ are the same mapping:

$$(T^{-1})^{-1} = T \tag{9-181}$$

EXAMPLE. The set U of linear homogeneous polynomials in x, y, z is a vector space and the mapping $T(ax + by + cz) = (a + b + c, b, c)$ is a one-to-one mapping of U onto V_3 with $T^{-1}(\alpha, \beta, \gamma) = (\alpha - \beta - \gamma)x + \beta y + \gamma z$.

PROBLEMS

1. Let S, T, M, N be the linear mappings of V_3 into V_3 given in Table 9-2.
   (a) Determine the rank and kernel of S and of T.
   (b) Same as part (a) for the mappings M and N.
   (c) Describe the mappings: $S + T$ and $S + M$ by giving their values at a basis for V_3. Also determine their range and kernel.
   (d) Same as part (c) for the mappings: $S + N$, $T + M$, $M + T + S$, $N - S$.
   ‡(e) Express S, T, M, and N as linear combinations of $T_{11}, \ldots, T_{33}$, where

   $$T_{ij}(a_1\mathbf{e}_1 + a_2\mathbf{e}_2 + a_3\mathbf{e}_3) = a_i\mathbf{e}_j, \qquad (i = 1, 2, 3, j = 1, 2, 3)$$

   (f) Find linear mappings A and B of rank 3 from V_3 into V_3 such that rank $(A + B) = 2$.
   (g) Find linear mappings A and B of rank 3 from V_3 into V_3 such that rank $(A + B) = 1$.

Table 9-2

	$\mathbf{e}_1$	$\mathbf{e}_2$	$\mathbf{e}_3$
S	$\mathbf{e}_1 - \mathbf{e}_2$	$\mathbf{e}_2 - 5\mathbf{e}_3$	$\mathbf{e}_1 - \mathbf{e}_2 + \mathbf{e}_3$
T	$4\mathbf{e}_1 - 2\mathbf{e}_2 + \mathbf{e}_3$	$\mathbf{e}_1 - 2\mathbf{e}_2 + 5\mathbf{e}_3$	$3\mathbf{e}_1 - 4\mathbf{e}_3$
M	$\mathbf{e}_2 + \mathbf{e}_3$	$3\mathbf{e}_3$	$\mathbf{0}$
N	$\mathbf{e}_1 + \mathbf{e}_2 + 3\mathbf{e}_3$	$-2\mathbf{e}_2 + 6\mathbf{e}_3$	$4\mathbf{e}_3$

2. Let S, T, M, N be the linear mappings of V_3 into V_3 given by Table 9-2 and let T_1, T_2, T_3, T_4 be the linear mappings of V_4 into V_3 given by Table 9-1 of the preceding problem set.
   (a) Describe the mappings TS, ST, TT, TTT by giving their values at a basis for V_3, and determine their rank.

(b) Same as part (a) for the mappings: MT_1, MMT_2, STT_3.

(c) A vector **x** is said to be *fixed* by a mapping T of V into V if $T(\mathbf{x}) = \mathbf{x}$. Find the fixed vectors for each of the mappings S, T, M, N.

3. Let S, T, M, N be as in Problem 1. (a) Determine whether S has an inverse and, if so, describe this inverse by giving its values for the standard basis of V_3.

(b) Proceed as in (a) for the mapping T. (c) Proceed as in (a) for M.

(d) Proceed as in (a) for N.

4. For the vector space $\mathcal{P}$ of all polynomials, let D denote the derivative, H the indefinite integral: $H(a_0 + \cdots + a_n x^n) = a_0 x + \cdots + [a_n/(n+1)]x^{n+1}$, and X the mapping: $X(a_0 + \cdots + a_n x^n) = a_0 x + \cdots + a_n x^{n+1}$. Then D, H, X are linear mappings from $\mathcal{P}$ to $\mathcal{P}$. Let J denote the definite integral from 0 to 1; for $j = 0, 1, \ldots$, let F_j denote the mapping: $F_j(a_0 + \cdots + a_n x^n) = a_j$. Then J and the F_j are linear mappings from $\mathcal{P}$ to R, which can also be viewed as linear mappings from $\mathcal{P}$ into $\mathcal{P}$. Prove (a), . . . , (f) and evaluate (g), . . . , (r).

(a) $F_{j-1}D = jF_j$.

(b) $XD \neq DX$ (Hence, the product of linear mappings need not commute).

(c) $DH = I =$ identity mapping.

(d) $HD = I - F_0$.

(e) $HD \neq DH$.

(f) $JF_j = F_j$, $(j = 0, 1, 2, \ldots)$.

(g) Kernel D,

(h) Kernel JD,

(i) Kernel F_j,

(j) Kernel $[DH - HD]$,

(k) Range $[DH - HD]$,

(l) Kernel $[XD - DX]$,

(m) Range $[XD - DX]$,

(n) Kernel $[F_jDXX]$,

(o) Kernel $[F_jDXH]$,

(p) Kernel $[DXD]$,

(q) Kernel $[F_0DXD]$,

(r) Kernel $[JDXD]$.

‡5. Prove: if neither U nor V is the zero space and one is infinite dimensional, then the set of linear mappings of U into V is an infinite dimensional vector space.

‡6. Let S be a linear mapping of U into V, and let T be a linear mapping of V into W. Prove:

(a) If S is onto, then rank $(TS) =$ rank T.

(b) If T is one-to-one, then rank $(TS) =$ rank S.

(c) If TS is one-to-one, then S is one-to-one.

(d) If TS is onto, then T is onto.

‡7. Find examples of linear mappings S, T such that TS is defined, $T \neq O$, $S \neq O$, and $TS = O$.

‡8. Let S, T be linear mappings of U into V.

(a) Show that Kernel $(S + T) \supset$ Kernel $S \cap$ Kernel T.

(b) Give an example where Kernel $(S + T) =$ Kernel $S \cap$ Kernel T.

(c) Give an example where equality does not hold true in the equation of part (b).

(d) Show that there exist S, T with rank $S =$ rank $T =$ min (dim U, dim V) such that rank $(S + T)$ can be any value from 0 up to min (dim U, dim V).

9. Complete the proof of Theorem 26.

‡10. Let S be a linear mapping of U into V and let T be a linear mapping of V into W. Prove:

(a) Kernel (TS) = S-pre-image of (Kernel $T \cap$ Range S).

(b) Range $(TS) = T(\text{Range } S)$.

(c) rank $(TS) \leq \min$ (rank T, rank S). [*Hint.* If Z is a subspace of V, then $\dim T(Z) \leq \dim Z$.]

(d) null $S \leq$ null $(TS) \leq$ null S + null T.

‡11. Give examples where null $T <$ null (TS), null $T =$ null (TS), null $T >$ null (TS).

9-19 LINEAR TRANSFORMATIONS ON A VECTOR SPACE

Thus far we have considered linear mappings from one vector space U into a vector space V. It may happen that U and V are the *same* vector space, so that we have a linear mapping of U into U. We call such a mapping a *linear transformation on U*. Thus a linear transformation on a vector space is a linear mapping of that vector space into itself. (The words transformation and mapping are sometimes used in other senses, but we shall consistently use them only as here indicated.)

EXAMPLE 1

(a) The derivative D is a linear transformation on the vector space $\mathcal{P}$ of all polynomials.

(b) The derivative is *not* a linear transformation on $\mathcal{C}^{(1)}(-\infty, \infty)$. For example, $D\left(\int_0^x |t|\, dt\right) = |x|$, a function which is not differentiable at $x = 0$ and so is not in $\mathcal{C}^{(1)}(-\infty, \infty)$.

(c) Let $\mathcal{A}$ be the set of all real analytic functions on $(-1, 1)$; that is, f is in $\mathcal{A}$ if f is the sum of a power series $\Sigma a_n x^n$ converging for $-1 < x < 1$. If f is in $\mathcal{A}$, then so is $f', f'', \ldots$ (see Section 8-17). Hence, $D, D^2, D^3, \ldots$ are all linear transformations on $\mathcal{A}$.

(d) The mapping $T(f) = f(x) \sin x$ is a linear transformation on $\mathcal{C}^{(1)}(-\infty, \infty)$.

EXAMPLE 2. Let U be the set of all vectors in the plane. Then the following are linear transformations on U.

(a) $T_c(\mathbf{u}) = c\mathbf{u}$, c a fixed real number.

(b) $J(\mathbf{u}) = J(a\mathbf{i} + b\mathbf{j}) = a\mathbf{i} - b\mathbf{j}$.

(c) $K_\alpha(\mathbf{u}) = \mathbf{v}$, where $\mathbf{v} = \mathbf{0}$ if $\mathbf{u} = \mathbf{0}$ and otherwise $\mathbf{v}$ has the same length as $\mathbf{u}$ and the directed angle from $\mathbf{u}$ to $\mathbf{v}$ is α, where α is fixed. For $\alpha = \pi/2$, $K_\alpha(\mathbf{u}) = \mathbf{u}^\perp$.

When $c = 1$, $T_c = I$ (the identity); when $c > 1$, T_c is called a *dilation;* when $0 < c < 1$, T_c is called a *contraction;* when $c = -1$, T_c is called a *reflection in the origin*. The mapping J is called a *reflection in the x-axis*. The mapping

K_α is called a *rotation* through angle α (see Figure 9-20). We note that $K_\pi = T_{-1}$.

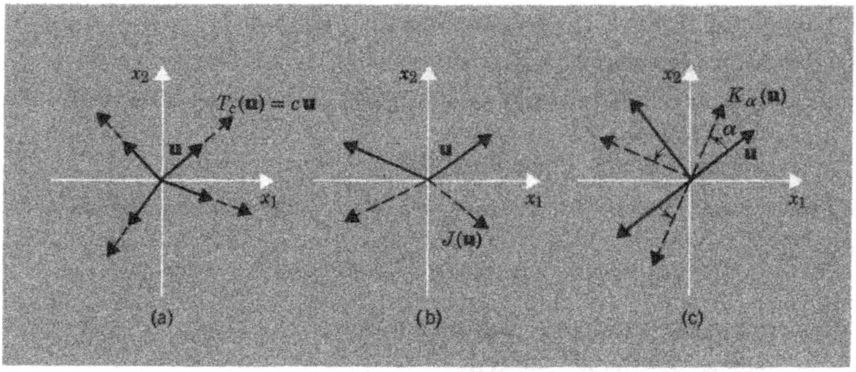

Figure 9-20 (*a*) Dilation; (*b*) reflection; (*c*) rotation.

If S and T are linear transformations on a vector space U, then $S + T$ and cS are defined and are also linear transformations on U (see Section 9-16). Hence, the set of *all* linear transformations on the vector space U is a vector space. We denote this set by $Lt(U)$. Furthermore, by Theorem 25, if U has finite dimension, then dim $Lt(U) = [\dim U]^2$.

Since the domain and target space of a linear transformation are the same vector space, it follows (see Section 9-17) that, if S and T are linear transformations on U, then ST and TS are both defined and are both linear transformations on U. The linear transformations ST and TS need not be equal. For example, in the notations of Example 2 above,

$$K_{\pi/2}J(\mathbf{i}) = K_{\pi/2}(\mathbf{i}) = \mathbf{j}, \qquad JK_{\pi/2}(\mathbf{i}) = J(\mathbf{j}) = -\mathbf{j}$$

Hence, $K_{\pi/2}J \neq JK_{\pi/2}$. If $ST = TS$, we say that S and T *commute*. The linear transformations K_α and T_c of Example 2 can be shown to commute.

If U is of dimension 1, then each pair of linear transformations on U commute (Problem 6 below). However, if dim $U \geq 2$, then there exist pairs of linear transformations on U which do not commute. For let $\{\mathbf{u}_1, \mathbf{u}_2, \ldots\}$ be a basis for U. Let $T(\mathbf{u}_2) = \mathbf{u}_1$, $T(\mathbf{u}_j) = \mathbf{0}$ if $j \neq 2$ and let $S(\mathbf{u}_1) = \mathbf{u}_2$, $S(\mathbf{u}_j) = \mathbf{0}$ for $j \geq 2$. Then $ST \neq TS$, since $ST(\mathbf{u}_1) = S(\mathbf{0}) = \mathbf{0}$, while $TS(\mathbf{u}_1) = T(\mathbf{u}_2) = \mathbf{u}_1 \neq \mathbf{0}$.

The zero transformation O commutes with all linear transformations on U.

The Identity Transformation. For each vector space U, we denote by I_U, or simply by I (if the space U is clearly understood), the linear transformation on U such that

$$I(\mathbf{u}) = \mathbf{u} \qquad \text{for all } \mathbf{u} \text{ in } U$$

For each linear transformation T on U, we then have

$$IT = T = TI \tag{9-190}$$

Thus, I commutes with all linear transformations on U and, relative to multiplication of linear transformations, it behaves as does the real number 1 for the ordinary multiplication of numbers. We call I the *identity transformation.*

Rules for Multiplication of Linear Transformations. By Theorem 26 (Section 9-17), multiplication of linear transformations obeys certain rules. We collect them here: *for all S, T, R in $Lt(U)$ and each scalar c,*

$$\begin{aligned} &S(T+R) = ST + SR, \qquad (S+T)R = SR + TR \\ &c(ST) = S(cT) = (cS)T, \qquad S(RT) = (SR)T, \qquad SO = OS = O \end{aligned} \tag{9-191}$$

The Algebra of Linear Transformations on a Vector Space. The set $Lt(U)$ of all linear transformations on U forms a vector space in which we have a multiplication obeying the rules (9-191). A vector space having such a multiplication is called an *algebra*. We have met other examples of algebras: for example, the real numbers, the complex numbers, the set $\mathcal{C}[a, b]$ of all continuous functions on $[a, b]$, and the set $\mathcal{P}$ of all polynomials—all of these are examples of algebras. The set $\mathcal{P}_m$ of all polynomials of degree at most m is *not* an algebra, since it is not closed under multiplication. An algebra has many properties in common with a number system; as we just observed, the real and complex number systems are algebras; in particular, an algebra is closed under addition and multiplication. However, an algebra need not have all the properties of a number system. For example, products need not commute, and in some algebras the product of two nonzero elements is zero. This is illustrated by the algebra of linear transformations of the vector space V_2. Here the transformation T: $T(x_1, x_2) = (x_2, 0)$ has the property that $T^2(\mathbf{x}) = \mathbf{0}$ for all $\mathbf{x}$; that is, $TT = O$, yet $T \neq O$. Also, an algebra need not contain a multiplicative identity I satisfying (9-190).

For algebras we have the concept of *subalgebra*, analogous to that of subspace for a vector space. A subalgebra of an algebra is a subset which forms an algebra by itself (with the same operations). As for vector spaces, a subalgebra of an algebra can be characterized as a subset of the algebra which is closed under the operations concerned: here, addition, multiplication by scalars, *and* multiplication of two elements of the algebra. An example of a subalgebra is $\mathcal{P}$, as a subset of $\mathcal{C}(-\infty, \infty)$. Also, in any vector space U having a nonzero element $\mathbf{u}$, the set of all transformations T on U such that $T(\mathbf{u})$ is a scalar times $\mathbf{u}$ forms a subalgebra B of $Lt(U)$. For if S and T are in B, then $S(\mathbf{u}) = a\mathbf{u}$, $T(\mathbf{u}) = b\mathbf{u}$, so that $(S + T)\mathbf{u} = (a + b)\mathbf{u}$, $(cS)(\mathbf{u}) = (ca)\mathbf{u}$, and $(ST)(\mathbf{u}) = (ab)\mathbf{u}$.

9-20 POLYNOMIALS IN A LINEAR TRANSFORMATION

With the aid of multiplication, we can define nonnegative powers of a linear transformation T on U:

$$T^0 = I, \quad T^1 = T, \quad T^2 = TT, \quad \ldots, \quad T^k = T(T^{k-1}), k \geq 1 \tag{9-200}$$

Each such power of T is again in $Lt(U)$.

EXAMPLE 1. Let S be the linear transformation on V_2 such that $S(\mathbf{e}_1) = \mathbf{e}_2$, $S(\mathbf{e}_2) = \mathbf{0}$. Then $S^0 = I$, $S^1 = S$, $S^2 = O$, $S^3 = O, \ldots$

EXAMPLE 2. Let M be the linear transformation on V_2 such that $M(\mathbf{e}_1) = \mathbf{e}_1$, $M(\mathbf{e}_2) = \mathbf{0}$. Then $M^0 = I$, $M^1 = M$, $M^2 = M$, $M^3 = M, \ldots$

EXAMPLE 3. Let N be the linear transformation on $\mathcal{C}(-\infty, \infty)$ such that $N(f) = xf(x)$, for all f in $C(-\infty, \infty)$. Then $N^0 = I$, $N^1 = N$, $N^2(f) = N(xf) = x^2f(x)$, $N^3(f) = x^3f(x)$, and, in general, $N^m(f)$ is the function $x^m f(x)$.

If $T^k = O$ for some positive integer k, we say that T is *nilpotent*. If $T^2 = T$, so that $T^k = T$ for all positive integers k, we say that T is *idempotent*. Example 1 illustrates a nilpotent transformation, Example 2 illustrates an idempotent transformation. The linear transformation of Example 3 is neither nilpotent nor idempotent. Clearly, the zero-transformation O is both nilpotent and idempotent, and no other transformation is both; for if $T^k = O$ for some k and $T^2 = T$, then $T = T^2 = \cdots = T^k = O$.

The powers of a linear transformation T satisfy the usual rules for positive exponents:

$$T^kT^l = T^lT^k = T^{l+k} \tag{9-201}$$

$$(T^k)^l = T^{lk} \tag{9-202}$$

The rules (9-201) and (9-202) can be proved by induction.

With the aid of addition, multiplication by scalars and formation of powers, we can now form *polynomials in the linear transformation T*:

$$a_kT^k + a_{k-1}T^{k-1} + \cdots + a_1T + a_0I$$

(I could be replaced by T^0 in the last term). Each such "polynomial in T" is again a linear transformation on U; that is, it is in $Lt(U)$.

Now let $g(x) = a_kx^k + a_{k-1}x^{k-1} + \cdots + a_1x + a_0$ be an ordinary polynomial whose coefficients are from the set of scalars for U (real or complex). With $g(x)$ and the particular transformation T in $Lt(U)$ we associate the linear transformation $a_kT^k + a_{k-1}T^{k-1} + \cdots + a_1T + a_0I$. We write

$$g(T) = a_kT^k + a_{k-1}T^{k-1} + \cdots + a_1T + a_0I$$

to indicate the relationship. The notation suggests that we are evaluating g at the value $x = T$, and it is simply a convenient way to describe the particular linear transformation $a_kT^k + \cdots + a_1T + a_0I$ obtained from T and a certain set of scalars $a_k, \ldots, a_0$ (given in that order).

THEOREM 27. *Let U be a vector space, let T be a linear transformation on U. Let g and h be polynomials, with coefficients of the same type as the scalars in U; let c denote such a scalar. Let $g(x) + h(x) = p(x)$, $g(x)h(x) = q(x)$, $cg(x) = r(x)$. Then*

$$g(T) + h(T) = p(T), \qquad g(T)h(T) = q(T), \qquad cg(T) = r(T)$$

Thus addition and multiplication of polynomials in T, and multiplication of such a polynomial by a scalar can be carried out exactly as for ordinary *polynomials. Furthermore, two polynomials in T are commuting linear transformations.*

PROOF. Let $g(x) = a_k x^k + \cdots + a_0$, $h(x) = b_l x^l + \cdots + b_0$. Then, as defined above,

$$g(T) = a_k T^k + \cdots + a_0 I, \qquad h(T) = b_l T^l + \cdots + b_0 I$$

If m is the larger of k and l, by adding terms with zero coefficient if necessary we can write $g(T) = a_m T^m + \cdots + a_0 I$, and $h(T) = b_m T^m + \cdots + b_0 I$. Then

$$\begin{aligned} g(T) + h(T) &= (a_m T^m + \cdots + a_0 I) + (b_m T^m + \cdots + b_0 I) \\ &= (a_m + b_m)T^m + \cdots + (a_1 + b_1)T + (a_0 + b_0)I \end{aligned}$$

But by algebra $p(x) = g(x) + h(x) = (a_m + b_m)x^m + \cdots + (a_1 + b_1)x + (a_0 + b_0)$ in exactly the same way. Hence, $g(T) + h(T) = p(T)$.

The proof for the rules $g(T)h(T) = q(T)$, $cg(T) = r(T)$ is similar.

Since $g(x)h(x) = h(x)g(x)$, we have $g(T)h(T) = h(T)g(T)$. Thus two polynomials in T always commute.

Remark. Each polynomial in T is a specific linear transformation on U, obtained from a specific transformation T. It may thus happen that two different polynomials in x lead to polynomials in T which are equal: that is, are the same linear transformation. For example, if $T^2 = T$ (T idempotent), then for this particular T we have $g(T) = h(T)$ with $g(x) = x^2$, $h(x) = x$. In general, $g(T)$ may equal O for certain choices of T [just as $g(a)$ may equal 0 for certain choices of the scalar $x = a$].

The results of Theorem 27 show that the set of all polynomials in a linear transformation T form an algebra.

Differential Operators. We consider the vector space $\mathcal{C}^{(\infty)}$ for a fixed interval. The derivative D is a linear transformation on $C^{(\infty)}$ and, hence, we can form polynomials in D such as $D + I$, $5D^2 + 3D + 2I$. Here it is customary to replace the I by 1 and to write simply $D + 1$, $5D^2 + 3D + 2$. By the definitions above,

$$(5D^2 + 3D + 2)f = 5D^2 f + 3Df + 2f = 5f'' + 3f' + 2f$$

for every f in $C^{(\infty)}$. These polynomials in D are called *differential operators with constant coefficients.*

By algebra, we have the identity $x^2 - 1 = (x + 1)(x - 1)$ and, therefore, we have the identity

$$D^2 - 1 = (D + 1)(D - 1)$$

For example, let $f(x) = \sin 3x$. Then

$$\begin{aligned} (D^2 - 1)f &= f'' - f = -10 \sin 3x, \\ (D + 1)(D - 1)f &= (D + 1)[(D - 1)f] = (D + 1)(f' - f) \\ &= (D + 1)(3 \cos 3x - \sin 3x) \\ &= (3 \cos 3x - \sin 3x)' + 3 \cos 3x - \sin 3x = -10 \sin 3x \end{aligned}$$

PROBLEMS

1. Let T be the linear transformation on V_2 such that $T(\mathbf{e}_1) = (1, 2)$, $T(\mathbf{e}_2) = (3, 1)$, so that $T(x, y) = (x + 3y, 2x + y)$.
   (a) Find $T^2(\mathbf{e}_1)$ and $T^2(\mathbf{e}_2)$ and, hence, obtain $T^2(x, y)$; then show that $T^2 = 2T + 5I$.
   (b) From the result of part (a) show that $T^4 = 4T^2 + 20T + 25I$.
   (c) From the results of parts (a) and (b) show that $T^4 = 28T + 45I$ and, hence, find $T^4(3, 2)$ and $T^4(-1, 7)$.
   (d) From the previous results show that $T^3 = 9T + 10I$ and, hence, find $T^3(5, 1)$ and $T^3(0, 6)$.
2. Let T be the linear transformation on V_2 such that $T(\mathbf{e}_1) = (1, 1)$ and $T(\mathbf{e}_2) = (0, 1)$, so that $T(x, y) = (x, x + y)$.
   (a) Find $T^2(\mathbf{e}_1)$ and $T^2(\mathbf{e}_2)$ and, hence, obtain $T^2(x, y)$. Then show that $T^2 - 2T + I = O$ and that $(T - I)^2 = O$, but that $T - I \neq O$.
   (b) From the results of part (a) show that $(T - I)^4 = O$ and that $T^4 = 4T^2 - 4T + I$.
   (c) From the results of parts (a) and (b) show that $T^4 = 4T - 3I$ and find $T^4(4, -2)$ and $T^4(1, 4)$.
   (d) From the previous results show that $T^3 = 3T - 2I$ and evaluate $T^3(5, 7)$ and $T^3(1, 4)$.
3. Let T be a linear transformation on V_3 such that $T(\mathbf{e}_1) = \mathbf{e}_2$, $T(\mathbf{e}_2) = \mathbf{e}_3$, $T(\mathbf{e}_3) = \mathbf{e}_1$.
   (a) Find $T^2(\mathbf{e}_i)$ and $T^3(\mathbf{e}_i)$ for $i = 1, 2, 3$ and, hence, show that $T^3 = I$.
   (b) Verify that $T^{-1} = T^2$.
   (c) Interpret T geometrically.
4. Let T be a linear transformation of V_3 such that $T(\mathbf{e}_1) = \mathbf{e}_2 + \mathbf{e}_3$, $T(\mathbf{e}_2) = \mathbf{e}_3 + \mathbf{e}_1$, $T(\mathbf{e}_3) = \mathbf{e}_1 + \mathbf{e}_2$.
   (a) Find $T^2(\mathbf{e}_i)$ and $T^3(\mathbf{e}_i)$ for $i = 1, 2, 3$ and, hence, show that $T^3 = 3T + 2I$.
   (b) Verify that $T^{-1} = \frac{1}{2}(T^2 - 3I)$.
5. Determine whether the following linear transformations are nilpotent, idempotent, or neither.
   (a) $T(x, y) = (-x, -y)$, (b) $T(x, y) = (x, 0)$,
   (c) $T(x, y) = (0, x)$, (d) $T(x, y, z) = (z, x, y)$,
   (e) $T(x, y, z) = (y + z, z, 0)$, (f) $T(x, y, z) = (x, 0, z)$.
6. Show that if U is one-dimensional and S, T are linear transformations on U, then $ST = TS$.
7. Let U be a one-dimensional complex vector space, so that the vectors of U can be represented as complex numbers z. Let S, T be linear transformations on U defined by $S(z) = e^{i\pi/4}z$, $T(z) = iz$. Show that
   (a) $T^2 = -I$ (b) $T^4 = I$ (c) $S^2 = T$ (d) $STST = -T$
   (e) $S^4 = -I$ (f) $S^8 = I$ (g) $ST = TS = S^3$
8. (a) Show that the derivative D is a linear transformation on $\mathcal{P}_h$, the vector space

of polynomials of degree at most h, and find the kernel and range of each of the transformations D, D^2, D^h, D^{h+1}.

(b) Verify that each of the identities:

$$\text{(i) } (D^2 + 3D + 2)f = (D + 1)(D + 2)f,$$
$$\text{(ii) } (D^3 + 1)f = (D + 1)(D^2 - D + 1)f$$

is true for $f(x) = x^k$, where k is a positive integer.

9. (a) Let U, V be finite dimensional vector spaces and let T be a linear mapping of U into V. Prove: If dim U = dim V and T is onto then T is one-to-one. If dim U = dim V and T is one-to-one, then T is onto. [*Hint.* Use Theorem 22.]

(b) Let T be a linear transformation on a finite dimensional space U. Prove: if T is one-to-one, then T is onto. If T is onto, then T is one-to-one.

(c) Show that the conclusions in part (a) are false if dim U = dim $V = \infty$.

(d) Show that the conclusions in part (a) are false if dim $U \neq$ dim V.

(e) Show that the conclusions in part (b) are false if dim $U = \infty$.

‡10. Let S and T be linear transformations on a vector space U such that $ST = TS$. Prove the following:

(a) {Kernel S + Kernel T} $\subset$ Kernel ST.

(b) If Kernel $S \cap$ Kernel T is the zero space V_0 and null S is finite, then T maps Kernel S one-to-one onto itself.

(c) Under the assumptions of part (b), {Kernel S + Kernel T} = Kernel ST.

(d) If U is infinite-dimensional, with basis $\{\mathbf{u}_0, \mathbf{u}_1, \ldots, \mathbf{u}_n, \ldots\}$ and $S(\mathbf{u}_0) = \mathbf{u}_1$, $S(\mathbf{u}_{2i-1}) = \mathbf{u}_{2i+1}$, $S(\mathbf{u}_{2i}) = \mathbf{0}$ for $i = 1, 2, \ldots$ and $T(\mathbf{u}_0) = \mathbf{u}_2$, $T(\mathbf{u}_{2i-1}) = \mathbf{0}$, $T(\mathbf{u}_{2i}) = \mathbf{u}_{2i+2}$ for $i = 1, 2, \ldots$, then $ST = TS$ and Kernel $ST \neq$ {Kernel S + Kernel T}. Thus in part (c) the assumption that null S (or, alternatively, null T) be finite, is necessary to obtain the conclusion.

(e) Show that on $\mathcal{P}_3$, Kernel $D^2 \neq$ {Kernel D + Kernel D} = Kernel D. Hence for part (c) the first assumption of part (b) is necessary.

9-21 NONSINGULAR LINEAR TRANSFORMATIONS

A linear transformation T on a vector space U is said to be *nonsingular* if T is one-to-one and onto. The identity I and the mappings cI ($c \neq 0$), called *scalar mappings*, are nonsingular linear transformations. Linear transformations which are not nonsingular are called *singular* linear transformations.

EXAMPLE 1. (a) The mapping $T(x, y) = (x + y, y)$ is a nonsingular linear transformation on V_2, since $T(x, y) = (0, 0)$ if, and only if, $x = y = 0$ (and, hence, T is one-to-one) and $T(a - b, b) = (a, b)$ (and, hence, T is onto).

(b) The mapping $T(f) = e^x f(x)$ is a nonsingular linear transformation on $\mathcal{C}(-\infty, \infty)$, since $e^x f(x) \equiv 0$ if, and only if, $f(x) \equiv 0$, so that Kernel T is the zero function and T is one-to-one; also $T(e^{-x} f) = f$, so that T is onto.

(c) The derivative D is singular on the space $\mathcal{C}^{(\infty)}$, since Kernel D equals Span (1), which is not the zero space and, hence, D is not one-to-one.

Let U have finite dimension n, and let T be a linear transformation on U. Then by Theorem 22 in Section 9-14, we have

$$n = \dim U = \text{null } T + \text{rank } T \tag{9-210}$$

We conclude that *a linear transformation on U is one-to-one, if and only if, it is an onto mapping.* For T is one-to-one precisely when Kernel $T = \{\mathbf{0}\}$ or, equivalently, when null $T = 0$; and T is onto exactly when Range $T = U$ or, equivalently, rank $T = \dim U$. But (9-210) shows that null $T = 0$ exactly when rank $T = \dim U$. Accordingly, *every one-to-one transformation on a finite-dimensional vector space is nonsingular, and every transformation of that space onto itself is nonsingular.* This result does not hold true for infinite-dimensional vector spaces. For example, D maps $\mathcal{C}^{(\infty)}$ onto itself but is not one-to-one.

Nonsingular linear transformations on U are important because they are one-to-one, onto mappings and so have inverses and the discussion in Section 9-18 is applicable. Hence, *if T is a nonsingular linear transformation on U, then its inverse T^{-1} exists and is also a nonsingular linear transformation on U. Furthermore,*

$$TT^{-1} = I, \qquad T^{-1}T = I, \qquad (T^{-1})^{-1} = T \tag{9-211}$$

The following theorem helps characterize nonsingular linear transformations.

THEOREM 28. *Let S, T be linear transformations on U.*

(a) *If U is finite dimensional and $TS = I$, then T and S are nonsingular and $T = S^{-1}$, $S = T^{-1}$.*

(b) *If U is not necessarily finite dimensional and*

$$TS = I = ST$$

then T and S are nonsingular and $T = S^{-1}$, $S = T^{-1}$.

PROOF. (a) Let U be finite-dimensional and let S, T be linear transformations on U such that $TS = I$. Then $S(\mathbf{u}) = \mathbf{0}$ implies $T[S(\mathbf{u})] = T(\mathbf{0}) = \mathbf{0}$, so that $(TS)(\mathbf{u}) = \mathbf{0}$. But $TS = I$. Hence, $S(\mathbf{u}) = \mathbf{0}$ implies $\mathbf{u} = \mathbf{0}$ and, therefore, the kernel of S consists of $\mathbf{0}$ alone. Therefore, S is one-to-one and, therefore, S is nonsingular. Consequently, S^{-1} exists and

$$T = TI = T(SS^{-1}) = (TS)S^{-1} = IS^{-1} = S^{-1}$$

Accordingly, T is also nonsingular and $T^{-1} = (S^{-1})^{-1}$. But, by (9-211), $(S^{-1})^{-1} = S$ and hence $T^{-1} = S$.

(b) For general U, let S, T be linear transformations on U such that $TS = I = ST$. Then as in the proof of (a) we show that both S and T are one-to-one. If Range $S \neq U$, then choose $\mathbf{u}_1$ not in Range S. Since T is one-to-one, it follows that $T(\mathbf{u}_1)$ is not in T(Range S) = Range (TS). This is impossible, since $TS = I$. Hence, Range $S = U$ and S is nonsingular. Similarly, T is nonsingular. As in the proof of (a) we conclude that $S = T^{-1}$, $T = S^{-1}$.

COROLLARY OF THEOREM 28. *If S, T are nonsingular linear transformations on U, then so are ST and cT for each nonzero scalar c. Furthermore, $(ST)^{-1} = T^{-1}S^{-1}$ and $(cT)^{-1} = c^{-1}T^{-1}$.*

PROOF. We have $(ST)(T^{-1}S^{-1}) = S(TT^{-1})S^{-1} = SIS^{-1} = SS^{-1} = I$ and, similarly, $(T^{-1}S^{-1})(ST) = I$. Hence, by Theorem 28, ST is nonsingular and its inverse is $T^{-1}S^{-1}$. The proof for cT is similar.

We can extend the result to products of more than two transformations:

$$(STP)^{-1} = P^{-1}T^{-1}S^{-1}, \qquad (STPQ)^{-1} = Q^{-1}P^{-1}T^{-1}S^{-1}, \qquad \ldots$$

The proof is similar. Notice that *the inverse of a product of linear transformation is the product of the inverses in reverse order.*

We note in particular that, if T is nonsingular, then so also are $T^2, T^3, \ldots, T^m, \ldots$. Consequently, T^m is never O (except for the trivial case when U is V_0) and T is never nilpotent.

Let T be a nonsingular transformation on U and let S be a linear transformation on U. Then the equation $TX = S$ has a unique solution for X in the algebra $Lt(U)$. For clearly $T^{-1}S$ is a solution and, if X and X_1 are solutions, then $O = T^{-1}O = T^{-1}(TX - TX_1) = T^{-1}TX - T^{-1}TX_1 = X - X_1$. Therefore, the solution is unique. The equation $YT = S$ also has a unique solution for Y, namely, ST^{-1}. Since ST^{-1} and $T^{-1}S$ need not be equal, the two equations need not have the same solution.

Negative Powers of a Nonsingular Linear Transformation. In the preceding section we defined nonnegative integral powers of a linear transformation T on U. If T is nonsingular, we can also define negative integral powers: T^{-1} is the inverse of T and, for $m > 1$, T^{-m} is defined to be $(T^{-1})^m$. By the corollary to Theorem 28, $(T^{-1})^m = (T^m)^{-1}$, so that T^{-m} is the inverse of T^m. Furthermore, in general, if T is nonsingular and k, l are integers (positive, negative or 0), then

$$T^kT^l = T^lT^k = T^{l+k}, \qquad (T^k)^l = T^{kl} \tag{9-212}$$

The proof is left as an exercise (Problem 6 below).

EXAMPLE 2. If $T(x, y) = (x + y, 2x - y)$, show that T is a nonsingular linear transformation on V_2 and determine T^{-1}.

Solution. Clearly T is a linear transformation. It is nonsingular, since V_2 is finite-dimensional and T has kernel $\{\mathbf{0}\}$; for if $T(x, y) = \mathbf{0}$, then $x + y = 0$ and $2x - y = 0$ and, accordingly, $x = y = 0$.

Now let $T(x, y) = (x', y')$, so that $x' = x + y$, $y' = 2x - y$. If we solve these equations for x and y, we obtain $x = (\frac{1}{3})(x' + y')$, $y = (\frac{1}{3})(2x' - y')$. The inverse transformation T^{-1} takes (x', y') to (x, y) or (on interchanging primed and unprimed letters), $T^{-1}(x, y) = (\frac{1}{3})(x + y, 2x - y)$.

For this example, we could also have found the result by observing that $T^2 = 3I$, so that, by Theorem 28, we must have $T^{-1} = (\frac{1}{3})T$. The following theorem generalizes this idea.

THEOREM 29. *Let $g(x)$ be a polynomial with nonzero constant term $g(0)$ and let $p(x)$ be the polynomial such that $g(x) = g(0)(1 - xp(x))$. Let T be a linear transformation on U such that $g(T) = O$. Then T is nonsingular and $T^{-1} = p(T)$.*

PROOF. Let $g(x) = c_0 + c_1x + \cdots + c_kx^k$, so that $c_0 = g(0)$ is the constant term of g and, by assumption, $c_0 \neq 0$. Now let

$$\begin{aligned} h(x) = \frac{g(x)}{c_0} &= 1 + b_1x + \cdots + b_kx^k \\ &= 1 - x(-b_1 - b_2x - \cdots - b_kx^{k-1}) \\ &= 1 - xp(x) \end{aligned}$$

with $p(x) = -b_1 - b_2x - \cdots - b_kx^{k-1}$. Then also $h(T) = O$, so that $I + b_1T + \cdots + b_kT^k = O$, or

$$I = TS = ST, \qquad \text{where} \qquad S = -b_1I - b_2T - \cdots - b_kT^{k-1} = p(T)$$

Therefore $T^{-1} = p(T)$, as asserted.

9-22 THE MINIMAL POLYNOMIAL OF A LINEAR TRANSFORMATION

The preceding theorem raises the question: given a linear transformation T, are there nonzero polynomials $g(x)$ such that $g(T) = O$? If dim U is infinite, the answer is: not necessarily; but if dim U is finite, the answer is yes.

THEOREM 30. *Let U be an n-dimensional vector space and let T be a linear transformation on U. Then there is a nonzero polynomial $g(x)$ such that $g(T) = O$.*

PROOF. By Theorem 25, $Lt(U)$ is a vector space of dimension n^2. Hence, each set of $n^2 + 1$ elements from $Lt(U)$ is linearly dependent. In particular, $I, T, T^2, \ldots, T^{n^2}$ is a linearly dependent set. Hence, there exist scalars a_0, $a_1, \ldots, a_{n^2}$, not all 0, such that

$$a_0I + a_1T + \cdots + a_{n^2}T^{n^2} = O$$

Accordingly, the polynomial $g(x) = a_{n^2}x^{n^2} + \cdots + a_1x + a_0$ is nonzero and $g(T) = O$.

Remark. Theorem 30 can be improved. We can always find a nonzero polynomial $g(x)$ of degree at most n satisfying the conclusion of the theorem. We shall prove this in Chapter 10 for linear transformations on V_n.

Definition. Let T be a linear transformation on U. A nonzero polynomial $g(x)$ is called a *minimal polynomial for T* if $g(T) = O$ and there is no nonzero polynomial $q(x)$ of degree less than that of g for which $q(T) = O$.

Once we know that $q(T) = O$ for some nonzero polynomial q, we know that T must have a minimal polynomial. For we simply choose $g(x)$ of degree as small as possible so that $g(T) = O$. In particular, if dim $U = n$, then T has a minimal polynomial of degree at most n^2 (or, as in the Remark, of degree at most n).

If $g(x)$ is a minimal polynomial for T, then $p(x)$ is a minimal polynomial for T if, and only if, $p(x) = cg(x)$ for a nonzero scalar c (Problem 7 below).

EXAMPLE 1. Show that, if a linear transformation T has a minimal polynomial $g(x)$ whose constant term is 0, then T is singular.

Solution. Let $g(x) = c_k x^k + \cdots + c_1 x$. Then

$$O = g(T) = TS, \qquad \text{where} \qquad S = c_k T^{k-1} + \cdots + c_1 I$$

If T is nonsingular, then T^{-1} exists and, hence,

$$O = T^{-1}(TS) = S$$

But then $q(T) = O$ for $q(x) = c_k x^{k-1} + \cdots + c_1$, so that $g(x)$ is not minimal. This contradicts our assumption. Thus T is singular.

EXAMPLE 2. Determine whether the linear transformation $T(x, y) = (x + 2y, -3x + y)$ is nonsingular and, if so, find its inverse.

Solution. According to the remark above, T has a minimal polynomial of degree at most 2. If the minimal polynomial for T were of degree 1, then T would be a scalar multiple of I, which is not true. Hence, we seek one of degree 2. We compute T^2 and find that

$$T^2(x, y) = (-5x + 4y, -6x - 5y) = -7(x, y) + (2x + 4y, -6x + 2y)$$

Therefore, $T^2 = -7I + 2T$ and $g(x) = x^2 - 2x + 7$ is a minimal polynomial for T. It follows from Theorem 29 that T is nonsingular and we find that $T^{-1} = (\frac{1}{7})(2I - T)$.

PROBLEMS

1. Show that each of the following mappings is nonsingular and find its inverse:
   (a) $T(x, y) = (x, 2y)$
   (b) $T(x, y) = (3x + y, 5x + 2y)$
   (c) $T(x, y, z) = (x + y, y + z, z)$
   (d) $T(x, y, z) = (x, x - y, y - z)$
   (e) $T(x, y, z) = (y, x + z, y - z)$
   (f) $T(x, y, z) = (x + y, y + 2z, z + x + y)$
2. Let $T(x_1, x_2, x_3, x_4) = (0, x_1, x_2 + 2x_1, x_3 + 2x_2 + 3x_1)$. Prove:
   (a) $T^4 = O$.
   (b) $I - T$ is nonsingular.
   (c) $I + T + T^2 + T^3 = (I - T)^{-1}$.
   (d) $I + T$ is nonsingular.
   (e) $I + 2T$ is nonsingular. (*Hint.* Factor $I - 16T^4$.)
   (f) For λ nonnegative, $I - \lambda T$ is nonsingular.
3. Let $U = \text{Span}\,(e^x, e^{2x}, \ldots, e^{nx}, \ldots)$. Prove the following:
   (a) The derivative D is a nonsingular linear transformation on U and the integral $\int_{-\infty}^{x} f(t)\,dt$ is its inverse.
   (b) $I + D$ is a nonsingular linear transformation on U.

(c) $I - D$ is a singular linear transformation on U.

(d) For λ a nonnegative integer, $\lambda I - D$ is singular.

(e) For λ a nonnegative integer greater than 1, $I - \lambda D$ is nonsingular.

4. Let $U = \text{Span}(\sin x, \cos x, \sin 2x, \cos 2x, \ldots)$.

(a) Show that D is a nonsingular linear transformation on U.

(b) Determine the inverse of D.

(c) Show that $I + D$, $I - D$ and $I - D^2$ are nonsingular linear transformations on U.

(d) Show that $I + D^2$ is a singular linear transformation on U.

(e) Discuss the singularity or nonsingularity of $I + \lambda D$ for λ an integer.

(f) Discuss the singularity or nonsingularity of $I + \lambda D^2$ for λ an integer.

5. Let A be a linear transformation with $x^2 - 5x + 6$ as minimal polynomial. Prove:

(a) $A^3 = 19A - 30I$. (b) $A^4 = 65A - 114I$.

(c) Each polynomial in A of degree $2, 3, \ldots$ is equal to a polynomial in A of degree 1.

(d) For $n = 2, 3, 4, \ldots, A^n = s_nA + t_nI$, where $s_{n+1} = 5s_n + t_n$, $t_{n+1} = -6s_n$, $s_2 = 5$, $t_2 = -6$. [*Hint.* Use induction.]

(e) The set of all real polynomials in A is a real vector space of dimension 2.

6. (a) Prove the rule $T^{k+l} = T^kT^l = T^lT^k$ in (9-212). [*Hint.* It is sufficient to prove $T^{k+l} = T^kT^l$. By (9-201) in Section 9-20, we have (i) $T^{k+l} = T^kT^l = T^lT^k$ for $k \geq 0$, $l \geq 0$. From these relations deduce the relations (for $k \geq 0$, $l \geq 0$):

(ii) $T^{k+l}T^{-k} = T^l$, (iii) $T^{-k-l}T^k = T^{-l}$, (iv) $T^{-k-l} = T^{-k}T^{-l}$

Show that (i), . . . , (iv) give the desired rule for all combinations of positive and negative exponents.]

(b) Prove the rule $(T^k)^l = T^{kl}$ in (9-212). [*Hint.* If $l > 0$, $(T^k)^l = T^k \cdots T^k$ and we can apply the result of (a); if $l < 0$, write $l = -m$, show that $(T^k)^l = (T^{-k})^m$ and apply the previous result; if $l = 0$, verify the rule directly.]

7. Let $g(x)$ be a minimal polynomial for T. Prove:

(a) $p(x) = cg(x)$ is also a minimal polynomial for T, provided that $c \neq 0$.

(b) If $p(x)$ is a minimal polynomial for T, then $p(x) = cg(x)$ for some $c \neq 0$. [*Hint.* p and g must have the same degree k, so that $p(x) = d_kx^k + \cdots$, and $g(x) = c_kx^k + \cdots$. Now consider $q(x) = p(x) - (d_k/c_k)g(x)$.]

9-23 EIGENVECTORS AND EIGENVALUES

Let T be a linear transformation on a vector space U. For some nonzero vector $\mathbf{u}$ in U it may happen that

$$T(\mathbf{u}) = \lambda \mathbf{u}$$

for an appropriate scalar λ. When this happens, we call $\mathbf{u}$ an *eigenvector* of T and we call λ an *eigenvalue* for T. We also say that $\mathbf{u}$ is an eigenvector associated with the eigenvalue λ. If $\mathbf{u}$ is an eigenvector associated with λ, then for $c \neq 0$, the vectors $c\mathbf{u}$ are also eigenvectors associated with λ. More generally, we can state: *the vector* $\mathbf{0}$ *together with all eigenvectors of* T *associated with a particular eigenvalue* λ *form a subspace* K_λ *of* U. For we have already seen that K_λ is closed under scalar multiplication, and we note that K_λ is also closed under addition, since if $T(\mathbf{u}_1) = \lambda\mathbf{u}_1$, $T(\mathbf{u}_2) = \lambda\mathbf{u}_2$, then $T(\mathbf{u}_1 + \mathbf{u}_2) = T(\mathbf{u}_1) + T(\mathbf{u}_2) = \lambda\mathbf{u}_1 + \lambda\mathbf{u}_2 = \lambda(\mathbf{u}_1 + \mathbf{u}_2)$. We also note in passing that if 0 is an eigenvalue for T, then $K_0 = \text{Kernel } T$. If 0 is not an eigenvalue for T, then Kernel $T = \{\mathbf{0}\}$ and, hence, T is a one-to-one linear transformation.

Eigenvectors and eigenvalues are important because they give considerable information about the behavior of a particular linear transformation T. For example, if T is a linear transformation on V_2, regarded as the set of all vectors in the plane, and if $\mathbf{u}$ is an eigenvector for T, associated with the eigenvalue λ, then T takes the line $\{t\mathbf{u}\}$ into itself. If $\lambda > 1$ T stretches vectors along this line in the ratio λ to 1; if $\lambda < 1$ it shrinks them in this ratio; if $\lambda = 1$, it leaves them unchanged; if $\lambda = 0$, it shrinks each vector to a point (the origin); if $\lambda < 0$ it reverses directions and stretches or shrinks.

EXAMPLE 1. Let $T(\mathbf{u}) = c\mathbf{u}$ in V_2, with $c > 1$. Then T is a *dilation* on V_2. Here c is the only eigenvalue for T, and *all* nonzero vectors are eigenvectors associated with c.

EXAMPLE 2. In V_2, let $J(a\mathbf{i} + b\mathbf{j}) = a\mathbf{i} - b\mathbf{j}$, so that J is a reflection in the x-axis. Clearly, $\mathbf{i}$ is an eigenvector associated with the eigenvalue 1. If λ is any eigenvalue, then

$$a\mathbf{i} - b\mathbf{j} = J(a\mathbf{i} + b\mathbf{j}) = \lambda(a\mathbf{i} + b\mathbf{j})$$

or $a = \lambda a$, $-b = \lambda b$. These equations hold true only if $\lambda = 1$ and $b = 0$ or if $\lambda = -1$ and $a = 0$. Thus the eigenvalues are ± 1; the eigenvectors associated with 1 are the nonzero scalar multiples of $\mathbf{i}$; those associated with -1 are the nonzero scalar multiples of $\mathbf{j}$.

EXAMPLE 3. In V_2, let $B(\mathbf{u}) = \mathbf{u}^{\perp}$, so that $B(a\mathbf{i} + b\mathbf{j}) = -b\mathbf{i} + a\mathbf{j}$. From the geometric meaning of B, it is clear that $B(\mathbf{u})$ can never be a scalar multiple of $\mathbf{u}$ (for $\mathbf{u}$ nonzero). Hence, the linear transformation B has no eigenvalues.

EXAMPLE 4. Find the eigenvalues and eigenvectors for $T(x, y) = (2x + 2y, x + 3y)$.

Solution. Let λ be an eigenvalue. Then for some x and y, not both 0, we have $(2x + 2y, x + 3y) = (\lambda x, \lambda y)$ or

$$(2 - \lambda)x + 2y = 0 \qquad \text{and} \qquad x + (3 - \lambda)y = 0$$

This is thus a set of homogeneous linear equations with a nontrivial solution. Hence, we must have (Section 0-9)

$$\begin{vmatrix} 2 - \lambda & 2 \\ 1 & 3 - \lambda \end{vmatrix} = 0$$

or $4 - 5\lambda + \lambda^2 = 0$ or $\lambda = 4$ or 1. The eigenvectors associated with 4 are those (x, y) such that $-2x + 2y = 0$ and $x - y = 0$. Thus the eigenvectors associated with eigenvalue 4 are the nonzero scalar multiples of (1, 1). Similarly, we find that the eigenvectors associated with the eigenvector 1 are the nonzero scalar multiples of (−2, 1). The vectors along the line $\{t(1, 1)\}$ are stretched in the ratio 4 to 1, those along the line $\{t(-2, 1)\}$ are left unchanged. One says that T is a *shearing* of the plane (Figure 9-21).

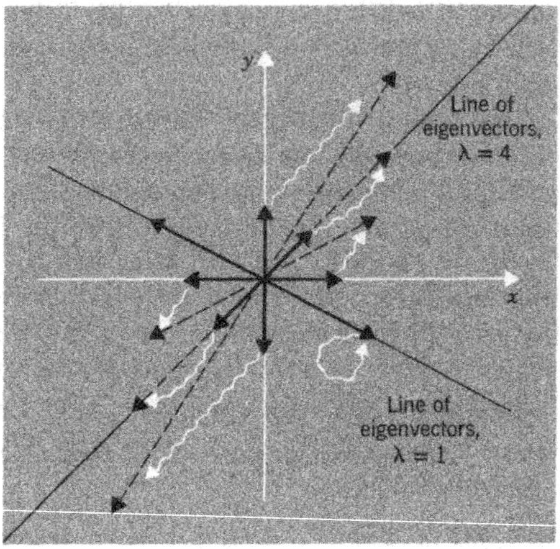

Figure 9-21 Shearing of the plane.

EXAMPLE 5. The derivative is a linear transformation on $C^{(\infty)}$ having each real number as an eigenvalue, since $D(e^{\lambda x}) = \lambda e^{\lambda x}$. The derivative is also a linear transformation on the vector space $\mathcal{P}$ of polynomials; but, on $\mathcal{P}$, D has only 0 as an eigenvalue.

THEOREM 31. *If* $\lambda_1, \ldots, \lambda_k$ *are k distinct eigenvalues of T and* $\mathbf{u}_1, \ldots, \mathbf{u}_k$ *are associated eigenvectors* ($\mathbf{u}_1$ *with* λ_1, $\mathbf{u}_2$ *with* $\lambda_2, \ldots$), *then* $\mathbf{u}_1, \ldots, \mathbf{u}_k$ *are linearly independent.*

PROOF. We prove the assertion by induction. It is true for $k = 1$, since $\mathbf{u}_1$ cannot be $\mathbf{0}$. Let us suppose it true for a particular k and seek to prove it true for $k + 1$. Let $\mathbf{u}_1, \ldots, \mathbf{u}_{k+1}$ be eigenvectors associated with the distinct eigenvalues $\lambda_1, \ldots, \lambda_{k+1}$. If

$$c_1\mathbf{u}_1 + \cdots + c_{k+1}\mathbf{u}_{k+1} = \mathbf{0}$$

then we apply T to both sides to obtain

$$c_1\lambda_1\mathbf{u}_1 + \cdots + c_{k+1}\lambda_{k+1}\mathbf{u}_{k+1} = \mathbf{0}$$

We multiply the previous equation by λ_1 and subtract from this equation to obtain

$$c_2(\lambda_2 - \lambda_1)\mathbf{u}_2 + \cdots + c_{k+1}(\lambda_{k+1} - \lambda_1)\mathbf{u}_{k+1} = \mathbf{0}$$

By the induction assumption, $\mathbf{u}_2, \ldots, \mathbf{u}_{k+1}$ are linearly independent. Therefore,

$$c_2(\lambda_2 - \lambda_1) = 0, \qquad \ldots, \qquad c_{k+1}(\lambda_{k+1} - \lambda_1) = 0$$

Since the λ's are distinct, these equations imply

$$c_2 = 0, \quad \ldots, \quad c_{k+1} = 0$$

Accordingly, $c_1\mathbf{u}_1 = \mathbf{0}$ and, since $\mathbf{u}_1 \neq \mathbf{0}$, $c_1 = 0$. Therefore, $\mathbf{u}_1, \ldots, \mathbf{u}_{k+1}$ are linearly independent. Thus the theorem is proved.

COROLLARY. *Let T be a linear transformation on a vector space U of finite dimension $n > 0$. Let T have n distinct eigenvalues $\lambda_1, \ldots, \lambda_n$ with corresponding eigenvectors $\mathbf{u}_1, \ldots, \mathbf{u}_n$. Then $\{\mathbf{u}_1, \ldots, \mathbf{u}_n\}$ forms a basis for U.*

PROBLEMS

1. For each of the following linear transformations of V_2 or V_3 find the eigenvalues and associated eigenvectors:
   (a) $T(x, y) = (2x, 3y)$ (b) $T(x, y) = (x, y)$
   (c) $T(x, y) = (4x, 2x + 3y)$ (d) $T(x, y) = (x, 2x - y)$
   (e) $T(x, y) = (3x + y, 6x + 2y)$ (f) $T(x, y) = (2x - y, x + 4y)$
   (g) $T(x, y, z) = (x + 2y - 2z, 2y + 4z, 3z)$
2. For the linear transformation D^2 show that each positive real λ is an eigenvalue and $c_1e^{\sqrt{\lambda}x} + c_2e^{-\sqrt{\lambda}x}$ are associated eigenvectors; 0 is an eigenvalue and $c_1 + c_2x$ are associated eigenvectors; each negative real λ is an eigenvalue and $a \sin \sqrt{-\lambda}x + b \cos \sqrt{-\lambda}x$ are associated eigenvectors.
3. Consider the operator $T = xD$ as a linear transformation on $\mathcal{C}^{(\infty)}(0, \infty)$; T assigns to each function $y = f(x)$ the function $xf'(x)$. Find the eigenvalues and associated eigenvectors of T.
4. Let U be the vector space of all real functions on $(-\infty, \infty)$. To each f in U assign $Tf = g$ where $g(x) = f(x + 1)$. Show that 1 is an eigenvalue of T and describe the corresponding eigenvectors.
5. Show that if T has the eigenvalue λ with associated eigenvector $\mathbf{u}$, then T^2 has the eigenvalue λ^2 with associated eigenvector $\mathbf{u}$.
6. Show that if T is nonsingular and has eigenvalue λ with associated eigenvector $\mathbf{u}$, then $\lambda \neq 0$ and T^{-1} has eigenvalue $1/\lambda$ with associated eigenvector $\mathbf{u}$.
7. Show that if T has the eigenvalue λ with associated eigenvector $\mathbf{u}$ and $p(x) = a_0 + a_1x + \cdots + a_nx^n$ is a polynomial then the linear transformation $p(T)$ has $p(\lambda)$ as an eigenvalue with $\mathbf{u}$ as the associated eigenvector.
8. Prove: a linear transformation T on a finite dimensional vector space is singular if, and only if, 0 is an eigenvalue for T.
9. Prove: if T is a linear transformation on a finite-dimensional space U, then T has at most n eigenvalues, where $n = \dim U$.

10
MATRICES AND DETERMINANTS

10-1 MATRICES

By a *matrix* we mean a rectangular array of the form

$$\begin{pmatrix} a_{11} & a_{12} & \cdots & a_{1n} \\ a_{21} & a_{22} & \cdots & a_{2n} \\ \vdots & & & \vdots \\ a_{m1} & a_{m2} & \cdots & a_{mn} \end{pmatrix} \tag{10-10}$$

In this text the a_{ij} shall be either numbers or functions, and in this chapter they shall be real numbers unless we specify otherwise. The following are examples of matrices:

$$A = \begin{pmatrix} 5 & 6 \\ 2 & -1 \end{pmatrix} \qquad B = \begin{pmatrix} 3 & 1 & 2 \\ 0 & 5 & 7 \end{pmatrix} \qquad C = \begin{pmatrix} \pi \\ 2 \end{pmatrix} \qquad D = (\sqrt{3}, -1, 5, e)$$

In particular, we note that each vector of V_n is a matrix. We call each a_{ij} in (10-10) an *entry* of the matrix. The first subscript of a_{ij} always designates the row in which the entry appears and the second subscript designates the column. Thus, a_{ij} appears in the ith row and the jth column. We shall sometimes say that a_{ij} appears in the ijth *spot* or *position* in the matrix. We shall often denote a matrix (10-10) by the shorthand (a_{ij}) or by a single capital letter, such as $A, B, \ldots$, as in the examples above. If the matrix has m rows and n columns, then we call it an m by n (or $m \times n$) matrix. The number of rows and number of columns need not be equal. When they are equal, we speak of a *square matrix of order n;* this is illustrated by matrix A above, of order 2. Two matrices are said to be *equal* when they have the same size (the same number of rows and the same number of columns) and all pairs of corresponding entries are equal.

Matrices arise in many branches of mathematics and its applications. Numerical tables (for example, tables of logarithms and tables of square roots) are usually arranged in rectangular arrays. The record of test grades for a class forms such an array; each row indicates the performance of a particular student; each column gives the performance of the class on a particular examination or paper. Here we shall concentrate on the role matrices play in the study of vector spaces. We shall see that they also play an important part in geometry and in the calculus of functions of several variables.

Row Vectors and Column Vectors. Let A be an m by n matrix. Then each row of A can be regarded as a vector of V_n. For example, for the matrix

$$A = \begin{pmatrix} 3 & 1 & 2 \\ 0 & 5 & 7 \end{pmatrix}$$

the first row of A is the vector $(3, 1, 2)$ of V_3; the second row of A is the vector $(0, 5, 7)$ of V_3. We call these vectors the *row vectors* of the matrix. For an m by n matrix there are m row vectors, each of which is a vector of V_n.

We can also regard the columns of A as vectors in V_m, which happen to be written vertically. For the matrix A above, the first column is the vector $(3, 0)$ of V_2, written vertically. Similarly, the second and third columns are the vectors $(1, 5)$ and $(2, 7)$ of V_2, written vertically. In general, we call these vectors the *column vectors* of the matrix. For an m by n matrix there are n column vectors, each of which is a vector in V_m.

For many operations on matrices and vectors it is important to know whether the vectors are written horizontally or vertically. In most cases, the context will make this clear. However, when there is any doubt, we shall make an appropriate remark. For column vectors, we shall also write (in order to save space in printing) col $\mathbf{v}$ or col $(v_1, \ldots, v_m)$ for the vector written vertically.

We remark that each vector of V_n, written horizontally, is itself a matrix, a 1 by n matrix. Each vector of V_n, written vertically, is also a matrix, an n by 1 matrix.

10-2 MATRICES AND LINEAR MAPPINGS OF V_n INTO V_m

To illustrate how the theory of vector spaces leads to matrices, we consider a linear mapping T from V_2 to V_2. Here it will be convenient to write the vectors as column vectors. Hence, our standard basis is formed of the vectors $\mathbf{e}_1 = \text{col}\,(1, 0) = \begin{pmatrix} 1 \\ 0 \end{pmatrix}$ and $\mathbf{e}_2 = \text{col}\,(0, 1) = \begin{pmatrix} 0 \\ 1 \end{pmatrix}$. We know from Section 9-11 that the linear mapping T is completely determined by its values at $\mathbf{e}_1$ and $\mathbf{e}_2$. Let

$$T(\mathbf{e}_1) = a\mathbf{e}_1 + c\mathbf{e}_2, \qquad T(\mathbf{e}_2) = b\mathbf{e}_1 + d\mathbf{e}_2 \tag{10-20}$$

Now an arbitrary vector $\mathbf{x}$ of V_2 can be written as $\mathbf{x} = \text{col}\,(x_1, x_2) = x_1\mathbf{e}_1 + x_2\mathbf{e}_2$. Hence, since T is linear,

$$\begin{aligned} T(\mathbf{x}) &= T(x_1\mathbf{e}_1 + x_2\mathbf{e}_2) = x_1T(\mathbf{e}_1) + x_2T(\mathbf{e}_2) \\ &= x_1(a\mathbf{e}_1 + c\mathbf{e}_2) + x_2(b\mathbf{e}_1 + d\mathbf{e}_2) \\ &= (ax_1 + bx_2)\mathbf{e}_1 + (cx_1 + dx_2)\mathbf{e}_2 \end{aligned}$$

Therefore, $T(\mathbf{x}) = \mathbf{y} = \text{col}\,(y_1, y_2)$, where

$$\begin{aligned} y_1 &= ax_1 + bx_2 \\ y_2 &= cx_1 + dx_2 \end{aligned} \tag{10-21}$$

We call these equations the *coordinate equations* of T. We can write the equations as one (vertical) vector equation

$$\begin{pmatrix} y_1 \\ y_2 \end{pmatrix} = T\begin{pmatrix} x_1 \\ x_2 \end{pmatrix} = \begin{pmatrix} ax_1 + bx_2 \\ cx_1 + dx_2 \end{pmatrix} \tag{10-22}$$

From the equations (10-21) we can extract the 2 by 2 matrix

$$A = \begin{pmatrix} a & b \\ c & d \end{pmatrix} \tag{10-23}$$

We call A the *matrix of the linear mapping* T.

Thus each linear mapping T of V_2 into V_2 gives rise to a 2 by 2 matrix. But conversely, for each 2 by 2 matrix A, given by (10-23), we obtain a linear mapping T whose matrix is A—namely, the one whose values at the basis vectors are given by (10-20). We call T the *linear mapping determined by* A. Hence, we can pass freely back and forth from linear mapping to matrix; we have established a one-to-one correspondence between the set of linear mappings of V_2 into V_2 and the set of 2 by 2 matrices.

The situation we have illustrated prevails in general. Let T be a linear mapping of V_n into V_m (where m and n need not be equal). Let $\mathbf{e}_1, \ldots, \mathbf{e}_n$ be the standard basis vectors of V_n, written as column vectors [so that $\mathbf{e}_1 = \text{col}\,(1, 0, \ldots, 0)$] and let $\boldsymbol{\epsilon}_1, \ldots, \boldsymbol{\epsilon}_m$ similarly denote the standard basis vectors of V_m [$\boldsymbol{\epsilon}_1 = \text{col}\,(1, 0, \ldots, 0)$]. The linear mapping T is completely determined by its values at $\mathbf{e}_1, \ldots, \mathbf{e}_n$. We can write

$$T(\mathbf{e}_1) = a_{11}\boldsymbol{\epsilon}_1 + \cdots + a_{m1}\boldsymbol{\epsilon}_m, \ldots, T(\mathbf{e}_n) = a_{1n}\boldsymbol{\epsilon}_1 + \cdots + a_{mn}\boldsymbol{\epsilon}_m \tag{10-24}$$

Then, if $\mathbf{x}$ is an arbitrary vector of V_n, we have

$$\begin{aligned} T(\mathbf{x}) &= T(x_1\mathbf{e}_1 + \cdots + x_n\mathbf{e}_n) = x_1T(\mathbf{e}_1) + \cdots + x_nT(\mathbf{e}_n) \\ &= x_1(a_{11}\boldsymbol{\epsilon}_1 + \cdots + a_{m1}\boldsymbol{\epsilon}_m) + \cdots + x_n(a_{1n}\boldsymbol{\epsilon}_1 + \cdots + a_{mn}\boldsymbol{\epsilon}_m) \\ &= (a_{11}x_1 + a_{12}x_2 + \cdots + a_{1n}x_n)\boldsymbol{\epsilon}_1 + \cdots + (a_{m1}x_1 + \cdots + a_{mn}x_n)\boldsymbol{\epsilon}_m \\ &= y_1\boldsymbol{\epsilon}_1 + \cdots + y_m\boldsymbol{\epsilon}_m = \mathbf{y} \end{aligned}$$

Thus we have

$$T\begin{pmatrix} x_1 \\ x_2 \\ \vdots \\ x_n \end{pmatrix} = \begin{pmatrix} a_{11}x_1 + \cdots + a_{1n}x_n \\ a_{21}x_1 + \cdots + a_{2n}x_n \\ \vdots \\ a_{m1}x_1 + \cdots + a_{mn}x_n \end{pmatrix} = \begin{pmatrix} y_1 \\ y_2 \\ \vdots \\ y_m \end{pmatrix} \tag{10-25}$$

or equivalently, we have the *coordinate equations of* T:

$$\begin{aligned} y_1 &= a_{11}x_1 + \cdots + a_{1n}x_n \\ y_2 &= a_{21}x_1 + \cdots + a_{2n}x_n \\ &\vdots \\ y_m &= a_{m1}x_1 + \cdots + a_{mn}x_n \end{aligned} \tag{10-26}$$

As with the mappings of V_2 into V_2, the linear mapping T determines and is determined by the equations (10-25) or (10-26). Furthermore, the equations (10-25) or (10-26) are completely determined by the matrix

$$A = \begin{pmatrix} a_{11} & \cdots & a_{1n} \\ \vdots & & \vdots \\ a_{m1} & \cdots & a_{mn} \end{pmatrix} \tag{10-27}$$

Hence, again we have a one-to-one correspondence between the linear mappings of V_n into V_m and the set of m by n matrices. Given either the linear mapping T or the matrix A, we have a prescribed procedure for going from one to the other. We call A the *matrix of the linear mapping* T; we call T the *linear mapping determined by* A.

Remark 1. The correspondence between matrices and linear mappings, as just described, depends on our choice of bases for V_n and V_m. If one or the other basis is changed, we obtain a different correspondence (see Section 10-21).

Remark 2. From (10-24) it is clear that the first column of A is given by $T(\mathbf{e}_1)$ written as a column vector; similarly, the second column is $T(\mathbf{e}_2), \ldots,$ the nth column is $T(\mathbf{e}_n)$.

EXAMPLE 1. Let T be the linear mapping of V_2 into V_3 such that

$$T\begin{pmatrix} 1 \\ 0 \end{pmatrix} = \begin{pmatrix} 1 \\ -1 \\ 0 \end{pmatrix}, \quad T\begin{pmatrix} 0 \\ 1 \end{pmatrix} = \begin{pmatrix} 2 \\ 6 \\ 3 \end{pmatrix}$$

Then T has the matrix $\begin{pmatrix} 1 & 2 \\ -1 & 6 \\ 0 & 3 \end{pmatrix}$ and the coordinate equations

$$\begin{aligned} y_1 &= x_1 + 2x_2 \\ y_2 &= -x_1 + 6x_2 \\ y_3 &= 3x_2 \end{aligned}$$

We can also write

$$\begin{pmatrix} y_1 \\ y_2 \\ y_3 \end{pmatrix} = T\begin{pmatrix} x_1 \\ x_2 \end{pmatrix} = \begin{pmatrix} x_1 + 2x_2 \\ -x_1 + 6x_2 \\ 3x_2 \end{pmatrix}$$

EXAMPLE 2. The linear mapping S corresponding to the matrix $\begin{pmatrix} 2 & 1 & 5 \\ 1 & -1 & 3 \end{pmatrix}$ is a linear mapping of V_3 into V_2 such that

$$S\begin{pmatrix} x_1 \\ x_2 \\ x_3 \end{pmatrix} = \begin{pmatrix} 2x_1 + x_2 + 5x_3 \\ x_1 - x_2 + 3x_3 \end{pmatrix}$$

and, in particular,

$$S(\mathbf{e}_1) = \begin{pmatrix} 2 \\ 1 \end{pmatrix}, \qquad S(\mathbf{e}_2) = \begin{pmatrix} 1 \\ -1 \end{pmatrix}, \qquad S(\mathbf{e}_3) = \begin{pmatrix} 5 \\ 3 \end{pmatrix}$$

PROBLEMS

1. Give the information requested for the following m by n matrices.

$$A = (a_{ij}) = \begin{pmatrix} 2 & 1 \\ 1 & 3 \end{pmatrix} \qquad B = (b_{ij}) = \begin{pmatrix} 1 & 3 & 5 \\ 1 & -1 & 0 \end{pmatrix}$$

$$C = (c_{ij}) = \begin{pmatrix} 3 & 1 \\ 5 & 2 \\ 7 & 3 \end{pmatrix} \qquad D = (d_{ij}) = \begin{pmatrix} 1 & 0 & 0 & 4 \\ 2 & -1 & 0 & 5 \\ 0 & 3 & 1 & 2 \end{pmatrix}$$

(a) The values of m and n for A and B.

(b) The values of m and n for C and D.

(c) The values $a_{11}, a_{22}, a_{21}, b_{23}, b_{13}$.

(d) The values $c_{12}, c_{21}, c_{32}, d_{24}, d_{31}, d_{33}$.

(e) The row vectors of A and B.

(f) The row vectors of C and D.

(g) The column vectors of A and B.

(h) The column vectors of C and D.

2. Let $\mathbf{e}_1, \ldots, \mathbf{e}_n$ be the standard basis of V_n and $\boldsymbol{\epsilon}_1, \ldots, \boldsymbol{\epsilon}_m$ the standard basis for V_m (written as column vectors). Find the coordinate equations and the matrix corresponding to the linear mapping T from the information given:

(a) $m = 2$, $n = 2$, $T(\mathbf{e}_1) = \boldsymbol{\epsilon}_1 - \boldsymbol{\epsilon}_2$, $T(\mathbf{e}_2) = \boldsymbol{\epsilon}_1 + 2\boldsymbol{\epsilon}_2$

(b) $m = 2$, $n = 3$, $T(\mathbf{e}_1) = \boldsymbol{\epsilon}_1 + 2\boldsymbol{\epsilon}_2$, $T(\mathbf{e}_2) = \boldsymbol{\epsilon}_1 - \boldsymbol{\epsilon}_2$, $T(\mathbf{e}_3) = 3\boldsymbol{\epsilon}_1 + \boldsymbol{\epsilon}_2$

(c) $m = 3$, $n = 2$, $T(\mathbf{e}_1) = \boldsymbol{\epsilon}_1 + 2\boldsymbol{\epsilon}_2 - \boldsymbol{\epsilon}_3$, $T(\mathbf{e}_2) = \boldsymbol{\epsilon}_1 - 3\boldsymbol{\epsilon}_2 + 5\boldsymbol{\epsilon}_3$

(d) $m = 3$, $n = 3$, $T(\mathbf{e}_1) = 3\boldsymbol{\epsilon}_1 + 2\boldsymbol{\epsilon}_2 + 4\boldsymbol{\epsilon}_3$, $T(\mathbf{e}_2) = \boldsymbol{\epsilon}_3$, $T(\mathbf{e}_3) = \boldsymbol{\epsilon}_1 + \boldsymbol{\epsilon}_2$

(e) $m = 2$, $n = 2$, $T(\mathbf{e}_1 + \mathbf{e}_2) = \boldsymbol{\epsilon}_1 - 2\boldsymbol{\epsilon}_2$, $T(\mathbf{e}_1 - \mathbf{e}_2) = 2\boldsymbol{\epsilon}_1 + 3\boldsymbol{\epsilon}_2$

3. For each of the following matrices, let T be the linear mapping of V_n into V_m determined by the matrix. Find $T(\mathbf{e}_1)$, $T(\mathbf{e}_2)$, $T(\mathbf{e}_1 + \mathbf{e}_2)$, $T(2\mathbf{e}_2)$, $T(\mathbf{e}_1 - 3\mathbf{e}_2)$.

(a) matrix A of Problem 1.

(b) matrix B of Problem 1.

(c) matrix C of Problem 1.

(d) Matrix D of Problem 1.

10-3 MATRICES AS LINEAR MAPPINGS

Since each m by n matrix A determines a unique linear mapping of V_n into V_m, it is natural to identify the matrix with the mapping. Thus, instead of speaking of "the linear mapping determined by A," we can speak simply of the "linear mapping A." The size of the matrix A determines the domain and target space of the linear mapping A; for if A is an m by n matrix, then the linear mapping is from V_n into V_m.

The linear mapping $A = (a_{ij})$ assigns $\mathbf{y}$ in V_m to the vector $\mathbf{x}$ in V_n, where $\mathbf{x}$ and $\mathbf{y}$ are related by the equations (10-25) or, equivalently, by (10-26). We now go one step further and interpret the operation of forming $\mathbf{y}$ from $\mathbf{x}$ as one of the *multiplication of* $\mathbf{x}$ *by* A. That is, we define the *product* $A\mathbf{x}$ of an m by n matrix $A = (a_{ij})$ and a column vector $\mathbf{x} = \text{col}\,(x_1, \ldots, x_n)$ to be the vector $\mathbf{y} = \text{col}\,(y_1, \ldots, y_m)$ of V_m such that

$$\begin{aligned} y_1 &= a_{11}x_1 + \cdots + a_{1n}x_n \\ &\;\vdots \\ y_m &= a_{m1}x_1 + \cdots + a_{mn}x_n \end{aligned}$$

We note that $A\mathbf{x}$ *is defined only when* A *has the same number of columns as* $\mathbf{x}$ *has entries.* Also *the product* $A\mathbf{x} = \mathbf{y}$ *has as many entries as* A *has rows.* We give a number of examples:

$$\begin{pmatrix} a & b \\ c & d \end{pmatrix}\begin{pmatrix} u \\ v \end{pmatrix} = \begin{pmatrix} au + bv \\ cu + dv \end{pmatrix}, \quad \begin{pmatrix} 1 & 2 \\ -1 & 3 \end{pmatrix}\begin{pmatrix} 1 \\ -1 \end{pmatrix} = \begin{pmatrix} -1 \\ -4 \end{pmatrix}, \quad \begin{pmatrix} 1 & 2 \\ 4 & 2 \end{pmatrix}\begin{pmatrix} -2 \\ 1 \end{pmatrix} = \begin{pmatrix} 0 \\ -6 \end{pmatrix}$$

$$\begin{pmatrix} a & b & c \\ d & e & f \end{pmatrix}\begin{pmatrix} u \\ v \\ w \end{pmatrix} = \begin{pmatrix} au + bv + cw \\ du + ev + fw \end{pmatrix}, \quad \begin{pmatrix} 3 & 1 & 5 \\ 2 & 4 & 7 \end{pmatrix}\begin{pmatrix} 0 \\ 2 \\ 1 \end{pmatrix} = \begin{pmatrix} 7 \\ 15 \end{pmatrix}$$

$$\begin{pmatrix} a & b & c \\ d & e & f \\ p & q & r \end{pmatrix}\begin{pmatrix} u \\ v \\ w \end{pmatrix} = \begin{pmatrix} au + bv + cw \\ du + ev + fw \\ pu + qv + rw \end{pmatrix}, \quad \begin{pmatrix} 1 & 2 & 3 \\ 0 & -1 & 2 \\ 4 & 1 & 0 \end{pmatrix}\begin{pmatrix} 1 \\ 2 \\ 1 \end{pmatrix} = \begin{pmatrix} 8 \\ 0 \\ 6 \end{pmatrix}$$

$$\begin{pmatrix} 1 & 0 & 7 & 1 \\ 2 & 1 & 0 & 2 \end{pmatrix}\begin{pmatrix} -1 \\ 1 \\ 2 \\ 3 \end{pmatrix} = \begin{pmatrix} 16 \\ 5 \end{pmatrix}, \quad \begin{pmatrix} 1 & 2 & 3 \\ 4 & 1 & 2 \\ -1 & 0 & 2 \\ 1 & 1 & 0 \end{pmatrix}\begin{pmatrix} 1 \\ -1 \\ 2 \end{pmatrix} = \begin{pmatrix} 5 \\ 7 \\ 3 \\ 0 \end{pmatrix}$$

Later we shall learn how to form products AB of arbitrary matrices (of sizes which match). The product $A\mathbf{x}$ which we have defined is a special case of general multiplication—the case in which B is a column vector. In all cases of matrix multiplication the *order* of the factors is very important; in general AB and BA are not the same—and one may be defined when the other is not. Here we consider only the order $A\mathbf{x}$, with the column vector to the right.

Now that we have defined the product $A\mathbf{x}$, we can state; *the linear mapping determined by matrix* A (or, simply, the linear mapping A) *has the equation*

$$\mathbf{y} = A\mathbf{x} \tag{10-30}$$

For our product $A\mathbf{x}$ was defined in accordance with the coordinate equations of the linear mapping. Since the mapping A is linear, we have the rules: *if* $\mathbf{x}, \mathbf{z}$ *are in* V_n *and* A *is an* m *by* n *matrix, then for every scalar* c

$$A(\mathbf{x} + \mathbf{z}) = A\mathbf{x} + A\mathbf{z}, \qquad A(c\mathbf{x}) = c(A\mathbf{x}) \tag{10-31}$$

EXAMPLE. The linear mapping $A = \begin{pmatrix} 1 & 2 \\ 2 & 3 \end{pmatrix}$ has the properties: $A\mathbf{e}_1 = \gamma_1 = \text{col}\ (1, 2)$, $A\mathbf{e}_2 = \gamma_2 = \text{col}\ (2, 3)$. As remarked previously, $A\mathbf{e}_1, \ldots$ are the successive column vectors of A. By the linearity, we now conclude that

$$A(5\mathbf{e}_1) = 5\gamma_1, \qquad A(-2\mathbf{e}_2) = -2\gamma_2, \qquad A(7\mathbf{e}_1 + 4\mathbf{e}_2) = 7\gamma_1 + 4\gamma_2$$

In general, $A(x_1\mathbf{e}_1 + x_2\mathbf{e}_2) = x_1\gamma_1 + x_2\gamma_2$. We remark that γ_1, γ_2 are linearly independent and, hence, are a basis for V_2. In particular, $x_1\gamma_1 + x_2\gamma_2 = \mathbf{0}$ if, and only if, $x_1 = x_2 = 0$. Hence, $A\mathbf{x} = \mathbf{0}$ if, and only if, $\mathbf{x} = \mathbf{0}$. Furthermore, each vector $\mathbf{y}$ in V_2 can be expressed as a linear combination of γ_1, γ_2; if $\mathbf{y} = a\gamma_1 + b\gamma_2$, we have $A(a\mathbf{e}_1 + b\mathbf{e}_2) = \mathbf{y}$. Hence, the linear mapping A is a one-to-one mapping of V_2 onto V_2.

10-4 KERNEL, RANGE, NULLITY, AND RANK OF A MATRIX

Since each m by n matrix A is now regarded as a linear mapping of V_n into V_m, we can apply familiar concepts for such mappings to the matrix A.

The *range* of A is the set of all $\mathbf{y}$ for which $A\mathbf{x} = \mathbf{y}$ for some $\mathbf{x}$. We saw in Section 9-12 that the range of a linear mapping T from V to W is the subspace of W spanned by the images $T(\mathbf{v}_1), \ldots, T(\mathbf{v}_n)$ of a basis of V. Here we can use $\mathbf{e}_1, \ldots, \mathbf{e}_n$ as a basis for V_n. The vectors $A\mathbf{e}_1, \ldots, A\mathbf{e}_n$ then span the range. But we saw in Section 10-2 that $A\mathbf{e}_1$ is the first column of A, $A\mathbf{e}_2$ is the second column, ..., $A\mathbf{e}_n$ is the nth column. Hence, *the range of a matrix* A *is the set of all linear combinations of the column vectors of* A.

If we denote the successive column vectors of A by $\gamma_1, \ldots, \gamma_n$, then we can write, for a general $\mathbf{x} = x_1\mathbf{e}_1 + \cdots + x_n\mathbf{e}_n$,

$$\begin{aligned} A\mathbf{x} &= A(x_1\mathbf{e}_1 + \cdots + x_n\mathbf{e}_n) \\ &= x_1(A\mathbf{e}_1) + \cdots + x_n(A\mathbf{e}_n) = x_1\gamma_1 + \cdots + x_n\gamma_n \end{aligned}$$

This equation shows clearly that, as $\mathbf{x}$ varies over V_n, the vectors $\mathbf{y} = A\mathbf{x}$ sweep out all linear combinations of $\gamma_1, \ldots, \gamma_n$.

EXAMPLE 1. The matrix A given below has the column vectors $\gamma_1, \gamma_2, \gamma_3$ shown:

$$A = \begin{pmatrix} 1 & 2 & 4 \\ 3 & 1 & 2 \\ 0 & 2 & 4 \end{pmatrix}, \qquad \gamma_1 = \begin{pmatrix} 1 \\ 3 \\ 0 \end{pmatrix}, \qquad \gamma_2 = \begin{pmatrix} 2 \\ 1 \\ 2 \end{pmatrix}, \qquad \gamma_3 = \begin{pmatrix} 4 \\ 2 \\ 4 \end{pmatrix}$$

We observe that $\gamma_3 = 2\gamma_2$, whereas γ_1, γ_2 are linearly independent. Hence, here the range of A consists of all vectors $\mathbf{y}$ in V_3 which are linear combinations of γ_1 and γ_2.

The rank of a linear mapping is the dimension of the range of that mapping. Hence, the *rank of a matrix* A, written rank A, is the number of vectors in a maximal linearly independent set of column vectors of A. For Example 1 above, we can find two linearly independent column vectors—namely, γ_1 and γ_2—but no more than two. Therefore, in the example, rank $A = 2$.

In Section 10-13 we shall prove that the rank of a matrix also equals the maximal number of linearly independent *row vectors* of A. For the matrix A of the example, the row vectors are $\mathbf{u}_1 = (1, 2, 4)$, $\mathbf{u}_2 = (3, 1, 2)$ and $\mathbf{u}_3 = (0, 2, 4)$. We verify that $\mathbf{u}_1$ and $\mathbf{u}_2$ are linearly independent but that scalars a, b, c, not all 0, can be found such that $a\mathbf{u}_1 + b\mathbf{u}_2 + c\mathbf{u}_3 = \mathbf{0}$; for example, $a = 6$, $b = -2$, $c = -5$. Hence, we again conclude that rank $A = 2$.

We shall also show (Section 10-13) that rank A can be found with the aid of determinants. We consider the square arrays obtained from A by deleting rows or columns or both or—if A is square—by possibly deleting nothing. The determinant of such a square array is called a *minor of* A. If a minor is an n by n array, we call n the order of the minor. The rank of A is 0 if every minor of A equals 0. Otherwise the rank equals the order of the minor of largest order which is not 0: briefly, the maximal order of a nonzero minor of A. For Example 1,

$$\begin{vmatrix} 1 & 2 & 4 \\ 3 & 1 & 2 \\ 0 & 2 & 4 \end{vmatrix} = 0 \quad \text{and} \quad \begin{vmatrix} 1 & 2 \\ 3 & 1 \end{vmatrix} \neq 0$$

Hence, the rank is again found to be 2.

Since the rank of a linear mapping cannot exceed the dimension of its domain or that of its target space, we have the rule: the rank of an m by n matrix is at most equal to the *smaller* of m and n (or, if $m = n$, at most equal to m). This is also clear from the determinant method of the preceding paragraph. When the rank of A *equals* the smaller of m and n (so that the rank is as large as it is permitted to be), we say that A has *maximal rank*.

The *kernel* of an m by n matrix $A = (a_{ij})$, denoted by Kernel A, is the set of all vectors $\mathbf{x}$ in V_n such that $A\mathbf{x} = \mathbf{0}$. Thus Kernel A consists of all vectors $\mathbf{x}$ whose coordinates satisfy the homogeneous linear equations

$$\begin{aligned} a_{11}x_1 + \cdots + a_{1n}x_n &= 0 \\ &\vdots \\ a_{m1}x_1 + \cdots + a_{mn}x_n &= 0 \end{aligned}$$

The kernel of A is a subspace of V_n. The *nullity* of A, denoted by null A, is the dimension of Kernel A.

We have thus the relations

$$\text{null } A = \dim(\text{Kernel } A), \qquad \text{rank } A = \dim(\text{Range } A) \qquad (10\text{-}40)$$

Also by Theorem 22 of Section 9-14, for every m by n matrix A,

$$\text{null } A + \text{rank } A = n \tag{10-41}$$

For Example 1 above, $n = 3$ and we found rank A to be 2. Consequently, null A should be 1. The linear equations to be considered are

$$\begin{aligned} x_1 + 2x_2 + 4x_3 &= 0 \\ 3x_1 + x_2 + 2x_3 &= 0 \\ 2x_2 + 4x_3 &= 0 \end{aligned}$$

We find readily that the solutions are the scalar multiples of $\mathbf{x} = (0, 2, -1)$ and, hence, the kernel of A is the one-dimensional space spanned by this vector. Thus, null $A = 1$, as found above.

EXAMPLE 2. Find the range, kernel, nullity, and rank of

$$B = \begin{pmatrix} 2 & 1 & 1 \\ -2 & 1 & 0 \\ 4 & 1 & 2 \\ 3 & 0 & 1 \end{pmatrix}$$

Solution. Let the column vectors of B be $\gamma_1, \gamma_2, \gamma_3$. Then we see at once that γ_2 and γ_3 are linearly independent (neither is a scalar multiple of the other). If $\gamma_1 = a\gamma_2 + b\gamma_3$, then $2 = a + b$, $-2 = a$, $4 = a + 2b$, $3 = b$. These equations are contradictory. Hence, $\gamma_1, \gamma_2, \gamma_3$ is a linearly independent set and rank $B = 3$. Since null B + rank $B = 3 = \dim V_3$, we see that null $B = 0$ and, hence, Kernel $B = \{\mathbf{0}\}$. Thus B is a one-to-one linear mapping of V_3 into V_4 whose range is the set of all linear combinations of the column vectors. Clearly, B is not an onto mapping.

THEOREM 1. *If Z is a subspace of V_n such that*

$$k = \dim Z \geq n - m$$

then Z is the kernel of an m by n matrix T of rank $n - k$. If W is a subspace of V_m such that

$$p = \dim W \leq n$$

then W is the range of some m by n matrix A.

This theorem is Theorem 23 (Section 9-14) phrased in the language of matrices.

Remark 1. It should be noted that the matrices T and A specified in Theorem 1 are not unique (see the remark in Section 9-15).

Remark 2. Theorem 1 shows that any subspace of V_m of dimension at most n is the range of some m by n matrix, and any subspace of V_n of dimension at least $n - m$ is the kernel of some m by n matrix. In particular, we note that each k dimensional subspace of V_n is the kernel of some linear mapping of V_n into V_{n-k}; that is, it is the kernel of some $n - k$ by n matrix. We also observe that there are linear mappings of V_n onto V_m if, and only if, $m \leq n$. Since a one-to-one linear mapping has kernel V_0, it follows from Theorem 1 that there are one-to-one linear mappings of V_n into V_m if, and only if, $m \geq n$. Hence, there are one-to-one linear mappings of V_n onto V_m if, and only if, $n = m$.

THEOREM 2. *Let L be a linear variety of V_n which is not a subspace of V_n and such that $n > k = \dim L \geq n - m$. Let $\mathbf{b} \neq 0$ be a fixed vector in V_m. Then there exists an m by n matrix T* of rank $n - k$ *such that the T-pre-image of* $\mathbf{b}$ *is L.*

This theorem is Theorem 24 (Section 9-14) phrased in the language of matrices.

Remark 3. It follows from Theorem 2 that, given a linear variety of dimension k in V_n and any nonzero vector $\mathbf{b}$ in V_{n-k}, there is an $(n - k)$ by n matrix whose pre-image of $\mathbf{b}$ is L.

10-5 IDENTITY MATRIX, SCALAR MATRIX, ZERO MATRIX, COMPLEX MATRICES

Let I_n be the square n by n matrix (δ_{ij}) with

$$\delta_{ii} = 1 \text{ for } i = 1, \ldots, n, \qquad \delta_{ij} = 0 \text{ for } i \neq j \tag{10-50}$$

(One calls δ_{ij} the Kronecker delta.) Thus

$$I_2 = \begin{pmatrix} 1 & 0 \\ 0 & 1 \end{pmatrix}, \qquad I_3 = \begin{pmatrix} 1 & 0 & 0 \\ 0 & 1 & 0 \\ 0 & 0 & 1 \end{pmatrix}$$

We call I_n the n by n *identity matrix*. We verify at once that it has the important property

$$I_n\mathbf{x} = \mathbf{x} \qquad \text{for every } \mathbf{x} \text{ in } V_n \tag{10-51}$$

Thus I_n is the identity mapping of V_n onto itself. When the context indicates the size of I_n, we can drop the subscript n and write simply I.

A square matrix A of form $(c\,\delta_{ij})$, with δ_{ij} as in (10-50), is said to be a *scalar matrix*. The following matrices are examples:

$$\begin{pmatrix} 2 & 0 \\ 0 & 2 \end{pmatrix}, \qquad \begin{pmatrix} -1 & 0 & 0 \\ 0 & -1 & 0 \\ 0 & 0 & -1 \end{pmatrix}$$

In particular, I_n is a scalar matrix. For $A = (c\,\delta_{ij})$, we verify that

$$A\mathbf{x} = c\mathbf{x} \qquad \text{for every } \mathbf{x} \text{ in } V_n \tag{10-52}$$

Thus A takes each $\mathbf{x}$ into the scalar multiple of $\mathbf{x}$ by c; this explains the name scalar matrix.

We denote by O_{mn} the m by n matrix having 0 at all entries. Here the context normally indicates the values of m and n, so that we write simply O. We have the general rule

$$O\mathbf{x} = \mathbf{0} \qquad \text{for every } \mathbf{x} \text{ in } V_n \tag{10-53}$$

Hence, O is the zero mapping of V_n into V_m. We call O the *zero matrix*.

A matrix $A = (a_{ij})$ such that $a_{ij} = 0$ where $i \neq j$ is said to be a *diagonal matrix*. Each scalar matrix is a diagonal matrix. For a square diagonal matrix,

$$A\mathbf{e}_j = a_{jj}\mathbf{e}_j \qquad \text{for } j = 1, 2, \ldots, n$$

The *principal diagonal* of a square n by n matrix $A = (a_{ij})$ is the vector $(a_{11}, \ldots, a_{nn})$.

Complex Matrices. We can allow the entries in our matrices to be complex numbers. We then speak of a complex matrix. The whole preceding discussion extends at once to complex matrices, provided that we replace the vector spaces V_n and V_m by $V_n{}^c$ and $V_m{}^c$, respectively.

PROBLEMS

1. Evaluate the following products:

(a) $\begin{pmatrix} 1 & 2 \\ 3 & -1 \end{pmatrix}\begin{pmatrix} 1 \\ 2 \end{pmatrix}$

(b) $\begin{pmatrix} 1 & 2 \\ 3 & -1 \end{pmatrix}\begin{pmatrix} -3 \\ 1 \end{pmatrix}$

(c) $\begin{pmatrix} 1 & 2 \\ 3 & -1 \end{pmatrix}\begin{pmatrix} 1 \\ 1 \end{pmatrix}$

(d) $\begin{pmatrix} 1 & 0 & 1 \\ 1 & 2 & 3 \\ -1 & 1 & 0 \end{pmatrix}\begin{pmatrix} 1 \\ 0 \\ 1 \end{pmatrix}$

(e) $\begin{pmatrix} 1 & 0 & 1 \\ 1 & 2 & 3 \\ -1 & 1 & 0 \end{pmatrix}\begin{pmatrix} 1 \\ 2 \\ 1 \end{pmatrix}$

(f) $\begin{pmatrix} 1 & 0 & 1 \\ 1 & 2 & 3 \\ -1 & 1 & 0 \end{pmatrix}\begin{pmatrix} -1 \\ 2 \\ -3 \end{pmatrix}$

(g) $\begin{pmatrix} 1 & 0 & 1 & 1 \\ 1 & 2 & 0 & 3 \\ 3 & -1 & 1 & 0 \end{pmatrix}\begin{pmatrix} 1 \\ 0 \\ 1 \\ 1 \end{pmatrix}$

(h) $\begin{pmatrix} 1 & 0 & 1 & 1 \\ 1 & 2 & 0 & 3 \\ 3 & -1 & 1 & 0 \end{pmatrix}\begin{pmatrix} 2 \\ 1 \\ -1 \\ 2 \end{pmatrix}$

(i) $\begin{pmatrix} 1 & 0 & 1 & 1 \\ 1 & 2 & 0 & 3 \\ 3 & -1 & 1 & 0 \end{pmatrix}\begin{pmatrix} 5 \\ 2 \\ -3 \\ -2 \end{pmatrix}$

(j) $\begin{pmatrix} 1 & 2 & 1 \\ 0 & 1 & 3 \\ 4 & 2 & 6 \\ 1 & 4 & 0 \end{pmatrix}\begin{pmatrix} 1 \\ 0 \\ 1 \end{pmatrix}$

(k) $\begin{pmatrix} 1 & 2 & 1 \\ 0 & 1 & 3 \\ 4 & 2 & 6 \\ 1 & 4 & 0 \end{pmatrix}\begin{pmatrix} 1 \\ 2 \\ 1 \end{pmatrix}$ (l) $\begin{pmatrix} 1 & 2 & 1 \\ 0 & 1 & 3 \\ 4 & 2 & 6 \\ 1 & 4 & 0 \end{pmatrix}\begin{pmatrix} -1 \\ 2 \\ -3 \end{pmatrix}$

2. Consider the matrix $A = \begin{pmatrix} 2 & 5 \\ 1 & 7 \end{pmatrix}$ as a linear mapping.

(a) Write the corresponding coordinate equations.

(b) Find $A\mathbf{x}$ for $\mathbf{x} = (1, 0)$, $\mathbf{x} = (0, 1)$ and $\mathbf{x} = (3, 2)$.

(c) Show that if $\mathbf{x} = (x_1, x_2)$, then $A\mathbf{x} = x_1(2, 1) + x_2(5, 7)$.

(d) From the result of (c) show that A has Span $\{(2, 1), (5, 7)\}$ as range.

3. Let $\mathbf{u} = \text{col}\ (3, 5)$, $\mathbf{v} = \text{col}\ (2, 7)$, $A = \begin{pmatrix} 3 & 0 \\ 0 & 3 \end{pmatrix}$. Evaluate:

(a) $A\mathbf{u}$ (b) $A(A\mathbf{u})$ (c) $A(A[A\mathbf{u}])$ (d) $I\mathbf{v}$

(e) $2(A\mathbf{v}) - 6(I\mathbf{v})$ (f) $O\mathbf{v}$ (g) $A(0\mathbf{u})$ (h) $O(I\mathbf{v})$

4. Let $\mathbf{u} = \text{col}\ (1 + i, 1 - i)$, $\mathbf{v} = \text{col}\ (i, 3)$, $A = \begin{pmatrix} 1 & i \\ 0 & 2 \end{pmatrix}$, $B = \begin{pmatrix} i & 1 \\ 1 & -i \end{pmatrix}$.

(a) Evaluate $A\mathbf{u}$. (b) Evaluate $B\mathbf{v}$.

(c) Find the range of A. (d) Find the kernel of B.

(e) Evaluate $A(B[\mathbf{u}])$. (f) Evaluate $B[A(\mathbf{u})]$.

5. Find the range, rank, kernel, and nullity of the following matrices appearing in Problem 1. (a) in 1(a), (b) in 1(d), (c) in 1(g), (d) in 1(j).

6. If T is a linear mapping of U into V, the pre-image (or T-pre-image) of the vector $\mathbf{v}$ in V is the set of all $\mathbf{u}$ in U such that $T(\mathbf{u}) = \mathbf{v}$ (see Section 9-13). Find the pre-image of the vector (1, 1, 1) for the matrices in parts (d) and (g) of Problem 1.

7. Find the kernel of each of the following matrices:

(a) $\begin{pmatrix} 1 & 2 & 3 & 4 \\ -1 & 0 & 2 & -3 \\ 3 & 1 & -6 & 2 \\ 4 & 7 & 7 & 8 \end{pmatrix}$, (b) $\begin{pmatrix} 1 & 4 & -1 & 2 & 0 \\ 0 & 3 & 2 & -1 & 1 \\ 1 & 2 & 3 & -3 & 4 \\ 2 & 3 & 0 & 0 & 3 \end{pmatrix}$,

(c) $\begin{pmatrix} 6 & 2 & 0 & 7 \\ 4 & 3 & 1 & 1 \\ 0 & 2 & 1 & 0 \\ 2 & 1 & 0 & 6 \\ 8 & 4 & 1 & 2 \end{pmatrix}$, (d) $\begin{pmatrix} 1 & 2 & 0 & 1 \\ 4 & -1 & 2 & 3 \\ 2 & 1 & -2 & 3 \\ 7 & 2 & 0 & 9 \\ 0 & -6 & 4 & 1 \end{pmatrix}$.

8. Decide which of the matrices below has (1, 2, 3) in its range.

(a) $\begin{pmatrix} 1 & 2 & 3 & 4 \\ -1 & 0 & 2 & -3 \\ 3 & 1 & -6 & 2 \end{pmatrix}$, (b) $\begin{pmatrix} 2 & 1 & -1 & 4 \\ 1 & 2 & 0 & 5 \\ 1 & -1 & -1 & -1 \end{pmatrix}$

9. For each matrix below find the pre-image of (1, 2).

(a) $\begin{pmatrix} 3 & 2 \\ 6 & 4 \end{pmatrix}$, (b) $\begin{pmatrix} 1 & 2 \\ 2 & 1 \end{pmatrix}$, (c) $\begin{pmatrix} 1 & 2 & 3 \\ 4 & 2 & -1 \end{pmatrix}$, (d) $\begin{pmatrix} 2 & 1 & 0 & 1 & 4 \\ 6 & 3 & 0 & 3 & 12 \end{pmatrix}$.

10. Let A be an m by n matrix. Prove:

(a) If A is a one-to-one linear mapping (of V_n into V_m) then $m \geq n$ and the column vectors of A are linearly independent.

(b) If A is an onto linear mapping, then $m \leq n$ and the column vectors of A span V_m.

(c) If A is one-to-one and onto, then $m = n$ and the column vectors of A are a basis for V_m.

11. For the specified m and Z find, when possible, an m by 3 matrix having Z as kernel.

(a) $m = 2$, $Z = \text{Span}\,((1, 1, 0))$.

(b) $m = 2$, $Z = \text{Span}\,((1, 2, 3), (1, 0, 1))$.

(c) $m = 3$, $Z = \text{Span}\,((1, 1, 1))$.

(d) $m = 1$, $Z = \text{Span}\,((1, 1, 1))$.

(e) $m = 2$, $Z = V_0$.

(f) $m = 4$, $Z = \text{Span}\,((1, 2, 0), (-1, -1, 1))$.

12. For each L and **b** find a 2 by 3 matrix having L as the pre-image of **b**.

(a) $L = \{(1, 0, 0) + \text{Span}\,(1, 1, 0)\}$, $\mathbf{b} = (1, 2)$

(b) $L = \{(1, 0, 1) + \text{Span}\,(1, 2, 3)\}$, $\mathbf{b} = (1, 0)$

10-6 LINEAR EQUATIONS

A common problem of algebra is the solution of simultaneous linear equations (Section 0-8). We here discuss them for the real case. A similar discussion is valid for the complex case. Typical examples are the following:

$$\text{(a)}\ \begin{cases} 2x - y = 5 \\ x + 2y = 3 \end{cases} \qquad \text{(b)}\ \begin{cases} x - y + z = 0 \\ 2x + y - z = 0 \\ x + 3y + 2z = 0 \end{cases}$$

$$\text{(c)}\ \begin{cases} 2x - y = 5 \\ x + 2y = 3 \\ 5x + 5y = 7 \end{cases} \qquad \text{(d)}\ \begin{cases} x - y + z = 1 \\ 5x + y + 3z = 2 \end{cases}$$

Examples (a) and (b) have the same number of unknowns as equations, example (c) has more equations than unknowns, while example (d) has less equations than unknowns.

In general, we are dealing with a system of m equations in n unknowns:

$$\begin{aligned} a_{11}x_1 + \cdots + a_{1n}x_n &= b_1 \\ &\vdots \\ a_{m1}x_1 + \cdots + a_{mn}x_n &= b_m \end{aligned} \qquad (10\text{-}60)$$

These equations are the same as the coordinate equations (10-26) except that $y_1, \ldots, y_m$ have been replaced by $b_1, \ldots, b_m$. Accordingly, we see that we are really studying a matrix equation

$$A\mathbf{x} = \mathbf{b} \tag{10-61}$$

where

$$A = \begin{pmatrix} a_{11} & \cdots & a_{1n} \\ \vdots & & \vdots \\ a_{m1} & \cdots & a_{mn} \end{pmatrix}, \qquad \mathbf{b} = \begin{pmatrix} b_1 \\ \vdots \\ b_m \end{pmatrix}$$

are given and fixed for the discussion. Since the matrix A is a linear mapping of V_n into V_m, our problem reduces to finding the vectors $\mathbf{x}$ whose image under the mapping A is the vector $\mathbf{b}$; or, equivalently, to finding all $\mathbf{x}$ such that $A\mathbf{x} = \mathbf{b}$. Thus we are seeking the *A-pre-image* of $\mathbf{b}$. Along with (10-60), we also study the homogeneous linear equations

$$a_{11}x_1 + \cdots + a_{1n}x_n = 0, \quad \ldots, \quad a_{m1}x_1 + \cdots + a_{mn}x_n = 0 \tag{10-62}$$

The solutions of these equations form the kernel of A: the set of all $\mathbf{x}$ such that $A\mathbf{x} = \mathbf{0}$.

The theory of linear mappings was discussed in detail in Sections 9-11 to 9-14. In Sections 10-3 and 10-4 we have seen how that theory can be applied to the matrix A as a linear mapping. We now use the theory in order to discuss the solutions of Equations (10-60). We use the notations of Sections 10-3 and 10-4; in particular, we let γ_j denote the jth column vector of A.

In the following paragraphs we list the main facts on the system of linear equations (10-60). These facts are simply restatements of results from the previous theory of linear mappings, as applied to the case at hand.

1. The range of A is a subspace Y of V_m of dimension $r = \text{rank } A$; r is at most equal to the smaller of m and n. The vectors of Y are the linear combinations of the column vectors $\gamma_1, \ldots, \gamma_n$; that is, $Y = \text{Span } \{\gamma_1, \ldots, \gamma_n\}$.

2. The equation $A\mathbf{x} = \mathbf{b}$ has a solution precisely when $\mathbf{b}$ is in Y; that is, precisely when $\mathbf{b}$ is a linear combination of $\gamma_1, \ldots, \gamma_n$.

3. The solutions of $A\mathbf{x} = \mathbf{0}$ form the kernel of A, which is a subspace K of V_n. The dimension k of K is the nullity of A (null A) and

$$0 \leq k \leq n, \qquad n - k \leq m, \qquad k + r = n$$

4. For $\mathbf{b}$ in Y, the solutions of $A\mathbf{x} = \mathbf{b}$ form a linear variety $L = \{\mathbf{x}^* + K\}$, where $\mathbf{x}^*$ is any one solution of $A\mathbf{x} = \mathbf{b}$.

5. For $\mathbf{b}$ in Y, the equation $A\mathbf{x} = \mathbf{b}$ has exactly one solution precisely when $k = 0$ or, equivalently, when $r = n$. By rule 3 this can occur only if $m \geq n$; that is, only if the number of equations is at least as large as the number of unknowns. One can also state: the equation $A\mathbf{x} = \mathbf{b}$ has exactly one solution precisely when (a) $\mathbf{b}$ is in Y and (b) the equation $A\mathbf{x} = \mathbf{0}$ has only the trivial solution $\mathbf{x} = \mathbf{0}$.

6. If A has less than maximal rank, then there are values of $\mathbf{b}$ for which the equation $A\mathbf{x} = \mathbf{b}$ has no solution; if $\mathbf{b}$ is such that the equation $A\mathbf{x} = \mathbf{b}$ has a solution, then the equation has infinitely many solutions.

7. Let A be of maximal rank. (a) If $m = n$, then the equation $A\mathbf{x} = \mathbf{b}$ has

a unique solution for every **b**. (b) If $m < n$ (less equations than unknowns), then for every **b** the equation $A\mathbf{x} = \mathbf{b}$ has infinitely many solutions. (c) If $m > n$ (more equations than unknowns), then for some choices of **b** the equation $A\mathbf{x} = \mathbf{b}$ has no solution; if **b** is such that the equation $A\mathbf{x} = \mathbf{b}$ has a solution, then that solution is unique.

Intuitive Discussion. One can gain an intuitive feeling for these results by considering the mapping A as a process of packing V_n into V_m. As usual, we can identify points of n-dimensional space with ordered n-tuples and, hence, with vectors in V_n. Thus our mapping A packs the points of n-dimensional space into the "box" V_m.

When $n = m$, we can fit V_n into V_m in many ways with minor distortion; that is, there are many one-to-one mappings A of V_n into V_n. Each such mapping takes $\mathbf{e}_1$ to $\gamma_1, \ldots, \mathbf{e}_n$ to γ_n, and $\gamma_1, \ldots, \gamma_n$ form a basis for V_n (see Figure 10-1 for the case $n = m = 3$). However, we can also pack V_n into V_m by collapsing V_n down to an r-dimensional space Y $(r < n)$. This is illustrated for $r = 2$, $n = m = 3$, in Figure 10-2. In the collapsing process of Figure 10-2, a whole line K through the origin is collapsed onto the origin of V_m. Furthermore, each line parallel to K also collapses onto a point. Accordingly, the pre-images of points in Y are lines (the lines which collapse to these points); for points not in Y the pre-image is empty. We could also fit V_n into V_m by squeezing V_n down to a line Y, as in Figure 10-3; in this case a whole plane K in V_n has to be squeezed to a point $(k = 2, r = 1)$. Finally, we can simply squeeze all of V_n to the vector $\mathbf{0}$ of V_m; in this case, $V_n = K$, so that $k = 3$ for the case $n = m = 3$; also Y coincides with the subspace formed of $\mathbf{0}$ alone and $r = 0$.

For $m > n$, we cannot possibly pack V_n into V_m to fill all of V_m. The most

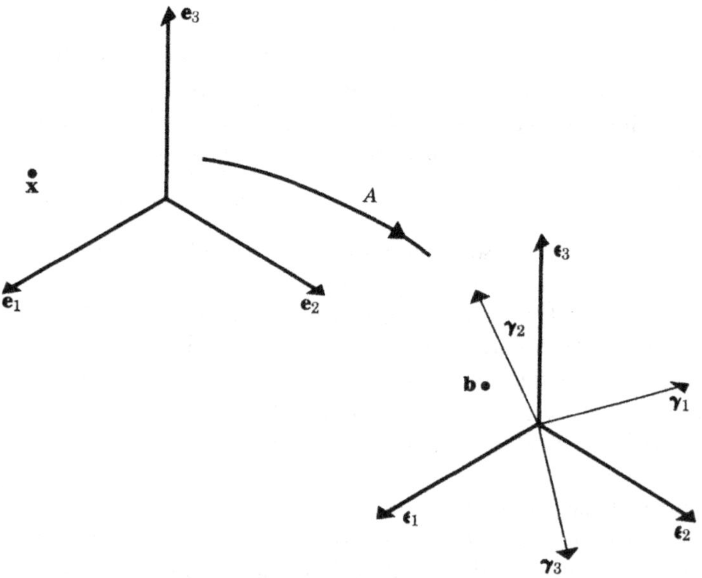

Figure 10-1 One-to-one mapping of V_n onto V_m, $n = m$ $(= 3)$.

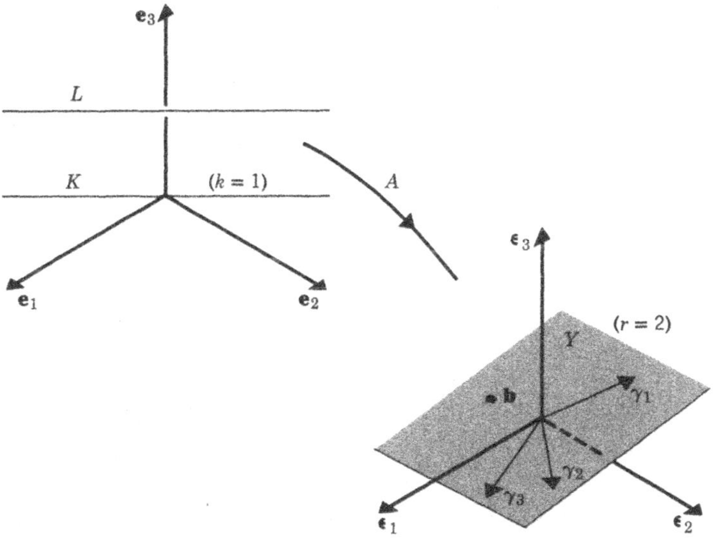

Figure 10-2 V_n mapped into V_m ($m = n = 3$) with rank 2, nullity 1.

room is taken up if V_n occupies an n-dimensional subspace Y of V_m, as shown in Figure 10-4 for $n = 2, m = 3$. If this is done, there are infinitely many vectors $\mathbf{b}$ not in Y (for which, accordingly, $A\mathbf{x} = \mathbf{b}$ is not solvable); when $\mathbf{b}$ is in Y the solution is unique ($k = 0, r = n$). We can always collapse V_n further, down to a subspace Y of $r = n - k$ dimensions; the smaller we squeeze it, the more vectors $\mathbf{b}$ for which $A\mathbf{x} = \mathbf{b}$ has no solution. The more we squeeze it, the higher the dimension k of the set K and the parallel linear varieties which are squeezed to a point of V_m.

For $n > m$, V_n is bigger than V_m, and we cannot pack V_n into V_m without

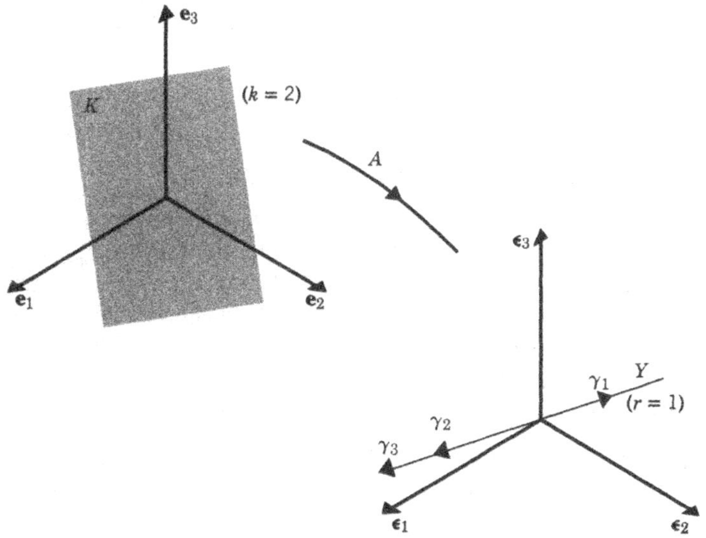

Figure 10-3 V_n mapped into V_m ($m = n = 3$) with rank 1, nullity 2.

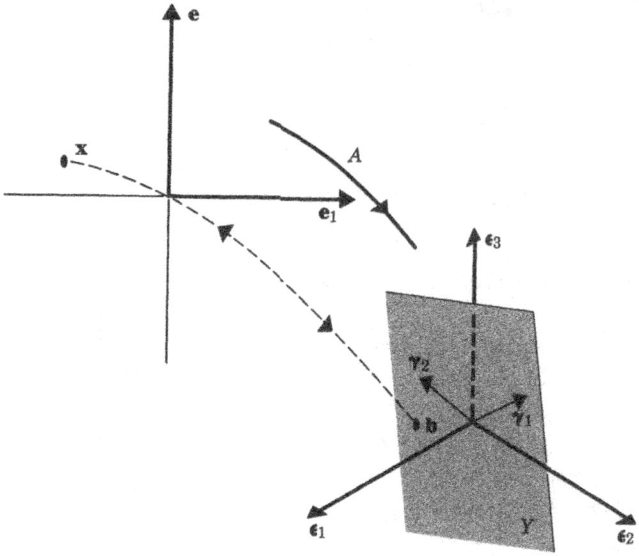

Figure 10-4 V_n mapped into V_m ($n = 2, m = 3$) with rank 2, nullity 0. Here $m > n$.

collapsing V_n. We must collapse at least enough that an $(n - m)$-dimensional subspace K of V_n is squeezed to **0.** If we collapse just this much, then V_n has been packed into V_m so as to fill all of V_m; $A\mathbf{x} = \mathbf{b}$ has a solution for every **b,** but the solutions are not unique, since each **b** comes from squeezing an $(n - m)$-dimensional linear variety to a point. If we squeeze even further, then K gets bigger (k increases), and we fill less and less of V_m (Y gets smaller, r decreases); the extreme case is when we squeeze V_n down to **0** in V_m; then K is all of V_n and Y is just **0.**

The Enlarged Matrix. If A is an m by n matrix and B is an m by p matrix, then we can form a new m by $(n + p)$ matrix by placing the array of B to the right of that of A:

$$\left(\begin{array}{ccc|ccc} a_{11} & \cdots & a_{1n} & b_{11} & \cdots & b_{1p} \\ \vdots & & \vdots & & & \vdots \\ a_{m1} & \cdots & a_{mn} & b_{m1} & \cdots & b_{mp} \end{array}\right)$$

We denote this new matrix by (A, B) and call it the *enlargement of A by B.* In the particular case when B is a column vector $\mathbf{b} = \text{col}\,(b_1, \ldots, b_m)$, then (A, B) becomes $(A, \mathbf{b})$, a matrix of m rows and $n + 1$ columns. This matrix can be used to give in one expression both the coefficients a_{ij} and the right-hand members for a set of simultaneous equations (10-60). As a matrix, $(A, \mathbf{b})$ has all the attributes: rank, nullity, and so on, attached to a matrix. We have the following useful rule to be added to our list of general rules:

8. *Equation* (10-61) *has a solution if, and only if, A and* $(A, \mathbf{b})$ *have the same rank.* The proof is left as an exercise (Problem 6 below).

Remarks on linear varieties. A linear variety L in V_n is given by $\{\mathbf{c} + K\}$, where K is a subspace of V_n, of dimension k. Let $\boldsymbol{\delta}_1, \ldots, \boldsymbol{\delta}_k$ be a basis for

K. We can use these vectors as column vectors to form an n by k matrix D:

$$D = \begin{pmatrix} d_{11} & \cdots & d_{1k} \\ \vdots & & \vdots \\ d_{n1} & \cdots & d_{nk} \end{pmatrix}, \qquad \boldsymbol{\delta}_i = \begin{pmatrix} d_{1i} \\ \vdots \\ d_{ni} \end{pmatrix}, \qquad i = 1, \ldots, k$$

Now the vectors of the linear variety L are all vectors of the form

$$\mathbf{x} = \mathbf{c} + t_1\boldsymbol{\delta}_1 + t_2\boldsymbol{\delta}_2 + \cdots + t_k\boldsymbol{\delta}_k$$

where $t_1, \ldots, t_k$ are arbitrary scalars or, in terms of coordinates, are all $(x_1, \ldots, x_n)$ such that

$$\begin{aligned} x_1 &= c_1 + d_{11}t_1 + d_{12}t_2 + \cdots + d_{1k}t_k \\ &\vdots \\ x_n &= c_n + d_{n1}t_1 + d_{n2}t_2 + \cdots + d_{nk}t_k \end{aligned} \tag{10-63}$$

Thus

$$\mathbf{x} = \mathbf{c} + D\mathbf{t} \tag{10-64}$$

where $\mathbf{t}$ is an arbitrary column vector in V_k. All vectors in (10-64) are written vertically. Equations (10-63) (or (10-64)) are termed a *parametric representation of the linear variety* L in terms of parameters $t_1, \ldots, t_k$. (When $k = 1$ and $n = 2$, we obtain the parametric equations of a line, as in Section 1-15.)

When $\mathbf{c} = \mathbf{0}$, $L = K$ and the linear variety is a subspace of V_n. Thus

$$\mathbf{x} = D\mathbf{t}$$

is the *parametric representation of a k-dimensional subspace* K in terms of parameters $t_1, \ldots, t_k$. When $k = n$, $\mathbf{t}$ is an arbitrary vector of V_n and the equation $\mathbf{x} = D\mathbf{t}$ is the same as $\mathbf{x} = t_1\boldsymbol{\delta}_1 + \cdots + t_n\boldsymbol{\delta}_n$, the representation of $\mathbf{x}$ in terms of a basis $\boldsymbol{\delta}_1, \ldots, \boldsymbol{\delta}_n$.

For general k, the equation $\mathbf{x} = D\mathbf{t}$ can be regarded as a linear mapping D of V_k into V_n whose range is K. Because $\boldsymbol{\delta}_1, \ldots, \boldsymbol{\delta}_k$ are a basis for K, the representation of $\mathbf{x}$ as $D\mathbf{t}$ is unique; that is, D is a one-to-one mapping (as in No. 5 above, with D of maximum rank, here equal to k).

We remark finally that, given a subspace Z of V_n of dimension k, there are systems of $n - k$ linear homogeneous equations in n unknowns whose solutions are the vectors in Z. Furthermore, if L is a k-dimensional linear variety of V_n, not a subspace of V_n, then L is the set of solutions of $n - k$ linear equations: $a_{i1}x_1 + a_{i2}x_2 + \cdots a_{in}x_n = b_i$, $i = 1, 2, \ldots, n - k$, where the matrix (a_{ij}) has rank $n - k$. These assertions follow from Theorems 1 and 2 in Section 10-4.

Results Formulated Without Vectors. Because of their importance for many applications, we restate some of the results for the case $m = n$ in other language:

THEOREM 3. *Let there be given the n equations in n unknowns*

$$a_{11}x_1 + \cdots + a_{1n}x_n = b_1, \quad \ldots, \quad a_{n1}x_1 + \cdots + a_{nn}x_n = b_n \tag{10-65}$$

with corresponding homogeneous equations

$$a_{11}x_1 + \cdots + a_{1n}x_n = 0, \quad \ldots, \quad a_{n1}x_1 + \cdots + a_{nn}x_n = 0 \tag{10-66}$$

If the homogeneous equations (10-66) *have only the trivial solution* $x_1 = 0, \ldots, x_n = 0$, *then the nonhomogeneous equations* (10-65) *have one, and only one, solution for each choice of* $b_1, \ldots, b_n$.

If the homogeneous equations have solutions other than the trivial solution, then there is a positive integer k such that the solutions of the homogeneous equation can be given in the form:

$$x_1 = d_{11}t_1 + \cdots + d_{1k}t_k, \quad \ldots, \quad x_n = d_{n1}t_1 + \cdots + d_{nk}t_k \qquad (10\text{-}67)$$

where, for each $j = 1, 2, \ldots, k$, $x_1 = d_{1j}, \ldots, x_n = d_{nj}$ *is a solution of* (10-66), $t_1, \ldots, t_k$ *are arbitrary scalars,* $1 \leq k \leq n$, *and* (10-67) *reduces to the trivial solution only for* $t_1 = 0, t_2 = 0, \ldots, t_k = 0$. *In this case, the nonhomogeneous equations are not generally solvable, but are solvable for special choices of* $b_1, \ldots, b_n$. *For each such special choice of* $b_1, \ldots, b_n$, *there are infinitely many solutions of* (10-65), *given by*

$$x_i = x_i^* + d_{i1}t_1 + d_{i2}t_2 + \cdots + d_{ik}t_k, \qquad i = 1, \ldots, n$$

where $x_1 = x_1^*, \ldots, x_n = x_n^*$ *is one solution and the* d_{ij}, t_j *and k are as in* (10-67).

Remark. As pointed out in Section 10-4 and proved in Section 10-14, the rank of a matrix can be computed by determinants. From the determinant rule it follows that an n by n matrix has maximal rank n precisely when $\det A \neq 0$, where $\det A$ denotes the determinant formed from the square array (a_{ij}). Thus the two cases in Theorem 3 correspond to the case when $\det A \neq 0$ (homogeneous equations have only the trivial solution) and the case when $\det A = 0$ (homogeneous equations have non-trivial solutions).

Some Examples. In the preceding discussion we have considered the existence of solutions of linear equations, but we have not considered ways of actually determining the solutions. We now give several examples that illustrate methods for finding solutions. The technique used is essentially the elimination of variables. This technique can be much improved by matrix methods; such refinements will be discussed in Section 10-16. One can also use determinants (as shown in Sections 0-9 and 10-13); however, except for small values of m and n, determinants are not convenient calculation tools.

EXAMPLE 1. $3x - 2y = 4, 4x - 3y = 5$. Here, $m = n = 2$. The corresponding homogeneous equations are

$$3x - 2y = 0, \qquad 4x - 3y = 0$$

Multiplying the first by 4, the second by -3, and adding, gives $y = 0$, so that $x = 0$. Thus the homogeneous equations have only the trivial solution $x = 0$, $y = 0$, and the nonhomogeneous equations must have a unique solution. Multiplying the nonhomogeneous equations by the same scalars and adding, gives $y = 1$, from which we find $x = 2$. Thus $x = 2$, $y = 1$ is the only solution.

In vector space language, we have found that the mapping by the matrix $A = \begin{pmatrix} 3 & -2 \\ 4 & -3 \end{pmatrix}$ has kernel consisting of $\mathbf{0}$ alone, so that A has nullity k equal to 0 and rank r equal to 2. Thus, as in rule 5, A provides a one-to-one mapping of V_2 onto V_2, and $A\mathbf{x} = \mathbf{b} = \text{col}\ (4, 5)$ has a unique solution for $\mathbf{x}$; we found $\mathbf{x} = \text{col}\ (2, 1)$.

EXAMPLE 2.

$$4x_1 - x_2 - x_3 = 2, \qquad 2x_1 + 2x_2 - 3x_3 = 31, \qquad 6x_1 + x_2 - 4x_3 = 33$$

Here the homogeneous equations are

$$4x_1 - x_2 - x_3 = 0, \qquad 2x_1 + 2x_2 - 3x_3 = 0, \qquad 6x_1 + x_2 - 4x_3 = 0$$

We eliminate x_2 from the first and second equations to find $10x_1 - 5x_3 = 0$ or $x_3 = 2x_1$. We replace x_3 by $2x_1$ in the first equation to obtain $4x_1 - x_2 - 2x_1 = 0$ or $x_2 = 2x_1$. We replace x_2 by $2x_1$, x_3 by $2x_1$ in the second and third equations to obtain $0 = 0$. Thus all three homogeneous equations are satisfied for $x_2 = 2x_1, x_3 = 2x_1$, no matter how x_1 is chosen. We write $x_1 = t$, $x_2 = 2t$, $x_3 = 2t$ and, thereby, give all solutions of the homogeneous equation in terms of the arbitrary scalar t.

We now know that the given nonhomogeneous equations either have no solution or else have infinitely many solutions. We try to find a solution with $x_3 = 0$, so that the first two nonhomogeneous equations become $4x_1 - x_2 = 2$, $2x_1 + 2x_2 = 31$. By elimination we find easily $x_1 = 7/2, x_2 = 12$. These values satisfy the third equation, with $x_3 = 0$. Hence, $x_1 = x_1^* = 7/2$, $x_2 = x_2^* = 12$, $x_3 = x_3^* = 0$ is a particular solution of the nonhomogeneous equations, and all solutions are given by $x_1 = (7/2) + t, x_2 = 12 + 2t, x_3 = 2t$.

In vector-space language, we have found that the matrix

$$A = \begin{pmatrix} 4 & -1 & -1 \\ 2 & 2 & -3 \\ 6 & 1 & -4 \end{pmatrix}$$

has kernel K equal to Span ((1, 2, 2)), so that K is one-dimensional, A has nullity k equal to 1 and rank r equal to $3 - 1 = 2$. Thus rule 4 applies. The range of A is a two-dimensional subspace Y of V_3, spanned by the column vectors $\gamma_1 = (4, 2, 6)$, $\gamma_2 = (-1, 2, -1)$, $\gamma_3 = (-1, -3, -4)$, only two of which are linearly independent (any two, as one easily verifies); the vector $\mathbf{b} = (2, 31, 33)$ happens to lie in this subspace, so that the equation $A\mathbf{x} = \mathbf{b}$ has solutions. These solutions form the one-dimensional linear variety $L = \{\mathbf{x}^* + K\}$, where $\mathbf{x}^*$ can be chosen as $(7/2, 12, 0)$.

EXAMPLE 3. $\quad x_1 - 2x_2 = 3,\ 2x_1 - 4x_2 = 1,\ 3x_1 - 6x_2 = 7.$
Here the homogeneous equations

$$x_1 - 2x_2 = 0, \qquad 2x_1 - 4x_2 = 0, \qquad 3x_1 - 6x_2 = 0$$

are all equivalent. We can take x_2 arbitrary—that is, set $x_2 = t$ and find $x_1 = 2t$; thus the solutions are $x_1 = 2t, x_2 = t$. We again know that the nonhomogeneous equations, if solvable, cannot have a unique solution (by rule 6). Elimination of x_2 from the first two equations gives $0 = -5$, so that the equations are inconsistent and no solution can be found.

In vector-space language, we have found that the matrix

$$A = \begin{pmatrix} 1 & -2 \\ 2 & -4 \\ 3 & -6 \end{pmatrix}$$

has kernel $K = \text{Span}\,((2, 1))$, so that the nullity k is 1 and A has rank $2 - 1 = 1$. Here A maps V_2 into V_3, and its range is a one-dimensional subspace Y of V_3; Y is spanned by $\gamma_1 = (1, 2, 3)$, for example. The vector $\mathbf{b} = (3, 1, 7)$ is not linearly dependent on γ_1 and, hence, does not lie in Y, so that $A\mathbf{x} = \mathbf{b}$ has no solution.

EXAMPLE 4. $x_1 + 3x_2 - x_3 + x_4 = 1$, $x_1 + 3x_2 + x_3 + 2x_4 = 5$. The corresponding homogeneous equations are

$$x_1 + 3x_2 - x_3 + x_4 = 0, \qquad x_1 + 3x_2 + x_3 + 2x_4 = 0$$

Here we have only two equations in four unknowns; hence, we try to solve for two of the unknowns in terms of the other two. We try to solve for x_1, x_2 by writing the equations as

$$x_1 + 3x_2 = x_3 - x_4, \qquad x_1 + 3x_2 = -x_3 - 2x_4$$

But if we eliminate x_1, we also eliminate x_2, so that our attempt fails. We try other combinations and find in particular that we can solve for x_1 and x_3:

$$x_1 - x_3 = -3x_2 - x_4, \qquad x_1 + x_3 = -3x_2 - 2x_4$$

Elimination gives $2x_1 = -6x_2 - 3x_4$, $2x_3 = 0x_2 - x_4$. Thus we can choose x_2, x_4 arbitrarily and express x_1, x_3 in terms of them. We avoid fractions if we take $x_2 = 2t_1$, $x_4 = 2t_2$ so that $2x_1 = -12t_1 - 6t_2$, $2x_3 = -2t_2$ and, finally,

$$x_1 = -6t_1 - 3t_2, \qquad x_2 = 2t_1, \qquad x_3 = -t_2, \qquad x_4 = 2t_2$$

gives all solutions of the homogeneous equations. For the nonhomogeneous equations we seek a particular solution by taking two of the unknowns equal to 0. The choice $x_3 = 0$, $x_4 = 0$ leads to contradictory equations for x_1, x_2. The choice $x_2 = 0$, $x_4 = 0$ leads to $x_1 - x_3 = 1$, $x_1 + x_3 = 5$ and, hence, $x_1 = 3$, $x_2 = 2$. Thus $x_1 = x_1^* = 3$, $x_2 = x_2^* = 0$, $x_3 = x_3^* = 2$, $x_4 = x_4^* = 0$ is a particular solution and

$$x_1 = 3 - 6t_1 - 3t_2, \qquad x_2 = 2t_1, \qquad x_3 = 2 - t_2, \qquad x_4 = 2t_2$$

gives all solutions of the nonhomogeneous equations.

In vector-space language, we have a mapping

$$A = \begin{pmatrix} 1 & 3 & -1 & 1 \\ 1 & 3 & 1 & 2 \end{pmatrix}$$

of V_4 into V_2. The kernel K of A is the two-dimensional subspace spanned by $(-6, 2, 0, 0)$ and $(-3, 0, -1, 2)$, so that A has nullity k equal to 2 and rank

$r = 4 - 2 = 2$. Since A has rank 2, its range Y is a two-dimensional subspace of V_2 and, hence, is all of V_2. Therefore, for each **b**, $A\mathbf{x} = \mathbf{b}$ has solutions; for each **b** the solutions form a linear variety $\{\mathbf{x}^* + K\}$ in V_4. For $\mathbf{b} = (1.5)$, we find a choice of $\mathbf{x}^*$ to be $(3, 0, 2, 0)$. For this example, rule 7(b) applies.

Complex Case. We have emphasized the case of real vector spaces. All results and methods extend at once to complex vector spaces. In each case, we deal with a complex matrix A mapping V_n^c into V_m^c.

It is important to know that if a system of linear equations with real coefficients has solutions in the complex numbers, then that system has solutions in the real numbers. For if $a_1, \ldots, a_n, b_1, \ldots, b_n, c_1, \ldots, c_m$ are real and $\operatorname{col}(a_1 + b_1 i, \ldots, a_n + b_n i) = \mathbf{a} + i\mathbf{b}$ is a solution of $A\mathbf{x} = \mathbf{c} = \operatorname{col}(c_1, \ldots, c_n)$, then $A(\mathbf{a} + i\mathbf{b}) = A\mathbf{a} + iA\mathbf{b} = \mathbf{c} = \mathbf{c} + i\mathbf{0}$; and therefore $A\mathbf{a} = \mathbf{c}$, $A\mathbf{b} = \mathbf{0}$. Thus **a** is a real solution of $A\mathbf{x} = \mathbf{c}$.

PROBLEMS

1. Solve where possible and interpret the results in vector-space language:

(a) $\begin{cases} x - 2y = 0, \\ 2x - 4y = 0. \end{cases}$ (b) $\begin{cases} x + y = 0, \\ 3x - y = 0. \end{cases}$ (c) $\begin{cases} x - y = 1, \\ 2x - 4y = 3. \end{cases}$

(d) $\begin{cases} x - y = 5, \\ 2x + y = 4. \end{cases}$ (e) $\begin{cases} x_1 + 3x_2 - 2x_3 = 1, \\ 3x_1 - x_2 - x_3 = 2, \\ x_1 + x_2 - x_3 = 0. \end{cases}$ (f) $\begin{cases} x_1 + x_2 - x_3 = 1, \\ 4x_1 + 2x_2 - 3x_3 = 3, \\ 2x_1 - x_2 + x_3 = 2. \end{cases}$

(g) $\begin{cases} x_1 - 2x_2 = 2, \\ 2x_1 + 3x_2 = 4, \\ 5x_1 - x_2 = 10. \end{cases}$ (h) $\begin{cases} 2x_1 - x_2 = 4, \\ x_1 + x_2 = 1, \\ 3x_1 - x_2 = 1. \end{cases}$ (i) $\begin{cases} x_1 + x_2 - x_3 + x_4 = 3, \\ x_1 + x_2 - x_3 - 2x_4 = 0. \end{cases}$

(j) $\begin{cases} x_1 + 2x_2 - 3x_3 + x_4 = 1, \\ 2x_1 - x_2 + x_3 - x_4 = 2, \\ 4x_1 + 3x_2 - 5x_3 + x_4 = 4. \end{cases}$ (k) $3x_1 - 2x_2 + 3x_3 - x_4 + x_5 = 0.$

(l) $x_1 + 3x_2 - x_3 + x_4 - x_5 = 0.$

2. Let A be an m by n matrix with nullity k and rank r. Let **b** be an m by 1 column vector such that $(A, \mathbf{b})$ has rank r_1. For each of the following sets of values, state whether it can arise and, if it can, give the dimension of the linear variety of solutions of $A\mathbf{x} = \mathbf{0}$ and of $A\mathbf{x} = \mathbf{b}$:

(a) $m = 2, n = 2, r = 2, k = 0, r_1 = 2$;

(b) $m = 2, n = 2, r = 1, k = 1, r_1 = 2$;

(c) $m = 3, n = 2, r = 3, k = 0, r_1 = 3$;

(d) $m = 3, n = 2, r = 2, k = 0, r_1 = 2$;

(e) $m = 3, n = 2, r = 1, k = 0, r_1 = 2$;

(f) $m = 3, n = 2, r = 0, k = 2, r_1 = 2$;

(g) $m = 3, n = 3, r = 3, k = 0, r_1 = 2$;

(h) $m = 3, n = 3, r = 3, k = 0, r_1 = 3$;

(i) $m = 3, n = 4, r = 4, k = 0, r_1 = 4$;

(j) $m = 3, n = 4, r = 2, k = 1, r_1 = 2$;

(k) $m = 3, n = 4, r = 3, k = 1, r_1 = 4$;

(l) $m = 3, n = 4, r = 2, k = 2, r_1 = 3$;

(m) $m = 3, n = 4, r = 1, k = 3, r_1 = 2$;

(n) $m = 3, n = 4, r = 0, k = 4, r_1 = 0$;

(o) $m = 7, n = 2, r = 4, k = 3, r_1 = 4$;

(p) $m = 5, n = 8, r = 6, k = 2, r_1 = 6$;

(q) $m = 11, n = 3, r = 3, k = 8, r_1 = 3$;

(r) $m = 9, n = 10, r = 10, k = 0, r_1 = 10$;

(s) $m = 8, n = 13, r = 7, k = 6, r_1 = 8$;

(t) $m = 15, n = 12, r = 9, k = 3, r_1 = 9$.

3. With the notations of Problem 2, find a matrix A and a vector $\mathbf{b}$ to fit the following cases and find all solutions of the corresponding equation $A\mathbf{x} = \mathbf{b}$:

(a) $m = 2, n = 2, r = 2, k = 0, r_1 = 2$.

(b) $m = 2, n = 2, r = 1, k = 1, r_1 = 2$.

(c) $m = 3, n = 3, r = 2, k = 1, r_1 = 2$.

(d) $m = 3, n = 3, r = 2, k = 1, r_1 = 3$.

4. With the notations of Problem 2, find matrices A, $\mathbf{b}$ to fit the following cases:

(a) $m = 4, n = 3, r = 1, k = 2, r_1 = 2$;

(b) $m = 3, n = 5, r = 2, k = 3, r_1 = 2$;

(c) $m = 5, n = 2, r = 1, k = 1, r_1 = 2$;

(d) $m = 5, n = 3, r = 2, k = 1, r_1 = 2$.

5. Solve where possible and express the results in the language of complex vector spaces:

(a) $\begin{cases} (2 - i)x_1 + (3 + i)x_2 = 0, \\ (3 + i)x_1 + (2 + 4i)x_2 = 0. \end{cases}$

(b) $\begin{cases} (1 + i)x_1 + (1 - i)x_2 = 2i, \\ (1 - i)x_1 + (1 + i)x_2 = 1. \end{cases}$

(c) $\begin{cases} x_1 + ix_2 - x_3 = 1, \\ 2ix_1 - x_2 + 3ix_3 = 0, \\ (1 + 5i)x_1 - (3 - i)x_2 - (1 - 5i)x_3 = 3 - i, \end{cases}$

(d) $\begin{cases} x_1 + (2 - i)x_2 - x_3 = 1, \\ x_1 - (1 + i)x_2 - x_3 = 1 - i. \end{cases}$

6. Prove: $A\mathbf{x} = \mathbf{b}$ has a solution if, and only if, rank A = rank $(A, \mathbf{b})$. [*Hint:* If there is a solution, $\mathbf{b}$ is in Range A.]

10-7 ADDITION OF MATRICES, SCALAR TIMES MATRIX

We have now seen that the m by n matrices are linear mappings of V_n into V_m. As in Chapter 9 (Section 9-16), the linear mappings of one vector space into another can be added and multiplied by scalars and, under these operations, they form a vector space. Accordingly, we must be able to add two m by n matrices and multiply each such matrix by a scalar, and these operations must obey the basic laws for vector spaces.

The sum $A + B$ of two m by n matrices $A = (a_{ij})$ and $B = (b_{ij})$ is the mapping such that $(A + B)\mathbf{x} = A\mathbf{x} + B\mathbf{x}$ for every $\mathbf{x}$ in V_n. If we write

$A + B = C = (c_{ij})$, then for $j = 1, \ldots, n$,

$$C\mathbf{e}_j = (A + B)\mathbf{e}_j = A\mathbf{e}_j + B\mathbf{e}_j$$

Hence, the jth column vector of C is the sum of the jth column vectors of A and B. This in turn implies that $c_{ij} = a_{ij} + b_{ij}$ for all i and j. We conclude:

The sum of two m by n matrices $A = (a_{ij})$ and $B = (b_{ij})$ is the m by n matrix $C = (c_{ij})$, where

$$c_{ij} = a_{ij} + b_{ij} \qquad \text{for } i = 1, \ldots, m, j = 1, \ldots, n \tag{10-70}$$

Thus to add two matrices of the same size, we simply add corresponding entries.

EXAMPLE 1 $$\begin{pmatrix} 2 & 1 & 3 \\ 3 & 4 & 5 \end{pmatrix} + \begin{pmatrix} 1 & 1 & 0 \\ 2 & 7 & 4 \end{pmatrix} = \begin{pmatrix} 3 & 2 & 3 \\ 5 & 11 & 9 \end{pmatrix}$$

Warning. We only add two matrices having the same number of rows and the same number of columns; the sum again has the same number of rows and columns.

If we regard $A = (a_{ij})$ as a linear mapping and if c is a scalar, then $cA = D = (d_{ij})$ is the linear mapping such that $D\mathbf{x} = c(A\mathbf{x})$ for every $\mathbf{x}$ in V_n. Hence, in particular, $D\mathbf{e}_j = c(A\mathbf{e}_j)$. Therefore, the jth column of D is c times the jth column vector of A, for $j = 1, \ldots, n$ and, accordingly, $d_{ij} = ca_{ij}$ for all i and j. We conclude:

A scalar c times an m by n matrix $A = (a_{ij})$ yields the m by n matrix $D = (d_{ij})$, where

$$d_{ij} = ca_{ij} \qquad \text{for } i = 1, \ldots, m, j = 1, \ldots, n \tag{10-71}$$

Thus to multiply a matrix by a scalar c, we simply multiply each entry by c.

EXAMPLE 2 $$3\begin{pmatrix} 5 & 8 \\ 6 & -1 \end{pmatrix} = \begin{pmatrix} 15 & 24 \\ 18 & -3 \end{pmatrix}$$

We remarked above that the linear mappings of V_n into V_m form a vector space. Therefore, we now conclude that the operations of the addition of matrices and of the multiplication of matrices by scalars obey the laws of Section 9-1:

$$\begin{aligned} &A + B = B + A, \quad (A + B) + C = A + (B + C), \quad A + O = A \\ &A + (-A) = O, \quad a(bA) = (ab)A, \quad (a + b)A = aA + bA \\ &a(A + B) = aA + aB, \quad 1A = A, \quad -A = (-1)A \end{aligned} \tag{10-72}$$

We have here used the fact that, since $Ox = \mathbf{0}$ for every x, O (that is, O_{mn}) must be the zero of our vector space of matrices.

We denote by $\mathfrak{M}_{mn}$ the vector space of all (real) m by n matrices. (There is a complex counterpart, which we denote by $\mathfrak{M}_{mn}{}^c$.)

Remark. The vector space $\mathfrak{M}_{mn}$ has dimension mn. This follows from Theorem 25 of Section 9-16. Also we can verify directly that the $m \times n$

matrices having one entry equal to 1 and all other entries equal to 0 form a basis for $\mathfrak{M}_{mn}$. For example,

$$E_{11} = \begin{pmatrix} 1 & 0 \\ 0 & 0 \end{pmatrix} \quad E_{12} = \begin{pmatrix} 0 & 1 \\ 0 & 0 \end{pmatrix} \quad E_{21} = \begin{pmatrix} 0 & 0 \\ 1 & 0 \end{pmatrix} \quad E_{22} = \begin{pmatrix} 0 & 0 \\ 0 & 1 \end{pmatrix}$$

form a basis for $\mathfrak{M}_{22}$, the vector space of all 2 by 2 matrices, since

$$a_{11}E_{11} + a_{12}E_{12} + a_{21}E_{21} + a_{22}E_{22} = \begin{pmatrix} a_{11} & a_{12} \\ a_{21} & a_{22} \end{pmatrix}$$

and this linear combination of $E_{11}, \ldots, E_{22}$ can reduce to $O = \begin{pmatrix} 0 & 0 \\ 0 & 0 \end{pmatrix}$ only when $a_{11} = 0, \ldots, a_{22} = 0$. In general, a basis for $\mathfrak{M}_{mn}$ is given by the analogous matrices $E_{11}, E_{12}, \ldots, E_{mn}$ and

$$A = (a_{ij}) = a_{11}E_{11} + a_{12}E_{12} + \cdots + a_{mn}E_{mn}$$

10-8 MULTIPLICATION OF MATRICES

Multiplication of linear mappings is defined through composition (Section 9-17). If T maps U into V and S maps V into W, then $R = ST = S \circ T$ maps U into W and

$$R(\mathbf{u}) = S[T(\mathbf{u})]$$

for each $\mathbf{u}$ in U. Therefore, we can multiply two matrices, of appropriate sizes. If $A = (a_{ij})$ is an m by p matrix and $B = (b_{ij})$ is a p by n matrix, then B maps V_n into V_p and A maps V_p into V_m (Figure 10-5), so that $C = AB$ maps V_n into V_m and, therefore, $C = (c_{ij})$ is the m by n matrix such that

$$C\mathbf{x} = A(B\mathbf{x}) \qquad \text{for every } \mathbf{x} \text{ in } V_n$$

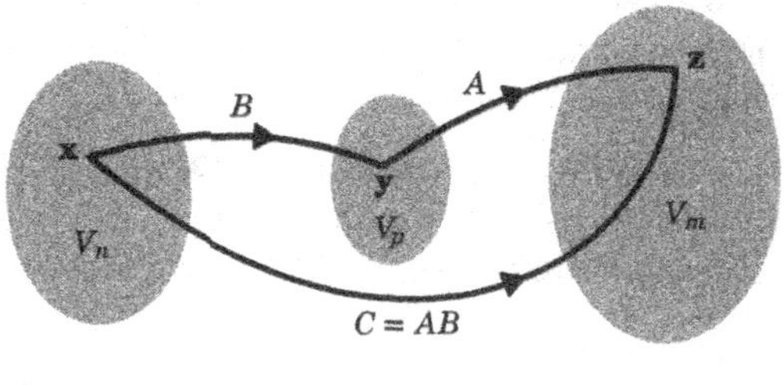

Figure 10-5 The product of matrices as composition of mappings.

To find the columns of C, we seek $C\mathbf{e}_j = A(B\mathbf{e}_j)$. But $B\mathbf{e}_j$ is the jth column of B. Hence, *the jth column of $C = AB$ is A times the jth column of B.*

Now the jth column of B is the vector col $(b_{1j}, b_{2j}, \ldots, b_{pj})$. Hence, the jth column of C is the column vector

$$\begin{pmatrix} a_{11}b_{1j} + a_{12}b_{2j} + \cdots + a_{1p}b_{pj} \\ a_{21}b_{1j} + a_{22}b_{2j} + \cdots + a_{2p}b_{pj} \\ \vdots \\ a_{m1}b_{1j} + a_{m2}b_{2j} + \cdots + a_{mp}b_{pj} \end{pmatrix}$$

and the ith element in that column is

$$c_{ij} = a_{i1}b_{1j} + a_{i2}b_{2j} + \cdots + a_{ip}b_{pj} \tag{10-80}$$

Equation (10-80) gives the basic rule for computing $C = AB$ from A and B. We can write it in concise form as follows:

$$C = AB = (c_{ij}), \quad c_{ij} = \sum_{k=1}^{p} a_{ik}b_{kj}, \; i = 1, \ldots, m, \; j = 1, \ldots, n \tag{10-81}$$

In practice it is simpler to think of the formation of the columns of C from the columns of B by multiplication by A.

EXAMPLE 1

$$\begin{pmatrix} 1 & 2 \\ 3 & 4 \end{pmatrix}\begin{pmatrix} 2 & 0 & -1 \\ 4 & 1 & 2 \end{pmatrix} = \begin{pmatrix} 2+8 & 0+2 & -1+4 \\ 6+16 & 0+4 & -3+8 \end{pmatrix} = \begin{pmatrix} 10 & 2 & 3 \\ 22 & 4 & 5 \end{pmatrix}$$

Here we obtain the columns of the product as follows:

$$\begin{pmatrix} 1 & 2 \\ 3 & 4 \end{pmatrix}\begin{pmatrix} 2 \\ 4 \end{pmatrix} = \begin{pmatrix} 10 \\ 22 \end{pmatrix}, \quad \begin{pmatrix} 1 & 2 \\ 3 & 4 \end{pmatrix}\begin{pmatrix} 0 \\ 1 \end{pmatrix} = \begin{pmatrix} 2 \\ 4 \end{pmatrix}, \quad \begin{pmatrix} 1 & 2 \\ 3 & 4 \end{pmatrix}\begin{pmatrix} -1 \\ 2 \end{pmatrix} = \begin{pmatrix} 3 \\ 5 \end{pmatrix}$$

Remark. It is to be emphasized that the product AB can be formed only when the sizes of the two matrices match; the width of A must equal the height of B, as in Figure 10-6. In particular, AB may be defined whereas BA is not defined (Problem 5 below).

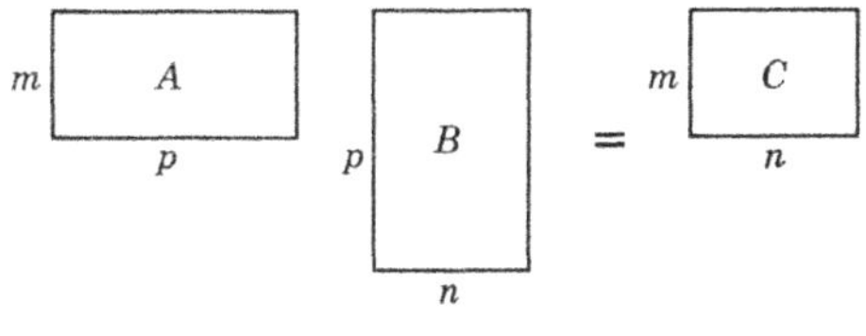

Figure 10-6 The multiplication of matrices.

EXAMPLE 2

$$\begin{pmatrix} a & b \\ c & d \\ e & f \end{pmatrix}\begin{pmatrix} \alpha & \beta & \gamma & \delta \\ \epsilon & \rho & \sigma & \tau \end{pmatrix} = \begin{pmatrix} a\alpha + b\epsilon & a\beta + b\rho & a\gamma + b\sigma & a\delta + b\tau \\ c\alpha + d\epsilon & c\beta + d\rho & c\gamma + d\sigma & c\delta + d\tau \\ e\alpha + f\epsilon & e\beta + f\rho & e\gamma + f\sigma & e\delta + f\tau \end{pmatrix}$$

Observe that the first factor is 3×2, the second is 2×4, so that the product is defined and is 3×4.

EXAMPLE 3. Let A be an $m \times n$ matrix. Find AI and IA.

Solution. For the product AI to have meaning, I must be the n by n identity matrix I_n, whose column vectors are the basis vectors $\mathbf{e}_1, \ldots, \mathbf{e}_n$. Hence, A times the first column is just the first column of A, A times the second column is the second column of A, and so on. Therefore, $AI = A$. Also, IA has meaning if I is m by m. Since $I\mathbf{x} = \mathbf{x}$ for every vector in V_m, we conclude that $IA = A$ also. Therefore

$$AI = A \qquad \text{and} \qquad IA = A \tag{10-82}$$

for the proper choice of I in each case.

EXAMPLE 4. Evaluate $(cI)A$ and $A(cI)$, where c is a scalar. We observe that cI is the scalar matrix

$$\begin{pmatrix} c & 0 & \cdots & 0 \\ 0 & c & \cdots & 0 \\ \vdots & \vdots & & \vdots \\ 0 & 0 & \cdots & c \end{pmatrix}$$

By reasoning like that for Example 3, we find that

$$(cI)A = cA \qquad \text{and} \qquad A(cI) = cA \tag{10-83}$$

for the proper choice of I in each case.

We remark also that

$$OA = O, \qquad AO = O \tag{10-84}$$

for a proper choice of O in each of the four cases in which it occurs.

The multiplication of matrices is a special case of the multiplication of linear mappings. Therefore, by Theorem 26 of Section 9-17, we conclude that, whenever the various sums and products are defined, one has the rules:

$$C(A + B) = CA + CB \tag{10-85}$$

$$(C + D)A = CA + DA \tag{10-86}$$

$$E(CA) = (EC)A \tag{10-87}$$

$$C(kA) = k(CA) \tag{10-88}$$

Remark. We have been led to the rules for addition and multiplication of matrices and for a scalar times a matrix by interpreting matrices as linear mappings. One can also regard these rules simply as definitions. For example, one simply *defines* $A + B$ to be the matrix $(a_{ij} + b_{ij})$ and cA to be the matrix (ca_{ij}). One can then establish all the properties we have listed [such as (10-85) to (10-88)] by direct algebraic manipulation.

Matrices sometimes arise whose elements are not real (or complex) numbers, but which can be added and multiplied, and multiplied by scalars. For such matrices the interpretation as a linear mapping may lose meaning. However, we can still use the purely algebraic definition of $A + B$, cA and AB, and then verify that all the algebraic laws still hold true.

10-9 THE TRANSPOSE

The *transpose* of the m by n matrix $A = (a_{ij})$ is the n by m matrix $E = (e_{ji})$, where $e_{ji} = a_{ij}$, $1 \le i \le m$, $1 \le j \le n$. We write $E = A'$. The following are examples:

$$A = \begin{pmatrix} a & b \\ c & d \end{pmatrix}, \qquad A' = \begin{pmatrix} a & c \\ b & d \end{pmatrix}$$

$$B = \begin{pmatrix} a & b & c \\ d & e & f \end{pmatrix}, \qquad B' = \begin{pmatrix} a & d \\ b & e \\ c & f \end{pmatrix}$$

In general, one has

$$A = \begin{pmatrix} a_{11} & a_{12} & \cdots & a_{1n} \\ a_{21} & a_{22} & \cdots & a_{2n} \\ \vdots & \vdots & & \vdots \\ a_{m1} & a_{m2} & \cdots & a_{mn} \end{pmatrix}, \qquad A' = \begin{pmatrix} a_{11} & a_{21} & \cdots & a_{m1} \\ a_{12} & a_{22} & \cdots & a_{m2} \\ \vdots & \vdots & & \vdots \\ a_{1n} & a_{2n} & \cdots & a_{mn} \end{pmatrix}$$

THEOREM 4. *Let A, D be m by p matrices, let B be a p by n matrix, let k be a scalar. Then*

$$\text{(i)}\ (A + D)' = A' + D' \qquad \text{(ii)}\ (kA)' = kA'$$

$$\text{(iii)}\ (AB)' = B'A' \qquad \text{(iv)}\ (A')' = A$$

PROOF. The proofs for parts (i), (ii), and (iv) are left as exercises (Problem 9 below). We give the proof of (iii) here.

We have $A = (a_{ij})$, $B = (b_{jk})$, $i = 1, \ldots, m$, $j = 1, \ldots, p$, $k = 1, \ldots, n$. Also, $AB = C = (c_{ik})$, where

$$c_{ik} = a_{i1}b_{1k} + a_{i2}b_{2k} + \cdots + a_{ip}b_{pk}$$

Next let $E = (e_{ji})$ be A', so that $e_{ji} = a_{ij}$ and let $F = (f_{kj})$ be B', so that $f_{kj} = b_{jk}$. Then F is n by p, E is p by m, so that we can form the product $FE = G = (g_{ki})$:

$$\begin{aligned} g_{ki} &= f_{k1}e_{1i} + f_{k2}e_{2i} + \cdots + f_{kp}e_{pi} = b_{1k}a_{i1} + b_{2k}a_{i2} + \cdots + b_{pk}a_{ip} \\ &= a_{i1}b_{1k} + a_{i2}b_{2k} + \cdots + a_{ip}b_{pk} = c_{ik} \end{aligned}$$

Thus $g_{ki} = c_{ik}$ and, accordingly, $G = C'$ or

$$FE = B'A' = (AB)'$$

as asserted.

Remark. The correspondence $A \to A'$ can be considered as a mapping of $\mathfrak{M}_{mn}$ into $\mathfrak{M}_{nm}$. Distinct matrices of $\mathfrak{M}_{mn}$ have distinct transposes, so that the correspondence is one-to-one [for if $A' = B'$, then $(A')' = (B')'$, so that

$A = B$ by (iv)]. Every n by m matrix C is the transpose of an m by n matrix [namely, C' by (iv) again]. Hence, the mapping $A \to A'$ is a one-to-one mapping of $\mathfrak{M}_{mn}$ onto $\mathfrak{M}_{nm}$. By (i) and (ii) the mapping is linear.

PROBLEMS

1. Let the following matrices be given:

$$A = \begin{pmatrix} 1 & 4 \\ 0 & 3 \end{pmatrix}, \quad B = \begin{pmatrix} -1 & 0 \\ 1 & 2 \end{pmatrix}, \quad C = \begin{pmatrix} 2 & 1 \\ -1 & 0 \end{pmatrix},$$

$$D = \begin{pmatrix} 1+i & 2 \\ 1 & 1-i \end{pmatrix}, \quad E = \begin{pmatrix} 5 & 6 & 2 \\ 0 & 1 & 0 \end{pmatrix}, \quad F = \begin{pmatrix} 1 \\ 3 \end{pmatrix},$$

$$G = \begin{pmatrix} 1 \\ 2 \\ 4 \end{pmatrix}, \quad H = \begin{pmatrix} 1 \\ -1 \\ 2 \end{pmatrix}, \quad J = \begin{pmatrix} 1 & 0 & 7 \\ 2 & 1 & 2 \\ 1 & 4 & 3 \end{pmatrix}, \quad K = \begin{pmatrix} 1 & 1 & 1 \\ 0 & 1 & 1 \\ 0 & 0 & 1 \end{pmatrix}.$$

Evaluate the expression requested:

(a) $A + B$ (b) $A + C$ (c) $3A$ (d) $4D$ (e) $G + H$
(f) $2G - H$ (g) $3A + B - C$ (h) $J + K$ (i) AB (j) BA
(k) AE (l) BF (m) EG (n) JG (o) JK
(p) $J(G + H)$ (q) J^2 (r) K^2 (s) K^3 (t) K^4

2. Which of the possible pairs of the following matrices have a product, and in which order?

$$A = \begin{pmatrix} 1 & 2 & 3 \\ 0 & 7 & 0 \\ 1 & 0 & 1 \end{pmatrix}, \quad B = \begin{pmatrix} 1 & 2 \\ 6 & 7 \\ 8 & 0 \end{pmatrix}, \quad C = \begin{pmatrix} 0 & 1 & 0 & 1 & 0 \\ 2 & 0 & 2 & 0 & 1 \end{pmatrix},$$

$$D = \begin{pmatrix} 1 & 1 & 0 \\ 0 & -1 & 2 \\ 2 & 2 & 0 \\ 1 & -1 & 0 \\ 0 & 0 & 1 \end{pmatrix}, \quad E = (0, 1, 2), \quad F = (1, 0, 2, 3, -1),$$

$$G = \begin{pmatrix} 7 & -1 \\ 0 & 2 \\ 1 & 1 \end{pmatrix}, \quad H = \begin{pmatrix} 2 & 1 & 1 & 1 & 0 \\ 1 & 0 & 0 & 0 & 1 \\ 0 & 1 & 2 & 1 & 0 \end{pmatrix}, \quad J = \begin{pmatrix} 1 & 0 & 0 & 0 & 1 \\ 0 & 1 & 0 & 1 & 0 \\ 0 & 0 & 1 & 0 & 0 \end{pmatrix}.$$

3. Evaluate:

(a)
$$\begin{pmatrix} 1 & 0 & x \\ 2x & -1 & 3 \\ x^2 & 0 & 4 \end{pmatrix}^2 \left\{ \begin{pmatrix} 1 & 0 \\ x & 2 \\ 1 & x^2 \end{pmatrix} + 3 \begin{pmatrix} x & 0 \\ 0 & x \\ 1 & 1 \end{pmatrix} \right\},$$

(b) $\begin{pmatrix} \sin x & \cos x \\ -1 & \tan x \\ 1 & 0 \end{pmatrix} \begin{pmatrix} e^x & x \\ -1 & \cos x \end{pmatrix}$.

(c) $\begin{pmatrix} 4 & 2 & \sin x & 1 \\ 0 & 1 & x & x^2 \\ e^x & 0 & 0 & \ln x \\ 1 & 0 & -1 & x \end{pmatrix} \begin{pmatrix} 0 & 2 & x \\ 1 & 0 & 1 \\ 1 & x & 1 \\ -\sin x & -1 & 0 \end{pmatrix} \begin{pmatrix} \cos x & 0 \\ 1 & 2x \\ x & \sin x \end{pmatrix}$.

4. For the matrices of Problem 1 find the expression requested:

(a) A' (b) B' (c) E' (d) F'

(e) AB' (f) $A'B'$ (g) $F'A$ (h) $G'J$

5. (a) If A is not a square matrix ($m \neq n$), then $A^2 = AA$ is not defined but AA' and $A'A$ are defined. Explain.

(b) Show that if $B = AA'$, then $B' = B$.

6. Let E_{st} be a 4 by 3 matrix with 1 in the st spot, and 0 elsewhere (here $1 \leq s \leq 4$, $1 \leq t \leq 3$). Let E^*_{ij} be a 3 by 3 matrix with 1 in the i, j spot and 0 elsewhere (here $1 \leq i \leq 3$, $1 \leq j \leq 3$). Evaluate:

(a) $E_{11}E^*_{11}$ (b) $E_{12}E^*_{12}$ (c) $E_{23}E^*_{32}$

(d) $E_{11}E^*_{22}$ (e) $E_{13}E^*_{32}$ (f) $E_{13}E^*_{22}$

(g) Formulate a general rule for when $E_{st}E^*_{ij} = O$.

(h) Is it true that $E_{st}E^*_{tj} = E_{sj}$?

(i) Formulate and prove a general rule for the product of an m by p matrix E_{st} and a p by n matrix E^*_{ij}, where E_{st} has 1 in the st spot and E^*_{ij} has 1 in the ij spot, and all other entries are 0.

7. (a) If $A = \begin{pmatrix} xy & -x^2 \\ y^2 & -xy \end{pmatrix}$, where x, y are arbitrary real numbers not both 0, show that $A^2 = O$, yet $A \neq O$.

(b) Prove: if $A^2 = O$, then there exist x, y such that $A = \pm\begin{pmatrix} xy & -x^2 \\ y^2 & -xy \end{pmatrix}$. Assume that A is 2×2.

(c) Show that if $X = \begin{pmatrix} 0 & x & y \\ 0 & 0 & z \\ 0 & 0 & 0 \end{pmatrix}$, then $X^3 = O$.

(d) Find a 3 by 3 matrix B such that B is not of form X above and $B \neq O$, $B^2 \neq O$, $B^3 = O$.

8. (a) Prove the rules (10-72) directly from (10-70) and (10-71), regarded as definitions of the addition of matrices and of scalar times a matrix.

‡(b) Prove the rules (10-85) to (10-88) directly from (10-81).

9. Prove the following parts of Theorem 4 and illustrate each by an example:

(a) Part (i) (b) Part (ii) (c) Part (iv)

10. (a) Find matrices A, B such that AB is defined, but BA is not defined.
    (b) When are both AB and BA defined?
    (c) Find matrices A, B such that AB and BA are defined but are not equal.
    (d) Find matrices A, B such that $AB = BA$.

10-10 PARTITIONING A MATRIX

In the matrix

$$A = \left(\begin{array}{cc|cc} 3 & 4 & 5 & 2 \\ 6 & 1 & 3 & 2 \\ \hline 1 & 2 & 5 & 6 \end{array}\right)$$

we have inserted broken lines to suggest that A can be thought of as a 2 by 2 array of matrices:

$$A = \begin{pmatrix} A_{11} & A_{12} \\ A_{21} & A_{22} \end{pmatrix}$$

where

$$A_{11} = \begin{pmatrix} 3 & 4 \\ 6 & 1 \end{pmatrix}, \qquad A_{12} = \begin{pmatrix} 5 & 2 \\ 3 & 2 \end{pmatrix}, \qquad A_{21} = (1, 2), \qquad A_{22} = (5, 6)$$

We say that the 3×4 matrix A has been partitioned into a $(2 + 1) \times (2 + 2)$ array. A similar procedure is applicable to an $m \times n$ matrix A. By writing $m = m_1 + m_2$ and $n = n_1 + n_2$, we partition A into a $(m_1 + m_2) \times (n_1 + n_2)$ array and obtain an array as follows

$$\left(\begin{array}{c|c} m_1 \times n_1 & m_1 \times n_2 \\ \hline m_2 \times n_1 & m_2 \times n_2 \end{array}\right)$$

Partitioning is useful for multiplication. Let A be m by p and let B be p by n, so that AB is defined. If we now partition A and B, splitting p *in the same way for both*, A as $(m_1 + m_2) \times (p_1 + p_2)$, B as $(p_1 + p_2) \times (n_1 + n_2)$, then

$$A = \left(\begin{array}{c|c} A_{11} & A_{12} \\ \hline A_{21} & A_{22} \end{array}\right), \qquad B = \left(\begin{array}{c|c} B_{11} & B_{12} \\ \hline B_{21} & B_{22} \end{array}\right)$$

$$AB = \left(\begin{array}{c|c} A_{11}B_{11} + A_{12}B_{21} & A_{11}B_{12} + A_{12}B_{22} \\ \hline A_{21}B_{11} + A_{22}B_{21} & A_{21}B_{12} + A_{22}B_{22} \end{array}\right) \tag{10-100}$$

Here A_{11} is $m_1 \times p_1$, B_{11} is $p_1 \times n_1$, so that $A_{11}B_{11}$ is defined and is $m_1 \times n_1$; A_{12} is $m_1 \times p_2$, B_{21} is $p_2 \times n_1$, so that $A_{12}B_{21}$ is defined and is $m_1 \times n_1$. Thus the first block of the array for AB is well defined. Similarly, one verifies that the others are well defined and that the right side of (10-100) is a matrix partitioned as an $(m_1 + m_2) \times (n_1 + n_2)$ array. Finally, direct computation shows that this array is a partitioned form of AB, so that (10-100) is correct. We illustrate this with an example.

EXAMPLE 1

$$\left(\begin{array}{cc|ccc} 1 & 2 & 4 & 6 & 7 \\ 0 & 1 & 1 & 2 & 0 \\ \hline 0 & 2 & 3 & 6 & -1 \end{array}\right)\left(\begin{array}{c|cc} 0 & -1 & 2 \\ 2 & 3 & 1 \\ \hline 1 & 4 & 5 \\ 2 & 2 & 0 \\ 0 & 7 & 6 \end{array}\right)$$

$$= \left(\begin{array}{c|c} \begin{pmatrix} 1 & 2 \\ 0 & 1 \end{pmatrix}\begin{pmatrix} 0 \\ 2 \end{pmatrix} + \begin{pmatrix} 4 & 6 & 7 \\ 1 & 2 & 0 \end{pmatrix}\begin{pmatrix} 1 \\ 2 \\ 0 \end{pmatrix} & \begin{pmatrix} 1 & 2 \\ 0 & 1 \end{pmatrix}\begin{pmatrix} -1 & 2 \\ 3 & 1 \end{pmatrix} + \begin{pmatrix} 4 & 6 & 7 \\ 1 & 2 & 0 \end{pmatrix}\begin{pmatrix} 4 & 5 \\ 2 & 0 \\ 7 & 6 \end{pmatrix} \\ \hline (0 \;\; 2)\begin{pmatrix} 0 \\ 2 \end{pmatrix} + (3 \;\; 6 \; -1)\begin{pmatrix} 1 \\ 2 \\ 0 \end{pmatrix} & (0 \;\; 2)\begin{pmatrix} -1 & 2 \\ 3 & 1 \end{pmatrix} + (3 \;\; 6 \; -1)\begin{pmatrix} 4 & 5 \\ 2 & 0 \\ 7 & 6 \end{pmatrix} \end{array}\right)$$

$$= \left(\begin{array}{c|c} \begin{pmatrix} 4 \\ 2 \end{pmatrix} + \begin{pmatrix} 16 \\ 5 \end{pmatrix} & \begin{pmatrix} 5 & 4 \\ 3 & 1 \end{pmatrix} + \begin{pmatrix} 77 & 62 \\ 8 & 5 \end{pmatrix} \\ \hline 4 + 15 & (6 \;\; 2) + (17 \;\; 9) \end{array}\right) = \left(\begin{array}{c|cc} 20 & 82 & 66 \\ 7 & 11 & 6 \\ \hline 19 & 23 & 11 \end{array}\right)$$

One usually chooses to partition a matrix when for appropriate n_1, n_2, m_1, m_2, some of the parts of the partitioning are the O-matrix. Thus one would usually partition as indicated:

$$\left(\begin{array}{cc|ccc} 1 & 3 & 4 & 2 & 1 \\ 0 & 1 & 2 & 1 & 0 \\ \hline 0 & 0 & 0 & 2 & 3 \\ 0 & 0 & 1 & 0 & 0 \end{array}\right)$$

If $A = \begin{pmatrix} A_{11} & O \\ O & A_{22} \end{pmatrix}$, then we say that A is the *direct sum of* A_{11}, A_{22}.

Of course, one can partition a matrix into more than 4 submatrices; for example, we might have

$$\begin{pmatrix} A_{11} & A_{12} & A_{13} \\ A_{21} & A_{22} & A_{23} \\ A_{31} & A_{32} & A_{33} \end{pmatrix}$$

If $A_{ij} = O$ for $i \neq j$, the partitioned matrix would now be called a direct sum of A_{11}, A_{22}, and A_{33}.

Two matrices are partitioned so that the partitions are compatible, say, $(m_1 + m_2 + m_3) \times (p_1 + p_2 + p_3)$ and $(p_1 + p_2 + p_3) \times (n_1 + n_2 + n_3)$, then they multiply in the standard way, for example,

$$\begin{pmatrix} A_{11} & A_{12} & A_{13} \\ A_{21} & A_{22} & A_{23} \\ A_{31} & A_{32} & A_{33} \end{pmatrix} \begin{pmatrix} B_{11} & B_{12} & B_{13} \\ B_{21} & B_{22} & B_{23} \\ B_{31} & B_{32} & B_{33} \end{pmatrix} = \begin{pmatrix} \Sigma A_{1j}B_{j1} & \Sigma A_{1j}B_{j2} & \Sigma A_{1j}B_{j3} \\ \Sigma A_{2j}B_{j1} & \Sigma A_{2j}B_{j2} & \Sigma A_{2j}B_{j3} \\ \Sigma A_{3j}B_{j1} & \Sigma A_{3j}B_{j2} & \Sigma A_{3j}B_{j3} \end{pmatrix}$$

The idea of partitioning a matrix is of value, since it often permits us to study the properties of a matrix by examining the properties of the smaller matrices that constitute the partition.

Addition and Multiplication by Scalars for Partitioned Matrices. Let two m by n matrices A and B be partitioned in the same way [both $(m_1 + m_2) \times (n_1 + n_2)$]. Then we can add them by adding corresponding blocks; also cA is obtained by multiplying each block by c; that is,

$$A + B = \begin{pmatrix} A_{11} & A_{12} \\ A_{21} & A_{22} \end{pmatrix} + \begin{pmatrix} B_{11} & B_{12} \\ B_{21} & B_{22} \end{pmatrix} = \begin{pmatrix} A_{11} + B_{11} & A_{12} + B_{12} \\ A_{21} + B_{21} & A_{22} + B_{22} \end{pmatrix}$$

$$cA = c\begin{pmatrix} A_{11} & A_{12} \\ A_{21} & A_{22} \end{pmatrix} = \begin{pmatrix} cA_{11} & cA_{12} \\ cA_{21} & cA_{22} \end{pmatrix}$$

In the case of addition, the process indicated yields a new m by n matrix each entry of which is the sum of the corresponding entries of A and B; hence, the equality holds true. There is a similar reasoning for cA.

10-11 THE ALGEBRA OF SQUARE MATRICES

In Section 9-19 we studied linear mappings of a vector space V into itself and called such mappings *linear transformations on V*. If we choose V to be V_n, then our mapping is given by a matrix, an n by n matrix (or a square matrix of order n). Thus the study of linear transformations of V_n is equivalent to the study of square matrices of order n.

If A and B are square matrices of order n and c is a scalar then, as above, $A + B$ and cA are defined and are square matrices of order n. Furthermore, as in Section 9-19, the product AB is defined and is again a square matrix of order n. Thus the set of all square matrices of order n is closed under the three operations: addition, multiplication by scalars, and matrix multiplication. The first two operations obey familiar rules: $A + B = B + A$, $c(A + B) = cA + cB, \ldots$. We summarize by saying that the set in question forms a vector space; this is simply the vector space $\mathfrak{M}_{nn}$, a special case of the vector space $\mathfrak{M}_{mn}$ discussed in Section 10-7. From the results of that section, we conclude that $\mathfrak{M}_{nn}$ is a vector space of dimension n^2. But we know more about this vector space. It has a multiplication, obeying the rules

$$\begin{aligned} c(ST) &= (cS)T = S(cT), & (ST)R &= S(TR) \\ S(T + R) &= ST + SR, & (S + T)R &= (SR + TR) \end{aligned} \qquad (10\text{-}110)$$

Furthermore, the n by n identity matrix $I = I_n$ and the n by n matrix O are in $\mathfrak{M}_{nn}$, and for every n by n matrix S

$$SI = S = IS, \qquad SO = O = OS \tag{10-111}$$

In Section 9-19 we described a vector space having a multiplication satisfying the rules specified in (10-110) as an *algebra*. Hence, the square matrices of order n form an algebra. We shall also denote this algebra by $\mathfrak{M}_{nn}$; it is really the same as the algebra of linear transformations on V_n: the algebra $Lt(V_n)$.

We point out, as was done in Section 9-19, that there is in general no commutative law for multiplication *in* $\mathfrak{M}_{nn}$ (the case $n = 1$ is an exception; $\mathfrak{M}_{11}$ is the same as the real number system R). For example, if

$$A = \begin{pmatrix} 0 & 1 \\ 0 & 0 \end{pmatrix}, \qquad B = \begin{pmatrix} 0 & 0 \\ 1 & 0 \end{pmatrix}$$

then

$$AB = \begin{pmatrix} 1 & 0 \\ 0 & 0 \end{pmatrix} \neq \begin{pmatrix} 0 & 0 \\ 0 & 1 \end{pmatrix} = BA$$

But then, for $n > 2$, say $n = 2 + r$, the partitioned matrices

$$\begin{pmatrix} A & O \\ O & I_r \end{pmatrix}, \qquad \begin{pmatrix} B & O \\ O & I_r \end{pmatrix}$$

are n by n and do not commute.

In applications of matrices, one is sometimes given a square matrix M of order n and asked to find $\mathcal{E}_M$, the set of all matrices which commute with M: that is, the set of all matrices A in $\mathfrak{M}_{nn}$ such that $AM = MA$. The set $\mathcal{E}_M$ contains O and is closed under all three operations discussed above. That O is in $\mathcal{E}_M$ follows from the equations $OM = MO = O$. If A and B are in $\mathcal{E}_M$, then $AM = MA$ and $BM = MB$, so that $(A + B)M = AM + BM = MA + MB = M(A + B)$, $(cA)M = c(AM) = c(MA) = M(cA)$, $(AB)M = A(BM) = A(MB) = (AM)B = (MA)B = M(AB)$. Therefore, $\mathcal{E}_M$ is a vector space and, since it is closed under multiplication, it is also an algebra, a subalgebra of $\mathfrak{M}_{nn}$. (We recall that a subalgebra is a nonempty subset of an algebra which is closed under all three operations; see Section 9-19).

EXAMPLE 1. Find the matrices that commute with $A = \begin{pmatrix} 0 & 1 \\ 0 & 0 \end{pmatrix}$.

Solution.

$$\begin{pmatrix} 0 & 1 \\ 0 & 0 \end{pmatrix}\begin{pmatrix} a & b \\ c & d \end{pmatrix} = \begin{pmatrix} c & d \\ 0 & 0 \end{pmatrix}, \qquad \begin{pmatrix} a & b \\ c & d \end{pmatrix}\begin{pmatrix} 0 & 1 \\ 0 & 0 \end{pmatrix} = \begin{pmatrix} 0 & a \\ 0 & c \end{pmatrix}$$

For these to be equal we need $c = 0$, $a = d$. Thus the matrices that commute with A are of the form $\begin{pmatrix} a & b \\ 0 & a \end{pmatrix}$, where a and b are arbitrary numbers. As we just saw above, the set of all such matrices forms a subalgebra of $\mathfrak{M}_{22}$. This could be shown directly. Clearly this subalgebra is a vector space of dimension 2, with basis $\begin{pmatrix} 1 & 0 \\ 0 & 1 \end{pmatrix}, \begin{pmatrix} 0 & 1 \\ 0 & 0 \end{pmatrix}$.

Since square matrices of order n can be multiplied, we can form nonnegative powers of each matrix: $A^0 = I, A^1 = A, A^2 = AA, A^3, \ldots$. As for linear transformations in general, we can then form polynomials in a square matrix A:

$$a_0 I + a_1 A + \cdots + a_k A^k \tag{10-112}$$

If $g(x) = a_0 + a_1 x + \cdots + a_k x^k$, we write $g(A)$ for the expression (10-112). Occasionally, one sees expressions like (10-112) in which the I does not appear next to a_0; in all such cases the expression should be treated as though the I were present, since we cannot add numbers and matrices.

Let a particular n by n matrix be given. There may exist two different polynomials $g(x)$, $h(x)$ such that $g(A) = h(A)$. For example, if $A = \begin{pmatrix} 1 & 0 \\ 0 & 0 \end{pmatrix}$, then $A^2 = A = A^3$, yet x, x^2, x^3 are three different polynomials. Since $\mathfrak{M}_{nn}$ is of dimension n^2, it follows (as in Section 9-22) that for each n by n matrix there are scalars $a_0, a_1, \ldots, a_{n^2}$, not all 0, such that

$$a_0 I + a_1 A + \cdots + a_{n^2} A^{n^2} = O$$

It should be noted that here $a_i \neq 0$ for, at least, one $i \geq 1$. Hence, for each square matrix A, there exist polynomials $p(x)$ of positive degree such that $p(A) = O$. But then there must be polynomials $m(x)$ of minimum positive degree such that $m(A) = O$. We call such polynomials, *minimal polynomials for* A. All minimal polynomials for a matrix A must be nonzero scalar multiples of one another (see Problem 7 following Section 9-22), and so there is exactly *one* minimal polynomial for A having 1 as coefficient of the highest degree term. We shall speak of that one polynomial as *the* minimal polynomial for A. It can be shown (see Section 10-21) that the minimal polynomial for an n by n matrix is of degree at most n. The minimal polynomial for O is x, for I is $x - 1$, for scalar matrices cI is $x - c$. If A has the minimal polynomial $x - c$, then $A = cI$. Hence, all matrices other than cI have a minimal polynomial of degree at least 2. Different matrices can have the same minimal polynomial.

THEOREM 5. *If $p(x)$ is a nonzero polynomial such that $p(A) = O$, then $p(x)$ is divisible by $m(x)$, the minimal polynomial for A.*

PROOF. We know that $p(x)$ has degree at least equal to that of $m(x)$. If we divide $p(x)$ by $m(x)$ by long division, we obtain

$$p(x) = m(x)q(x) + r(x)$$

where $r(x)$ has degree less than the degree of $m(x)$. On evaluating these polynomials at A, we obtain

$$O = p(A) = m(A)q(A) + r(A) = Oq(A) + r(A) = r(A)$$

Since $m(x)$ is the minimal polynomial for A, it follows that $r(x)$ must be the zero polynomial; thus $p(x) = m(x)q(x)$ and the theorem is proved.

Theorem 5 shows that if we find one polynomial $p(x)$ which equals O at A, then we can find the minimal polynomial for A by examining the various factors of the polynomial $p(x)$.

A square matrix A is said to be *nilpotent* if $A^k = O$ for some positive integer k. It, therefore, follows that the minimal polynomial for a nilpotent matrix is of the form x^m. The converse is also true. A square matrix A is said to be *idempotent* if $A^2 = A$. Except for O, I, all idempotent matrices have $x^2 - x$ as the minimal polynomial (Problem 8). For every c, the matrix $\begin{pmatrix} 1 & c \\ 0 & 0 \end{pmatrix}$ is idempotent. Thus we see that an infinity of matrices can have the same minimal polynomial.

EXAMPLE 2. Find the minimal polynomial for $A = \begin{pmatrix} 1 & 1 & 0 \\ 0 & 1 & 0 \\ 0 & 0 & 1 \end{pmatrix}$.

Solution. We observe that $A^2 = \begin{pmatrix} 1 & 2 & 0 \\ 0 & 1 & 0 \\ 0 & 0 & 1 \end{pmatrix} = 2A - I$. Since A is not a scalar matrix its minimal polynomial is at least of degree 2 and, hence, $x^2 - 2x + 1$ is the minimal polynomial for A.

PROBLEMS

1. Let $A = \begin{pmatrix} 2 & 2 \\ 3 & 2 \end{pmatrix}$, $B = \begin{pmatrix} 1 & 1 \\ 0 & 1 \end{pmatrix}$, $C = \begin{pmatrix} 1 \\ 0 \end{pmatrix}$, $D = (2, -1)$, $E = \begin{pmatrix} 3 \\ 1 \end{pmatrix}$.

   Find each of the products requested with the aid of the rule (10-100) or its generalization and check by direct multiplication:

   (a) $\begin{pmatrix} A & O \\ O & B \end{pmatrix}\begin{pmatrix} B & O \\ O & I \end{pmatrix}$, (b) $\begin{pmatrix} A & B \\ B & A \end{pmatrix}\begin{pmatrix} A & O \\ I & B \end{pmatrix}$,

   (c) $\begin{pmatrix} A & C \\ D & O \end{pmatrix}\begin{pmatrix} E & I \\ O & D \end{pmatrix}$, (d) $\begin{pmatrix} A & O & O \\ O & B & O \\ O & O & O \end{pmatrix}\begin{pmatrix} I \\ O \\ B \end{pmatrix}$.

2. Let X be the set of all direct sums of two 3 by 3 matrices; that is, X is the set

of all matrices of the form $\begin{pmatrix} A & O \\ O & B \end{pmatrix}$, where A, B are 3 by 3 matrices. Prove:

(a) X is a subset of the set of all 6 by 6 matrices.

(b) X is a vector space of dimension 18.

(c) X is closed under multiplication.

(d) X is a subalgebra of $\mathfrak{M}_{66}$.

3. Let $f(x) = x^2 - x + 3$, $g(x) = x^3 - 2x + 1$, $A = \begin{pmatrix} 3 & 1 \\ 2 & 0 \end{pmatrix}$, $B = \begin{pmatrix} 1 & -2 \\ 3 & 4 \end{pmatrix}$.

Evaluate: (a) $f(A)$ (b) $f(B)$ (c) $f(I)$ (d) $f(O)$
(e) $g(A)$ (f) $g(B)$ (g) $f(A + B)$ (h) $g(AB)$

4. For each of the following choices of matrix A show that A can be expressed as $J + N$, where J is idempotent and N is nilpotent (recall that O is both nilpotent and idempotent:

(a) $A = \begin{pmatrix} 1 & 0 \\ 0 & 1 \end{pmatrix}$, (b) $A = \begin{pmatrix} 0 & 0 \\ 1 & 0 \end{pmatrix}$, (c) $A = \begin{pmatrix} 1 & 0 \\ 0 & 0 \end{pmatrix}$,

(d) $A = \begin{pmatrix} 1 & 0 \\ 1 & 1 \end{pmatrix}$, (e) $A = \begin{pmatrix} 1 & 0 \\ 1 & 0 \end{pmatrix}$, (f) $A = \begin{pmatrix} 1 & 0 & 0 \\ 0 & 1 & 0 \\ 1 & 0 & 0 \end{pmatrix}$,

(g) $A = (a_{ij})$, $1 \leq i \leq n$, $1 \leq j \leq n$, where $a_{ii} = 0$ or 1 for $i = 1, \ldots, n$, $a_{ij} = 0$ for $1 \leq i < j \leq n$.

5. Prove: if A is a square matrix, then

(a) $A^3 - I = (A - I)(A^2 + A + I)$ (b) $A^4 - I = (A^2 + I)(A + I)(A - I)$

(c) $3A^2 - 2A - I = (3A + I)(A - I)$

(d) $A^k - I = (A - I)(A^{k-1} + A^{k-2} + \cdots + I)$ $(k = 2, 3, \ldots)$

6. Let $A = \begin{pmatrix} -3 & 2 \\ -15 & 8 \end{pmatrix}$, $B = \begin{pmatrix} -4 & 2 \\ -15 & 7 \end{pmatrix}$, $p(x, y) = x^2 - xy + 2y^2$, and

$q(x, y) = x^2 - y^2 - 2y - 1$.

(a) Show that A and B commute. (b) Evaluate $p(A, B)$.

(c) Evaluate $q(A, B)$.

(d) Prove that, in general, if A, B are n by n matrices and $AB = BA$, then

$$(A + B)^2 = A^2 + 2AB + B^2, \qquad A^2 - B^2 = (A + B)(A - B)$$

and verify these relations for A, B as given.

7. Diagonal matrices were defined in Section 10-5. Prove the following results for square diagonal matrices:

(a) A diagonal matrix is the direct sum of 1 by 1 matrices.

(b) All n by n diagonal matrices commute.

(c) The only diagonal matrices that commute with all matrices A are scalar matrices.

(d) The set of all n by n diagonal matrices form an algebra of dimension n.

(e) The direct sum of two diagonal matrices is diagonal.

(f) If D is diagonal so is its transpose D'.

8. (a) Prove: the minimal polynomial of an idempotent matrix not O or I is $x^2 - x$.

(b) Find all 2 by 2 idempotent matrices.

9. Prove: (a) A matrix and its transpose have the same minimal polynomial.

(b) If A is a direct sum of B and C, then A' is a direct sum of B' and C'.

(c) If A is a direct sum of B and C, then rank A = rank B + rank C.

(d) If A is the direct sum of B and C, then A is nilpotent if, and only if, B and C are nilpotent.

10. Let A be the direct sum of an m by m matrix B and a q by q matrix C. Prove:

(a) If $p(x)$ is a polynomial, then $p(A)$ is the direct sum of $p(B)$ and $p(C)$.

(b) The minimal polynomial for A is the polynomial of least degree divisible by both the minimal polynomial for B and the minimal polynomial for C. In particular, the minimal polynomial for A is a divisor of the product of the minimal polynomials for B and C.

11. Prove: if $c_1, \ldots, c_k$ are the distinct numbers appearing on the diagonal of a diagonal matrix D, then the minimal polynomial for D is $(x - c_1)(x - c_2) \cdots (x - c_k)$.

12. Find the minimal polynomials for the following matrices. (*Hint.* Sometimes the results of Problems 9, 10, and 11 can be useful.) Here $c \neq 0$, $d \neq 0$.

(a) $A = \begin{pmatrix} 2 & 0 \\ 0 & 1 \end{pmatrix}$, (b) $B_c = \begin{pmatrix} 1 & c \\ 0 & 1 \end{pmatrix}$, (c) $C_c = \begin{pmatrix} 1 & 0 \\ c & 1 \end{pmatrix}$, (d) $D = \begin{pmatrix} 1 & 0 & 0 \\ 0 & 2 & 0 \\ 0 & 0 & 3 \end{pmatrix}$.

(e) E = direct sum of B_c and C_c, (f) F = direct sum of B_c and B_d,

(g) G = direct sum of B_c and C_d, (h) H = direct sum of A and D,

(i) J = direct sum of D and D, (j) K = direct sum of C_1 and D.

13. An m by n matrix $A = (a_{ij})$ is said to be *upper triangular* if $a_{ij} = 0$ for $i > j$. A is said to be *lower triangular* if $a_{ij} = 0$ for $i < j$. Thus in Problem 12, B_c is upper triangular and C_c is lower triangular. Prove:

(a) The set of m by n upper triangular matrices form a vector space of dimension $mn - \frac{1}{2}(m^2 - m)$ if $m \leq n$ and of dimension $\frac{1}{2}(n^2 + n)$ if $n \leq m$.

(b) A is upper triangular if, and only if, A' is lower triangular.

(c) The set of upper triangular n by n matrices is a subalgebra of $\mathfrak{M}_{nn}$.

(d) Diagonal matrices are the only square matrices which are both upper and lower triangular. (Discuss the situation for rectangular matrices.)

(e) A triangular square matrix with 0 at all diagonal entries is nilpotent.

(f) A triangular n by n matrix with c at all diagonal entries has a minimal polynomial which is a divisor of $(x - c)^n$.

14. Find the minimal polynomial for the following matrices. (*Hint.* Problem 13 will be useful.)

(a) $A = \begin{pmatrix} 0 & 2 & 3 \\ 0 & 0 & 4 \\ 0 & 0 & 0 \end{pmatrix}$, (b) $B = \begin{pmatrix} -1 & 2 & 3 \\ 0 & -1 & 4 \\ 0 & 0 & 1 \end{pmatrix}$, (c) $C = \begin{pmatrix} 2 & 0 & 0 \\ 3 & 2 & 0 \\ 1 & 5 & 2 \end{pmatrix}$.

(d) $D =$ direct sum of A and B, (e) $E =$ direct sum of B and $2I_3$,
(f) $F =$ direct sum of B and C, (g) $G =$ direct sum of B and B.

10-12 NONSINGULAR MATRICES

A *nonsingular matrix* M is a square (say, n by n) matrix which maps V_n onto itself (see Section 9-21). Hence, Range $M = V_n$, so that M has rank n. And, therefore, the column vectors of M are linearly independent and, hence, a basis for V_n. Conversely, if M has rank n, then M is nonsingular. It follows from relation (10-41) that a nonsingular matrix is also a one-to-one mapping of V_n into itself. A linear mapping that is both one-to-one and onto has an inverse; hence, to each nonsingular matrix M there is a unique matrix denoted by M^{-1} such that

$$MM^{-1} = I = M^{-1}M$$

Also, if $MS = I$, then $S = M^{-1}$; and if $SM = I$, then $S = M^{-1}$ (Theorem 28, Section 9-21). The matrix M^{-1} is also one-to-one and onto and so is also nonsingular. The n by n nonsingular matrices are exactly those matrices in $\mathfrak{M}_{nn}$ which have an inverse in $\mathfrak{M}_{nn}$. If M and N are nonsingular matrices, then MN and NM are nonsingular matrices, since they have inverses. In particular, the inverse of MN is $N^{-1}M^{-1}$, since

$$(MN)(N^{-1}M^{-1}) = M(NN^{-1})M^{-1} = MIM^{-1} = I$$

Similarly, the inverse of NM is $M^{-1}N^{-1}$. Thus we see that products of nonsingular matrices are nonsingular. A nonzero scalar multiple of a nonsingular matrix is also nonsingular; the inverse of cM is $c^{-1}M^{-1}$. However, the sum of two nonsingular matrices need not be nonsingular; for example, I and $-I = (-1)I$ are nonsingular but their sum is O.

As in the remark following Theorem 3 in Section 10-6, *M is nonsingular precisely when* $\det M \neq 0$, (see Theorem 15, Section 10-13).

EXAMPLE 1. The matrix $M = \begin{pmatrix} a & b \\ c & d \end{pmatrix}$ is nonsingular if, and only if, $ad - bc \neq 0$, and in that case

$$M^{-1} = \frac{1}{ad - bc}\begin{pmatrix} d & -b \\ -c & a \end{pmatrix} \tag{10-120}$$

PROOF. If $ad - bc \neq 0$, then (10-120) is a nonzero matrix and direct computation shows that it is the inverse of M and, hence, M is nonsingular.

If M is nonsingular, then the column vectors of M are linearly independent and, hence, at least one of b and d is nonzero. If $ad - bc = 0$ and $d \neq 0$, then $(c/d)(b, d) = (a, c)$; while if $ad - bc = 0$ and $b \neq 0$, then $(a/b)(b, d) =$

(a, c). Hence, it is impossible for M to be nonsingular and for $ad - bc = 0$ (see also Section 1-12).

EXAMPLE 2. Find the inverse of the matrix $A = \begin{pmatrix} 1 & 1 & 2 \\ 0 & 2 & 1 \\ 1 & 4 & 0 \end{pmatrix}$

Solution. The reader can verify that the columns vectors of A are linearly independent and, therefore, that A is nonsingular. Let

$$A^{-1} = \begin{pmatrix} x_1 & x_2 & x_3 \\ y_1 & y_2 & y_3 \\ z_1 & z_2 & z_3 \end{pmatrix}$$

Then $A^{-1}A = \begin{pmatrix} x_1 + x_3 & x_1 + 2x_2 + 4x_3 & 2x_1 + x_2 \\ y_1 + y_3 & y_1 + 2y_2 + 4y_3 & 2y_1 + y_2 \\ z_1 + z_3 & z_1 + 2z_2 + 4z_3 & 2z_1 + z_2 \end{pmatrix} = \begin{pmatrix} 1 & 0 & 0 \\ 0 & 1 & 0 \\ 0 & 0 & 1 \end{pmatrix}.$

Thus x_1, x_2, x_3 is a common solution for the equations

$$x_1 + x_3 = 1, \qquad x_1 + 2x_2 + 4x_3 = 0, \qquad 2x_1 + x_2 = 0$$

Solving, we find that $x_1 = \frac{4}{7}$, $x_2 = -\frac{8}{7}$, $x_3 = \frac{3}{7}$ is the solution of this set of linear equations. Similarly, y_1, y_2, y_3 is the solution of

$$y_1 + y_3 = 0, \qquad y_1 + 2y_2 + 4y_4 = 1, \qquad 2y_1 + y_2 = 0$$

and z_1, z_2, z_3 is the solution of

$$z_1 + z_3 = 0, \qquad z_1 + 2z_2 + 4z_3 = 0, \qquad 2z_1 + z_2 = 1$$

Solving, we find that $y_1 = -\frac{1}{7}$, $y_2 = \frac{2}{7}$, $y_3 = \frac{1}{7}$, $z_1 = \frac{2}{7}$, $z_2 = \frac{3}{7}$, $z_3 = -\frac{2}{7}$. Hence

$$A^{-1} = \tfrac{1}{7}\begin{pmatrix} 4 & -8 & 3 \\ -1 & 2 & 1 \\ 2 & 3 & -2 \end{pmatrix}$$

The situation illustrated in Example 2 holds true in general. If $M = (m_{ij})$ is a nonsingular n by n matrix, then the jth row vector $(x_1, x_2, \ldots, x_n)$ of the matrix M^{-1} is the unique solution of the system of linear equations:

$$m_{i1}x_1 + \cdots + m_{in}x_n = \delta_{ij}, \qquad i = 1, 2, \ldots, n$$

where δ_{ij} is the Kronecker delta: $\delta_{ij} = 1$, if $i = j$; $\delta_{ij} = 0$ otherwise.

When n is large, the task of finding the inverse of an n by n nonsingular matrix can be quite tedious. Here, and in the problems, we discuss a few techniques that can simplify the task. We also discuss other methods in Sections 10-13 and 10-16.

(1) The inverse of the n by n matrix $M = (m_{ij})$ can be found by solving

one set of simultaneous linear equations:

$$\sum_{j=1}^{n} m_{ij}x_j = y_i, \qquad i = 1, \ldots, n$$

For these equations are equivalent to $M\mathbf{x} = \mathbf{y}$ and, hence, to $\mathbf{x} = M^{-1}\mathbf{y}$. If one solves the equations for $x_1, \ldots, x_n$, one obtains

$$x_i = \sum_{j=1}^{n} b_{ij}y_j \qquad (i = 1, \ldots, n)$$

and M^{-1} is the matrix (b_{ij}).

For Example 2, above, one must solve

$$x_1 + x_2 + 2x_3 = y_1, \qquad 2x_2 + x_3 = y_2, \qquad x_1 + 4x_2 = y_3$$

By elimination, one finds easily that

$$x_1 = \tfrac{1}{7}(4y_1 - 8y_2 + 3y_3), \qquad x_2 = \tfrac{1}{7}(-y_1 + 2y_2 + y_3),$$
$$x_3 = \tfrac{1}{7}(2y_1 + 3y_2 - 2y_3)$$

From these expressions one can read off the matrix A^{-1} as given above.

(2) *A matrix M is nonsingular if, and only if, its minimal polynomial m(x) has a nonzero constant term. If* $m(x) = xh(x) + c$, *with* $c \neq 0$, *then* $M^{-1} = -c^{-1}h(M)$ (see Theorem 29 in Section 9-21). Thus, if M is a square matrix with $x^3 - x + 2$ as minimal polynomial, then $m(x) = x(x^2 - 1) + 2$, so that $h(x) = x^2 - 1$ and $c = 2$; thus $M^{-1} = (-\frac{1}{2})(M^2 - I)$. Since all idempotent matrices except O, I have $x^2 - x$ as minimal polynomial, we see that I is the only nonsingular idempotent matrix.

(3) *If A is the direct sum of nonsingular matrices B and C, then* A^{-1} *is the direct sum of* B^{-1} *and* C^{-1}; *that is,*

$$\begin{pmatrix} B & O \\ O & C \end{pmatrix}^{-1} = \begin{pmatrix} B^{-1} & O \\ O & C^{-1} \end{pmatrix}$$

One easily verifies that this is the inverse of A. Hence, *the direct sum of nonsingular matrices is nonsingular.*

(4) *If A is nonsingular so is its transpose A′, and*

$$(A')^{-1} = (A^{-1})' \qquad (10\text{-}121)$$

To prove this, we observe that $AA^{-1} = I$ and, hence, that $(AA^{-1})' = I'$. But $I' = I$, and $(XY)' = Y'X'$. Hence, $(A^{-1})'A' = I$ and this establishes (10-121).

Remark 1. As a consequence of (4), we see that the column vectors of a square matrix are linearly independent if, and only if, the row vectors are linearly independent.

Remark 2. Each rectangular matrix A can be expressed uniquely as a product: $A = LDU$, where L is lower triangular, D is diagonal, U is upper triangular, and the diagonal entries for L and U are all 1 (for definitions see Problem 13, Section 10-11). The matrices L and U are all nonsingular, and there is a simple algorithm for determining their entries and those of D from those of A, and there are other simple algorithms for determining the entries in L^{-1}, D^{-1} (if it exists), U^{-1} from those for L, D, and U. If D^{-1} exists, we can then determine $A^{-1} = U^{-1}D^{-1}L^{-1}$. These algorithms are easily adaptable for use on a computer and can be used quite effectively to compute inverses of matrices having order as large as 50. For a detailed discussion of these methods see Faddeev and Faddeeva: *Computational Methods in Linear Algebra* (Freeman, San Francisco, 1963).

PROBLEMS

1. Show that each of the following matrices is nonsingular and find the inverse of each.

(a) $\begin{pmatrix} 1 & 0 \\ 0 & 2 \end{pmatrix}$, (b) $\begin{pmatrix} 3 & 1 \\ 5 & 2 \end{pmatrix}$, (c) $\begin{pmatrix} 1 & 2 & 3 \\ 2 & 0 & -1 \\ 4 & 2 & 2 \end{pmatrix}$,

(d) $\begin{pmatrix} 1 & -1 & 2 \\ 2 & 0 & 2 \\ 3 & 1 & 3 \end{pmatrix}$, (e) $\begin{pmatrix} 1 & 2 & 3 \\ 0 & 1 & 4 \\ 0 & 0 & 1 \end{pmatrix}$, (f) $\begin{pmatrix} 1 & \alpha & \beta \\ 0 & 1 & \gamma \\ 0 & 0 & 1 \end{pmatrix}$,

(g) $\begin{pmatrix} 1 & 0 & 0 \\ \alpha & 1 & 0 \\ \beta & \gamma & 1 \end{pmatrix}$, (h) $\begin{pmatrix} 1 & i \\ i & 1 \end{pmatrix}$, (i) $\begin{pmatrix} 1+2i & i \\ 1 & 1-2i \end{pmatrix}$.

2. Prove: (a) If $d_1, d_2, \ldots, d_n$ are different from 0, then

$$\begin{pmatrix} d_1 & 0 & \cdots & 0 \\ 0 & d_2 & \cdots & 0 \\ \vdots & \vdots & & \vdots \\ 0 & 0 & \cdots & d_n \end{pmatrix}^{-1} = \begin{pmatrix} d_1^{-1} & 0 & \cdots & 0 \\ 0 & d_2^{-1} & \cdots & 0 \\ \vdots & \vdots & & \vdots \\ 0 & 0 & \cdots & d_n^{-1} \end{pmatrix}$$

   (b) Prove: In a diagonal matrix, if any $d_i = 0$, then the matrix is singular.

   (c) A square triangular matrix is nonsingular if, and only if, all its diagonal entries are nonzero. The inverse of a nonsingular upper (lower) triangular matrix is upper (lower) triangular. [For definitions see Problem 13, Section 10-11.]

   (d) If T is upper triangular and nonsingular and AT is upper triangular, then A is upper triangular.

3. (a) ... (j) Find the inverse of each matrix in Problem 12, following Section 10-11, which is nonsingular.

4. (a) ... (g) Find the inverse of each matrix in Problem 14, following Section 10-11, which is nonsingular.

5. An n by n matrix P is said to be a *permutation matrix* if its column vectors are $\mathbf{e}_1, \ldots, \mathbf{e}_n$ in some order. Prove:

   (a) A permutation matrix is nonsingular.

(b) The row vectors of a permutation matrix are $\mathbf{e}_1, \ldots, \mathbf{e}_n$ in some order.

(c) The transpose of a permutation matrix is a permutation matrix.

(d) The set of permutation matrices is closed under multiplication.

(e) If P is a permutation matrix, then $P' = P^{-1}$.

(f) The only triangular permutation matrix is I. (See Problem 13, following Section 10-11.)

(g) The direct sum of B and C is a permutation matrix if, and only if, B and C are permutation matrices.

(h) If P is a permutation matrix, then the columns of AP are those of A, but possibly in a different order, and the rows of PA are those of A but possibly in a different order.

6. Prove: if $A = \begin{pmatrix} B & O \\ C & D \end{pmatrix}$ is a nonsingular matrix and if B and D are square matrices, then B and D are nonsingular and $A^{-1} = \begin{pmatrix} B^{-1} & O \\ -D^{-1}CB^{-1} & D^{-1} \end{pmatrix}$.

7. Find the inverse of each of the following matrices:

(a) $\begin{pmatrix} 0 & 1 & 0 & 0 & 0 \\ 1 & 0 & 0 & 0 & 0 \\ 2 & 3 & 0 & 1 & 0 \\ 1 & 0 & 0 & 0 & 1 \\ -1 & 1 & 1 & 0 & 0 \end{pmatrix}$, (b) $\begin{pmatrix} 1 & 1 & 1 & 0 & 0 \\ 0 & 1 & 1 & 1 & 0 \\ 0 & 0 & 1 & 1 & 1 \\ 0 & 0 & 0 & 1 & 1 \\ 0 & 0 & 0 & 0 & 1 \end{pmatrix}$, (c) $\begin{pmatrix} 1 & 2 & 3 & 4 & 5 \\ 0 & 1 & 2 & 3 & 4 \\ 0 & 0 & 1 & 2 & 3 \\ 0 & 0 & 0 & 1 & 2 \\ 0 & 0 & 0 & 0 & 1 \end{pmatrix}$

8. Show: $\begin{pmatrix} 1 & 0 & 0 \\ a & 1 & 0 \\ b & c & 1 \end{pmatrix}^{-1} = \begin{pmatrix} 1 & 0 & 0 \\ -a & 1 & 0 \\ ac - b & -c & 1 \end{pmatrix}$

9. Evaluate $\begin{pmatrix} 1 & 0 & 0 & 0 \\ a & 1 & 0 & 0 \\ b & c & 1 & 0 \\ d & e & f & 1 \end{pmatrix}^{-1}$ (*Hint:* Problems 6 and 8 should be useful.)

10. Find the matrix solutions for the equations:

(a) $\begin{pmatrix} 1 & 2 & 3 \\ 0 & 1 & 4 \\ 0 & 0 & 2 \end{pmatrix} X = \begin{pmatrix} 1 & 0 & 3 \\ 4 & 1 & 1 \\ 5 & 1 & 4 \end{pmatrix}$, (b) $\begin{pmatrix} 1 & 0 & 1 \\ 0 & 2 & 4 \\ 0 & 0 & 3 \end{pmatrix} X = \begin{pmatrix} 1 & 0 & 1 \\ 2 & 1 & 3 \\ -1 & 4 & 1 \end{pmatrix}$

10-13 DETERMINANTS

In Section 0-9 the idea of determinants is discussed and illustrated. Here we review the definition and the main properties of determinants, derive some new properties, and relate determinants to matrices in several important ways. In order to avoid interrupting the development, several of the more technical proofs are postponed to the next section.

Definition of determinant. A determinant can be thought of as a (real or complex) polynomial obtained by a special process. A determinant of order 1 is just the polynomial $p(x) = x$. A determinant of order 2 is

$$\begin{vmatrix} a & b \\ c & d \end{vmatrix} = ad - bc$$

This is a polynomial of degree 2 in the variables a, b, c, d. A determinant of order n $(n \geq 2)$ is defined inductively by the rule:

$$\begin{vmatrix} a_{11} & a_{12} & \cdots & a_{1n} \\ a_{21} & a_{22} & \cdots & a_{2n} \\ \vdots & \vdots & & \vdots \\ a_{n1} & a_{n2} & \cdots & a_{nn} \end{vmatrix} = a_{11}\,\Delta_{11} - a_{12}\,\Delta_{12} + \cdots + (-1)^{n+1}a_{1n}\,\Delta_{1n} \quad (10\text{-}130)$$

Here Δ_{ij}, the *minor* of a_{ij}, is the determinant obtained from the given one by deleting the row and column containing a_{ij}. The rule (10-130) is called *expansion of the determinant by minors of the first row.*

By means of the rule (10-130) a determinant of order n is expressed in terms of determinants of order $n - 1$, and then in terms of determinants of order $n - 2$, and so on, until ultimately determinants of order 2 (or even of order 1) are obtained and evaluated as above. From the definition it follows that *a determinant of order n is a polynomial of degree n in the n^2 variables* $a_{11}, \ldots, a_{nn}$. By induction, one can show that a determinant of order n is the sum of $n!$ terms. Since the variables in a determinant are arranged in a square array, we can associate with them a square n by n matrix $A = (a_{ij})$ in the variables a_{ij}. When the a_{ij} are given real or complex values, the determinant assigns a value (real or complex) to each such matrix and, hence, we write

$$\begin{vmatrix} a_{11} & \cdots & a_{1n} \\ \vdots & & \vdots \\ a_{n1} & \cdots & a_{nn} \end{vmatrix} = \det A = \det(a_{ij})$$

Thus for each n, det A is a real- (or complex-) valued function whose domain is $\mathfrak{M}_{nn}$ (or $\mathfrak{M}^c_{nn}$). It will be seen that several properties of determinants can be stated most simply in terms of matrices. Throughout the following discussion A, B will denote n by n square matrices.

Expression of det A as sum of signed products of elements. Given a square matrix $A = (a_{ij})$, we can select n elements from A in such a way that one element has been selected from each row of A, and one element from each column of A. For example, from the first row, we select the i_1-*st* element, from the second row the i_2-*nd* element, and so on, so that $i_1, \ldots, i_n$ contain no repetitions and, hence, form a *permutation* of $1, \ldots, n$. (Permutations are discussed in Section 0-21.) A particular case, for $n = 5$, is

$$a_{13}, \quad a_{24}, \quad a_{32}, \quad a_{45}, \quad a_{51}$$

Here $i_1 = 3$, $i_2 = 4$, $i_3 = 2$, $i_4 = 5$, $i_5 = 1$. We now multiply the elements selected and insert a + or − sign before the product, according to a basic Rule of Signs. The rule can be described in terms of permutations. Here we give a graphical version, due to G. D. Birkhoff and used by P. R. Rider in his "*College Algebra*" (Macmillan, N.Y., 1940).

Rule of Signs. Place a point at each a_{ij} in the square array and, for those a_{ij} used in the product, join all the corresponding points by line segments. In Figure 10-7 this is done for the product $a_{13} \ldots a_{51}$ cited above. We now count the number of segments with positive slope; 6 in the case of the figure. When that number is even, as in the figure, the sign is $+$; when the number is odd, the sign is $-$.

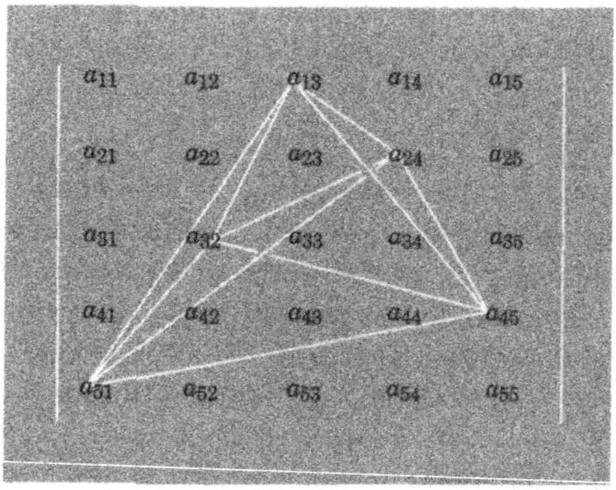

Figure 10-7 Rule of signs.

THEOREM 6. *Let* $A = (a_{ij})$ *be an n by n square matrix. Then* det A *is the sum of all products, each containing exactly one element from each row and each column of A, and each prefaced by a* $+$ *or* $-$ *sign in accordance with the Rule of Signs.*

The proof is given in Section 10-14. From the theory of permutations (Section 0-21) one can verify that the sum has $n!$ terms.

EXAMPLE 1

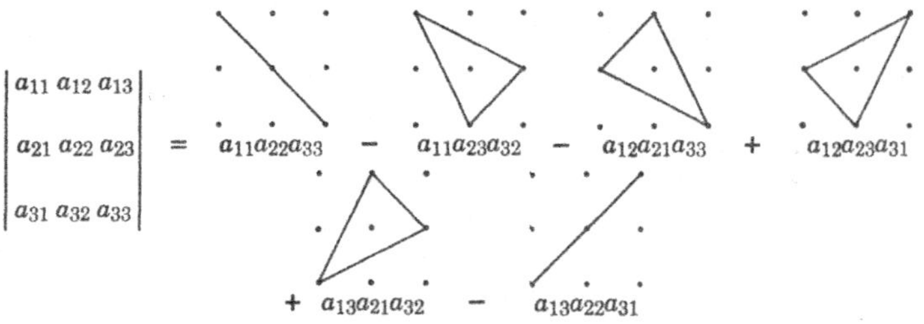

Here we have indicated the slopes of the segments above each term. The first term has no segments with positive slope, the second has 1, the third has 1, and so on. Thus one obtains the signs $+, -, -, \ldots$ as shown.

THEOREM 7. $\det A' = \det A$.

PROOF. The rule is a consequence of Theorem 6. For if we form A' and then form the products as above, we again get the same terms except perhaps for the + and −. But A' is obtained from A by reflecting in the principal diagonal ($a_{11}, a_{22}, \ldots$) and, hence, segments of positive slope go into segments of positive slope, and those of negative slope go into ones of negative slope (Problem 16 below). The figure giving the slopes for the term $a_{1i_1} \cdots a_{ni_n}$ in A is the reflection about the principal diagonal of the figure giving the slopes for the term $a_{i_1 1}a_{i_2 2} \cdots a_{i_n n}$ in A' (see Figure 10-8) and, hence, these two terms have the same sign. Hence, the signs of the various terms are unchanged and the theorem follows.

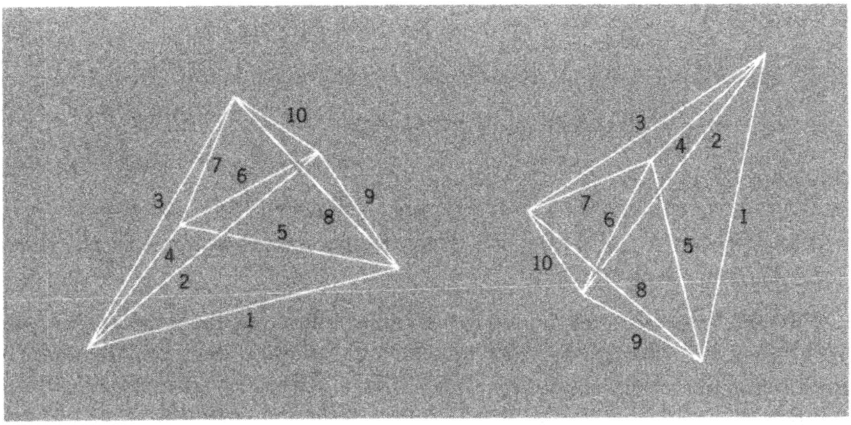

Figure 10-8 The effect of transposing the matrix.

The rule of Theorem 7 is often described by saying that *interchange of rows and columns does not affect the value of a determinant.* The theorem shows that if we prove a property for the rows of a determinant, then the same property must hold true for the columns.

THEOREM 8. *Let the square matrix B be obtained from the square matrix A by interchanging two rows (or two columns). Then* $\det B = -\det A$.

The theorem follows from the Rule of Signs; a proof is given in the next section.

EXAMPLE 2

$$\begin{vmatrix} 1 & 2 & 3 \\ 0 & 1 & 2 \\ 1 & 0 & 0 \end{vmatrix} = 1, \qquad \begin{vmatrix} 1 & 0 & 0 \\ 0 & 1 & 2 \\ 1 & 2 & 3 \end{vmatrix} = -1, \qquad \begin{vmatrix} 2 & 1 & 3 \\ 1 & 0 & 2 \\ 0 & 1 & 0 \end{vmatrix} = -1$$

THEOREM 9. *Let* Δ_{ij} *be the minor of* a_{ij} *in* $A = (a_{ij})$. *Let* $A_{ij} = (-1)^{i+j}\Delta_{ij}$. *Then for each fixed i and j,*

$$\begin{aligned} \det A &= a_{i1}A_{i1} + a_{i2}A_{i2} + \cdots + a_{in}A_{in} \\ &= a_{1j}A_{1j} + a_{2j}A_{2j} + \cdots + a_{nj}A_{nj} \end{aligned} \qquad (10\text{-}131)$$

Thus a determinant can be expanded by minors using any row or column, with a corresponding assignment of + and − signs.

The theorem follows from Theorem 8; a proof is given in the next section.

The number $A_{ij} = (-1)^{i+j}\,\Delta_{ij}$ is called the *cofactor of the element* a_{ij} of matrix A. The matrix $(A_{ij})'$ (note the transpose) is called the *adjoint of* A and denoted by adj A. Note that (10-130) is the case $i = 1$ of (10-131). It is very convenient to be able to expand by any row or column and not to be restricted to the first row.

EXAMPLE 3

$$\begin{vmatrix} a_{11} & a_{12} & a_{13} \\ a_{21} & a_{22} & a_{23} \\ a_{31} & a_{32} & a_{33} \end{vmatrix} = a_{12}A_{12} + a_{22}A_{22} + a_{32}A_{32}$$
$$= -a_{12}\,\Delta_{12} + a_{22}\,\Delta_{22} - a_{32}\,\Delta_{32}$$

THEOREM 10. *Let 2 rows (or columns) of a square matrix A be the same (as vectors). Then* det $A = 0$.

PROOF. Interchanging 2 rows (or columns) which are the same, on the one hand, does not change the determinant and, on the other, by Theorem 8, multiplies the determinant by -1. Hence, $\det A = -\det A$ and, thus, $\det A = 0$.

An application. If A_{21}, A_{22}, A_{23} are the cofactors of the second row of a 3 by 3 matrix A, then

$$a_{11}A_{21} + a_{12}A_{22} + a_{13}A_{23} = 0$$

since the left-hand side is the expansion by cofactors of the second row of the determinant:

$$\begin{vmatrix} a_{11} & a_{12} & a_{13} \\ a_{11} & a_{12} & a_{13} \\ a_{31} & a_{32} & a_{33} \end{vmatrix}$$

and this determinant has two rows the same. Similarly, $a_{31}A_{21} + a_{32}A_{22} + a_{33}A_{23} = 0$. More generally: *if* $k \neq l$ *and* $A_{k1}, A_{k2}, \ldots, A_{kn}$ *are the cofactors of the* k*th row of the n by n matrix* A, *then* $a_{l1}A_{k1} + \cdots + a_{ln}A_{kn} = 0$. For, as above, the left-hand side is the expansion of an n by n matrix whose lth and kth rows are the same.

THEOREM 11. *Let matrix B be obtained from matrix A by multiplying all elements of one row (or column) of A by a scalar c. Then* $\det B = c \det A$.

PROOF. Expand by minors of the chosen row or column. Each term is multiplied by c. Hence, the determinant is multiplied by c.

EXAMPLE 4

(a) $$\begin{vmatrix} 3 & 2 & 3 \\ 0 & 1 & 2 \\ 3 & 0 & 0 \end{vmatrix} = 3\begin{vmatrix} 1 & 2 & 3 \\ 0 & 1 & 2 \\ 1 & 0 & 0 \end{vmatrix},$$ (b) $$\begin{vmatrix} 3 & 6 & 9 \\ 0 & 3 & 6 \\ 3 & 0 & 0 \end{vmatrix} = 27\begin{vmatrix} 1 & 2 & 3 \\ 0 & 1 & 2 \\ 1 & 0 & 0 \end{vmatrix} = 27$$

In general, if A is n by n, then $\det(cA) = c^n \det A$.

THEOREM 12. *Let the kth row vector (column vector) of A be expressed as a sum of two vectors* **a**$'$, **a**$''$. *Then*

$$\det A = \det A' + \det A''$$

where A', A'' are obtained from A by replacing the kth row (column) vector by **a**$'$, **a**$''$, *respectively.*

EXAMPLE 5

$$\begin{vmatrix} a_{11} & a'_{12} + a''_{12} & a_{13} \\ a_{21} & a'_{22} + a''_{22} & a_{23} \\ a_{31} & a'_{32} + a''_{32} & a_{33} \end{vmatrix} = \begin{vmatrix} a_{11} & a'_{12} & a_{13} \\ a_{21} & a'_{22} & a_{23} \\ a_{31} & a'_{32} & a_{33} \end{vmatrix} + \begin{vmatrix} a_{11} & a''_{12} & a_{13} \\ a_{21} & a''_{22} & a_{23} \\ a_{31} & a''_{32} & a_{33} \end{vmatrix}$$

PROOF OF THEOREM 12. This is proved in the same way as Theorem 11, by expanding by minors of the chosen row (or column). A typical term in the expansion of det A by minors of the kth row is

$$(-1)^{k+j}(a'_{kj} + a''_{kj})\,\Delta_{kj} = (-1)^{k+j}a'_{kj}\,\Delta_{kj} + (-1)^{k+j}a''_{kj}\,\Delta_{kj}$$

which is the sum of a term from the expansion of det A' and a term from the expansion of det A''.

THEOREM 13. *Let matrix B be obtained from matrix A by adding to the kth row (column) vector c times the lth row (column) vector, $k \neq l$. Then* $\det B = \det A$.

EXAMPLE 6

$$\begin{vmatrix} a_{11} & a_{12} & a_{13} \\ a_{21} & a_{22} & a_{23} \\ a_{31} & a_{32} & a_{33} \end{vmatrix} = \begin{vmatrix} a_{11} & a_{12} + ca_{13} & a_{13} \\ a_{21} & a_{22} + ca_{23} & a_{23} \\ a_{31} & a_{32} + ca_{33} & a_{33} \end{vmatrix}$$

$$= \begin{vmatrix} a_{11} & a_{12} & a_{13} \\ a_{21} & a_{22} & a_{23} \\ a_{31} & a_{32} & a_{33} \end{vmatrix} + \begin{vmatrix} a_{11} & ca_{13} & a_{13} \\ a_{21} & ca_{23} & a_{23} \\ a_{31} & ca_{33} & a_{33} \end{vmatrix}$$

$$= \begin{vmatrix} a_{11} & a_{12} & a_{13} \\ a_{21} & a_{22} & a_{23} \\ a_{31} & a_{32} & a_{33} \end{vmatrix} + c\begin{vmatrix} a_{11} & a_{13} & a_{13} \\ a_{21} & a_{23} & a_{23} \\ a_{31} & a_{33} & a_{33} \end{vmatrix}$$

and the last determinant is 0 by Theorem 10.

PROOF OF THEOREM 13. As in Example 6, one can apply Theorems 11 and 12 to write $\det B = \det A + c \det A''$, where A'' is obtained from A by replacing the kth row (column) by the lth. In A'' two rows (columns) are equal; hence, by Theorem 10, $\det A'' = 0$ and the rule is proved.

Theorem 13 is particularly useful when evaluating determinants. For if a matrix $A = (a_{ij})$ has $a_{IJ} \neq 0$, we can use Theorem 13 repeatedly, adding certain multiples of the Ith row to each of the other rows to obtain a matrix B where $\det A = \det B$ and the Jth column of B has 0 in all places except the Ith place. If we then expand B by minors of the Jth column, we find that

$$\det A = \det B = (-1)^{I+J} b_{IJ}\, \Delta_{IJ}$$

and we have reduced our problem to the evaluation of a determinant of smaller order.

EXAMPLE 7

$$\begin{vmatrix} 1 & 4 & 0 & 4 \\ 3 & 2 & 1 & 0 \\ 4 & -1 & -2 & 2 \\ 5 & 1 & -3 & 1 \end{vmatrix} = \begin{vmatrix} -19 & 0 & 12 & 0 \\ 3 & 2 & 1 & 0 \\ -6 & -3 & 4 & 0 \\ 5 & 1 & -3 & 1 \end{vmatrix}$$

$$= \begin{vmatrix} -19 & 0 & 12 \\ 3 & 2 & 1 \\ -6 & -3 & 4 \end{vmatrix} = \begin{vmatrix} -55 & -24 & 0 \\ 3 & 2 & 1 \\ -18 & -11 & 0 \end{vmatrix}$$

$$= -\begin{vmatrix} -55 & -24 \\ -18 & -11 \end{vmatrix} = -(605 - 432) = -173.$$

THEOREM 14. $\det(AB) = (\det A)(\det B)$.

We give a proof here for $n = 2$, reserving the general case to the next section.

$$\det(AB) = \det\left\{\begin{pmatrix} a_{11} & a_{12} \\ a_{21} & a_{22} \end{pmatrix}\begin{pmatrix} b_{11} & b_{12} \\ b_{21} & b_{22} \end{pmatrix}\right\} = \begin{vmatrix} a_{11}b_{11} + a_{12}b_{21} & a_{11}b_{12} + a_{12}b_{22} \\ a_{21}b_{11} + a_{22}b_{21} & a_{21}b_{12} + a_{22}b_{22} \end{vmatrix}$$

$$= \begin{vmatrix} a_{11}b_{11} & a_{11}b_{12} \\ a_{21}b_{11} + a_{22}b_{21} & a_{21}b_{12} + a_{22}b_{22} \end{vmatrix} + \begin{vmatrix} a_{12}b_{21} & a_{12}b_{22} \\ a_{21}b_{11} + a_{22}b_{21} & a_{21}b_{12} + a_{22}b_{22} \end{vmatrix}$$

$$= a_{11}\begin{vmatrix} b_{11} & b_{12} \\ a_{21}b_{11} + a_{22}b_{21} & a_{21}b_{12} + a_{22}b_{22} \end{vmatrix} + a_{12}\begin{vmatrix} b_{21} & b_{22} \\ a_{21}b_{11} + a_{22}b_{21} & a_{21}b_{12} + a_{22}b_{22} \end{vmatrix}$$

$$= a_{11}\begin{vmatrix} b_{11} & b_{12} \\ a_{22}b_{21} & a_{22}b_{22} \end{vmatrix} + a_{12}\begin{vmatrix} b_{21} & b_{22} \\ a_{21}b_{11} & a_{21}b_{12} \end{vmatrix} = a_{11}a_{22}\begin{vmatrix} b_{11} & b_{12} \\ b_{21} & b_{22} \end{vmatrix} + a_{12}a_{21}\begin{vmatrix} b_{21} & b_{22} \\ b_{11} & b_{12} \end{vmatrix}$$

$$= a_{11}a_{22}(b_{11}b_{22} - b_{12}b_{21}) - a_{12}a_{21}(-b_{21}b_{12} + b_{22}b_{11})$$
$$= a_{11}a_{22} \det B - a_{12}a_{21} \det B = (\det B)(a_{11}a_{22} - a_{12}a_{21}) = (\det A)(\det B).$$

The proof has been so arranged as to set the pattern for the general case.

Minors of a matrix. We have used the term *minor* only for determinants or for square matrices. For a general m by n matrix by deleting certain rows and columns one can leave a square array, hence, a square matrix; the determinant of such a square matrix is said to be a minor of the given m by n matrix. For example, given the 3 by 4 matrix,

$$C = \begin{pmatrix} a & b & c & d \\ e & f & g & h \\ p & q & r & s \end{pmatrix}$$

one can obtain minors of order 3 by deleting one of the columns, of order 2 by deleting one row and two columns, of order 1 by deleting two rows and three columns. Thus

$$\begin{vmatrix} a & b & d \\ e & f & h \\ p & q & s \end{vmatrix}, \quad \begin{vmatrix} b & c \\ q & r \end{vmatrix}, \quad e$$

are minors of the matrix C. For a square matrix A, $\det A$ is also considered to be a minor of A (the "major minor").

THEOREM 15. *Matrix A is nonsingular if, and only if,* $\det A \neq 0$. *If A is nonsingular, then*

$$\det A^{-1} = \frac{1}{\det A} \tag{10-132}$$

PROOF. If A is nonsingular, then A has an inverse A^{-1} and $A \cdot A^{-1} = I$ so that by Theorem 14

$$(\det A)(\det A^{-1}) = \det I = 1$$

Hence, $\det A \neq 0$ and (10-132) follows.

Conversely, let $\det A \neq 0$. We seek the kernel of A; that is, the solutions of $A\mathbf{x} = \mathbf{0}$, or the solutions of the n homogeneous equations

$$\begin{cases} a_{11}x_1 + \cdots + a_{1n}x_n = 0 \\ a_{21}x_1 + \cdots + a_{2n}x_n = 0 \\ \vdots \\ a_{n1}x_1 + \cdots + a_{nn}x_n = 0 \end{cases}$$

We multiply the first equation by A_{11}, the second by $A_{21}, \ldots$ and add. In the resulting equation, x_1 has coefficient

$$a_{11}A_{11} + a_{21}A_{21} + \cdots + a_{n1}A_{n1} = \det A$$

(expansion by minors of the first column); x_2 has coefficient

$$a_{12}A_{11} + a_{22}A_{21} + \cdots + a_{n2}A_{n1} = \begin{vmatrix} a_{12} & a_{12} & \cdots & a_{1n} \\ a_{22} & a_{22} & \cdots & a_{2n} \\ \vdots & \vdots & & \vdots \\ a_{n2} & a_{n2} & \cdots & a_{nn} \end{vmatrix}$$

and this is 0, since two columns are the same. Similarly, the coefficients of $x_3, \ldots, x_n$ are 0. Thus the equation reads $(\det A)x_1 = 0$ and, since $\det A \neq 0$, we conclude that $x_1 = 0$. Similarly, $x_2 = 0, \ldots, x_n = 0$. Thus Kernel A consists of **0** alone and A is one-to-one, hence, nonsingular.

THEOREM 16. *Let C be an m by n matrix. If $C = O$, let $\rho = 0$. If $C \neq O$, let ρ be the positive integer such that there is at least one minor of C of order ρ which is not 0 and such that all minors of C of order greater than ρ, if there are any, are 0. Then C has rank ρ.*

The proof is given in the next section. We here give an example. We recall first that the rank of C is the dimension of the range of C, or equivalently it is the dimension of the subspace of V_m spanned by the n column vectors of C.

EXAMPLE 8. Let $C = \begin{pmatrix} 3 & 4 & 2 & 11 \\ 1 & 2 & 0 & 5 \\ 2 & 1 & 3 & 4 \end{pmatrix}$. The minors of order 3 are

$$\begin{vmatrix} 3 & 4 & 2 \\ 1 & 2 & 0 \\ 2 & 1 & 3 \end{vmatrix}, \quad \begin{vmatrix} 3 & 4 & 11 \\ 1 & 2 & 5 \\ 2 & 1 & 4 \end{vmatrix}, \quad \begin{vmatrix} 3 & 2 & 11 \\ 1 & 0 & 5 \\ 2 & 3 & 4 \end{vmatrix}, \quad \begin{vmatrix} 4 & 2 & 11 \\ 2 & 0 & 5 \\ 1 & 3 & 4 \end{vmatrix}$$

These are all found to be 0. However, the minor $\begin{vmatrix} 3 & 4 \\ 1 & 2 \end{vmatrix}$ of order 2 is not 0. Hence, $\rho = 2$ and C has rank 2. Thus the 4 column vectors (3, 1, 2), (4, 2, 1), (2, 0, 3), (11, 5, 4) span a 2-dimensional subspace of V_3. The first two of these are linearly independent and, hence, the subspace is Span ((3, 1, 2), (4, 2, 1)). We verify that the other two column vectors lie in this subspace:

$$(2, 0, 3) = 2(3, 1, 2) - (4, 2, 1), \qquad (11, 5, 4) = (3, 1, 2) + 2(4, 2, 1)$$

Theorem 16 includes the first part of Theorem 15 as a special case. For if C is a square n by n matrix, then $\rho = n$ precisely when $\det C \neq 0$; that is, C has rank n, precisely when $\det C \neq 0$. But, as pointed out in Section 10-12, C has rank n precisely when C is nonsingular.

COROLLARY 1 TO THEOREM 16. *If C is an m by n matrix then C and C' have the same rank.*

PROOF. Each minor of C of order s is the determinant of a matrix of the form

$$E = \begin{pmatrix} c_{i_1j_1} & \cdots & c_{i_1j_s} \\ \vdots & & \vdots \\ c_{i_sj_1} & \cdots & c_{i_sj_s} \end{pmatrix}$$

where $1 \leq i_1 < i_2 < \cdots < i_s \leq m$, $1 \leq j_1 < j_2 < \cdots < j_s \leq n$. Now the minors of C' are just the determinants of the matrices E', where E is as above. Since $\det E = \det E'$ (Theorem 7), we can conclude that if ρ is defined as in Theorem 16, then the ρ for E and the ρ for E' are the same. It then follows from Theorem 16 that C and C' have the same rank.

COROLLARY 2 TO THEOREM 16. *If C is an m by n matrix, then the rank of C is the maximal number of linearly independent row vectors of C.*

PROOF. By Corollary 1, above, rank C = rank C'. But rank C' is the maximal number of linearly independent column vectors of C', which is the maximal number of linearly independent row vectors of C.

THEOREM 17 (*Cramer's Rule*). *If $\det A \neq 0$, the linear equations*

$$a_{i1}x_1 + a_{i2}x_2 + \cdots + a_{in}x_n = b_i \qquad (i = 1, \ldots, n) \tag{10-133}$$

have a unique solution:

$$x_i = \frac{D_i}{\det A} \qquad (i = 1, \ldots, n)$$

where D_i is the determinant of the matrix obtained from A by replacing the ith column by $\text{col}\,(b_1, b_2, \ldots, b_n)$.

PROOF. By Theorem 15, A is nonsingular, so that the equation $A\mathbf{x} = \mathbf{b}$ —that is, the given set (10-133) of simultaneous equations—has a unique solution. To find the solution, we multiply the first equation by A_{11}, the second by $A_{21}, \ldots$ and add, as in the proof of Theorem 15. We obtain $(\det A)x_1$ on the left and

$$b_1A_{11} + b_2A_{21} + \cdots + b_nA_{n1} = \begin{vmatrix} b_1 & a_{12} & \cdots & a_{1n} \\ b_2 & a_{22} & \cdots & a_{2n} \\ \vdots & \vdots & & \vdots \\ b_n & a_{n2} & \cdots & a_{nn} \end{vmatrix} = D_1$$

on the right. Hence, $(\det A)x_1 = D_1$, so that $x_1 = D_1/\det A$. Similarly, $x_2 = D_2/\det A, \ldots, x_n = D_n/\det A$.

THEOREM 18. *Let $A = (a_{ij})$ be nonsingular. Then the inverse of A is $C = (c_{ij})$, where*

$$c_{ij} = \frac{A_{ji}}{\det A} = \frac{(-1)^{j+i}\,\Delta_{ji}}{\det A}$$

Thus

$$A^{-1} = \frac{1}{\det A}\,\text{adj}\,A \tag{10-134}$$

PROOF. Let γ_k be the kth column vector of C. Then from the equation $AC = I$ we conclude that $A\gamma_k$ equals the kth column of I; that is, $A\gamma_k = \mathbf{e}_k = \text{col}\,(0, 0, \ldots, 1, \ldots, 0)$ (here 1 is in the kth position); or, for k fixed and $i = 1, \ldots, n$,

$$\sum_{j=1}^{n} a_{ij}c_{jk} = \delta_{ik} = \begin{cases} 0 & \text{for } i \neq k \\ 1 & \text{for } i = k \end{cases}$$

Hence, we can solve for $c_{1k}, \ldots, c_{nk}$ by Cramer's rule:

$$c_{ik} \det A = \begin{vmatrix} a_{11} & \cdots & 0 & \cdots & a_{1n} \\ \vdots & & \vdots & & \vdots \\ a_{k1} & \cdots & 1 & \cdots & a_{kn} \\ \vdots & & \vdots & & \vdots \\ a_{n1} & \cdots & 0 & \cdots & a_{nn} \end{vmatrix}$$

where the exceptional column is the ith and 1 is the ki-entry. If we expand this determinant by minors of the ith column, we obtain $1 \cdot A_{ki} = (-1)^{k+i}\Delta_{ki}$. Thus $c_{ik} = (1/\det A)A_{ki}$ and the theorem is proved.

EXAMPLE 9 $$\begin{pmatrix} a & b \\ c & d \end{pmatrix}^{-1} = \frac{1}{ad - bc}\begin{pmatrix} d & -b \\ -c & a \end{pmatrix}$$

The rule (10-134) is a simple one for the inverse of a matrix and is quite useful in proving theorems. However, it is not a very practical rule for the actual computation of the inverse of matrices of even moderate size. For example, if we were to use (10-134) to compute the inverse of a 10 by 10 matrix we would need to evaluate 101 determinants—100 of order 9 and one of order 10. In Section 10-16 we shall give another method for finding the inverse of a nonsingular matrix which involves fewer numerical calculations (see also Section 10-12). Similarly, Cramer's rule is not a very practical rule for determining the solution of linear equations when the number of equations exceeds 5 or 6—usually the elimination of unknowns is a faster method.

THEOREM 19. *If* $\det A = 0$ *and the equations* $\sum_{j=1}^{n} a_{ij}x_j = b_i$ $(i = 1, \ldots, n)$ *have a solution, then* $D_1 = 0, \ldots, D_n = 0$ *(with* D_i *as in Theorem 17). If* $\det A = 0$, $D_1 = 0, \ldots, D_n = 0$ *and, at least, one minor of A of order* $n - 1$ *is not 0, then the equations have infinitely many solutions, forming a one-dimensional linear variety in* V_n.

More generally, the given linear equations have solutions precisely when the two matrices

$$A = \begin{pmatrix} a_{11} & \cdots & a_{1n} \\ \vdots & & \vdots \\ a_{n1} & \cdots & a_{nn} \end{pmatrix}, \qquad B = (A, \mathbf{b}) = \begin{pmatrix} a_{11} & \cdots & a_{1n} & b_1 \\ \vdots & & \vdots & \vdots \\ a_{n1} & \cdots & a_{nn} & b_n \end{pmatrix}$$

have the same rank r and then the solutions form a linear variety of dimension $n - r$.

PROOF. The statement in the second part of the theorem is the rule 8 of Section 10-6. If $\det A = 0$, then A has rank less than n. Hence, if solutions exist, B has rank less than n. In particular, $D_1, \ldots, D_n$ (each equal to plus or minus a minor of B) must be 0.

Now suppose $\det A = 0$ and that a minor of order $n - 1$ of A is not 0. Then A has rank $n - 1$. If also $D_1 = \cdots = D_n = 0$, then, since $\det A$, $\pm D_1, \ldots, \pm D_n$ are the minors of B of order n, B has rank less than n and, because the nonzero minor of A is a nonzero minor of B, B must have rank $n - 1$. Thus rank A = rank $B = r = n - 1$ and solutions exist, forming a one-dimensional linear variety.

Remark. The statement in the second paragraph of the theorem is applicable to general systems of m equations in n unknowns (see Section 10-6). Determinants help here in finding the ranks of A and B, as in Theorem 16. Determinants are also useful in finding the solutions. The following example shows this; the same example is treated without determinants as Example 4 in Section 10-6.

EXAMPLE 10. $x_1 + 3x_2 - x_3 + x_4 = 1$, $x_1 + 3x_2 + x_3 + 2x_4 = 5$. Thus we are to solve $A\mathbf{x} = \mathbf{b}$, where

$$A = \begin{pmatrix} 1 & 3 & -1 & 1 \\ 1 & 3 & 1 & 2 \end{pmatrix}, \qquad B = (A, \mathbf{b}) = \begin{pmatrix} 1 & 3 & -1 & 1 & 1 \\ 1 & 3 & 1 & 2 & 5 \end{pmatrix}$$

Clearly the minor of order 2 of A formed by the first and second columns is 0, all others are not 0. Hence, A has rank 2 and, since B can have no larger rank, B also has rank 2. To find the solutions, we select a nonzero minor of A of order 2 and write the equations so that this minor is the "determinant of the coefficients"; for example,

$$\begin{aligned} x_1 - x_3 &= 1 - 3x_2 - x_4 \\ x_1 + x_3 &= 5 - 3x_2 - 2x_4 \end{aligned}$$

Now we regard these equations as two simultaneous equations in x_1, x_3. We solve, by elimination or by Cramer's rule, and obtain x_1, x_3 in terms of x_2, x_4. Setting $x_2 = 2t_1$, $x_4 = 2t_2$ as in Section 10-6, we obtain our linear variety of solutions:

$$x_1 = 3 - 6t_1 - 3t_2, \qquad x_2 = 2t_1, \qquad x_3 = 2 - t_2, \qquad x_4 = 2t_2$$

‡10-14 PROOFS OF THEOREMS ON DETERMINANTS

In this section we give the proofs of five theorems of the preceding section for which proofs were not provided.

PROOF OF THEOREM 6 (*Expansion by Rule of Signs*). We use induction. For $n = 2$,

$$\begin{vmatrix} a & b \\ c & d \end{vmatrix} = \overset{\searrow}{ad} - \overset{\swarrow}{bc}$$

and the rule is verified. Let us assume that the expansion is true for determinants of order $n-1$ and let $A=(a_{ij})$ be an n by n matrix. We expand det A by minors of the first row as in the definition:

$$\det A = a_{11}\,\Delta_{11} - a_{12}\,\Delta_{12} + a_{13}\,\Delta_{13} - \cdots. \tag{10-140}$$

By our induction assumption, Δ_{11} equals a sum of signed products of elements, in each case formed by selecting an element from each of the 2nd, 3rd, . . . , nth rows, and columns of A. If we multiply each term by a_{11}, we obtain a sum of signed products of n factors, one each from the 1st, 2nd, . . . , nth rows and columns of A. Similar remarks apply to $-a_{12}\,\Delta_{12}$, $a_{13}\,\Delta_{13}$, ... Furthermore, each selection of one element from each row and column of A is represented by precisely one term in (10-140), after expansion of Δ_{11}, Δ_{12}, . . . , and multiplying by a_{11}, $-a_{12}$, ... We need only verify that the Rule of Signs is obeyed. For the terms coming from $a_{11}\,\Delta_{11}$, introduction of the factor a_{11} cannot add any segments of positive slope (see Figure 10-9). Hence, these terms should keep the same signs as in the expansion of Δ_{11}, as they do in (10-140). For each term coming from $-a_{12}\,\Delta_{12}$, there is one new segment of positive slope, as in Figure 10-9. Hence, this term should reverse sign, as the $-$ in front forces it to do. Similarly, for $a_{13}\,\Delta_{13}$ there are 2 new segments of positive slope, hence, no change in sign, and so on. Therefore, the Rule of Signs holds true for determinants of nth order, and the theorem is proved.

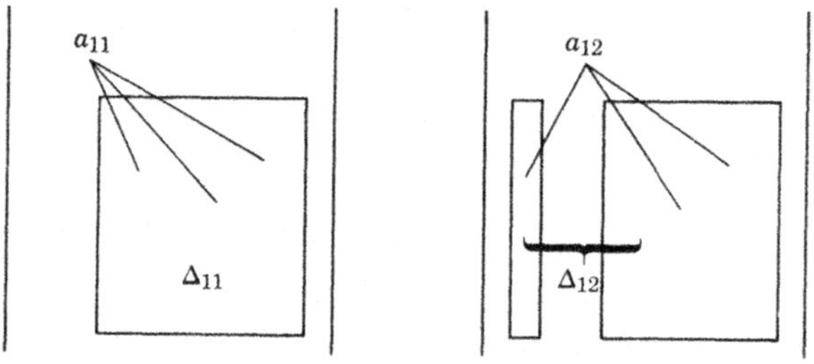

Figure 10-9 Proof of expansion by rule of signs.

Remark. As noted at the beginning of Section 10-13, we can write each term of the expansion as

$$\pm a_{1i_1}a_{2i_2}\cdots a_{ni_n}$$

where $(i_1, \ldots, i_n)$ is a permutation of $(1, \ldots, n)$. One can verify that our Rule of Signs assigns a $+$ or $-$ according as the permutation is even or odd; that is, obtained from $(1, \ldots, n)$ by an even or odd number of interchanges of pairs. For a discussion of determinants based on permutations, see Birkhoff and MacLane: *Survey of Modern Algebra* (Macmillan, N.Y., 1969).

PROOF OF THEOREM 8 (*Interchange of Two Rows or Columns*). By Theorem 7 any theorem about rows is applicable to columns. Hence, it is sufficient to consider only interchange of two rows. After such an interchange, expansion by the Rule of Signs gives us the same terms as before, except perhaps for signs. In a particular product in the expansion of the determinant, let a, b be the elements whose rows are interchanged. The interchange reverses the slope of the segment connecting a, b, and, hence, causes one change in sign. For each element of the product whose row and column are between those of a and b (such as c in Figure 10-10), there are two changes of sign, hence, no net effect; for each other element there is no change in the number of segments with positive slope. Hence, in all, one change of sign is needed. This applies to every term, so that the theorem is proved.

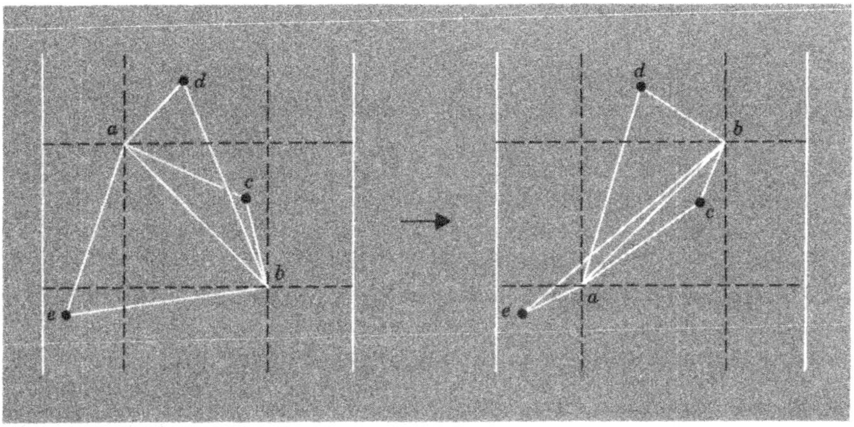

Figure 10-10 The effect of interchanging two rows.

PROOF OF THEOREM 9 (*Expansion by Minors of Any Row or Column*). As for Theorem 8, we need only consider the case of a row. To expand det A by minors of the ith row, for $i > 1$, we can successively interchange the ith and $(i-1)$st rows, the $(i-1)$st and $(i-2)$nd rows, and so on, until we have moved the original ith row up to become the first row. There are $i-1$ interchanges of rows here, hence, by Theorem 8, $i-1$ changes of sign. For each element of the original ith row, the process described does not affect the orders of rows and columns not containing the element; hence, each element has the same minor in its new position in the first row as in its old position. For the new position expansion (10-140) is the same as the expansion by minors of the ith row for the original position, except for the $i-1$ changes in sign. For the element a_{ij} one has, therefore, a factor $(-1)^{i-1}$ from the shift of rows and a factor $(-1)^{j-1}$ from the expansion rule in (10-140). Thus, in all, one has the factor $(-1)^{i+j-2} = (-1)^{i+j}$. Accordingly, the rule is proved.

PROOF OF THEOREM 14 (*Determinant of AB*). We use induction. For $n = 2$, the rule is proved in the preceding section. We assume the rule proved for matrices or order $n-1$ and assume A, B to be n by n matrices. We then imitate the proof for $n = 2$:

$$\det(AB) = \begin{vmatrix} a_{11}b_{11} + \cdots + a_{1n}b_{n1} & a_{11}b_{12} + \cdots & \cdots & a_{11}b_{1n} + \cdots + a_{1n}b_{nn} \\ a_{21}b_{11} + \cdots + a_{2n}b_{n1} & a_{21}b_{12} + \cdots & \cdots & a_{21}b_{1n} + \cdots + a_{2n}b_{nn} \\ \vdots & \vdots & \vdots & \vdots \\ a_{n1}b_{11} + \cdots + a_{nn}b_{n1} & a_{n1}b_{12} + \cdots & \cdots & a_{n1}b_{1n} + \cdots + a_{nn}b_{nn} \end{vmatrix}$$

Clearly we can express the first row vector of AB as a sum of n vectors: $a_{11}(b_{11}, b_{12}, \ldots, b_{1n}) + \cdots + a_{1n}(b_{n1}, \ldots, b_{nn})$. Hence, by Theorems 11 and 12, det $(AB) =$

$$a_{11}\begin{vmatrix} b_{11} & b_{12} & \cdots & b_{1n} \\ a_{21}b_{11} + \cdots + a_{2n}b_{n1} & a_{21}b_{12} + \cdots + a_{2n}b_{n2} & \cdots & a_{21}b_{1n} + \cdots + a_{2n}b_{nn} \\ \vdots & \vdots & \vdots & \vdots \\ a_{n1}b_{11} + \cdots + a_{nn}b_{n1} & a_{n1}b_{12} + \cdots + a_{nn}b_{n2} & \cdots & a_{n1}b_{1n} + \cdots + a_{nn}b_{nn} \end{vmatrix}$$

$$+ a_{12}\begin{vmatrix} b_{21} & \cdots & b_{2n} \\ a_{21}b_{11} + a_{22}b_{21} + \cdots + a_{2n}b_{n1} & \cdots & a_{21}b_{1n} + a_{22}b_{2n} + \cdots + a_{2n}b_{nn} \\ \vdots & \vdots & \vdots \\ a_{n1}b_{11} + a_{n2}b_{21} + \cdots + a_{nn}b_{n1} & \cdots & a_{n1}b_{1n} + a_{n2}b_{2n} + \cdots + a_{nn}b_{nn} \end{vmatrix}$$

$$+ \cdots + a_{1n}\begin{vmatrix} b_{n1} & \cdots & b_{nn} \\ a_{21}b_{11} + \cdots + a_{2n}b_{n1} & \cdots & a_{21}b_{1n} + \cdots + a_{2n}b_{nn} \\ \vdots & \vdots & \vdots \\ a_{n1}b_{11} + \cdots + a_{nn}b_{n1} & \cdots & a_{n1}b_{1n} + \cdots + a_{nn}b_{nn} \end{vmatrix}$$

In the first determinant here we multiply the first row by a_{21} and subtract from the second row; this eliminates the terms in $b_{11}, b_{12}, \ldots$, from the second row. We multiply the first row by a_{31} and subtract from the third; this eliminates the terms in $b_{11}, b_{12}, \ldots$ from the third row. We proceed similarly throughout: in each determinant we eliminate the terms in the elements of the first row from the other rows, as suggested by the slanting broken lines as above. We then expand each of the resulting determinants by minors of the first row. For the first determinant the minor of b_{11} is seen to be the determinant of the matrix $A_{(1,1)}B_{(1,1)}$, where $A_{(1,1)}$ is the $(n-1)$ by $(n-1)$ matrix obtained from A by striking the first row and first column of A, and $B_{(1,1)}$ is the $(n-1)$ by $(n-1)$ matrix obtained from B by striking the first row and first column of B. Then

$$\det A_{(1,1)} = \Delta_{11} = \text{minor of } a_{11} \text{ in } A$$
$$\det B_{(1,1)} = H_{11} = \text{minor of } b_{11} \text{ in } B$$

By the induction assumption for $(n-1)$ by $(n-1)$ matrices,

$$\det(A_{(1,1)}B_{(1,1)}) = \det A_{(1,1)} \det B_{(1,1)} = \Delta_{11}H_{11}$$

Thus the minor of b_{11} in the first determinant of the expansion of det (AB) is $\Delta_{11}H_{11}$. Similarly, the minor of b_{12} in this first determinant is det $(A_{(1,1)}B_{(1,2)}) = \Delta_{11}H_{12}$, where $B_{(1,2)}$ is the matrix obtained from B by striking the first row and second column of B and, hence, det $B_{(1,2)} = H_{12}$, the minor of b_{12} in B. In like fashion, we find that the minor of b_{13} in the first determinant is $\Delta_{11}H_{13}, \ldots$, and the minor of b_{1n} in the first determinant is $\Delta_{11}H_{1n}$. Similarly, the minor of b_{21} in the second determinant is $\Delta_{12}H_{21}$, the minor of b_{22} in the second determinant is $\Delta_{12}H_{22}, \ldots$, and so on. Accordingly, det (AB) equals

$$a_{11}(b_{11}\,\Delta_{11}H_{11} - b_{12}\,\Delta_{11}H_{12} + \cdots) + a_{12}(b_{21}\,\Delta_{12}H_{21} - b_{22}\,\Delta_{12}H_{22} + \cdots) + \cdots + a_{1n}(b_{n1}\,\Delta_{1n}H_{n1} - b_{n2}\,\Delta_{1n}H_{n2} + \cdots)$$
$$= a_{11}\,\Delta_{11}(b_{11}H_{11} - b_{12}H_{12} + \cdots) - a_{12}\,\Delta_{12}(-b_{21}H_{21} + b_{22}H_{22} - \cdots) + \cdots + (-1)^{n+1}a_{1n}\,\Delta_{1n}((-1)^{n+1}b_{n1}H_{n1} + (-1)^{n+2}b_{n2}H_{n2} + \cdots)$$

Now $b_{11}H_{11} - b_{12}H_{12} + \cdots = \det B$, by expansion by minors of the first row, $-b_{21}H_{21} + b_{22}H_{22} - \cdots = \det B$, by expansion by minors of the second row, and so on. Hence,

$$\begin{aligned}\det(AB) &= a_{11}\,\Delta_{11}\det B - a_{12}\,\Delta_{12}\det B + \cdots + (-1)^{n+1}a_{1n}\,\Delta_{1n}\det B\\ &= (a_{11}\,\Delta_{11} - a_{12}\,\Delta_{12} + \cdots + (-1)^{n+1}a_{1n}\,\Delta_{1n})(\det B)\\ &= (\det A)(\det B)\end{aligned}$$

Thus the induction is complete and the theorem is proved.

PROOF OF THEOREM 16 (*Rank of an m by n Matrix*). Let C have rank r. We must show that $r = \rho$. If $C = O$, both r and ρ are 0. Hence, we may assume that $C \neq O$, whence both r and ρ are positive. If $r = m$, there are no minors of C of order greater than r and, hence, $\rho \leq r$. Now suppose that $r < m$. Since C has rank r, every set of s column vectors, with $s > r$, is linearly dependent. Let $s > r$ columns have numbers $j_1, \ldots, j_s$. Then there are scalars $k_{j_1}, k_{j_2}, \ldots, k_{j_s}$, not all 0, such that

$$k_{j_1}c_{ij_1} + k_{j_2}c_{ij_2} + \cdots + k_{j_s}c_{ij_s} = 0, \qquad (i = 1, \ldots, m)$$

If $m \geq s$, it follows that the first s equations

$$c_{ij_1}x_1 + c_{ij_2}x_2 + \cdots + c_{ij_s}x_s = 0 \qquad (i = 1, \ldots, s)$$

have a nontrivial solution $\mathbf{x} = (k_{j_1}, k_{j_2}, \ldots, k_{j_s})$; that is, the (square) matrix of coefficients is singular and, hence, by Theorem 15, its determinant is 0:

$$\begin{vmatrix} c_{1j_1} & c_{1j_2} & \cdots & c_{1j_s} \\ \vdots & \vdots & & \vdots \\ c_{sj_1} & c_{sj_2} & \cdots & c_{sj_s} \end{vmatrix} = 0$$

Similarly, if $i_1, i_2, \ldots, i_s$ are distinct numbers between 1 and m, the equations

$$c_{ij_1}x_1 + \cdots + c_{ij_s}x_s = 0 \qquad (i = i_1, i_2, \ldots, i_s)$$

also have the same nontrivial solution and, hence, the determinant of their coefficients is 0. It thus follows that all minors of C of order s are 0. Hence, $\rho \leq r$.

If $\rho = \min(m, n)$, then $r \leq \rho$, since $r = \operatorname{rank} C \leq \min(m, n)$. Now suppose that $1 \leq \rho < \min(m, n)$. Then there is a minor of C of order ρ which is not 0, and every minor of C of order $s > \rho$ is 0. Permuting rows and columns of C affects neither r nor ρ. Hence, without loss of generality, we can assume that

$$K^* = \begin{vmatrix} c_{11} & \cdots & c_{1\rho} \\ \vdots & & \vdots \\ c_{\rho 1} & \cdots & c_{\rho\rho} \end{vmatrix} \neq 0, \qquad \begin{vmatrix} c_{11} & \cdots & c_{1\rho} & c_{1s} \\ \vdots & & \vdots & \vdots \\ c_{\rho 1} & \cdots & c_{\rho\rho} & c_{\rho s} \\ c_{h1} & \cdots & c_{h\rho} & c_{hs} \end{vmatrix} = 0$$

for $h > \rho$ and $s > \rho$. For $h \leq \rho$ or $s \leq \rho$, the second determinant is 0 in any case, since two rows or two columns are the same. Hence, for any s and h, the second determinant is 0. We keep s fixed ($s > \rho$) and expand the second determinant by the minors of the last row. The cofactor of c_{hs} is $K_0 = \pm K^* \neq 0$, and we denote the cofactors of $c_{h1}, \ldots, c_{h\rho}$ by $K_1, K_2, \ldots, K_\rho$. None of these cofactors change as we change h. Thus

$$K_1 c_{h1} + \cdots + K_\rho c_{h\rho} + K_0 c_{hs} = 0, \qquad h = 1, \ldots, m$$

Since $K_0 \neq 0$, these equations state that the sth column of C is expressible as a linear combination of the first ρ columns of C. This is true for $s = \rho + 1$, $\rho + 2, \ldots, n$. Hence, the dimension of range C is at most ρ; that is, $r \leq \rho$. Since we proved in all cases that $r \leq \rho$ and $\rho \leq r$, we conclude that $\rho = r$.

‡10-15 FURTHER REMARKS ON DETERMINANTS

For simplicity we emphasize the real case here. As earlier, $\mathfrak{M}_{nn}$ denotes the set of n by n real matrices, with the usual operations on these matrices.

The Determinant as a Real Function on $\mathfrak{M}_{nn}$. Our very notation $\det A$ indicates that the determinant is a function of matrices; in particular, it assigns a real number to each matrix in $\mathfrak{M}_{nn}$. If we write $\Phi(A)$ instead of $\det A$ for the value, then the function Φ has the four properties:

Ia. $\Phi(AB) = \Phi(A)\Phi(B)$ for all A, B in $\mathfrak{M}_{nn}$.
Ib. $\Phi(cA) = c^n\Phi(A)$ for each real c and each A in $\mathfrak{M}_{nn}$.
Ic. $\Phi(A) = -1$ for each matrix A in $\mathfrak{M}_{nn}$ obtained from I by interchanging two rows.
Id. Φ is continuous in the following sense: for each A in $\mathfrak{M}_{nn}$ and each $\epsilon > 0$ there is a $\delta > 0$ such that $|\Phi(B) - \Phi(A)| < \epsilon$ for all B in $\mathfrak{M}_{nn}$ such that $|b_{ij} - a_{ij}| < \delta$ for $i = 1, \ldots, n$, $j = 1, \ldots, n$.

Property Ia is Theorem 14. Property Ib follows at once from Theorem 11. Property Ic follows from Theorem 8. Property Id can be proved from the fact that $\Phi(A)$ is a polynomial in the elements a_{ij}, and a polynomial is a continuous function (see Chapter 12).

It is a remarkable fact that the determinant is the only function Φ on $\mathfrak{M}_{nn}$ with the four properties stated. In particular, knowledge that Φ has these four properties permits one to deduce that Φ has all the other properties of determinants.

The Determinant as a Function of Row Vectors. A real-valued function on $\mathfrak{M}_{nn}$ can be written as $F(\mathbf{u}_1, \ldots, \mathbf{u}_n)$, where $\mathbf{u}_1, \ldots, \mathbf{u}_n$ are the successive row vectors of each matrix; the domain of F is then the set of all ordered n-tuples of vectors in V_n. In particular, we could write $\det A$ as $\det(\mathbf{u}_1, \ldots, \mathbf{u}_n)$, where $\mathbf{u}_i = (a_{i1}, \ldots, a_{in})$. For this particular function $F(\mathbf{u}_1, \ldots, \mathbf{u}_n) = \det(\mathbf{u}_1, \ldots, \mathbf{u}_n)$ we have the following properties:

IIa. F is linear in each argument; that is, for each i, F is linear in $\mathbf{u}_i$ for fixed $\mathbf{u}_1, \ldots, \mathbf{u}_{i-1}, \mathbf{u}_{i+1}, \ldots, \mathbf{u}_n$:

$$F(\mathbf{u}_1, \ldots, \mathbf{u}_{i-1}, a\mathbf{u}_i' + b\mathbf{u}_i'', \mathbf{u}_{i+1}, \ldots, \mathbf{u}_n)$$
$$= aF(\mathbf{u}_1, \ldots, \mathbf{u}_i', \ldots, \mathbf{u}_n) + bF(\mathbf{u}_1, \ldots, \mathbf{u}_i'', \ldots, \mathbf{u}_n)$$

IIb. $F(\mathbf{u}_1, \ldots, \mathbf{u}_n) = 0$ whenever $\mathbf{u}_i = \mathbf{u}_j$ for some pair of distinct indices i, j.

IIc. $F(\mathbf{e}_1, \ldots, \mathbf{e}_n) = 1$, where $\mathbf{e}_1 = (1, 0, \ldots, 0)$, $\mathbf{e}_2 = (0, 1, 0, \ldots, 0), \ldots$.

Here Property IIa follows from Theorems 11 and 12, Property IIb is Theorem 10, Property IIc is just $\det I = 1$.

A function F, having domain as above, and having property IIa is called a *linear form on* V_n; a function F with both properties IIa and IIb is called an *alternating linear form on* V_n. It is a remarkable fact that each alternating linear form on V_n must be a constant k times $\det(\mathbf{u}_1, \ldots, \mathbf{u}_n) = \det A$. It follows at once that an alternating linear form on V_n satisfying IIc must have $k = 1$, that is, it reduces to $\det A$. Thus $\det A$ is an alternating linear form on its row vectors such that $\det I = 1$. For a proof of these assertions, see D. J. Lewis: *Introduction to Algebra* (Harper and Row, N.Y. 1965), or S. Lang: *A Second Course in Calculus* (Addison Wesley, Reading, Mass., 1964).

There is a similar discussion of $\det A$ as an alternating linear form on its column vectors.

Determinants and Volume. In Section 1-12 we saw that $\pm \begin{vmatrix} u_x & u_y \\ v_x & v_y \end{vmatrix}$ was the area of the parallelogram whose sides, properly directed, represented $\mathbf{u} = u_x\mathbf{i} + u_y\mathbf{j}$ and $\mathbf{v} = v_x\mathbf{i} + v_y\mathbf{j}$. It can be shown that

$$\pm \begin{vmatrix} u_1 & u_2 & u_3 \\ v_1 & v_2 & v_3 \\ w_1 & w_2 & w_3 \end{vmatrix}$$

is the volume of a parallelepiped in 3-dimensional space whose edges, properly directed, represent $\mathbf{u} = (u_1, u_2, u_3)$, $\mathbf{v} = (v_1, v_2, v_3)$, $\mathbf{w} = (w_1, w_2, w_3)$; here all components refer to a fixed Cartesian coordinate system. There is an analogous interpretation of a determinant of order n as a signed n-dimensional volume of an n-dimensional parallelepiped. Properties IIa, b, c have natural interpretations in terms of volume. For details, see Sections 11-15 and 11-16.

PROBLEMS

1. Give the proper sign to be given to each of the following terms obtained from expansion of $\det A = \det(a_{ij})$:

(a) $a_{12}a_{21}$, (b) $a_{13}a_{21}a_{32}$, (c) $a_{12}a_{23}a_{34}a_{41}$,
(d) $a_{14}a_{21}a_{32}a_{43}$, (e) $a_{15}a_{24}a_{33}a_{42}a_{51}$, (f) $a_{14}a_{25}a_{31}a_{42}a_{53}$,
(g) $a_{17}a_{26}a_{35}a_{44}a_{53}a_{62}a_{71}$, (h) $a_{11}a_{22}a_{33} \cdots a_{nn}$.

2. Evaluate the following determinants:

(a) $\begin{vmatrix} 3 & 5 \\ 7 & 2 \end{vmatrix}$, (b) $\begin{vmatrix} -1 & -3 \\ -4 & -12 \end{vmatrix}$, (c) $\begin{vmatrix} 1 & 0 & 0 \\ 2 & 3 & 0 \\ 5 & 6 & 7 \end{vmatrix}$,

(d) $\begin{vmatrix} 5 & 2 & 1 \\ 0 & 3 & 4 \\ 0 & 0 & -8 \end{vmatrix}$, (e) $\begin{vmatrix} 1 & 0 & 1 \\ 0 & 1 & 1 \\ 1 & 1 & 0 \end{vmatrix}$, (f) $\begin{vmatrix} 1 & -4 & -5 \\ 4 & 2 & -6 \\ 5 & 6 & 3 \end{vmatrix}$,

(g) $\begin{vmatrix} 1 & 0 & 2 & 0 \\ 3 & 5 & 1 & 2 \\ 4 & 0 & 6 & 4 \\ 3 & 0 & 1 & 7 \end{vmatrix}$, (h) $\begin{vmatrix} 4 & 0 & 1 & 2 \\ 3 & 2 & 0 & 4 \\ 5 & -1 & -2 & 3 \\ 4 & 1 & 1 & 0 \end{vmatrix}$,

(i) $\begin{vmatrix} 1 & 2 & 3 & 6 \\ 0 & 4 & 5 & 1 \\ 3 & -2 & 0 & 1 \\ -1 & 1 & 2 & 3 \end{vmatrix}$, (j) $\begin{vmatrix} 1 & i & 1 & -i \\ i & 1 & i & -i \\ 1+i & 1-i & 1-i & -1+i \\ 1-i & 1+i & 1+i & -1-i \end{vmatrix}$,

(k) $\begin{vmatrix} x & x+1 & x+2 & x+3 \\ 0 & 1 & x & x+2 \\ x & 1 & 0 & 1 \\ 0 & 0 & x+3 & 3 \end{vmatrix}$, (l) $\begin{vmatrix} \sin\theta\cos\varphi & \cos\theta\cos\varphi & \sin\varphi \\ -\cos\theta\cos\varphi & \sin\theta\cos\varphi & \sin\varphi \\ -\cos\varphi & -\cos\varphi & 1 \end{vmatrix}$,

(m) $\begin{vmatrix} x & y & 0 & 0 & 0 \\ 0 & x & y & 0 & 0 \\ 0 & 0 & x & y & 0 \\ 0 & 0 & 0 & x & y \\ 0 & 0 & 0 & 0 & y \end{vmatrix}$, (n) $\begin{vmatrix} 1 & x & x & x & x \\ 0 & 2 & x & x & x \\ 0 & 0 & 3 & x & x \\ 0 & 0 & 0 & 4 & x \\ 0 & 0 & 0 & 0 & 5 \end{vmatrix}$, (o) $\begin{vmatrix} 1 & x & x & x & x \\ y & 2 & x & x & x \\ y & y & 3 & x & x \\ y & y & y & 4 & x \\ y & y & y & y & 5 \end{vmatrix}$

3. Prove by induction that the expansion of a determinant of order n by the Rule of Signs has $n!$ terms.

4. Let $n = p + q$ and let $A = \begin{pmatrix} B & C \\ D & E \end{pmatrix}$, where B is a p by p matrix and E is q by q. Prove:

(a) If $p < q$, and $E = O$, then det $A = 0$. (*Hint.* Consider the last q column vectors of A.)

(b) If $p = q$ and $C = O$, then det A = det B det E.

(c) If $C = O$ and $D = O$, so that A is the direct sum of B and E, then det A = det B det E.

5. Let $A = (a_{ij})$ be an n by n matrix and let $a_{1n} = 1$, $a_{2,n-1} = 1, \ldots, a_{n1} = 1$, $a_{ij} = 0$ otherwise. Show that det $A = (-1)^{n(n-1)/2}$.

6. Let $a_1, \ldots, a_5$ be distinct real numbers. Find all solutions of the equation

for x:

$$\begin{vmatrix} 1 & x & x^2 & \cdots & x^5 \\ 1 & a_1 & a_1^2 & \cdots & a_1^5 \\ \vdots & \vdots & \vdots & & \vdots \\ 1 & a_5 & a_5^2 & \cdots & a_5^5 \end{vmatrix} = 0$$

(*Hint.* The determinant is a polynomial in x of degree 5.)

7. Evaluate:

(a) $\begin{vmatrix} 1 & 1 & 1 & 1 & \cdots & 1 \\ 1 & x+1 & 1 & 1 & \cdots & 1 \\ 1 & 1 & x+2 & 1 & \cdots & 1 \\ \vdots & \vdots & \vdots & \vdots & & \vdots \\ 1 & 1 & 1 & 1 & \cdots & x+n \end{vmatrix}$ (b) $\begin{vmatrix} a_0 & -1 & 0 & 0 & \cdots & 0 & 0 \\ a_1 & x & -1 & 0 & \cdots & 0 & 0 \\ a_2 & 0 & x & -1 & \cdots & 0 & 0 \\ \vdots & \vdots & \vdots & \vdots & & \vdots & \vdots \\ a_{n-1} & 0 & 0 & \cdots & & x & -1 \\ a_n & 0 & 0 & \cdots & & 0 & x \end{vmatrix}$

8. Evaluate the Vandermonde determinants:

(a) $\begin{vmatrix} 1 & 1 \\ x_1 & x_2 \end{vmatrix}$, (b) $\begin{vmatrix} 1 & 1 & 1 \\ x_1 & x_2 & x_3 \\ x_1^2 & x_2^2 & x_3^2 \end{vmatrix}$, (c) $\begin{vmatrix} 1 & 1 & 1 & 1 \\ x_1 & x_2 & x_3 & x_4 \\ x_1^2 & x_2^2 & x_3^2 & x_4^2 \\ x_1^3 & x_2^3 & x_3^3 & x_4^3 \end{vmatrix}$

9. Show that, for the general Vandermonde determinant,

$$\begin{vmatrix} 1 & 1 & \cdots & 1 \\ x_1 & x_2 & \cdots & x_n \\ x_1^2 & x_2^2 & \cdots & x_n^2 \\ \vdots & \vdots & & \vdots \\ x_1^{n-1} & x_2^{n-1} & \cdots & x_n^{n-1} \end{vmatrix} = \prod_{1 \le i < j \le n} (x_j - x_i)$$

where the symbol on the right denotes the product of all $(x_j - x_i)$ for which i, j satisfy the stated inequalities. [*Hint:* Use induction. For the step from $n - 1$ to n let the nth order determinant be considered as a function f of x_n; show that f is a polynomial of degree $n - 1$ in x_n such that $f(x_1) = 0, \ldots, f(x_{n-1}) = 0$, and conclude that $f(x_n) = k(x_n - x_1) \cdots (x_n - x_{n-1})$, where k is a Vandermonde determinant of order $n - 1$. Now use the induction hypothesis.]

10. Determine whether the given matrix is singular or nonsingular:

(a) $\begin{pmatrix} 12 & 15 \\ 16 & 20 \end{pmatrix}$, (b) $\begin{pmatrix} 2 & 0 & 1 \\ 3 & 0 & 5 \\ 1 & 2 & 7 \end{pmatrix}$, (c) $\begin{pmatrix} 2 & 1 & 0 & 1 \\ 3 & 5 & 1 & 0 \\ 1 & -3 & -1 & 1 \\ 1 & 0 & 7 & 4 \end{pmatrix}$,

(d) $\begin{pmatrix} 1 & 1 & 0 & 1 \\ 2 & 1 & 1 & 0 \\ 0 & 0 & 1 & 2 \\ 3 & 1 & 2 & 1 \end{pmatrix}$, (e) $\begin{pmatrix} 2 & -1 & 3 & 4 \\ 0 & 2 & -2 & 3 \\ 1 & 1 & -3 & -6 \\ 0 & -1 & 7 & 19 \end{pmatrix}$, (f) $\begin{pmatrix} 6 & 3 & 1 & 2 \\ -2 & 5 & 1 & 3 \\ 4 & 1 & 7 & -5 \\ 0 & 2 & 1 & 0 \end{pmatrix}$

11. Use determinants to determine whether the following sets of vectors are linearly dependent:

(a) (1, 2, 2), (2, 1, 2), (2, 2, 1)

(b) (1, 2, 3, 4), (−6, 7, 1, 0), (0, −3, 2, 1), (8, −2, 4, 5)

(c) (2, −1, 2, 1), (3, −1, 1, 3), (1, −2, 2, 2), (0, 1, 2, 3)

(d) (7, 8, 1, 2, 3), (4, 0, 2, −1, 8), (0, 0, 2, 3, 1), (6, 7, −1, 2, −3), (1, 2, 3, 0, 1)

(e) (2, 0, 5, 1), (3, 1, 7, 2), (0, −2, 1, −1)

(f) (4, 1, 2, 2), (3, 5, 1, 2), (1, 0, 3, 4)

12. Determine the rank of the matrix:

(a) $\begin{pmatrix} 3 & 1 & 4 \\ 2 & 2 & 3 \end{pmatrix}$, (b) $\begin{pmatrix} 10 & 12 & 16 \\ 25 & 30 & 40 \end{pmatrix}$, (c) $\begin{pmatrix} 2 & 1 & 8 & 5 \\ -1 & 0 & -3 & -1 \\ 1 & 2 & 7 & 5 \end{pmatrix}$,

(d) $\begin{pmatrix} 4 & 1 & 7 & 3 \\ 2 & 3 & 1 & 0 \\ 1 & 5 & -3 & 2 \end{pmatrix}$, (e) $\begin{pmatrix} 1 & 0 & 7 & 9 \\ 4 & 3 & 2 & 7 \\ 1 & -1 & 0 & 3 \\ 2 & -1 & 3 & 8 \end{pmatrix}$, (f) $\begin{pmatrix} 1 & 2 & 3 & -5 & -1 \\ 4 & 1 & 5 & -6 & 10 \\ 2 & 0 & 2 & -2 & 6 \end{pmatrix}$

13. Find the inverse matrix if it exists:

(a) $\begin{pmatrix} 3 & 1 \\ 2 & 4 \end{pmatrix}$, (b) $\begin{pmatrix} 5 & 3 \\ 7 & 2 \end{pmatrix}$, (c) $\begin{pmatrix} 1 & 2 & 2 \\ 3 & 1 & 0 \\ -1 & 2 & 3 \end{pmatrix}$, (d) $\begin{pmatrix} 5 & 1 & 3 \\ 2 & 2 & 1 \\ 1 & 0 & 1 \end{pmatrix}$

14. Determine whether solutions exist and, if so, the dimension of the linear variety of solutions:

(a) $\begin{cases} x_1 + x_2 - x_3 = 4 \\ 2x_1 + x_2 + x_3 = 3 \\ x_1 + 2x_2 - x_3 = 9 \end{cases}$ (b) $\begin{cases} 3x_1 - x_2 + x_3 = 3 \\ 2x_1 + 2x_2 + 3x_3 = 1 \\ 11x_1 - 13x_2 - 7x_3 = 2 \end{cases}$

(c) $\begin{cases} x_1 + x_2 - x_3 + 2x_4 = 3 \\ x_1 - x_2 + x_3 - 2x_4 = 5 \end{cases}$ (d) $\begin{cases} 2x_1 - x_2 + x_3 - x_4 = 1 \\ x_1 + x_2 - x_3 + x_4 = 2 \\ 4x_1 - 5x_2 + 5x_3 - 5x_4 = -1 \end{cases}$

(e) $\begin{cases} 2x_1 + x_2 = 3 \\ 3x_1 - x_2 = 5 \\ x_1 + 3x_2 = 1 \end{cases}$ (f) $\begin{cases} x_1 - x_2 + 2x_3 = 3 \\ 3x_1 + x_2 = 2 \\ 2x_1 - 6x_2 - 10x_3 = 13 \\ 9x_1 - x_2 + 6x_3 = 4 \end{cases}$

15. Prove by determinants: if matrix A is nonsingular, then so is A^k for $k = 0, \pm 1, \pm 2, \ldots$

16. Prove: reflection of a line L in the xy-plane in the line $y = -x$ does not change the sign of the slope of L. [*Hint.* Show that for such a reflection the point (a, b) becomes the point $(-b, -a)$.]

17. (a) Show that

$$\det\left\{\begin{pmatrix} a_1 & a_2 & a_3 \\ b_1 & b_2 & b_3 \end{pmatrix}\begin{pmatrix} c_1 & d_1 \\ c_2 & d_2 \\ c_3 & d_3 \end{pmatrix}\right\} = \sum_{1 \le i < j \le 3} (a_ib_j - a_jb_i)(c_id_j - c_jd_i)$$

(b) Show that $\left(\sum_1^3 a_i^2\right)\left(\sum_1^3 b_i^2\right) - \left(\sum_1^3 a_i b_i\right)^2 = \sum_{1 \leq i < j \leq 3} (a_i b_j - a_j b_i)^2.$

[*Hint.* Let the matrices in part (a) be transposes of one another.]

(c) Show that $\left(\sum_1^3 a_i^2\right)\left(\sum_1^3 b_i^2\right) \geq \left(\sum_1^3 a_i b_i\right)^2.$

(d) Generalize the ideas in parts (a), (b), (c).

18. (a) Let $f_k(x) = a_k x^2 + b_k x + c_k$ $(k = 1, 2, 3, 4)$. Let x_1, x_2, x_3, x_4 be real numbers. Show that

$$\begin{vmatrix} f_1(x_1) & f_2(x_1) & f_3(x_1) & f_4(x_1) \\ \vdots & \vdots & \vdots & \vdots \\ f_1(x_4) & f_2(x_4) & f_3(x_4) & f_4(x_4) \end{vmatrix} = 0$$

[*Hint.* The polynomials of degree at most 2 form a vector space of dimension 3. Hence, $f_1(x), \ldots, f_4(x)$ must be linearly dependent.]

(b) Generalize (a) by showing that if $f_1, \ldots, f_{n+2}$ are polynomials of degree at most n and $x_1, \ldots, x_{n+2}$ are real numbers, then $\det A = 0$, where $A = (a_{ij})$ and $a_{ij} = f_j(x_i)$.

19. (*The problem of interpolation*). Let $x_1, \ldots, x_{n+1}$ be distinct real numbers, let $c_1, \ldots, c_{n+1}$ be given real numbers. Show that there is exactly one polynomial $f(x) = a_0 x^n + \cdots + a_{n-1}x + a_n$ such that $f(x_i) = c_i$ for $i = 1, \ldots, n + 1$. [*Hint.* Show that the conditions lead to an equation $A\mathbf{a} = \mathbf{c}$ for $\mathbf{a}$, where A is an $(n + 1)$ by $(n + 1)$ matrix, $\mathbf{a} = \operatorname{col}(a_0, a_1, \ldots, a_n)$, $\mathbf{c} = \operatorname{col}(c_1, \ldots, c_{n+1})$. Show that A has kernel consisting of $\mathbf{0}$ alone by considering the stated problem for the case $c_1 = 0, \ldots, c_{n+1} = 0$. Another way is to note that $\det A$ is a Vandermonde determinant and use the result of Problem 9.]

20. Write out the proof of Theorem 14 for the case $n = 3$, following the method of Section 10-14.

†10-16 THE METHOD OF ELIMINATION

In Section 10-6 we discussed and proved a variety of theorems regarding solutions of a system of m linear equations in n unknowns. There, we saw that the solutions $x_1, \ldots, x_n$ for the system

$$a_{11}x_1 + \cdots + a_{1n}x_n = b_1, \quad \ldots, \quad a_{m1}x_1 + \cdots + a_{mn}x_n = b_m \qquad (10\text{-}160)$$

correspond to the vector solutions $\mathbf{x} = (x_1, \ldots, x_n)$ of the matrix equation

$$A\mathbf{x} = \mathbf{b}, \qquad A = (a_{ij}) \qquad (10\text{-}161)$$

We saw (rule 8 of Section 10-6) that a necessary and sufficient condition that the equations (10-160) have a solution is that

$$\operatorname{rank} A = \operatorname{rank}(A, \mathbf{b})$$

However, no systematic method was given for determining the solutions. We also saw (Theorem 17, Section 10-13) that when A is nonsingular the equations

(10-160) have a unique solution $x_1, \ldots, x_n$ where

$$x_j = \frac{\det B_j}{\det A}, \qquad (j = 1, \ldots, n)$$

and B_j is a matrix that agrees with A except that the jth column vector of B_j is $\mathbf{b}$. From the result just stated one can deduce the following result for the case $n > m$. If

$$A_m = \begin{pmatrix} a_{11} & \cdots & a_{1m} \\ \vdots & & \vdots \\ a_{m1} & \cdots & a_{mm} \end{pmatrix}$$

is nonsingular, then the solutions of (10-160) are $x_1, \ldots, x_n$, where $x_{m+1}, \ldots, x_n$ can be chosen arbitrarily and then

$$x_j \cdot \det A_m = \det B_j - x_{m+1} \det C_j^{(m+1)} - \cdots - x_n \det C_j^{(n)}, \qquad j = 1, \ldots, m$$

and here B_j differs from A_m in having $\mathbf{b}$ as its jth column, and the $C_j^{(l)}$ differ from A_m in having the lth column of A as the jth column. While these formulae are extremely beautiful and are quite useful in proofs, they are not very practical when m, n are large, since they entail the time-consuming evaluation of many determinants.

A more practical method for solving linear equations is *Gauss's method of elimination*. In this method one replaces the given system by a second system, where the two systems have the same solutions and the second system is of the form

$$A^*\mathbf{x} = \mathbf{b}^* \tag{10-162}$$

where A^* is a partitioned matrix of the form

$$A^* = \begin{pmatrix} A_{11} & A_{12} & \cdots & A_{1l} & A_{1,l+1} \\ O & A_{22} & \cdots & A_{2l} & A_{2,l+1} \\ \vdots & \vdots & & \vdots & \vdots \\ O & \cdots & & A_{ll} & A_{l,l+1} \\ O & \cdots & & O & O \end{pmatrix} \tag{10-163}$$

and each A_{jj} is of the form (I, G_{jj}). An example of such a matrix A^* is the following:

$$\left(\begin{array}{cc|cc|ccc} 1 & 2 & 3 & 0 & 7 & 6 & -1 \\ \hline 0 & 0 & 1 & 0 & 2 & 4 & 1 \\ 0 & 0 & 0 & 1 & 1 & 0 & 2 \\ \hline 0 & 0 & 0 & 0 & 1 & 0 & 0 \\ 0 & 0 & 0 & 0 & 0 & 1 & 2 \\ \hline 0 & 0 & 0 & 0 & 0 & 0 & 0 \end{array}\right)$$

The equation (10-162) has no solution if, for some j, the jth row of A^* is $\mathbf{0}$ and the jth entry of $\mathbf{b}^*$ is not 0. However, if the jth coordinate of $\mathbf{b}^*$ is 0 whenever the jth row of A^* is $\mathbf{0}$, then we can easily read off the solutions of equation (10-162) (see Example 1 below). The rank of A equals the number of nonzero rows in A^*.

Finding the solutions of (10-160) is equivalent to finding the set of solutions of the equations $L_1 = 0, \ldots, L_m = 0$, where

$$L_1 = \sum_{j=1}^{n} a_{1j}x_j - b_1, \qquad \ldots, \qquad L_m = \sum_{j=1}^{n} a_{mj}x_j - b_m$$

Here, $L_1, \ldots, L_m$ are polynomials in $x_1, \ldots, x_n$ of at most first degree. We say such a system of polynomials is equivalent to a second such system:

$$L_1^* = \sum_{j=1}^{n} a_{1j}^*x_j - b_j^*, \ldots, \qquad L_m^* = \sum_{j=1}^{n} a_{mj}^*x_j - b_m^*$$

if we can go from the first to the second system by repeated application of the following procedures: (a) interchanging the order of the polynomials, (b) replacing a polynomial by a nonzero multiple of itself, (c) replacing a polynomial L_i by $L_i + aL_j$, where a is a number and $i \neq j$. It can be easily shown that if $L_1, \ldots, L_m$ is equivalent to $L_1^*, \ldots, L_m^*$, then $L_1^*, \ldots, L_m^*$ is equivalent to $L_1, \ldots, L_m$. Also if $L_1, \ldots, L_m$ is equivalent to $L_1^*, \ldots, L_m^*$ and $L_1^*, \ldots, L_m^*$ is equivalent to $L_1', \ldots, L_m'$, then $L_1, \ldots, L_m$ is equivalent to $L_1', \ldots, L_m'$. Clearly, if one system is obtained from another by one of the operations (a), (b), or (c), the two systems have the same set of solutions. Hence, equivalent systems have the same solutions. Gauss's elimination method is the process of successive applications of the operations (a), (b), (c) to the given system to obtain an equivalent system with coefficient matrix of the form (10-163). These operations are very simple and quickly performed; furthermore, it is easy to program a machine to perform them.

We now illustrate the elimination procedure. We can view the x_i as place markers and, therefore, they can be omitted. Thus we have the array

$$\begin{array}{cccc|c} a_{11} & a_{12} & \cdots & a_{1n} & b_1 \\ \vdots & \vdots & & \vdots & \vdots \\ a_{m1} & a_{m2} & \cdots & a_{mn} & b_m \end{array}$$

The operations (a), (b), (c) listed above correspond to the following operations on the array: (a) interchanging rows, (b) multiplying a row by a non-zero scalar, (c) adding a multiple of one row to another. We may suppose all variables are present and hence no column vector of A is $\mathbf{0}$. Consequently some $a_{j1} \neq 0$. Then we can interchange rows, and suppose that $a_{11} \neq 0$. We perform operation (b) by dividing the first row of the array by a_{11} to obtain

$$\begin{array}{cccc|c} 1 & a_{12}' & \cdots & a_{1n}' & b_1' \\ a_{21} & a_{22} & \cdots & a_{2n} & b_2 \\ \vdots & \vdots & & \vdots & \vdots \\ a_{m1} & a_{m2} & \cdots & a_{mn} & b_m \end{array}$$

Next we subtract appropriate multiples of the first row from each of the remaining rows to get an array where col $(1, 0, \ldots, 0) = \mathbf{e}_1$ is the first column. Thus we have

$$\begin{array}{cccc|c} 1 & a'_{12} & \cdots & a'_{1n} & b'_1 \\ 0 & a'_{22} & \cdots & a'_{2n} & b'_2 \\ \vdots & \vdots & & \vdots & \vdots \\ 0 & a'_{m2} & \cdots & a'_{mn} & b'_m \end{array}$$

where $a'_{ij} = a_{ij} - a'_{1j}a_{i1}$, $b'_i = b_i - b'_1 a_{i1}$ for $i \geq 2$. If the second column of this new array is not a scalar multiple of the first, some $a'_{j2} \neq 0$. We then interchange rows so that we can assume $a'_{22} \neq 0$. Next we divide the second row by a'_{22} and subtract appropriate multiples of the second row from the other rows to obtain an array whose second column is $\mathbf{e}_2$. If the second column is a scalar multiple of the first, then we do nothing with regard to the second column. Thus we have passed to a new equivalent system whose first column vector is $\mathbf{e}_1$ and the second column vector is either $\mathbf{e}_2$, or it is a scalar multiple of $\mathbf{e}_1$. We next examine the third column of this new array. If it is a linear combination of the first two columns, we do nothing, but if it is not, then we can pass to an equivalent system with the same first and second columns and with the third column either $\mathbf{e}_3$ (if $\mathbf{e}_1$, $\mathbf{e}_2$ are the first two columns) or $\mathbf{e}_2$ (if the first two columns are linearly dependent). One continues the process through all n columns of the coefficient matrix. The end result is an array of the form (10-163).

EXAMPLE 1. Find all solutions of the system of equations

$$\begin{aligned} x_1 + 3x_2 + 2x_3 + 3x_4 - 7x_5 &= 14 \\ 2x_1 + 6x_2 + x_3 - 2x_4 + 5x_5 &= -2 \\ x_1 + 3x_2 - x_3 \qquad\quad + 2x_5 &= -1 \end{aligned}$$

Solution. We form the array associated with these equations and carry out the elimination process, putting each new array under its predecessor. The work is shown in Table 10-1. Then our solutions are given by the equations

$$\begin{aligned} x_1 &= 1 - 3x_2 - x_5 \\ x_2 &= \qquad x_2 \\ x_3 &= 2 \qquad\quad + x_5 \\ x_4 &= 3 \qquad\quad + 2x_5 \\ x_5 &= \qquad\qquad x_5 \end{aligned}$$

They clearly form a 2-dimensional linear variety, with x_2, x_5 as parameters.

If we need to solve several systems of linear equations all with the same coefficient matrix, say

$$\mathbf{A}\mathbf{x} = \mathbf{b}, \qquad \mathbf{A}\mathbf{x} = \mathbf{c}, \qquad \ldots, \qquad \mathbf{A}\mathbf{x} = \mathbf{d}$$

we can apply the elimination process to all the problems at once. We consider the array

$$\begin{array}{ccc|cccc} a_{11} & \cdots & a_{1n} & b_1 & c_1 & \cdots & d_1 \\ \vdots & & \vdots & \vdots & & & \vdots \\ a_{m1} & \cdots & a_{mn} & b_m & c_m & \cdots & d_m \end{array}$$

Table 10-1

1	3	2	3	−7	14
2	6	1	−2	5	−2
1	3	−1	0	2	−1
1	3	2	3	−7	14
0	0	−3	−8	19	−30
0	0	−3	−3	9	−15
1	3	2	3	−7	14
0	0	1	1	−3	5
0	0	−3	−8	19	−30
1	3	0	1	−1	4
0	0	1	1	−3	5
0	0	0	−5	10	−15
1	3	0	0	1	1
0	0	1	0	−1	2
0	0	0	1	−2	3

We then eliminate variables on the left as before to get the array

$$A^* \mid \mathbf{b}^* \qquad \mathbf{c}^* \qquad \cdots \qquad \mathbf{d}^*$$

where A^* is of the form (10-163). We can then solve the original equations by solving

$$A^*\mathbf{x} = \mathbf{b}^*, \qquad \ldots, \qquad A^*\mathbf{x} = \mathbf{d}^*$$

each in turn.

Computation of the Inverse of a Nonsingular Matrix. In Section 10-13 we saw that for a nonsingular matrix A,

$$A^{-1} = (\det A)^{-1} \operatorname{adj} A \tag{10-164}$$

where $\operatorname{adj} A = (A_{ij})'$ and A_{ij} is $(-1)^{i+j}$ times the minor Δ_{ij} of A. Clearly, (10-164) suggests a procedure for determining A^{-1}, but it is not a practical one when the size of A is much beyond 4 or 5. To compute the inverse of a 10 by 10 matrix using only (10-164) would require more than 350 million operations of addition, multiplication, and division—a rather large number even for a high-speed computing machine. Thus, for very practical reasons, we seek methods for determining the inverse of a matrix that is both quick and simple.

One such method which is quite effective for matrices of moderate size (say $n \leq 10$) is based on the method of elimination. Recall that if A is nonsingular, then the jth column of A^{-1} is the unique solution of the equation

$A\mathbf{x} = \mathbf{e}_j$. Hence, we can find A^{-1} by starting with the array

$$(A \mid I)$$

and using the method of elimination on A. Since the columns of A constitute a linearly independent set, the elimination process ends with an equivalent system having I on the left, say, $(I \mid X)$. If γ_j is the jth column of X, then $I\mathbf{x} = \gamma_j$ has γ_j as solution, and hence so does the equivalent system $A\mathbf{x} = \mathbf{e}_j$. Hence, $AX = I$, so that $X = A^{-1}$. Thus if we apply the elimination process to $(A \mid I)$, we obtain

$$(I \mid A^{-1})$$

This procedure is quick and easy and is quite suitable for computer calculation for reasonable-sized matrices.

EXAMPLE 1. Determine

$$\begin{pmatrix} 1 & 1 & 0 \\ 3 & 0 & 2 \\ 3 & 1 & 1 \end{pmatrix}^{-1}$$

Solution. We apply the method of the preceding paragraph in Table 10-2.

Table 10-2

1	1	0	1	0	0
3	0	2	0	1	0
3	1	1	0	0	1
1	1	0	1	0	0
0	−3	2	−3	1	0
0	−2	1	−3	0	1
1	0	$\frac{2}{3}$	0	$\frac{1}{3}$	0
0	1	$-\frac{2}{3}$	1	$-\frac{1}{3}$	0
0	0	$-\frac{1}{3}$	−1	$-\frac{2}{3}$	1
1	0	0	−2	−1	2
0	1	0	3	1	−2
0	0	1	3	2	−3

Thus

$$\begin{pmatrix} 1 & 1 & 0 \\ 3 & 0 & 2 \\ 3 & 1 & 1 \end{pmatrix}^{-1} = \begin{pmatrix} -2 & -1 & 2 \\ 3 & 1 & -2 \\ 3 & 2 & -3 \end{pmatrix}$$

Remark. In determining A^{-1} we have been solving the matrix equation $AX = I$. We can also use the elimination method to solve $AX = B$, where

A is an n by n nonsingular matrix and B is an n by m matrix. For if we start with the array

$$(A \mid B)$$

and apply the elimination process to A, we obtain the array

$$(I \mid A^{-1}B)$$

To solve $XA = B$, we can first solve $A'X' = B'$ for X', and then $X = (X')'$.

PROBLEMS

For each of the following systems of equations, find the complete set of solutions, if any exist.

1. $\begin{cases} x_1 - 3x_2 + x_3 - x_4 = 7 \\ 2x_1 + x_2 + x_4 = 0 \\ 3x_2 - x_3 + 5x_4 = -6 \end{cases}$

2. $\begin{cases} x_1 + 3x_2 - 4x_3 + 12x_4 + 2x_5 = 5 \\ 2x_1 + x_2 - x_3 + x_4 - x_5 = 9 \\ -x_1 - x_2 + x_3 + 6x_4 - 6x_5 = 19 \end{cases}$

3. $\begin{cases} x_1 + x_2 - 3x_3 - x_4 = 4 \\ 2x_1 - x_2 + x_3 - 4x_4 = 4 \\ x_1 + 2x_2 - 2x_3 + 2x_4 = 3 \\ 7x_1 + 6x_2 - 12x_3 - 5x_4 = 19 \end{cases}$

4. $\begin{cases} x_1 - 2x_2 + x_3 = 1 \\ 2x_1 + 3x_2 - x_3 = 6 \\ -x_1 - x_2 + 4x_3 = 1 \\ 2x_1 - x_2 + 4x_3 = 7 \end{cases}$

5. $\begin{cases} 2x_1 - x_2 + 3x_3 + 5x_4 = 5 \\ x_1 + x_2 - x_3 - 2x_4 = -4 \\ x_2 + 7x_3 + 3x_4 = 2 \\ 2x_1 + 4x_2 + 2x_3 + x_4 = 3 \end{cases}$

6. $\begin{cases} x_1 + 2x_2 + 3x_4 + x_5 = 6 \\ 3x_1 - 2x_2 + x_4 - x_5 = 0 \\ x_1 + 3x_2 - 3x_4 + 2x_5 = 1 \\ 4x_1 - x_2 + 16x_4 - x_5 = 13 \end{cases}$

‡7. Prove:

(a) If $L_1, \ldots, L_m$ is equivalent to $L_1^*, \ldots, L_m^*$, then $L_1^*, \ldots, L_m^*$ is equivalent to $L_1, \ldots, L_m$. [*Hint.* Reason that it is sufficient to prove this for the case in which only one of the steps (a), (b), or (c) is used in going from the first system to the second.]

(b) If $L_1, \ldots, L_m$ is equivalent to $L_1^*, \ldots, L_m^*$ and $L_1^*, \ldots, L_m^*$ is equivalent to $L_1', \ldots, L_m'$, then $L_1, \ldots, L_m$ is equivalent to $L_1', \ldots, L_m'$.

(c) If $L_1, \ldots, L_m$ is equivalent to $L_1^*, \ldots, L_m^*$, then the solution set for $L_1 = \cdots = L_m = 0$ is the same as the solution set for $L_1^* = \cdots = L_m^* = 0$.

8. Determine the solutions for each of the three systems.

(a) $\begin{cases} 3x + 4y = 7 \\ 3x + 3.9999y = 6.9988 \end{cases}$

(b) $\begin{cases} 3x + 4y = 7 \\ 3x + 4.0001y = 7.0001 \end{cases}$

(c) $\begin{cases} 3x + 4y = 7 \\ 3x + 3.9999y = 7.0004 \end{cases}$

These are examples of equations where the entries of the coefficient matrix A and the constant vector $\mathbf{b}$ differ very little yet the solutions differ a great deal. This difficulty often occurs when det A is small compared with the largest $|a_{ij}|$.

9. (a), (b), (c) Find the inverses of each of the matrices of parts (a), (b), and (c) of Problem 7 following Section 10-12.

10. Solve the matrix equations:

(a) $\begin{pmatrix} 1 & 1 & -1 \\ 2 & 1 & 0 \\ 1 & -1 & 1 \end{pmatrix} X = \begin{pmatrix} 1 & -1 & 3 \\ 4 & 3 & 2 \\ 1 & -2 & 5 \end{pmatrix}$ (b) $X \begin{pmatrix} 1 & 1 & -1 \\ 2 & 1 & 0 \\ 1 & -1 & 1 \end{pmatrix} = \begin{pmatrix} 1 & -1 & 3 \\ 4 & 3 & 2 \\ 1 & -2 & 5 \end{pmatrix}$

11. For each of the following matrices use the method of elimination to determine its rank: (a) The matrix of Problem 10(d) following Section 10-15. (b) The matrix of Problem 12(d) following Section 10-15.

†10-17 MATRICES OF FUNCTIONS

Up till now we have considered matrices whose entries were real or complex numbers. We can also consider matrices that have real functions as entries:

$$A(t) = \begin{pmatrix} a_{11}(t) & \cdots & a_{1n}(t) \\ \vdots & & \vdots \\ a_{m1}(t) & \cdots & a_{mn}(t) \end{pmatrix} \tag{10-170}$$

In Section 9-19 we remarked that the set of real valued functions on an interval form an algebra; more specifically, they form an algebra for which the multiplication is commutative. Hence, we can form sums and products of such functions, and as long as we do not divide, they behave as do the numbers. Since adding two matrices, multiplying two matrices, and multiplying a matrix by a scalar are all carried out by forming sums and products of the entries of the matrices, it follows that these same procedures for combining matrices can be carried out for matrices with real valued functions as entries. Furthermore, such rules as $A + B = B + A$, $(AB)C = A(BC)$, $A(B + C) = (AB + AC)$, $(A + B)C = AC + BC$, $IA = A$ must continue to hold true for matrices of functions.

A matrix such as (10-170), where the $a_{ij}(t)$ are defined on an interval J, can be thought of as a function from J into $\mathfrak{M}_{mn}$; it is the function that assigns to any t_0 in J the matrix $A(t_0)$, where the entries in $A(t_0)$ are the values of the functions a_{ij} at t_0. Conversely, given a function F from J to $\mathfrak{M}_{mn}$, the values in the ij entry of F as t ranges over J determine a real function on J which we denote by $f_{ij}(t)$. Hence, $F(t) = (f_{ij}(t))$. Thus functions from J into $\mathfrak{M}_{mn}$ determine and are determined by matrices of functions. (When $m = 1$ and $n = 2$, our statement reduces to the familiar fact that a vector function $\mathbf{F}(t)$, whose values are vectors in the plane, is equivalent to a pair of scalar functions: $\mathbf{F}(t) = f(t)\mathbf{i} + g(t)\mathbf{j}$.)

Now suppose that the entries in the matrix (10-170) are continuous functions on an interval J. Then a small change in t causes only a small change Δ_{ij} in the entries a_{ij} and

$$A(t + \Delta t) = (a_{ij}(t + \Delta t)) = (a_{ij}(t) + \Delta_{ij}) = (a_{ij}(t)) + (\Delta_{ij}) = A(t) + (\Delta_{ij})$$

Thus $A(t + \Delta t) - A(t)$ is a matrix with very small entries whose values tend to 0 as Δt tends to 0. In analogy with real functions and vector functions, we say that $A(t)$ is *continuous in t*. In Chapter 12 we shall study the calculus of such matrix valued functions. We have seen that a 2 by 2 matrix maps V_2, that is, the x_1x_2-plane, into itself. If we have a continuous 2 by 2 matrix valued function, we then have a continuously varying deformation of the plane. This can best be seen by watching the image of the square S: $|x_1| \leq 1$, $|x_2| \leq 1$. A 2 by 2 matrix takes the square S into a parallelogram. Thus a continuous matrix function continuously varies the parallelogram (which is the image of the square S). The area of the parallelogram at time t is the absolute value of the determinant of $A(t)$. The parallelogram collapses into a line segment or a point when the matrix $A(t)$ is singular.

Remark. We do not need to restrict ourselves to matrices with numbers or real valued functions as entries. If the entries come from an algebra with a commutative multiplication, all the algebraic theory goes through. Thus we can speak of matrices with complex functions as entries, of matrices with polynomials as entries, of matrices with entries that are polynomials in the derivative D.

The definition of a determinant also only involves forming sums and products and, hence, we can speak of the determinant of square matrices having functions as entries. If $a_{ij}(t)$ are functions defined on an interval J, then

$$\det(a_{ij}(t)) = q(t) \tag{10-171}$$

is a real valued function defined on J. If all the $a_{ij}(t)$ are continuous on J, then $q(t)$ is continuous on J, since sums and products of continuous functions are continuous. We can thus apply the theory of the calculus to the real function $q(t)$. For example, if J is a closed interval, then $q(t)$ has a maximum value on J. If the $a_{ij}(t)$ are differentiable, we can also differentiate $q(t)$. We have the theorem:

THEOREM 20. *If the functions $a_{ij}(t)$ are differentiable at t, then at this value*

$$\frac{d}{dt}\begin{vmatrix} a_{11}(t) & \cdots & a_{1n}(t) \\ \vdots & & \vdots \\ a_{n1}(t) & \cdots & a_{nn}(t) \end{vmatrix} = \begin{vmatrix} a'_{11}(t) & a_{12}(t) & \cdots & a_{1n}(t) \\ \vdots & & & \vdots \\ a'_{n1}(t) & a_{n2}(t) & \cdots & a_{nn}(t) \end{vmatrix}$$

$$+ \begin{vmatrix} a_{11}(t) & a'_{12}(t) & \cdots & a_{1n}(t) \\ \vdots & & & \vdots \\ a_{n1}(t) & a'_{n2}(t) & \cdots & a_{nn}(t) \end{vmatrix} + \cdots + \begin{vmatrix} a_{11}(t) & \cdots & a'_{1n}(t) \\ \vdots & & \vdots \\ a_{n1}(t) & \cdots & a'_{nn}(t) \end{vmatrix}$$

PROOF. Since $q(t) = \Sigma \pm a_{1i_1}a_{2i_2} \cdots a_{ni_n}$, where the sum is over all permutations $i_1, \ldots, i_n$ of $1, 2, \ldots, n$, we have, by repeated application of the rule for differentiating products,

$$q'(t) = \sum \pm \{a'_{1i_1}a_{2i_2} \cdots a_{ni_n} + a_{1i_1}a'_{2i_2} \cdots a_{ni_n} + \cdots + a_{1i_1}a_{2i_2} \cdots a'_{ni_n}\}$$

If for each s we collect all terms containing a factor of the form a'_{si_s}, we obtain

$$\begin{vmatrix} a_{11} & a_{12} & \cdots & a'_{1s} & \cdots & a_{1n} \\ \vdots & & & \vdots & & \vdots \\ a_{n1} & a_{n2} & \cdots & a'_{ns} & \cdots & a_{nn} \end{vmatrix}$$

and, hence, we obtain the theorem.

PROBLEMS

1. For each of the matrix functions listed below describe the varying image of the square S: $|x_1| \le 1$, $|x_2| \le 1$ as t ranges from 0 to infinity.

(a) $A(t) = \begin{pmatrix} 1 & t \\ 0 & 1 \end{pmatrix}$, (b) $B(t) = \begin{pmatrix} t & 0 \\ 0 & t \end{pmatrix}$, (c) $C(t) = \begin{pmatrix} t & 0 \\ 0 & 1 \end{pmatrix}$

(d) $D(t) = \begin{pmatrix} \sin t & \cos t \\ -\cos t & \sin t \end{pmatrix}$, (e) $E(t) = \begin{pmatrix} t\sin t & t\cos t \\ -t\cos t & t\sin t \end{pmatrix}$

2. (a) . . . (e) Find the maximum of the determinant of each matrix in Problem 1 for t on the interval $-1 \le t \le 1$.

3. Prove: a square matrix with polynomial entries which is nonsingular at one value of the variable is nonsingular at all but finitely many values of the variable.

4. For each of the following matrices determine the values of t for which the matrix is singular.

(a) $\begin{pmatrix} 1 & t \\ 0 & 1 \end{pmatrix}$, (b) $\begin{pmatrix} 2+t & 1 \\ 2 & 1-t \end{pmatrix}$, (c) $\begin{pmatrix} 3-t & t^2 \\ 4 & 2+t \end{pmatrix}$, (d) $\begin{pmatrix} \sin t & \cos t \\ -\cos t & \sin t \end{pmatrix}$

(e) $\begin{pmatrix} \sin^2 t & 1 \\ 1 & 4\cos^2 t \end{pmatrix}$, (f) $\begin{pmatrix} e^t & 3e^{2t} \\ 2e^t & 4e^{2t} \end{pmatrix}$, (g) $\begin{pmatrix} te^t & e^{-t} \\ 2e^{2t} & 2t \end{pmatrix}$

†10-18 EIGENVALUES, EIGENVECTORS, CHARACTERISTIC POLYNOMIAL OF A MATRIX

Since an n by n matrix A is a linear transformation on V_n, we can speak of its eigenvalues and its eigenvectors (see Section 9-23). If λ is a scalar and $\mathbf{u}$ is a nonzero vector in V_n such that

$$A\mathbf{u} = \lambda\mathbf{u} \tag{10-180}$$

then λ is an *eigenvalue* of A and $\mathbf{u}$ is an *eigenvector* of A. Since $(\lambda I)\mathbf{u} = \lambda\mathbf{u}$, for all vectors $\mathbf{u}$ in V_n, we see that (10-180) is equivalent to the equation:

$$(A - \lambda I)\mathbf{u} = \mathbf{0} \tag{10-181}$$

Thus the eigenvectors of A associated with the eigenvalue λ are the nonzero vectors in the kernel of the matrix $A - \lambda I$. We denote this kernel by K_λ. We also note that a scalar λ is an eigenvalue of A if, and only if, $A - \lambda I$ is a singular matrix. The matrix $A - \lambda I$ is singular exactly when $\det(A - \lambda I) = 0$ (see Theorem 15 in Section 10-13). On expressing this determinant in terms of its entries, we see that it is a polynomial of degree n in λ:

$$\det(A - \lambda I) = (-1)^n\lambda^n + \cdots + \det A$$

We denote this polynomial by $\chi_A(\lambda)$ and call it the *characteristic polynomial of* A. Thus *the eigenvalues of an n by n matrix A are the zeros of the characteristic polynomial of* A and, hence, there can be no more than n distinct eigenvalues for the matrix A.

EXAMPLE 1. Determine the eigenvalues and the eigenvectors of the matrix

$$A = \begin{pmatrix} 1 & 2 & -2 \\ 1 & 0 & 2 \\ 1 & 1 & 1 \end{pmatrix}$$

Solution. The characteristic polynomial of A is

$$\chi_A(x) = \det(A - xI) = -x^3 + 2x^2 + x - 2 = -(x-1)(x+1)(x-2)$$

Hence, the eigenvalues of A are 1, -1, and 2. To find the eigenvectors associated with these eigenvalues, we must find the kernels of the matrices: $A - I$, $A + I$ and $A - 2I$. Now

$$A - I = \begin{pmatrix} 0 & 2 & -2 \\ 1 & -1 & 2 \\ 1 & 1 & 0 \end{pmatrix}, \quad A + I = \begin{pmatrix} 2 & 2 & -2 \\ 1 & 1 & 2 \\ 1 & 1 & 2 \end{pmatrix},$$

$$A - 2I = \begin{pmatrix} -1 & 2 & -2 \\ 1 & -2 & 2 \\ 1 & 1 & -1 \end{pmatrix}$$

Clearly, $A - I$ is of rank 2 and $(-1, 1, 1)$ is in its Kernel. Hence, $K_1 = \text{Span}(-1, 1, 1)$. The matrices $A + I$ and $A - 2I$ are also of rank 2 and we observe that $K_{-1} = \text{Span}(1, -1, 0)$, $K_2 = \text{Span}(0, 1, 1)$. The eigenvectors associated with each eigenvalue λ are the nonzero vectors in K_λ.

EXAMPLE 2. A real matrix need not have any real eigenvalues. For example $A = \begin{pmatrix} -1 & -3 \\ 1 & -1 \end{pmatrix}$ has no real eigenvalues, since its characteristic polynomial is $x^2 + 2x + 4 = (x + 1)^2 + 3$, which clearly has no real zeros. On the other hand, a complex matrix always has complex eigenvalues, since by the Fundamental Theorem of Algebra (see Section 0-18) every polynomial of positive degree with complex coefficients has a complex zero. Now we can always view a real matrix as a complex matrix and can find its complex eigenvalues. In that case the eigenvectors corresponding to the complex eigenvalues will generally be complex vectors. For the matrix A we would have $-1 \pm \sqrt{3}\,i$ as its complex eigenvalues and $K_{-1+\sqrt{3}i} = \text{Span}(\sqrt{3}\,i, 1)$ and $K_{-1-\sqrt{3}i} =$

Span $(\sqrt{3}i, -1)$. It is important, when studying matrices, to consider their complex eigenvalues.

THEOREM 21. *Let A be an n by n matrix with real or complex entries. Then A has at most n distinct eigenvalues. If* $\lambda_1, \ldots, \lambda_k$ *are the distinct eigenvalues of A with corresponding eigenvectors* $\mathbf{u}_1, \ldots, \mathbf{u}_k$, *then the set* $\{\mathbf{u}_1, \ldots, \mathbf{u}_k\}$ *is linearly independent.*

COROLLARY 1. *If A is an n by n matrix with n distinct eigenvalues* $\lambda_1, \ldots, \lambda_n$ *with corresponding eigenvectors* $\mathbf{u}_1, \ldots, \mathbf{u}_n$, *then* $\{\mathbf{u}_1, \ldots, \mathbf{u}_n\}$ *is a basis for* V_n.

This is a restatement of Theorem 31 and its corollary, as in Section 9-23.

COROLLARY 2. *If* $\lambda_1, \ldots, \lambda_k$ *are distinct eigenvalues of a matrix A, then* $K_{\lambda_1} \cap \{K_{\lambda_2} + K_{\lambda_3} + \cdots + K_{\lambda_k}\} = \mathbf{0}$.

PROOF. Suppose that Corollary 2 is false. Then there exist a nonzero vector $\mathbf{u}_1$ in K_{λ_1} and vectors $\mathbf{u}_2$ in $K_{\lambda_2}, \ldots, \mathbf{u}_k$ in K_{λ_k} such that $\mathbf{u}_1 = \mathbf{u}_2 + \cdots + \mathbf{u}_k$. If we omit those $\mathbf{u}_i$ that are zero, we obtain a set of eigenvectors corresponding to distinct eigenvalues which, by Theorem 21, must be a linearly independent set; thus it is impossible to have a relation $\mathbf{u}_1 = \mathbf{u}_2 + \cdots + \mathbf{u}_k$. Therefore, we have a contradiction and Corollary 2 is proved.

COROLLARY 3. *If* $\lambda_1, \ldots, \lambda_k$ *are all the distinct eigenvalues of an n by n matrix A, then*

$$n \geq \dim \{K_{\lambda_1} + \cdots + K_{\lambda_k}\} = \dim K_{\lambda_1} + \cdots + \dim K_{\lambda_k}$$

PROOF. Since $\{K_{\lambda_1} + \cdots + K_{\lambda_k}\}$ is a subspace of V_n, we must have $n \geq \dim \{K_{\lambda_1} + \cdots + K_{\lambda_k}\}$. We prove the second part of the Corollary by induction. We assert: if $\lambda_1, \ldots, \lambda_h$ are distinct eigenvalues of A, then

$$\dim \{K_{\lambda_1} + \cdots + K_{\lambda_h}\} = \dim K_{\lambda_1} + \cdots + \dim K_{\lambda_h}$$

This result is trivially true for $h = 1$. Now let us suppose that it is true for h and let us prove it to be true for $h + 1$. By Theorem 14 in Section 9-9 we have

$$\begin{aligned}&\dim \{K_{\lambda_1} + \cdots + K_{\lambda_h} + K_{\lambda_{h+1}}\}\\ &= \dim \{K_{\lambda_1} + \cdots + K_{\lambda_h}\} + \dim K_{\lambda_{h+1}} - \dim K_{\lambda_{h+1}} \cap \{K_{\lambda_1} + \cdots + K_{\lambda_h}\}\end{aligned}$$

But Corollary 2 implies that $\dim K_{\lambda_{h+1}} \cap \{K_{\lambda_1} + \cdots + K_{\lambda_h}\} = 0$ and, hence, we obtain the result.

Remark. The matrix $A = \begin{pmatrix} 0 & 1 & 2 \\ 0 & 0 & 3 \\ 0 & 0 & 0 \end{pmatrix}$ has only the one eigenvalue 0, and all its eigenvectors are scalar multiples of $\mathbf{e}_1$. Thus we see that $K_0 = \text{Span}(\mathbf{e}_1)$ and $\dim K_0 < 3$. Similarly, the matrix $B = \begin{pmatrix} 0 & 1 & 2 \\ 0 & 0 & 3 \\ 0 & 0 & 1 \end{pmatrix}$ has 0 and 1 as eigenvalues and $K_0 = \text{Span}(\mathbf{e}_1)$, $K_1 = \text{Span}[(5, 3, 1)]$. Thus $\dim\{K_0 + K_1\} = 2 < 3$.

We now make several observations regarding eigenvalues. Proofs of most of these statements are left as exercises for the reader.

(a) A matrix and its transpose have the same characteristic equation, and so have the same eigenvalues.

(b) If A is a direct sum of B and C, then $\chi_A(x) = \chi_B(x)\chi_C(x)$ and, hence, the set of distinct eigenvalues of A is the union of the set of eigenvalues for B and the set of eigenvalues for C.

(c) If λ is an eigenvalue for A, then λ^2 is an eigenvalue for A^2. For if $A\mathbf{u} = \lambda\mathbf{u}$, then $A^2\mathbf{u} = A(A\mathbf{u}) = A(\lambda\mathbf{u}) = \lambda(A\mathbf{u}) = \lambda(\lambda\mathbf{u}) = \lambda^2\mathbf{u}$.

More generally, we have

(d) If $p(x)$ is a polynomial and λ is an eigenvalue for A, then $p(\lambda)$ is an eigenvalue for $p(A)$.

(e) A matrix is nonsingular if, and only if, its eigenvalues are all nonzero. If A is nonsingular, then the eigenvalues of A^{-1} are the reciprocals of the eigenvalues for A.

(f) If $A = M^{-1}BM$, then A and B have the same eigenvalues.

Eigenspaces. Let an n by n matrix A be given. A subspace W of V_n is said to be an *invariant set for* A or to be *invariant under* A if $A\mathbf{x}$ is in W for each vector $\mathbf{x}$ in W. Thus V_n and V_0 are invariant spaces for A.

If λ is an eigenvalue of A, then the space K_λ is invariant under A. For if $\mathbf{x}$ is in K_λ, then $(A - \lambda I)\mathbf{x} = \mathbf{0}$, or $A\mathbf{x} = \lambda\mathbf{x}$. But $\lambda\mathbf{x}$ is in K_λ, since K_λ is a subspace and so closed under scalar multiplication.

We now generalize the idea of the subspace $K_\lambda = \text{Kernel}(A - \lambda I)$. Let W_λ be the set of all vectors $\mathbf{u}$ in V_n such that $(A - \lambda I)^r\mathbf{u} = \mathbf{0}$, for some nonnegative integer r. Clearly, W_λ contains K_λ, and it is easily seen that W_λ is closed under multiplication by scalars. If $(A - \lambda I)^r\mathbf{u} = \mathbf{0}$, then $(A - \lambda I)^s\mathbf{u} = \mathbf{0}$ for all $s \geq r$. Thus, if $\mathbf{u}_1, \mathbf{u}_2$ are in W_λ, there exist r and t so that $(A - \lambda I)^r\mathbf{u}_1 = \mathbf{0}$, $(A - \lambda I)^t\mathbf{u}_2 = \mathbf{0}$ and, hence, with $s = \max(r, t)$, we have $(A - \lambda I)^s\mathbf{u}_1 = \mathbf{0}$, $(A - \lambda I)^s\mathbf{u}_2 = \mathbf{0}$ and, consequently, $(A - \lambda I)^s(\mathbf{u}_1 + \mathbf{u}_2) = \mathbf{0}$. Thus W_λ is also closed under addition and, therefore, W_λ *is a subspace of* V_n. Now $(A - \lambda I)^r$ is a polynomial in A and, hence, it commutes with A. If $\mathbf{x}$ is in W_λ, then $(A - \lambda I)^r\mathbf{x} = \mathbf{0}$ and hence, $(A - \lambda I)^r(A\mathbf{x}) = A(A - \lambda I)^r\mathbf{x} = A\mathbf{0} = \mathbf{0}$. It follows that W_λ *is invariant under* A. The spaces W_λ are called *eigenspaces.*

Just as for the subspaces K_λ we can prove: if $\lambda_1, \ldots, \lambda_k$ are distinct eigenvalues for a matrix A, and if $\mathbf{u}_1, \ldots, \mathbf{u}_k$ are nonzero vectors from $W_{\lambda_1}, \ldots, W_{\lambda_k}$, respectively, then $\mathbf{u}_1, \ldots, \mathbf{u}_k$ is a linearly independent set. Hence, as in Corollary 3 of Theorem 21, we have

$$n \geq \dim\{W_{\lambda_1} + \cdots + W_{\lambda_k}\} = \dim W_{\lambda_1} + \cdots + \dim W_{\lambda_k}$$

Indeed, we can show that if $\lambda_1, \ldots, \lambda_k$ are all the distinct real eigenvalues of A, and these are all the eigenvalues of A, then

$$V_n = \{W_{\lambda_1} + \cdots + W_{\lambda_k}\}$$

If we let $m_i = \dim W_{\lambda_i}$, then m_i is exactly the multiplicity of the zero λ_i for the polynomial $\chi_A(x)$:

$$\chi_A(x) = (\lambda_1 - x)^{m_1}(\lambda_2 - x)^{m_2} \cdots (\lambda_k - x)^{m_k}$$

There are analogous statements for the complex case with V_n replaced by $V_n{}^c$. See D. J. Lewis, *Introduction to Algebra*, Harper and Row, New York, 1965.

PROBLEMS

1. Find the eigenvalues and corresponding eigenvectors of the following real matrices:

(a) $\begin{pmatrix} 2 & 0 \\ 0 & 3 \end{pmatrix}$, (b) $\begin{pmatrix} 1 & 0 \\ 0 & 1 \end{pmatrix}$, (c) $\begin{pmatrix} 1 & 0 \\ 2 & 3 \end{pmatrix}$,

(d) $\begin{pmatrix} 0 & 4 \\ 1 & 0 \end{pmatrix}$, (e) $\begin{pmatrix} 1 & 5 \\ -1 & 2 \end{pmatrix}$, (f) $\begin{pmatrix} 3 & 1 \\ 6 & 2 \end{pmatrix}$,

(g) $\begin{pmatrix} 2 & -1 \\ 1 & 4 \end{pmatrix}$, (h) $\begin{pmatrix} 3 & -1 \\ 9 & -3 \end{pmatrix}$, (i) $\begin{pmatrix} 1 & 2 & -2 \\ 0 & 2 & 4 \\ 0 & 0 & 3 \end{pmatrix}$,

2. Find the eigenvalues and corresponding eigenvectors of the following complex matrices:

(a) $\begin{pmatrix} 1+i & 1 \\ 1 & 1-i \end{pmatrix}$, (b) $\begin{pmatrix} 1 & i \\ i & 1 \end{pmatrix}$

3. For each of the following matrices find the eigenvalues and the subspaces K_λ and W_λ. [*Hint*. As above, $\dim W_\lambda$ is the multiplicity of λ.]

(a) $\begin{pmatrix} 1 & 2 & 3 \\ 0 & 1 & 4 \\ 0 & 0 & 2 \end{pmatrix}$, (b) $\begin{pmatrix} 0 & 1 & 2 \\ 0 & 0 & 3 \\ 0 & 0 & 1 \end{pmatrix}$, (c) $\begin{pmatrix} 0 & 0 & 1 \\ 0 & 0 & 2 \\ 0 & 0 & 0 \end{pmatrix}$,

(d) $\begin{pmatrix} 1 & 1 & -3 \\ 0 & -2 & -6 \\ 0 & 0 & 2 \end{pmatrix}$, (e) $\begin{pmatrix} 0 & -1 & 0 \\ 0 & -1 & 1 \\ 0 & -1 & 1 \end{pmatrix}$, (f) $\begin{pmatrix} 1 & 2 & 0 \\ 0 & 1 & 2 \\ 2 & 0 & 1 \end{pmatrix}$

4. (a), (b), (d), (e), (f). Prove the rules for eigenvalues stated and not proved in the text. [*Hint for* (b). See Problem 4-c following Section 10-15.]

5. (a) Show that if **u** is an eigenvector corresponding to the eigenvalue λ of A, then **u** need not be an eigenvector for A'.

   (b) Discuss the relationship between the eigenvectors of A and those of $p(A)$, where $p(x)$ is a polynomial.

   (c) If M is a nonsingular matrix, discuss the relationship between the eigenvectors of A and those of MAM^{-1}.

6. Prove: if $\lambda_1, \ldots, \lambda_k$ are distinct eigenvalues for A, $\mathbf{0} \neq \mathbf{u}_1$ is in $W_{\lambda_1}, \ldots, \mathbf{0} \neq \mathbf{u}_k$ is in W_{λ_k}, then $\{\mathbf{u}_1, \ldots, \mathbf{u}_k\}$ is a linearly independent set.

‡10-19 MATRIX REPRESENTATIONS OF A LINEAR MAPPING

In this section and the following two sections, we consider the effect on the preceding theory of *change from one basis to another one* in the vector spaces considered. The analysis here is more difficult, and we present only the main facts, in a number of cases without proof. The reader desiring to acquire a full grasp of the subject matter should read more advanced books and take appropriate courses. It is hoped that the abridged discussion given here will encourage him to do so.

In Section 10-2 we saw that the m by n matrices arose naturally in the study of linear mappings of V_n into V_m. We now show that this same situation prevails for finite dimensional spaces in general.

Let $\{\mathbf{u}_1, \ldots, \mathbf{u}_n\}$ be a basis for a finite dimensional space U and let $\{\mathbf{v}_1, \ldots, \mathbf{v}_m\}$ be a basis for a finite dimensional space V. We fix the order in which we arrange the elements in the bases, so that we are actually considering *ordered bases* for U and V. A linear mapping T from U into V is completely determined by its action on a basis for U, say

$$\begin{aligned} T(\mathbf{u}_1) &= a_{11}\mathbf{v}_1 + \cdots + a_{m1}\mathbf{v}_m \\ &\vdots \\ T(\mathbf{u}_n) &= a_{1n}\mathbf{v}_1 + \cdots + a_{mn}\mathbf{v}_m \end{aligned} \tag{10-190}$$

We can then assign to the linear mapping T the matrix

$$A = \begin{pmatrix} a_{11} & \cdots & a_{1n} \\ \vdots & & \vdots \\ a_{m1} & \cdots & a_{mn} \end{pmatrix} \tag{10-191}$$

We call the matrix A a *matrix representation* of T determined by the ordered bases $\{\mathbf{u}_1, \ldots, \mathbf{u}_n\}$ and $\{\mathbf{v}_1, \ldots, \mathbf{v}_m\}$.

EXAMPLE 1. Let U be $\mathcal{P}_3$, the vector space of polynomials of degree at most 3 and let V be $\mathcal{P}_2$, then the derivative D is a linear mapping of U into V. Let $\{1, x, x^2, x^3\}$ be an ordered basis for U and let $\{1, x, x^2\}$ be an ordered basis for V. Then the matrix representation of D determined by these bases is

$$\begin{pmatrix} 0 & 1 & 0 & 0 \\ 0 & 0 & 2 & 0 \\ 0 & 0 & 0 & 3 \end{pmatrix}$$

If we change the bases or if we change the order in which we arrange the basis elements, we obtain a different matrix representation. If in Example 1 we replace the basis 1, x, x^2 for V by the basis 1, $x-1$, x^2-x, then D is represented by the matrix

$$\begin{pmatrix} 0 & 1 & 2 & 3 \\ 0 & 0 & 2 & 3 \\ 0 & 0 & 0 & 3 \end{pmatrix}$$

Let us now fix an ordered basis for U and one for V. Then corresponding to each linear mapping T from U into V is its matrix representation A determined by the fixed choice of bases. This correspondence is one-to-one, since if T and S are linear mappings having the same matrix representation (determined by the same ordered bases), then T and S agree on a basis of U and so are the same mapping. Also, given an m by n matrix (10-191), it is the matrix representation of a linear mapping of U into V, namely, the linear mapping (10-190). Thus for fixed sets of ordered bases the correspondence

$$T \to A \qquad (A \text{ the representation of } T)$$

is a one-to-one mapping of the set $\mathcal{L}$ of all linear mappings of U into V onto the set $\mathfrak{M}_{mn}$ of m by n matrices. This correspondence is a linear mapping of the vector space $\mathcal{L}$ onto $\mathfrak{M}_{mn}$. For if $T \to A$ and $S \to B$, then

$$T + S \to A + B \qquad \text{and} \qquad cT \to cA \tag{10-192}$$

Furthermore, if we form the correspondence

$$\mathbf{u} = x_1\mathbf{u}_1 + \cdots + x_n\mathbf{u}_n \to \mathbf{x} = (x_1, \ldots, x_n)$$

between U and V_n, and the correspondence

$$\mathbf{v} = y_1\mathbf{v}_1 + \cdots + y_m\mathbf{v}_m \to \mathbf{y} = (y_1, \ldots, y_m)$$

between V and V_m (as in Theorem 11, Section 9-8) then

$$T(\mathbf{u}) \to A\mathbf{x} \tag{10-193}$$

This last relation permits us to determine Kernel T by finding Kernel A and then determining the subspace of U which corresponds to Kernel A (see Figure 10-11). Similarly, other properties of the linear mapping T can be determined by first finding the corresponding result for the matrix representing T.

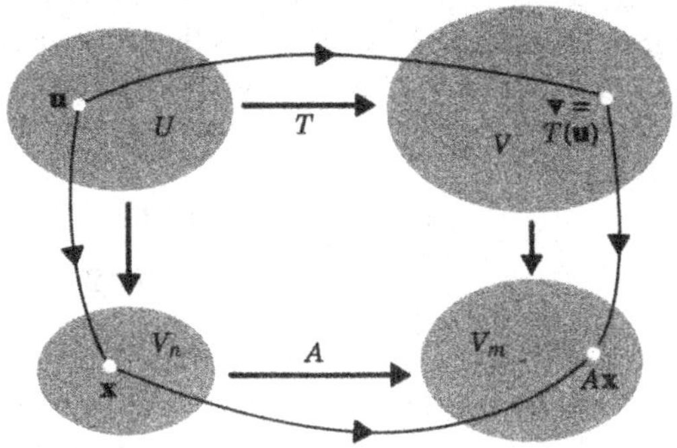

Figure 10-11 The matrix representation of a linear mapping.

‡10-20 JORDAN MATRICES

In this section we shall tacitly assume that all matrices appearing are real and have only real eigenvalues; analogous results hold for complex matrices with V_n replaced by $V_n{}^c$.

Since linear transformations on finite dimensional vector space V are linear mappings of V into V, it follows that linear transformations have a matrix representation—by *square* matrices. Since for linear transformations the domain and the target space of the mapping are the same, we can use the same ordered basis of V for the domain space and for the target space, and we shall always do so when studying matrix representations of linear transformations. Thus, if $\{\mathbf{u}_1, \ldots, \mathbf{u}_n\}$ is an ordered basis for V and T is a linear transformation on V such that

$$\begin{aligned} T(\mathbf{u}_1) &= a_{11}\mathbf{u}_1 + \cdots + a_{n1}\mathbf{u}_n \\ &\vdots \\ T(\mathbf{u}_n) &= a_{1n}\mathbf{u}_1 + \cdots + a_{nn}\mathbf{u}_n \end{aligned} \tag{10-200}$$

then

$$A = \begin{pmatrix} a_{11} & \cdots & a_{1n} \\ \vdots & & \vdots \\ a_{n1} & \cdots & a_{nn} \end{pmatrix} \tag{10-201}$$

is a matrix representation for T.

As for linear mappings, for a fixed choice of ordered basis, there is a one-to-one correspondence between the linear transformations on V and the n by n matrices which represent them:

$$T \to A$$

This correspondence is such that if $T \to A$ and $S \to B$, then

$$T + S \to A + B, \qquad TS \to AB, \qquad cT \to cA \tag{10-202}$$

It follows from the relations (10-202) that if $T \to A$, then T and A have the same minimal polynomial. Hence, T is nonsingular exactly when A is nonsingular (or equivalently, when $\det A \neq 0$).

If also we map U onto V_n by the assignment

$$\mathbf{u} = x_1\mathbf{u}_1 + \cdots + x_n\mathbf{u}_n \to \mathbf{x} = (x_1, \ldots, x_n)$$

as in Section 10-19, then

$$T(\mathbf{u}) \to A\mathbf{x} \tag{10-203}$$

From this last relation, we see that T and A have the same eigenvalues. If we let $K_\lambda^{(T)}$ be the set of vectors $\mathbf{u}$ in V such that $(T - \lambda I)\mathbf{u} = \mathbf{0}$, then $K_\lambda^{(T)}$ corresponds to the space $K_\lambda = K_\lambda^{(A)}$ for the matrix A. Similarly, we define the eigenspaces for T: $W_\lambda^{(T)}$ is the set of vectors $\mathbf{u}$ in V such that there is an integer r such that $(T - \lambda I)^r\mathbf{u} = \mathbf{0}$. Then the correspondence is such that $K_\lambda^{(T)}$ has image $K_\lambda^{(A)}$ and $W_\lambda^{(T)}$ has image $W_\lambda^{(A)}$.

We saw in Section 10-18 that, given a fixed matrix A with $\lambda_1, \ldots, \lambda_k$ as its distinct eigenvalues, then

$$V_n = \{W_{\lambda_1}^{(A)} + \cdots + W_{\lambda_k}^{(A)}\}$$

Hence, our correspondence tells us that

$$V = \{W_{\lambda_1}^{(T)} + \cdots + W_{\lambda_k}^{(T)}\}$$

and the subspaces $W_{\lambda_i}^{(T)}$ are invariant under T. It should be remarked that the subspaces $W_{\lambda_i}^{(T)}$ depend only on T.

Now let us choose ordered bases for the subspaces $W_{\lambda_1}^{(T)}, \ldots, W_{\lambda_k}^{(T)}$, say $\mathbf{u}_{11}, \ldots, \mathbf{u}_{1m_1}$ for $W_{\lambda_1}^{(T)}, \ldots, \mathbf{u}_{k1}, \ldots, \mathbf{u}_{km_k}$ for $W_{\lambda_k}^{(T)}$. Together, and in this order, they give us an ordered basis for the vector space V. Since $W_{\lambda_i}^{(T)}$ is invariant under T, we find that

$$T(\mathbf{u}_{ij}) = a_{j1}^{(i)}\mathbf{u}_{i1} + \cdots + a_{jm_i}^{(i)}\mathbf{u}_{jm_i} \qquad (j = 1, \ldots, m_i)$$

Hence, for this choice of basis, the matrix representation for T is a partitioned matrix of the form

$$A = \begin{pmatrix} A_1 & O & \cdots & O \\ O & A_2 & \cdots & O \\ \vdots & \vdots & & \vdots \\ O & O & \cdots & A_k \end{pmatrix} \tag{10-204}$$

where the A_i are m_i by m_i matrices. We note in particular, that if T has n distinct eigenvalues, then $K_{\lambda_i}^{(T)} = W_{\lambda_i}^{(T)}$ $(i = 1, \ldots, n)$ and these spaces are one-dimensional. Thus *if T has n distinct eigenvalues, then T has a diagonal matrix representation with its eigenvalues as the entries along the diagonal.*

When the spaces $W_{\lambda_i}^{(T)}$ are of dimension greater than 1 we can choose the bases for $W_{\lambda_i}^{(T)}$ so that the submatrices have a very simple form. Let $W = W_{\lambda_i}^{(T)}$. Choose any nonzero vector $\mathbf{u}_1$ in W. If $(T - \lambda_i I)\mathbf{u}_1 \neq \mathbf{0}$, we let $\mathbf{u}_2 = (T - \lambda_i I)\mathbf{u}_1$, (the vectors $\mathbf{u}_1, \mathbf{u}_2$ are linearly independent, see Problem 12); but if $(T - \lambda_i I)\mathbf{u}_1 = \mathbf{0}$, we let $\mathbf{u}_2$ be any vector in W linearly independent of $\mathbf{u}_1$. If $\dim W > 2$, we consider $(T - \lambda_i I)\mathbf{u}_2$. If it is $\mathbf{0}$, we let $\mathbf{u}_3$ be a vector in W not in Span $(\mathbf{u}_1, \mathbf{u}_2)$; but if $(T - \lambda_i I)\mathbf{u}_2 \neq \mathbf{0}$, we let $\mathbf{u}_3 = (T - \lambda_i I)\mathbf{u}_2$ and show that $\mathbf{u}_1, \mathbf{u}_2, \mathbf{u}_3$ are a linearly independent set. We continue in this way until we obtain a basis for W. Relative to this chosen basis we have

$$\begin{aligned} T(\mathbf{u}_1) &= \lambda_i\mathbf{u}_1 + \epsilon_1\mathbf{u}_2 \\ T(\mathbf{u}_2) &= \lambda_i\mathbf{u}_2 + \epsilon_2\mathbf{u}_3 \\ &\vdots \\ T(\mathbf{u}_{m-1}) &= \lambda_i\mathbf{u}_{m-1} + \epsilon_{m-1}\mathbf{u}_m \\ T(\mathbf{u}_m) &= \lambda_i\mathbf{u}_m \end{aligned}$$

where the ϵ_i are either 0 or 1 and $m = \dim W$. When we have chosen bases for each of the $W_{\lambda_i}^{(T)}$ in this way we get a matrix representation for T of the form (10-204) where the submatrices A_i have the form

$$A_i = \begin{pmatrix} \lambda_i & \epsilon_1 & 0 & \cdots & 0 & 0 \\ 0 & \lambda_i & \epsilon_2 & & 0 & 0 \\ \vdots & \vdots & \vdots & & \vdots & \vdots \\ 0 & 0 & 0 & & \lambda_i & \epsilon_{m_i-1} \\ 0 & 0 & 0 & \cdots & 0 & \lambda_i \end{pmatrix} \tag{10-205}$$

Matrices of the form (10-204), where the A_i have the form (10-205), are called *Jordan matrices.* Thus *each linear transformation on a finite dimensional vector space has a matrix representation which is a Jordan matrix.*

The Jordan matrices of order 2 are of the form:

$$\begin{pmatrix} \lambda & 0 \\ 0 & \mu \end{pmatrix}, \quad \begin{pmatrix} \lambda & 0 \\ 0 & \lambda \end{pmatrix}, \quad \begin{pmatrix} \lambda & 1 \\ 0 & \lambda \end{pmatrix}$$

where $\lambda \neq \mu$ are arbitrary numbers. Every linear transformation on a 3-dimensional vector space has a Jordan matrix of one (and only one) of the forms:

$$\text{(a)} \begin{pmatrix} \lambda & 0 & 0 \\ 0 & \mu & 0 \\ 0 & 0 & \sigma \end{pmatrix}, \quad \text{(b)} \begin{pmatrix} \lambda & 0 & 0 \\ 0 & \lambda & 0 \\ 0 & 0 & \mu \end{pmatrix}, \quad \text{(c)} \begin{pmatrix} \lambda & 1 & 0 \\ 0 & \lambda & 0 \\ 0 & 0 & \mu \end{pmatrix},$$

$$\text{(d)} \begin{pmatrix} \lambda & 0 & 0 \\ 0 & \lambda & 0 \\ 0 & 0 & \lambda \end{pmatrix}, \quad \text{(e)} \begin{pmatrix} \lambda & 1 & 0 \\ 0 & \lambda & 0 \\ 0 & 0 & \lambda \end{pmatrix}, \quad \text{(f)} \begin{pmatrix} \lambda & 1 & 0 \\ 0 & \lambda & 1 \\ 0 & 0 & \lambda \end{pmatrix}$$

where λ, μ, σ are distinct numbers.

It should be observed that the product of Jordan matrices need not be a Jordan matrix. Thus if we choose a basis for V so that, relative to that basis, T has a Jordan matrix representation, we cannot expect that every other linear transformation on V will also have a Jordan matrix representation relative to that one chosen basis. To get a Jordan matrix representation, we must choose a special basis for each linear transformation T.

We saw in Problem 10 following Section 10-11 that the minimal polynomial for a matrix A which is the direct sum of matrices $A_1, \ldots, A_k$ was a factor of the product of the minimal polynomials for the A_j. If A_i is of the form (10-205), then $A_i - \lambda_i I$ is nilpotent and its minimal polynomial is a factor of $(x - \lambda_i)^{m_i}$. Thus the minimal polynomial for a Jordan matrix of the form (10-204) where the A_i are of the form (10-205) is a factor of $(x - \lambda_1)^{m_1}(x - \lambda_2)^{m_2} \cdots (x - \lambda_k)^{m_k}$. This polynomial is the characteristic polynomial for the given matrix. Thus, if $\chi_A(x)$ is the characteristic polynomial for the Jordan matrix A, then $\chi_A(A) = O$, and the minimal polynomial is of degree at most n (the order of the matrix). Since a linear transformation T has the same minimal polynomial as any of its matrix representations, and since each T has a matrix representation that is a Jordan matrix, we can conclude that *each linear transformation on a vector space of dimension n has its minimal polynomial of degree at most n.*

The minimal polynomials for the three 2 by 2 Jordan matrices listed above are $(x - \lambda)(x - \mu)$, $(x - \lambda)$, $(x - \lambda)^2$, respectively.

‡10-21 SIMILAR MATRICES

Let A be an n by n matrix. Then A is itself a linear transformation on V_n and, hence, A has matrix representations. We note in particular that A is the matrix representation for A determined by the ordered basis $\{\mathbf{e}_1, \ldots, \mathbf{e}_n\}$. Let

M be another matrix representation for A determined by the ordered basis $\{\mathbf{u}_1, \ldots, \mathbf{u}_n\}$. We now show how M and A are related.

Since $\mathbf{u}_1, \ldots, \mathbf{u}_n$ are in V_n, they can be expressed as linear combinations of the basis $\mathbf{e}_1, \ldots, \mathbf{e}_n$, say

$$\begin{aligned} \mathbf{u}_1 &= p_{11}\mathbf{e}_1 + \cdots + p_{n1}\mathbf{e}_n \\ &\vdots \\ \mathbf{u}_n &= p_{1n}\mathbf{e}_1 + \cdots + p_{nn}\mathbf{e}_n \end{aligned} \tag{10-210}$$

But $\mathbf{u}_1, \ldots, \mathbf{u}_n$ are also a basis for V_n and, therefore, we can express the $\mathbf{e}_i$ in terms of $\mathbf{u}_1, \ldots, \mathbf{u}_n$, say

$$\begin{aligned} \mathbf{e}_1 &= q_{11}\mathbf{u}_1 + \cdots + q_{n1}\mathbf{u}_n \\ &\vdots \\ \mathbf{e}_n &= q_{1n}\mathbf{u}_1 + \cdots + q_{nn}\mathbf{u}_n \end{aligned}$$

Simple computation shows that $(p_{ij})(q_{jk}) = I$. Hence, $P = (p_{ij})$ and $Q = (q_{jk})$ are nonsingular matrices and $P^{-1} = Q$. To determine M, we need to find the m_{ij} such that

$$\begin{aligned} A\mathbf{u}_1 &= m_{11}\mathbf{u}_1 + \cdots + m_{n1}\mathbf{u}_n \\ &\vdots \\ A\mathbf{u}_n &= m_{1n}\mathbf{u}_1 + \cdots + m_{nn}\mathbf{u}_n \end{aligned}$$

We proceed to compute

$$\begin{aligned} A\mathbf{u}_j &= A(p_{1j}\mathbf{e}_1 + \cdots + p_{nj}\mathbf{e}_n) = p_{1j}A\mathbf{e}_1 + \cdots + p_{nj}A\mathbf{e}_n \\ &= p_{1j}(a_{11}\mathbf{e}_1 + \cdots + a_{n1}\mathbf{e}_n) + \cdots + p_{nj}(a_{1n}\mathbf{e}_1 + \cdots + a_{nn}\mathbf{e}_n) \\ &= (a_{11}p_{1j} + \cdots + a_{1n}p_{nj})\mathbf{e}_1 + \cdots + (a_{n1}p_{1j} + \cdots + a_{nn}p_{nj})\mathbf{e}_n \\ &= (a_{11}p_{1j} + \cdots + a_{1n}p_{nj})(q_{11}\mathbf{u}_1 + \cdots + q_{n1}\mathbf{u}_n) \\ &\qquad + \cdots + (a_{n1}p_{1j} + \cdots + a_{nn}p_{nj})(q_{1n}\mathbf{u}_1 + \cdots + q_{nn}\mathbf{u}_n) \\ &= \left(\sum_{t=1}^{n}\sum_{k=1}^{n} q_{1t}a_{tk}p_{kj}\right)\mathbf{u}_1 + \cdots + \left(\sum_{t=1}^{n}\sum_{k=1}^{n} q_{nt}a_{tk}p_{kj}\right)\mathbf{u}_n \end{aligned}$$

Thus we see that the matrix M representing A is

$$M = QAP = P^{-1}AP = QAQ^{-1} \tag{10-211}$$

We say that the matrix A is *similar* to the matrix B if there is a nonsingular matrix P such that $A = PBP^{-1}$. We note that A is similar to itself. If A is similar to B, then B is similar to A. If A is similar to B and B is similar to C, then A is similar to C; for if $A = PBP^{-1}$ and $B = QCQ^{-1}$, then $A = (PQ)C(PQ)^{-1}$. Thus the property of being "similar" divides the set of all n by n matrices into classes; each class consists of all matrices similar to a particular matrix. We call these classes *similarity classes*.

From (10-211) it follows that if M is a matrix representation of A, then A and M belong to the same similarity class. The converse also holds true. If A and M are in the same similarity class, then $A = PMP^{-1}$ and we can form the basis (10-210) using the matrix P. Relative to this basis M is a matrix representation for A. Thus the matrices in a similarity class are just those matrices that can be matrix representations of one another.

We note in particular that it follows from Section 9-20 that each square matrix is similar to a Jordan matrix. It can be shown that there is "essentially" only one Jordan matrix in any similarity class. For full details on results, see Lewis, *Introduction to Algebra*, Harper and Row, N.Y., 1965, or Faddeev and Faddeeva, *Computational Methods in Linear Algebra*, Freeman, San Francisco, 1963.

PROBLEMS

1. Find the matrix representing D, where
   (a) D is the derivative on $\mathcal{P}_5 = U = V$ and $\mathbf{v}_1 = \mathbf{u}_1 = 1$, $\mathbf{v}_2 = \mathbf{u}_2 = x, \ldots, \mathbf{v}_6 = \mathbf{u}_6 = x^5$.
   (b) D is the derivative on $\mathcal{P}_5 = U = V$, $\mathbf{u}_1 = 1, \ldots, \mathbf{u}_6 = x^5$, $\mathbf{v}_1 = 1$, $\mathbf{v}_2 = 1 + x$, $\mathbf{v}_3 = 1 + x + x^2, \ldots, \mathbf{v}_6 = 1 + x + \cdots + x^5$.
   (c) D is the derivative, $U = \mathcal{P}_5$, $V = \mathcal{P}_4$, $\mathbf{u}_1 = 1, \ldots, \mathbf{u}_6 = x^5$, $\mathbf{v}_j = jx^{j-1}$ $(j = 1, \ldots, 5)$.
2. Let T be the linear mapping of V_4 into V_3 such that

$$T(1, 0, 0, 0) = (2, 3, 6), \qquad T(0, 1, 0, 0) = (1, 2, 0)$$
$$T(0, 0, 1, 0) = (-1, 2, -3), \qquad T(0, 0, 0, 1) = (0, 2, -1)$$

   (a) Find the matrix corresponding to T if $\{(1, 0, 0, 0), \ldots, (0, 0, 0, 1)\}$ is the basis for V_4 and $\{(1, 0, 0), (0, 1, 0), (0, 0, 1)\}$ is the basis for V_3.
   (b) Find the matrix corresponding to T if $\{(1, 1, 0, 0), (0, 1, 1, 0), (0, 0, 1, 1), (1, 0, 0, 1)\}$ is the basis for V_4 and $\{(1, 0, 0), (0, 1, 0), (0, 0, 1)\}$ is the basis for V_3.
   (c) Find the matrix corresponding to T if $\{(1, 1, 0, 0), (0, 1, 1, 0), (0, 0, 1, 1), (1, 0, 0, 1)\}$ is the basis for V_4 and $\{(1, 1, 0), (0, 1, 1), (1, 0, 1)\}$ is the basis for V_3.
   (d) Find the kernel of T.
   (e) Let $\{\mathbf{u}_1, \mathbf{u}_2, \mathbf{u}_3\}$ be a basis for Kernel T, and extend to a basis for V_4. If we use this basis for V_4, what can be said about the matrix for T, regardless of the choice of basis for V_3?
3. (a) Prove that the correspondences (10-192) hold true.
   (b) Prove that the correspondence (10-193) holds true.
   (c) Prove that the correspondences (10-202) hold true.
4. Let A be a matrix representation for T. Prove:
   (a) A and T have the same minimal polynomial.
   (b) A and T have the same eigenvalues.
5. Let A, B be similar matrices. Prove:
   (a) For each integer $k \geq 0$, A^k and B^k are similar.
   (b) For each scalar c, cA and cB are similar.

(c) If $p(x)$ is a polynomial, then $p(A)$ and $p(B)$ are similar.

(d) A' and B' are similar.

6. Determine the minimal polynomial and the characteristic polynomial for each of the 3 by 3 Jordan matrices (a), . . . , (f) listed in Section 10-20 and show that no two have the same minimal polynomial and the same characteristic polynomial.

7. Let A and B be similar matrices. Prove:

(a) If B is nonsingular, then A is nonsingular and A^{-1}, B^{-1} are similar.

(b) $\det A = \det B$.

(c) A and B have the same characteristic polynomial.

(d) A and B have the same eigenvalues, but not necessarily the same eigenvectors.

(e) A and B have the same minimal polynomial.

8. Let

$$A = \begin{pmatrix} 1 & 1 & 0 \\ 0 & 1 & 0 \\ 0 & 0 & 2 \end{pmatrix}, \quad B = \begin{pmatrix} 1 & 0 & 0 \\ 0 & 2 & 1 \\ 0 & 0 & 2 \end{pmatrix}, \quad C = \begin{pmatrix} 1 & 0 & 0 \\ 0 & 2 & 0 \\ 0 & 0 & 2 \end{pmatrix}$$

(a) Show that these matrices have the same eigenvalues.

(b) Show that the characteristic polynomials for A and B are different but that the characteristic polynomials for B and C are the same.

(c) Determine the minimal polynomial for each of these matrices.

(d) Show that no two of these matrices are similar.

(e) Show that AB and AC have the same characteristic polynomial and the same minimal polynomial.

(f) Show that AB, AC are similar by finding a nonsingular matrix M such that $MABM^{-1} = AC$.

(g) Show that C, C^2 are not similar.

9. For each matrix below, find a Jordan matrix similar to it.

(a) $\begin{pmatrix} 1 & 2 & 0 \\ 0 & 2 & 0 \\ -1 & -2 & -1 \end{pmatrix}$, (b) $\begin{pmatrix} -7 & 12 & 2 \\ 3 & 4 & 0 \\ 2 & 0 & 2 \end{pmatrix}$, (c) $\begin{pmatrix} 13 & 16 & 16 \\ -5 & -7 & -6 \\ -6 & -8 & -7 \end{pmatrix}$,

(d) $\begin{pmatrix} 1 & 2 & 3 & 4 \\ 0 & 1 & 2 & 3 \\ 0 & 0 & 1 & 2 \\ 0 & 0 & 0 & 1 \end{pmatrix}$, (e) $\begin{pmatrix} 0 & 1 & 0 & 1 \\ 1 & 0 & 0 & 0 \\ 0 & 0 & 1 & 1 \\ 0 & 0 & 0 & 1 \end{pmatrix}$

10. Let $\chi(x)$ denote the characteristic polynomial and $m(x)$ denote the minimal polynomial. Exhibit all Jordan matrices for which

(a) $\chi(x) = (x-1)^5$, $m(x) = (x-1)^3$. (b) $\chi(x) = (x-1)^5$, $m(x) = (x-1)^2$.

(c) $\chi(x) = (x-1)^3(x-2)(x+3) = m(x)$.

(d) $m(x) = x^2 - 1$ and the matrix has order 3.

(e) $m(x) = x^2 - 1$ and the matrix has order 7.

(f) $m(x) = (x - 1)^2(x + 2)$ and the matrix has order 5.

(g) $m(x) = x^2 + 1$ and the matrix has order 4.

11. Prove: if $\chi(x)$ is the characteristic polynomial for a matrix A, then $\chi(A) = O$. [*Hint:* This result holds true when A is a Jordan matrix; see second last paragraph of Section 10-20.]

12. Let λ be an eigenvalue for a linear transformation T. Let W_λ be the corresponding eigenspace. Prove:

(a) If $\mathbf{u}$ is in W_λ and $\mathbf{v} = (T - \lambda I)\mathbf{u} \neq \mathbf{0}$, then $\{\mathbf{u}, \mathbf{v}\}$ is a linearly independent set.

(b) If $\mathbf{u}$ is in W_λ and $(T - \lambda I)^2\mathbf{u} \neq \mathbf{0}$, then $\{\mathbf{u}, (T - \lambda I)\mathbf{u}, (T - \lambda I)^2\mathbf{u}\}$ is a linearly independent set.

11
LINEAR EUCLIDEAN GEOMETRY

INTRODUCTION

In this chapter we develop geometry in 3-dimensional space. At several points, especially in the exercises, we also indicate how one can develop geometry in n-dimensional space.

In the spirit of this text, geometry will be developed on the basis of linear algebra. This means that we take the point of view of *analytic geometry* and describe points, lines, and other geometric objects by vectors, coordinates, equations, and inequalities. For example, the inner product of two vectors $\mathbf{u}, \mathbf{v}$ in V_3 will be defined by the algebraic formula $\mathbf{u} \cdot \mathbf{v} = u_1v_1 + u_2v_2 + u_3v_3$, and it will then be shown that angles can be defined in such a way that $\mathbf{u} \cdot \mathbf{v} = \|\mathbf{u}\| \, \|\mathbf{v}\| \cos \varphi$, where $\|\mathbf{u}\|$, $\|\mathbf{v}\|$ are the magnitudes of the vectors and φ is $\sphericalangle(\mathbf{u}, \mathbf{v})$ (that is, φ is the angle between $\mathbf{u}$ and $\mathbf{v}$).

It will be seen that the theorems of geometry are deducible from results of linear algebra and, in many cases, are mere restatements of them.

An alternative approach to geometry is by means of axioms and postulates, as has been traditional in plane and solid geometry ever since the work of Euclid more than 2000 years ago. For geometry of n-dimensional space, with $n > 3$, the axiomatic procedure becomes more complicated, and we are hampered by failure of "geometric intuition," whereas the algebraic procedure generalizes very simply. The study of geometry in spaces of dimension more than 3 has become increasingly important for mathematics and its applications and, therefore, it is essential to seek the simplest way to develop the theory. Linear algebra provides a tool exceptionally suited to this purpose.

Matrices appear at a number of points in the chapter. In particular, we recall that the *rank* of a matrix A equals the maximum number of linearly independent column (or row) vectors; it also is the largest integer r such that A has a nonzero minor of order r. If A is an m by n matrix of rank r, then the solutions of $A\mathbf{u} = \mathbf{0}$ (the kernel of A) form a subspace of V_n of dimension $n - r$; the set of solutions is also the solution set of the associated set of m linear equations in n unknowns (see Section 10-6).

11-1 INNER PRODUCT AND NORM IN V_3

Let V_3 be as usual the vector space of all ordered triples of real numbers, with the usual addition and multiplication by scalars. For vectors in V_3 we use the familiar notation $\mathbf{u} = (u_1, u_2, u_3)$, $\mathbf{v} = (v_1, v_2, v_3), \ldots$. We also write (x_1, x_2, x_3), (a, b, c) or (x, y, z) for such vectors. The index notation is preferred for general formulas and for generalizations to V_n. Occasionally, one writes $\mathbf{u} = (u_x, u_y, u_z)$ and calls u_x the x-component of $\mathbf{u}$, u_y the y-component, u_z the z-component. However, this notation is less desirable, since it can be confused with the one for partial derivatives (Chapter 12).

Basis Vectors. In V_3 one has a natural basis formed of the three vectors

$$\mathbf{i} = (1, 0, 0) \qquad \mathbf{j} = (0, 1, 0), \qquad \mathbf{k} = (0, 0, 1)$$

Each vector $\mathbf{u}$ of V_3 is expressible as a linear combination of these basis vectors: $\mathbf{u} = (u_1, u_2, u_3) = u_1\mathbf{i} + u_2\mathbf{j} + u_3\mathbf{k}$ (see Figure 11-1).

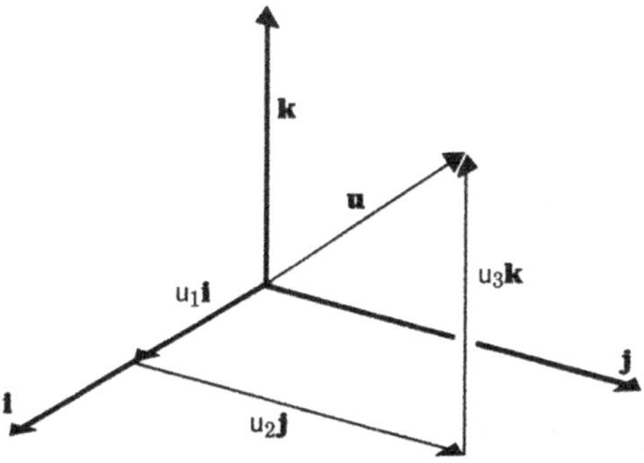

Figure 11-1 The basis for V_3.

Inner Product. In V_3 we now define an *inner product* (also called *dot product* and, occasionally, *scalar product*): the inner product of $\mathbf{u} = (u_1, u_2, u_3)$ and $\mathbf{v} = (v_1, v_2, v_3)$ is the real number $\mathbf{u} \cdot \mathbf{v}$, where

$$\mathbf{u} \cdot \mathbf{v} = u_1v_1 + u_2v_2 + u_3v_3$$

For example, $(3, 5, 6) \cdot (2, 1, -4) = 6 + 5 - 24 = -13$.

THEOREM 1. *The inner product has the properties: for all* $\mathbf{u}$, $\mathbf{v}$ *in* V_3 *and all scalars* a,

$$\left.\begin{aligned} \mathbf{u} \cdot \mathbf{v} &= \mathbf{v} \cdot \mathbf{u}, \\ \mathbf{u} \cdot (\mathbf{v} + \mathbf{w}) &= \mathbf{u} \cdot \mathbf{v} + \mathbf{u} \cdot \mathbf{w}, \\ (a\mathbf{u}) \cdot \mathbf{v} &= a(\mathbf{u} \cdot \mathbf{v}), \end{aligned}\right\} \tag{11-10}$$

$$\mathbf{u} \cdot \mathbf{u} \geq 0, \textit{ with equality if, and only if, } \mathbf{u} = \mathbf{0}; \tag{11-11}$$

$$\mathbf{i} \cdot \mathbf{i} = \mathbf{j} \cdot \mathbf{j} = \mathbf{k} \cdot \mathbf{k} = 1, \qquad \mathbf{i} \cdot \mathbf{j} = \mathbf{j} \cdot \mathbf{k} = \mathbf{k} \cdot \mathbf{i} = 0; \tag{11-12}$$

$$\mathbf{u} \cdot \mathbf{0} = 0 \tag{11-13}$$

The proofs are left as exercises (Problem 3 below).

Norm. In V_3 we assign a norm $\|\mathbf{u}\|$ to each vector $\mathbf{u}$ as follows:

$$\|\mathbf{u}\| = \sqrt{\mathbf{u}\cdot\mathbf{u}} = \sqrt{u_1^2 + u_2^2 + u_3^2}$$

(The notation $\|\mathbf{u}\|$, rather than $|\mathbf{u}|$, is preferred for advanced work.)

THEOREM 2. *The norm has the properties: for all* $\mathbf{u}, \mathbf{v}$ *in* V_3 *and all scalars* a,

$$\|\mathbf{u}\| \geq 0, \text{ with equality if, and only if, } \mathbf{u} = \mathbf{0}; \tag{11-14}$$

$$\|a\mathbf{u}\| = |a|\,\|\mathbf{u}\|; \tag{11-15}$$

$$|\mathbf{u}\cdot\mathbf{v}| \leq \|\mathbf{u}\|\,\|\mathbf{v}\| \text{ (Schwarz inequality)}, \tag{11-16}$$

where equality holds if, and only if, $\mathbf{u}, \mathbf{v}$ *are linearly dependent;*

$$\|\mathbf{u}+\mathbf{v}\| \leq \|\mathbf{u}\| + \|\mathbf{v}\| \text{ (Minkowski inequality)}, \tag{11-17}$$

where equality holds if, and only if, either $\mathbf{u} = k\mathbf{v}$ *or* $\mathbf{v} = k\mathbf{u}$ *with* $k \geq 0$.

PROOF. Rule (11-14) follows at once from (11-11). To obtain Rule (11-15) we write $\|a\mathbf{u}\|^2 = (a\mathbf{u})\cdot(a\mathbf{u}) = a^2(\mathbf{u}\cdot\mathbf{u})$ by (11-10), and then take square roots.

To prove the Schwarz inequality (11-16), we first obtain an equivalent inequality by squaring both sides. In terms of components, this is

$$(u_1v_1 + u_2v_2 + u_3v_3)^2 \leq (u_1^2 + u_2^2 + u_3^2)(v_1^2 + v_2^2 + v_3^2)$$

If we multiply out and move all terms to the right side, we again obtain an equivalent inequality:

$$\begin{aligned} 0 \leq u_1^2v_2^2 - 2u_1u_2v_1v_2 + u_2^2v_1^2 + u_1^2v_3^2 - 2u_1u_3v_1v_3 \\ + u_3^2v_1^2 + u_2^2v_3^2 - 2u_2u_3v_2v_3 + u_3^2v_2^2 \end{aligned}$$

or

$$0 \leq (u_1v_2 - u_2v_1)^2 + (u_1v_3 - u_3v_1)^2 + (u_2v_3 - u_3v_2)^2 \tag{11-18}$$

But the last inequality is true, since all terms on the right are squared. Hence, the two previous inequalities are true and (11-16) is proved. It is also clear from (11-18) that equality occurs in (11-16) if, and only if, $u_1v_2 - u_2v_1 = 0$, $u_1v_3 - u_3v_1 = 0$ and $u_2v_3 - u_3v_2 = 0$; but this last condition is equivalent to the condition that the matrix

$$\begin{pmatrix} u_1 & u_2 & u_3 \\ v_1 & v_2 & v_3 \end{pmatrix}$$

have rank less than 2 and, hence, to the condition that $\mathbf{u}, \mathbf{v}$ be linearly dependent. Therefore, in (11-16) equality occurs precisely when $\mathbf{u}, \mathbf{v}$ are linearly dependent.

PROOF OF MINKOWSKI INEQUALITY. We have by (11-14) and (11-16)

$$\begin{aligned} 0 \leq \|\mathbf{u}+\mathbf{v}\|^2 &= (\mathbf{u}+\mathbf{v})\cdot(\mathbf{u}+\mathbf{v}) = \|\mathbf{u}\|^2 + 2\mathbf{u}\cdot\mathbf{v} + \|\mathbf{v}\|^2 \\ &\leq \|\mathbf{u}\|^2 + 2\|\mathbf{u}\|\,\|\mathbf{v}\| + \|\mathbf{v}\|^2 = (\|\mathbf{u}\| + \|\mathbf{v}\|)^2 \end{aligned}$$

Since $\|\mathbf{u} + \mathbf{v}\|$ and $\|\mathbf{u}\| + \|\mathbf{v}\|$ are nonnegative, we obtain (11-17). The proof of the conditions for equality is left as an exercise (Problem 10 below).

11-2 UNIT VECTORS, ANGLE BETWEEN VECTORS

A vector of norm 1 is called a *unit vector*. Thus $\mathbf{i} = (1, 0, 0)$ is a unit vector, as are $\mathbf{j}$ and $\mathbf{k}$. Every nonzero vector $\mathbf{v}$ can be written as the positive scalar $\|\mathbf{v}\|$ times a unit vector $\mathbf{u} = \mathbf{v}/\|\mathbf{v}\|$:

$$\mathbf{v} = \|\mathbf{v}\| \left(\frac{1}{\|\mathbf{v}\|} \mathbf{v} \right) = \|\mathbf{v}\| \mathbf{u}$$

Angle Between Vectors. Let $\mathbf{u}$, $\mathbf{v}$ be nonzero vectors. By the Schwarz inequality (11-16)

$$|\mathbf{u} \cdot \mathbf{v}| \leq \|\mathbf{u}\|\,\|\mathbf{v}\| \qquad \text{or} \qquad -\|\mathbf{u}\|\,\|\mathbf{v}\| \leq \mathbf{u} \cdot \mathbf{v} \leq \|\mathbf{u}\|\,\|\mathbf{v}\|$$

or

$$-1 \leq \frac{\mathbf{u} \cdot \mathbf{v}}{\|\mathbf{u}\|\,\|\mathbf{v}\|} \leq 1$$

Hence, we can find an angle φ (in radians) so that

$$\cos \varphi = \frac{\mathbf{u} \cdot \mathbf{v}}{\|\mathbf{u}\|\,\|\mathbf{v}\|}, \qquad 0 \leq \varphi \leq \pi \tag{11-20}$$

Because of the restriction $0 \leq \varphi \leq \pi$, the angle φ is unique. We call it *the angle between* $\mathbf{u}$ *and* $\mathbf{v}$ (more precisely, the radian measure of the angle between $\mathbf{u}$ and $\mathbf{v}$), and write:

$$\varphi = \measuredangle(\mathbf{u}, \mathbf{v})$$

From (11-20) it follows that

$$\mathbf{u} \cdot \mathbf{v} = \|\mathbf{u}\|\,\|\mathbf{v}\| \cos \varphi$$

as in Section 1-10; in fact, this is the reason for the definition (11-20).

We do not define $\measuredangle(\mathbf{u}, \mathbf{v})$ if $\mathbf{u}$ or $\mathbf{v}$ is $\mathbf{0}$.

Orthogonal Vectors. If $\mathbf{u} \cdot \mathbf{v} = 0$, we say that $\mathbf{u}$, $\mathbf{v}$ are *orthogonal* (or *perpendicular*) vectors. It follows from (11-12) that the basis vectors $\mathbf{i}$, $\mathbf{j}$, $\mathbf{k}$ are pairwise orthogonal. From (11-20) it follows that two nonzero vectors $\mathbf{u}$, $\mathbf{v}$ are orthogonal precisely when $\measuredangle(\mathbf{u}, \mathbf{v}) = \pi/2$. In addition, by our definition, for any vector $\mathbf{v}$, the vectors $\mathbf{0}$, $\mathbf{v}$ are always orthogonal.

Parallel Vectors. We say that $\mathbf{u}$, $\mathbf{v}$ are *parallel* if they are linearly dependent: $\mathbf{u} = k\mathbf{v}$ or $\mathbf{v} = k\mathbf{u}$. If $\mathbf{u}$ and $\mathbf{v}$ are nonzero vectors, then we find that (11-20) gives $\cos \varphi = \pm 1$ (according as $k > 0$ or $k < 0$), so that $\varphi = 0$ or π. In fact, by the condition for equality in the Schwarz inequality, $\mathbf{u}$, $\mathbf{v}$ are linearly dependent precisely when $\mathbf{u} \cdot \mathbf{v} = \pm\|\mathbf{u}\|\,\|\mathbf{v}\|$, so that $\cos \varphi = \pm 1$ and $\varphi = 0$ or π. By our definition, we also say that $\mathbf{0}$, $\mathbf{v}$ ($\mathbf{v}$ arbitrary) are parallel vectors.

Vectors Having the Same Direction. Two nonzero vectors **u**, **v** are said to have the *same direction* if $\mathbf{u} = k\mathbf{v}$ (or $\mathbf{v} = k\mathbf{u}$) with $k > 0$. This again corresponds to a case of equality in (11-16); since here $\mathbf{u} \cdot \mathbf{v} > 0$, we have $\mathbf{u} \cdot \mathbf{v} = \|\mathbf{u}\| \, \|\mathbf{v}\|$ and (11-20) gives $\varphi = 0$. Thus **u**, **v** have the same direction precisely when $\measuredangle(\mathbf{u}, \mathbf{v}) = 0$. When **u**, $-\mathbf{v}$ have the same direction, we say that **u**, **v** have *opposite directions*. This is clearly the case when $\measuredangle(\mathbf{u}, \mathbf{v}) = \pi$.

Component of One Vector in the Direction of Another. Let **v** be a nonzero vector. Then for each vector **u**, we write $\text{comp}_v\, \mathbf{u}$ for the inner product of **u** with a unit vector in the direction of **v**. Hence,

$$\text{comp}_v\, \mathbf{u} = \mathbf{u} \cdot \frac{\mathbf{v}}{\|\mathbf{v}\|} = \frac{\mathbf{u} \cdot \mathbf{v}}{\|\mathbf{v}\|} \tag{11-21}$$

If **u** is a nonzero vector, then $\mathbf{u} \cdot \mathbf{v} = \|\mathbf{u}\| \, \|\mathbf{v}\| \cos \varphi$ and (11-21) becomes

$$\text{comp}_v\, \mathbf{u} = \frac{\|\mathbf{u}\| \, \|\mathbf{v}\| \cos \varphi}{\|\mathbf{v}\|} = \|\mathbf{u}\| \cos \varphi$$

as suggested in Figure 11-2. When $\mathbf{u} = \mathbf{0}$ or when **u** is orthogonal to **v**, then $\text{comp}_v\, \mathbf{u} = 0$. For $\mathbf{v} = \mathbf{i}$ and $\mathbf{u} = u_1\mathbf{i} + u_2\mathbf{j} + u_3\mathbf{k}$, we obtain

$$\text{comp}_i\, \mathbf{u} = \mathbf{u} \cdot \mathbf{i} = (u_1\mathbf{i} + u_2\mathbf{j} + u_3\mathbf{k}) \cdot \mathbf{i} = u_1 \tag{11-22}$$

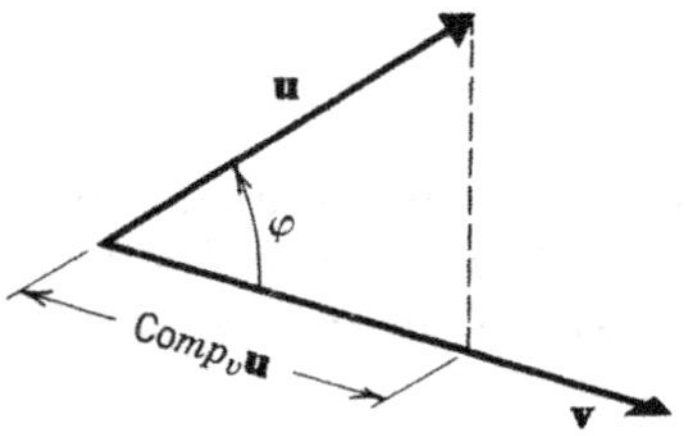

Figure 11-2 The component of u in direction of v.

Hence, u_1 is the component of **u** in the direction of **i** (or, as we often say, u_1 is the x-component of **u**, the component of **u** in the direction of the x-axis). Similarly,

$$\text{comp}_j\, \mathbf{u} = \mathbf{u} \cdot \mathbf{j} = u_2, \qquad \text{comp}_k\, \mathbf{u} = \mathbf{u} \cdot \mathbf{k} = u_3 \tag{11-22$'$}$$

Euclidean Vector Spaces. The inner product and related norm we have introduced in V_3 are called the *Euclidean inner product* and *Euclidean norm*, since they form the basis for Euclidean geometry in 3-dimensional space. We call V_3, with this inner product and norm, a *3-dimensional Euclidean vector space*. Throughout the rest of this book, V_3 will always be assumed to have this norm and inner product.

‡11-3 EUCLIDEAN VECTOR SPACE OF DIMENSION n

In the vector space V_n we introduce the inner product and norm by the definitions:

$$\mathbf{u} \cdot \mathbf{v} = u_1v_1 + \cdots + u_nv_n \tag{11-30}$$

$$\|\mathbf{u}\| = \sqrt{\mathbf{u} \cdot \mathbf{u}} \tag{11-31}$$

From the definition (11-30) one can then verify (11-10), (11-11), (11-13), and the analogue of (11-12):

$$\mathbf{e}_i \cdot \mathbf{e}_j = 1 \text{ for } i = j, \quad \mathbf{e}_i \cdot \mathbf{e}_j = 0 \text{ for } i \neq j$$

where $\mathbf{e}_1 = (1, 0, \ldots, 0)$, $\mathbf{e}_2 = (0, 1, \ldots, 0), \ldots, \mathbf{e}_n = (0, \ldots, 0, 1)$. It follows from (11-11) that (11-31) is always meaningful. One can then prove that Theorem 2 is valid for V_n. The proofs are left as exercises (Problem 10 below). The vector space V_n with this norm and inner product is called *n-dimensional Euclidean vector space.* Throughout this book V_n will be assumed to have this norm and inner product. One can define unit vectors, the angle between two vectors, orthogonal or parallel vectors, vectors having the same direction and components as in Section 11-2. In particular, $\mathbf{e}_1, \ldots, \mathbf{e}_n$ are unit vectors, and for $\mathbf{u} = (u_1, \ldots, u_n)$,

$$\text{comp}_{\mathbf{e}_i} \mathbf{u} = \mathbf{u} \cdot \mathbf{e}_i = u_i \qquad (i = 1, \ldots, n)$$

PROBLEMS

1. Let $\mathbf{u} = (3, 4, 2)$, $\mathbf{v} = (5, 0, 5)$, $\mathbf{w} = (2, -1, -3)$. Evaluate
   (a) $\mathbf{u} \cdot \mathbf{v}$, (b) $\mathbf{v} \cdot \mathbf{w}$, (c) $\|\mathbf{u}\|$, (d) $\|\mathbf{u} + \mathbf{v}\|$,
   (e) $(3\mathbf{u}) \cdot \mathbf{w}$, (f) $(2\mathbf{u} + 5\mathbf{v}) \cdot \mathbf{u}$, (g) $|\mathbf{v} \cdot \mathbf{w}|$,
   (h) $\|(\mathbf{u} \cdot \mathbf{v})\mathbf{u}\|$, (i) $\measuredangle(\mathbf{u}, \mathbf{v})$, (j) $\measuredangle(\mathbf{u}, \mathbf{v} + \mathbf{w})$.
2. With $\mathbf{u}$ and $\mathbf{v}$ as in Problem 1, verify: (a) the Schwarz inequality (11-16), (b) the Minkowski inequality (11-17).
3. Prove the following parts of Theorem 1:
   (a) (11-10) (b) (11-11) (c) (11-12) (d) (11-13)
4. *Direction angles, direction cosines.* Let $\mathbf{u} = (u_1, u_2, u_3)$ be a nonzero vector in V_3. The angles $\alpha = \measuredangle(\mathbf{u}, \mathbf{i})$, $\beta = \measuredangle(\mathbf{u}, \mathbf{j})$, $\gamma = \measuredangle(\mathbf{u}, \mathbf{k})$ are called the *direction angles* of $\mathbf{u}$; the numbers $l = \cos\alpha$, $m = \cos\beta$, $n = \cos\gamma$ are called the *direction cosines* of $\mathbf{u}$. Prove (a)–(d) and solve (e)–(f).
   (a) $l = (\mathbf{u} \cdot \mathbf{i})/\|\mathbf{u}\|$, $m = (\mathbf{u} \cdot \mathbf{j})/\|\mathbf{u}\|$, $n = (\mathbf{u} \cdot \mathbf{k})/\|\mathbf{u}\|$
   (b) $\mathbf{u} = \|\mathbf{u}\| (\cos\alpha\, \mathbf{i} + \cos\beta\, \mathbf{j} + \cos\gamma\, \mathbf{k})$
   (c) $l^2 + m^2 + n^2 = 1$
   (d) If $\mathbf{u}$ is a unit vector, then u_1, u_2, u_3 are the direction cosines of $\mathbf{u}$.
   (e) With $\mathbf{v}$ as in Problem 1, find the direction angles and cosines of $\mathbf{v}$.
   (f) With $\mathbf{w}$ as in Problem 1, find the direction angles and cosines of $\mathbf{w}$.
5. Let

$$A = \begin{pmatrix} \dfrac{1}{\sqrt{2}} & \dfrac{1}{\sqrt{2}} & 0 \\[2ex] \dfrac{-1}{2} & \dfrac{1}{2} & \dfrac{1}{\sqrt{2}} \\[2ex] \dfrac{1}{2} & \dfrac{-1}{2} & \dfrac{1}{\sqrt{2}} \end{pmatrix}$$

Prove:

(a) Regarded as a vector in V_3, each row of A is a unit vector, and each column of A is a unit vector.

(b) The row vectors of A are pairwise orthogonal.

(c) The column vectors of A are pairwise orthogonal.

Remark. A matrix with properties (a), (b), (c) is called an *orthogonal matrix* (see Section 11-21).

6. Prove: (a) if **u**, **v** are orthogonal, then $\|\mathbf{u} + \mathbf{v}\|^2 = \|\mathbf{u}\|^2 + \|\mathbf{v}\|^2$.

(b) If $\|\mathbf{u} + \mathbf{v}\|^2 = \|\mathbf{u}\|^2 + \|\mathbf{v}\|^2$, then **u**, **v** are orthogonal.

(*Note.* These assertions are a form of the Pythagorean Theorem and its converse.)

7. Prove: (a) $\|\mathbf{u} - \mathbf{v}\|^2 = \|\mathbf{u}\|^2 + \|\mathbf{v}\|^2 - 2\mathbf{u} \cdot \mathbf{v}$,

(b) $\|\mathbf{u} - \mathbf{v}\|^2 = \|\mathbf{u}\|^2 + \|\mathbf{v}\|^2 - 2\|\mathbf{u}\|\,\|\mathbf{v}\| \cos \varphi$, $\varphi = \measuredangle(\mathbf{u}, \mathbf{v})$.

(*Note.* These relations show that inner product and angle in V_3 are expressible in terms of norms. Identity (b) is a form of the Law of Cosines.)

8. Let $\mathbf{v}_1, \mathbf{v}_2, \mathbf{v}_3$ be nonzero vectors which are pairwise orthogonal. Prove:

(a) $\{\mathbf{v}_1, \mathbf{v}_2, \mathbf{v}_3\}$ is a linearly independent set. (*Hint.* Let $c_1\mathbf{v}_1 + c_2\mathbf{v}_2 + c_3\mathbf{v}_3 = \mathbf{0}$ and take the inner product of both sides with $\mathbf{v}_1, \mathbf{v}_2, \mathbf{v}_3$ in turn.)

(b) If **v** is in Span $(\mathbf{v}_1, \mathbf{v}_2, \mathbf{v}_3)$ and is orthogonal to $\mathbf{v}_1, \mathbf{v}_2, \mathbf{v}_3$, then **v** is **0**.

Remark. The result (b) shows that there can be at most 3 nonzero pairwise orthogonal vectors in V_3.

9. Let **a, b, c** be three unit vectors which are pairwise orthogonal. The result of Problem 8(a) shows that **a, b, c** form a basis for V_3, so that we can write $\mathbf{u} = u_a\mathbf{a} + u_b\mathbf{b} + u_c\mathbf{c}$ for each **u** in V_3. Prove:

(a) $u_a = \mathbf{u} \cdot \mathbf{a}$, $u_b = \mathbf{u} \cdot \mathbf{b}$, $u_c = \mathbf{u} \cdot \mathbf{c}$

(b) $\mathbf{u} \cdot \mathbf{v} = u_a v_a + u_b v_b + u_c v_c$ for all **u**, **v**

(c) $\|\mathbf{u}\| = \sqrt{u_a^2 + u_b^2 + u_c^2}$

Remark. The results (b), (c) show that there is nothing special about the basis **i, j, k.** We can develop the *same* inner product and norm using any basis of pairwise orthogonal unit vectors. We call such a basis an *orthonormal basis.*

‡10. On the basis of the definitions of Section 11-3, prove the following in V_n:

(a) (11-10) (b) (11-11) (c) $\mathbf{e}_i \cdot \mathbf{e}_j = 1$ for $i = j$, $\mathbf{e}_i \cdot \mathbf{e}_j = 0$ otherwise

(d) **u**, **v** are linearly independent if, and only if, $(\mathbf{v} \cdot \mathbf{v})\mathbf{u} - (\mathbf{u} \cdot \mathbf{v})\mathbf{v} \neq \mathbf{0}$

(e) (11-13) (f) (11-14) (g) (11-15)

(h) $\|(\mathbf{v} \cdot \mathbf{v})\mathbf{u} - (\mathbf{u} \cdot \mathbf{v})\mathbf{v}\|^2 = \|\mathbf{v}\|^2\{\|\mathbf{v}\|^2\|\mathbf{u}\|^2 - (\mathbf{u} \cdot \mathbf{v})^2\}$

(i) The Schwarz inequality (11-16) [*Hint.* First verify it for $\mathbf{v} = \mathbf{0}$. For $\mathbf{v} \neq \mathbf{0}$, apply (h).]

(j) If **u**, **v** are linearly dependent, then equality holds true in (11-16).

(k) If equality holds true in (11-16), then **u**, **v** are linearly dependent. [*Hint.* Use (h) and (d).]

(l) (11-17)

(m) If $\mathbf{u} = k\mathbf{v}$, with $k \geq 0$, then equality holds true in (11-17).

(n) If equality holds true in (11-17), then $\mathbf{u} = k\mathbf{v}$ or $\mathbf{v} = k\mathbf{u}$ with $k \geq 0$.

‡11. Let $\mathcal{C}$ be the vector space of all functions continuous on a given interval $[a, b]$. For f and g in $\mathcal{C}$, one defines the inner product: (f, g) and norm: $\|f\|$ by the equations:

$$(f, g) = \int_a^b f(x)g(x)\,dx, \qquad \|f\| = (f, f)^{1/2}$$

Prove the rules analogous to the following:

(a) (11-10) (b) (11-11) (c) (11-13)

(d) (11-14) (e) (11-15) (f) (11-16) [*Hint.* The proof of the Schwarz inequality suggested in Problem 10 (i) is applicable to $\mathcal{C}$.]

(g) Show from the Schwarz inequality that for all f, g in $\mathcal{C}$,

$$\left[\int_a^b f(x)g(x)\,dx\right]^2 \leq \left(\int_a^b [f(x)]^2\,dx\right)\left(\int_a^b [g(x)]^2\,dx\right)$$

(h) By taking $g(x) \equiv 1$ in (g), prove that for every f in $\mathcal{C}$ the absolute value of the average of f is less than or equal to the root mean square of f (see Section 4-24).

11-4 POINTS, VECTORS, DISTANCE, LINES IN 3-DIMENSIONAL EUCLIDEAN SPACE R^3

We are now prepared to begin our description of the geometry of 3-dimensional space R^3. The points of the space are the ordered triples of real numbers (x, y, z). We write $P = (x, y, z)$ and say that point P has *coordinates* x, y, z. The point $O = (0, 0, 0)$ is called the *origin* (see Figure 11-3). Two points P_1: (x_1, y_1, z_1) and P_2: (x_2, y_2, z_2) are the same (or equal) if, and only if, corresponding coordinates are equal. The x-axis consists of all points $(x, 0, 0)$, the y-axis of all points $(0, y, 0)$, the z-axis of all points $(0, 0, z)$; the xy-plane consists of all points $(x, y, 0)$, and the yz and xz-planes are defined similarly. One refers

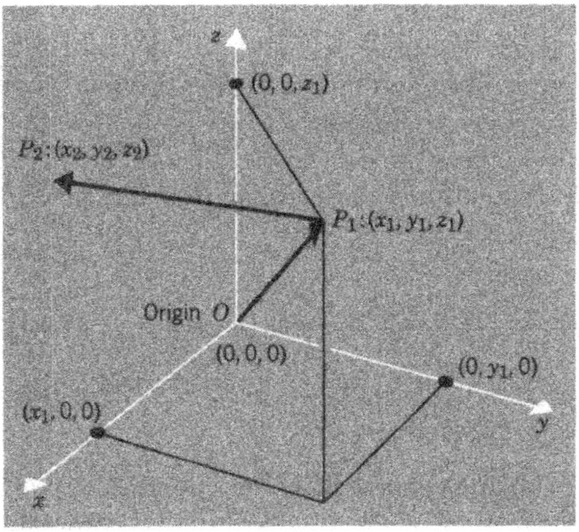

Figure 11-3 The points in 3-dimensional space R^3.

to these axes and planes as *coordinate axes* and *coordinate planes*. Space is divided into eight *octants*, according to the signs of the coordinates; the *first octant* consists of all points (x, y, z) for which $x \geq 0$, $y \geq 0$ and $z \geq 0$.

Throughout this chapter we shall consider various geometric objects in R^3. Often the object will appear as the graph of an equation in x, y, and z; that is, as the set of all points satisfying the equation. In some cases the object will be given by two equations; it is then the intersection of the corresponding graphs (as in Section 0-6). Occasionally, it will be described by several inequalities—as for the first octant at the close of the preceding paragraph. It is important to be able to sketch rough graphs in all these cases. We do not discuss the techniques for making such sketches (a subject considered in books on descriptive geometry), but try to suggest through examples how they can be made. By attempting such sketches, one can improve one's intuitive grasp of relationships in space. In difficult cases, one may wish to make a 3-dimensional model of the object.

Vector Determined by Two Points. The vector from P_1: (x_1, y_1, z_1) to P_2: (x_2, y_2, z_2), denoted by $\overrightarrow{P_1P_2}$, is the vector $(x_2 - x_1)\mathbf{i} + (y_2 - y_1)\mathbf{j} + (z_2 - z_1)\mathbf{k}$, which is the vector $(x_2 - x_1, y_2 - y_1, z_2 - z_1)$ of V_3:

$$\overrightarrow{P_1P_2} = (x_2 - x_1, y_2 - y_1, z_2 - z_1)$$

It follows that $P = Q$ if, and only if, $\overrightarrow{PQ} = \mathbf{0}$. As a particular case of our definition, for $P = (x, y, z)$, we have

$$\overrightarrow{OP} = x\mathbf{i} + y\mathbf{j} + z\mathbf{k}$$

We can now write, for P_1 and P_2 as above,

$$\overrightarrow{P_1P_2} = (x_2, y_2, z_2) - (x_1, y_1, z_1) = \overrightarrow{OP_2} - \overrightarrow{OP_1} \tag{11-40}$$

From (11-40) it follows that

$$\overrightarrow{QP} = \overrightarrow{OP} - \overrightarrow{OQ} = -\overrightarrow{PQ}$$

Remark. Both vectors and points are here represented by triples of numbers, and one might expect this to lead to confusion. However, it will be seen that the context always makes clear whether a point or a vector is meant. Furthermore, the notation provides a certain flexibility in passing from point P to vector $\overrightarrow{OP}$, both of which are represented by the same triple. From a more profound point of view, we can say that points and vectors are merely two different ways of viewing the same objects; if we consider all our vectors as directed line segments $\overrightarrow{OP}$, with O fixed, then the vectors are given by their end points P—the rest of the arrow is redundant.

We note one other consequence of (11-40):

$$\overrightarrow{PQ} + \overrightarrow{QS} = \overrightarrow{PS} \tag{11-41}$$

For $\overrightarrow{PQ} + \overrightarrow{QS} = (\overrightarrow{OQ} - \overrightarrow{OP}) + (\overrightarrow{OS} - \overrightarrow{OQ}) = \overrightarrow{OS} - \overrightarrow{OP} = \overrightarrow{PS}$. If we take $S = P$, we get the rule $\overrightarrow{PQ} = -\overrightarrow{QP}$ noted above.

Distance Between Two Points. We now define the distance from P_1: (x_1, y_1, z_1) to P_2: (x_2, y_2, z_2) to be the number $d(P_1,P_2) = \|\overrightarrow{P_1P_2}\|$. Accordingly,

$$d(P_1, P_2) = \sqrt{(x_2 - x_1)^2 + (y_2 - y_1)^2 + (z_2 - z_1)^2} \qquad (11\text{-}42)$$

The introduction of distance is the crucial step in developing Euclidean geometry in space. As in plane geometry, knowledge of distance permits one to find angles, to tell when lines are perpendicular, and to find lengths, areas, volumes. For this reason, one formalizes the process as follows:

Definition of Euclidean 3-Dimensional Space R^3. Three-dimensional Euclidean space is the set of all ordered triples of real numbers, with distance defined by Equation (11-42). One denotes this space by R^3.

The definition generalizes to *n-dimensional Euclidean space* R^n ($n = 1, 2, 3, \ldots$). Points become ordered n-tuples $(x_1, \ldots, x_n)$ and distance becomes

$$d(P', P'') = \left\{\sum_{i=1}^{n} (x_i'' - x_i')^2\right\}^{1/2}$$

The vectors in R^3 are vectors from the Euclidean vector space V_3, with norm and inner product as in Section 11-1. To emphasize this, we shall write: a vector of R^3 (just as we referred to "vectors in the plane" in Chapter 1). (From the more profound point of view mentioned above, R^3 can be considered to be the same as the vector space in question.)

THEOREM 3. *For all P, Q, S in R^3 one has*

$$d(P, Q) = d(Q, P); \qquad (11\text{-}43)$$

$$d(P, Q) \geq 0;\ d(P, Q) = 0 \textit{ if, and only if, } P = Q; \qquad (11\text{-}44)$$

$$d(P, S) \leq d(P, Q) + d(Q, S) \textit{ (triangle inequality)}. \qquad (11\text{-}45)$$

PROOF. The assertions are consequences of properties of the norm, by virtue of the relation $d(P, Q) = \|\overrightarrow{PQ}\|$. For (11-43) we use the fact that $\|\overrightarrow{PQ}\| = \|-\overrightarrow{QP}\| = \|\overrightarrow{QP}\|$ by (11-15); (11-44) is an immediate consequence of (11-11). The inequality (11-45) follows from (11-41) and (11-17):

$$d(P, S) = \|\overrightarrow{PS}\| = \|\overrightarrow{PQ} + \overrightarrow{QS}\| \leq \|\overrightarrow{PQ}\| + \|\overrightarrow{QS}\| = d(P, Q) + d(Q, S)$$

Straight Lines. A set L of points in R^3 is said to be a *straight line* (or, more simply, a line) if the set of all vectors $\overrightarrow{OP}$, for P in L, forms a one-dimensional linear variety of V_3. Thus the vectors (x, y, z) form a one-dimensional linear variety exactly when the points (x, y, z) form a line. We say that each point P in L is a point of the line L, or that P lies on L, and that L passes through P.

Let L be a straight line and let $\{\mathbf{v}_0 + t\mathbf{u}\}$ be the corresponding linear variety (with $\mathbf{u} \neq \mathbf{0}$). Then L consists of all P: (x, y, z) for which

$$\overrightarrow{OP} = \mathbf{v}_0 + t\mathbf{u}, \qquad -\infty < t < \infty \qquad (11\text{-}46)$$

or, in components, with $\mathbf{v}_0 = (x_0, y_0, z_0)$ and $\mathbf{u} = (a, b, c)$,

$$x = x_0 + at, \qquad y = y_0 + bt, \qquad z = z_0 + ct \qquad (11\text{-}46')$$

We call (11-46) a *vector equation of the line* and (11-46′) a set of *parametric equations of the line.*

When $t = 0$, $\overrightarrow{OP} = \mathbf{v}_0 = (x_0, y_0, z_0)$, so that P is the point P_0: (x_0, y_0, z_0), and we can write (11-46) as

$$\overrightarrow{OP} = \overrightarrow{OP_0} + t\mathbf{u}$$

or, since $\overrightarrow{P_0P} = \overrightarrow{OP} - \overrightarrow{OP_0}$, as

$$\overrightarrow{P_0P} = t\mathbf{u}$$

Thus, as t varies, P moves along the line and $\overrightarrow{P_0P}$ becomes all scalar multiples of $\mathbf{u}$; for $t = 1$, $\overrightarrow{P_0P} = \mathbf{u}$, for $t = 2$, $\overrightarrow{P_0P} = 2\mathbf{u}$, and so on (see Figure 11-4).

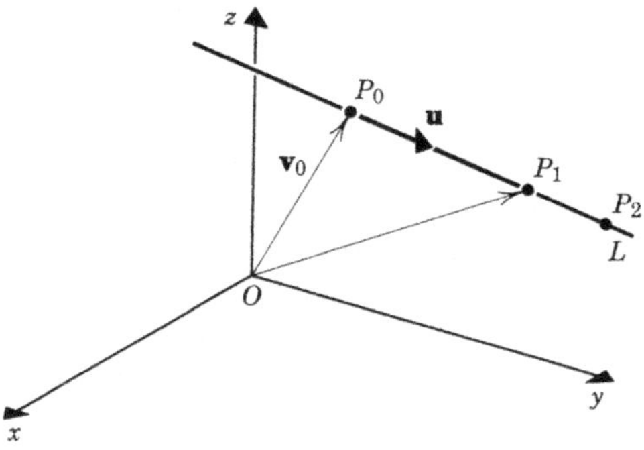

Figure 11-4 Line L in space.

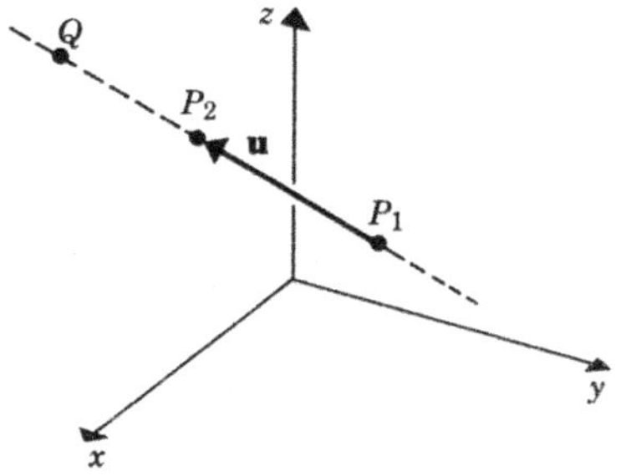

Figure 11-5 Line through two points.

The scalar multiples of $\mathbf{u}$ form the base space of the linear variety; we call this set the *base space of the line L;* it is a subspace of V_3. The same base space can also be represented as all multiples of $\mathbf{w}$, where $\mathbf{w}$ is a nonzero multiple of $\mathbf{u}$. We can also start at a different point, say P_1: (x_1, y_1, z_1) on L. The points P such that

$$\overrightarrow{OP} = \overrightarrow{OP_1} + t\mathbf{w}, \qquad -\infty < t < \infty$$

sweep out the same line L. Our geometric discussion is equivalent to the algebraic statement that $\{\mathbf{p} + \{t\mathbf{u}\}\}$ and $\{\mathbf{q} + \{t\mathbf{w}\}\}$ represent the same linear variety precisely when $\mathbf{u}$, $\mathbf{w}$ are nonzero multiples of each other (base space the same) and $\mathbf{q} - \mathbf{p}$ is a multiple of $\mathbf{u}$ ($\mathbf{q}$ is contained in $\{\mathbf{p} + \{t\mathbf{u}\}\}$); see Theorem 6 and its corollary in Section 9-5 (also see Section 1-14).

A line L is uniquely determined by any two distinct points P_1, P_2 on L. For $\overrightarrow{P_1P_2}$ must be in the base space and, hence,

$$\overrightarrow{OP} = \overrightarrow{OP_1} + t\,\overrightarrow{P_1P_2} \tag{11-47}$$

must be a vector equation of L. Thus (11-47) is our formula for the *equation of a line through two given points.* The corresponding parametric equations are

$$x = x_1 + t(x_2 - x_1), \quad y = y_1 + t(y_2 - y_1), \quad z = z_1 + t(z_2 - z_1) \qquad (11\text{-}47')$$

These formulas show clearly that the vectors $\mathbf{u}$ appearing in the different vector equations of a line L are the vectors "along L," that is, vectors $\overrightarrow{P_1P_2}$ for P_1 and P_2 on L, as in Figure 11-5. The components a, b, c of $\mathbf{u}$ are sometimes termed a set of *direction numbers or direction components* of L. If a, b, c is one set of direction numbers, then all sets are given by ka, kb, kc for $k \neq 0$. One can think of a, b, c as a triple proportion: $a:b:c$. This proportion is the 3-dimensional analogue of slope in the plane.

EXAMPLE 1. Find equations of the line L through P_1: (1, 2, 3) and P_2: (5, 0, 6), and determine whether Q: (9, −2, 9) is on the line.

Solution. By (11-47′) the equations for L are

$$x = 1 + t(5 - 1) = 1 + 4t, \qquad y = 2 - 2t, \qquad z = 3 + 3t$$

Hence, $\mathbf{u} = (4, -2, 3)$ is a vector along the line. The point Q is on the line if, and only if, $\overrightarrow{P_1Q}$ is a multiple of $\mathbf{u}$. Since $\overrightarrow{P_1Q} = (8, -4, 6)$, we ask whether

$$(8, -4, 6) = k(4, -2, 3)$$

The answer is yes, with $k = 2$. Therefore, Q is on the line (see Figure 11-5).

EXAMPLE 2. Show that the lines L_1: $\overrightarrow{OP} = (3, 4, 6) + t(2, 1, 1)$ and L_2: $\overrightarrow{OP} = (4, 2, 7) + t(3, -1, 2)$ intersect and find the point of intersection.

Solution. The parameter values need not be the same for both lines at the point of intersection. Hence, we seek t_1, t_2 so that $(3, 4, 6) + t_1(2, 1, 1)$ equals $(4, 2, 7) + t_2(3, -1, 2)$ or, in components,

$$3 + 2t_1 = 4 + 3t_2, \qquad 4 + t_1 = 2 - t_2, \qquad 6 + t_1 = 7 + 2t_2$$

We find that these equations are satisfied for precisely one pair: $t_1 = -1$, $t_2 = -1$. Hence, the lines meet at just one point P_0 and $\overrightarrow{OP_0} = (3, 4, 6) - (2, 1, 1) = (1, 3, 5)$; that is, P_0 is (1, 3, 5).

Parallel Lines. Two lines

$$L_1\colon \overrightarrow{OP} = \mathbf{a} + t\mathbf{u}, \qquad L_2\colon \overrightarrow{OP} = \mathbf{b} + t\mathbf{v}$$

are said to be parallel ($L_1 \parallel L_2$ or $L_2 \parallel L_1$) if $\mathbf{u}, \mathbf{v}$ are parallel—that is, are linearly dependent. Thus $L_1 \parallel L_2$ when the set of vectors along L_1 is the same as the set of vectors along L_2: namely, the set of all vectors $t\mathbf{u}$ or $t\mathbf{v}$ (Figure 11-6); or, in other language, $L_1 \parallel L_2$ when the corresponding linear varieties have the same base space $W = \text{Span}(\mathbf{u}) = \text{Span}(\mathbf{v})$. The definition is satisfied when L_1 coincides with L_2; we still call L_1 parallel to L_2 in this case but, for clarity, say that L_1, L_2 are *coincident parallel lines*.

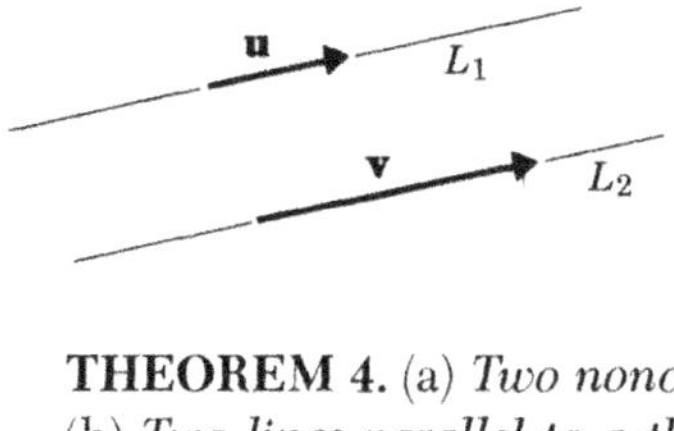

Figure 11-6 Parallel lines.

THEOREM 4. (a) *Two noncoincident parallel lines have no common point.* (b) *Two lines parallel to a third line are parallel to each other.* (c) *Through each point of R^3 there is one, and only one, line parallel to a given line.*

The proofs are left as exercises (Problem 10 below).

EXAMPLE 3. In the plane two lines having no point of intersection must be parallel. In 3-dimensional space this is not true. For example, the lines

$$L_1\colon \overrightarrow{OP} = (1, 0, 0) + t(0, 1, 0), \qquad L_2\colon \overrightarrow{OP} = (0, 0, 1) + t(1, 0, 0)$$

have no point in common and are not parallel (Figure 11-7). That the lines

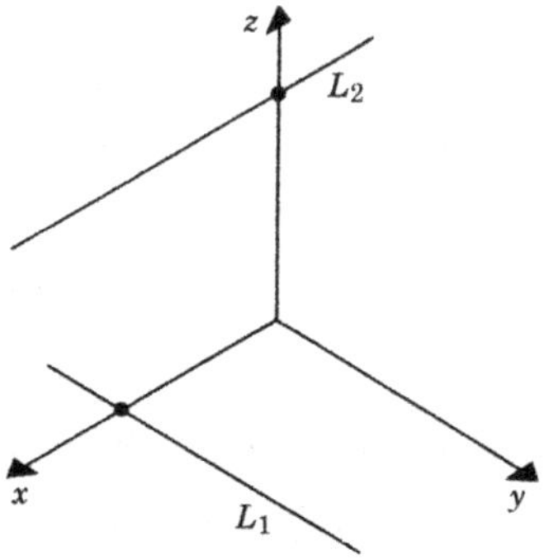

Figure 11-7 Skew lines.

are not parallel follows from the fact that $\mathbf{u} = (0, 1, 0)$ and $\mathbf{v} = (1, 0, 0)$ are linearly independent. To seek a point of intersection, we relabel the parameter on L_2 as τ and try to find t and τ so that

$$(1, 0, 0) + t(0, 1, 0) = (0, 0, 1) + \tau(1, 0, 0)$$

or $1 + 0t = 0 + \tau, 0 + t = 0\tau, 0 + 0t = 1 + 0\tau$. The first and last equations are contradictory, so that there is no point of intersection.

Two lines which do not intersect and are not parallel are called *skew lines*.

Distance along a line, line segments. On a line $L\colon \overrightarrow{OP} = \mathbf{a} + t\mathbf{u}$, the values of t serve as coordinates, so that L appears just as the number axis in Section 0-1. However, the distance from the "origin", where $t = 0$, to the position P corresponding to a general value of t, is

$$\|(\mathbf{a} + t\mathbf{u}) - (\mathbf{a} + 0\mathbf{u})\| = \|t\mathbf{u}\| = |t|\,\|\mathbf{u}\|$$

Hence if $\mathbf{u}$ is taken to be a unit vector, then the distance is precisely $|t|$; that is, t is a true coordinate measuring "directed distance" along L. To emphasize this, we usually denote the parameter by s in this case:

$$\overrightarrow{OP} = \mathbf{a} + s\mathbf{u} \qquad (\mathbf{u} = \text{unit vector})$$

Then s is the arc length along the path, exactly as in Section 4-28.

For a general line $L\colon \overrightarrow{OP} = \mathbf{a} + t\mathbf{u}$, we define a *line segment on* L to be the portion of L for which t lies in a closed interval, say $[t_1, t_2]$. Thus a line segment has vector equation

$$\overrightarrow{OP} = \mathbf{a} + t\mathbf{u}, \qquad t_1 \le t \le t_2$$

A line segment is the range of the function $\mathbf{f}(t) = \mathbf{a} + t\mathbf{u}$, having $[t_1, t_2]$ as its domain. Thus a line segment is the image of a finite closed interval. When the arc length parameter s is used, the length of the segment is just $|s_2 - s_1|$.

Otherwise, it is $|t_2 - t_1| \|\mathbf{u}\|$, by the same reasoning as above. To find the midpoint of the segment, or to divide it in a given ratio, one need only carry out the desired operation with respect to t (or s) in its interval.

Angles. The angle between two straight lines

$$L_1\colon \overrightarrow{OP} = \mathbf{a} + t\mathbf{u}, \qquad L_2\colon \overrightarrow{OP} = \mathbf{b} + t\mathbf{v}, \qquad -\infty < t < \infty$$

is defined to be the angle φ between $\mathbf{u}$ and $\mathbf{v}$ or its supplement $\pi - \varphi = \measuredangle(\mathbf{u}, -\mathbf{v})$. In each case, one must specify which value is chosen. When $\varphi = \pi/2$, we say that L_1, L_2 are *orthogonal* or *perpendicular*. We tend to avoid speaking of the angle between nonintersecting lines, but the definition is always meaningful.

We can also refer to angles between a line and a vector, to orthogonality of a line and a vector, to a vector parallel to a line. Thus $L\colon \overrightarrow{OP} = \mathbf{a} + t\mathbf{u}$ is orthogonal to the vector $\mathbf{v}$ if $\mathbf{u}, \mathbf{v}$ are orthogonal.

‡11-5 LINES IN n-DIMENSIONAL EUCLIDEAN SPACE

The theory of Section 11-4 extends to n-dimensional Euclidean space R^n. The points are the ordered n-tuples $(x_1, \ldots, x_n)$, the origin O is the point $(0, \ldots, 0)$. The vector from $P\colon (x_1, \ldots, x_n)$ to $Q\colon (y_1, \ldots, y_n)$ is the vector $\mathbf{u} = (y_1 - x_1, \ldots, y_n - x_n)$. In particular, $\overrightarrow{OP} = (x_1, \ldots, x_n)$. The rules $\overrightarrow{P_1P_2} = \overrightarrow{OP_2} - \overrightarrow{OP_1}$ and $\overrightarrow{PQ} + \overrightarrow{QS} = \overrightarrow{PS}$ remain valid, as does Theorem 3, with the same proofs. A straight line L in R^n is a set of points such that the corresponding vectors form a one-dimensional linear variety $\{\mathbf{v} + \{t\mathbf{u}\}\}$ of V_n. Hence, the line has vector equation $\overrightarrow{OP} = \mathbf{v} + t\mathbf{u}$, and corresponding parametric equations $x_i = v_i + tu_i$, $i = 1, \ldots, n$. As in R^3, lines in R^n are parallel if they have the same base space and Theorem 4 remains valid in R^n, with essentially the same proofs.

The concepts of distance along a line, line segments, angle between lines, and orthogonal lines can also be treated as in Section 11-4.

PROBLEMS

(Throughout these problems assume that $-\infty < t < \infty$.)

1. Determine whether L_1 and L_2 are parallel or coincident. Also graph the lines.
   (a) $L_1\colon \overrightarrow{OP} = (1, 2, 2) + t(3, 5, 7)$; $L_2\colon \overrightarrow{OP} = (7, 12, 16) + t(6, 10, 14)$.
   (b) $L_1\colon \overrightarrow{OP} = (3, 4, 2) + t(1, 0, 7)$; $L_2\colon \overrightarrow{OP} = (6, 8, 4) + t(2, 0, 14)$.
2. In each case find a vector equation of the line L through P_1, P_2 and determine whether Q is on L:
   (a) $P_1 = (3, 5, 1)$, $P_2 = (2, 0, 7)$, $Q = (0, -10, 19)$.
   (b) $P_1 = (2, 0, 5)$, $P_2 = (-1, 0, 0)$, $Q = (4, 2, 3)$.
   (c) $P_1 = (4, -1, 5)$, $P_2 = (6, 2, -3)$, $Q = (0, 1, 2)$.
3. Graph and determine whether the given lines intersect or are skew and, if they

intersect, find the point of intersection:

(a) L_1: $\overrightarrow{OP} = (3, 5, 1) + t(1, 0, 2)$; L_2: $\overrightarrow{OP} = (1, 2, 5) + t(4, 3, 2)$.

(b) L_1: line through (2, 3, 3) and (6, 1, 7); L_2: line through (6, 2, 8) and (7, 2, 9).

(c) L_1: $x=-2+3t$, $y=-4+5t$, $z=3-2t$; L_2: $x=5+2t$, $y=1$, $z=3+t$.

(d) L_1: the line through P_1 and Q_1; L_2: the line through P_2 and Q_2, where $\overrightarrow{P_1Q_1}$, $\overrightarrow{P_2Q_2}$, $\overrightarrow{P_1Q_2}$ are known to be linearly independent.

4. Graph and determine whether the following sets of points in R^3 are *collinear*, that is, lie on a line.

(a) (2, 1, 4), (4, 4, −1), (6, 7, −6); (b) (1, 2, 3), (−4, 2, 1), (1, 1, 2);

(c) (1, 2, 3), (5, −4, 7), (3, −1, 5); (d) (7, 3, 2), (6, 0, −4), (7, 3, 2).

5. Let four lines be given: L_1: $\overrightarrow{OP} = (1, -1, 2) + t(1, 2, 3)$; L_2: through (7, 2, −1) and (4, −4, 0); L_3: through (1, 1, 4) with all direction numbers equal; L_4: through (1, 1, 1) parallel to the line $\overrightarrow{OP} = t(7, -4, -3)$. Test for orthogonality:

(a) L_1, L_2 (b) L_1, L_3 (c) L_1, L_4

(d) L_2, L_3 (e) L_2, L_4 (f) L_3, L_4

6. Find the acute angle between the following pairs of lines of Problem 5:

(a) L_1, L_2 (b) L_1, L_3 (c) L_1, L_4 (d) L_2, L_4

7. The *direction angles* of a line L: $\overrightarrow{OP} = \mathbf{a} + t\mathbf{u}$ are defined to be the direction angles of $\mathbf{u}$ or those of $-\mathbf{u}$ (see Problem 4 following Section 11-3); for each choice, the cosines of the angles are said to form a set of *direction cosines* of L. The unit sphere in R^3 is the set of all (x, y, z) for which $x^2 + y^2 + z^2 = 1$. Show that the two sets of direction cosines of a line L are the two sets of coordinates of the points at which a line parallel to L through O meets the unit sphere.

8. Let the line L through P_1: (2, 5, 6) and P_2: (1, 2, 2) be given.

(a) Find the midpoint of the segment P_1P_2 joining P_1 to P_2.

(b) Divide the segment P_1P_2 into three equal parts.

(c) Find all points P on L such that $\|\overrightarrow{P_1P}\| = 3\|\overrightarrow{P_2P}\|$.

9. Let $A = (a_{ij})$ be a 2 by 3 matrix of rank 2. Show that the solutions of

$$A \operatorname{col}(x, y, z) = \operatorname{col}(k_1, k_2)$$

that is, the solutions of the equations

$$a_{11}x + a_{12}y + a_{13}z = k_1, \qquad a_{21}x + a_{22}y + a_{23}z = k_2$$

form a line in space.

10. Prove the following parts of Theorem 4: (a) Part (a), (b) Part (b), (c) Part (c).

‡11. Let L_1: $\overrightarrow{OP} = \mathbf{u} + t\mathbf{v}$, L_2: $\overrightarrow{OP} = \mathbf{z} + t\mathbf{w}$ be two lines in R^n. Show that L_1 and L_2 coincide if, and only if, the matrix

$$\begin{pmatrix} v_1 & v_2 & \cdots & v_n \\ w_1 & w_2 & \cdots & w_n \\ z_1 - u_1 & z_2 - u_2 & \cdots & z_n - u_n \end{pmatrix}$$

has rank 1.

‡12. In R^n let three points P_k: $(x_{k1}, \ldots, x_{kn})$ $(k = 1, 2, 3)$ be given. Show that the three points lie on a line (are collinear) if, and only if,

$$\operatorname{rank}\begin{pmatrix} x_{11} & \cdots & x_{1n} & 1 \\ x_{21} & \cdots & x_{2n} & 1 \\ x_{31} & \cdots & x_{3n} & 1 \end{pmatrix} \leq 2$$

‡13. Let $A = (a_{ij})$ be an $n - 1$ by n matrix of rank $n - 1$. Let $\mathbf{k}$ be a vector of V_{n-1}. Show that the solutions of the equation $A\mathbf{x} = \mathbf{k}$ form a line in R^n.

11-6 THE CROSS PRODUCT (VECTOR PRODUCT)

In this section we introduce a new multiplication of vectors in R^3, the cross product or vector product of $\mathbf{u}$ and $\mathbf{v}$, denoted by $\mathbf{u} \times \mathbf{v}$ and defined by equation (11-61) below. Before proceeding to this definition, we discuss intuitively the geometric ideas leading to the cross product.

Orientation of Space. In introducing coordinate axes in space one has a choice of a right-handed system, as in Figure 11-8*a*, or a left-handed system, as in Figure 11-8*b*. For the first case the directions of the x-, y-, and z-axes can be approximated by the thumb, index finger, and middle finger, respectively, of the right hand; for the second case, the right hand cannot be used, but one can use the left hand.

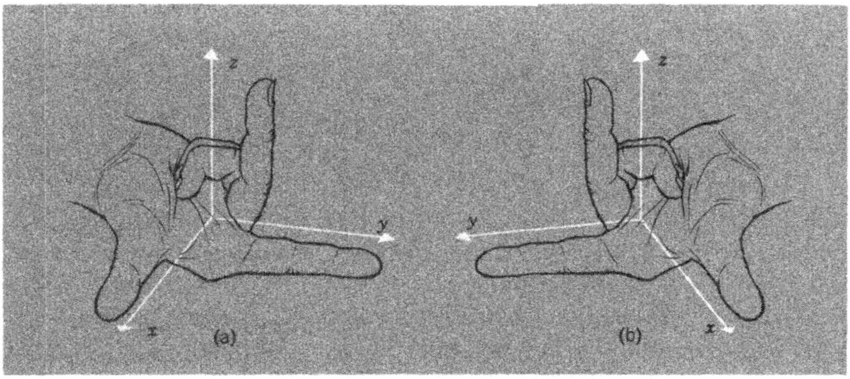

Figure 11-8 (*a*) Right-handed and (*b*) left-handed coordinate system.

Once axes have been chosen, one says that space has been *oriented*. The vectors $\mathbf{i}, \mathbf{j}, \mathbf{k}$, along the axes, then serve as a standard to which all other triples of vectors can be compared. One says that a triple of linearly independent vectors $\mathbf{u}, \mathbf{v}, \mathbf{w}$ *in that order* have orientation similar to $\mathbf{i}, \mathbf{j}, \mathbf{k}$ if one can gradually change the directions of the vectors $\mathbf{u}, \mathbf{v}, \mathbf{w}$, without ever making them linearly dependent, so that they finally take on the directions of $\mathbf{i}, \mathbf{j}, \mathbf{k}$, respectively. If this is not possible, we say that the triple $\mathbf{u}, \mathbf{v}, \mathbf{w}$ has orientation opposite to **i**, **j**, **k**. When **u**, **v**, **w** are linearly dependent, no statement is made on orientation. One can now group all triples of vectors having orientation similar to **i**, **j**, **k** in one class, which we call the class of *positively oriented triples,* or simply the class of positive triples. Similarly, the triples having orientation opposite to **i**, **j**, **k** form the class of *negative triples.*

We remark that **i**, **j**, **k** itself is a positive triple, as are **j**, **k**, **i** and **k**, **i**, **j**. For one can rotate the triple **i**, **j**, **k** about the line through the origin and (1, 1, 1) to gradually permute the vectors. However, one cannot go from **i**, **j**, **k** to **j**, **i**, **k**; this would be like going from a right-handed system to a left-handed system. In general, if **u**, **v**, **w** is a positive triple, then **w**, **u**, **v** and **v**, **w**, **u** are also positive, whereas **v**, **u**, **w** and **w**, **v**, **u** and **u**, **w**, **v** are negative. Hence, it is only the "cyclic order" that counts.

We arrive at an algebraic test for a positive triple by the following reasoning. If **u**, **v**, **w** have the same directions as **i**, **j**, **k**, then $\mathbf{u} = a\mathbf{i}$, $\mathbf{v} = b\mathbf{j}$, $\mathbf{w} = c\mathbf{k}$ for *positive* scalars a, b, c. The determinant whose rows are **u**, **v**, and **w** is then

$$\begin{vmatrix} a & 0 & 0 \\ 0 & b & 0 \\ 0 & 0 & c \end{vmatrix} = abc$$

and, hence, is positive. If we now gradually change **u**, **v**, and **w**, then the determinant will change continuously, but as long as **u**, **v**, **w** never become linearly dependent, the determinant cannot become 0 and, hence, *remains positive.* Since by such a process we can eventually reach an arbitrary positive triple, it follows that for every positive triple the corresponding determinant must be positive. Thus **u**, **v**, **w** *is a positive triple precisely when*

$$\begin{vmatrix} u_1 & u_2 & u_3 \\ v_1 & v_2 & v_3 \\ w_1 & w_2 & w_3 \end{vmatrix} > 0 \tag{11-60}$$

and, similarly, **u**, **v**, **w** *is a negative triple when the determinant is negative.* We have arrived at this criterion by an intuitive reasoning. For what follows, we shall however regard the criterion as our definition, and shall base all future discussion and proofs on it. In particular, we are thus *defining* a triple **u**, **v**, **w** to be positive when (11-60) holds true.

Idea of the Cross Product. Now let space be oriented and let **u** and **v** be two vectors in space. It then appears evident that we can always find a nonzero vector **q** orthogonal to both **u** and **v**, as in Figure 11-9. In fact, if **q** is one such vector, then all vectors $k\mathbf{q}$ are also orthogonal to **u** and to **v**, so that there are infinitely many nonzero vectors orthogonal to the two given vectors. We want to pick out just *one* of these in a natural way. To do so, we first fix the length of **q** by requiring that $\|\mathbf{q}\|$ equal the area of the parallelogram whose edges, properly directed, represent **u** and **v**. This requirement still leaves two choices. For if **q** is one such vector, then $-\mathbf{q}$ is another. To settle the matter, we choose **q** so that **u**, **v**, **q** *is a positive triple.*

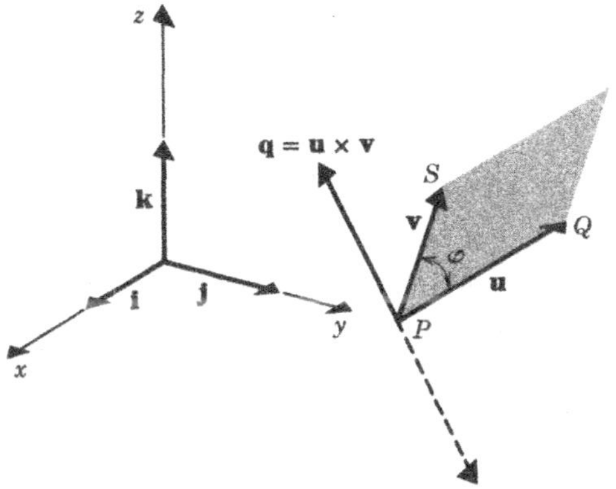

Figure 11-9 The formation of the cross product.

The vector **q** is then uniquely determined by the three conditions, and we denote it by **u** × **v**. Thus **q** = **u** × **v** is the unique vector which has the three properties:

(a) **q** is orthogonal to **u** and to **v**,
(b) ‖**q**‖ equals the area of the parallelogram of edges **u**, **v**,
(c) **u**, **v**, **q** is a positive triple.

When **u** and **v** are linearly dependent, we would naturally assign area 0 to the parallelogram, so that, in this case, we take **q** to be **0**.

With this intuitive geometric background, we seek an algebraic formula for **q** = **u** × **v**. We write **q** = (x, y, z). Then the requirement that **q** be orthogonal to **u** and to **v** is equivalent to **u** · **q** = 0 and **v** · **q** = 0 or

$$u_1x + u_2y + u_3z = 0$$
$$v_1x + v_2y + v_3z = 0$$

These are two simultaneous equations for x, y and z. Now if **u**, **v** are linearly independent, the matrix

$$B = \begin{pmatrix} u_1 & u_2 & u_3 \\ v_1 & v_2 & v_3 \end{pmatrix}$$

has rank 2, and we obtain solutions forming a one-dimensional subspace of V_3 (see Section 10-6). Thus the vectors orthogonal to **u** and **v** are all the vectors of form c**q**, where **q** is one nonzero vector orthogonal to **u** and **v**. If, for example, $u_1v_2 - u_2v_1 \neq 0$, then we can solve for x and y in terms of z. We find that

$$\begin{vmatrix} u_1 & u_2 \\ v_1 & v_2 \end{vmatrix} x = \begin{vmatrix} u_2 & u_3 \\ v_2 & v_3 \end{vmatrix} z, \qquad \begin{vmatrix} u_1 & u_2 \\ v_1 & v_2 \end{vmatrix} y = \begin{vmatrix} u_3 & u_1 \\ v_3 & v_1 \end{vmatrix} z$$

Hence, a particular solution is

$$x = \begin{vmatrix} u_2 & u_3 \\ v_2 & v_3 \end{vmatrix}, \qquad y = \begin{vmatrix} u_3 & u_1 \\ v_3 & v_1 \end{vmatrix}, \qquad z = \begin{vmatrix} u_1 & u_2 \\ v_1 & v_2 \end{vmatrix}$$

In fact, we shall see that, regardless of the rank of B, the vector

$$\mathbf{q} = \begin{vmatrix} u_2 & u_3 \\ v_2 & v_3 \end{vmatrix}\mathbf{i} + \begin{vmatrix} u_3 & u_1 \\ v_3 & v_1 \end{vmatrix}\mathbf{j} + \begin{vmatrix} u_1 & u_2 \\ v_1 & v_2 \end{vmatrix}\mathbf{k} \tag{11-61}$$

is orthogonal to $\mathbf{u}$ and to $\mathbf{v}$, and has all the desired properties. Vectors of the form $c\mathbf{q}$ are also orthogonal to $\mathbf{u}$ and $\mathbf{v}$, but only for $c = 1$ does one satisfy all three conditions mentioned above. We are thus led to the definition:

Definition. The *cross product* (vector product) of $\mathbf{u}$, $\mathbf{v}$ in that order, is the vector $\mathbf{q}$ defined by (11-61). We write $\mathbf{q} = \mathbf{u} \times \mathbf{v}$. Thus

$$\mathbf{u} \times \mathbf{v} = \left(\begin{vmatrix} u_2 & u_3 \\ v_2 & v_3 \end{vmatrix}, \begin{vmatrix} u_3 & u_1 \\ v_3 & v_1 \end{vmatrix}, \begin{vmatrix} u_1 & u_2 \\ v_1 & v_2 \end{vmatrix} \right) \tag{11-61'}$$

THEOREM 5. *Let $\mathbf{u}$, $\mathbf{v}$, $\mathbf{w}$ be vectors in V_3, let c be a scalar. Then the following rules hold true:*

$$\left.\begin{array}{l} \mathbf{u} \times \mathbf{v} = \mathbf{0}, \textit{ if } \mathbf{u}, \mathbf{v} \textit{ are linearly dependent,} \\ \mathbf{u} \times \mathbf{v} \neq \mathbf{0}, \textit{ if } \mathbf{u}, \mathbf{v} \textit{ are linearly independent.} \end{array}\right\} \tag{11-62}$$

$$\mathbf{u} \times \mathbf{v} = -\mathbf{v} \times \mathbf{u}. \tag{11-63}$$

$$\mathbf{w} \cdot (\mathbf{u} \times \mathbf{v}) = \begin{vmatrix} w_1 & w_2 & w_3 \\ u_1 & u_2 & u_3 \\ v_1 & v_2 & v_3 \end{vmatrix} \tag{11-64}$$

$$\mathbf{u} \cdot (\mathbf{u} \times \mathbf{v}) = 0, \qquad \mathbf{v} \cdot (\mathbf{u} \times \mathbf{v}) = 0 \tag{11-65}$$

$$\mathbf{u} \times (\mathbf{w} + \mathbf{v}) = (\mathbf{u} \times \mathbf{w}) + (\mathbf{u} \times \mathbf{v}) \tag{11-66}$$

$$(c\mathbf{u}) \times \mathbf{v} = c(\mathbf{u} \times \mathbf{v}) = \mathbf{u} \times (c\mathbf{v}) \tag{11-67}$$

$$\|\mathbf{u} \times \mathbf{v}\| = \|\mathbf{u}\| \, \|\mathbf{v}\| \sin \varphi, \text{ where } \varphi = \measuredangle(\mathbf{u}, \mathbf{v}) \tag{11-68}$$

If $\mathbf{u}$, $\mathbf{v}$ are linearly independent, then $\mathbf{u}$, $\mathbf{v}$, $\mathbf{u} \times \mathbf{v}$ are linearly independent and form a positive triple.

PROOF. If $\mathbf{u}$, $\mathbf{v}$ are linearly dependent, then the matrix B, defined above, has rank less than 2, so that all three components of $\mathbf{u} \times \mathbf{v}$ are 0. If $\mathbf{u}$, $\mathbf{v}$ are linearly independent, then B has rank 2 and at least one component of $\mathbf{u} \times \mathbf{v}$ is nonzero. Thus (11-62) is proved.

To form $\mathbf{v} \times \mathbf{u}$, we interchange the rows in all three determinants in (11-61'); hence, all the signs are reversed and (11-63) follows.

To prove (11-64) one expands the determinant by minors of the first row to obtain

$$w_1 \begin{vmatrix} u_2 & u_3 \\ v_2 & v_3 \end{vmatrix} + w_2 \begin{vmatrix} u_3 & u_1 \\ v_3 & v_1 \end{vmatrix} + w_3 \begin{vmatrix} u_1 & u_2 \\ v_1 & v_2 \end{vmatrix} = \mathbf{w} \cdot (\mathbf{u} \times \mathbf{v})$$

(after changing sign and interchanging columns in the second determinant).

The relation (11-65) follows immediately from (11-64), since a determinant is zero whenever two rows are equal. Relations (11-66) and (11-67) also follow from properties of determinants (Theorems 12 and 11 of Section 10-13).

To prove (11-68) we observe that

$$\begin{aligned}\|\mathbf{u}\times\mathbf{v}\|^2 &= (u_2v_3 - u_3v_2)^2 + (u_1v_2 - u_2v_1)^2 + (u_3v_1 - u_1v_3)^2\\ &= (u_1{}^2 + u_2{}^2 + u_3{}^2)(v_1{}^2 + v_2{}^2 + v_3{}^2) - (u_1v_1 + u_2v_2 + u_3v_3)^2\\ &= \|\mathbf{u}\|^2\|\mathbf{v}\|^2 - \|\mathbf{u}\|^2\|\mathbf{v}\|^2\cos^2\varphi\\ &= \|\mathbf{u}\|^2\|\mathbf{v}\|^2(1-\cos^2\varphi) = \|\mathbf{u}\|^2\|\mathbf{v}\|^2\sin^2\varphi\end{aligned}$$

We now take square roots, observing that $\sin\varphi \geq 0$, since $0 \leq \varphi \leq \pi$.

Now let **u**, **v** be linearly independent. Then $\mathbf{u}\times\mathbf{v} \neq \mathbf{0}$ by (11-62). To test for linear independence of **u**, **v**, $\mathbf{u}\times\mathbf{v}$, we form the determinant whose rows are $\mathbf{u}\times\mathbf{v}$, **u**, **v**. By (11-64) this determinant equals $(\mathbf{u}\times\mathbf{v})\cdot(\mathbf{u}\times\mathbf{v}) = \|\mathbf{u}\times\mathbf{v}\|^2 > 0$. Therefore $\mathbf{u}\times\mathbf{v}$, **u**, **v** are linearly independent and, in the order stated, form a positive triple. Hence, also, **u**, **v**, $\mathbf{u}\times\mathbf{v}$ form a positive triple.

Remark 1. From (11-68) we conclude that, for $\overrightarrow{PQ} = \mathbf{u}$, $\overrightarrow{PS} = \mathbf{v}$,

$$\|\overrightarrow{PQ}\times\overrightarrow{PS}\| = \|\overrightarrow{PQ}\|\,\|\overrightarrow{PS}\|\sin\varphi$$

Hence, as in Figure 11-9, we conclude that $\|\mathbf{u}\times\mathbf{v}\|$ is the area of a parallelogram whose sides, properly directed, represent **u** and **v**; a full justification of this interpretation of $\|\mathbf{u}\times\mathbf{v}\|$ is given in Section 11-15 below.

Remark 2. To compute a vector product, one can write the formula (11-61′) in the following symbolic form:

$$\mathbf{u}\times\mathbf{v} = \begin{vmatrix}\mathbf{i} & \mathbf{j} & \mathbf{k}\\ u_1 & u_2 & u_3\\ v_1 & v_2 & v_3\end{vmatrix} = \begin{vmatrix}u_2 & u_3\\ v_2 & v_3\end{vmatrix}\mathbf{i} - \begin{vmatrix}u_1 & u_3\\ v_1 & v_3\end{vmatrix}\mathbf{j} + \begin{vmatrix}u_1 & u_2\\ v_1 & v_2\end{vmatrix}\mathbf{k}$$

$$= \begin{vmatrix}u_2 & u_3\\ v_2 & v_3\end{vmatrix}\mathbf{i} + \begin{vmatrix}u_3 & u_1\\ v_3 & v_1\end{vmatrix}\mathbf{j} + \begin{vmatrix}u_1 & u_2\\ v_1 & v_2\end{vmatrix}\mathbf{k} \qquad (11\text{-}61'')$$

Thus the determinant is to be expanded by minors of the first row. For example,

$$(2, 1, 2)\times(3, 0, 5) = \begin{vmatrix}\mathbf{i} & \mathbf{j} & \mathbf{k}\\ 2 & 1 & 2\\ 3 & 0 & 5\end{vmatrix} = \begin{vmatrix}1 & 2\\ 0 & 5\end{vmatrix}\mathbf{i} - \begin{vmatrix}2 & 2\\ 3 & 5\end{vmatrix}\mathbf{j} + \begin{vmatrix}2 & 1\\ 3 & 0\end{vmatrix}\mathbf{k}$$

and the value $5\mathbf{i} - 4\mathbf{j} - 3\mathbf{k} = (5, -4, -3)$ is obtained. The minus sign before the second small determinant can be avoided by interchanging columns. We deduce the following useful vector products:

$$\mathbf{i}\times\mathbf{j} = \begin{vmatrix}\mathbf{i} & \mathbf{j} & \mathbf{k}\\ 1 & 0 & 0\\ 0 & 1 & 0\end{vmatrix} = \mathbf{k}, \qquad \mathbf{j}\times\mathbf{k} = \mathbf{i}, \qquad \mathbf{k}\times\mathbf{i} = \mathbf{j}$$

while

$$\mathbf{j} \times \mathbf{i} = -\mathbf{i} \times \mathbf{j} = -\mathbf{k}, \quad \mathbf{k} \times \mathbf{j} = -\mathbf{i}, \quad \mathbf{i} \times \mathbf{k} = -\mathbf{j}$$
$$\mathbf{i} \times \mathbf{i} = \mathbf{0}, \quad \mathbf{j} \times \mathbf{j} = \mathbf{0}, \quad \mathbf{k} \times \mathbf{k} = \mathbf{0}$$

by the rules (11-62), (11-63). We can now also write, by (11-66) and (11-67):

$$\begin{aligned}(2, 1, 2) \times (3, 0, 5) &= (2\mathbf{i} + \mathbf{j} + 2\mathbf{k}) \times (3\mathbf{i} + 5\mathbf{k}) \\ &= 6\mathbf{i} \times \mathbf{i} + 10\mathbf{i} \times \mathbf{k} + 3\mathbf{j} \times \mathbf{i} + \cdots \\ &= \mathbf{0} - 10\mathbf{j} - 3\mathbf{k} + \cdots = 5\mathbf{i} - 4\mathbf{j} - 3\mathbf{k}\end{aligned}$$

Remark 3. The rule (11-63) indicates that the vector product is not commutative. It is also not associative. For example, $\mathbf{i} \times (\mathbf{i} \times \mathbf{j}) = \mathbf{i} \times \mathbf{k} = -\mathbf{j}$, $(\mathbf{i} \times \mathbf{i}) \times \mathbf{j} = \mathbf{0} \times \mathbf{j} = \mathbf{0}$.

11-7 TRIPLE PRODUCTS

We saw in (11-64) that $\mathbf{w} \cdot (\mathbf{u} \times \mathbf{v})$ equals the determinant whose rows are $\mathbf{w}$, $\mathbf{u}$, and $\mathbf{v}$. We call $\mathbf{w} \cdot (\mathbf{u} \times \mathbf{v})$ a *scalar triple product of* $\mathbf{w}$, $\mathbf{u}$, *and* $\mathbf{v}$. Here the parentheses are not needed, since $\mathbf{w} \cdot \mathbf{u} \times \mathbf{v}$ can only be interpreted as $\mathbf{w} \cdot (\mathbf{u} \times \mathbf{v})$ [the expression $(\mathbf{w} \cdot \mathbf{u}) \times \mathbf{v}$ has no meaning, since $\mathbf{w} \cdot \mathbf{u}$ is not a vector]. Since the inner product is commutative,

$$\mathbf{w} \cdot \mathbf{u} \times \mathbf{v} = \mathbf{u} \times \mathbf{v} \cdot \mathbf{w} \tag{11-70}$$

If we interchange two rows of the determinant in (11-64), we multiply the value by -1. Thus $\mathbf{w} \cdot \mathbf{u} \times \mathbf{v} = -\mathbf{u} \cdot \mathbf{w} \times \mathbf{v}$. If we repeatedly interchange, we deduce that

$$\begin{aligned}\mathbf{w} \cdot \mathbf{u} \times \mathbf{v} &= -\mathbf{u} \cdot \mathbf{w} \times \mathbf{v} = \mathbf{u} \cdot \mathbf{v} \times \mathbf{w} \\ &= \mathbf{v} \cdot \mathbf{w} \times \mathbf{u} = -\mathbf{v} \cdot \mathbf{u} \times \mathbf{w} = -\mathbf{w} \cdot \mathbf{v} \times \mathbf{u}\end{aligned}$$

If we combine these results with (11-70), we see that one, in fact, can obtain only two values $\pm c$ for a scalar triple product formed of $\mathbf{u}$, $\mathbf{v}$, and $\mathbf{w}$ in some order and for some choice of position of dot and cross. Interchanging dot and cross has no effect, and only the cyclic order counts. Thus

$$\begin{aligned}\mathbf{w} \cdot \mathbf{u} \times \mathbf{v} &= \mathbf{w} \times \mathbf{u} \cdot \mathbf{v} = \mathbf{u} \cdot \mathbf{v} \times \mathbf{w} = \mathbf{u} \times \mathbf{v} \cdot \mathbf{w} \\ &= \mathbf{v} \cdot \mathbf{w} \times \mathbf{u} = \mathbf{v} \times \mathbf{w} \cdot \mathbf{u} = c \\ \mathbf{w} \cdot \mathbf{v} \times \mathbf{u} &= \mathbf{w} \times \mathbf{v} \cdot \mathbf{u} = \mathbf{v} \cdot \mathbf{u} \times \mathbf{w} = \mathbf{v} \times \mathbf{u} \cdot \mathbf{w} \\ &= \mathbf{u} \cdot \mathbf{w} \times \mathbf{v} = \mathbf{u} \times \mathbf{w} \cdot \mathbf{v} = -c\end{aligned} \tag{11-71}$$

If $\mathbf{u}$, $\mathbf{v}$, $\mathbf{w}$ form a positive triple, then c is positive and, hence, $-c$ is negative. If $\mathbf{u}$, $\mathbf{v}$, $\mathbf{w}$ are linearly dependent, then the determinant whose rows are $\mathbf{u}$, $\mathbf{v}$ and $\mathbf{w}$ must equal 0, so that all the scalar triple products in (11-71) are 0.

It will be seen in Section 11-15 that $\mathbf{w} \cdot \mathbf{u} \times \mathbf{v}$ equals $\pm V$, where V is the volume of the parallelepiped whose edges, properly directed, represent the three vectors. The $+$ sign is used when the vectors form a positive triple, the $-$ sign when they form a negative triple. In the case of a positive triple, the angle β between $\mathbf{w}$ and $\mathbf{u} \times \mathbf{v}$ is acute and $\|\mathbf{w}\| \cos \beta$, the component of $\mathbf{w}$ in the direction of $\mathbf{u} \times \mathbf{v}$, is the altitude of the parallelepiped. But $\mathbf{w} \cdot \mathbf{u} \times \mathbf{v} = \mathbf{u} \times \mathbf{v} \cdot \mathbf{w} = (\|\mathbf{u} \times \mathbf{v}\|)(\|\mathbf{w}\| \cos \beta) =$ area of base times altitude (See Figure 11-10).

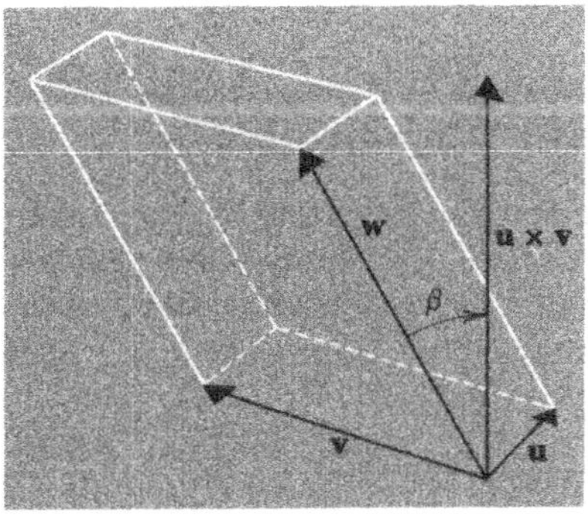

Figure 11-10 Scalar triple product as volume.

One can also form a triple product $\mathbf{u} \times (\mathbf{v} \times \mathbf{w})$. Here, the parentheses are needed, since both $\mathbf{u} \times (\mathbf{v} \times \mathbf{w})$ and $(\mathbf{u} \times \mathbf{v}) \times \mathbf{w}$ have meaning and are, in general, unequal. These triple products are called *vector triple products;* in each case, the value is a vector. For these products one has the identities:

$$\mathbf{u} \times (\mathbf{v} \times \mathbf{w}) = (\mathbf{u} \cdot \mathbf{w})\mathbf{v} - (\mathbf{u} \cdot \mathbf{v})\mathbf{w} \qquad (11\text{-}72)$$

$$(\mathbf{u} \times \mathbf{v}) \times \mathbf{w} = (\mathbf{w} \cdot \mathbf{u})\mathbf{v} - (\mathbf{w} \cdot \mathbf{v})\mathbf{u} \qquad (11\text{-}73)$$

The right-hand side, for both identities, can be described by the words: (outer dot remote) adjacent minus (outer dot adjacent) remote.

To prove (11-72) one can reason as follows. Let $\mathbf{v}$ and $\mathbf{w}$ be fixed. Then the left side of (11-72) assigns to each vector $\mathbf{u}$ of V_3 a vector $T(\mathbf{u}) = \mathbf{u} \times (\mathbf{v} \times \mathbf{w})$ of V_3. Furthermore, T is a linear transformation, since

$$\begin{aligned} T(c_1\mathbf{u}_1 + c_2\mathbf{u}_2) &= (c_1\mathbf{u}_1 + c_2\mathbf{u}_2) \times (\mathbf{v} \times \mathbf{w}) \\ &= (c_1\mathbf{u}_1) \times (\mathbf{v} \times \mathbf{w}) + (c_2\mathbf{u}_2) \times (\mathbf{v} \times \mathbf{w}) \\ &= c_1[\mathbf{u}_1 \times (\mathbf{v} \times \mathbf{w})] + c_2[\mathbf{u}_2 \times (\mathbf{v} \times \mathbf{w})] \\ &= c_1T(\mathbf{u}_1) + c_2T(\mathbf{u}_2) \end{aligned}$$

Similarly, for fixed $\mathbf{v}$ and $\mathbf{w}$, the right side of (11-72) assigns to each vector $\mathbf{u}$ a vector $S(\mathbf{u}) = (\mathbf{u} \cdot \mathbf{w})\mathbf{v} - (\mathbf{u} \cdot \mathbf{v})\mathbf{w}$ of V_3. The mapping S is also a linear transformation on V_3 (Problem 6). To show that T and S are the same linear transformation, we need only show that they agree on the basis vectors $\mathbf{i}, \mathbf{j}, \mathbf{k}$ (see Section 9-11). But

$$\begin{aligned} T(\mathbf{i}) &= \mathbf{i} \times (\mathbf{v} \times \mathbf{w}) \\ &= \mathbf{i} \times [(v_2w_3 - v_3w_2)\mathbf{i} + (v_3w_1 - v_1w_3)\mathbf{j} + (v_1w_2 - v_2w_1)\mathbf{k}] \\ &= (v_3w_1 - v_1w_3)\mathbf{k} + (v_1w_2 - v_2w_1)(-\mathbf{j}) \\ S(\mathbf{i}) &= (\mathbf{i} \cdot \mathbf{w})\mathbf{v} - (\mathbf{i} \cdot \mathbf{v})\mathbf{w} \\ &= w_1(v_1\mathbf{i} + v_2\mathbf{j} + v_3\mathbf{k}) - v_1(w_1\mathbf{i} + w_2\mathbf{j} + w_3\mathbf{k}) \\ &= (v_3w_1 - v_1w_3)\mathbf{k} + (v_1w_2 - v_2w_1)(-\mathbf{j}) \end{aligned}$$

Hence, $T(\mathbf{i}) = S(\mathbf{i})$. Similarly, $T(\mathbf{j}) = S(\mathbf{j})$ and $T(\mathbf{k}) = S(\mathbf{k})$. Therefore, $T = S$ and the identity is proved.

The identity (11-73) can be proved in the same way. However, once we have (11-72), we can give a simpler proof. We write $(\mathbf{u} \times \mathbf{v}) \times \mathbf{w} = -\mathbf{w} \times (\mathbf{u} \times \mathbf{v})$ and now apply the identity (11-72) to the last expression, to obtain the right side of (11-73).

EXAMPLE 1. Let $\mathbf{u}$ and $\mathbf{v}$ be linearly independent. Find a nonzero vector $\mathbf{p}$ which is a linear combination of $\mathbf{u}$ and $\mathbf{v}$ and is orthogonal to $\mathbf{u}$.

Solution. The vectors $\mathbf{u}, \mathbf{v}$ are given as linearly independent. The vector $\mathbf{u} \times \mathbf{v}$ is orthogonal to both $\mathbf{u}$ and $\mathbf{v}$. Also, $\mathbf{u}, \mathbf{v}, \mathbf{u} \times \mathbf{v}$ are linearly independent. The vector $\mathbf{p} = (\mathbf{u} \times \mathbf{v}) \times \mathbf{u}$ is then a nonzero vector orthogonal to both $\mathbf{u} \times \mathbf{v}$ and $\mathbf{u}$; $\mathbf{p}$ can be expressed as a linear combination of $\mathbf{u}, \mathbf{v}$, and $\mathbf{u} \times \mathbf{v}$, but the coefficient of $\mathbf{u} \times \mathbf{v}$ is 0. In fact, by (11-73),

$$\mathbf{p} = (\mathbf{u} \times \mathbf{v}) \times \mathbf{u} = (\mathbf{u} \cdot \mathbf{u})\mathbf{v} - (\mathbf{u} \cdot \mathbf{v})\mathbf{u}$$

In geometric language, $\mathbf{u} \times \mathbf{v}$ is perpendicular to the plane of $\mathbf{u}$ and $\mathbf{v}$, $\mathbf{p}$ is perpendicular to a perpendicular to that plane, and, hence, lies in the plane again (see Figure 11-11).

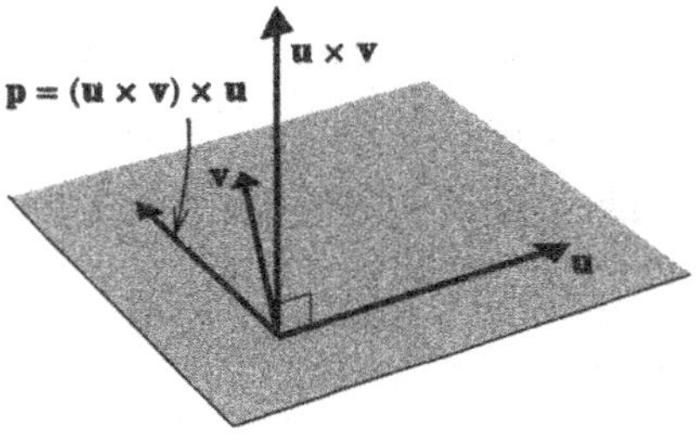

Figure 11-11 Example 1.

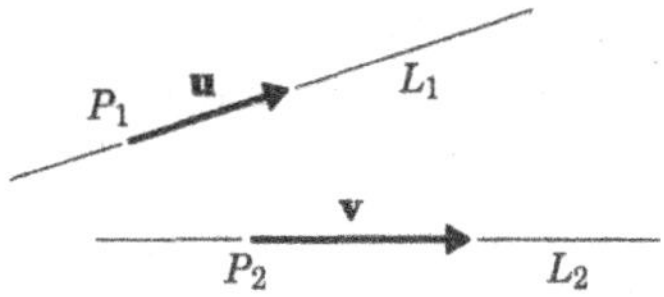

Figure 11-12 Relation between two lines.

EXAMPLE 2. Prove: $(\mathbf{u} \times \mathbf{v}) \cdot (\mathbf{v} \times \mathbf{w}) = (\mathbf{u} \cdot \mathbf{v})(\mathbf{v} \cdot \mathbf{w}) - (\mathbf{v} \cdot \mathbf{v})(\mathbf{u} \cdot \mathbf{w})$.

Solution. We first regard the left side as a scalar triple product $\mathbf{u} \times \mathbf{v} \cdot \mathbf{p}$ and interchange the dot and cross to obtain $\mathbf{u} \cdot \mathbf{v} \times \mathbf{p} = \mathbf{u} \cdot \mathbf{v} \times (\mathbf{v} \times \mathbf{w})$. But by (11-72)

$$\mathbf{u} \cdot \mathbf{v} \times (\mathbf{v} \times \mathbf{w}) = \mathbf{u} \cdot [(\mathbf{v} \cdot \mathbf{w})\mathbf{v} - (\mathbf{v} \cdot \mathbf{v})\mathbf{w}] = (\mathbf{u} \cdot \mathbf{v})(\mathbf{v} \cdot \mathbf{w}) - (\mathbf{u} \cdot \mathbf{w})(\mathbf{v} \cdot \mathbf{v})$$

Hence, the identity is proved.

11-8 APPLICATION OF THE CROSS PRODUCT TO LINES IN SPACE

Let two straight lines in space be given (Figure 11-12):

$$L_1\colon \overrightarrow{OP} = \overrightarrow{OP_1} + t\mathbf{u}, \qquad L_2\colon \overrightarrow{OP} = \overrightarrow{OP_2} + t\mathbf{v}, \qquad -\infty < t < \infty$$

First let $\mathbf{u} \times \mathbf{v} = \mathbf{0}$. Then $\mathbf{u}, \mathbf{v}$ are linearly dependent and the lines must be parallel, possibly coincident. To decide whether they coincide or not, one can form $\overrightarrow{P_1P_2} = \overrightarrow{OP_2} - \overrightarrow{OP_1}$. If $\overrightarrow{P_1P_2} \times \mathbf{u} = \mathbf{0}$, then P_2 must lie on L_1 and the lines must coincide. If $\overrightarrow{P_1P_2} \times \mathbf{u} \neq \mathbf{0}$, then P_2 is not on L_1 and the lines are parallel, but not coincident.

Next let $\mathbf{u} \times \mathbf{v} \neq \mathbf{0}$, so that $\mathbf{u}, \mathbf{v}$ are linearly independent. Then the lines cannot be parallel and consequently they must either intersect at a single point Q or be skew and not intersect at all. To find Q, one must be able to find t and τ so that

$$\overrightarrow{OP_1} + t\mathbf{u} = \overrightarrow{OP_2} + \tau\mathbf{v} \tag{11-80}$$

Then the common value of both sides is $\overrightarrow{OQ}$. But (11-80) is equivalent to

$$\overrightarrow{P_1P_2} = t\mathbf{u} - \tau\mathbf{v}$$

Hence, the lines intersect if, and only if, $\overrightarrow{P_1P_2}$ is linearly dependent on $\mathbf{u}$ and $\mathbf{v}$. Since $\mathbf{u}, \mathbf{v}$ are linearly independent, this condition is equivalent to the condition that the determinant whose rows are $\overrightarrow{P_1P_2}$, $\mathbf{u}, \mathbf{v}$ be equal to 0; that is, to the condition

$$\overrightarrow{P_1P_2} \cdot \mathbf{u} \times \mathbf{v} = 0 \tag{11-81}$$

If $\overrightarrow{P_1P_2} \cdot \mathbf{u} \times \mathbf{v} \neq 0$, then the lines cannot intersect. In fact, the condition $\overrightarrow{P_1P_2} \cdot \mathbf{u} \times \mathbf{v} \neq 0$ alone implies that $\mathbf{u} \times \mathbf{v} \neq \mathbf{0}$ and that the lines are skew.

We summarize our conclusions.

Parallel, distinct: $\mathbf{u} \times \mathbf{v} = \mathbf{0}$, $\overrightarrow{P_1P_2} \times \mathbf{u} \neq \mathbf{0}$

Coincident: $\mathbf{u} \times \mathbf{v} = \mathbf{0}$, $\overrightarrow{P_1P_2} \times \mathbf{u} = \mathbf{0}$

Intersecting: $\mathbf{u} \times \mathbf{v} \neq \mathbf{0}$, $\overrightarrow{P_1P_2} \cdot \mathbf{u} \times \mathbf{v} = 0$

Skew: $\overrightarrow{P_1P_2} \cdot \mathbf{u} \times \mathbf{v} \neq 0$

EXAMPLE 1. The lines L_1: $\overrightarrow{OP} = (2, 3, 5) + t(1, 5, 6)$ and L_2: $\overrightarrow{OP} = (1, 0, -7) + t(3, 1, 2)$ are not parallel, since $\mathbf{u} = (1, 5, 6)$ and $\mathbf{v} = (3, 1, 2)$ are linearly independent (so that $\mathbf{u} \times \mathbf{v} \neq \mathbf{0}$). Here, $\overrightarrow{P_1P_2} = (1, 0, -7) - (2, 3, 5) = (-1, -3, -12)$ and

$$\overrightarrow{P_1P_2} \cdot \mathbf{u} \times \mathbf{v} = \begin{vmatrix} -1 & -3 & -12 \\ 1 & 5 & 6 \\ 3 & 1 & 2 \end{vmatrix} = -116 \neq 0$$

Hence, the lines are skew.

The cross product can be used to write the *equation of a line*. For example, if $\mathbf{u} \neq \mathbf{0}$, the equation

$$\overrightarrow{P_1P} \times \mathbf{u} = \mathbf{0} \tag{11-82}$$

is satisfied by all P such that $\overrightarrow{P_1P}$ is a scalar multiple of $\mathbf{u}$. Thus (11-82) is an equation for the line L: $\overrightarrow{OP} = \overrightarrow{OP_1} + t\mathbf{u}$. Similarly, if $P_1 \neq P_2$,

$$\overrightarrow{P_1P} \times \overrightarrow{P_1P_2} = \mathbf{0} \tag{11-83}$$

is an equation for the line through P_1 and P_2.

EXAMPLE 2. Find the equation of a line L through $(0, 0, 0)$ perpendicular to the lines L_1 and L_2 of Example 1.

Solution. The vectors along L are the vectors $t\mathbf{w}$, where $\mathbf{w}$ is a nonzero vector orthogonal to both $\mathbf{u} = (1, 5, 6)$ and $\mathbf{v} = (3, 1, 2)$. Hence, we can take

$$\mathbf{w} = \mathbf{u} \times \mathbf{v} = \begin{vmatrix} \mathbf{i} & \mathbf{j} & \mathbf{k} \\ 1 & 5 & 6 \\ 3 & 1 & 2 \end{vmatrix} = 4\mathbf{i} + 16\mathbf{j} - 14\mathbf{k}$$

The line L has the equation $\overrightarrow{OP} = \mathbf{0} + t\mathbf{w} = t(4, 16, -14)$ or, as in (11-82),

$$\overrightarrow{OP} \times (4, 16, -14) = \mathbf{0}$$

‡11-9 THE CROSS PRODUCT IN V_n

The cross product can be generalized to V_n, where $n \geq 2$. One might expect the generalization to consist of assigning to each pair of vectors $\mathbf{u}, \mathbf{v}$ a vector $\mathbf{u} \times \mathbf{v}$ orthogonal to $\mathbf{u}$ and $\mathbf{v}$. However, a sensible way of making such an assignment has been found for $n = 3$ only. The difficulty arises from the fact that, for $n = 4$, for example, there are *too many* vectors orthogonal to $\mathbf{u}$ and $\mathbf{v}$ and one has no way of choosing a particular one with geometric significance.

The successful generalization consists in assigning to each set of $n - 1$ vectors $\mathbf{v}_1, \ldots, \mathbf{v}_{n-1}$ of V_n a vector $\mathbf{w}$ orthogonal to each of $\mathbf{v}_1, \ldots, \mathbf{v}_{n-1}$. The formula found is a natural generalization of (11-61″):

$$\mathbf{v}_1 \times \mathbf{v}_2 \times \cdots \times \mathbf{v}_{n-1} = \begin{vmatrix} v_{11} & \cdots & v_{1n} \\ \vdots & & \vdots \\ v_{n-1,1} & \cdots & v_{n-1,n} \\ \mathbf{e}_1 & \cdots & \mathbf{e}_n \end{vmatrix} \tag{11-90}$$

Here $\mathbf{v}_i = (v_{i1}, \ldots, v_{in})$, so that $\mathbf{v}_1, \ldots, \mathbf{v}_{n-1}$ are the row vectors of the determinant. The determinant is to be expanded by minors of the last row.

For $n = 2$ there is only one vector $\mathbf{v}_1 = a\mathbf{i} + b\mathbf{j}$ and the determinant in (11-90) becomes

$$\begin{vmatrix} a & b \\ \mathbf{i} & \mathbf{j} \end{vmatrix} = -b\mathbf{i} + a\mathbf{j} = \mathbf{v}_1^{\dashv}$$

Thus our left-turn operation (Section 1-12) is the 2-dimensional form of the cross product. For $n = 3$, (11-90) is easily seen to agree with (11-61″), even though the determinant is written in a different form.

We leave to the exercises (Problem 12 below) a proof that (11-90) defines a cross product with properties analogous to those of $\mathbf{u} \times \mathbf{v}$ in V_3. The relationship to volume in R^n is considered in Section 11-16 below.

PROBLEMS

1. Given the points A: $(3, 1, -2)$, B: $(1, 3, 4)$, C: $(2, 0, 1)$, D: $(1, 3, 5)$, evaluate:
   (a) $\overrightarrow{AB} \times \overrightarrow{AC}$ (b) $\overrightarrow{AC} \times (2\overrightarrow{BC})$ (c) $\overrightarrow{AB} \cdot (\overrightarrow{BC} \times \overrightarrow{CD})$
   (d) $(\overrightarrow{AB} \times \overrightarrow{BC}) \cdot \overrightarrow{CD}$ (e) $\overrightarrow{AB} \cdot [(2\overrightarrow{AB} + \overrightarrow{BC}) \times \overrightarrow{AD}]$
   (f) $\overrightarrow{BC} \cdot (\overrightarrow{CD} \times \overrightarrow{CB})$ (g) $\overrightarrow{AB} \times (\overrightarrow{BC} \times \overrightarrow{CD})$
   (h) $(\overrightarrow{AB} \times \overrightarrow{BC}) \times \overrightarrow{CD}$
   (i) $(\overrightarrow{AB} \times \overrightarrow{AC}) \times (\overrightarrow{AC} \times \overrightarrow{AD})$
   (j) $(\overrightarrow{AB} \times \overrightarrow{AC}) \times (\overrightarrow{AB} \times \overrightarrow{AD})$
2. Find an equation for line L satisfying the conditions stated:
   (a) L contains $(1, 0, 1)$ and L is orthogonal to both of the lines L_1: $\overrightarrow{OP} = (1, 2, -3) + t(-1, 1, 2)$, L_2: $\overrightarrow{OP} = (-1, 0, 6) + t(2, 4, -3)$, $-\infty < t < \infty$.
   (b) L contains $(3, 5, 2)$ and is perpendicular to the line segments $\overrightarrow{AB}$ and $\overrightarrow{AC}$, with A, B, C as in Problem 1.
3. Determine whether the lines L_1, L_2 are coincident, parallel and noncoincident, intersecting, or skew:
   (a) L_1: $\overrightarrow{OP} = \mathbf{i} + t(\mathbf{i} + \mathbf{j})$, L_2: $\overrightarrow{OP} = 2\mathbf{i} - \mathbf{j} - t(\mathbf{i} + \mathbf{j})$
   (b) L_1: $\overrightarrow{OP} = (6, 6, 6) + t(3, 1, 4)$, L_2: $\overrightarrow{OP} = (5, 7, 4) + t(1, 1, 1)$
   (c) L_1: $\overrightarrow{OP} = (1, 2, -1) + t(1, 0, 1)$, L_2: $\overrightarrow{OP} = (3, 0, 5) + t(0, 2, 2)$
   (d) L_1: $\overrightarrow{OP} = (4, 1, 8) + t(4, 2, 6)$, L_2: $\overrightarrow{OP} = (0, -1, 2) + t(6, 3, 9)$
4. Let L_1, L_2 be nonparallel lines. Let L_3 and L_4 both be perpendicular to L_1 and to L_2. Prove: in R^3, $L_3 \parallel L_4$.
5. Prove: $\mathbf{u} \cdot \mathbf{v} = 0$ if, and only if, $\|\mathbf{u} \times \mathbf{v}\| = \|\mathbf{u}\|\,\|\mathbf{v}\|$.
6. (a) Prove (11-66). (b) Prove (11-67).
   (c) Show that, for fixed $\mathbf{v}$, the equation $\mathbf{q} = \mathbf{u} \times \mathbf{v}$ defines a linear transformation $\mathbf{q} = T(\mathbf{u})$ of V_3. Represent this transformation by a matrix for the particular case $\mathbf{v} = (2, 3, 1)$.
   (d) For the linear transformation T of part (c) find a matrix representing T for the case of a general fixed vector $\mathbf{v} = (v_1, v_2, v_3)$.
   (e) Show that, for fixed $\mathbf{v}$ and $\mathbf{w}$, the equation $S(\mathbf{u}) = (\mathbf{u} \cdot \mathbf{w})\mathbf{v} - (\mathbf{u} \cdot \mathbf{v})\mathbf{w}$ defines a linear transformation of V_3.
   (f) Represent the transformation S of part (e) by a matrix for the special case $\mathbf{v} = (1, 0, 2)$, $\mathbf{w} = (0, -1, 1)$.
7. Prove the identities:
   (a) $\mathbf{u} \times (\mathbf{v} \times \mathbf{w}) + \mathbf{v} \times (\mathbf{w} \times \mathbf{u}) + \mathbf{w} \times (\mathbf{u} \times \mathbf{v}) = \mathbf{0}$
   (b) $(\mathbf{u} \times \mathbf{v}) \cdot (\mathbf{w} \times \mathbf{z}) = \begin{vmatrix} \mathbf{u} \cdot \mathbf{w} & \mathbf{u} \cdot \mathbf{z} \\ \mathbf{v} \cdot \mathbf{w} & \mathbf{v} \cdot \mathbf{z} \end{vmatrix}$
   (c) $\|\mathbf{u} \times \mathbf{v}\|^2 = \begin{vmatrix} \mathbf{u} \cdot \mathbf{u} & \mathbf{u} \cdot \mathbf{v} \\ \mathbf{v} \cdot \mathbf{u} & \mathbf{v} \cdot \mathbf{v} \end{vmatrix}$
   (d) If $\mathbf{u} + \mathbf{v} + \mathbf{w} = \mathbf{0}$, then $\mathbf{u} \times \mathbf{v} = \mathbf{v} \times \mathbf{w} = \mathbf{w} \times \mathbf{u}$.

(e) If $\mathbf{u} + \mathbf{v} + \mathbf{w} = \mathbf{0}$ and $\mathbf{u}, \mathbf{v}$ are linearly independent, then one has the Law of Sines:

$$\frac{\|\mathbf{u}\|}{\sin \measuredangle(\mathbf{v}, \mathbf{w})} = \frac{\|\mathbf{v}\|}{\sin \measuredangle(\mathbf{u}, \mathbf{w})} = \frac{\|\mathbf{w}\|}{\sin \measuredangle(\mathbf{u}, \mathbf{v})}$$

8. Determine whether the given ordered triples are positive or negative, or linearly dependent. In each case make a sketch.
   (a) $\mathbf{k}, \mathbf{j}, \mathbf{i}$ (b) $\mathbf{j}, \mathbf{i}, \mathbf{k}$ (c) $\mathbf{i} + \mathbf{k}, \mathbf{j} + \mathbf{i}, \mathbf{k} + \mathbf{j}$
   (d) $\mathbf{i}, \mathbf{i} + \mathbf{j}, \mathbf{i} + \mathbf{k}$ (e) $\mathbf{i} + \mathbf{j}, \mathbf{k}, 2\mathbf{i} + 2\mathbf{j} + \mathbf{k}$

9. In mechanics one defines the moment of a force about a point O as the vector product $\overrightarrow{OP_0} \times \mathbf{F}$, where P_0 is the point of application of the force and $\mathbf{F}$ is the vector representing the force (Figure 11-13) in appropriate units. The line of action of the force is the line L: $\overrightarrow{OP} = \overrightarrow{OP_0} + t\mathbf{F}$, $-\infty < t < \infty$.

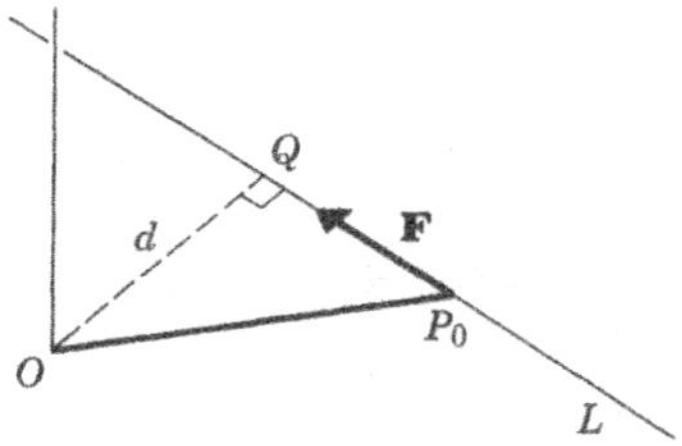

Figure 11-13 Moment of force.

   (a) Show that the moment of the force $\mathbf{F}$ depends only on the line of action L and not on the particular point P_0 of L at which the force is applied.
   (b) Show that the magnitude (norm) of the moment equals $d\|\mathbf{F}\|$, where d is the length of a segment OQ perpendicular to L at Q.

10. In mechanics one considers a rigid body rotating about an axis L at angular velocity ω (in appropriate units). One assigns to this motion an *angular velocity vector* $\mathbf{w}$, whose norm is ω, along L, and shows that the velocity of each particle P of the body is $\mathbf{v}_P = \mathbf{w} \times \overrightarrow{OP}$, where O is a point on the axis L (see Figure 11-14).
   (a) Show that $\mathbf{v}_P$ is unaffected by the choice of O on L.
   (b) Show that if OP is perpendicular to L, then $\|\mathbf{v}_P\| = r\omega$, where $r = \|\overrightarrow{OP}\|$.
   (c) Show the velocity pattern in the xy-plane for $\mathbf{w} = \mathbf{k}$, by sketching a number of vectors $\mathbf{v}_P$ at points P in the xy-plane.

11. Let L_1: $\overrightarrow{OP} = \mathbf{a} + t\mathbf{u}$, L_2: $\overrightarrow{OP} = \mathbf{b} + \tau\mathbf{v}$ be skew lines. It can then be shown [part (d) below] that there is a unique line L meeting L_1 and L_2 and perpendicular to both. One calls L the *common perpendicular* of L_1, L_2. Let L meet L_1 at P_1, L_2 at P_2.
   (a) Show that, if Q_1 is a point on L_1 and Q_2 is a point on L_2, then $\overrightarrow{Q_1Q_2} \cdot \mathbf{u} \times \mathbf{v} = \overrightarrow{P_1P_2} \cdot \mathbf{u} \times \mathbf{v}$. [*Hint.* Write $\overrightarrow{Q_1Q_2} = \overrightarrow{Q_1P_1} + \overrightarrow{P_1P_2} + \overrightarrow{P_2Q_2}$.]

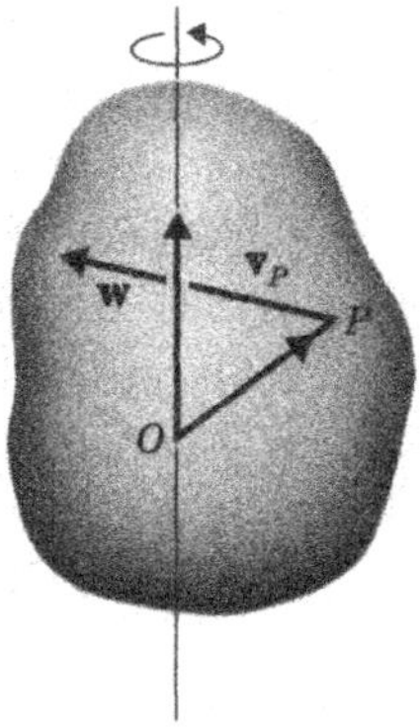

Figure 11-14 Angular velocity vector.

   (b) From the result of (a) show that

$$d(P_1, P_2) = \frac{|\overrightarrow{Q_1Q_2} \cdot \mathbf{u} \times \mathbf{v}|}{\|\mathbf{u} \times \mathbf{v}\|}$$

(c) From the result of (b) show that $d(P_1, P_2) \leq d(Q_1, Q_2)$, with equality only when $\overrightarrow{Q_1Q_2}$ is perpendicular to $\mathbf{u}$ and to $\mathbf{v}$ (so that, by uniqueness of the common perpendicular, $Q_1 = P_1$ and $Q_2 = P_2$). Hence, the formula in (b) gives the shortest distance between the skew lines.

‡(d) Prove the existence and uniqueness of the common perpendicular L by showing that, for $\overrightarrow{OP_1} = \mathbf{a} + t_1\mathbf{u}$, $\overrightarrow{OP_2} = \mathbf{b} + \tau_2\mathbf{v}$, the conditions $\overrightarrow{P_1P_2} \cdot \mathbf{u} = 0$, $\overrightarrow{P_1P_2} \cdot \mathbf{v} = 0$ lead to simultaneous linear equations for t_1, τ_2, for which the determinant of coefficients is

$$\begin{vmatrix} (\mathbf{u} \cdot \mathbf{u}) & -(\mathbf{v} \cdot \mathbf{u}) \\ (\mathbf{u} \cdot \mathbf{v}) & -(\mathbf{v} \cdot \mathbf{v}) \end{vmatrix}$$

Now use the identity of Problem 7c and the fact that L_1, L_2 are skew.

(e) Find the shortest distance between the skew lines L_1: $\overrightarrow{OP} = (1, 0, 1) + t(1, 1, 2)$ and L_2: $\overrightarrow{OP} = (0, 1, 2) + \tau(1, 0, 2)$ and graph.

‡12. Let $\mathbf{v}_1, \ldots, \mathbf{v}_{n-1}$ be vectors in V_n and let $\mathbf{w} = \mathbf{v}_1 \times \mathbf{v}_2 \times \cdots \times \mathbf{v}_{n-1}$ as defined by (11-90). Prove the following:

(a) $\mathbf{v}_1 \times \cdots \times \mathbf{v}_{n-1} \cdot \mathbf{u} = \det A$, where A is the matrix whose row vectors are $\mathbf{v}_1, \ldots, \mathbf{v}_{n-1}, \mathbf{u}$.

(b) $\mathbf{w} \cdot \mathbf{v}_1 = 0, \ldots, \mathbf{w} \cdot \mathbf{v}_{n-1} = 0$

(c) $\mathbf{v}_2 \times \mathbf{v}_1 \times \mathbf{v}_3 \times \cdots \times \mathbf{v}_{n-1} = -\mathbf{v}_1 \times \mathbf{v}_2 \times \cdots \times \mathbf{v}_{n-1}$; and, in general, the interchange of two $\mathbf{v}$'s multiplies the cross product by -1.

(d) If $\mathbf{v}_j = \mathbf{p}_j + \mathbf{q}_j$, then

$$\begin{aligned} \mathbf{v}_1 \times \cdots \times \mathbf{v}_{n-1} = \mathbf{v}_1 \times \cdots \times \mathbf{v}_{j-1} \times \mathbf{p}_j \times \mathbf{v}_{j+1} \times \cdots \times \mathbf{v}_{n-1} \\ + \mathbf{v}_1 \times \cdots \times \mathbf{v}_{j-1} \times \mathbf{q}_j \times \mathbf{v}_{j+1} \times \cdots \times \mathbf{v}_{n-1} \end{aligned}$$

(e) $\mathbf{w} = \mathbf{0}$ if, and only if, $\mathbf{v}_1, \ldots, \mathbf{v}_{n-1}$ are linearly dependent.

(f) If $\mathbf{v}_1, \ldots, \mathbf{v}_{n-1}$ are linearly independent, then $\mathbf{v}_1, \ldots, \mathbf{v}_{n-1}, \mathbf{w}$ are linearly independent and form a positive n-tuple (that is, $d > 0$, where d is the value of the determinant whose rows are $\mathbf{v}_1, \ldots, \mathbf{v}_{n-1}, \mathbf{w}$); furthermore, the vectors orthogonal to $\mathbf{v}_1, \ldots, \mathbf{v}_{n-1}$ are given by $c\mathbf{w}$ for $-\infty < c < \infty$.

11-10 PLANES IN R^3

A set of points (x, y, z) in R^3 is said to form a *plane* if the corresponding vectors form a 2-dimensional linear variety of V_3. By the results of Sections 9-5 and 10-6, it follows that every linear equation

$$Ax + By + Cz + D = 0 \qquad (A, B, C \text{ not all } 0) \tag{11-100}$$

represents a plane in R^3, and every plane in R^3 can be so represented.

EXAMPLE 1. $2x + y + z - 4 = 0$. This plane is shown graphically in Figure 11-15. We note that when $z = 0$, the equation becomes $2x + y - 4 = 0$; thus the plane meets the xy-plane along the line $2x + y - 4 = 0$, which we call the *xy-trace* of the plane. Similarly, the *xz-trace* is the line $2x + z - 4 = 0$ in the xz-plane, the *yz-trace* is the line $y + z - 4 = 0$ in the yz-plane. For

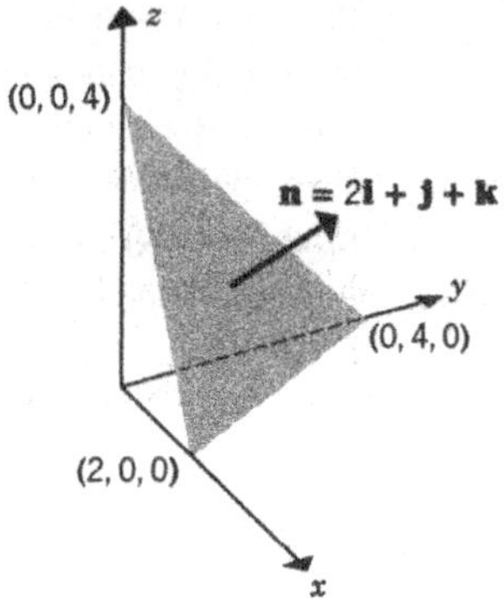

Figure 11-15 The plane $2x + y + z - 4 = 0$.

y and z equal to 0, the equation becomes $2x - 4 = 0$; that is, the plane meets the x-axis at $(2, 0, 0)$; we say that the plane has *x-intercept* 2; similarly, the *y-intercept* is 4, the *z-intercept* is 4.

Different Equations for a Given Plane. The linear equation $Ax + By + Cz + D = 0$, representing a given plane, is not unique, since we obtain an equivalent equation by multiplying by a nonzero number. However, this is the only freedom permitted. *If*

$$A_1x + B_1y + C_1z + D_1 = 0, \qquad A_2x + B_2y + C_2z + D_2 = 0 \quad (11\text{-}101)$$

are equations of the same plane, then the coefficients must be proportional; that is, in vector language, the two vectors of V_4:

$$(A_1, B_1, C_1, D_1), \qquad (A_2, B_2, C_2, D_2)$$

must be linearly dependent, or the matrix

$$\begin{pmatrix} A_1 & B_1 & C_1 & D_1 \\ A_2 & B_2 & C_2 & D_2 \end{pmatrix}$$

must have rank $r = 1$. For we know that the simultaneous equations (11-101) have solutions. Hence, as in Section 10-6, both matrices

$$\begin{pmatrix} A_1 & B_1 & C_1 \\ A_2 & B_2 & C_2 \end{pmatrix}, \qquad \begin{pmatrix} A_1 & B_1 & C_1 & D_1 \\ A_2 & B_2 & C_2 & D_2 \end{pmatrix}$$

must have the same rank. Since the solutions form a linear variety of dimension 2, the rank of the first matrix must be $3 - 2 = 1$ (Section 10-4).

EXAMPLE 2. $3x - 2y - z + 4 = 0$ and $6x - 4y - 2z + 8 = 0$ are two equations for the same plane, since the coefficients are proportional, or

$$\text{rank} \begin{pmatrix} 3 & -2 & -1 & 4 \\ 6 & -4 & -2 & 8 \end{pmatrix} = 1$$

Now let a plane H be given by (11-100) and let P_0: (x_0, y_0, z_0) be a point in H. Then

$$Ax_0 + By_0 + Cz_0 + D = 0$$

If we subtract this equation from (11-100), we obtain the equivalent equation

$$A(x - x_0) + B(y - y_0) + C(z - z_0) = 0 \tag{11-102}$$

If we write $\mathbf{n} = A\mathbf{i} + B\mathbf{j} + C\mathbf{k}$, then Equation (11-102) can be written as follows:

$$\mathbf{n} \cdot \overrightarrow{P_0P} = 0 \tag{11-103}$$

where P: (x, y, z) is an arbitrary point of H. Hence, the vector $\mathbf{n} = A\mathbf{i} + B\mathbf{j} + C\mathbf{k}$ is orthogonal to each vector joining two points of H. We call each vector having this property (of being orthogonal to each vector joining two points of H) a *normal vector of H*. Therefore, $\mathbf{n}$ is a normal vector of H (see Figure 11-16).

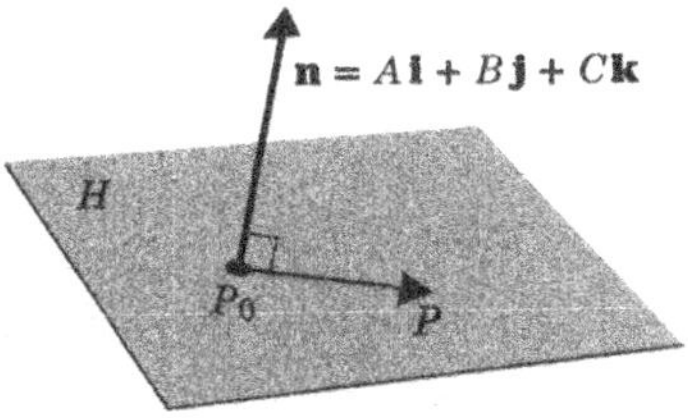

Figure 11-16 Plane and normal vector.

For the plane of Example 1, $2\mathbf{i} + \mathbf{j} + \mathbf{k}$ is thus a normal vector, as shown in Figure 11-15.

Now let $\mathbf{n}'$ be an arbitrary nonzero normal vector of H and let P_0 be a point of H. Then the equation $\mathbf{n}' \cdot \overrightarrow{P_0P} = 0$, when expanded, is a linear equation and, hence, represents a plane H'. But, since $\mathbf{n}'$ is a normal vector of H, the equation $\mathbf{n}' \cdot \overrightarrow{P_0P} = 0$ is satisfied by every point P: (x, y, z) in H. Hence, H' includes H and, since H' and H are both 2-dimensional linear varieties, H' must coincide with H. Therefore, for each normal vector $\mathbf{n}'$ of H and each point P_0 of H, H has the equation $\mathbf{n}' \cdot \overrightarrow{P_0P} = 0$. From our discussion of the different forms of the equation of H, it follows that $\mathbf{n}'$ must be a scalar multiple of the normal vector $\mathbf{n} = A\mathbf{i} + B\mathbf{j} + C\mathbf{k}$ and, conversely, every scalar multiple of $A\mathbf{i} + B\mathbf{j} + C\mathbf{k}$ is a normal vector.

In terms of subspaces of V_3, we can say that the vectors normal to H form a one-dimensional subspace U, and the vectors joining points of H are the vectors orthogonal to each nonzero vector in U. We shall see below that the vectors joining points of H form a subspace of V_3, which is the base space of H as a 2-dimensional linear variety.

Plane Through Three Points. Let P_0, P_1, P_2 be three points of space which are not collinear (do not lie on a line). Then there is a unique plane H containing the three points. For if H is such a plane and $\mathbf{n}$ is a normal vector of H, then $\mathbf{n} \cdot \overrightarrow{P_0P_1} = 0$ and $\mathbf{n} \cdot \overrightarrow{P_0P_2} = 0$, so that $\mathbf{n}$ is a scalar multiple of $\overrightarrow{P_0P_1} \times \overrightarrow{P_0P_2}$. In particular, $\overrightarrow{P_0P_1} \times \overrightarrow{P_0P_2}$ is itself a normal vector of H and is nonzero, since $\overrightarrow{P_0P_1}$, $\overrightarrow{P_0P_2}$ are linearly independent (because the points are not collinear). Therefore, the equation

$$\overrightarrow{P_0P_1} \times \overrightarrow{P_0P_2} \cdot \overrightarrow{P_0P} = 0 \tag{11-104}$$

is an equation for H. But this equation is satisfied for $P = P_0$, $P = P_1$, and $P = P_2$, since for each such choice of P the vectors $\overrightarrow{P_0P_1}$, $\overrightarrow{P_0P_2}$, $\overrightarrow{P_0P}$ are linearly dependent. Therefore, Equation (11-104) does describe a plane H passing through the three given points and H is the only such plane.

Equation (11-104) can be written in terms of a determinant:

$$\begin{vmatrix} x - x_0 & y - y_0 & z - z_0 \\ x_1 - x_0 & y_1 - y_0 & z_1 - z_0 \\ x_2 - x_0 & y_2 - y_0 & z_2 - z_0 \end{vmatrix} = 0 \qquad (11\text{-}105)$$

EXAMPLE 3. Find an equation for the plane through P_0: $(1, 1, 1)$, P_1: $(2, 1, -1)$ and P_2: $(1, 3, -1)$.

Solution. By (11-105), the equation is

$$\begin{vmatrix} x - 1 & y - 1 & z - 1 \\ 1 & 0 & -2 \\ 0 & 2 & -2 \end{vmatrix} = 0$$

or, after simplification, $2x + y + z - 4 = 0$.

Vector Equation of a Plane. Let a plane H be given. Since the vectors $\overrightarrow{OP}$, for P in H, fill a 2-dimensional linear variety, we can give a *vector equation* for H:

$$\overrightarrow{OP} = \mathbf{a} + t\mathbf{u} + \tau\mathbf{v}, \quad -\infty < t < \infty, \quad -\infty < \tau < \infty \qquad (11\text{-}106)$$

where $\mathbf{a}, \mathbf{u}, \mathbf{v}$ are given vectors and $\mathbf{u}, \mathbf{v}$ form a basis for the base space W of H. Thus the vectors $\overrightarrow{OP}$, for P in H, form the set $\{\mathbf{a} + W\}$. For $t = 0$ and $\tau = 0$ in (11-106), $\overrightarrow{OP} = \mathbf{a}$, so that we can write $\mathbf{a} = \overrightarrow{OP_0} = (x_0, y_0, z_0)$, where P_0 is a point of H, as in Figure 11-17. Hence, (11-106) can be written as

$$\overrightarrow{OP} = \overrightarrow{OP_0} + t\mathbf{u} + \tau\mathbf{v}$$

or as

$$\overrightarrow{P_0P} = t\mathbf{u} + \tau\mathbf{v} \qquad (11\text{-}106')$$

(where we always understand $-\infty < t < \infty$, $-\infty < \tau < \infty$). Now for $t = 1$, $\tau = 0$ in (11-106′), we have $\overrightarrow{P_0P} = \mathbf{u}$. Thus $\mathbf{u}$ can be interpreted as $\overrightarrow{P_0P_1}$, where P_1 is a point in the plane H (see Figure 11-17). Similarly, $\mathbf{v}$ can be interpreted as $\overrightarrow{P_0P_2}$, where P_2 is a point in the plane H. As Figure 11-17 shows, (11-106′) then instructs us to obtain all points of the plane H by forming all linear

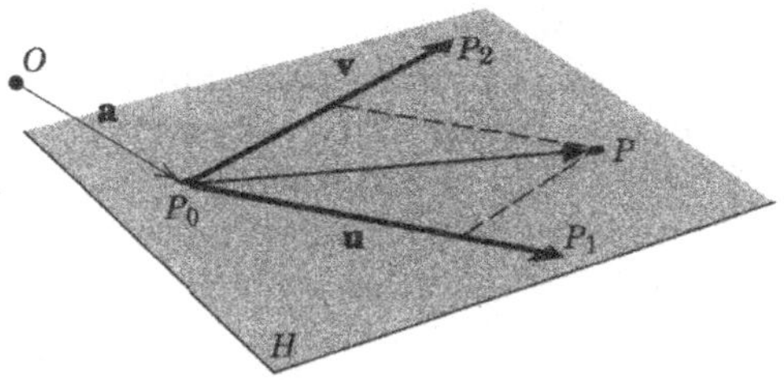

Figure 11-17 Vector equation of plane.

combinations of **u** and **v**. We can now write (11-106′) in the form:

$$\overrightarrow{P_0P} = t\overrightarrow{P_0P_1} + \tau\overrightarrow{P_0P_2} \tag{11-107}$$

If P_0, P_1, P_2 are given noncollinear points (that is, points not lying on a line), then $\overrightarrow{P_0P_1}$, $\overrightarrow{P_0P_2}$ are linearly independent and (11-107) provides us with a *vector equation of the unique plane through the three points.*

EXAMPLE 4. Find the vector equation of the plane containing P_0: (1, 1, 1), P_1: (2, 1, −1), P_2: (1, 3, −1) and determine whether the point (3, 5, 2) is in the plane.

Solution. Here $\overrightarrow{P_0P_1} = (1, 0, -2)$, $\overrightarrow{P_0P_2} = (0, 2, -2)$. Hence, by (11-107), the vector equation is

$$\overrightarrow{P_0P} = t(1, 0, -2) + \tau(0, 2, -2)$$

To test whether the point (3, 5, 2) is in the plane, we set $P = (3, 5, 2)$, so that $\overrightarrow{P_0P} = (3, 5, 2) - (1, 1, 1) = (2, 4, 1)$ and try to determine t and τ so that

$$(2, 4, 1) = t(1, 0, -2) + \tau(0, 2, -2)$$

or, on taking components,

$$2 = t, \qquad 4 = 2\tau, \qquad 1 = -2t - 2\tau$$

We see at once that these equations are contradictory, so that (3, 5, 2) is not in the plane.

In general, we can write $\overrightarrow{OP} = (x, y, z)$, $\mathbf{a} = (x_0, y_0, z_0)$ and, on taking components, (11-106) becomes

$$\begin{aligned} x &= x_0 + u_1t + v_1\tau \\ y &= y_0 + u_2t + v_2\tau \\ z &= z_0 + u_3t + v_3\tau \end{aligned} \tag{11-108}$$

These are called *parametric equations* of the plane H. For Example 4, $\mathbf{a} = \overrightarrow{OP_0} = (1, 1, 1)$ and the parametric equations are

$$x = 1 + t, \qquad y = 1 + 2\tau, \qquad z = 1 - 2t - 2\tau$$

From Vector Equation to Linear Equation. Given a vector equation of a plane H, one can obtain a corresponding linear equation by writing the parametric equations (11-108) and then eliminating t and τ. The cross product provides a simple way of doing this. From (11-106′), we proceed by forming the inner product of both sides with $\mathbf{u} \times \mathbf{v}$. Since $\mathbf{u} \times \mathbf{v} \cdot \mathbf{u} = 0$ and $\mathbf{u} \times \mathbf{v} \cdot \mathbf{v} = 0$, we obtain

$$\mathbf{u} \times \mathbf{v} \cdot \overrightarrow{P_0P} = 0$$

or

$$\begin{vmatrix} x - x_0 & y - y_0 & z - z_0 \\ u_1 & u_2 & u_3 \\ v_1 & v_2 & v_3 \end{vmatrix} = 0 \tag{11-109}$$

If the determinant is expanded, we obtain a linear equation of form (11-100).

EXAMPLE 5. Find a linear equation for the plane of Example 3.

Solution. Here $(x_0, y_0, z_0) = (1, 1, 1)$, $\mathbf{u} = (1, 0, -2)$, $\mathbf{v} = (0, 2, -2)$ and the plane is

$$\begin{vmatrix} x-1 & y-1 & z-1 \\ 1 & 0 & -2 \\ 0 & 2 & -2 \end{vmatrix} = 0$$

From Linear Equation to Vector Equation. Given a linear equation $Ax + By + Cz + D = 0$, we can find a vector equation of the plane in question by simply finding three noncollinear points P_0, P_1, P_2 in the plane and then writing out the equation (11-107). For example, for the plane $2x + y + z - 4 = 0$ of Example 1, we can choose P_0 as $(1, 1, 1)$ P_1 as $(2, 1, -1)$, P_2 as $(1, 3, -1)$; the three points are noncollinear and in the plane, as one verifies directly. (We can obtain as many such points as we wish by choosing x, y arbitrarily and then finding the proper z.) From these points we obtain $\mathbf{a} = (1, 1, 1)$, $\mathbf{u} = \overrightarrow{P_0P_1} = (1, 0, -2)$, $\mathbf{v} = \overrightarrow{P_0P_2} = (0, 2, -2)$ and, hence, obtain

$$\overrightarrow{OP} = (1, 1, 1) + t(1, 0, -2) + \tau(0, 2, -2)$$

as a vector equation of the plane.

Another procedure is simply to solve the equation for one letter, say z, in terms of the other two, and then treat the other two as parameters. For example, from $2x + y + z - 4 = 0$, we obtain $z = 4 - 2x - y$ and then

$$x = t, \qquad y = \tau, \qquad z = 4 - 2t - \tau$$

and, hence,

$$(x, y, z) = (t, \tau, 4 - 2t - \tau) = (0, 0, 4) + t(1, 0, -2) + \tau(0, 1, -1)$$

Different Vector Equations for a Given Plane. For the vector equation (11-106), we are free to take $\mathbf{a} = (x_0, y_0, z_0)$ as the position vector of any point in H, and $\mathbf{u}, \mathbf{v}$ as any basis for W, the base space of H. For by the theory of Section 9-5, $\{\mathbf{a} + W\}$ and $\{\mathbf{b} + V\}$ represent the same linear variety if, and only if, $W = V$ and $\mathbf{b}$ is contained in $\{\mathbf{a} + W\}$—that is, $\mathbf{b} - \mathbf{a}$ is in W.

EXAMPLE 6. Let two sets of parametric equations be given: $x = 1 + t$, $y = 1 + 2\tau$, $z = 1 - 2t - 2\tau$ and $x = t$, $y = \tau$, $z = 4 - 2t - \tau$. Determine whether they represent the same plane.

Solution. Here we have the vector equations

$$\overrightarrow{OP} = (1, 1, 1) + t(1, 0, -2) + \tau(0, 2, -2)$$

$$\overrightarrow{OP} = (0, 0, 4) + t(1, 0, -2) + \tau(0, 1, -1)$$

We observe that Span $[(1, 0, -2), (0, 2, -2)] =$ Span $[(1, 0, -2), (0, 1, -1)]$ $= W$. Thus the base spaces are the same. Here $\mathbf{b} - \mathbf{a} = (-1, -1, 3)$, and we ask whether

$$(-1, -1, 3) = t(1, 0, -2) + \tau(0, 2, -2)$$

or

$$-1 = t, \qquad -1 = 2\tau, \qquad 3 = -2t - 2\tau$$

can be satisfied. The values $t = -1$, $\tau = -\frac{1}{2}$ satisfy all three equations. Hence $\mathbf{b} - \mathbf{a}$ is in W and the linear varieties coincide. Therefore, the planes coincide.

PROBLEMS

1. Graph the plane, showing traces and intercepts:

(a) $2x + 3y + z = 6$ (b) $2x + y - z = 2$

(c) $x - y = 1$ (d) $2x + z = 3$

(e) $x = 0$ (f) $z = 1$

2. Determine whether the two given equations represent the same plane:

(a) $6x + 8y - 4z - 10 = 0$ and $9x + 12y - 6z - 15 = 0$

(b) $6x + 12y - 42z - 12 = 0$ and $7x + 14y - 49z - 7 = 0$

3. (a) . . . (f) Find a nonzero normal vector for each plane of Problem 1.

4. Find a linear equation for the plane satisfying the given conditions and graph:

(a) Contains the point $(1, 3, 2)$, has the normal vector $2\mathbf{i} - \mathbf{j} + 5\mathbf{k}$.

(b) Contains the point $(4, 0, 7)$, has the normal vector $4\mathbf{j} - \mathbf{k}$.

(c) Contains the points $(2, 1, -1)$, $(1, 1, 0)$, $(2, 0, 0)$.

(d) Contains the points $(1, 3, 1)$, $(2, 1, 0)$, $(1, 7, 2)$.

5. Find a vector equation for a plane satisfying the given conditions:

(a) Contains points $(0, 0, 0)$, $(1, 1, 1)$, $(3, 1, 4)$.

(b) Contains points $(2, 1, 2)$, $(3, 4, 2)$, $(3, 3, 5)$.

(c) Contains the point Q: $(1, 2, 1)$ and the line L: $\overrightarrow{OP} = (3, 1, 5) + t(2, 0, 1)$.

(d) Contains the point Q: $(3, 5, 2)$ and the line L: $\overrightarrow{OP} = t(2, 2, 1)$.

(e) Contains lines L_1: $\overrightarrow{OP} = (2, 1, 2) + t(5, 6, 7)$ and L_2: $\overrightarrow{OP} = (2, 1, 2) + t(1, 0, 0)$.

(f) Contains lines L_1: $\overrightarrow{OP} = (1, 3, 5) + t(1, 0, 2)$ and L_2: $\overrightarrow{OP} = (1, 3, 5) + t(4, 1, 7)$.

6. Find a linear equation for the plane with given vector equation or parametric equations:

(a) $\overrightarrow{OP} = \mathbf{i} + t(\mathbf{i} + 2\mathbf{j} + \mathbf{k}) + \tau(2\mathbf{i} + \mathbf{k})$

(b) $\overrightarrow{OP} = (1, 0, 3) + t(1, 0, 0) + \tau(0, 1, 0)$

(c) $x = 2 + t + \tau$, $y = 1 - t$, $z = 2 - 3t + \tau$

(d) $x = 1 - t - \tau$, $y = 2 + t + 3\tau$, $z = -1 + t + \tau$

7. Find a vector equation of the plane with given linear equation:

(a) $x + 3y - z = 5$ (b) $2x - y + z = 2$

(c) $2(x - 1) + 7(y - 3) - (z - 1) = 0$ (d) $z = 1$

8. Determine whether the given vector equations represent the same plane.

(a) $\overrightarrow{OP} = (1, 1, 2) + t(3, 0, 5) + \tau(1, 2, 2)$ and $\overrightarrow{OP} = (4, 1, 6) + t(3, 0, 5) + \tau(2, 4, 4)$

(b) $\overrightarrow{OP} = \mathbf{i} + t\mathbf{j} + \tau\mathbf{k}$, and $\overrightarrow{OP} = \mathbf{i} + \mathbf{j} + \mathbf{k} + t(\mathbf{j} + \mathbf{k}) + \tau(\mathbf{j} - \mathbf{k})$

9. Find a nonzero normal vector for each plane of Problem 6.

11-11 RELATIONS BETWEEN LINES AND PLANES

Let a plane H be given with base space W, and let H have vector equation

$$\overrightarrow{P_0P} = t\mathbf{u} + \tau\mathbf{v}, \qquad -\infty < t < \infty,\ -\infty < \tau < \infty \tag{11-110}$$

The vectors $t\mathbf{u} + \tau\mathbf{v}$ of the base space W are called vectors *along* H or *in* H. As (11-110) shows, each of these vectors equals a vector joining two points of H. Conversely, if P', P'' are points of H, then

$$\overrightarrow{P_0P'} = t'\mathbf{u} + \tau'\mathbf{v}, \qquad \overrightarrow{P_0P''} = t''\mathbf{u} + \tau''\mathbf{v}$$

so that

$$\overrightarrow{P'P''} = \overrightarrow{P_0P''} - \overrightarrow{P_0P'} = (t'' - t')\mathbf{u} + (\tau'' - \tau')\mathbf{v}$$

and, accordingly, $\overrightarrow{P'P''}$ is also in W. *Thus the vectors of W are the vectors joining pairs of points of H.*

A straight line L: $\overrightarrow{OP} = \overrightarrow{OP_1} + t\mathbf{w}$ is said to be *parallel* to a plane H: $\overrightarrow{OP} = a + t\mathbf{u} + \tau\mathbf{v}$ if the vectors along L are also vectors along H; that is, if $\mathbf{w}$ is contained in the base space W of H. We here allow for the possibility that L meets H. But if it does, then *L must be contained in H.* For if $\overrightarrow{OP_1} + t_1\mathbf{w}$ is in H for a particular t_1, and L is parallel to H, then

$$\overrightarrow{OP_1} + t_1\mathbf{w} = \mathbf{a} + s_1\mathbf{u} + \tau_1\mathbf{v}$$

for some s_1, τ_1 and, hence, $\overrightarrow{OP_1} = \mathbf{a} + \mathbf{w}'$, where $\mathbf{w}'$ is an element of W. Thus the equation of L can be written:

$$\overrightarrow{OP} = \mathbf{a} + \mathbf{w}' + t\mathbf{w}$$

and, hence, $\overrightarrow{OP} - \mathbf{a}$ is in W for every point of L; that is, every point P on L is also in H. The reasoning just given shows, in particular, that if a line L meets a plane H in two distinct points, then L must be contained in H.

Let a line L: $\overrightarrow{OP} = \overrightarrow{OP_1} + t\mathbf{w}$ and a plane H with nonzero normal vector $\mathbf{n}$ be given. Then L is parallel to H if, and only if, $\mathbf{w} \cdot \mathbf{n} = 0$. For L is parallel to H if, and only if, $\mathbf{w}$ is in the base space W of H, and we know that W consists of all vectors orthogonal to $\mathbf{n}$.

EXAMPLE 1. Show that the line L: $\overrightarrow{OP} = (1, 3, 2) + t(2, -1, 4)$ is parallel to the plane H: $3x + 2y - z - 5 = 0$ and determine whether L lies in H.

Solution. Here $2\mathbf{i} - \mathbf{j} + 4\mathbf{k}$ is orthogonal to the normal vector $\mathbf{n} = 3\mathbf{i} + 2\mathbf{j} - \mathbf{k}$, since $2 \cdot 3 - 1 \cdot 2 - 4 = 0$. Therefore, L is parallel to H. The point $(1, 3, 2)$ is in L but does not satisfy the equation $3x + 2y - z - 5 = 0$. Hence, L does not meet H at all.

A line L and a plane H are said to be *perpendicular* or *orthogonal* when the vectors along L are orthogonal to the vectors along H; that is, when the vectors along L are parallel to a nonzero normal vector of H.

EXAMPLE 2. Find the equation of a line L through $(3, 1, 2)$ orthogonal to the plane $H: x - y + z = 2$.

Solution. Here $\mathbf{i} - \mathbf{j} + \mathbf{k}$ is a nonzero normal vector to H. Hence, this is a vector along L, and L has the equation $\overrightarrow{OP} = (3, 1, 2) + t(1, -1, 1)$ (see Figure 11-18).

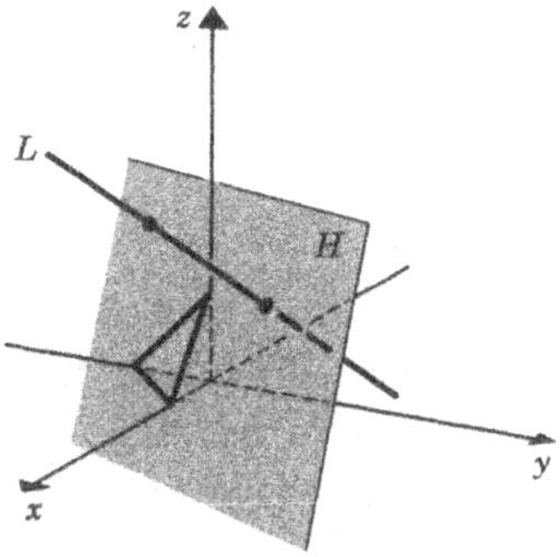

Figure 11-18 Line perpendicular to a plane.

EXAMPLE 3. Find the equation of a plane H through $(1, 0, 5)$ and orthogonal to the line L: $x = 1 - t$, $y = 2 + 2t$, $z = 3 + 5t$.

Solution. Here, $-\mathbf{i} + 2\mathbf{j} + 5\mathbf{k}$ is a nonzero vector along L, hence, a normal vector to H. Therefore, H has the equation:

$$-(x - 1) + 2(y - 0) + 5(z - 5) = 0$$

Remark. As these examples show, in general, there is a unique line through a given point orthogonal to a given plane, and there is a unique plane through a given point orthogonal to a given line (see Problems 6 and 7 below).

A line L that is not parallel to a plane H must intersect H in just one point. We illustrate this by an example and leave the general proof as an exercise (Problem 6 below).

EXAMPLE 4. Find the point of intersection of the line L: $\overrightarrow{OP} = (1, 2, 3) + t(1, 0, 5)$ and the plane H: $x - y + z = 5$.

Solution. We write parametric equations for L: $x = 1 + t, y = 2, z = 3 + 5t$ and substitute in the equation for H to obtain $1 + t - 2 + 3 + 5t = 5$, and find $t = \frac{1}{2}$. Thus $(\frac{3}{2}, 2, \frac{11}{2})$ is the point of intersection.

Remark 1. In the example, the vector $\mathbf{i} + 5\mathbf{k}$ along L is not orthogonal to the vector $\mathbf{i} - \mathbf{j} + \mathbf{k}$ normal to H. Hence, L is not parallel to H.

Remark 2. When L is orthogonal to H: $\overrightarrow{OP} = \mathbf{a} + t\mathbf{u} + \tau\mathbf{v}$, L cannot be parallel to H, since $\mathbf{u}, \mathbf{v}, \mathbf{u} \times \mathbf{v}$ are linearly independent. Hence, L meets H in a unique point. In particular, for each point P_0, one can construct a line L through P_0 perpendicular to a given plane H, and L meets H at a single point P_1, called the *foot of the perpendicular from P_0 to H.*

11-12 RELATIONS BETWEEN TWO PLANES

Two planes

$$H_1\colon A_1x + B_1y + C_1z + D_1 = 0, \quad H_2\colon A_2x + B_2y + C_2z + D_2 = 0 \quad (11\text{-}120)$$

are said to be *parallel* if they have the same base space W. It follows from Section 11-10 that H_1 and H_2 are parallel if, and only if, they have the same normal vectors; that is, if, and only if,

$$\mathbf{n}_1 = A_1\mathbf{i} + B_1\mathbf{j} + C_1\mathbf{k}, \qquad \mathbf{n}_2 = A_1\mathbf{i} + B_2\mathbf{j} + C_2\mathbf{k}$$

are linearly dependent.

We introduce the two matrices

$$S = \begin{pmatrix} A_1 & B_1 & C_1 \\ A_2 & B_2 & C_2 \end{pmatrix}, \qquad Q = \begin{pmatrix} A_1 & B_1 & C_1 & D_1 \\ A_2 & B_2 & C_2 & D_2 \end{pmatrix}$$

It then follows that H_1 and H_2 are parallel precisely when S has rank 1. If Q also has rank 1, then H_1 and H_2 coincide, as pointed out in Section 11-10. If S has rank 1 and Q has rank 2, then by the theory of Section 10-6, the simultaneous equations (11-120) have no solutions and H_1, H_2 are *parallel and nonintersecting*.

EXAMPLE 1. $2x + 3y - z - 2 = 0$, $4x + 6y - 2z - 1 = 0$. Here $\mathbf{n}_1 = 2\mathbf{i} + 3\mathbf{j} - \mathbf{k}$ and $\mathbf{n}_2 = 4\mathbf{i} + 6\mathbf{j} - 2\mathbf{k} = 2\mathbf{n}_1$, so that the planes are parallel. However, the coefficients are not proportional; that is, the matrix $\begin{pmatrix} 2 & 3 & -1 & -2 \\ 4 & 6 & -2 & -1 \end{pmatrix}$ has rank 2. Hence, the planes are parallel and nonintersecting (Figure 11-19).

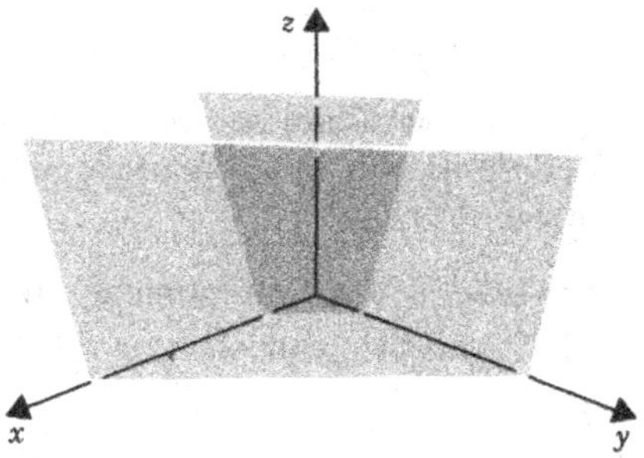

Figure 11-19 Parallel nonintersecting planes.

Two nonparallel planes must intersect in a line L. For if H_1 and H_2 are nonparallel, then matrix S has rank 2 and, as in Section 10-6, the solutions of the simultaneous equations (11-120) form a one-dimensional linear variety—that is, a line L. Let $\mathbf{w}$ be a nonzero vector along L. Then $\mathbf{w} \cdot \mathbf{n}_1 = 0$, since L is in H_1, and $\mathbf{w} \cdot \mathbf{n}_2 = 0$, since L is in H_2. Hence, $\mathbf{w}$ must be a scalar multiple of $\mathbf{n}_1 \times \mathbf{n}_2$. In particular, $\mathbf{n}_1 \times \mathbf{n}_2$ is itself a vector along L (Figure 11-20).

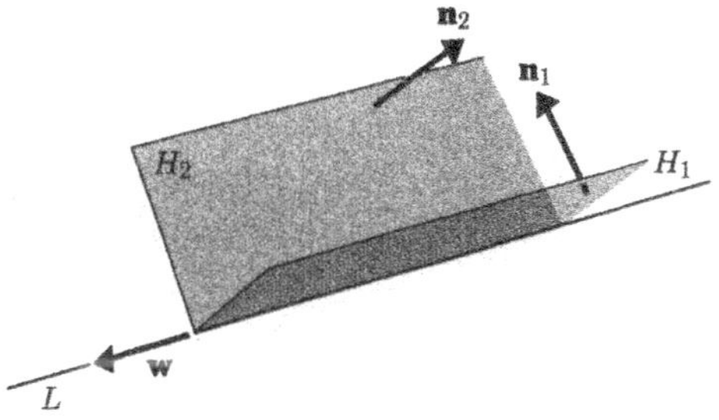

Figure 11-20 Planes intersecting in a line.

Conversely, *every line L can be represented as the intersection of two planes.* To show this, we write parametric equations for L:

$$x = x_1 + at, \qquad y = y_1 + bt, \qquad z = z_1 + ct$$

Not all of a, b, c can be 0. If, for example, $c \neq 0$, then we can eliminate t between the first and last equations, and also between the second and last equations to obtain the linear equations:

$$cx - az = cx_1 - az_1, \qquad cy - bz = cy_1 - bz_1$$

These equations represent two planes H_1, H_2 with respective normal vectors $\mathbf{n}_1 = (c, 0, -a)$ and $\mathbf{n}_2 = (0, c, -b)$, which are linearly independent. Therefore, H_1 and H_2 intersect in a line. But every point of the given line L satisfies both linear equations and, hence, L lies in both H_1 and H_2. Accordingly, H_1 and H_2 must intersect in L.

EXAMPLE 2. Find a vector equation for the line L of intersection of the planes

$$x + y + z - 2 = -0, \qquad 2x + 3y - z - 4 = 0$$

Solution. A vector along L is

$$\mathbf{n}_1 \times \mathbf{n}_2 = (\mathbf{i} + \mathbf{j} + \mathbf{k}) \times (2\mathbf{i} + 3\mathbf{j} - \mathbf{k}) = -4\mathbf{i} + 3\mathbf{j} + \mathbf{k}$$

We need only find a point on L. Since there are two equations in three unknowns, we can choose one unknown arbitrarily. For example, we take $z = 0$ and solve the two simultaneous equations $x + y - 2 = 0$, $2x + 3y - 4 = 0$ to obtain $x = 2$, $y = 0$. Hence $(2, 0, 0)$ is on L, and L has the equation $\overrightarrow{OP} = (2, 0, 0) + t(-4, 3, 1)$.

EXAMPLE 3. Represent L as the intersection of two planes if L has the vector equation $\overrightarrow{OP} = (1, 2, 1) + t(1, -1, 3)$.

Solution. Here $x = 1 + t$, $y = 2 - t$, $z = 1 + 3t$. Hence, $x + y = 3$, $3x - z = 2$ are two planes which intersect in the given line L. The two planes are graphed in Figure 11-21. The z-axis is parallel to the first plane, the y-axis is parallel to the second.

Remark. A line L can be given in many ways as the intersection of a pair of planes. If H_1: $\mathbf{n}_1 \cdot \overrightarrow{P_0P} = 0$, H_2: $\mathbf{n}_2 \cdot \overrightarrow{P_0P} = 0$ is one such pair and $\mathbf{u}, \mathbf{v}$ are

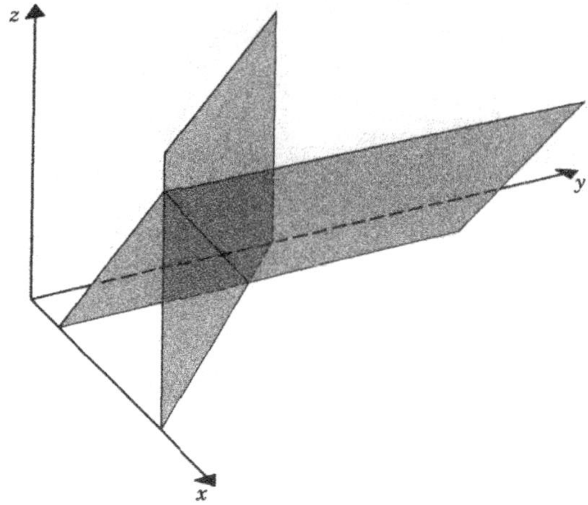

Figure 11-21 Example 3.

a linearly independent pair in Span ($\mathbf{n}_1$, $\mathbf{n}_2$), then H_3: $\mathbf{u} \cdot \overrightarrow{P_0P} = 0$ and H_4: $\mathbf{v} \cdot \overrightarrow{P_0P} = 0$ is another such pair, and every such pair is of this form (Problem 9 below).

‡11-13 HYPERPLANES AND LINEAR MANIFOLDS IN R^n

The lines and planes of R^3 have their counterpart in R^n: the linear manifolds of R^n. A set in R^n is a *k-dimensional linear manifold* if the corresponding vectors form a k-dimensional linear variety of V_n. Thus much of the theory of linear manifolds is simply a consequence of the linear algebra of linear varieties.

A one-dimensional linear manifold of R^n is a straight line (Section 11-5); a 2-dimensional linear manifold of R^n is called a *plane*. An $(n-1)$-dimensional linear manifold of R^n is called a *hyperplane*. If $\mathbf{a} = (a_1, \ldots, a_n)$, is a nonzero vector, then the solutions of the linear equation

$$a_1x_1 + \cdots + a_nx_n + b = 0 \tag{11-130}$$

form a linear variety of dimension $n - 1$ and, hence, define a hyperplane H. Conversely, every hyperplane H can be represented as the set of solutions $(x_1, \ldots, x_n)$ of a linear equation (11-130) (Section 10-4).

More generally, a linear manifold of dimension k can be defined by a vector equation

$$\overrightarrow{OP} = \overrightarrow{OP_1} + t_1\mathbf{w}_1 + \cdots + t_k\mathbf{w}_k$$

where $\mathbf{w}_1, \ldots, \mathbf{w}_k$ are linearly independent, or by $n - k$ linear equations

$$\sum_{j=1}^{n} a_{ij}x_j = 0, \qquad i = 1, \ldots, n-k$$

where $A = (a_{ij})$ has rank $n - k$ (see Remark 2, Section 10-4). Thus a k-dimensional linear manifold is the intersection of $n - k$ hyperplanes.

PROBLEMS

1. Find a linear equation for a plane satisfying the stated condition:
   (a) Contains P: $(6, 4, -2)$ and is orthogonal to the line through the points $(7, -2, 3)$ and $(1, 4, -5)$.
   (b) Contains P: $(1, 2, -3)$, is orthogonal to the y-axis.
   (c) Contains P: $(1, 2, 3)$, is orthogonal to the line $x = t$, $y = 2 - 2t$, $z = 1 + 3t$.
   (d) Contains the point $(1, 0, 1)$, is orthogonal to the line $\overrightarrow{OP} = (1, 0, 3) + t(2, 1, 4)$.
   (e) Contains P: $(-1, 2, 3)$ and the line $x = 1 - t$, $y = 2$, $z = 3 - t$.
   (f) Contains P: $(2, 0, 1)$, is parallel to the plane $2x - y + z = 2$.
   (g) Contains P: $(7, -1, 2)$, is parallel to the plane $x = 1 + t$, $y = 1 - \tau$, $z = 2 + t + 2\tau$.
   (h) Contains P: $(1, 0, 0)$, is orthogonal to the line $x - y + z = 1$, $2x + y + z = 2$.
   (i) Contains $(1, 2, 1)$, and the lines $\overrightarrow{OP} = (1, 5, 3) + t(1, 0, 2)$, $\overrightarrow{OP} = (7, 5, 9) + t(3, 1, 4)$ are parallel to it.
   (j) Contains $(3, 0, 4)$, and the lines $\overrightarrow{OP} = (0, 0, 1) + t(1, 0, 1)$, $\overrightarrow{OP} = (3, 0, 1) + t(0, 1, 0)$ are parallel to it.
2. Find the line in R^3 such that
   (a) It contains $(-2, 1, 3)$ and is orthogonal to the plane $2x + 3y + z = 1$;
   (b) It contains $(0, 1, 1)$ and is orthogonal to each of the lines $\overrightarrow{OP} = (1, 3, 5) + t(2, 0, 3)$, $\overrightarrow{OP} = (2, 1, 2) + t(7, 5, 1)$;
   (c) It lies in the plane $x - y + z = 7$, contains $(7, 0, 0)$ and is orthogonal to the line $x = 1 + t$, $y = 1 + 2t$, $z = 1 + 3t$;
   (d) It lies in the plane $2x - y + 3z = 5$, meets and is orthogonal to the line $x = t$, $y = -1 + t$, $z = 2 + t$.
3. Determine whether line L intersects plane H in a single point and, if it does, find the point of intersection:
   (a) L: $\overrightarrow{OP} = (1, 2, 2) + t(3, 5, 0)$, H: $x - y + z = 2$
   (b) L: $\overrightarrow{OP} = (3, 2, 2) + t(1, 2, 4)$, H: $2x + 2y - z = 0$
4. Determine whether the two given planes intersect in a line and, if they do, find a vector equation for the line:
   (a) $2x + y - z = 1$, $3x + 2y + z = 4$
   (b) $x + 2y + 3z = 1$, $2x + 4y + 6z = 1$
   (c) $3x + 6y + 3z = 27$, $2x + 4y + 2z = 14$
   (d) $2x + 2y + z = 0$, $x - y = 0$
5. Represent the given line as the intersection of two planes:
   (a) $\overrightarrow{OP} = (1, 2, 2) + t(3, 0, 5)$ (b) $x = 1 - t$, $y = 2 + t$, $z = 3$
6. Prove the following theorems:
   (a) If line L is not parallel to plane H, then L intersects H in exactly one point.
   (b) Planes perpendicular to the same line are parallel.
   (c) There is one and only one plane through a given point Q parallel to a given plane H_1.

(d) Two planes parallel to a third plane are parallel to each other.

(e) A line parallel to a second line which is parallel to a given plane is parallel to the plane.

(f) Two lines perpendicular to the same plane are parallel.

(g) There is one and only one line through a given point orthogonal to a given plane.

7. Prove that there is one and only one plane H in R^3

(a) Containing a given line L and a given point Q not on L.

(b) Containing two given nonparallel intersecting lines L_1, L_2.

(c) Containing two distinct parallel lines L_1, L_2.

(d) Through a given point Q and such that H is parallel to L_1 and to L_2, where L_1, L_2 are given nonparallel lines.

(e) Through a given point Q and orthogonal to a given line L.

8. Let line L have equation $\overrightarrow{OP} = \mathbf{a} + t\mathbf{v}$, where $\mathbf{v} = (v_1, v_2, v_3)$ and none of v_1, v_2, v_3 is 0. Show that (x_1, x_2, x_3) is on L if, and only if,

$$\frac{x_1 - a_1}{v_1} = \frac{x_2 - a_2}{v_2} = \frac{x_3 - a_3}{v_3}$$

[*Note.* These equations are called *symmetric equations* for L. From the equations we get the equations of two planes intersecting in L by using the first equality and then the second:

$$v_2(x_1 - a_1) - v_1(x_2 - a_2) = 0, \qquad v_3(x_2 - a_2) - v_2(x_3 - a_3) = 0$$

From the symmetric equations we can recover the parametric equations by letting t be the common value of the 3 ratios and, hence, obtaining $x_1 - a_1 = v_1 t$ and so on. The symmetric equations can be used even if some (at most 2) of the v_i are zero, providing that one then understands that each numerator corresponding to a zero denominator must also be zero.]

‡9. Prove the assertions in the remark at the end of Section 11-12.

‡10. Show by example that in R^4 two 2-dimensional linear manifolds (planes) need not intersect, even though they are not parallel (that is, have different base spaces).

‡11. Describe the intersection (if any) of the linear manifolds in R^5: $\overrightarrow{OP} = (2, 1, 0, 0, 0) + t(3, 5, 2, 1, 4) + \tau(1, 2, 2, 0, 1)$, $x_1 + x_2 + x_3 + x_4 + x_5 = 1$.

‡12. Prove: in R^n, if a k-dimensional linear manifold and an m-dimensional linear manifold intersect, then the intersection is a linear manifold of dimension at least $k + m - n$.

11-14 OTHER CARTESIAN COORDINATE SYSTEMS IN R^3

Thus far we have used only one coordinate system in R^3; we call this a Cartesian coordinate system because the coordinate axes are perpendicular and the same unit of distance is used on each. Another Cartesian coordinate system can be obtained in R^3 by choosing a new origin and/or changing the directions of the axes. These directions are given by corresponding unit vectors.

In general, three nonzero vectors of V_3 which are pairwise orthogonal are called an *orthogonal set* of vectors; such vectors are necessarily linearly independent (Problem 8 following Section 11-3). Hence, they form a basis, called an *orthogonal basis*, of V_3. When the three vectors are unit vectors, they are said to form an *orthonormal basis*.

We remark that there are many orthogonal bases for V_3. In particular, from a given nonzero vector $\mathbf{v}_1$ one can obtain vectors $\mathbf{v}_2, \mathbf{v}_3$ to complete an orthogonal basis. For one need only choose a vector $\mathbf{w}$ such that $\mathbf{w}, \mathbf{v}_1$ are linearly independent. Then $\mathbf{v}_1$ and $\mathbf{v}_2 = \mathbf{w} \times \mathbf{v}_1$ are linearly independent, and $\mathbf{v}_1$ is orthogonal to $\mathbf{v}_2$. Then we take $\mathbf{v}_3$ to be $\mathbf{v}_1 \times \mathbf{v}_2$ and have an orthogonal basis $\mathbf{v}_1$, $\mathbf{v}_2$, $\mathbf{v}_3$ (Figure 11-22). The vectors $\mathbf{i}' = \mathbf{v}_1/\|\mathbf{v}_1\|$, $\mathbf{j}' = \mathbf{v}_2/\|\mathbf{v}_2\|$, $\mathbf{k}' = \mathbf{v}_3/\|\mathbf{v}_3\|$ then form an orthonormal basis.

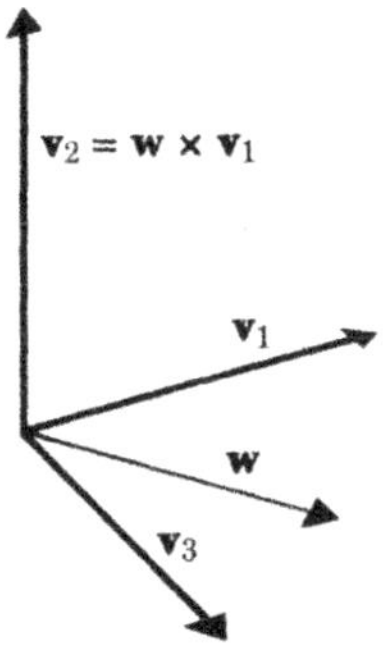

Figure 11-22 Formation of orthogonal basis.

Now let $\mathbf{i}', \mathbf{j}', \mathbf{k}'$ be an orthonormal basis for V_3 and let O' be a point of R^3 which we select as new origin. We can now represent each vector $\mathbf{v}$ of V_3 in terms of this basis:

$$\mathbf{v} = v_1'\mathbf{i}' + v_2'\mathbf{j}' + v_3'\mathbf{k}' \tag{11-140}$$

In particular, for each point P of R^3 we can write

$$\overrightarrow{O'P} = x'\mathbf{i}' + y'\mathbf{j}' + z'\mathbf{k}' \tag{11-141}$$

to obtain new Cartesian coordinates for P (see Figure 11-23).

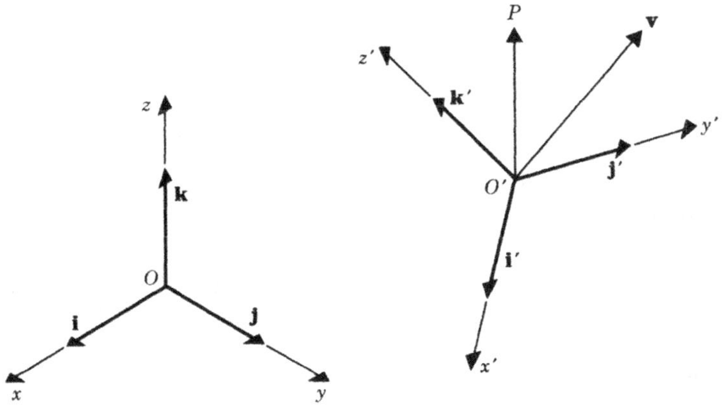

Figure 11-23 New coordinates in space.

THEOREM 6. *Let* $\mathbf{i}', \mathbf{j}', \mathbf{k}'$ *be an orthonormal basis for* V_3 *and let each vector* $\mathbf{v}$ *of* V_3 *be assigned a new set of components by Equation* (11-140). *Let* O' *be a point of* R^3 *and let a new* $x'y'z'$ *coordinate system be introduced in* R^3 *by Equation* (11-141). *Then for all* $\mathbf{v}, \mathbf{w}$ *in* V_3 *and all scalars* c,

(a) $\mathbf{v} + \mathbf{w} = (v_1' + w_1')\mathbf{i}' + (v_2' + w_2')\mathbf{j}' + (v_3' + w_3')\mathbf{k}'$

(b) $c\mathbf{v} = (cv_1')\mathbf{i}' + (cv_2')\mathbf{j}' + (cv_3')\mathbf{k}'$

(c) $\mathbf{v} \cdot \mathbf{w} = v_1'w_1' + v_2'w_2' + v_3'w_3'$

(d) $\|\mathbf{v}\| = \sqrt{v_1'^2 + v_2'^2 + v_3'^2}$

and, provided that $\mathbf{i}', \mathbf{j}', \mathbf{k}'$ *is a positive triple,*

(e) $\mathbf{v} \times \mathbf{w} = \begin{vmatrix} \mathbf{i}' & \mathbf{j}' & \mathbf{k}' \\ v_1' & v_2' & v_3' \\ w_1' & w_2' & w_3' \end{vmatrix}$

For each pair of points P_1: (x_1', y_2', z_3'), P_2: (x_1', y_2', z_3') *(new coordinates) one has*

(f) $\overrightarrow{P_1P_2} = (x_2' - x_1')\mathbf{i}' + (y_2' - y_1')\mathbf{j}' + (z_2' - z_1')\mathbf{k}'$

(g) $d(P_1, P_2) = \|\overrightarrow{P_1P_2}\| = \sqrt{(x_2' - x_1')^2 + (y_2' - y_1')^2 + (z_2' - z_1')^2}$

PROOF. Statements (a) and (b) follow from basic algebraic rules. For (c) we write $\mathbf{v}$ as in (11-140) and write $\mathbf{w}$ similarly. We then find that

$$\mathbf{v} \cdot \mathbf{w} = v_1'w_1'\mathbf{i}' \cdot \mathbf{i}' + v_1'w_2'\mathbf{i}' \cdot \mathbf{j}' + \cdots$$

Since $\mathbf{i}' \cdot \mathbf{i}' = 1$, $\mathbf{i}' \cdot \mathbf{j}' = 0, \ldots$, (c) follows; and (d) is an immediate consequence. For (e) we write similarly

$$\mathbf{v} \times \mathbf{w} = v_1'w_1'\mathbf{i}' \times \mathbf{i}' + v_1'w_2'\mathbf{i}' \times \mathbf{j}' + \cdots$$

Since $\mathbf{i}', \mathbf{j}', \mathbf{k}'$ is a positive triple, $\mathbf{i}' \times \mathbf{j}' \cdot \mathbf{k}' > 0$. But $\mathbf{i}' \times \mathbf{j}'$ must be parallel to $\mathbf{k}'$ and must be a unit vector. Hence, $\mathbf{i}' \times \mathbf{j}' = \mathbf{k}'$. Evaluating the other cross products in the same way, we find that $\mathbf{v} \times \mathbf{w} = (v_2'w_3' - v_3'w_2')\mathbf{i}' + \cdots$ as in (e). For P_1, P_2 as given, we have

$$\begin{aligned}\overrightarrow{P_1P_2} &= \overrightarrow{O'P_2} - \overrightarrow{O'P_1} \\ &= x_2'\mathbf{i}' + y_2'\mathbf{j}' + z_2'\mathbf{k}' - (x_1'\mathbf{i}' + y_1'\mathbf{j}' + z_1')\mathbf{k}'\end{aligned}$$

and one obtains the expression in (f). Finally, (g) follows from (f) and (d).

Discussion. This theorem shows that R^3 is "homogeneous": all points are the same. Any point O_1 can be used as origin, with any orthonormal basis, to determine coordinates. (Thus all directions are the same: space is "isotropic.") The expressions for norms, inner products, and so on, are the same in the new coordinates as in the old. We can develop the theory of lines and planes as above, with exactly the same formulas. (For formulas using the cross product, we must be sure we have not changed orientation.)

Because of these results, in any particular problem we can choose the origin and coordinate axes as desired to best fit the problem.

THEOREM 7. *Each line L in R^3 can serve as the x-axis of a Cartesian coordinate system in R^3. Each plane H in R^3 can serve as the xy-plane of a Cartesian coordinate system in R^3.*

PROOF. Let O' be a point on L, $\mathbf{v}_1$ a vector along L. Then, as remarked above, we can choose an orthonormal basis $\mathbf{i}', \mathbf{j}', \mathbf{k}'$ such that $\mathbf{i}' = \mathbf{v}_1/\|\mathbf{v}_1\|$ and, hence, $\mathbf{i}'$ is along L. With this basis, we define our coordinates by (11-141), and L is the x'-axis as desired. Also, the x'-axis has as direction that of the given vector $\mathbf{v}_1$. (After introducing the new coordinates, one may wish to forget the old ones; in this case, one can drop the primes on the new coordinates.)

For the case of a plane H, we first select a point O' in H and a basis $\mathbf{u}$, $\mathbf{v}$ for the base space of H. We then take $\mathbf{v}_1 = \mathbf{u}$, $\mathbf{v}_3 = \mathbf{u} \times \mathbf{v}$ and $\mathbf{v}_2 = \mathbf{v}_3 \times \mathbf{u}$ (Figure 11-24) to obtain an orthogonal basis. Also $\mathbf{v}_1$ and $\mathbf{v}_2$ are in the base space of H and, in fact, $\mathbf{v}_1, \mathbf{v}_2, \mathbf{v}_3$ is a positive triple (see Problem 10 below). The unit vectors $\mathbf{i}' = \mathbf{v}_1/\|\mathbf{v}_1\|$, $\mathbf{j}' = \mathbf{v}_2/\|\mathbf{v}_2\|$, $\mathbf{k}' = \mathbf{v}_3/\|\mathbf{v}_3\|$ then provide the desired basis, and (11-141) defines new coordinates for which H is the $x'y'$-plane (and one may now wish to drop the primes).

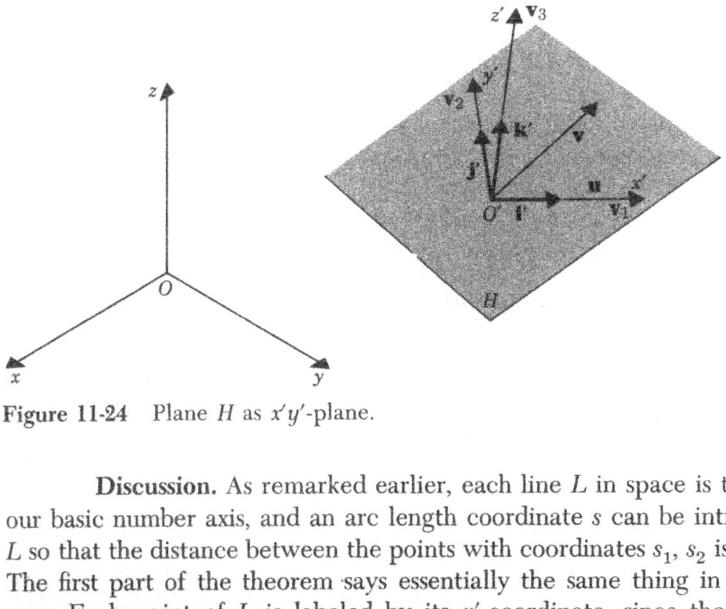

Figure 11-24 Plane H as $x'y'$-plane.

Discussion. As remarked earlier, each line L in space is the same as our basic number axis, and an arc length coordinate s can be introduced on L so that the distance between the points with coordinates s_1, s_2 is $|s_2 - s_1|$. The first part of the theorem says essentially the same thing in a different way. Each point of L is labeled by its x'-coordinate, since the y' and z'-coordinates are 0; x' serves as an arc length coordinate, since the distance between $(x_1', 0, 0)$ and $(x_2', 0, 0)$ is

$$\sqrt{(x_2' - x_1')^2} = |x_2' - x_1'|$$

For the case of a plane, our results show that each plane in space is really the same as the xy-plane of 2-dimensional geometry, as in Chapter 1. Each point can be located by coordinates (x', y') (z' being 0 in H), and distances are given by the formula

$$d((x_1', y_1', 0), (x_2', y_2', 0)) = \sqrt{(x_2' - x_1')^2 + (y_2' - y_1')^2}$$

as usual in the plane. All the other properties of plane geometry can be recovered from this information (as is implied by Chapter 1). Hence, each plane in space is the same as the plane of plane Euclidean geometry.

In particular, to establish a plane-geometric property of a plane in space, one can without loss of generality assume that the plane is the xy-plane in space, that the x-axis is along a specified line in that plane, and that the origin is a chosen point on that line.

EXAMPLE 1. Show that the perpendicular is the shortest distance from a point to a plane.

Solution. We are given a plane H and a point P_1. We want to show that $d(P_1, P_0) < d(P, Q)$, where P_0 is the foot of the perpendicular from P_1 to H and Q is any other point of H. We choose coordinates so that H is the xy-plane and the origin O is P_0. Then P_1 must be on the z-axis, P_1: $(0, 0, c)$. Each point Q in H has coordinates $(x, y, 0)$. Hence, $d(P_1, Q) = (x^2 + y^2 + c^2)^{1/2} \geq |c| = d(P_1, P_0)$, with equality only for $x = 0$ and $y = 0$—that is, for Q at P_0. Thus the assertion is proved.

EXAMPLE 2. Show that parallel planes are everywhere equidistant.

Solution. Let H_1 and H_2 be distinct parallel planes. We want to show that the shortest distance from P_2 to H_1 is the same for all points P_2 in H_2. We choose coordinates so that H_1 is the xy-plane and, hence, has equation $z = 0$. Therefore, H_2 has equation $z = c$, for some c. Hence, a general point P_2 of H_2 has coordinates (x_2, y_2, c) and $(x_2, y_2, 0)$ is the foot of the perpendicular from P_2 to H_1. Therefore, by Example 1, the shortest distance from P_2 to H_1 is $(c^2)^{1/2} = |c|$ and this is the same for all choices of P_2 in H_2.

Remark. Further information on new coordinate systems is given in Sections 11-21 and 11-22 below.

11-15 LENGTHS, AREAS AND VOLUMES IN R^3

Thus far in this chapter we have assigned length only to line segments. By attaching several line segments in sequence (see Figure 11-25), we obtain

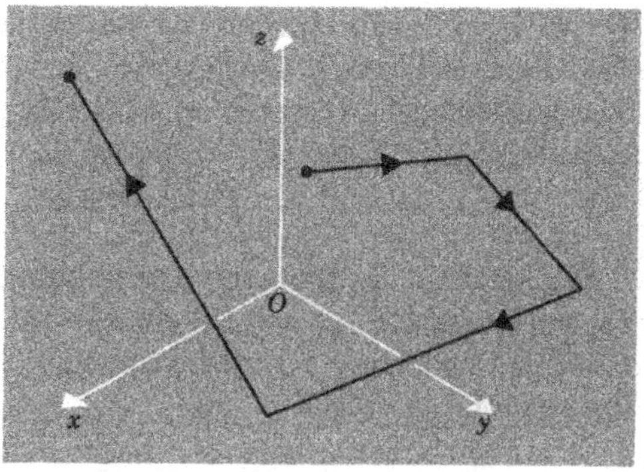

Figure 11-25 Broken line—polygonal path.

a *broken line* or *polygonal path*. Its length can be defined as the sum of the lengths of the several segments. A limiting process as in Section 4-27 then permits us to assign length to a more general path by an appropriate integral.

For area we first consider planar figures; that is, figures lying in a plane. As in Section 11-14, each plane can be assumed to be the xy-plane and then, as in Chapter 4, plane geometry (or calculus) can be applied to find the area. In particular, for a parallelogram whose sides, properly directed, are the vectors $\mathbf{u}$, $\mathbf{v}$, we know from plane geometry that the area is $\|\mathbf{u}\|\ \|\mathbf{v}\| \sin\theta$, where $\theta = \measuredangle(\mathbf{u}, \mathbf{v})$. Hence, as in Section 11-6,

$$\|\mathbf{u} \times \mathbf{v}\| = \|\mathbf{u}\|\ \|\mathbf{v}\| \sin\theta = \text{area of parallelogram}$$

From this formula and plane geometry we obtain the area of a triangle in space:

$$\tfrac{1}{2}\|\mathbf{u} \times \mathbf{v}\| = \tfrac{1}{2}\|\mathbf{u}\|\ \|\mathbf{v}\| \sin\theta = \text{area of triangle}$$

where $\mathbf{u}$, $\mathbf{v}$ are two sides of the triangle, properly directed. Now

$$\|\mathbf{u}\|^2\ \|\mathbf{v}\|^2 \sin^2\theta = \|\mathbf{u}\|^2\ \|\mathbf{v}\|^2\ (1 - \cos^2\theta) = \|\mathbf{u}\|^2\ \|\mathbf{v}\|^2 - (\mathbf{u}\cdot\mathbf{v})^2$$

and, hence, we have the formula:

$$(\text{area of parallelogram})^2 = \begin{vmatrix} \mathbf{u}\cdot\mathbf{u} & \mathbf{u}\cdot\mathbf{v} \\ \mathbf{u}\cdot\mathbf{v} & \mathbf{v}\cdot\mathbf{v} \end{vmatrix} = 4(\text{area of triangle})^2$$

Thus the area of the parallelogram with sides $\mathbf{u} = (1, 2, 3)$, $\mathbf{v} = (1, -1, 0)$ is

$$\begin{vmatrix} 14 & -1 \\ -1 & 2 \end{vmatrix}^{1/2} = \sqrt{27}$$

By combining planar figures, as we did above for line segments, we obtain nonplanar surfaces: for example, the surface of a cube. The area of such a surface is defined as the sum of the areas of the planar parts of which it is composed. Limiting processes, as in Section 7-8, then permit one to assign area meaningfully to more general surfaces in spaces. (This turns out to be extraordinarily subtle to carry out. See Section 13-9.)

Volumes of Figures in Space. The theory of volumes of solids is normally developed in solid geometry and leads to the formulas $V = \frac{1}{3}bh$ for volume of a tetrahedron and $V = bh$ for a parallelepiped (b = area of base, h = altitude). We do not attempt to explore this theory here, but merely point out several vector formulas for volume, formulas which are exceptionally convenient for a systematic development of the theory.

Let a tetrahedron be given, with vertices P_1, P_2, P_3, P_4, as in Figure 11-26. We refer to the solid as the tetrahedron $P_1P_2P_3P_4$. We can compute the volume as $\frac{1}{3}bh$, where b is the area of the triangular base $P_1P_2P_3$ and h is the altitude (perpendicular distance from P_4 to the plane of the base). We write $\mathbf{u} = \overrightarrow{P_1P_2}$, $\mathbf{v} = \overrightarrow{P_1P_3}$. Then the area of the base is $\frac{1}{2}\|\mathbf{u} \times \mathbf{v}\|$ and the altitude h is $\|\overrightarrow{P_4P_0}\|$, where P_0 is the foot of the perpendicular from P_4 to the plane of the base.

Hence, the volume is

$$V = \frac{1}{3}\frac{1}{2}\|\mathbf{u} \times \mathbf{v}\| \, \|\overrightarrow{P_4P_0}\|$$

Now $\mathbf{u} \times \mathbf{v}$ is a vector normal to the base, as is $\overrightarrow{P_4P_0}$. If they are similarly directed, as we can (and shall) assume for proper numbering of the vertices, then we can write

$$V = \frac{1}{6}\mathbf{u} \times \mathbf{v} \cdot \overrightarrow{P_4P_0} \tag{11-150}$$

For the inner product of two vectors $\mathbf{p}$, $\mathbf{q}$, is $\|\mathbf{p}\| \, \|\mathbf{q}\| \cos \phi$ and this reduces to $\|\mathbf{p}\| \, \|\mathbf{q}\|$ when $\phi = 0$.

Now on the right side of (11-150) we can replace $\overrightarrow{P_4P_0}$ by $\overrightarrow{P_4P_0'}$, where P_0' is an arbitrary point in the plane of P_1, P_2, P_3. For $\overrightarrow{P_4P_0'} = \overrightarrow{P_4P_0} + \overrightarrow{P_0P_0'}$ and, hence,

$$\begin{aligned}\mathbf{u} \times \mathbf{v} \cdot \overrightarrow{P_4P_0'} &= \mathbf{u} \times \mathbf{v} \cdot (\overrightarrow{P_4P_0} + \overrightarrow{P_0P_0'}) \\ &= \mathbf{u} \times \mathbf{v} \cdot \overrightarrow{P_4P_0} + \mathbf{u} \times \mathbf{v} \cdot \overrightarrow{P_0P_0'}\end{aligned}$$

But the last term is 0, since $\mathbf{u}$, $\mathbf{v}$, $\overrightarrow{P_0P_0'}$ are linearly dependent. Therefore, P_0 can be replaced by P_0' in (11-150). We can also interpret the vector $\frac{1}{2}\mathbf{u} \times \mathbf{v}$ as a vector normal to the base $P_1P_2P_3$ and having magnitude equal to the area of the base. We call such a vector an *area vector* and denote it here by $\mathbf{n}$. Since $\mathbf{u} \times \mathbf{v}$ and $\overrightarrow{P_4P_0}$ have the same direction, $\mathbf{n}$ is an "outer normal"—that is, $\mathbf{n}$ points to the exterior of the tetrahedron. We can now write (11-150) as

$$V = \frac{1}{3}\mathbf{n} \cdot \overrightarrow{P_4P_0'} \tag{11-150'}$$

We next select a point Q inside the tetrahedron and use Q as a vertex to subdivide the tetrahedron into four smaller tetrahedra $QP_1P_2P_3$, $QP_2P_3P_4$, $QP_3P_4P_1$, $QP_4P_1P_2$ (Figure 11-26). We find the volume of each by the formula

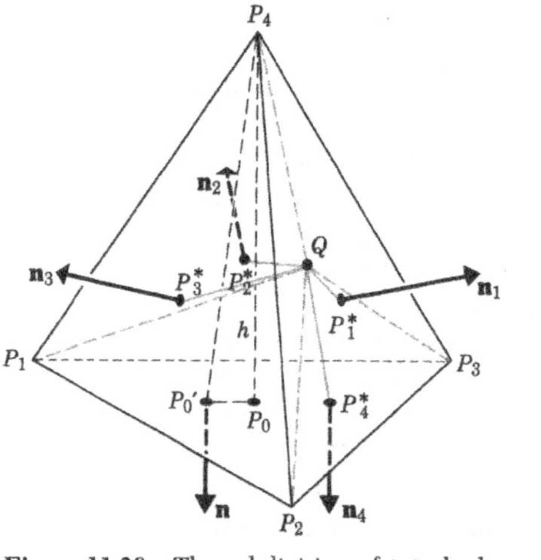

Figure 11-26 The subdivision of tetrahedron.

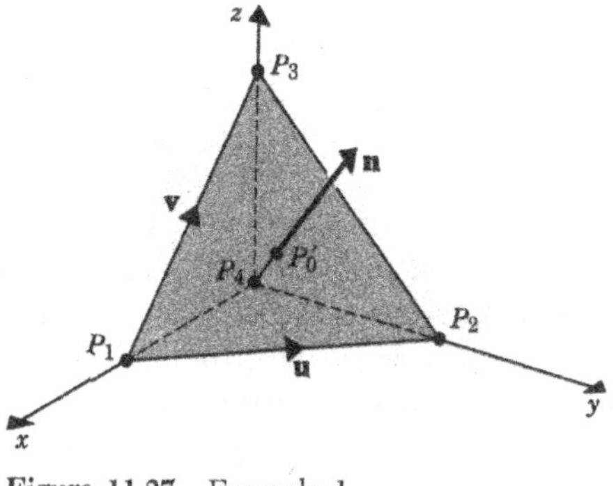

Figure 11-27 Example 1.

analogous to (11-150′) and add to obtain the volume of V:

$$V = \frac{1}{3}\sum_{i=1}^{4} \overrightarrow{QP_i^*} \cdot \mathbf{n}_i \tag{11-151}$$

Here we have numbered the triangular faces from 1 to 4, chosen a point P_i^* in the plane of the ith face and a corresponding area vector $\mathbf{n}_i$, as in Figure 11-26.

Equation (11-151) is an exceptionally useful expression for the volume of the tetrahedron. The point Q can be chosen as desired—one, in fact, can permit it to be any point in space (Problem 12 below). For $Q = P_4$, (11-151) reduces to (11-150′), since for $Q = P_4$ all terms in (11-151) except the one for the face $P_1P_2P_3$ reduce to 0.

EXAMPLE 1. For the tetrahedron with vertices P_1: (1, 0, 0), P_2: (0, 1, 0), P_3: (0, 0, 1), P_4: (0, 0, 0) we can apply (11-150′) with $\mathbf{u} = \mathbf{j} - \mathbf{i}$, $\mathbf{v} = \mathbf{k} - \mathbf{i}$, $\mathbf{n} = \frac{1}{2}\mathbf{u} \times \mathbf{v} = \frac{1}{2}(\mathbf{i} + \mathbf{j} + \mathbf{k})$, and P_0' as $(\frac{1}{3}, \frac{1}{3}, \frac{1}{3})$. (For this point lies in the plane $x + y + z = 1$ of the vertices P_1, P_2, P_3; see Figure 11-27.) Hence,

$$V = \frac{1}{3} \cdot \frac{1}{2}(\mathbf{i} + \mathbf{j} + \mathbf{k}) \cdot \left(\frac{1}{3}\mathbf{i} + \frac{1}{3}\mathbf{j} + \frac{1}{3}\mathbf{k}\right) = \frac{1}{6}$$

We can also use (11-151) with Q at (1, 1, 1), $\mathbf{n}_1 = -\frac{1}{2}\mathbf{i}$, $\mathbf{n}_2 = -\frac{1}{2}\mathbf{j}$, $\mathbf{n}_3 = -\frac{1}{2}\mathbf{k}$, $\mathbf{n}_4 = \mathbf{n}$ as above, and $P_1^* = P_2^* = P_3^* = (0, 0, 0)$, $P_4^* = (0, 0, 1)$. Then $\overrightarrow{QP_1^*} = \overrightarrow{QP_2^*} = \overrightarrow{QP_3^*} = -\mathbf{i} - \mathbf{j} - \mathbf{k}$, $\overrightarrow{QP_4^*} = -\mathbf{i} - \mathbf{j}$, and we find that

$$V = \frac{1}{3}(\overrightarrow{QP_1^*} \cdot \mathbf{n}_1 + \cdots) = \frac{1}{3}[(-\mathbf{i} - \mathbf{j} - \mathbf{k}) \cdot (-\tfrac{1}{2}\mathbf{i}) + \cdots] = \frac{1}{6}$$

The formula (11-151) can be extended to a solid whose boundary is an arbitrary polyhedron of m faces. As suggested in Figure 11-28, one selects a reference point Q anywhere in space, a point P_i^* in each face, and an area vector pointing to the exterior of the solid. Then the volume is given by

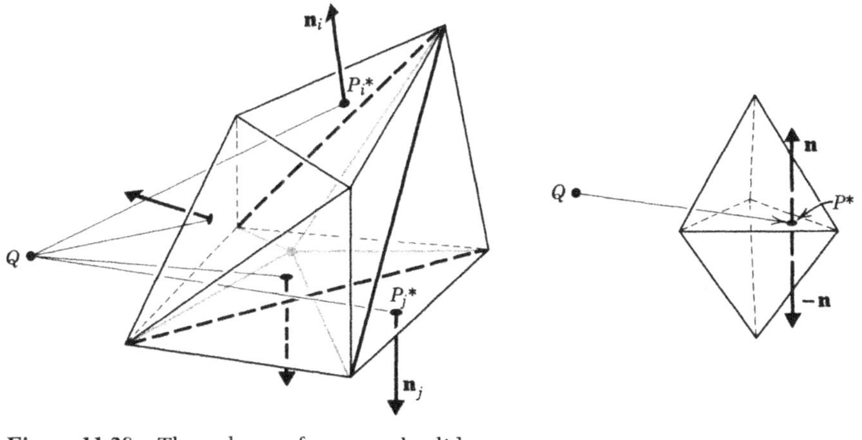

Figure 11-28 The volume of a general solid.

$$V = \frac{1}{3} \sum_{i=1}^{m} \overrightarrow{QP_i^*} \cdot \mathbf{n}_i \tag{11-152}$$

We can deduce this formula from (11-151) by subdividing the solid into tetrahedra as suggested, applying (11-151) to each tetrahedron, and by adding the results. When one adds, the expressions $\overrightarrow{QP^*} \cdot \mathbf{n}$ coming from "inner" faces of tetrahedra cancel, since where two tetrahedra have such a face in common, the two normals are equal and opposite and the points P^* can be taken as the same for both; thus one has two terms $\overrightarrow{QP^*} \cdot \mathbf{n} + \overrightarrow{QP^*}(-\mathbf{n}) = 0$ (see detail enlarged in Figure 11-28). Thus one is left with the contributions from the exterior faces alone. For those faces of the original solid that are composed of several triangles, one can choose the same point P^* for all and then get several terms $\overrightarrow{QP^*} \cdot \mathbf{n}_1' + \overrightarrow{QP^*} \cdot \mathbf{n}_2' + \cdots + \overrightarrow{QP^*} \cdot \mathbf{n}_k'$, where $\mathbf{n}_1', \ldots, \mathbf{n}_k'$ are area vectors for the triangular faces. But $\mathbf{n}_1' + \cdots + \mathbf{n}_k' = \mathbf{n}$, where $\mathbf{n}$ is an area vector for the whole face. Hence, the sum reduces to (11-152).

It is important to note that (11-152) is a formula for the volume of a polyhedral solid that does not refer to a subdivision into tetrahedra. In deriving (11-152), we had to know and use that such a subdivision was possible, but the end result does not depend on the subdivision chosen.

From the general formula (11-152) one can deduce the familiar formulas for volume of a parallelepiped, a prism, and a pyramid. For example, for a parallelepiped as in Figure 11-29, one can introduce edge vectors $\mathbf{u}, \mathbf{v}, \mathbf{w}$

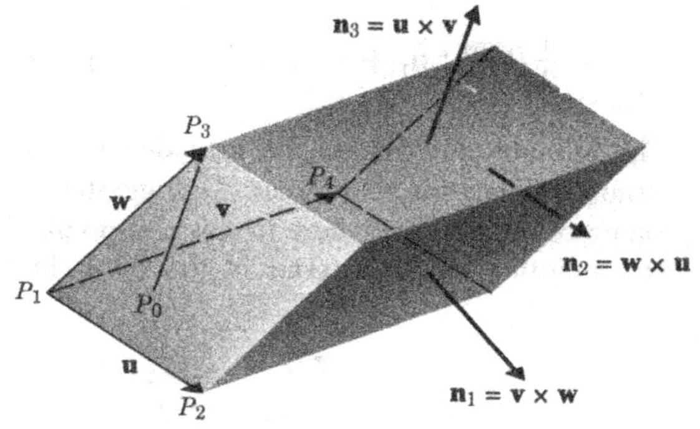

Figure 11-29 The volume of a parallelepiped.

as shown, forming a positive triple. Then the area vectors are cross products of pairs of these vectors, as shown. If one takes Q at P_1, then only the terms in the three area vectors shown are different from 0 and

$$V = \frac{1}{3}(\mathbf{u} \cdot \mathbf{n}_1 + \mathbf{v} \cdot \mathbf{n}_2 + \mathbf{w} \cdot \mathbf{n}_3)$$

$$= \frac{1}{3}(\mathbf{u} \cdot \mathbf{v} \times \mathbf{w} + \mathbf{v} \cdot \mathbf{w} \times \mathbf{u} + \mathbf{w} \cdot \mathbf{u} \times \mathbf{v})$$

Since all three scalar triple products are equal, we deduce the valuable formula

$$V = \mathbf{u} \cdot \mathbf{v} \times \mathbf{w} = \begin{vmatrix} u_1 & u_2 & u_3 \\ v_1 & v_2 & v_3 \\ w_1 & w_2 & w_3 \end{vmatrix} \qquad (11\text{-}153)$$

for *volume of a parallelepiped*. This formula shows that a determinant of order 3 can be interpreted as volume, just as a determinant of order 2 can be interpreted as area. (If the edges **u, v, w** form a negative triple, then the volume becomes $-\mathbf{u} \cdot \mathbf{v} \times \mathbf{w}$.)

If we let P_0 be the foot of the perpendicular from P_3 to the plane of $P_1P_2P_4$, as in Figure 11-29, then $\mathbf{w} = \overrightarrow{P_1P_0} + \overrightarrow{P_0P_3}$, and (11-153) gives

$$\begin{aligned} V &= \mathbf{u} \times \mathbf{v} \cdot (\overrightarrow{P_1P_0} + \overrightarrow{P_0P_3}) = \mathbf{u} \times \mathbf{v} \cdot \overrightarrow{P_1P_0} + \mathbf{u} \times \mathbf{v} \cdot \overrightarrow{P_0P_3} \\ &= \mathbf{u} \times \mathbf{v} \cdot \overrightarrow{P_0P_3} \end{aligned}$$

since **u**, **v**, and $\overrightarrow{P_1P_0}$ are linearly dependent. Also $\mathbf{u} \times \mathbf{v}$ and $\overrightarrow{P_0P_3}$ have the same direction. Therefore

$$V = \|\mathbf{u} \times \mathbf{v}\| \, \|\overrightarrow{P_0P_3}\|$$

or the volume is base times altitude, as expected.

The rule (11-152) can be regarded as a very practical way for finding volumes of complicated solids of the type considered. Indeed, if one were faced with the problem of finding the volume of a huge piece of metal with triangular faces, one would find the rule an exceptionally convenient way to provide the numerical value; one need only measure areas of the faces and determine the appropriate vectors.

By passage to the limit, as in Chapter 7, or Chapter 13, one can find volumes of solids bounded by curved surfaces. Thus the rule: volume of a cylinder equals base times altitude can be obtained as the limiting case of the rule for prisms; a similar statement applies to pyramids and cones. The sphere is most easily treated as a solid of revolution, as in Section 7-5.

The rule (11-152) itself in the limiting cases becomes a surface integral

$$\frac{1}{3}\iint r_n \, dA$$

where r_n is the component of $\mathbf{r} = \overrightarrow{OP}$ along the "outer normal" **n**. This is discussed in books on advanced calculus (see also Section 13-15). This formula is, in fact, the 3-dimensional analogue of the area formula (7-40).

Remark. Solids in space are often defined by *inequalities*, referring to a given coordinate system. We remark that, if $\mathbf{n} = A\mathbf{i} + B\mathbf{j} + C\mathbf{k}$ is not **0**, then the *linear inequality*

$$A(x - x_0) + B(y - y_0) + C(z - z_0) \geq 0$$

is satisfied by all points P: (x, y, z) such that $\overrightarrow{P_0P} \cdot \mathbf{n} \geq 0$; that is, by all P such that $\overrightarrow{P_0P}$ has a positive or 0 component in the direction of the normal **n** to the plane H: $\mathbf{n} \cdot \overrightarrow{P_0P} = 0$. Hence, the inequality describes a "half-space,"

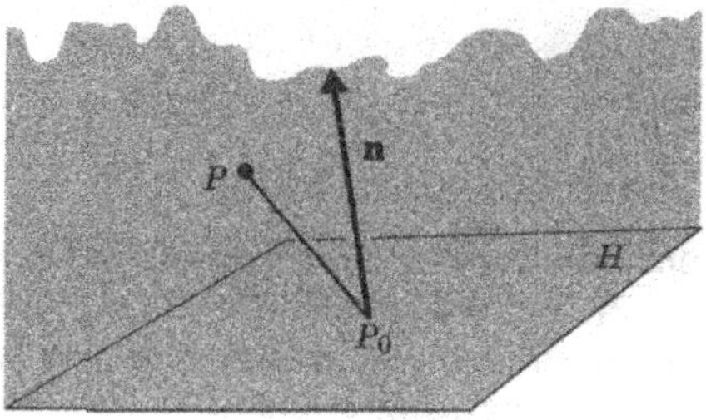

Figure 11-30 Half-space.

the portion of space on one side of the plane H, as suggested in Figure 11-30. Points satisfying a double linear inequality

$$a \leq Ax + By + Cz \leq b$$

form the intersection of two half-spaces and form a region between two parallel planes $Ax + By + Cz = a$, $Ax + By + Cz = b$. The inequalities $0 \leq x \leq a$, $0 \leq y \leq b$, $0 \leq z \leq c$ describe a rectangular parallelepiped of edges a, b, c. A general parallelepiped is described by inequalities of form

$$a_\alpha \leq A_\alpha x + B_\alpha y + C_\alpha z \leq b_\alpha, \qquad \alpha = 1, 2, 3$$

where $a_\alpha < b_\alpha$, and the vectors $A_\alpha \mathbf{i} + B_\alpha \mathbf{j} + C_\alpha \mathbf{k}$ are linearly independent. A general tetrahedron can be described by four inequalities:

$$A_\alpha x + B_\alpha y + C_\alpha z + D_\alpha \geq 0, \qquad \alpha = 1, \ldots, 4$$

A discussion of these representations is left to the exercises (Problems 13 and 14 below).

‡11-16 NEW COORDINATES AND VOLUME IN R^n

The introduction of new Cartesian coordinates in R^n follows the pattern of Section 11-14. However, the formation of an orthonormal basis from a given nonorthogonal basis requires a special construction (the Gram-Schmidt process, see Problem 15 below).

The discussion of volume in Section 11-15 also generalizes to R^n. We must now refer to n-volume, or n-dimensional volume. We here mention only one formula, that for the n-volume V of an n-dimensional parallelotope (generalization of parallelogram and parallelepiped to n-dimensions): if the edge vectors $\mathbf{v}_1, \mathbf{v}_2, \ldots, \mathbf{v}_n$ form a positive n-tuple, then

$$V = \mathbf{v}_1 \times \mathbf{v}_2 \times \cdots \times \mathbf{v}_{n-1} \cdot \mathbf{v}_n = \begin{vmatrix} v_{11} & \cdots & v_{1n} \\ \vdots & & \vdots \\ v_{n1} & \cdots & v_{nn} \end{vmatrix} \qquad (11\text{-}160)$$

PROBLEMS

1. With the aid of the vectors $\mathbf{v}_1$ and $\mathbf{w}$ given, find an orthonormal basis for V_3: $\mathbf{i}'$, $\mathbf{j}'$, $\mathbf{k}'$, such that $\mathbf{i}' = \mathbf{v}_1/\|\mathbf{v}_1\|$:
(a) $\mathbf{v}_1 = (2, 2, 1)$, $\mathbf{w} = (1, 1, 1)$ (b) $\mathbf{v}_1 = (1, 1, 1)$, $\mathbf{w} = (1, 0, 0)$

2. Let line L be parallel to plane H and not meet H. Let c be the perpendicular distance from one point on L to H. Show that, for every point on L, its perpendicular distance to H is c.

3. Let P_1 be a point in plane H_1, P_1Q_2 a segment meeting H_1 only at P_1, Q_1 the foot of the perpendicular from Q_2 to H_1. Show that for every point S_1 other than P_1 in H_1 one has $\measuredangle(\overrightarrow{P_1S_1}, \overrightarrow{P_1Q_2}) \geq \measuredangle(\overrightarrow{P_1Q_1}, \overrightarrow{P_1Q_2})$, with equality only if $\overrightarrow{P_1S_1} = k\overrightarrow{P_1Q_1}$ with $k > 0$. [*Hint.* By the reasoning of Section 11-14, one can assume coordinate axes chosen so that H_1 is the xy-plane, P_1 is the origin, Q_1 is on the positive x-axis. Show that Q_2 must then be in the xz-plane. Now find the cosines of the angles in question.]

4. Prove: the medians of a tetrahedron meet at a point $\frac{3}{4}$ of the way from each vertex to the opposite face. [*Hint.* Each median joins a vertex to the centroid (intersection of medians) of the opposite face. By the reasoning of Section 11-14 one can assume without loss of generality that the four vertices of the tetrahedron are $(0, 0, 0)$, $(a, 0, 0)$, $(b, c, 0)$, (d, e, f) with $a > 0$, $c > 0$, $f > 0$.]

5. Let $P_1 = (1, 0, 0)$, $P_2 = (0, 1, 1)$, $P_3 = (3, 2, 2)$, $P_4 = (6, 3, 3)$, $P_5 = (2, 3, 4)$. Find the length of the broken line:

   (a) $P_1P_2P_3$, (b) $P_1P_3P_5$, (c) $P_3P_4P_5$, (d) $P_1P_2P_3P_4P_5$.

6. With $P_1, \ldots, P_5$ as in Problem 5, find the area of the triangle:

   (a) $P_1P_2P_3$, (b) $P_1P_3P_5$, (c) $P_2P_3P_4$, (d) $P_1P_4P_5$.

7. Graph and find the surface area of the solid described.

   (a) The tetrahedron of vertices $(0, 0, 0)$, $(1, 0, 0)$, $(0, 1, 0)$, $(0, 0, 1)$.

   (b) The tetrahedron of vertices P_1, P_2, P_3, P_5 as given in Problem 5.

   (c) The pyramid of vertices $(\pm 1, 0, 0)$, $(0, \pm 1, 0)$, $(0, 0, 1)$.

   (d) The cube of vertices $(\pm 1, \pm 1, \pm 1)$.

8. Show by (11-150′) that the volume of the tetrahedron with vertices (x_1, y_1, z_1), $\ldots, (x_4, y_4, z_4)$ is $|D|/6$, where

$$D = \begin{vmatrix} x_2 - x_1 & y_2 - y_1 & z_2 - z_1 \\ x_3 - x_1 & y_3 - y_1 & z_3 - z_1 \\ x_4 - x_1 & y_4 - y_1 & z_4 - z_1 \end{vmatrix}$$

9. With $P_1, \ldots, P_5$ as in Problem 5 find the volume of the tetrahedron:

   (a) $P_1P_2P_4P_5$, (b) $P_1P_2P_3P_4$.

10. Let $\mathbf{u}$, $\mathbf{v}$ be a basis for the 2-dimensional subspace W of V_3. Let $\mathbf{v}_1 = \mathbf{u}$, $\mathbf{v}_3 = \mathbf{u} \times \mathbf{v}$ and $\mathbf{v}_2 = \mathbf{v}_3 \times \mathbf{u}$ as in the proof of Theorem 7. Show that $\mathbf{v}_1$, $\mathbf{v}_2$ are a basis for W and that $\mathbf{v}_1$, $\mathbf{v}_2$, $\mathbf{v}_3$ form a positive triple.

‡11. Let the tetrahedron $P_1P_2P_3P_4$ be given, let $\mathbf{u} = \overrightarrow{P_1P_2}$, $\mathbf{v} = \overrightarrow{P_1P_3}$, $\mathbf{w} = \overrightarrow{P_1P_4}$ and let $\mathbf{u}$, $\mathbf{v}$, $\mathbf{w}$ be a *negative* triple.

   (a) Show that $\overrightarrow{P_4P_1} \cdot \mathbf{u} \times \mathbf{v} > 0$ and, hence, the correct choice of the area vector for the face $P_1P_2P_3$ is $\frac{1}{2}\mathbf{u} \times \mathbf{v}$.

   (b) Show that the other area vectors should be chosen as $\frac{1}{2}\overrightarrow{P_1P_3} \times \overrightarrow{P_1P_4}$, $\frac{1}{2}\overrightarrow{P_2P_1} \times \overrightarrow{P_2P_4}$ and $\frac{1}{2}\overrightarrow{P_4P_3} \times \overrightarrow{P_4P_2}$.

   (c) Show that the sum of all four area vectors is $\mathbf{0}$.

‡12. Show from the result of Problem 11(c) that the expression $\Sigma\, \overrightarrow{QP_i^*} \cdot \mathbf{n}_i$ in (11-151)

has the same value for all choices of Q and then take $Q = P_4$ to conclude that (11-150′) implies (11-151). [*Remark.* This reasoning shows that if we define the volume of the tetrahedron by (11-150), then (11-151) follows. Then if we define the volume of a general solid bounded by a polyhedron by first subdividing it into tetrahedra, we obtain the value (11-152), which does not depend on the subdivision chosen.]

13. A parallelepiped can be defined algebraically as the set of all points P for which

$$(^*) \qquad \overrightarrow{OP} = \overrightarrow{OQ} + s_1\mathbf{v}_1 + s_2\mathbf{v}_2 + s_3\mathbf{v}_3$$

holds true, where Q: (x_0, y_0, z_0) is a given point, $\mathbf{v}_1, \mathbf{v}_2, \mathbf{v}_3$ are three given linearly independent vectors, and $0 \leq s_1 \leq 1$, $0 \leq s_2 \leq 1$, $0 \leq s_3 \leq 1$.

(a) Graph the parallelepiped $\overrightarrow{OP} = \mathbf{i} + \mathbf{j} + s_1\mathbf{i} + s_2(\mathbf{i} + \mathbf{j}) + s_3(\mathbf{i} + \mathbf{j} + \mathbf{k})$.

‡(b) Let A be the matrix whose columns are $\mathbf{v}_1, \mathbf{v}_2, \mathbf{v}_3$. Show that $(^*)$ can be written as follows: $(x, y, z) = (x_0, y_0, z_0) + A\mathbf{s}$, where A is nonsingular. Let $B = A^{-1} = (b_{ij})$, $\mathbf{c} = B(x_0, y_0, z_0)$ and show that for each point (x, y, z) of the parallelepiped one has $s_\alpha = b_{\alpha 1}x + b_{\alpha 2}y + b_{\alpha 3}z - c_\alpha$, for $\alpha = 1, 2, 3$ and, hence, (x, y, z) is in the parallelepiped if, and only if,

$$(^{**}) \qquad 0 \leq b_{\alpha 1}x + b_{\alpha 2}y + b_{\alpha 3}z - c_\alpha \leq 1, \qquad \alpha = 1, 2, 3$$

‡(c) Represent the parallelepiped of part (a) by inequalities of form $(^{**})$.

14. A tetrahedron can be defined algebraically as the set of all points P for which

$$(^{***}) \qquad \overrightarrow{OP} = s_1\overrightarrow{OP_1} + s_2\overrightarrow{OP_2} + s_3\overrightarrow{OP_3} + s_4\overrightarrow{OP_4}$$

where P_1, P_2, P_3, P_4 are four given points not in a plane, $s_\alpha \geq 0$ for $\alpha = 1, \ldots, 4$ and $s_1 + s_2 + s_3 + s_4 = 1$. The numbers $s_1, \ldots, s_4$ are called *barycentric coordinates* of P. They can be regarded as masses placed at the vertices $P_1, \ldots, P_4$ which force the center of mass to lie at P.

(a) Graph the tetrahedron with vertices P_1: $(0, 0, 0)$, P_2: $(1, 0, 0)$, P_3: $(1, 1, 0)$, and P_4: $(1, 1, 1)$ and show the points for which $(s_1, \ldots, s_4) = (0, 0, 0, 1)$, $(0, 0, \frac{1}{2}, \frac{1}{2})$, $(\frac{1}{3}, \frac{1}{3}, \frac{1}{3}, 0)$, $(\frac{1}{4}, \frac{1}{4}, \frac{1}{4}, \frac{1}{4})$.

‡(b) For a given tetrahedron $(^{***})$ let A be the 4 by 4 matrix whose columns are the vectors $(x_1, y_1, z_1, 1), \ldots, (x_4, y_4, z_4, 1)$. Show that $\operatorname{col}(x, y, z, 1) = A \operatorname{col}(s_1, s_2, s_3, s_4)$ and that A is nonsingular. Let $B = A^{-1}$, so that $s_\alpha = b_{\alpha 1}x + b_{\alpha 2}y + b_{\alpha 3}z + b_{\alpha 4}$ for $\alpha = 1, 2, 3, 4$ and, hence, show that (x, y, z) is in the tetrahedron if, and only if,

$$(^{****}) \qquad b_{\alpha 1}x + b_{\alpha 2}y + b_{\alpha 3}z + b_{\alpha 4} \geq 0 \qquad \text{for } \alpha = 1, \ldots, 4$$

‡(c) Represent the tetrahedron of part (a) in the form $(^{****})$.

‡15. Let $\mathbf{w}_1, \ldots, \mathbf{w}_n$ be a basis for V_n. Show that an orthogonal basis $\mathbf{v}_1, \ldots, \mathbf{v}_n$ for V_n can be found such that $\mathbf{v}_1 = \mathbf{w}_1$, $\mathbf{v}_2 = c_{21}\mathbf{w}_1 + \mathbf{w}_2, \ldots$, and, in general, $\mathbf{v}_k$ is a linear combination of $\mathbf{w}_1, \ldots, \mathbf{w}_k$ in which the coefficient of $\mathbf{w}_k$ is 1.

‡16. (a) From (11-160) show that, if $\mathbf{v}_1, \ldots, \mathbf{v}_{n-1}$ are linearly independent, then $\|\mathbf{v}_1 \times \cdots \times \mathbf{v}_{n-1}\|$ equals the $(n-1)$-volume of the $(n-1)$-dimensional parallelotope of edge vectors $\mathbf{v}_1, \ldots, \mathbf{v}_{n-1}$. [*Hint.* Assume a new basis for V_n chosen so that $\mathbf{v}_1, \ldots, \mathbf{v}_{n-1}$ are all perpendicular to $\mathbf{e}_n$.]

(b) From (11-160) show that the n-volume of the n-dimensional parallelotope equals base times altitude.

‡17. (a) Let $U = (u_{ij})$ be a nonsingular n by n matrix. Let $\mathbf{u}_1, \ldots, \mathbf{u}_n$ be the row vectors of U. Let B be the matrix $(\mathbf{u}_i \cdot \mathbf{u}_j)$. Prove that $\det U = \pm\sqrt{\det B}$. [*Hint.* Verify that $UU' = B$, and use the rule $\det CD = \det C \det D$.]

(b) Show that the n-dimensional parallelotope whose edges are the vectors $\mathbf{u}_1, \ldots, \mathbf{u}_n$ of part (a) has volume $\sqrt{\det B}$.

Remark. The result of Part (b) provides a formula for the n-volume of an n-dimensional parallelotope in R^m, $m \geq n$. For the case of a one-dimensional parallelotope in R^2—that is, a line segment P_1P_2 in the plane, the formula reduces to $\sqrt{\mathbf{u} \cdot \mathbf{u}}$, where $\mathbf{u} = (x_2 - x_1)\mathbf{i} + (y_2 - y_1)\mathbf{j}$ and, hence, to $\sqrt{(x_2 - x_1)^2 + (y_2 - y_1)^2}$, the familiar expression for distance in the plane. The formula can be regarded as a very general form of the *Pythagorean theorem.*

11-17 LINEAR MAPPINGS OF R^3 INTO R^3

In this section we shall use index notation for coordinates. Here it will be convenient to think of two different spaces R^3: one with coordinates (x_1, x_2, x_3), one with coordinates (y_1, y_2, y_3), as in Figure 11-31. Let f be a

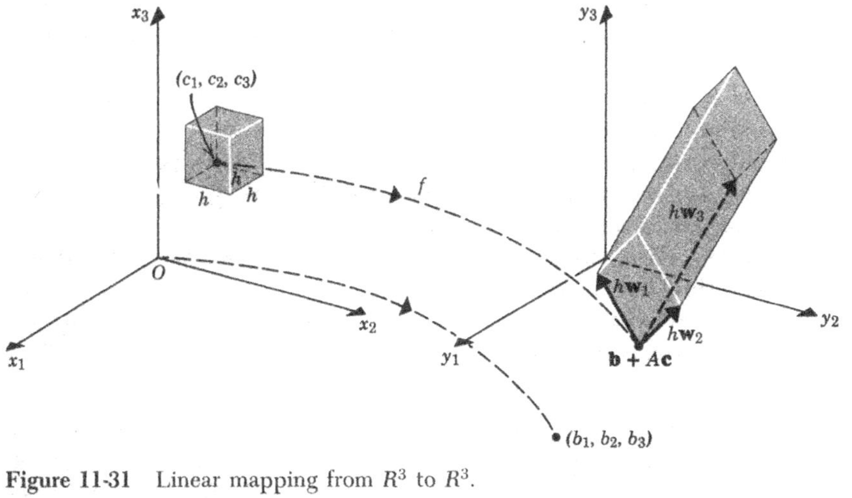

Figure 11-31 Linear mapping from R^3 to R^3.

mapping of the first space into the second. The mapping f is said to be *linear* if f is given by linear equations:

$$y_i = b_i + \sum_{j=1}^{3} a_{ij}x_j, \qquad i = 1, 2, 3 \tag{11-170}$$

Thus here the term "linear mapping" is used in a somewhat different sense from that of Chapter 9; see the comment following Theorem 16 in Section 9-11.

EXAMPLE 1. The mapping f given by the equations

$$y_1 = 1 + 2x_1 + x_2 - x_3, \qquad y_2 = 3 - x_1 + x_2 + 2x_3, \qquad y_3 = x_1 + x_3$$

is linear. This mapping assigns to $(x_1, x_2, x_3) = (0, 0, 0)$ the value $(y_1, y_2, y_3) = (1, 3, 0)$, to $(1, 2, 3)$ the value $(2, 10, 4)$.

We can write the equations (11-170) in matrix notation:

$$\mathbf{y} = \mathbf{b} + A\mathbf{x}$$

where $\mathbf{x}$, $\mathbf{b}$ and $\mathbf{y}$ are now column vectors and A is the 3 by 3 matrix (a_{ij}). We can also write

$$(y_1, y_2, y_3) =$$
$$(b_1, b_2, b_3) + x_1(a_{11}, a_{21}, a_{31}) + x_2(a_{12}, a_{22}, a_{32}) + x_3(a_{13}, a_{23}, a_{33})$$

Now let P be the point (y_1, y_2, y_3) and let Q be the point (b_1, b_2, b_3) of the y-space. Let $\mathbf{w}_i$ be the vector (a_{1i}, a_{2i}, a_{3i}) (the ith column of A). Then our equation reads

$$\overrightarrow{OP} = \overrightarrow{OQ} + x_1\mathbf{w}_1 + x_2\mathbf{w}_2 + x_3\mathbf{w}_3 \tag{11-171}$$

Now as (x_1, x_2, x_3) varies over R^3 (the x-space), the point P: (y_1, y_2, y_3) varies in the y-space. If $\mathbf{w}_1, \mathbf{w}_2, \mathbf{w}_3$ are linearly independent, then they form a basis for vectors in space and every vector $\overrightarrow{OP}$ can be represented in the form (11-171), and in just one way; that is, the mapping f is one-to-one and the range of f is all of R^3. If $\mathbf{w}_1, \mathbf{w}_2, \mathbf{w}_3$ are linearly dependent, then (11-171) still describes a linear variety of R^3; that is, a plane or a line (or a point, if all the $\mathbf{w}_i = \mathbf{0}$); the dimension of the linear variety is dim Span $(\mathbf{w}_1, \mathbf{w}_2, \mathbf{w}_3)$. Now if $\mathbf{w}_1, \mathbf{w}_2, \mathbf{w}_3$ are linearly independent, then as (x_1, x_2, x_3) varies over a plane (or line) in the x-space, the corresponding point P: (y_1, y_2, y_3) sweeps out a plane (or line) in the y-space, and the correspondence must remain one-to-one (Problem 5 below).

In general, we call the matrix $A = (a_{ij})$ the *matrix* of the linear mapping f. In terms of notations introduced in Chapter 12, A can also be called the *Jacobian matrix* of the mapping f.

The relation between f and the matrix A becomes even closer if we choose a new origin at Q in the y-space and choose new axes at this origin with the same directions as the old ones (translation of axes). If we denote the new coordinates of P by (z_1, z_2, z_3) then

$$\begin{aligned}\overrightarrow{QP} &= z_1\mathbf{e}_1 + z_2\mathbf{e}_2 + z_3\mathbf{e}_3 = \overrightarrow{OP} - \overrightarrow{OQ}\\ &= (y_1 - b_1)\mathbf{e}_1 + (y_2 - b_2)\mathbf{e}_2 + (y_3 - b_3)\mathbf{e}_3\end{aligned}$$

and, hence (as in the plane, Section 6-5),

$$z_1 = y_1 - b_1, \qquad z_2 = y_2 - b_2, \qquad z_3 = y_3 - b_3$$

In the new coordinates, the mapping f is given by the equations:

$$z_i = \sum_{j=1}^{3} a_{ij}x_j \qquad \text{or} \qquad \mathbf{z} = A\mathbf{x}$$

One can now apply the theory of linear mappings of V_n into V_n (Sections 10-2 to 10-4).

In general, the mapping f is one-to-one precisely when A is nonsingular (vectors $\mathbf{w}_1$, $\mathbf{w}_2$, $\mathbf{w}_3$ linearly independent) and, in that case, the inverse mapping is given by $\mathbf{x} = A^{-1}\mathbf{z}$, that is, by

$$\mathbf{x} = A^{-1}(\mathbf{y} - \mathbf{b}) = A^{-1}\mathbf{y} - A^{-1}\mathbf{b}$$

Thus the inverse mapping is a linear mapping of the y-space onto the x-space.

Now let f be a one-to-one mapping and let us consider a cube of edge h in the x-space: $c_i \leq x_i \leq c_i + h$ ($i = 1, 2, 3$); the cube has volume h^3. We can describe the cube as the set of all points (x_1, x_2, x_3) for which $x_i = c_i + s_ih$, where $0 \leq s_i \leq 1$. Then by (11-171) the corresponding points P: (y_1, y_2, y_3) are the set of all P for which

$$\overrightarrow{OP} = \overrightarrow{OQ} + x_1\mathbf{w}_1 + x_2\mathbf{w}_2 + x_3\mathbf{w}_3, \qquad x_i = c_i + s_ih, \qquad 0 \leq s_i \leq 1$$

or

$$\overrightarrow{OP} = \overrightarrow{OQ} + c_1\mathbf{w}_1 + c_2\mathbf{w}_2 + c_3\mathbf{w}_3 + s_1h_1\mathbf{w}_1 + s_2h\mathbf{w}_2 + s_3h\mathbf{w}_3$$

with $0 \leq s_i \leq 1$ for $i = 1, 2, 3$. As in Problem 13 following Section 11-16, the image of the cube is thus a parallelepiped in the y-space, with edge vectors $h\mathbf{w}_1$, $h\mathbf{w}_2$, $h\mathbf{w}_3$ (see Figure 11-31). Therefore, the volume of the image is

$$\pm h\mathbf{w}_1 \cdot h\mathbf{w}_2 \times h\mathbf{w}_3 = h^3|\mathbf{w}_1 \cdot \mathbf{w}_2 \times \mathbf{w}_3|$$

But $\mathbf{w}_1 \cdot \mathbf{w}_2 \times \mathbf{w}_3 = \det A$. Hence, the volume of the image is $|\det A|$ times the volume of the cube in the x-space. It can be shown that a similar relationship holds for tetrahedra and, hence, for all solid polyhedra: in fact, *all volumes of solids in the x-space are multiplied by the factor $|\det A|$ through the mapping f*. This shows the importance of the Jacobian matrix of a linear mapping.

We have here a very close parallel to the idea of *slope* for linear mappings of R^1 into R^1: that is, for the linear functions

$$y = mx + b$$

We know that m is the slope, or derivative, of the function, and that $|m|$ is the ratio of $|\Delta y|$ to $|\Delta x|$; that is, it is the ratio of the length of the image of a segment to the length of that segment. For our mappings (11-170): $\mathbf{y} = A\mathbf{x} + \mathbf{b}$, it is $|\det A|$ that gives the corresponding ratio of volumes, so that $|\det A|$ is like a 3-dimensional slope.

There is an analogous discussion for linear mappings of R^2 into R^2; that is of the plane into the plane. They are given by $\mathbf{y} = A\mathbf{x} + \mathbf{b}$, where A is now a 2 by 2 matrix. The same reasoning (applied now to squares and parallelograms) shows that $|\det A|$ is the ratio of the *area* of the image of a figure to the area of the figure itself.

‡11-18 LINEAR MAPPINGS OF R^n INTO R^m

The ideas of the preceding section generalize to linear mappings of an n-dimensional space R^n of coordinates $(x_1, \ldots, x_n)$, into an m-dimensional space R^m, of coordinates $(y_1, \ldots, y_m)$. A linear mapping f is given by equations

$$y_i = b_i + \sum_{j=1}^{n} a_{ij}x_j, \qquad i = 1, \ldots, m$$

or in matrix form by

$$\mathbf{y} = \mathbf{b} + A\mathbf{x}$$

For $n = m = 1$, we have the linear function $y = b + ax$. For $n = 1$, $m = 2$ we have the equations:

$$y_1 = b_1 + a_{11}x_1, \qquad y_2 = b_2 + a_{21}x_1, \qquad -\infty < x_1 < \infty$$

These are, except for notation, the same as the parametric equations of a line as studied in Chapter 1.

When $n = m$, A is a square matrix and the mapping f is one-to-one precisely when A is nonsingular. If f is one-to-one, then its inverse is the linear mapping

$$\mathbf{x} = A^{-1}(\mathbf{y} - \mathbf{b}) = A^{-1}\mathbf{y} - A^{-1}\mathbf{b}$$

Furthermore, each n-dimensional cube $a_i \leq x_i \leq a_i + h$ has as image an n-dimensional parallelotope whose n-volume is $h^n|\det A|$ and, in general, n-volumes are multiplied by $|\det A|$ in going from the x-space to the y-space via the mapping f.

When $n < m$, the mapping f is one-to-one when A has maximum rank n. In this case, the range of f is an n-dimensional linear manifold H in the y-space. Each n-dimensional cube $a_i \leq x_i \leq a_i + h$ $(i = 1, \ldots, n)$ corresponds to an n-dimensional parallelotope in H having n-volume $h^n \sqrt{\det B}$, where B is the n by n matrix $(\mathbf{w}_i \cdot \mathbf{w}_j)$, and $\mathbf{w}_i$ is the ith column vector of A. For $n = 1$ and $m = 2$, H is a straight line in the plane and $\sqrt{\det B}$ becomes $\sqrt{a_{11}^2 + a_{12}^2} = [(dy_1/dx_1)^2 + (dy_2/dx_1)^2]^{1/2}$ and the statement about volumes is equivalent to the familiar formula

$$ds = \sqrt{\left(\frac{dx}{dt}\right)^2 + \left(\frac{dy}{dt}\right)^2}\, dt$$

The proofs of the assertions made in this section are left as exercises (Problems 7 to 9 below).

PROBLEMS

1. A linear mapping f of R^3 into R^3 is given by the equations:

$$y_1 = 1 + x_1 - x_2 + x_3, \qquad y_2 = 2 + x_1 - x_3, \qquad y_3 = x_1 + x_2$$

(a) Find the images of the following points in the x-space: (0, 0, 0), (1, 0, 0), (1, 2, 1).

(b) Show that f is one-to-one.

(c) Find equations for the inverse mapping f^{-1}.

(d) Find the image of the x_1-axis, the x_2-axis, and the x_3-axis.

(e) Find the image of the plane: $x_1 - x_2 + x_3 = 0$, the plane $x_1 + x_2 = 0$, and the plane $x_1 + x_2 + x_3 = 1$.

(f) Find the image of the cube $0 \leq x_1 \leq h, 0 \leq x_2 \leq h, 0 \leq x_3 \leq h$, and find the volume of the image. Verify that this volume is $|\det A|h^3$.

2. (a) . . . (f). Carry out the steps of Problem 1 for the mapping given by the equations $y_1 = 3 + 2x_1 - 2x_2 + x_3$, $y_2 = 1 + 2x_2 + x_3$, $y_3 = -x_2$.

3. Let a linear map f of R^3 into R^3 be given by the equations:

$$y_1 = 2x_1 + x_2 + x_3, \quad y_2 = x_1 + 3x_2 + 2x_3, \quad y_3 = 3x_1 - x_2.$$

(a) Show that f is not one-to-one.

(b) Find the range of f.

(c) Find the image of the plane $4x_1 - 3x_2 - x_3 = 0$.

(d) Find the image of the line $x_1 = 2 + t$, $x_2 = 3 + 3t$, $x_3 = 1 - 5t$.

4. Let a linear mapping f of R^3 into R^3 be given by the equations:

$$y_1 = 1 + 3x_1 - x_2 + x_3, y_2 = 5 - 3x_1 + x_2 - x_3, \ y_3 = 2 + 6x_1 - 2x_2 + 2x_3.$$

(a) Show that f is not one-to-one.

(b) Find the range of f.

(c) Find the image of the plane $x_1 + x_2 + x_3 = 0$.

(d) Find the image of the line $x_1 = 1 + t, x_2 = 2 + 2t, x_3 = 2 + t$.

5. Let a one-to-one linear mapping f of R^3 into R^3 be given by equations (11-170).

(a) Show that the image of each line L: $x_i = k_i + h_i t$ ($i = 1, 2, 3$) in the x-space is a line in the y-space.

(b) Show that the image of each plane $\mathbf{x} = \mathbf{c} + t\mathbf{u} + \tau\mathbf{v}$ in the x-space is a plane in the y-space.

(c) Show that the image of each line segment $\mathbf{x} = \mathbf{k} + t\mathbf{u}$, $0 \leq t \leq 1$ in the x-space is a line segment in the y-space.

Remark. Parts (a) and (c) show that lines and segments correspond to lines and segments respectively; it is for this reason that one refers to f as a *linear* mapping. One can show that every mapping taking lines to lines is a linear mapping.

‡6. Let a linear mapping f of R^2 into R^3 be given by equations:

$$y_1 = 1 + x_1 + 2x_2, \qquad y_2 = 1 + 2x_1 - x_2, \qquad y_3 = 1 + x_1 + 2x_2$$

(a) Show that f is one-to-one.

(b) Find the image of the x_1-axis and of the x_2-axis.

(c) Find the image of the square $0 \leq x_1 \leq h, 0 \leq x_2 \leq h$ and find the area of the image. Verify that this area is h^2 times $\sqrt{\det B}$, as described in the text.

‡7. Let a linear mapping f of R^n into R^m be given, where $n \leq m$. Let f have the equation $\mathbf{y} = \mathbf{b} + A\mathbf{x}$. Show that f is one-to-one if, and only if, the kernel of A consists of $\mathbf{0}$ alone (and hence if, and only if, A has rank n).

‡8. Let $f: \mathbf{y} = \mathbf{b} + A\mathbf{x}$ be a one-to-one linear mapping of R^n into R^n. Show that each n-dimensional cube $a_i \leq x_i \leq a_i + h\ (i = 1, \ldots, n)$ has as image an n-dimensional parallelotope of n-volume $h^n |\det A|$.

‡9. Let f: $\mathbf{y} = \mathbf{b} + A\mathbf{x}$ be a one-to-one linear mapping of R^n into R^m.

(a) Show that $n \leq m$.

(b) Show that each n-dimensional cube $a_i \leq x_i \leq a_i + h$ has as image an n-dimensional parallelotope of n-volume $h^n \sqrt{\det B}$, where $B = (\mathbf{w}_i \cdot \mathbf{w}_j)$ and $\mathbf{w}_i$ is the ith column vector of A. (See Problem 17 following Section 11-16.)

11-19 SURFACES IN R^3

Surfaces in space are studied in Chapter 12. Here we consider only the special case of spheres, cylinders, and cones (see also Problem 5 at the end of this chapter).

A *sphere* S of radius $a > 0$ and center Q is the set of all points at distance a from Q: (x_0, y_0, z_0) (Figure 11-32). Hence, S has the equation:

$$(x - x_0)^2 + (y - y_0)^2 + (z - z_0)^2 = a^2 \qquad (11\text{-}190)$$

When expanded, this equation has the form:

$$Ax^2 + Ay^2 + Az^2 + Dx + Ey + Fz + G = 0 \qquad (11\text{-}191)$$

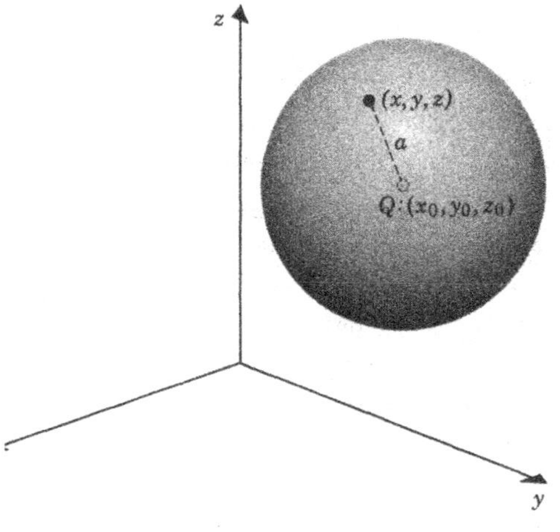

igure 11-32 Sphere.

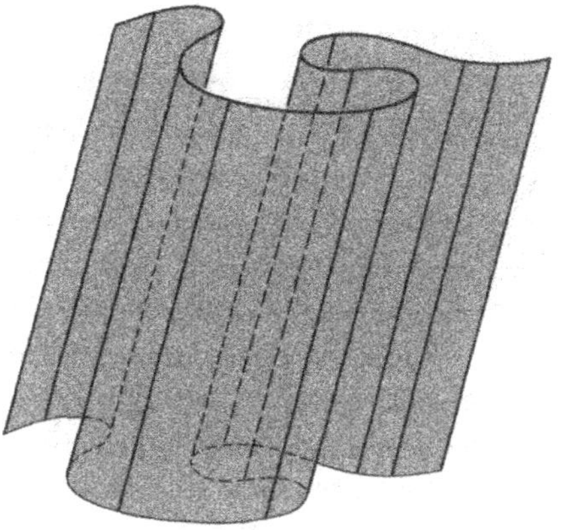

Figure 11-33 Cylindrical surface.

By completing the square, as for circles in the plane (Section 0-14), one can go from an equation (11-191), with $A \neq 0$, to an equation of form (11-190). However, a^2 may be 0, so that the graph reduces to the point Q, or a^2 may be negative, in which case the graph is empty.

A *cylindrical surface* (or simply, a cylinder) can be defined generally as the union of all lines (rulings) passing through a given curve (generating curve) and parallel to a given line (Figure 11-33). We here consider only the case in which the generating curve lies in a plane and the rulings are perpendicular to that plane. We can choose coordinates so that the generating curve is the xy-plane and the rulings are parallel to the z-axis. If the generating curve is a circle $(x - x_0)^2 + (y - y_0)^2 = a^2$, then the surface is called a *right circular cylinder*. We remark that the points of the surface are those points (x, y, z) for which (x, y) is on the circle and z is *arbitrary*. Hence, the equation

$$(x - x_0)^2 + (y - y_0)^2 = a^2$$

is itself an equation for the cylinder as a set in xyz-space. The cylinder $(x - 1)^2 + (y - 1)^2 = 1$ is shown in Figure 11-34. Similarly, the equation $2x^2 + y^2 = 1$, which describes an ellipse in the xy-plane, is also the equation of an *elliptical* cylinder in space. In general, if a curve in the xy-plane has an equation in x and y, then that same equation has as graph a cylindrical surface in space, with rulings parallel to the z-axis. Thus any equation in x and y represents a cylindrical surface in space. Similarly, an equation in which y or x is missing represents a cylindrical surface, with rulings parallel to the y-axis or x-axis, respectively.

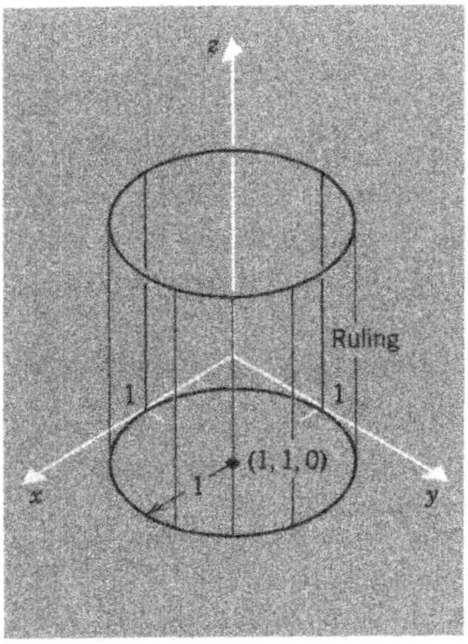

Figure 11-34 Right circular cylinder.

Cones. A cone (or conical surface) in space can be described as a surface formed by the union of all lines (*rulings*) passing through a given point (the *vertex*) and meeting a given curve (*generating curve*). We here consider only the case when the generating curve lies in a plane and the vertex is not in the plane. We can always choose coordinates so that the vertex is the origin O and the generating curve lies in a plane $z = k$, where k is a nonzero constant.

For example, let the generating curve be the circle $x^2 + y^2 = a^2$ in the plane $z = k$. Then the cone is the union of all lines L through O and a point (x_1, y_1, k), where $x_1{}^2 + y_1{}^2 = a^2$. The points of L are all points (x, y, z) with $x = tx_1$, $y = ty_1$, $z = tk$ and $x_1{}^2 + y_1{}^2 = a^2$ (see Figure 11-35). Hence, $x^2 + y^2 = t^2(x_1{}^2 + y_1{}^2) = a^2t^2 = z^2a^2/k^2$. Therefore, each point (x, y, z) of

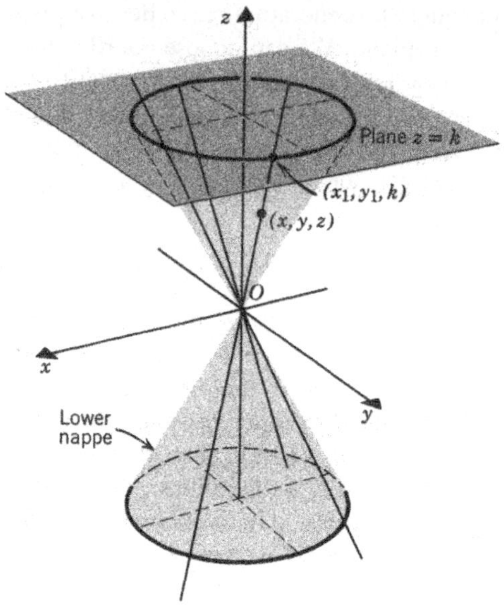

Figure 11-35 Right circular cone.

the cone satisfies the equation:

$$k^2x^2 + k^2y^2 - a^2z^2 = 0 \tag{11-192}$$

Conversely, if a point (x, y, z) satisfies this equation, then either $z \neq 0$, in which case the point lies on the line through O and (x_1, y_1, k), where $x_1 = xk/z$ and $y_1 = yk/z$, so that $x_1{}^2 + y_1{}^2 = a^2$, or else $z = 0$, in which case $x = 0$ and $y = 0$, so that the point is O. Therefore, (11-192) is the equation of the cone in question, called a *right circular cone.*

By similar reasoning, we show that the *elliptical cone* generated by the ellipse $2x^2 + y^2 = 1$ in the plane $z = 1$ has the equation:

$$2x^2 + y^2 - z^2 = 0 \tag{11-193}$$

By reasoning in general as we did in the circular case, we conclude that for a generating curve with given equation in x and y in the plane $z = k$, one obtains the equation of the cone with vertex at O by replacing x by kx/z, y by ky/z in the equation of the curve; this procedure does not take care of the origin O, but usually that difficulty is taken care of by multiplying the equation by a suitable power of z.

We remark that the equations (11-192) and (11-193) are *homogeneous:* that is, if (x, y, z) satisfies the equation, then so does (cx, cy, cz) for every scalar c. We can say: the set V of vectors $\overrightarrow{OP}$: (x, y, z) to points P in the surface is closed under multiplication by scalars. (Thus V is like a subspace, except that we do not require that V be closed under addition.) Since each conical surface with vertex O is a union of lines through O, it is clear that every equation of a conical surface must be homogeneous. (Conversely, each homogeneous equation describes a conical surface; however, the generating curve need not be planar, and may not even be a curve in the usual sense). A homogeneous *algebraic* equation can be recognized as an equation of the form:

polynomial in x, y, and z equals 0, where the polynomial is homogeneous—that is, all terms are of the same degree (degree of $ax^l y^m z^n$ is $l + m + n$). For example, (11-192) and (11-193) are homogeneous algebraic equations, with all terms of degree 2.

Nappes of a Cone. For a cone generated by curve C in the plane $z = k$ with vertex O, one refers to the part of the cone for which $z \geq 0$ as one *nappe* of the cone; and to the part for which $z \leq 0$ as the other nappe (see Figure 11-35).

PROBLEMS

1. Describe the nature of the surface having the equation given and graph:
   (a) $x^2 + y^2 + z^2 + 2x + 4y - 9z + 13 = 0$
   (b) $x^2 + y^2 + z^2 - 2y + 2z = 0$
   (c) $x^2 + y^2 + z^2 - 2x - 4y - 4z + 9 = 0$
   (d) $x^2 + y^2 + z^2 - 4y - 2z + 6 = 0$
   (e) $x^2 + y^2 = 4$ (f) $x^2 + y^2 = 9$ (g) $3x^2 + 2y^2 = 6$
   (h) $4x^2 + y^2 = 4$ (i) $y^2 - 4x + 8 = 0$ (j) $x^2 + 3y - 9 = 0$
   (k) $xy = 1$ (l) $3x^2 + xy + 3y^2 = 1$ (m) $x^2 + y^2 - 4z^2 = 0$
   (n) $x^2 + y^2 = 9z^2$ (o) $x^2 - y^2 = z^2$ (p) $x^2 - z^2 = y^2$
   (q) $4x^2 + 3y^2 = 12z^2$ (r) $9x^2 + z^2 = y^2$
2. Find an equation of the cone with vertex O and given generating curve:
   (a) $x^2 + y^2 = 4$ in plane $z = 1$ (b) $x^2 + y^2 = 1$ in plane $z = 5$
   (c) $x^4 + y^4 = 1$ in plane $z = 2$ (d) $x^2 + 3xy + y^2 = 1$ in plane $z = 1$
   (e) $2y^2 + z^2 = 4$ in plane $x = 1$ (f) $x^2 + 4z^2 = 1$ in plane $y = -1$
   (g) $x + y = 1$ in plane $z = 1$ (h) $x = 0$ in plane $z = 1$
3. Show that the given equation is homogeneous and, hence, represents a cone with vertex at O. Find a generating curve by considering the intersection of the surface with the plane $z = 1$.
   (a) $x^4 + y^4 - 3z^4 = 0$ (b) $x^2y + 3xy^2 + xz^2 + y^2z = 0$
   (c) $x^4 + x^2y^2 + y^4 - z^4 = 0$ (d) $x + y - z = 0$

   Remark. In (b) and (d) the generating curve suggested does not provide the whole surface; one loses the points where $z = 0$.

11-20 CYLINDRICAL AND SPHERICAL COORDINATES

In this section we give two generalizations of polar coordinates to 3-dimensional space.

Cylindrical Coordinates. Let $P = (x, y, z)$ be a point of R^3. Let $(x, y, 0)$ have polar coordinates (r, θ) in the xy-plane. Then the ordered triple (r, θ, z) forms the cylindrical coordinates of P in space. Accordingly, the

Cartesian coordinates (x, y, z) and the cylindrical coordinates (r, θ, z) of P are related by the equations:

$$x = r \cos \theta, \qquad y = r \sin \theta, \qquad z = z \tag{11-200}$$

One can solve for r, θ, z:

$$r = \sqrt{x^2 + y^2}, \qquad \theta = \tan^{-1} \frac{y}{x}, \qquad z = z \tag{11-201}$$

However, the equation for θ is ambiguous, and more information must be given (such as in which quadrant θ is to be chosen—see Section 5-4).

The graph of each equation $r = \text{const} = a$, where $a > 0$, is a right cylindrical surface. For $r = a$ is the same as $x^2 + y^2 = a^2$, in rectangular coordinates, and $x^2 + y^2 = a^2$ is the equation of a right circular cylinder (Section 11-19). It is for this reason that the name "cylindrical coordinates" is used. We note that $r = \sqrt{x^2 + y^2}$ is the shortest distance from (x, y, z) to the z-axis (Problem 6 below). Hence, the cylinder $r = a$ can be described as the locus of points at distance a from the z-axis.

Each locus $\theta = \text{constant} = \alpha$ meets the xy-plane in a ray and, since z is unrestricted, the locus $\theta = \alpha$ corresponds to the set of all (x, y, z) for which $(x, y, 0)$ is on the ray. This locus is a *half-plane in space* (Figure 11-36; see Problem 5 below).

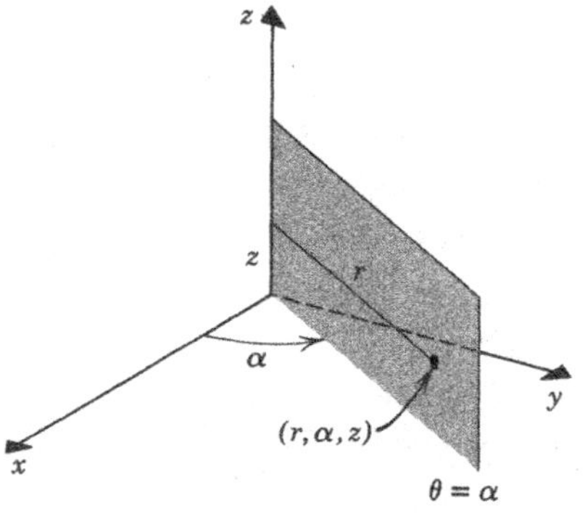

Figure 11-36 Cylindrical coordinates.

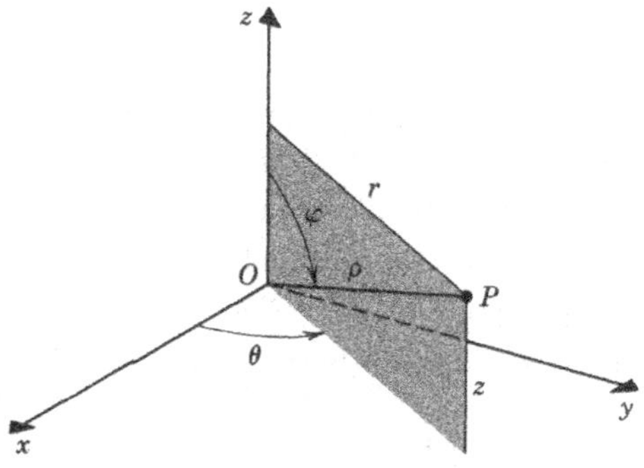

Figure 11-37 Spherical coordinates.

Spherical Coordinates. The spherical coordinates (ρ, φ, θ) of a point P: (x, y, z) in space are obtained from the cylindrical coordinates by the equations:

$$\theta = \theta, \qquad r = \rho \sin \varphi, \qquad z = \rho \cos \varphi \qquad (\rho \geq 0) \tag{11-202}$$

Thus, in effect, one is introducing polar coordinates (ρ, φ) in each half-plane: $\theta = \text{constant}$, with polar axis Oz, polar angle φ and distance ρ (Figure 11-37). We usually restrict φ to the interval $[0, \pi]$. From (11-200) and (11-202) we obtain the Cartesian coordinates (x, y, z):

$$x = \rho \sin \varphi \cos \theta, \qquad y = \rho \sin \varphi \sin \theta, \qquad z = \rho \cos \varphi \tag{11-203}$$

From (11-202) $r^2 + z^2 = \rho^2$, so that $x^2 + y^2 + z^2 = \rho^2$. Hence, the locus

ρ = constant = a (> 0) is a sphere with center at the origin. This is the reason for the term "spherical coordinates." On a sphere $\rho = a$, the loci φ = constant and θ = constant appear as circles of latitude and longitude (see Figure 11-38). However, latitude is usually measured from the equator, while φ (sometimes called "co-latitude") is measured from the North Pole.

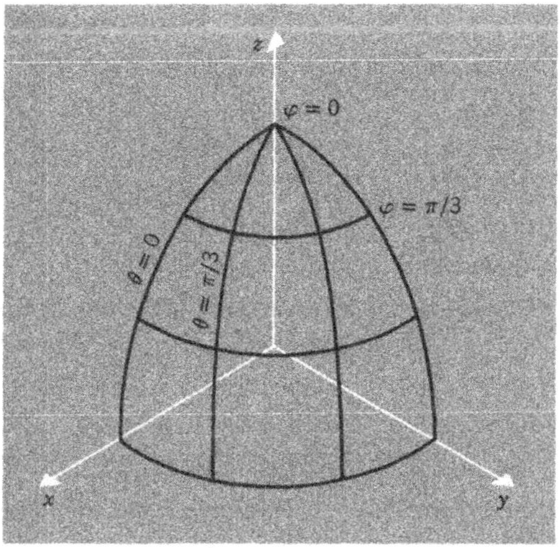

Figure 11-38 Spherical coordinates on a sphere.

Changing Equations from One Type of Coordinates to Another. A locus in space described by one or more equations in one type of coordinates can be described by equations in another type with the aid of Equations (11-200), (11-202), (11-203) above. One has to be careful when the angles φ or θ appear other than in trigonometric functions.

EXAMPLE 1. Describe the locus of the equation $x^2 + 6xy + y^2 + z^2 = 1$ in cylindrical and spherical coordinates.

Solution. By (11-200) one has in cylindrical coordinates

$$r^2 \cos^2 \theta + 6r^2 \cos \theta \sin \theta + r^2 \sin^2 \theta + z^2 = 1$$

or

$$r^2 + 3r^2 \sin 2\theta + z^2 = 1$$

By (11-202) this becomes

$$\rho^2 \sin^2 \varphi + 3\rho^2 \sin^2 \varphi \sin 2\theta + \rho^2 \cos^2 \varphi = 1$$

or

$$\rho^2 + 3\rho^2 \sin^2 \varphi \sin 2\theta = 1$$

EXAMPLE 2. Find the Cartesian-coordinate equation and the cylindrical coordinate equation for the set of points whose spherical coordinates satisfy $\rho = 2 \cos \varphi$. Describe this set of points.

Solution. We observe that the equation is satisfied for $\rho = 0$ (at the origin, φ is ambiguous but can be chosen as $\pi/2$, for example). Hence the locus is the same as that of the equation $\rho^2 - 2\rho \cos \varphi = 0$. In cylindrical coordinates this becomes $r^2 + z^2 - 2z = 0$ and in Cartesian coordinates $x^2 + y^2 + z^2 - 2z = 0$ or $x^2 + y^2 + (z - 1)^2 = 1$. Thus the surface is a sphere of radius 1 with center at (0, 0, 1) in Cartesian coordinates.

PROBLEMS

1. Sketch and describe each of the following sets in R^3, as determined by conditions on the cylindrical coordinates:

   (a) $r = 2$ (b) $\theta = \pi/6$ (c) $z = -6$

   (d) $r = z$ (e) $r = 4 \cos \theta$ (f) $r^2 \sin^2 \theta = 4(1 - r^2)$

   (g) $r \leq 3$ (h) $0 \leq \theta \leq \pi/2$ (i) $0 \leq z \leq 2r$

   (j) $r = 3, \sin \theta = 1$ (k) $r \leq 3, z = 7$ (l) $\theta = -\pi/4, z = 1$

2. Sketch and describe each of the following sets in R^3, as determined by conditions on the spherical coordinates:

   (a) $\rho = 2$ (b) $\theta = \pi/6$ (c) $\varphi = \pi/3$

   (d) $\rho = \varphi$ (e) $\rho = 2 \sin \varphi$ (f) $\rho \leq 1, 0 \leq \theta \leq \pi/2$

   (g) $0 \leq \varphi \leq \pi/4$ (h) $\rho \sin \theta \sin \varphi = 2$ (i) $\theta = \varphi$

3. Find the spherical-coordinate equation and the cylindrical-coordinate equation for the set of points whose Cartesian coordinates satisfy

   (a) $x^2 + y^2 = 5$ (b) $3x - 4y + 5z = 1$ (c) $x^2 + y^2 + 2z^2 = 4$

   (d) $x^2 + y^2 = z^2$ (e) $x^2 + 2y^2 + 3z^2 = 6$ (f) $xy + yz + xz = 0$

   (g) $x^2 - z^2 = 4$ (h) $y^2 = 4z$

4. Find an expression for the distance between two points of R^3:

   (a) in cylindrical coordinates, (b) in spherical coordinates.

5. To analyze the locus $\theta = \alpha =$ constant, we introduce a new orthonormal basis $\mathbf{i}_1, \mathbf{j}_1, \mathbf{k}_1$, with $\mathbf{i}_1 = \cos \alpha \mathbf{i} + \sin \alpha \mathbf{j}$, $\mathbf{j}_1 = -\sin \alpha \mathbf{i} + \cos \alpha \mathbf{j}$, $\mathbf{k}_1 = \mathbf{k}$. Show that a point P is on the locus precisely when $\overrightarrow{OP} = r \cos \alpha \mathbf{i} + r \sin \alpha \mathbf{j} + z\mathbf{k}$, $r \geq 0$, or when $\overrightarrow{OP} = r\mathbf{i}_1 + z\mathbf{k}_1$, $r \geq 0$, so that the locus is the half-plane $r \geq 0$ in an rz-plane.

6. Prove that $r = \sqrt{x^2 + y^2}$ is the shortest distance from the point (x, y, z) to the z-axis.

11-21 CHANGE OF COORDINATES IN R^3

It is convenient here to use index notation (x_1, x_2, x_3) for coordinates and to denote our standard basis **i, j, k** by $\mathbf{e}_1, \mathbf{e}_2, \mathbf{e}_3$.

Now let a new orthonormal basis $\mathbf{e}'_1, \mathbf{e}'_2, \mathbf{e}'_3$ be chosen. We form the inner products

$$a_{ij} = \mathbf{e}_i \cdot \mathbf{e}'_j$$

For example, $a_{1j} = \mathbf{e}_1 \cdot \mathbf{e}'_j$, $a_{2j} = \mathbf{e}_2 \cdot \mathbf{e}'_j$, $a_{3j} = \mathbf{e}_3 \cdot \mathbf{e}'_j$ are the three components of $\mathbf{e}_j'$ with respect to the old basis $\mathbf{e}_1$, $\mathbf{e}_2$, $\mathbf{e}_3$, and we can write

$$\mathbf{e}_j' = a_{1j}\mathbf{e}_1 + a_{2j}\mathbf{e}_2 + a_{3j}\mathbf{e}_3$$

Since $\mathbf{e}_j'$ is a unit vector, it follows that

$$a_{ij} = \cos \measuredangle(\mathbf{e}_i, \mathbf{e}_j')$$

(see Figure 11-39) and

$$a_{1j}^2 + a_{2j}^2 + a_{3j}^2 = 1 \tag{11-210}$$

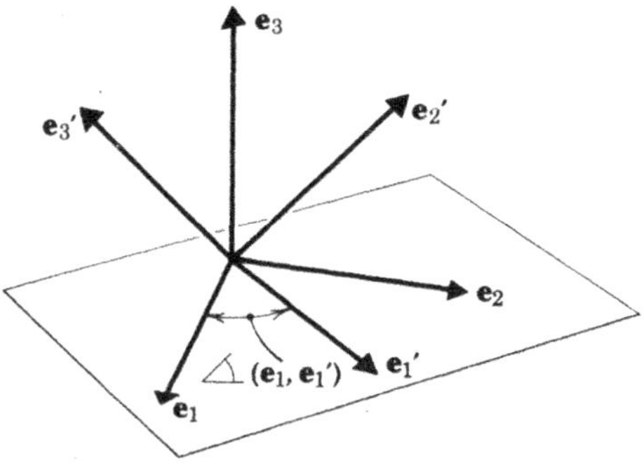

Figure 11-39 Old and new orthonormal bases in V_3.

Also, since $\mathbf{e}_1'$, $\mathbf{e}_2'$, $\mathbf{e}_3'$ are pairwise orthogonal, $\mathbf{e}_1' \cdot \mathbf{e}_2' = 0$, $\mathbf{e}_2' \cdot \mathbf{e}_3' = 0$, $\mathbf{e}_3' \cdot \mathbf{e}_1' = 0$ and, hence,

$$a_{1j}a_{1k} + a_{2j}a_{2k} + a_{3j}a_{3k} = 0, \qquad 1 \leq j < k \leq 3 \tag{11-211}$$

We can form the matrix:

$$A = (a_{ij}) = \begin{pmatrix} a_{11} & a_{12} & a_{13} \\ a_{21} & a_{22} & a_{23} \\ a_{31} & a_{32} & a_{33} \end{pmatrix}$$

The three columns of A are the three sets of components of $\mathbf{e}_1'$, $\mathbf{e}_2'$, $\mathbf{e}_3'$ with respect to the old basis. Equations (11-210) and (11-211) express the fact that these columns are mutually perpendicular unit vectors.

Now we can also use $\mathbf{e}_1'$, $\mathbf{e}_2'$, $\mathbf{e}_3'$ as basis. For example, the components of $\mathbf{e}_1$ with respect to this basis are $\mathbf{e}_1 \cdot \mathbf{e}_1' = a_{11}$, $\mathbf{e}_1 \cdot \mathbf{e}_2' = a_{12}$, $\mathbf{e}_1 \cdot \mathbf{e}_3' = a_{13}$. In general, the three rows of A are the three sets of components of $\mathbf{e}_1$, $\mathbf{e}_2$, $\mathbf{e}_3$ with respect to the new basis, and

$$\mathbf{e}_i = a_{i1}\mathbf{e}_1' + a_{i2}\mathbf{e}_2' + a_{i3}\mathbf{e}_3'$$

For the same reasons as above we now have

$$a_{i1}^2 + a_{i2}^2 + a_{i3}^2 = 1 \tag{11-210$'$}$$

$$a_{i1}a_{k1} + a_{i2}a_{k2} + a_{i3}a_{k3} = 0, \qquad 1 \leq i < k \leq 3 \tag{11-211$'$}$$

We can state simply: the rows of A are mutually perpendicular unit vectors, the columns of A are mutually perpendicular unit vectors. A real 3 by 3 matrix with these properties is called an *orthogonal matrix*.

THEOREM 8. *A real 3 by 3 matrix $A = (a_{ij})$ is orthogonal if, and only if,*

$$AA' = I$$

If A is orthogonal, the columns of A define a new orthonormal basis $\mathbf{e}'_1$, $\mathbf{e}'_2$, $\mathbf{e}'_3$ and $a_{ij} = \mathbf{e}_i \cdot \mathbf{e}'_j = \cos \measuredangle(\mathbf{e}_i, \mathbf{e}'_j)$.

Proof. We have

$$A = \begin{pmatrix} a_{11} & a_{12} & a_{13} \\ a_{21} & a_{22} & a_{23} \\ a_{31} & a_{32} & a_{33} \end{pmatrix}, \qquad A' = \begin{pmatrix} a_{11} & a_{21} & a_{31} \\ a_{12} & a_{22} & a_{32} \\ a_{13} & a_{23} & a_{33} \end{pmatrix}$$

Hence AA' equals

$$\begin{pmatrix} a_{11}^2 + a_{12}^2 + a_{13}^2 & a_{11}a_{21} + a_{12}a_{22} + a_{13}a_{23} & a_{11}a_{31} + a_{12}a_{32} + a_{13}a_{33} \\ a_{21}a_{11} + a_{22}a_{12} + a_{23}a_{13} & a_{21}^2 + a_{22}^2 + a_{23}^2 & a_{21}a_{31} + a_{22}a_{32} + a_{23}a_{33} \\ a_{31}a_{11} + a_{32}a_{12} + a_{33}a_{13} & a_{31}a_{21} + a_{32}a_{22} + a_{33}a_{23} & a_{31}^2 + a_{32}^2 + a_{33}^2 \end{pmatrix}$$

If A is orthogonal, then (11-210′) and (11-211′) hold true, so that

$$AA' = \begin{pmatrix} 1 & 0 & 0 \\ 0 & 1 & 0 \\ 0 & 0 & 1 \end{pmatrix} = I$$

Conversely, if $AA' = I$, then we must have $a_{11}^2 + a_{12}^2 + a_{13}^2 = 1$, $a_{11}a_{21} + a_{12}a_{22} + a_{13}a_{23} = 0, \ldots$, so that (11-210′) and (11-211′) hold true. Also $AA' = I$ implies that $A' = A^{-1}$, so that $A'A = A^{-1}A = I$. If we multiply A' by A and equate the result to I, then as above for AA', we obtain (11-210) and (11-211) (Problem 6 below). Hence, A is orthogonal.

When A is orthogonal, we can define $\mathbf{e}'_j$ as $a_{1j}\mathbf{e}_1 + a_{2j}\mathbf{e}_2 + a_{3j}\mathbf{e}_3$. Equations (11-210) and (11-211) then show that $\mathbf{e}'_1$, $\mathbf{e}'_2$, $\mathbf{e}'_3$ is an orthonormal basis, and $\mathbf{e}_i \cdot \mathbf{e}'_j = \mathbf{e}_i \cdot (a_{1j}\mathbf{e}_1 + a_{2j}\mathbf{e}_2 + a_{3j}\mathbf{e}_3) = a_{ij}$.

EXAMPLE 1. We take $\mathbf{e}'_3 = \mathbf{e}_3$ and obtain $\mathbf{e}'_1$, $\mathbf{e}'_2$ from $\mathbf{e}_1$, $\mathbf{e}_2$ by rotating through angle φ in the plane of $\mathbf{e}_1$, $\mathbf{e}_2$, as suggested in Figure 11-40. Thus $\mathbf{e}'_1 = \cos\varphi\, \mathbf{e}_1 + \sin\varphi\, \mathbf{e}_2$, $\mathbf{e}'_2 = -\sin\varphi\, \mathbf{e}_1 + \cos\varphi\, \mathbf{e}_2$, $\mathbf{e}'_3 = \mathbf{e}_3$, and

$$A = \begin{pmatrix} \cos\varphi & -\sin\varphi & 0 \\ \sin\varphi & \cos\varphi & 0 \\ 0 & 0 & 1 \end{pmatrix}$$

We see at once that each column (row) defines a unit vector, that each pair of column (row) vectors are orthogonal.

In general, we note that $\det A = \mathbf{e}'_1 \cdot \mathbf{e}'_2 \times \mathbf{e}'_3$ [see Theorem 5, Equation (11-64)] and, hence, $\det A > 0$ or < 0 according as the new basis is positively or negatively oriented. Since the $\mathbf{e}'_j$ are mutually perpendicular vectors, we have $\det A = \pm 1$.

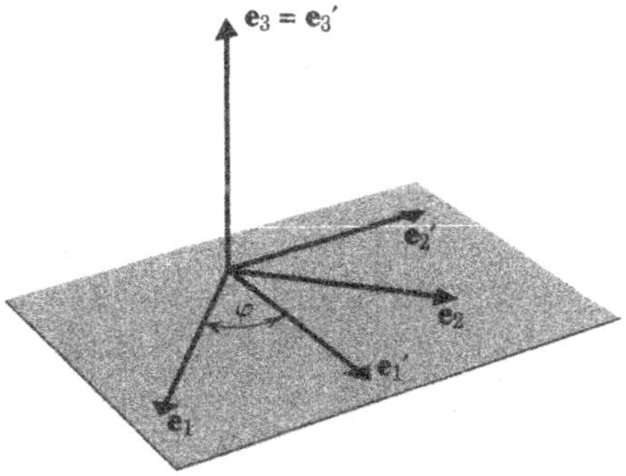

Figure 11-40 The rotation of axes.

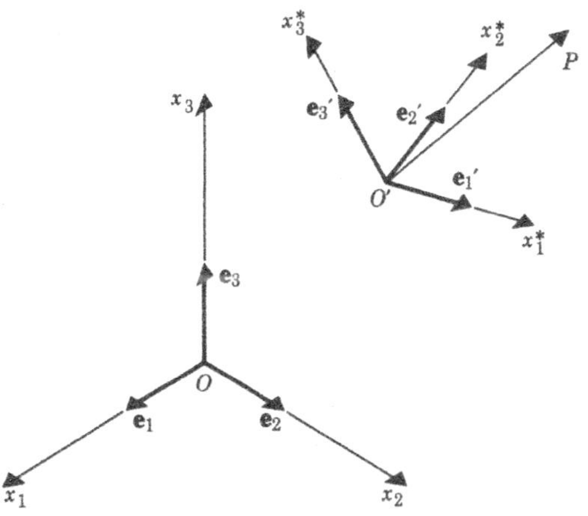

Figure 11-41 New coordinates in space

We now use our new orthonormal basis to define new coordinates in R^3. We also allow for a change of origin and denote by O': (b_1, b_2, b_3) the new origin. Then for each P: (x_1, x_2, x_3) new coordinates (x_1', x_2', x_3') are defined by the equation

$$\overrightarrow{OP} = \overrightarrow{OO'} + \overrightarrow{O'P}$$

or

$$x_1\mathbf{e}_1 + x_2\mathbf{e}_2 + x_3\mathbf{e}_3 = b_1\mathbf{e}_1 + b_2\mathbf{e}_2 + b_3\mathbf{e}_3 + x_1^*\mathbf{e}_1' + x_2^*\mathbf{e}_2' + x_3^*\mathbf{e}_3'$$

or

$$(x_1 - b_1)\mathbf{e}_1 + (x_2 - b_2)\mathbf{e}_2 + (x_3 - b_3)\mathbf{e}_3 = x_1^*\mathbf{e}_1' + x_2^*\mathbf{e}_2' + x_3^*\mathbf{e}_3'$$

(see Figure 11-41). If we form the inner product of both sides of the last equation with $\mathbf{e}_i$, we get expressions for the old coordinates in terms of the new:

$$x_i - b_i = a_{i1}x_1^* + a_{i2}x_2^* + a_{i3}x_3^*, \qquad i = 1, 2, 3 \qquad \text{(11-212)}$$

Similarly, if we form the inner product of both sides with $\mathbf{e}_j'$ we obtain the new coordinates in terms of the old:

$$x_j^* = a_{1j}(x_1 - b_1) + a_{2j}(x_2 - b_2) + a_{3j}(x_3 - b_3), \qquad j = 1, 2, 3 \quad \text{(11-213)}$$

The two sets of equations (11-212), (11-213) allow us to pass freely from old to new coordinates and back. We can write these equations in matrix form:

$$\mathbf{x} = A\mathbf{x}^* + \mathbf{b} \qquad \text{(11-212')}$$

$$\mathbf{x}^* = A'(\mathbf{x} - \mathbf{b}) = A'\mathbf{x} + \mathbf{c} \qquad \text{(11-213')}$$

where $\mathbf{x}$, $\mathbf{x}^*$, $\mathbf{b}$, $\mathbf{c}$ are the column vectors (x_1, x_2, x_3), (x_1^*, x_2^*, x_3^*), (b_1, b_2, b_3), (c_1, c_2, c_3) and

$$\mathbf{c} = -A'\,\mathbf{b}$$

When $x = \mathbf{0}$, $\mathbf{x}^* = \mathbf{c}$. Thus (c_1, c_2, c_3) are the new coordinates of the old origin.

EXAMPLE 2. The basis is unchanged but a new origin is taken at (1, 2, 3). Here $\mathbf{e}_i = \mathbf{e}_i'$ for $i = 1, 2, 3$, so that $A = I$ and Equations (11-212) and (11-213) become

$$x_i - b_i = x_i^* \qquad (i = 1, 2, 3), \qquad x_j^* = x_j - b_j \qquad (j = 1, 2, 3)$$

(both sets the same). Thus

$$x_1 - 1 = x_1^*, \qquad x_2 - 2 = x_2^*, \qquad x_3 - 3 = x_3^*$$

are the relations between new and old coordinates. Since the directions of the axes are unchanged, we speak of a *translation of axes* (see Section 6-5). The equation

$$(x_1 - 1)^2 + (x_2 - 2)^2 + (x_3 - 3)^2 = 9$$

becomes

$$x_1^{*2} + x_2^{*2} + x_3^{*2} = 9$$

From either form we recognize the locus of these equations as a sphere of radius 3, with center at the new origin.

EXAMPLE 3. The origin is unchanged but the basis is rotated as in Figure 11-40 with $\varphi = \pi/3$, so that

$$A = \begin{pmatrix} \frac{1}{2} & -\frac{\sqrt{3}}{2} & 0 \\ \frac{\sqrt{3}}{2} & \frac{1}{2} & 0 \\ 0 & 0 & 1 \end{pmatrix}$$

Hence, Equations (11-212) give

$$x_1 = \frac{1}{2}(x_1^* - \sqrt{3}\, x_2^*), \qquad x_2 = \frac{1}{2}(\sqrt{3}\, x_1^* + x_2^*), \qquad x_3 = x_3^*$$

We speak of a *rotation of axes*. The equation $2x_3^{\,2} = x_1^{\,2} - 2\sqrt{3}\, x_1x_2 - x_2^{\,2}$ becomes (after some calculation)

$$2x_3^{*2} = -2x_1^{*2} + 2x_2^{*2}$$

or

$$x_2^{*2} = x_1^{*2} + x_3^{*2}$$

One can verify that the surface is a right circular cone (Section 11-19).

EXAMPLE 4. We leave the origin unchanged but reverse the directions of the axes: $\mathbf{e}_1' = -\mathbf{e}_1$, $\mathbf{e}_2' = -\mathbf{e}_2$, $\mathbf{e}_3' = -\mathbf{e}_3$. Here

$$A = \begin{pmatrix} -1 & 0 & 0 \\ 0 & -1 & 0 \\ 0 & 0 & -1 \end{pmatrix} \qquad \det A = -1$$

We have reversed orientation. Equations (11-213) give $x_j^* = -x_j$ for $j = 1$,

2, 3. We say that the axes have been subjected to a *reflection in the origin.*

It can be shown that the most general change of coordinates can be achieved by successive application of several simple changes like those of Examples 2, 3, 4. (See, for example, M. Hausner, *A Vector Space Approach to Geometry*, Prentice Hall, Englewood Cliffs, 1965 or N. Kuiper, *Linear Algebra and Geometry*, North Holland, Amsterdam, 1963.)

Although we have denoted our old origin, basis and coordinates by O, $\mathbf{e}_1$, $\mathbf{e}_2$, $\mathbf{e}_3$, (x_1, x_2, x_3), the analysis of change of coordinates applies to any two choices of coordinates in R^3: one can be called the "old" coordinates and denoted, say, by (x_1, x_2, x_3), the other called the "new" coordinates and denoted by (x_1^*, x_2^*, x_3^*). Relations of form (11-212) and (11-213) must then hold true.

‡11-22 CHANGE OF COORDINATES IN R^n

The ideas of the preceding section generalize with very little change to n-dimensional space. If a new orthonormal basis $\{\mathbf{e}_1', \ldots, \mathbf{e}_n'\}$ is introduced, then the matrix $A = (a_{ij}) = (\mathbf{e}_i \cdot \mathbf{e}_j')$ is orthogonal: its rows are mutually perpendicular unit vectors and its columns are mutually perpendicular unit vectors. Furthermore, A is orthogonal if, and only if, $AA' = I$. The proofs are just like those for $n = 3$.

One can introduce new coordinates with a new origin at O': $(b_1, \ldots, b_n)$ and new axes having the directions of $\mathbf{e}_1', \ldots, \mathbf{e}_n'$. Just as for $n = 3$, new and old coordinates are related by (11-212′) and (11-213′).

PROBLEMS

1. Verify that the following matrices are orthogonal:

(a) $\begin{pmatrix} \frac{2}{3} & \frac{2}{3} & \frac{1}{3} \\ -\frac{2}{3} & \frac{1}{3} & \frac{2}{3} \\ \frac{1}{3} & -\frac{2}{3} & \frac{2}{3} \end{pmatrix}$ (b) $\begin{pmatrix} \frac{2}{7} & \frac{3}{7} & \frac{6}{7} \\ \frac{3}{7} & -\frac{6}{7} & \frac{2}{7} \\ \frac{6}{7} & \frac{2}{7} & -\frac{3}{7} \end{pmatrix}$

2. Show that for every choice of θ, φ, ψ the matrix A is orthogonal where A is

$$\begin{pmatrix} \cos\varphi\cos\psi - \sin\varphi\sin\psi\cos\theta & \sin\varphi\cos\psi + \cos\varphi\sin\psi\cos\theta & \sin\psi\sin\theta \\ -\cos\varphi\sin\psi - \sin\varphi\cos\psi\cos\theta & -\sin\varphi\sin\psi + \cos\varphi\cos\psi\cos\theta & \cos\psi\sin\theta \\ \sin\varphi\sin\theta & -\cos\varphi\sin\theta & \cos\theta \end{pmatrix}$$

[One can verify that every orthogonal matrix A with $\det A = +1$ can be represented in this form.]

3. (a) Show that the planes $x_1 + 2x_2 + 2x_3 = 0$, $2x_1 + x_2 - 2x_3 = 0$, $2x_1 = 2x_2 + x_3 = 0$ may be used as the $x_2^*x_3^*$-, $x_1^*x_3^*$- and $x_1^*x_2^*$-planes of a new coordinate system in R^3. Find the matrix A such that $\mathbf{x} = A\mathbf{x}^*$.

(b) Find the equation for a change of coordinates so that $x_1 + x_2 + x_3 = 0$ becomes the $x_1^*x_2^*$-plane in the new system.

4. In R^3, let the axes be translated so that the new origin is (1, 2, 3).

(a) Find the new coordinates of the points whose old coordinates are (0, 0, 0), (0, 0, 1), (2, −1, 0), (3, 2, 1), (−1, −2, −3).

(b) Find the new equations for the points whose old coordinates satisfy $x_1 + x_2 + x_3 = 6$, $x_1^2 + x_2^2 = x_3^2$, $x_1^2 + x_2^2 + x_3^2 - 2x_1 - 4x_2 - 6x_3 = 12$.

5. A surface in R^3 which, for proper choice of Cartesian coordinates, has an equation of form $ax_1^2 + bx_2^2 + cx_3^2 = 1$ with $a > 0$, $b > 0$, $c > 0$, is called an *ellipsoid*; if $a > 0$, $b > 0$, $c < 0$, the surface is called a *hyperboloid of one sheet;* if $a > 0$, $b < 0$, $c < 0$, the surface is called a *hyperboloid of two sheets.* If the form is $dx_1^2 + ex_2^2 + fx_3 = 0$, with $d > 0$, $e > 0$, $f \neq 0$, the surface is called an *elliptic paraboloid;* when $d > 0$, $e < 0$, $f \neq 0$, the surface is called a *hyperbolic paraboloid.* By choosing appropriate new coordinates determine the type of each of the following surfaces:

(a) $x_1^2 + 4x_2^2 + x_3^2 + 2x_1 - 8x_2 + 4 = 0$

(b) $x_1^2 - 4x_2^2 - x_3^2 - 2x_1 + 8x_2 + 10 = 0$

(c) $2x_1^2 + 4x_2^2 - 4x_3^2 + 8x_3 - 8 = 0$

(d) $4x_1^2 + x_2^2 - x_3^2 + 4x_1 - 4x_3 + 1 = 0$

(e) $x_1x_2 - x_3 = 0$ [*Hint.* Rotate by 45° about the x_3-axis.]

6. With A as in the proof of Theorem 8, show that the condition $A'A = I$ gives Equations (11-210) and (11-211).

7. Let A be an orthogonal matrix and let **x**, **y** be vectors of V_3. Prove that

$$(A\mathbf{x}) \cdot (A\mathbf{y}) = \mathbf{x} \cdot \mathbf{y}$$

(a) by interpreting **x**, **y** and A**x**, A**y** as new and old coordinates, respectively, of two points P_1, P_2 of R^3 (so that $\mathbf{x} \cdot \mathbf{y} = \overrightarrow{OP_1} \cdot \overrightarrow{OP_2}$);

(b) by direct calculation.

8. From the result of Problem 7 prove that, if A is orthogonal, then for every **x** in V_3

$$\|A\mathbf{x}\|^2 = \|\mathbf{x}\|^2$$

9. Let A, B be orthogonal 3 by 3 matrices. Prove:

(a) AB is orthogonal, (b) A^{-1} is orthogonal.

‡10. Let $A = (a_{ij})$ be an n by n orthogonal matrix. Let **u** and **v** be vectors of V_n, written as column vectors, hence, as n by 1 *matrices.*

(a) Show that $\mathbf{u} \cdot \mathbf{v} = \mathbf{u}'\mathbf{v}$ (matrix multiplication on the right!)

(b) Show that $\mathbf{u} \cdot \mathbf{v} = (A\mathbf{u}) \cdot (A\mathbf{v})$ [*Hint.* Use the results of part (a) to write the right side as $(A\mathbf{u})'(A\mathbf{v})$.]

(c) Show that $\|A\mathbf{u}\| = \|\mathbf{u}\|$

12
DIFFERENTIAL CALCULUS OF FUNCTIONS OF SEVERAL VARIABLES

INTRODUCTION

We began our study of the calculus with the straight line. Throughout our study, the line and the related linear function have played a central role. The derivative of a function of one variable is closely related to a linear function—that function whose graph is the tangent line to the graph of the given function.

The calculus of functions of several variables is similarly based on "linear objects"—lines, planes, and their generalizations to higher dimensions. In the preceding chapters we have developed the algebra and geometry needed to work with these linear objects. Now we proceed to relate them to the calculus. We shall see that appropriate derivatives (partial derivatives) can be introduced, through which we can find tangent lines and planes to the corresponding graphs. The set of derivatives itself forms a matrix, which we can regard as the matrix representing a *linear mapping*. This linear mapping can be regarded as a close approximation to the generally *nonlinear* mapping represented by our function (or functions) of several variables.

The relationship is suggested schematically in Figure 12-1. Here we have a nonlinear mapping f from R^3 to R^3. For such a mapping, each rectangular solid in the x-space corresponds to a curvilinear solid in the y-space; the planar faces of the rectangular solid correspond to curved surfaces in the y-space;

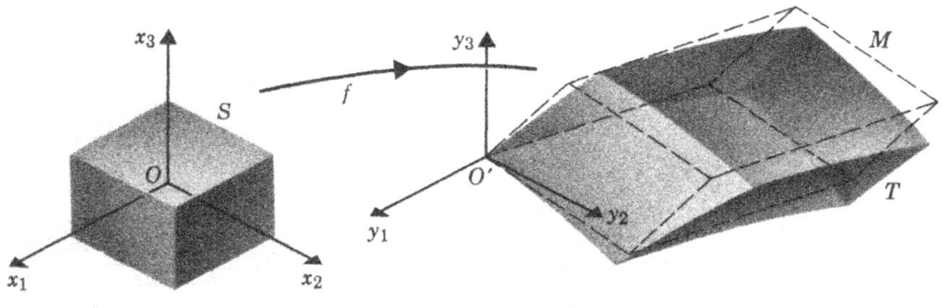

Figure 12-1 Linear mapping approximating nonlinear mapping.

the edges of the solid correspond to curves in the y-space. Through the calculus we shall obtain a linear mapping

$$y_i = \sum_{j=1}^{3} a_{ij}x_j \qquad (i = 1, 2, 3)$$

or, in matrix form,

$$\mathbf{y} = A\mathbf{x}$$

which closely approximates the given nonlinear mapping. In particular, the linear mapping assigns $f(O)$ to O'; we have assumed coordinates chosen so that f takes the origin in the x-space into the origin in the y-space. The linear mapping takes the rectangular solid to a parallelepiped in the y-space; the edges of the solid are mapped on the edges of the parallelepiped, which are lines *tangent* to the curves previously obtained; the faces of the solid correspond to parallelograms in planes *tangent* to the curved surfaces previously obtained. Thus we "linearize" our mapping f by replacing it by an approximating linear mapping.

It will be seen that vector and matrix notation greatly simplify the statement and understanding of results in the calculus of many variables. Throughout this chapter we shall gradually increase the emphasis on this notation.

12-1 SETS IN THE PLANE

For our study of functions of two variables we need several concepts concerning sets in the xy-plane.

By a *neighborhood*, of radius ρ, of a point P_0 we mean the set of all points P in the plane within distance ρ of P_0; that is, the set of all P such that $d(P, P_0) = \|\overrightarrow{P_0P}\| < \rho$. Here ρ is a positive number. Thus the neighborhood is formed of all points *inside* a circle (Figure 12-2).

If E is a set in the plane, then P_0 is said to be an *interior point* of E if one can find a neighborhood of P_0 lying wholly in E. The set of all interior points of E is called the *interior* of E. Thus if E is the set of all points inside

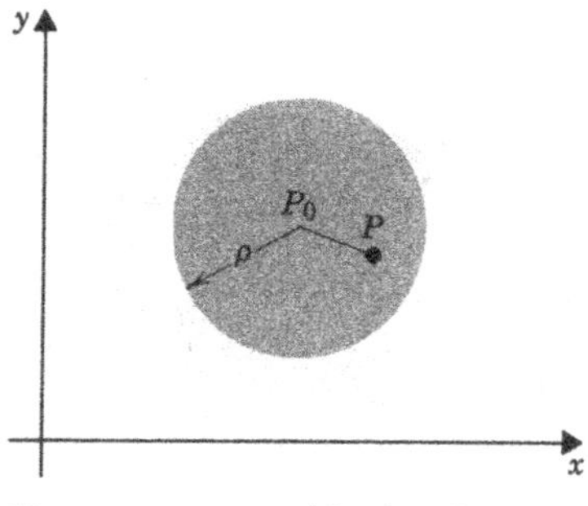

Figure 12-2 Neighborhood.

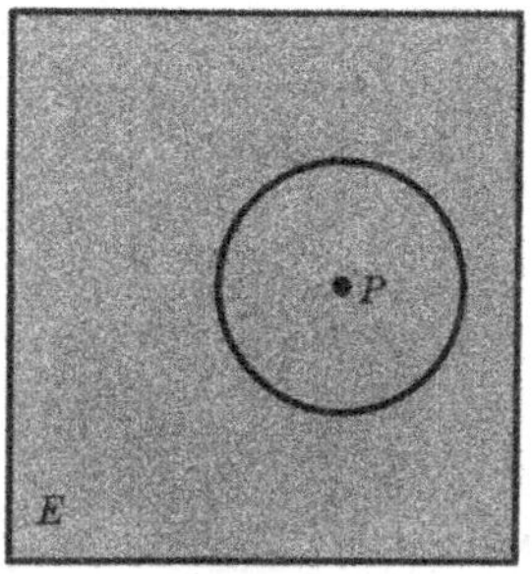

Figure 12-3 The interior of a square as an open set.

and on the edges of a square (Figure 12-3), then only the points inside the square are interior points of E.

A point P in the plane is said to be a *boundary point* of a set E provided each of the neighborhoods of P contain points of E and points not in E. The set of all boundary points of E is called the *boundary of E*. The boundary of the square of Figure 12-3 consists of the four edges.

A set E consisting solely of interior points is called an *open set*. For example, the interior of a square (Figure 12-3) is an open set, as is the half-plane: $x > 0$ (Figure 12-4); also each neighborhood (Figure 12-2) is an open set. For technical reasons, the empty set is also considered to be open.

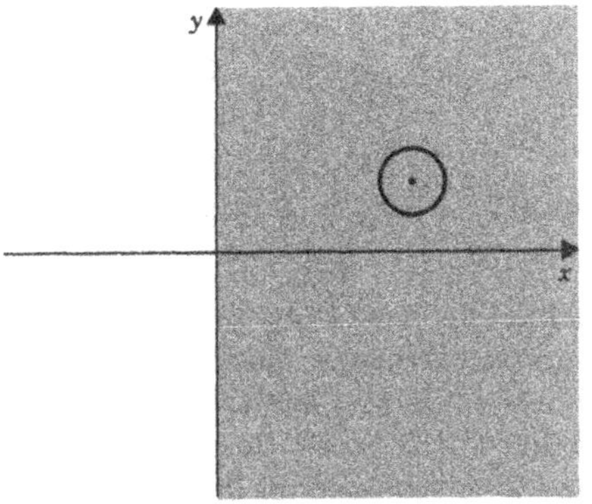

Figure 12-4 The half-plane $x > 0$ as an open set.

A set E is said to be *pathwise connected* if for each two points P_1, P_2 in E, one can find a path $\mathbf{r} = \mathbf{r}(t)$ ($\mathbf{r} = \overrightarrow{OP}$), $a \leq t \leq b$, lying entirely in E, such that $\mathbf{r}(a) = \overrightarrow{OP_1}$, $\mathbf{r}(b) = \overrightarrow{OP_2}$; that is, each two points in E can be joined by a path in E. Figure 12-5*a* shows a pathwise connected set; Figure 12-5*b* shows a set E formed of two pieces and, hence, not pathwise connected.

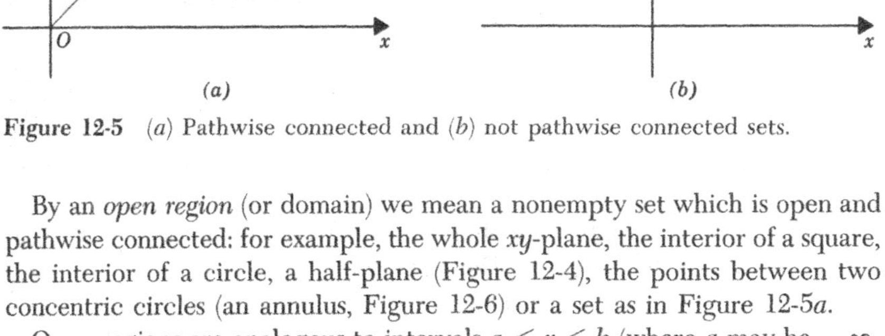

Figure 12-5 (*a*) Pathwise connected and (*b*) not pathwise connected sets.

By an *open region* (or domain) we mean a nonempty set which is open and pathwise connected: for example, the whole xy-plane, the interior of a square, the interior of a circle, a half-plane (Figure 12-4), the points between two concentric circles (an annulus, Figure 12-6) or a set as in Figure 12-5*a*.

Open regions are analogous to intervals $a < x < b$ (where a may be $-\infty$,

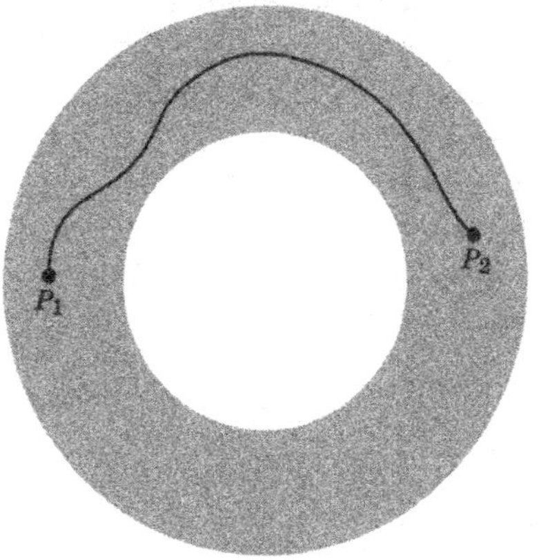

Figure 12-6 Annulus as an open region.

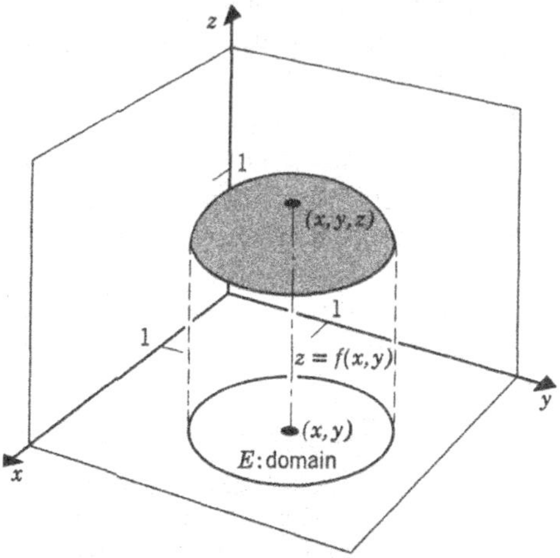

Figure 12-7 A function of two variables.

b may be ∞), and for most of this chapter our functions will be defined in open regions. The analogues of the closed intervals on the line are the *bounded closed regions* in the plane. For example, a circle plus interior is a bounded closed region, as is a square plus interior. The bounded closed regions are considered in Section 12-22 below. They are not needed for the earlier parts of this chapter.

Because of the analogies mentioned, we shall, in fact, on occasion refer to intervals (a, b) on the x-axis as open regions, and shall refer to closed intervals as bounded closed regions. Also a neighborhood of x_0 of radius ρ will be the open interval $(x_0 - \rho, x_0 + \rho)$. These definitions are consistent with those for the plane.

12-2 FUNCTIONS OF TWO VARIABLES

By a real function of two real variables (briefly, function of two variables) we mean a function f assigning a real number z to each ordered pair of real numbers (x, y) in a set E of such pairs. Thus the domain of f can be considered as a set in the xy-plane, and we shall generally consider it so. The range of f is a set of numbers on the z-axis.

Such a function can be represented graphically as in Figure 12-7. The graph consists of all triples (x, y, z) in space such that $z = f(x, y)$. Here $f(2, 1)$ is the value of z for $x = 2$, $y = 1$, $f(1, 3)$ is the value of z for $x = 1$, $y = 3$, $f(a, b)$ is the value of z for $x = a$, $y = b$, $f(x, y)$ is the value of z for given x and y.

Functions of two variables are often defined by equations: $z = 1 - x^2 - y^2$ defines $z = f(x, y)$ for all (x, y). The equations

$$z = \frac{1}{x} - \frac{1}{y}, \qquad u = \sqrt{v^2 - w^2}$$

define functions $g(x, y)$ and $F(v, w)$, the first for $x \neq 0$, $y \neq 0$, the second for $|v| \geq |w|$.

A function f of two variables is often conveniently represented by its *level curves*. They are the sets on which f has different constant values c. For example, if $z = f(x, y) = x^2 + y^2$, then $z = 1$ for the pairs (x, y) such that $x^2 + y^2 = 1$, $z = 2$ for $x^2 + y^2 = 2$, $z = 0$ for $x^2 + y^2 = 0$—that is, $z = 0$ only at $(0, 0)$, $z = -1$ for $x^2 + y^2 = -1$—that is, $z = -1$ for no real pair (x, y). The results are shown graphically in Figure 12-8.

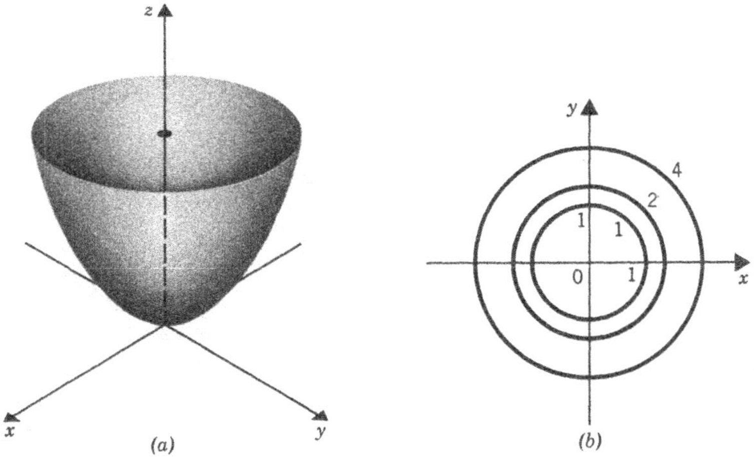

Figure 12-8 (*a*) The function $z = x^2 + y^2$ and (*b*) its level curves.

We can think of $z = f(x, y)$ as the altitude above sea level for certain points on the earth's surface (a small portion being approximately planar). Then our level curves correspond to *contour lines* in a topographical map, as in Figure 12-9. Practice with such maps makes it easy to visualize the 3-dimensional graph of the function, as in Figure 12-7.

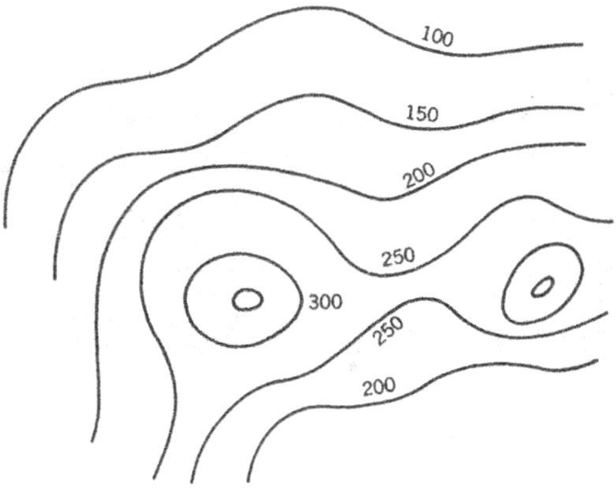

Figure 12-9 A topographical map.

For the function $z = xy$, the level curves are the curves $xy = c$, for different values of c. For $c = 0$, one obtains the two coordinate axes: $x = 0$ and $y = 0$;

for $c \neq 0$, the level curve is a hyperbola. The function and its level curves are graphed in Figure 12-10.

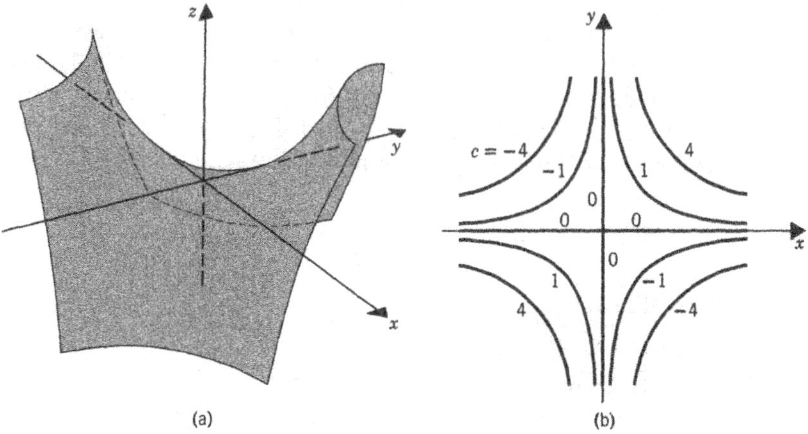

Figure 12-10 (*a*) The function $z = xy$ and (*b*) its level curves.

For the function $z = \sin(x - y)$ each level curve is defined by an equation $\sin(x - y) = c$. For each c between -1 and 1, inclusive, one thus obtains infinitely many parallel lines, as suggested in Figure 12-11. As the figure shows, the graph of the function is a wavy surface, like the surface of the ocean. In fact, the function itself is related to wave propagation. If one replaces y by "time" t, the function becomes $z = \sin(x - t)$ and describes the propagation of a wave along the x-axis (Problem 10 below).

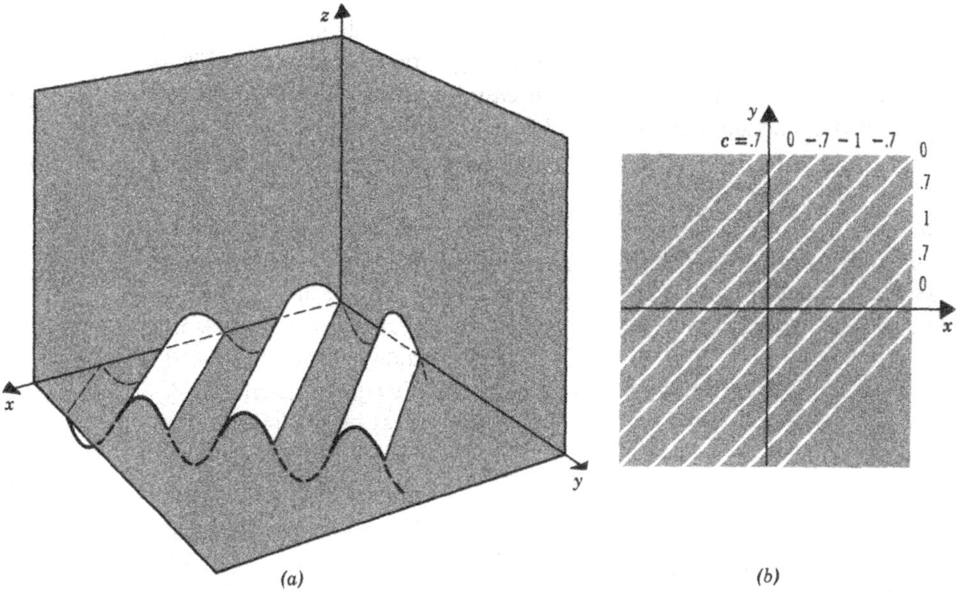

Figure 12-11 (*a*) The function $z = \sin(x - y)$ and (*b*) its level curves.

For the function $z = 1/(x^2 + y^2)$ (Figure 12-12), one must exclude (0, 0) from the domain of the function. In fact, as we approach the origin, the values of z approach ∞, so that we have a bad discontinuity (Section 12-7 below). Observe that the domain of the function is still an open region.

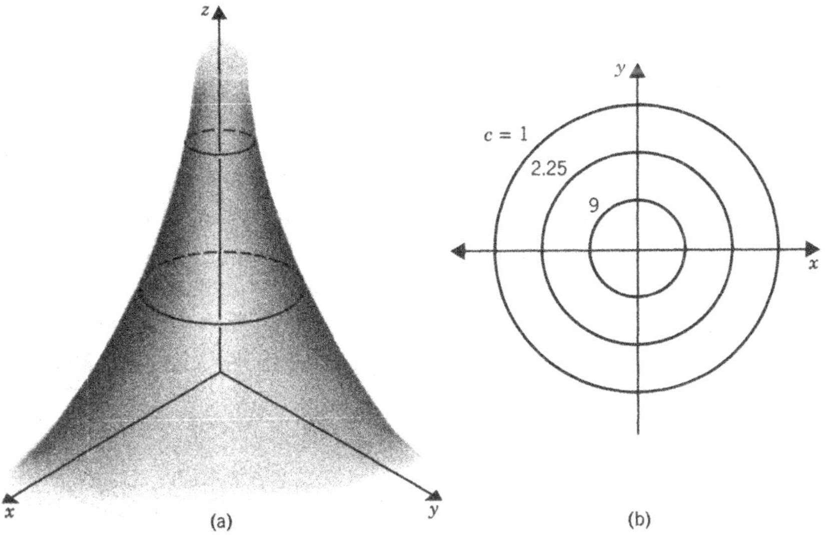

Figure 12-12 (*a*) The function $z = 1/(x^2 + y^2)$ and (*b*) its level curves.

Other Notations. Instead of writing $z = f(x, y)$, we can write $z = f(P)$, where P stands for the point (x, y) in the plane. For functions defined geometrically, this notation is preferable. For example, the function $z = x^2 + y^2$ can be written as $z = f(P) = \|\overrightarrow{OP}\|^2$, where O is the origin as usual.

We can also write $z = f(\mathbf{v})$, where $\mathbf{v} = x\mathbf{i} + y\mathbf{j}$. Here we are using the fact that we have a one-to-one correspondence between vectors and points (x, y) in the plane. For example, the function $z = 2x + 3y$ can be written as $z = (2\mathbf{i} + 3\mathbf{j}) \cdot \mathbf{v}$. The vector notation is especially useful for such linear functions, but also can be used generally and proves to be of very great value. It should be stressed that by using $z = f(\mathbf{v})$ instead of $z = f(x, y)$, we have replaced a function of *two* real variables by a function of *one* vector variable.

PROBLEMS

1. For each of the following sets in the xy-plane make a sketch and describe the interior. In each case the set consists of all (x, y) satisfying the given inequality or inequalities.

   (a) $x + y \geq 1$ (b) $x - y \leq 2$ (c) $|x| > 1$
   (d) $2x^2 + y^2 \geq 1$ (e) $x^2 + y^2 > 0$ (f) $(x - 1)^2(y - 2)^2 > 0$
   (g) $x - y > 1,\ 2x + y > 0,\ x + 3y > 2$
   (h) $2x + y > 5,\ -x + 4y > 0,\ 4x - y > 7$

2. For each of the following sets in the xy-plane make a sketch and state whether the set is open, pathwise connected, an open region. In each case the set consists of all (x, y) satisfying the given inequality or inequalities.
   (a) $x^2 + y^2 < 4$ (b) $3x^2 + 2y^2 \leq 1$
   (c) $[x^2 + y^2 - 1][(x - 2)^2 + y^2 - 1] < 0$ (d) $0 < y < x^2$
   (e) $r^2 < \cos\theta$ (in polar coordinates) (f) $r^2 \sin\theta \leq 1$
   (g) $e^x \cos y > 1$ (h) $x^2 + e^y > 1,\ y - \sin x < 0$
3. (a) Prove: the union of two open sets is open.
   (b) Prove: the intersection of two open sets is open.
   (c) Is the union of two open regions necessarily an open region?
   (d) Is the intersection of two open regions necessarily an open region?
4. If $z = f(x, y) = x^3 + 2y^3$ for all (x, y), evaluate
   (a) $f(1, 0)$ (b) $f(3, 2)$ (c) $f(a, b)$
   (d) $f(2x, 2y)$ (e) $f(-x, -y)$
5. Represent the function graphically by level curves, also by a sketch of a surface, as in Figures 12-8 and 12-10:
   (a) $z = x + y$
   (b) $z = y - 2x$
   (c) $z = y^2 - x$
   (d) $z = x + y^2$
   (e) $z = x^2 + 2y^2$
   (f) $z = \sqrt{1 - x^2 - y^2}$
   (g) $z = \sqrt{x^2 + y^2}$
   (h) $z = \sqrt{x^2 - y^2}$
   (i) $z = \sin(2x + y)$
   (j) $z = e^{-x^2-y^2}$
   (k) $z = \dfrac{1}{x^2 + 2y^2}$
   (l) $z = \dfrac{x^2 - y^2}{x^2 + y^2}$

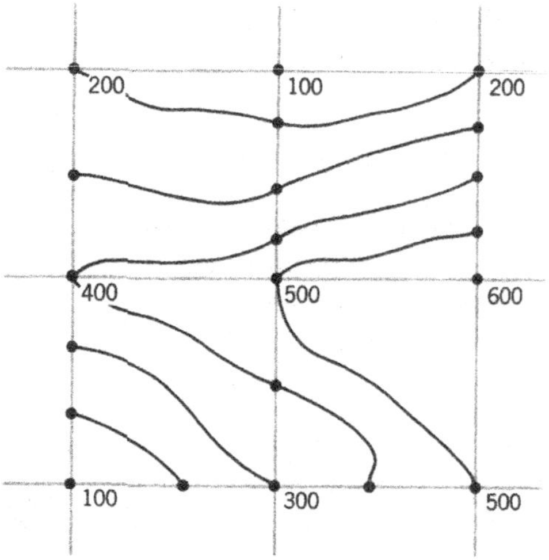

Figure 12-13 The preparation of a topographical map.

6. In preparing a topographical map, one generally has information concerning the altitude z only at certain points in the xy-plane. By linear interpolation along parallels to the axes, one can then estimate z at other points and, in particular, locate points at which z has standard values 100, 200, ... By joining the points at which $z = c$ by smooth curves, one obtains the contour lines. The process is illustrated in Figure 12-13. A similar procedure is used in drawing the lines of constant pressure (isobars) or constant temperature (isotherms) in a weather map. Use this method to sketch level curves of the function $z = f(x, y)$ from the values of z given in the table.

$y \backslash x$	0	1	2	3
0	0	1	4	9
1	2	3	6	11
2	8	9	12	12
3	18	19	22	27

7. Let $\mathbf{v}$ be a general vector in V_2, $\mathbf{a}$ a fixed vector of length 2. Let $z = f(\mathbf{v}) = \mathbf{a} \cdot \mathbf{v}$. Evaluate:

(a) $f(\mathbf{0})$ (b) $f(\mathbf{a})$ (c) $f(3\mathbf{a})$ (d) $f(\mathbf{v} + \mathbf{a})$ (e) $f(2\mathbf{v})$

8. Let $\mathbf{a}$, $\mathbf{b}$ be fixed vectors in V_2 such that $\mathbf{a} \cdot \mathbf{a} = 1$, $\mathbf{a} \cdot \mathbf{b} = 3$, $\mathbf{b} \cdot \mathbf{b} = 5$. Let $f(\mathbf{v}) = \mathbf{a} \cdot \mathbf{v} + \mathbf{b} \cdot \mathbf{v}$. Evaluate:

(a) $f(\mathbf{a})$ (b) $f(\mathbf{b})$ (c) $f(2\mathbf{a} - \mathbf{b})$ (d) $f(3\mathbf{a} + \mathbf{b})$

9. For each of the functions given, state whether the domain is open and whether it is an open region:

(a) $z = y + x^{-1}$ (b) $z = \ln |xy|$ (c) $z = \ln xy$ (d) $z = \sqrt{x + y^2}$

10. For the function $z = \sin (x - t)$ (see Figure 12-11), one sees the wave motion by considering z as a function of x for various fixed values of t. As t increases, one sees the wave moving along the x-axis. Graph z as a function of x for each of the following t-values: $0, \pi/8, \pi/4, 3\pi/8, \pi/2$, showing all curves in one graph in the xz-plane. From these curves state whether the wave is moving to the left or to the right, as t increases. A crude physical example is achieved by snapping a long string which is fastened at one end. Other examples are transmission of sound and of light.

12-3 FUNCTIONS OF THREE OR MORE VARIABLES

The concepts of neighborhood, interior, open set and open region generalize at once to 3-dimensional space R^3. For example, a *neighborhood* of P_0 of radius ρ consists of all P: (x, y, z) whose distance from P_0 is less than ρ. Accordingly, the neighborhood consists of all points inside a sphere of radius ρ. The following sets are open regions in R^3: all of R^3, all (x, y, z) for which $|x| < 1$, $|y| < 1$, $|z| < 1$ (interior of a cube), each neighborhood in R^3.

Functions of Three Variables. A function f of three variables assigns real numbers w to certain points (x, y, z) in 3-space. For example, $w = 2x - y + z$ for all (x, y, z). For this function f we have $f(1, 3, 2) = 2 - 3 + 2 = 1$, $f(1, 0, 0) = 2$, $f(a, b, c) = 2a - b + c$, $f(x, y, z) = 2x - y + z$.

A function f of three variables cannot be graphed as in Figure 12-7, for such a graph would require four dimensions; it can be projected onto a plane, as in Figure 12-7, but the results are difficult to visualize. One can draw graphs of $w = f(x, y, z)$ for different fixed values of z, for example; that is, graph $w = f(x, y, 0)$, $w = f(x, y, 1)$, $w = f(x, y, 2)$, . . . and so on. Such graphs together give some insight into the behavior of the function. The level curves of Figure 12-8 now become level surfaces, again more difficult to draw and to visualize. Thus for $w = x^2 + y^2 + z^2$, the level surfaces are concentric spheres, as in Figure 12-14.

For functions of three variables one can again use *geometric notation:* $w = f(P)$, where P is a point (x, y, z) of R^3, or *vector notation:* $\mathbf{w} = f(\mathbf{v})$, where $\mathbf{v} = x\mathbf{i} + y\mathbf{j} + z\mathbf{k}$ or $\mathbf{v}$ is the vector (v_1, v_2, v_3). We can also translate the discussion of sets into vector language. For example, a neighborhood of radius ρ of the vector $\mathbf{v}_0$ is the set of all vectors $\mathbf{v}$ such that $\|\mathbf{v} - \mathbf{v}_0\| < \rho$.

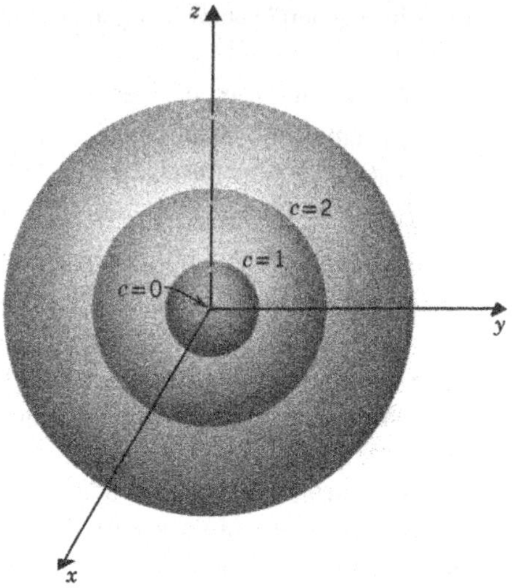

Figure 12-14 Level surfaces of $w = x^2 + y^2 + z^2$.

All of this discussion extends to R^n and functions defined on sets in R^n, that is, functions of n variables $x_1, \ldots, x_n$. We can write $f(x_1, \ldots, x_n)$ for such a function; or $f(P)$, where P is a point of R^n; or $f(\mathbf{v})$, where $\mathbf{v}$ is a vector of V_n. Through the vector notation, we always appear to be dealing with functions of one variable and, to a remarkable extent, the theory of functions of one real variable carries over to the theory of functions of one vector variable.

12-4 VECTOR FUNCTIONS

A pair of functions

$$u = f(x, y), \qquad v = g(x, y)$$

defined on a set E in the xy-plane can be considered as a mapping from the xy-plane to the uv-plane, as suggested in Figure 12-15. The map assigns a pair (u, v) to each (x, y) in E; or we can say it assigns a vector $\mathbf{w} = u\mathbf{i} + v\mathbf{j}$

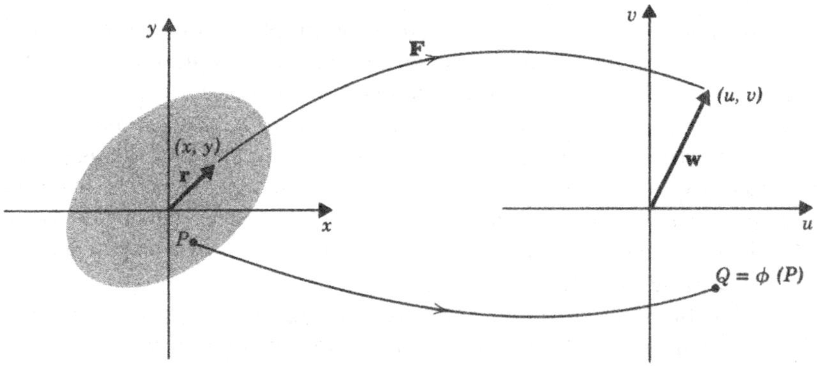

Figure 12-15 Mapping $u = f(x, y)$, $v = g(x, y)$ as a vector function.

to each vector $\mathbf{r} = x\mathbf{i} + y\mathbf{j}$. Hence, the mapping can be interpreted as a vector function:

$$\mathbf{w} = \mathbf{F}(\mathbf{r})$$

We considered many such vector functions in Chapters 9 and 10: namely, the linear mappings, given in matrix notation by

$$\mathbf{w} = A\mathbf{r}$$

One can also describe the mapping of Figure 12-15 in geometric language: $Q = \phi(P)$, where P is a point of the set E in the xy-plane, and Q is a point in the uv-plane. We call Q the *image* of P. For each set contained in E, the image of that set is the set of all points $\phi(P)$ for P in the set.

EXAMPLE. The equations

$$u = x^2 - y^2, \qquad v = 2xy$$

are equivalent to the vector function $\mathbf{F}$ given by

$$u\mathbf{i} + v\mathbf{j} = (x^2 - y^2)\mathbf{i} + (2xy)\mathbf{j}$$

or, in parenthesis notation, by

$$(u, v) = (x^2 - y^2, 2xy)$$

Thus $(x, y) = (0, 0)$ corresponds to $(u, v) = (0, 0)$, $(x, y) = (2, 1)$ corresponds to $(u, v) = (3, 4)$; correspondingly, we write $\mathbf{F}(0\mathbf{i} + 0\mathbf{j}) = 0\mathbf{i} + 0\mathbf{j}$, $\mathbf{F}(2\mathbf{i} + \mathbf{j}) = 3\mathbf{i} + 4\mathbf{j}$. For this mapping, in geometric language, we observe that the image of the origin is the origin; the image of the x-axis (where $y = 0$) is the set of all (u, v) for which $u = x^2$ and $v = 0$—that is, the positive u-axis.

A similar discussion applies to pairs of functions of three variables:

$$u = f(x, y, z), \qquad v = g(x, y, z)$$

Here we have a vector function $\mathbf{w} = \mathbf{F}(\mathbf{r})$, where $\mathbf{r}$ is the vector (x, y, z) and $\mathbf{w}$ is the vector (u, v). Thus we have a mapping from 3-dimensional space to 2-dimensional space (Figure 12-16).

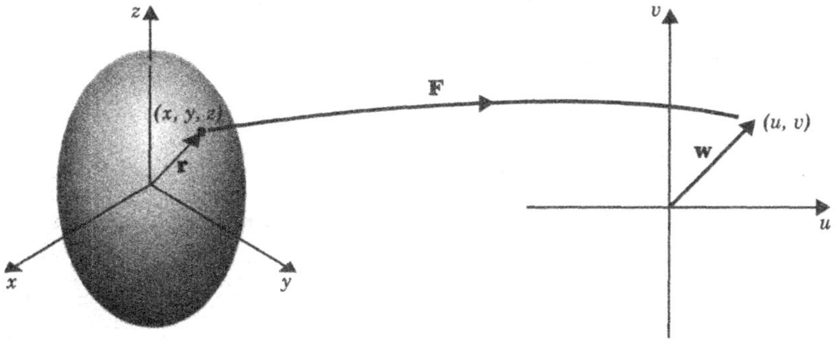

Figure 12-16 Vector function from 3-space to 2-space.

One can also consider pairs of functions of one variable:

$$x = f(t), \quad y = g(t), \qquad a \leq t \leq b \tag{12-40}$$

These are equivalent to a vector function $\mathbf{r} = \mathbf{F}(t)$, as in Section 3-10. Similarly, a triple of functions

$$x = f(t), \qquad y = g(t), \qquad z = h(t), \qquad a \leq t \leq b \tag{12-41}$$

are equivalent to a vector function $\mathbf{r} = \mathbf{F}(t)$, where now $\mathbf{r} = x\mathbf{i} + y\mathbf{j} + z\mathbf{k}$. The functions (12-40) generally represent paths in the xy-plane; the functions (12-41) generally represent paths in 3-dimensional space.

In general, we can consider any set of m functions of n variables:

$$u_1 = F_1(x_1, \ldots, x_n), \qquad \ldots, \qquad u_m = F_m(x_1, \ldots, x_n) \tag{12-42}$$

We usually assume these to be defined on the same set E of n-tuples $(x_1, \ldots, x_n)$. The set E can then be interpreted as a set of points in R^n; the corresponding m-tuples $(u_1, \ldots, u_m)$ form a set in R^m. Thus (12-42) can be thought of as a mapping f from R^n to R^m; we can then simply write $Q = f(P)$, where P is a point of E and Q is the corresponding point $(u_1, \ldots, u_m)$ in R^m.

As above, we can also consider $(x_1, \ldots, x_n)$ and $(u_1, \ldots, u_m)$ as sets of components of vectors. If we denote these vectors by $\mathbf{x}$ and $\mathbf{u}$, then (12-42) becomes

$$\mathbf{u} = \mathbf{F}(\mathbf{x}) \tag{12-42$'$}$$

a mapping $\mathbf{F}$ whose domain is a set in the vector space V_n and whose range is contained in the vector space V_m. Among these vector functions are the *linear mappings* studied in Chapters 9 and 10. They are given in matrix notation by

$$\mathbf{u} = A\mathbf{x}$$

or, in coordinates, by

$$u_i = \sum_{j=1}^{n} a_{ij}x_j, \qquad i = i, \ldots, m$$

Each such linear mapping is defined in all of V_n. Its range is a set in V_m, but need not be all of V_m.

12-5 MATRIX FUNCTIONS

The elements of a matrix $A = (a_{ij})$ may be functions of several variables. For example, we may have

$$A = \begin{pmatrix} x^2 - y^2 & 2x - 3y \\ x + 3y & 3xy \end{pmatrix}$$

The general case is

$$A = \begin{pmatrix} a_{11}(x_1, \ldots, x_n) & \cdots & a_{1p}(x_1, \ldots, x_n) \\ \vdots & & \vdots \\ a_{q1}(x_1, \ldots, x_n) & \cdots & a_{qp}(x_1, \ldots, x_n) \end{pmatrix}$$

We can abbreviate in several ways:

$$A = (a_{ij}(x_1, \ldots, x_n)) = (a_{ij}(\mathbf{x})) = A(\mathbf{x})$$

Normally all a_{ij} are defined in the same set E in R^n (or V_n). Then $A(\mathbf{x})$ becomes a function mapping E into the vector space of all q by p matrices (Section 10-7). In Section 10-17 we discussed such functions, when a_{ij} are functions of a single variable t.

12-6 OPERATIONS ON FUNCTIONS

Real-valued functions of a given type, with the same domain, may be combined as was done for functions of one variable (Section 2-3). Thus for functions of two variables $f(x, y)$, $g(x, y)$, $\ldots$, we may form $f + g$, $f - g$, fg, f/g. Similarly, vector functions can be combined with the allowable operations: $\mathbf{f} + \mathbf{g}$, $\mathbf{f} \cdot \mathbf{g}$, $\varphi\mathbf{f}$ (where φ is a scalar function), as can matrix functions. Composition of functions requires some matching of domains and ranges and hence makes sense only for certain combinations. Thus we cannot compose $f(x, y)$ and $\varphi(u, v)$. But we can compose $F(u, v)$ and the pair $u = f(x, y)$, $v = g(x, y)$ to form

$$F(f(x, y), g(x, y))$$

provided that f and g have the same domain and that the corresponding pairs (u, v) lie in the domain of F. In general, *we can compose* $\mathbf{F}(\mathbf{u})$ *and* $\mathbf{u} = \mathbf{f}(\mathbf{x})$ *if the range of* $\mathbf{f}$ *lies in the domain of* $\mathbf{F}$: *we then obtain* $\mathbf{F}(\mathbf{f}(\mathbf{x}))$, *or the composite* $\mathbf{F} \circ \mathbf{f}$. One can also compose matrix functions and vector functions: for example, compose $A(\mathbf{u})$ and $\mathbf{u} = \mathbf{f}(\mathbf{x})$ to obtain the new matrix function $A(\mathbf{f}(\mathbf{x}))$, under appropriate hypotheses.

We remark that, if $\mathbf{F}$ and $\mathbf{f}$ are linear mappings and $\mathbf{F} \circ \mathbf{f}$ is defined, then $\mathbf{F} \circ \mathbf{f}$ is a linear mapping. If $\mathbf{F}$ is represented by matrix A and $\mathbf{f}$ by matrix B, then $\mathbf{F} \circ \mathbf{f}$ is represented by the matrix AB (Section 10-8). This rule of linear algebra plays an important role in the calculus, as will be seen.

Zeros of Functions. As for functions of one variable, we say that (x_0, y_0) is a zero of $f(x, y)$ if $f(x_0, y_0) = 0$. In general, $\mathbf{b}$ is a zero of $\mathbf{f}(\mathbf{x})$ if $\mathbf{f}(\mathbf{b}) = \mathbf{0}$, and $\mathbf{b}$ is a zero of $A(\mathbf{x})$ if $A(\mathbf{b}) = O$. Note that for a function of two variables $f(x, y)$, the zeros of f together form the level curve $f(x, y) = 0$.

PROBLEMS

1. If $z = f(x, y, z) = (x - y)/(y - z)$, evaluate

(a) $f(2, 3, 0)$ (b) $f(0, 5, -2)$ (c) $f(2, 2, 1)$

(d) $(1, 2, 2)$ (e) $f(1, 1, 1)$ (f) $f(3, 2, 2)$

(g) $f(a, b, c)$ (h) $f(2x, 2y, 2z)$ (i) $f(-x, -y, -z)$

2. Describe the level surfaces of the functions:

   (a) $w = 2x - y + z$ (b) $w = x - y + z$

   (c) $w = f(x, y, z) = 2x - y$ (d) $w = f(x, y, z) = x^2 + y^2$

   (e) $w = 2x^2 + y^2 + z^2$ (f) $w = x^2 + y^2 - z^2$

3. Consider the equations $u = e^x \cos y$, $v = e^x \sin y$ as a mapping of the xy-plane into the uv-plane.

   (a) Find (u, v) for each of the following choices of the point (x, y): $(0, 0)$, $(0, \pi)$, $(0, \pi/2)$, $(1, 3\pi/4)$.

   (b) Find the image of the x-axis in the uv-plane.

   (c) Find the image of the y-axis.

   (d) Is the mapping one-to-one?

4. For vectors $\mathbf{u}$ in the plane, let $\mathbf{f}(\mathbf{u}) = [\mathbf{u} \cdot (\mathbf{i} + \mathbf{j})]\mathbf{u}$. Evaluate $\mathbf{f}(\mathbf{i})$, $\mathbf{f}(\mathbf{j})$, $\mathbf{f}(2\mathbf{i} - \mathbf{j})$, $\mathbf{f}(x\mathbf{i} + y\mathbf{j})$.

5. For vectors $\mathbf{u}$ in the plane let $\mathbf{f}(\mathbf{u}) = 3(\mathbf{u} \cdot \mathbf{u})\mathbf{i} + (\mathbf{u} \cdot \mathbf{u})^2\mathbf{j}$. Evaluate $\mathbf{f}(\mathbf{i})$, $\mathbf{f}(2\mathbf{i})$, $\mathbf{f}(\mathbf{j})$, $\mathbf{f}(\mathbf{i} - \mathbf{j})$, $\mathbf{f}(x\mathbf{i} + y\mathbf{j})$.

6. For vectors $\mathbf{u}$ in the plane let $\mathbf{f}(\mathbf{u}) = 3\mathbf{u} + 2\mathbf{u}^{\perp}$. Evaluate $\mathbf{f}(\mathbf{i})$, $\mathbf{f}(\mathbf{j})$, $\mathbf{f}(2\mathbf{i} + 3\mathbf{j})$, $\mathbf{f}(x\mathbf{i} + y\mathbf{j})$.

7. For real t, let $\mathbf{F}(t) = t^2\mathbf{i} + t^3\mathbf{j} + 3\mathbf{k}$. Evaluate $\mathbf{F}(0)$, $\mathbf{F}(1)$, $\mathbf{F}(-1)$.

8. For real t, let $\mathbf{F}(t) = \cos t\mathbf{i} + \sin t\mathbf{j} + t\mathbf{k}$. Evaluate $\mathbf{F}(0)$, $\mathbf{F}(\pi/2)$, $\mathbf{F}(\pi)$.

9. For vectors $\mathbf{v}$ in space, let $F(\mathbf{v}) = \mathbf{v} \cdot \mathbf{i} + \mathbf{v} \cdot \mathbf{v}$. Evaluate $F(\mathbf{i})$, $F(2\mathbf{i})$, $F(\mathbf{i} - \mathbf{j} + 2\mathbf{k})$, $F(x\mathbf{i} + y\mathbf{j} + z\mathbf{k})$.

10. For vectors $\mathbf{v}$ in space, let $F(\mathbf{v}) = \mathbf{v} \times \mathbf{i} \cdot (\mathbf{i} + \mathbf{j})$. Evaluate $F(\mathbf{i})$, $F(\mathbf{i} + \mathbf{j})$, $F(\mathbf{k})$, $F(x\mathbf{i} + y\mathbf{j} + z\mathbf{k})$.

11. For vectors $\mathbf{v}$ in V_2, let $\mathbf{F}(\mathbf{v}) = B\mathbf{v}$, where $B = \begin{pmatrix} 3 & 1 \\ 4 & 5 \end{pmatrix}$. Find $\mathbf{F}(\mathbf{i} + \mathbf{j})$, $\mathbf{F}(2\mathbf{i} - \mathbf{j})$, $\mathbf{F}(\mathbf{0})$, $\mathbf{F}(x\mathbf{i} + y\mathbf{j})$.

12. For each vector $\mathbf{u} = x\mathbf{i} + y\mathbf{j}$ in the plane, let

$$\mathbf{F}(\mathbf{u}) = x(\mathbf{i} + \mathbf{j} + 3\mathbf{k}) + y(\mathbf{i} - \mathbf{j} + 2\mathbf{k}) + 2\mathbf{i} + 5\mathbf{j} + \mathbf{k}$$

   (a) Evaluate $\mathbf{F}(\mathbf{0})$, $\mathbf{F}(\mathbf{i})$, $\mathbf{F}(\mathbf{j})$, $\mathbf{F}(3\mathbf{i} + 7\mathbf{j})$. (b) What is the range of $\mathbf{F}$?

13. For each vector $\mathbf{u} = x\mathbf{i} + y\mathbf{j} + z\mathbf{k}$ in V_3, let $\mathbf{F}(\mathbf{u}) = 2\mathbf{i} + \mathbf{j} + x(\mathbf{i} + \mathbf{j}) + y(2\mathbf{i} - \mathbf{k}) + z(\mathbf{i} + \mathbf{j} + \mathbf{k})$. Evaluate: $\mathbf{F}(\mathbf{i})$, $\mathbf{F}(\mathbf{j})$, $\mathbf{F}(\mathbf{k})$, $\mathbf{F}(\mathbf{i} - \mathbf{k})$. What is the range of $\mathbf{F}$?

14. For each real t let $A(t) = \begin{pmatrix} 2t^2 & t^3 \\ 0 & t \end{pmatrix}$. Evaluate $A(0)$, $A(1)$, $A(-1)$.

15. Let $f(x, y) = x - y$, $g(x, y) = xy$, $h(x, y) = x^2 + y^2$, $F(u, v) = u^2 - v$. Find expressions for each of the composite functions:

   (a) $F[f(x, y), g(x, y)]$ (b) $F[f(x, y), h(x, y)]$

   (c) $F[h(x, y), f(x, y)]$ (d) $F[f(x, y), f(x, y)]$

16. For vectors $\mathbf{u}$, $\mathbf{v}$ in the plane let $\mathbf{f}(\mathbf{u}) = (\mathbf{u} \cdot \mathbf{u})\mathbf{i} + (\mathbf{u} \cdot \mathbf{i})\mathbf{j}$, $\mathbf{g}(\mathbf{v}) = 2\mathbf{v} + 3\mathbf{v}^{\perp}$.

   (a) Find an expression for $\mathbf{g}[\mathbf{f}(\mathbf{u})]$. (b) Find an expression for $\mathbf{f}[\mathbf{g}(\mathbf{v})]$.

17. Find all zeros of the following functions:

   (a) $f(x, y) = x^2 + y^2$ (b) $f(x, y) = x - y$

(c) $f(x, y) = x^2 + y - 1$ (d) $f(x, y, z) = x^2 + y^2 + z^2 - 1$

(e) $\mathbf{f}(\mathbf{u}) = A\mathbf{u}$, for $\mathbf{u}$ in V_2 and $A = \begin{pmatrix} 1 & 3 \\ 2 & 6 \end{pmatrix}$.

12-7 LIMITS AND CONTINUITY

For functions of several variables, the concept of continuity is of greater importance than the theory of limits. Here we give a definition of continuity not referring to limits and deduce the principal properties of continuous functions. At the close of the section, we make some remarks on limits.

Let a function f of two variables be defined in an open region D of the xy-plane and let P: (x_0, y_0) be a point of D. Since D is open, we can then choose a $\delta_0 > 0$ so that the neighborhood of P_0 of radius δ_0 is contained in D (see Figure 12-17). The function f is said to be *continuous* at P_0 if the values

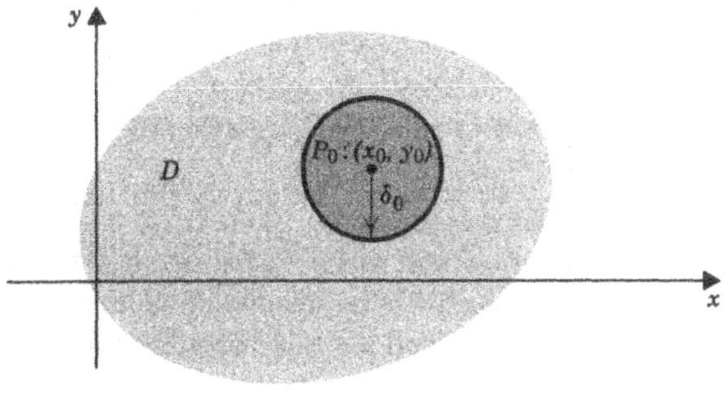

Figure 12-17 Definition of continuity.

of f can be made as close to $f(x_0, y_0)$ as desired by keeping (x, y) sufficiently close to (x_0, y_0); that is, if for each $\epsilon > 0$ we can choose a positive δ (less than δ_0) such that

$$|f(x, y) - f(x_0, y_0)| < \epsilon \qquad \text{for } (x - x_0)^2 + (y - y_0)^2 < \delta^2$$

We can write the definition in geometric notation: f is continuous at P_0 if

$$|f(P) - f(P_0)| < \epsilon \qquad \text{for } d(P, P_0) < \delta \tag{12-70}$$

Or we can say: for each $\epsilon > 0$ we can choose a neighborhood of P_0, of radius δ, within which $|f(P) - f(P_0)| < \epsilon$. We can also write the definition in vector notation: f is continuous at $\mathbf{v}_0$ if

$$|f(\mathbf{v}) - f(\mathbf{v}_0)| < \epsilon \qquad \text{for } \|\mathbf{v} - \mathbf{v}_0\| < \delta \tag{12-71}$$

The function f is said to be *continuous in D* if f is continuous at each point P of D. If P is in D and f is not continuous at P, then f is said to be *discontinuous* at P. The function f is also automatically discontinuous wherever the function is not defined.

EXAMPLE. The function $f(x, y) = x + y$ is continuous for all (x, y). For let a point (x_0, y_0) and a number $\epsilon > 0$ be given. Then

$$|x + y - (x_0 + y_0)| = |(x - x_0) + (y - y_0)| \leq |x - x_0| + |y - y_0| < \epsilon$$

if $(x - x_0)^2 + (y - y_0)^2 < \delta^2$, where $\delta = \epsilon/2$. For $(x - x_0)^2 + (y - y_0)^2 < \delta^2 = \epsilon^2/4$ implies $|x - x_0| < \epsilon/2$ and $|y - y_0| < \epsilon/2$, so that $|x - x_0| + |y - y_0| < (\epsilon/2) + (\epsilon/2) = \epsilon$.

We now can formulate the basic theorems on continuous functions, as in Section 2-7. The labeling of the theorems is in agreement with that section.

THEOREM B. *Let f and g be defined in the open region D of the xy-plane. If both f and g are continuous at (x_0, y_0), then so are $f \pm g$, $f \cdot g$, f/g—provided, for f/g, that $g(x_0, y_0) \neq 0$. If f and g are continuous in D, then so are $f \pm g$, $f \cdot g$, f/g—provided, for f/g, that g has no zeros in D.*

THEOREM C. *Let f and g be defined in the open region D of the xy-plane; let $F(u, v)$ be defined in an open region D_1 of the uv-plane. Let the composite function $F[f(x, y), g(x, y)]$ be defined in D. (a) If f and g are continuous at the point (x_0, y_0) of D and F is continuous at (u_0, v_0), where $f(x_0, y_0) = u_0$ and $g(x_0, y_0) = v_0$, then $F[f(x, y), g(x, y)]$ is also continuous at (x_0, y_0). (b) If f and g are continuous in D and F is continuous in D_1, then $F[f(x, y), g(x, y)]$ is continuous in D.*

‡PROOF OF THEOREM C. It suffices to prove (a), since (b) is a consequence of (a). We use geometric notation and write P_0 for (x_0, y_0), P for (x, y), Q_0 for (u_0, v_0), Q for (u, v). Let a positive ϵ be given and choose δ_1 so that $|F(Q) - F(Q)| < \epsilon$ for Q in the neighborhood of Q_0 of radius δ_1. We then choose ϵ_1 so small that the square open region:

$$|u - u_0| < \epsilon_1, \qquad |v - v_0| < \epsilon_1$$

is contained in the neighborhood of Q_0 of radius δ_1 (see Figure 12-18). We

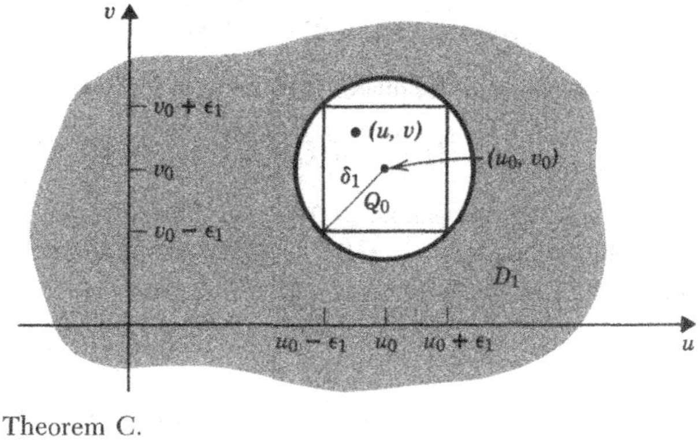

Figure 12-18 Proof of Theorem C.

next choose $\delta' > 0$ so small that $|f(P) - f(P_0)| < \epsilon_1$ for $d(P, P_0) < \delta'$ and choose $\delta'' > 0$ so small that $|g(P) - g(P_0)| < \epsilon_1$ for $d(P, P_0) < \delta''$. Finally, we let δ be the smaller of δ', δ''. For $d(P, P_0) < \delta$ we can then conclude that

$$|f(P) - f(P_0)| < \epsilon_1 \qquad \text{and} \qquad |g(P) - g(P_0)| < \epsilon_1$$

and, hence, that the point $(u, v) = (f(P), g(P))$ lies in the square region specified above, so that (u, v) also lies inside the neighborhood of Q_0 of radius δ_1. Therefore, $|F(u, v) - F(u_0, v_0)| < \epsilon$ or

$$|F[f(P), g(P)] - F[(f(P_0), g(P_0)]| < \epsilon$$

Accordingly, the composite function is continuous at P_0 and (a) is proved.‡

Now let $F(u, v) = u + v$. Then F is continuous for all (u, v), as in the example above. If we apply Theorem C to this case, we conclude that

$$F[f(x, y), g(x, y)] \equiv f(x, y) + g(x, y)$$

is continuous wherever both f and g are continuous. This gives the part of Theorem B referring to $f + g$. The assertions for $f - g$, fg, f/g are proved in the same way by taking $F(u, v) = u - v$, uv, u/v $(v \neq 0)$, respectively.

Remark. Theorem C can be applied to the case in which f and g depend on x alone. Thus, for example, $F[f(x), g(x)]$ is continuous for $a \leq x \leq b$ provided that f and g are continuous in $[a, b]$, $F(u, v)$ is defined and continuous in D_1, and $F[f(x), g(x)]$ is defined in $[a, b]$. (Here, f and g are not defined in an open region, but that condition is not actually needed for the proof of Theorem C; see the discussion of functions continuous on general sets below.)

The Intermediate Value Theorem also has an analogue.

THEOREM D. *Let $z = f(x, y)$ be defined and continuous in the open region D and let $f(x_1, y_1) = z_1$, $f(x_2, y_2) = z_2$, $z_1 \neq z_2$. Then for every number z_0 between z_1 and z_2 there is a point (x_0, y_0) of D for which $f(x_0, y_0) = z_0$.*

PROOF. Since D is an open region, D is pathwise connected. Hence, we can choose a path $x = x(t)$, $y = y(t)$, $a \leq t \leq b$, joining P_1 to P_2 in D as in Figure 12-19. By the remark above, $f(x(t), y(t))$ is continuous in $[a, b]$. Also, $f(x(a), y(a)) = f(x_1, y_1) = z_1$, $f(x(b), y(b)) = f(x_2, y_2) = z_2$. By the ordinary Intermediate Value Theorem there is a t_0 in $[a, b]$ such that $f(x(t_0), y(t_0)) = z_0$. Thus $(x_0, y_0) = (x(t_0), y(t_0))$ has the desired property.

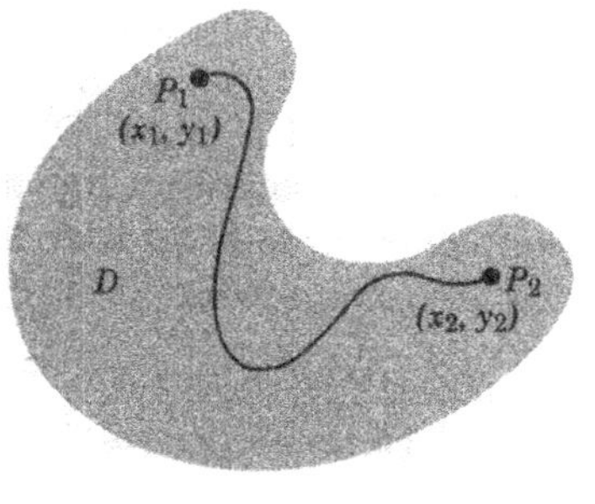

Figure 12-19 Intermediate Value Theorem.

Theorem E, concerning the inverse function, is discussed in Section 12-16 below.

Functions Defined on General Sets. The definition of continuity applies, with essentially no change, to a function f whose domain is not necessarily an open region; for example, a function whose domain is the closed square region: $|x| \leq 1$, $|y| \leq 1$; or a function defined only on a line segment. If f has domain E and P_0 is a point of E, then f is said to be continuous at P_0 if for each $\epsilon > 0$ one can find a $\delta > 0$ such that $|f(P) - f(P_0)| < \epsilon$ for every P in E such that $d(P, P_0) < \delta$. For example, $f(x, y) = \sqrt{x}\sqrt{y}$ is defined only for $x \geq 0$ and $y \geq 0$ and f is continuous at P_0: $(0, 0)$, as one can verify directly (see Problem 10 below). Here every neighborhood of P_0 contains points at which f is not defined, but we simply ignore such points in testing for continuity. The situation is analogous to that for a function of one variable defined on an interval $[a, b]$; here f is continuous at a if f is continuous to the *right* at a. Theorems B and C and their proofs remain valid, with the appropriate modifications, for continuous functions defined on general sets.

Existence of Maximum and Minimum. The basic Theorem F on maxima and minima for functions of one variable has its counterpart for functions of two variables: A function defined and continuous in a bounded closed region E has an absolute maximum and minimum in E. This theorem is discussed in Sections 12-22 and 12-25 below. We remark that there need be no absolute maximum or minimum for functions defined in an open region. For example, the function $f(x, y) = x + y$, $0 < x < 1$, $0 < y < 1$ has no absolute maximum or minimum in the given domain; however, the function $f(x, y) = x + y$, $0 \leq x \leq 1$, $0 \leq y \leq 1$ has its absolute maximum of 2 at $(1, 1)$ and its absolute minimum of 0 at $(0, 0)$. Thus the maximum or minimum can occur on the *boundary* of the domain.

We note one other property of continuous functions: *if f is continuous in the open region D, then for each real c the set of (x, y) in D for which $f(x, y) > c$ is an open set.* In other words, the conditions

$$f(x, y) > c, \qquad (x, y) \text{ in } D$$

characterize an open subset E of D. For let (x_0, y_0) be in E, so that $f(x_0, y_0) > c$, and we can write $f(x_0, y_0) = c + 2\epsilon$, with $\epsilon > 0$. For (x, y) in a sufficiently small neighborhood of (x_0, y_0) (in D), we have $|f(x, y) - f(x_0, y_0)| < \epsilon$ and, hence, clearly $f(x, y) > c + \epsilon$, so that (x, y) is in E. Hence, E is open. In a similar fashion the set characterized by two such inequalities:

$$f(x, y) > c_1, \qquad g(x, y) > c_2, \qquad (x, y) \text{ in } D \tag{12-72}$$

is open, as is the set characterized by any finite number of such inequalities. The sets, of course, may be empty (see Problem 14). We note that the set (12-72) is the intersection of the two sets: $f(x, y) > c_1$, (x, y) in D; $g(x, y) > c_2$, (x, y) in D.

Functions of More than Two Variables and Vector Functions. The preceding discussion extends at once to functions of more than two variables. In fact, the formulation of continuity based on (12-70) is applicable to functions of n variables, and the Theorems B and C can be restated and proved with obvious modifications for the general case.

The discussion also extends to a vector function $\mathbf{F}$. The basic definition (12-71) carries over, with f replaced by $\mathbf{F}$ and absolute value by norm. Theorem

B has an analogue for vector functions: *if* **F**, **G** *and* f *are continuous at* $\mathbf{v}_0$ *in* V_n, *then so are* $\mathbf{F} \pm \mathbf{G}$, $\mathbf{F} \cdot \mathbf{G}$ and $f\mathbf{F}$. Theorem C has an analogue: *if* **F** *and* **G** *are vector functions such that the composite function* $\mathbf{F} \circ \mathbf{G}$ *is defined at* $\mathbf{v}_0$, *if* **G** *is continuous at* $\mathbf{v}_0$ *and if* **F** *is continuous at* $\mathbf{u}_0 = \mathbf{G}(\mathbf{v}_0)$, *then* $\mathbf{F} \circ \mathbf{G}$ *is continuous at* $\mathbf{v}_0$. The proofs are similar to those given above. One can show (as in Section 3-10) that a vector function $\mathbf{F}(\mathbf{v}) = (F_1(\mathbf{v}), \ldots, F_m(\mathbf{v}))$ is continuous at $\mathbf{v}_0$ if, and only if, its components $F_1(\mathbf{v}), \ldots, F_m(\mathbf{v})$ are each continuous at $\mathbf{v}_0$. Then all theorems on properties of vector functions can be proved by reference to the appropriate theorems for (scalar) functions of several variables.

For a matrix function $A(\mathbf{x}) = (a_{ij}(\mathbf{x}))$, one can simply define continuity at $\mathbf{x}_0$ to mean continuity of all functions $a_{ij}(\mathbf{x})$ at $\mathbf{x}_0$. Then one immediately deduces that the sum and product of continuous matrix functions are continuous, also that the analogue of Theorem C holds true.

Limits. The definition of limit is the same as that of continuity, except that the function need not be defined at the value approached by the independent variable, and one never allows the independent variable to equal that value, in the definition. Thus for a function of n variables, defined in an open region D except perhaps at the point P_0 of D, one writes

$$\lim_{P \to P_0} f(P) = c$$

if, for each $\epsilon > 0$, one can find a $\delta > 0$ such that

$$|f(P) - c| < \epsilon \qquad \text{for } 0 < d(P, P_0) < \delta$$

There is an analogous definition for vector functions. One can also define infinite limits for scalar functions, as for functions of one variable. One can also verify that f is continuous at P_0 if, and only if,

$$\lim_{P \to P_0} f(P) = f(P_0)$$

The theorems on limits of sum, and so on, can be proved as usual for functions of several variables and for vector functions.

The study of limits for functions of several variables is much more complex than for functions of one variable. Examples of the complexity are given in Problem 8 below. As remarked earlier, we do not need such limits for the theory in this text.

Classes of Continuous Functions of Several Variables. Among the functions of two variables x, y are those which depend on x alone or on y alone: for example, $z = e^x$, $z = \cos y$. In general, if $z = f(x)$ and f, as a function of x, is continuous for $a < x < b$, then f, as a function of x and y, is continuous for $a < x < b$, $-\infty < y < \infty$; that is, f is continuous in a *strip region* as in Figure 12-20. This assertion follows at once from the definition of continuity. The level curves of such a function are the vertical lines $x = \text{const.}$

Starting with known continuous functions of one variable and repeatedly

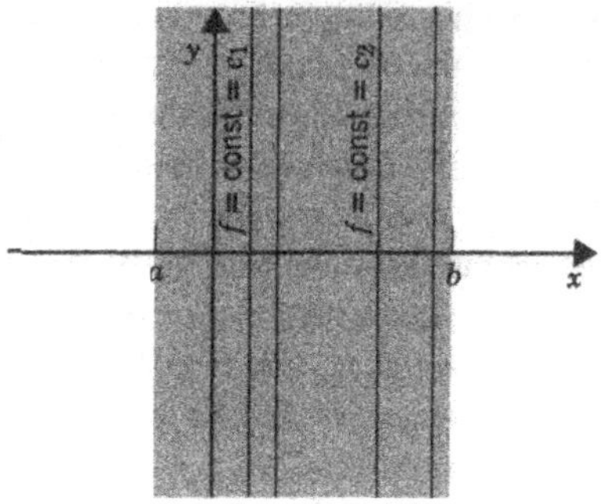

Figure 12-20 Function $z = f(x)$ as a function of x and y.

using the processes of Theorems B and C, we can construct new continuous functions of two or more variables; for example, $z = x^2 + y^3$, $z = e^x \cos y$, $z = \cos(x + y)$, $w = x_1^3 + x_2^3 + \cdots + x_n^3$. Among the important functions obtained in this way are the following:

Polynomials. For example, $z = 2 + x + 3y + x^2 - y^2 + x^3 + 5x^2y - y^4$
Rational functions. For example

$$w = \frac{xyz}{1 - x^2 - y^2 - z^2}.$$

Trigonometric polynomials. Sums of terms having the form $a \sin nx \cos my$, $a \cos nx \sin my$, $a \sin nx \sin my$, or $a \cos nx \cos my$, for example,

$$z = 2 + \sin x + 2 \cos y + 5 \sin 2x \cos 3y + 6 \sin 5x \cos 7y$$

The polynomials and trigonometric polynomials are continuous everywhere, the rational functions are continuous except at the zeros of the denominator.

Other functions of several variables are obtained by limit processes: for example,

$$f(x, y) = \lim_{n\to\infty} \left(\frac{1}{x^y} + \frac{1}{(x+1)^y} + \frac{1}{(x+2)^y} + \cdots + \frac{1}{(x+n)^y}\right)$$

can be shown to define a function (the generalized Zeta Function) for $0 < x \leq 1$, $y > 1$.

It follows from Theorem B that the collection of all functions continuous in a given open region D in the xy-plane forms a vector space, easily seen to be of infinite dimension. Other vector spaces of functions occur naturally: all polynomials in (x, y), all trigonometric polynomials in (x, y), all polynomials in (x, y) of degree at most n. The rational functions continuous in a given open region also form a vector space.

Effect of Fixing Variables. We remarked above that a function of one variable gives rise to a function of two variables. Conversely, from a function $F(x, y)$ we can, by fixing y at a particular value, obtain a function of x. Thus, if $F(x, y) = \cos xy$, then for $y = 2$, F becomes the function $\cos 2x$. If F is continuous at (x_0, y_0), then $F(x, y_0)$ is continuous in x at x_0 and $F(x_0, y)$ is continuous in y at y_0, as follows at once from the definition of continuity.

(The converse of this last statement is false; see Problem 13 below.) In general, from $F(x_1, \ldots, x_n)$ we obtain, by this process, functions such as

$$f(x_1) = F(x_1, x_2^0, \ldots, x_n^0)$$
$$f(x_1, x_2) = F(x_1, x_2, x_3^0, \ldots, x_n^0)$$

If F is continuous at $(x_1^0, \ldots, x_n^0)$, these are continuous at x_1^0, and (x_1^0, x_2^0), respectively.

PROBLEMS

1. From the results derived in the text, deduce where the following functions are continuous:

   (a) $z = x^2 + 3xy + y^2$ (b) $z = (x^2 + y^3)^5$ (c) $z = \dfrac{x^2 + y^2}{x - y}$

   (d) $z = \dfrac{2x + 3y}{x^2 + y^2}$ (e) $z = \ln(2x + 3y)$ (f) $z = \tan\dfrac{x}{y}$

   (g) $z = \mathrm{Sin}^{-1}\dfrac{x}{y}$ (h) $z = \ln(1 + e^x \cos y)$ (i) $w = x^2y^2 \ln|z|$

   (j) $w = x^{(y^2)}$ (k) $t^2\mathbf{i} + \ln t\,\mathbf{j} + e^{2t}\mathbf{k}$ (l) $t^{1/4}\mathbf{i} + t^{1/3}\mathbf{j} + t\mathbf{k}$

   (m) $(x^2 - y^2)\mathbf{i} + 2xy\mathbf{j}$ (n) $\ln(x + y)\,\mathbf{i} + \sin(x + 2y)\,\mathbf{j}$

   (o) $[\mathbf{u} \cdot (2\mathbf{i} + 3\mathbf{j} + \mathbf{k})]\mathbf{u}$, where $\mathbf{u} = x\mathbf{i} + y\mathbf{j} + z\mathbf{k}$

   (p) $(\ln\|\mathbf{u}\| + \mathbf{u} \cdot \mathbf{k})\mathbf{u}$, where $\mathbf{u}$ is as in (o)

   (q) $y = \dfrac{\mathbf{a} \cdot \mathbf{x}}{\|\mathbf{x}\|}$, a const. (r) $y = \dfrac{\mathbf{x} \cdot \mathbf{x}}{\mathbf{a} \cdot \mathbf{x}}$, a const., $\mathbf{a} \neq \mathbf{0}$.

   [In (q) and (r) $\mathbf{x}$ and $\mathbf{a}$ are vectors in R^n.]

2. Each of the following functions of (x, y) is to be evaluated with an error of at most 0.001 for $x = \sqrt{2}$, $y = \sqrt{3}$. Give a number of decimal places for x and y that will insure this accuracy.

   (a) $2x + 3y$ (b) $5x - y$ (c) $\dfrac{1}{x + y}$ (d) $\dfrac{x}{y}$

3. Prove from the definition that the function is continuous where stated:

   (a) $z = x - y$, all (x, y), (b) $z = c =$ const., all (x, y),

   (c) $z = x \cdot y$, all (x, y). [*Hint.* Write $x = x_0 + h$, $y = y_0 + k$, show that $|xy - x_0y_0|$ can be made less than ϵ by making $|h| < \frac{1}{3}$, $|hy_0| < \epsilon/3$, $|k| < \epsilon$ and $|kx_0| < \epsilon/3$, and that these conditions are satisfied for $h^2 + k^2 < \delta^2$, provided that δ is properly chosen.]

   (d) $z = x/y$, $y > 0$. [*Hint.* Make a geometric analysis of the level curves of z.]

4. (a) Can the point (x_0, y_0) of the Intermediate Value Theorem for $f(x, y)$ be unique?
   (b) Formulate and prove the theorem for a function of three variables.

5. (a) Let $F(u, v) = (u^2 + 2uv + v^3)/uv$, $f(x, y) = (x - y)^2 + 1$, $g(x, y) = x^2 - 3xy - 4y^2$. Determine open regions in the xy-plane for which the function $G(x, y) = F(f(x, y), g(x, y))$ is defined.

(b) Is the function $G(x, y)$ continuous in the open region $0 < 4y < x$?

6. (a) Let $F(u, v) = (u^3 - uv + v)/(uv - v)$, $f(x, y) = (\sin x)/(y - 1)$, $g(x, y) = x + 3y$. Determine the open regions in the xy-plane for which the function $G(x, y) = F(f(x, y), g(x, y))$ is defined.

   (b) Is the function $G(x, y)$, continuous at the points $(\pi/2, 1)$, $(\pi, 2)$, $(5\pi/2, 2)$, $(-3\pi, \pi)$? Explain.

7. (a) Prove: if $f(\mathbf{x})$ is continuous in an open region D of V_n, then for each real c, the set of $\mathbf{x}$ in D for which $f(\mathbf{x}) > c$ is an open set of V_n.

   (b) Would the same conclusion hold true for the $\mathbf{x}$ in D for which $f(\mathbf{x}) < c$?

8. Show that each function $z = f(x, y)$ has no limit as $(x, y) \to (0, 0)$:

   (a) $z = \dfrac{xy}{x^2 + y^2}$ (b) $z = \dfrac{x^2 - y^2}{x^2 + y^2}$ (c) $z = \dfrac{x + y}{x^2 + y^2}$

9. Let $z = f(x, y) = y(x^2 + y^2)/[y^2 + (x^2 + y^2)^2]$. Show that f has limit 0 as (x, y) approaches $(0, 0)$ on a ray $(x = at, y = bt)$, but f does not have limit 0 as $(x, y) \to (0, 0)$. (*Hint.* Find the level curve on which $f = \frac{1}{2}$.)

10. Prove from the definition of continuity that the function $f(x, y) = \sqrt{x}\sqrt{y}$, $x \geq 0$ and $y \geq 0$, is continuous at $(0, 0)$.

‡11. (a) Let f be a continuous mapping of an open region in R^n into R^m. Prove: for every open set E in R^m, the set of all points P for which $f(P)$ lies in E is an open set.

   (b) Prove the converse of the result of part (a): that is, if f is a mapping of an open region in R^n into R^m and, for every open set E in R^m, the set of all points P for which $f(P)$ lies in E is open, then f is continuous throughout its domain.

   (c) Let f be a continuous mapping of an open region in R^n into R^m. Is the image of every open set contained in the domain of f necessarily an open set in R^m? Show that, if the image of an open region is open, then the image is an open region.

12. (a) Prove the analogue of Theorem C for vector functions: if $\mathbf{F}(\mathbf{v})$ and $\mathbf{f}(\mathbf{u})$ are such that the composite function $\mathbf{F} \circ \mathbf{f}$ is defined in a neighborhood of $\mathbf{u}_0$, and $\mathbf{f}$ is continuous at $\mathbf{u}_0$ and $\mathbf{F}$ is continuous at $\mathbf{v}_0 = \mathbf{f}(\mathbf{u}_0)$, then $\mathbf{F} \circ \mathbf{f}$ is continuous at $\mathbf{u}_0$.

   (b) Show that Theorem C, as formulated and proved in Section 12-7, is a special case of part (a). (*Hint.* Use the fact that a vector function is continuous if, and only if, the component functions are continuous.)

‡13. (a) Let $f(x, y) = xy/(x^2 + y^2)$ for $(x, y) \neq (0, 0)$, $f(0, 0) = 0$. Show that $f(x, y_0)$ is continuous for all x, $f(x_0, y)$ is continuous for all y, but $f(x, y)$ is not continuous at $(0, 0)$. Here x_0 and y_0 are arbitrary real numbers.

   (b) Let $f(x, y) = \sin(x/y)$ for $y > 0$ and $0 \leq x \leq \pi y$, let f have value 0 for all other (x, y). Show that $f(x, y_0)$ is continuous for all x, $f(x_0, y)$ is continuous for all y, but f is discontinuous at $(0, 0)$.

14. Show that the set characterized by $1 - x^2 - y^2 > 0$, $x > 3$ is the empty set.

12-8 PARTIAL DERIVATIVES

Let a function $z = f(x, y)$ be given in an open region D of the xy-plane, and let (x_0, y_0) be a point of D. Then, for x sufficiently close to x_0, all points (x, y_0) are also in D (Figure 12-21). Thus we can consider $z = f(x, y_0)$ as a

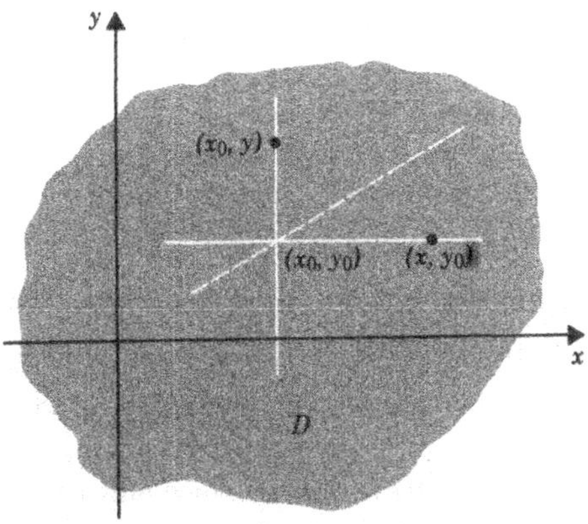

Figure 12-21 Partial derivative.

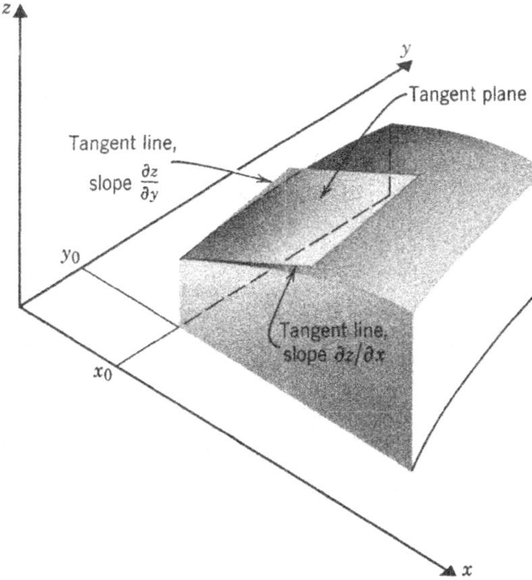

Figure 12-22 The geometric meaning of partial derivative.

function of x, in a small interval about x_0. The derivative at x_0 of this function of x (if the derivative exists) is called *the partial derivative of f with respect to x at* (x_0, y_0) and is denoted by one of the following:

$$f_x(x_0, y_0), \qquad \frac{\partial f}{\partial x}(x_0, y_0), \qquad z_x(x_0, y_0), \qquad \frac{\partial z}{\partial x}(x_0, y_0)$$

If we go back to the definition of the derivative as a limit, we can write

$$f_x(x_0, y_0) = \lim_{\Delta x \to 0} \frac{f(x_0 + \Delta x, y_0) - f(x_0, y_0)}{\Delta x} \tag{12-80}$$

We can interpret the partial derivative geometrically as a slope, as suggested in Figure 12-22. Here one considers a section of the surface $z = f(x, y)$ in a vertical plane $y = y_0$. In that plane the curve $z = f(x, y_0)$ (or more precisely, its tangent line) has slope $f_x(x_0, y_0)$ at x_0.

If the partial derivative of f with respect to x exists at all points of D, then that derivative becomes a new function in D, which we denote by f_x, $\partial f/\partial x$, z_x or $\partial z/\partial x$.

By considering z as a function of y, for fixed x, we obtain in similar fashion a partial derivative $f_y = \partial f/\partial y = z_y = \partial z/\partial y$. This also corresponds to a slope, as in Figure 12-22. To make clear which variable is held constant, one sometimes writes

$$\left(\frac{\partial z}{\partial x}\right)_y \quad \text{for} \quad \frac{\partial z}{\partial x}, \qquad \left(\frac{\partial z}{\partial y}\right)_x \quad \text{for} \quad \frac{\partial z}{\partial y}$$

To find the partial derivatives of a function given by an equation, one can apply the usual rules for functions of one variable, treating all independent variables but one as constants.

EXAMPLE 1. $z = x^3y$. Here $\frac{\partial z}{\partial x} = 3x^2y$, $\frac{\partial z}{\partial y} = x^3$.

EXAMPLE 2. $x^2 + y^2 + z^2 - 1 = 0$. Here z is given implicitly as a function of x and y. We could solve for z, but we can also differentiate implicitly:

$$2x + 2z\frac{\partial z}{\partial x} = 0 \qquad \text{hence} \qquad \frac{\partial z}{\partial x} = -\frac{x}{z}$$

$$2y + 2z\frac{\partial z}{\partial y} = 0 \qquad \text{hence} \qquad \frac{\partial z}{\partial y} = -\frac{y}{z}$$

These relations hold true for each function $z = f(x, y)$ differentiable with respect to x and y and satisfying the given equation.

From Figure 12-22 it is clear that the two partial derivatives $f_x(x_0, y_0)$ and $f_y(x_0, y_0)$ concern only the behavior of $z = f(x, y)$ along the two lines $y = y_0$, $x = x_0$. Thus the partial derivatives may exist even if f is defined only on the two lines. The partial derivatives appear to tell us nothing whatsoever about the behavior of f along an oblique line, such as the dashed line in Figure 12-21. It will be seen that, under further assumptions about f, we can use the partial derivatives to find the rate of change of z along such a line (the directional derivative, see Section 12-11 below).

Tangent Plane. We observed above that, in the plane $y = y_0$, the line with slope $f_x(x_0, y_0)$ is tangent to the section of our surface in the plane. There is a similar statement for the section of our surface in the plane $x = x_0$. The tangent plane to a surface at a point P_0 can be defined as the plane containing the tangent lines at P_0 to *all* curves in the surface passing through P_0. But if there is a tangent plane, then (since it is a plane) it is determined by just *two* lines in the plane. If we use the two tangent lines in the planes $y = y_0$ and $x = x_0$, then we are led to the following *equation for the tangent plane* at P_0:

$$z - z_0 = f_x(x_0, y_0)(x - x_0) + f_y(x_0, y_0)(y - y_0) \qquad (12\text{-}81)$$

Here $z_0 = f(x_0, y_0)$. For the equation (12-81) represents a plane passing through P_0: (x_0, y_0, z_0), and the plane contains the two tangent lines mentioned; for example, when $y = y_0$, the equation reduces to

$$z - z_0 = f_x(x_0, y_0)(x - x_0)$$

the equation of the tangent line to the section of the surface in the plane $y = y_0$. We shall see in Section 12-18 below that, when the partial derivatives of f are continuous, the plane (12-81) also contains the tangent lines at P to all other curves in the surface passing through P_0.

EXAMPLE 3. Let $z = x^2 + y^2$ (Figure 12-8). Then $\partial z/\partial x = 2x$, $\partial z/\partial y = 2y$; for $(x, y) = (3, 1)$, $z = 10$ and $\partial z/\partial x = 6$, $\partial z/\partial y = 2$. Therefore, the tangent plane at (3, 1, 10) is

$$z - 10 = 6(x - 3) + 2(y - 1) \qquad \text{or} \qquad 6x + 2y - z = 10$$

The Gradient Vector. Local Maxima and Minima. From the two partial derivatives of f at a point we can form a vector in V_2 called the *gradient vector* and denoted by grad f or ∇f:

$$\text{grad } f = \nabla f = \frac{\partial z}{\partial x}\mathbf{i} + \frac{\partial z}{\partial y}\mathbf{j}$$

When the partial derivatives exist in an open region, grad f becomes a vector function in that region. The symbol ∇f is pronounced "del f."

EXAMPLE 4. Let $f(x, y) = x^3 + 3xy^2$. Then grad $f = (3x^2 + 3y^2)\mathbf{i} + 6xy\mathbf{j}$.

We say that $f(x, y)$ has a *local maximum* at (x_0, y_0) if, in some neighborhood of (x_0, y_0), $f(x, y) \leqq f(x_0, y_0)$. Similarly, f has a *local minimum* at (x_0, y_0) if, in some neighborhood of (x_0, y_0), $f(x, y) \geqq f(x_0, y_0)$.

EXAMPLE 5. The function $z = x^2 + y^2$ has a local minimum at (0, 0). For $f(x, y) = x^2 + y^2 \geq 0 = f(0, 0)$ for all (x, y).

EXAMPLE 6. The function $z = e^{-x^2-y^2}$ has a local maximum at (0, 0). For, $f(x, y) \leq f(0, 0) = 1$ for all (x, y), since $e^{-u} < 1$ for $u > 0$.

THEOREM 1. *Let $f(x, y)$ be defined in the open region D and have a local maximum or minimum at (x_0, y_0). If f_x and f_y exist at (x_0, y_0), then*

$$f_x(x_0, y_0) = 0 \qquad \textit{and} \qquad f_y(x_0, y_0) = 0$$

That is, at a local maximum or minimum, grad f is **0**.

PROOF. Let f have a local maximum at (x_0, y_0). Then we can choose a neighborhood of (x_0, y_0) in which $f(x, y) \leq f(x_0, y_0)$. Then, as in forming the partial derivative above, we obtain $f(x, y_0)$, a function of x, defined in an interval $(x_0 - \delta, x_0 + \delta)$, in which $f(x, y_0) \leq f(x_0, y_0)$. Hence, the function of one variable $f(x, y_0)$ has a local maximum at x_0, at which its derivative with respect to x exists. But then that derivative must be 0; that is, $f_x(x_0, y_0) = 0$. Similarly, $f_y(x_0, y_0) = 0$. The case of a minimum is treated in the same way.

For Example 5 above, $\nabla f = 2x\mathbf{i} + 2y\mathbf{j}$. The only point at which ∇f is $\mathbf{0}$ is (0, 0). Therefore, this is the only point at which a local maximum or minimum can occur. We saw above that there is a local minimum at the point.

At a local maximum or minimum, the equation of the tangent plane (12-81) (if there is one) becomes

$$z = z_0$$

Thus the *tangent plane is horizontal at each local maximum or minimum.* However, as for functions of one variable, the fact that the tangent plane is horizontal does not ensure that we have a local maximum or minimum.

EXAMPLE 7. $z = xy$. Here, $\nabla f = y\mathbf{i} + x\mathbf{j}$ and a local maximum or minimum can occur only at (0, 0), where $\nabla f = \mathbf{0}$ and the tangent plane is horizontal. However, $f(0, 0) = 0$, $f(a, a) = a^2$, $f(a, -a) = -a^2$. We can make the points (a, a) and $(a, -a)$ as close to (0, 0) as we wish; f is positive at the former, negative at the latter. Hence, f has neither local maximum nor local minimum at (0, 0). The function is graphed in Figure 12-10; the graph is called a "saddle surface." At (0, 0) one has a formation like that at the highest point of a mountain pass.

Remark. The equation of the tangent plane (12-81) can be written in terms of the gradient vector as follows

$$z - z_0 = \nabla f \cdot (\mathbf{r} - \mathbf{r}_0)$$

where $\mathbf{r} = x\mathbf{i} + y\mathbf{j}$, $\mathbf{r}_0 = x_0\mathbf{i} + y_0\mathbf{j}$ and ∇f is evaluated at (x_0, y_0).

Generalizations. Partial derivatives can be defined in analogous fashion for functions of three or more variables. For a function of n variables $z = f(\mathbf{x})$ or $f(x_1, \ldots, x_n)$, one can write f_{x_k} for the partial derivative with respect to x_k, all the other variables being held constant; one can also abbreviate this as f_k. This notation (although less often used) has advantages even for functions of two variables: f_1 for f_x, f_2 for f_y, where $f(x, y)$ is given.

The gradient vector of $f(x_1, \ldots, x_n)$ is the vector of V_n:

$$\nabla f = \text{grad } f = (f_{x_1}, \ldots, f_{x_n}) = f_{x_1}\mathbf{e}_1 + \cdots + f_{x_n}\mathbf{e}_n$$

where $\mathbf{e}_1 = (1, 0, \ldots, 0)$, $\mathbf{e}_2 = (0, 1, \ldots, 0), \ldots$ Thus for $f(x, y, z)$

$$\nabla f = f_x\mathbf{i} + f_y\mathbf{j} + f_z\mathbf{k}$$

Local maxima and minima are defined for $f(x_1, \ldots, x_n)$ as above for $f(x, y)$, and Theorem 1 and its proof remain valid: *at a local maximum of* $f(x_1, \ldots, x_n)$, $\text{grad } f = \mathbf{0}$ (provided all partial derivatives exist at the point). The equation (12-81) has a generalization in terms of a "tangent hyperplane" (see Section 11-13), but we do not discuss this here.

PROBLEMS

1. Find $\partial z/\partial x$ and $\partial z/\partial y$:

(a) $z = 2x + 3y$ (b) $z = 5x - 7y$ (c) $z = x^2 - y^2$

(d) $z = 3x^2 + 4y^2$ (e) $z = 2x^2y$ (f) $z = xy^3$

(g) $z = e^{xy}\cos(x - 2y)$ (h) $z = \dfrac{x - y}{x + y}$ (i) $z = \ln(x^2 - 2xy - y^2)$

(j) $z = \log_x y$ (k) $z = \sqrt[x]{y}$ (l) $xyz + z^2 = 1$

(m) $x^3 + ye^z + z = 0$ (n) $z = x \sin xy$ (o) $\ln z + xye^y = 5$

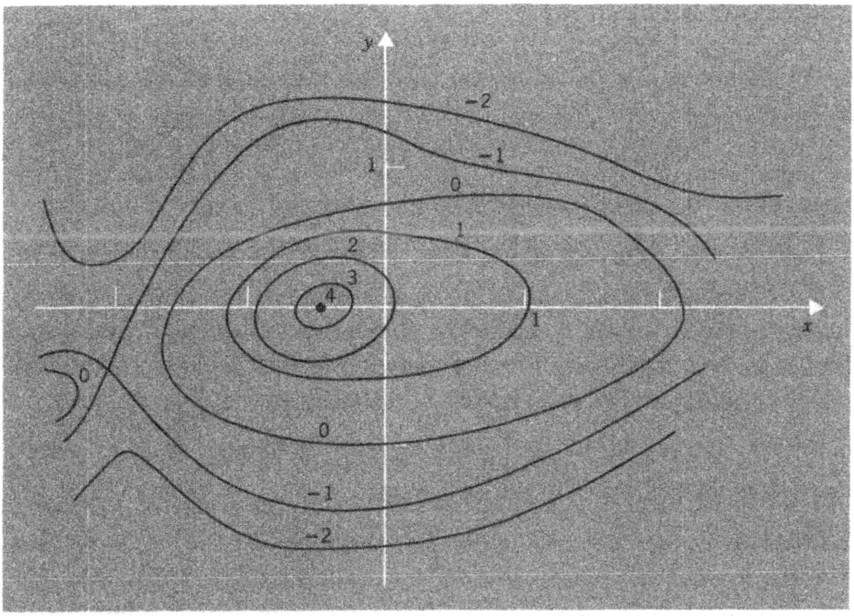

Figure 12-23 Problem 2.

2. From the level curves of f given in Figure 12-23 estimate the following partial derivatives:

   (a) $f_x(1, 0)$ (b) $f_y(1, 0)$ (c) $f_x(0, 1)$

   (d) $f_y(0, 1)$ (e) $f_x(-1, -1)$ (f) $f_y(-1, -1)$

3. Find the tangent plane at the point requested:

   (a) $z = x^2 - y^2$ at $(2, 1, 3)$ (b) $z = 3xy$ at $(1, 2, 6)$

   (c) $z = \ln(x + y)$ at $(2, -1, 0)$ (d) $z = e^x \cos y$ at $(0, \pi, -1)$

4. Find the gradient vector at the point specified:

   (a) $z = 2x^2 + y^2$ at $(1, 1)$ (b) $z = 3x^2 - y^2$ at $(2, 1)$

   (c) $z = \dfrac{x}{y}$ at $(1, 1)$ (d) $z = x^2e^y$ at $(1, 0)$

   (e) $z = \sin(x + 2y)$ at (x, y) (f) $z = \ln(x - 3y)$ at (x, y)

5. Locate all points where grad f is $\mathbf{0}$ and determine whether f has a local maximum or local minimum at each:

   (a) $f = 2x^2 + 4y^2$ (b) $f = x^2 + 3y^2$ (c) $f = \dfrac{1}{1 + x^2 + y^2}$

   (d) $f = 1 - x^2 - y^2$ (e) $f = x^2 - y^2$ (f) $f = 2x^2 - 3y^2$

   (g) $f = x^2 + 2xy + y^2$ (h) $f = (x^2 + y^2)^{1/3}$

6. Find the gradient vector for each function:

   (a) $f(x, y, z) = x^3y^2z$ (b) $f(x, y, z) = x^3y - y^2z^2$

   (c) $f(u, v, w) = \sin(u + 2v - 3w)$ (d) $f(u, v, w) = \ln(u^2 + v^2 - w^2)$

   (e) $f(x_1, \ldots, x_n) = x_1^2 + x_2^2 + \cdots + x_n^2$ (f) $f(x_1, \ldots, x_n) = x_1x_2 \cdots x_n$

12-9 THE DIFFERENTIAL

For a function $y = f(x)$ of one variable, the differential dy was introduced in Section 3-22 as the "linear approximation" to Δy. Thus $dy = m\,\Delta x$, when $\Delta y = m\,\Delta x + \Delta x\, p(\Delta x)$, where $p(\Delta x)$ is continuous at $\Delta x = 0$ and $p(0) = 0$. For example, we might have $\Delta y = m\,\Delta x + \Delta x^2$. We saw that m had to be the derivative at the point considered, so that we could write

$$dy = f'(x)\,\Delta x \qquad \text{or} \qquad dy = f'(x)\,dx$$

For functions of two variables we now seek an analogous linear approximation to the increment. Let $z = f(x, y)$ be given in the open region D and let (x_0, y_0) be a point of D. We now consider the change Δz in z when both x and y change, say by Δx, Δy from x_0, y_0, respectively. Thus we study

$$\Delta z = f(x_0 + \Delta x, y_0 + \Delta y) - f(x_0, y_0) \tag{12-90}$$

(see Figure 12-24). We consider an example:

EXAMPLE 1. Let $z = x^2 + xy^2 = f(x, y)$ and let $x_0 = 2$, $y_0 = 3$. Then $f(x_0, y_0) = f(2, 3) = 22$ and

$$\begin{aligned} f(x_0 + \Delta x, y_0 + \Delta y) &= f(2 + \Delta x, 3 + \Delta y) = (2 + \Delta x)^2 + (2 + \Delta x)(3 + \Delta y)^2 \\ &= 22 + 13\Delta x + 12\Delta y + (\Delta x)^2 + 6\Delta x\,\Delta y + 2(\Delta y)^2 + \Delta x(\Delta y)^2 \end{aligned}$$

Accordingly,

$$\begin{aligned} \Delta z &= f(2 + \Delta x, 3 + \Delta y) - f(2, 3) \\ &= 13\Delta x + 12\Delta y + (\Delta x)^2 + 6\Delta x\,\Delta y + 2(\Delta y)^2 + \Delta x(\Delta y)^2 \end{aligned}$$

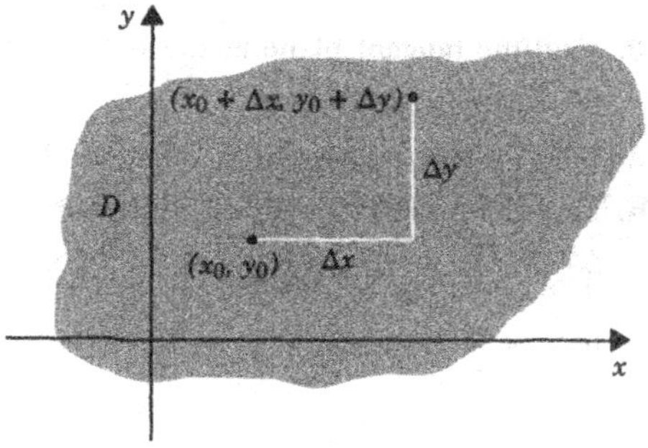

Figure 12-24 Increments for $f(x, y)$.

Thus in our example Δz equals a linear function $13\Delta x + 12\Delta y$ plus terms of higher degree. When Δx and Δy are very small, the linear terms dominate. For example, for $\Delta x = 0.1$, $\Delta y = 0.1$,

$$\Delta z = 13 \times 0.1 + 12 \times 0.1 + 0.01 + 6 \times 0.01 + 2 \times 0.01 + 0.001$$

The sum of the linear terms is 2.5. The sum of the others is 0.09. Thus we can write

$$\Delta z \sim 13\Delta x + 12\Delta y$$

where the approximation improves as Δx, Δy approach 0. We note that each of the higher degree terms can be written as Δx or Δy times an expression reducing to 0 for $\Delta x = 0$, $\Delta y = 0$. All these observations suggest our general definition:

Definition. The function $z = f(x, y)$ has a *differential* $dz = a\,\Delta x + b\,\Delta y = df$ at a particular point (x_0, y_0) if for $(\Delta x, \Delta y)$ in a neighborhood of $(0, 0)$,

$$\begin{aligned}\Delta z &= f(x_0 + \Delta x, y_0 + \Delta y) - f(x_0, y_0) \\ &= a\,\Delta x + b\,\Delta y + \Delta x\, p_1(\Delta x, \Delta y) + \Delta y\, p_2(\Delta x, \Delta y)\end{aligned} \tag{12-91}$$

where a, b are constants and $p_1(\Delta x, \Delta y)$, $p_2(\Delta x, \Delta y)$ are defined in a neighborhood of $(\Delta x, \Delta y) = (0, 0)$ and are continuous at $(0, 0)$ with value 0 at $(0, 0)$.

In our example, we can write

$$\Delta z = 13\Delta x + 12\Delta y + \Delta x(\Delta x + 6\Delta y) + \Delta y(2\Delta y + \Delta x\,\Delta y)$$

so that $a = 13$, $b = 12$, $p_1(\Delta x, \Delta y) = \Delta x + 6\Delta y$, $p_2(\Delta x, \Delta y) = 2\Delta x + \Delta x\,\Delta y$; by grouping the higher degree terms in other ways, we obtain other permissible choices of p_1 and p_2.

THEOREM 2. *If $z = f(x, y)$ has a differential $dz = a\,\Delta x + b\,\Delta y$ at (x_0, y_0), then $f_x(x_0, y_0)$ and $f_y(x_0, y_0)$ exist and*

$$f_x(x_0, y_0) = a, \qquad f_y(x_0, y_0) = b$$

so that

$$dz = f_x(x_0, y_0)\,\Delta x + f_y(x_0, y_0)\,\Delta y$$

PROOF. The definition of $\partial z/\partial x$ in (12-80) can be written as the limit as $\Delta x \to 0$ of $\Delta z/\Delta x$, for $\Delta y = 0$. For $\Delta y = 0$, (12-91) gives

$$\Delta z = a\,\Delta x + \Delta x\, p_1(\Delta x, 0)$$

$$\frac{\Delta z}{\Delta x} = a + p_1(\Delta x, 0) \qquad (\Delta x \neq 0)$$

If now $\Delta x \to 0$, then $p_1(\Delta x, 0) \to p_1(0, 0) = 0$. Hence

$$\lim_{\Delta x \to 0} \frac{\Delta z}{\Delta x} = \lim_{\Delta x \to 0} \frac{f(x_0 + \Delta x, y_0) - f(x_0, y_0)}{\Delta x} = a$$

Thus $f_x(x_0, y_0) = a$ and, similarly, $f_y(x_0, y_0) = b$.

Remark. The theorem shows that the differential, if it exists, is uniquely determined. If $dz = a\,\Delta x + b\,\Delta y$, then a must be $f_x(x_0, y_0)$ and b must be $f_y(x_0, y_0)$.

When f has a differential at (x_0, y_0), we say that f is *differentiable* at (x_0, y_0). When f has a differential at each point of the open region D, we say that "f is differentiable in D." In this case, we can write

$$df = \frac{\partial f}{\partial x}\Delta x + \frac{\partial f}{\partial y}\Delta y \qquad \text{or} \qquad dz = \frac{\partial z}{\partial x}\Delta x + \frac{\partial z}{\partial y}\Delta y$$

where $\partial f/\partial x$ and $\partial f/\partial y$ are functions of x and y. Other notations are used for the differential:

$$df = \frac{\partial f}{\partial x} h + \frac{\partial f}{\partial y} k, \qquad df = \frac{\partial f}{\partial x} dx + \frac{\partial f}{\partial y} dy$$

$$df = \frac{\partial f}{\partial x}(x - x_0) + \frac{\partial f}{\partial y}(y - y_0) \qquad (\text{at } (x_0, y_0))$$

Thus Δx, Δy can be replaced by other symbols. The use of dx, dy will be explained below.

We can also write the differential in vector notation:

$$dz = \nabla f \cdot d\mathbf{r}$$

where $d\mathbf{r} = dx\,\mathbf{i} + dy\,\mathbf{j}$.

When f has a differential at a point (x_0, y_0), we have information about the behavior of f in all directions from (x_0, y_0). We can write

$$f(x, y) \sim f(x_0, y_0) + a(x - x_0) + b(y - y_0) \tag{12-92}$$

Thus f is approximated, near (x_0, y_0), by a linear function, whose graph is a plane in space. We saw in Section 12-8 that this plane is the *tangent plane* to the surface $z = f(x, y)$ at (x_0, y_0, z_0). The relation (12-92) can be used to calculate values of f approximately.

EXAMPLE 2. Evaluate $(2.01)^{1.02}$. We let $z = x^y$. Then

$$dz = \frac{\partial z}{\partial x} dx + \frac{\partial z}{\partial y} dy = yx^{y-1}\,dx + x^y \ln x\, dy$$

We take $x_0 = 2$, $y_0 = 1$, $dx = 0.01$, $dy = 0.02$, so that

$$dz = dx + 2 \ln 2\, dy = 0.01 + 0.04 \ln 2 = 0.0377$$

and

$$z \sim 2 + 0.0377 = 2.0377$$

The exact value is found from tables to be 2.0383.

THEOREM 3. *If $z = f(x, y)$ has a differential at (x_0, y_0), then f is continuous at (x_0, y_0).*

PROOF. We can write (12-91) as follows:

$$f(x, y) - f(x_0, y_0) = a(x - x_0) + b(y - y_0) + (x - x_0)p_1(x - x_0, y - y_0) + (y - y_0)p_2(x - x_0, y - y_0)$$

Since p_1 and p_2 are continuous at (0, 0), it follows from Theorem C (Section 12-7) that the whole right side of the equation is continuous at (x_0, y_0). Since $f(x, y)$ equals $f(x_0, y_0)$ plus the right side, f must be continuous at (x_0, y_0).

THEOREM 4. *Let $z = f(x, y)$ be defined in the open region D and let $\partial z/\partial x$, $\partial z/\partial y$ be continuous in D. Then f is differentiable in D.*

‡PROOF. Let $f_x(x, y) = \phi(x, y)$, $f_y(x, y) = \psi(x, y)$, so that ϕ, ψ are continuous

in D. Then we can write

$$\begin{aligned} f(x_0 + h, y_0 + k) - f(x_0, y_0) \\ &= f(x_0 + h, y_0 + k) - f(x_0, y_0 + k) + f(x_0, y_0 + k) - f(x_0, y_0) \\ &= \int_{x_0}^{x_0+h} f_x(x, y_0 + k)\, dx + \int_{y_0}^{y_0+k} f_y(x_0, y)\, dy \\ &= \int_{x_0}^{x_0+h} \phi(x, y_0 + k)\, dx + \int_{y_0}^{y_0+k} \psi(x_0, y)\, dy \end{aligned} \tag{12-93}$$

Thus we have obtained the total increment in f as the sum of an increment due to a change in x and an increment due to a change in y (see Figure 12-25).

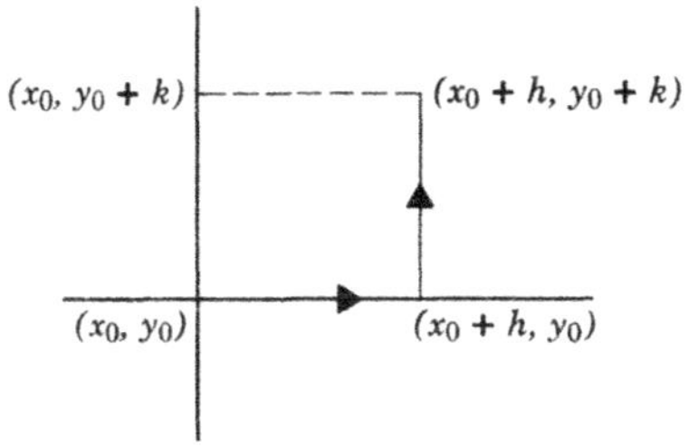

Figure 12-25 Proof of Theorem 4.

We evaluate each of these increments by integrating the corresponding partial derivative. Now for $h \neq 0$ the first term of the last expression is

$$\int_{x_0}^{x_0+h} \phi(x, y_0 + k)\, dx = h\left[\frac{1}{h}\int_{x_0}^{x_0+h} \phi(x, y_0 + k)\, dx\right] = hq_1(h, k)$$

where $q_1(h, k)$ is the *average* of ϕ on the segment from $(x_0, y_0 + k)$ to $(x_0 + h, y_0 + k)$. Now ϕ is continuous in D. Hence, for h and k sufficiently close to 0 we can ensure that the values of ϕ remain as close to $\phi(x_0, y_0)$ as desired. Hence, the same applies to the average, $q_1(h, k)$. We have not defined $q_1(h, k)$ for $h = 0$; we assign the value $\phi(x_0, y_0)$ for such points, and can then say that $q_1(h, k)$ is continuous at $(0, 0)$ with $q_1(0, 0) = \phi(x_0, y_0) = f_x(x_0, y_0)$. Hence, the first term to the right of the equals sign in (12-93) equals $hq_1(h, k)$ for all (h, k) in a neighborhood of $(0, 0)$. Similarly, the second term in (12-93) can be written as $kq_2(h, k)$, where $q_2(h, k)$ is continuous at $(0, 0)$ and $q_2(0, 0) = \psi(x_0, y_0) = f_y(x_0, y_0)$. Accordingly,

$$\begin{aligned} f(x_0 + h, y_0 + k) - f(x_0, y_0) &= hq_1(h, k) + kq_2(h, k) \\ &= h[f_x(x_0, y_0) + p_1(h, k)] + k\,[f_y(x_0, y_0) + p_2(h, k)] \end{aligned}$$

where $p_1(h, k)$ and $p_2(h, k)$ are continuous at $(0, 0)$ and have value 0 at $(0, 0)$. Thus, finally,

$$\begin{aligned} f(x_0 + h, y_0 + k) - f(x_0, y_0) \\ &= hf_x(x_0, y_0) + kf_y(x_0, y_0) + hp_1(h, k) + kp_2(h, k) \end{aligned}$$

and f has a differential at (x_0, y_0).‡

Remark. Mere existence of the partial derivatives (without continuity) does not insure differentiability of f; in fact, it does not even imply continuity of f! (See Problem 9 below.)

Theorem 4 allows us to conclude that many familiar functions are differentiable (except at obvious points). For example, the polynomial $z = 2x^2y - 3x^5y^7$ is differentiable for all (x, y), since $z_x = 4xy - 15x^4y^7$, $z_y = 2x^2 - 21x^5y^6$, and these derivatives are continuous everywhere. Thus $dz = (4xy - 15x^4y^7)\,dx + (2x^2 - 21x^5y^6)\,dy$. The following are further examples of differentials:

$$d(xy) = y\,dx + x\,dy \quad (\text{all } (x, y)), \qquad d(e^x \cos y) = e^x \cos y\,dx - e^x \sin y\,dy$$

$$d\left(\frac{y}{x}\right) = \frac{-y\,dx + x\,dy}{x^2} \quad (x \neq 0), \qquad d\left(\mathrm{Tan}^{-1}\frac{y}{x}\right) = \frac{-y\,dx + x\,dy}{x^2 + y^2} \quad (x \neq 0)$$

$$d(xy^3 \ln x) = y^3(1 + \ln x)\,dx + 3xy^2 \ln x\,dy \quad (x > 0)$$

We note that the differentials here given are the same as those obtained in elementary calculus, with x and y regarded as functions of t, for example. The reason for this is given in the next section.

THEOREM 5. *Let f be differentiable in the open region D and let $df \equiv 0$ in D. Then f is identically constant in D.*

PROOF. By Theorem 1, we must have $f_x \equiv 0$, $f_y \equiv 0$ in D. Hence, by one-variable calculus, f is constant along each line segment in D parallel to the x-axis or to the y-axis. Hence, in each circular neighborhood D_1 in D, f is constant. For if (x_1, y_1) and (x_2, y_2) are in D_1, then these two points can be joined by a broken line in D_1 on which f is constant: from (x_1, y_1) to (x_2, y_1) (parallel to the x-axis) and then from (x_2, y_1) to (x_2, y_2) parallel to the y-axis.

Now let P_1, P_2 be any two points of D. Since D is pathwise connected, we can join them by a path $\mathbf{r} = \overrightarrow{OP} = \mathbf{r}(t)$, $a \leq t \leq b$, with $\mathbf{r}(a) = \overrightarrow{OP_1}$, $\mathbf{r}(b) = \overrightarrow{OP_2}$. For each t_0, $a \leq t_0 \leq b$, the point P_0 with $\mathbf{r}(t_0) = \overrightarrow{OP_0}$ lies in D and, hence, lies in a circular neighborhood D_0 contained in D. As above, f is constant in D_0. By continuity of $\mathbf{r}(t)$, there is a $\delta_0 > 0$ such that $\mathbf{r}(t)$ lies in the neighborhood D_0 for $t_0 - \delta < t < t_0 + \delta$ (and t in $[a, b]$). It follows that $g(t) = f(\mathbf{r}(t)) = f(x(t), y(t))$ is constant for $t_0 - \delta < t < t_0 + \delta$. Hence $g'(t) \equiv 0$ in $[a, b]$, and thus $g(t) \equiv$ const. Accordingly, $f(\mathbf{r}(a)) = f(\mathbf{r}(b))$ or $f(P_1) = f(P_2)$. Thus f is constant in D (see Figure 12-26).

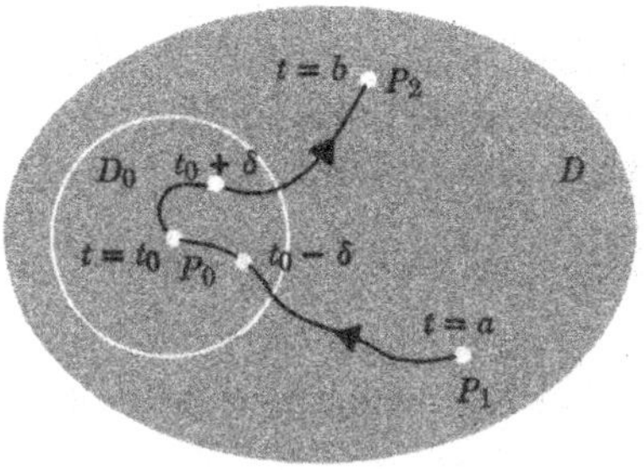

Figure 12-26 Proof of Theorem 5.

The concept of the differential extends at once to functions of more than two variables. For $w = f(x, y, z)$, for example,

$$dw = f_x\,dx + f_y\,dy + f_z\,dz$$

For "small" increments dx, dy and dz, dw approximates the increment $\Delta w = f(x + dx, y + dy, z + dz) - f(x, y, z)$. The analogues of Theorems 2, 3, 4, and 5 continue to hold true.

PROBLEMS

1. Find the differential of the given function:
   (a) $z = x^2 + 2y$ (b) $z = 3xy - y^2$
   (c) $z = xye^x$ (d) $z = xy^2e^{\cos x}$
   (e) $z = \dfrac{x - y}{x + y}$ (f) $z = \dfrac{x^2}{x - y^2}$
   (g) $w = x^2 + y^2 + z^2$ (h) $w = xyz$
   (i) $z = 3uv^2w^3$ (j) $z = e^u(v^2 - w^2)$
   (k) $z = \sin(uvw)$ (l) $z = e^u \ln(vw)$
2. A certain function $z = f(x, y)$ is known to have the following properties: $f(0, 0) = 1$, $f(1, 1) = 3$, $f(2, 3) = 5$, $f_x(0, 0) = 2$, $f_y(0, 0) = -1$, $f_x(1, 1) = 5$, $f_y(1, 1) = -2$, $f_x(2, 3) = 2$, $f_y(2, 3) = 0$. Evaluate approximately:
   (a) $f(0.1, 0)$ (b) $f(0, -0.2)$ (c) $f(1.8, 3)$
   (d) $f(2, 2.8)$ (e) $f(1, 1.1)$ (f) $f(0.7, 1)$
3. Find Δz at the point indicated and express Δz in the form (12-91):
   (a) $z = xy$ at $(2, 1)$ (b) $z = x^2y$ at $(2, 3)$
   ‡(c) $z = x/y$ at $(3, 2)$ ‡(d) $z = \dfrac{1}{x + y}$ at $(1, 1)$
4. Evaluate approximately:
   (a) $\sqrt{4.02} + \sqrt[3]{8.03}$ (b) $(1.04)^3(1.99)^5$
   (c) $\sin[1.99 \ln(1.03)]$ (d) $(\cos 0.3\pi)^{1.1}$
   (e) $(1.01)(9.98)(8.03)$ (f) $3.99^{1.01/2.03}$
5. (a) Interpret $z = xy$, for positive x and y, as the area of a rectangle and show by diagrams the way in which $dz = y\,\Delta x + x\,\Delta y$ approximates the area of the rectangle of sides $x + \Delta x$, $y + \Delta y$ minus the area of the rectangle of sides x, y.
   (b) Extend the idea of part (a) to $w = xyz$ as volume of a rectangular solid of sides x, y, z.
6. (a) Prove: if $F(x, y)$ and $G(x, y)$ have continuous partial derivatives in an open region D and $F_x(x, y) = G_x(x, y)$, $F_y(x, y) = G_y(x, y)$ in D, then $F(x, y) = G(x, y) + C$, where C is a constant.
   (b) Find all functions $z = F(x, y)$ such that $\partial z/\partial x = 2xy$, $\partial z/\partial y = x^2$ for all (x, y).
7. Let $f(x, y) = \sqrt{x^2 + 2xy + y^2}/(x + y)$. Show that $f_x \equiv 0$, $f_y \equiv 0$, but $f(1, 1) = 1$, $f(-1, -1) = -1$. Does this contradict Theorem 5?

8. On the basis of Theorem 4, show that f is differentiable where stated.
   (a) $f(x, y) = x^2y^5$, all (x, y) (b) $f(x, y) = \ln(x^2 + y^2)$, $(x, y) \neq (0, 0)$
   (c) $f(x, y) = \dfrac{e^{xy}}{x - y}$, $x \neq y$ (d) $f(x, y) = \dfrac{xy}{x^2 + y^2}$, $(x, y) \neq (0, 0)$
9. Let $z = f(x, y) = (xy)/(x^2 + y^2)$ for $(x, y) \neq (0, 0)$, $f(0, 0) = 0$.
   (a) Show that $\partial z/\partial x$ and $\partial z/\partial y$ exist for all (x, y).
   (b) Show that f is discontinuous at $(0, 0)$.

12-10 CHAIN RULES

For functions of one variable the chain rule

$$\frac{dy}{dx} = \frac{dy}{du}\frac{du}{dx}$$

is of much value. This rule has analogues for functions of several variables. For $z = f(x, y)$, where $x = x(t)$ and $y = y(t)$, the rule reads

$$\frac{dz}{dt} = \frac{\partial z}{\partial x}\frac{dx}{dt} + \frac{\partial z}{\partial y}\frac{dy}{dt} \tag{12-100}$$

or, in vector language,

$$\frac{dz}{dt} = \nabla f \cdot \frac{d\mathbf{r}}{dt} \tag{12-100$'$}$$

We illustrate the rule and then proceed to prove it:

EXAMPLE 1. Let $z = xy$, where x and y are functions of t. Then

$$\frac{dz}{dt} = y\frac{dx}{dt} + x\frac{dy}{dt}$$

This is just the rule for differentiating products of functions of one variable.

EXAMPLE 2. Let $z = \sin(2x + 3y)$, where x and y are functions of t. Then

$$\frac{dz}{dt} = 2\cos(2x + 3y)\frac{dx}{dt} + 3\cos(2x + 3y)\frac{dy}{dt} = (\cos(2x + 3y))\left(2\frac{dx}{dt} + 3\frac{dy}{dt}\right)$$

This result also follows from the rules of one-variable calculus.

THEOREM 6 (*A chain rule*). *Let $x(t)$ and $y(t)$ be functions which are differentiable at a particular t; let $z = f(x, y)$ be differentiable at the corresponding point (x, y). Then $z = g(t) = f[x(t), y(t)]$ is differentiable at the given t and*

$$\frac{dz}{dt} = \frac{\partial z}{\partial x}\frac{dx}{dt} + \frac{\partial z}{\partial y}\frac{dy}{dt} = \nabla f \cdot \frac{d\mathbf{r}}{dt}$$

$$dg = \frac{\partial z}{\partial x}dx + \frac{\partial z}{\partial y}dy$$

where $\partial z/\partial x$, $\partial z/\partial y$ are evaluated at $(x, y) = (x(t), y(t))$ and, in the last equation, $dx = x'(t)\,dt$, $dy = y'(t)\,dt$.

PROOF. Since f is differentiable at (x, y), at this point

$$\Delta z = \frac{\partial z}{\partial x}\Delta x + \frac{\partial z}{\partial y}\Delta y + \Delta x\, p_1 + \Delta y\, p_2$$

where p_1, p_2 are functions of Δx, Δy which are continuous for $\Delta x = 0$, $\Delta y = 0$ and have value 0 at this point. Hence, for $\Delta t \neq 0$

$$\frac{\Delta z}{\Delta t} = \frac{\partial z}{\partial x}\frac{\Delta x}{\Delta t} + \frac{\partial z}{\partial y}\frac{\Delta y}{\Delta t} + \frac{\Delta x}{\Delta t}p_1 + \frac{\Delta y}{\Delta t}p_2 \tag{12-101}$$

Here, we can consider $\Delta x = x(t + \Delta t) - x(t)$ to be expressed in terms of Δt, and Δy to be expressed in terms of Δt. Thus p_1 and p_2 become functions of Δt, continuous for $\Delta t = 0$ and equal to 0 for this value. Hence, if we let Δt approach 0 in (12-101), we obtain

$$g'(t) = \frac{dz}{dt} = \frac{\partial z}{\partial x}\frac{dx}{dt} + \frac{\partial z}{\partial y}\frac{dy}{dt} + \frac{dx}{dt}\cdot 0 + \frac{dy}{dt}\cdot 0$$

and (12-100) follows. The expression for dg given in the theorem follows on multiplication of both sides of (12-100) by dt.

The formula for the differential of $z = g(t) = f(x(t), y(t))$ can be written

$$dz = \frac{\partial z}{\partial x}dx + \frac{\partial z}{\partial y}dy \tag{12-102}$$

This is identical with the usual expression for $dz = df(x, y)$, as in Theorem 2. However, for $df(x, y)$ we consider dx, dy as arbitrary changes in x and in y: $dx = \Delta x$, $dy = \Delta y$. In (12-102) we are thinking of dx and dy as in Theorem 6: $dx = x'(t)\,dt$, $dy = y'(t)\,dt$; thus dx and dy are related. Since the formula (12-102) is the same in both cases, we conclude: the relationship between dz, dx, dy is the same for $z = f(x, y)$ and $dx = \Delta x$, $dy = \Delta y$ unrestricted, as for $z = f(x(t), y(t))$, and dz, dx, dy all expressed in terms of dt. It is for this reason that Δx, Δy can be replaced by dx, dy in the expression for dz.

A similar result holds true for differentials generally, as one can verify for the different kinds of relations: if an equation in differentials is correct when one variable is considered a function of the others, so that the differentials of the others are arbitrary increments, then the equation remains correct when all variables are expressed as functions of one or more new variables, and all differentials refer to these functions of the new variables. Theorem 7 below is a second illustration of this principle (Equation (12-105)).

EXAMPLE 3. Let $z = f(x(t), y(t))$, where $x = \cos t$, $y = \sin t$. Then

$$\frac{dz}{dt} = -f_x \sin t + f_y \cos t$$

THEOREM 7 (*Another chain rule*). *Let $z = F(u, v)$ be differentiable at*

(u_0, v_0). Let $u = f(x, y)$, $v = g(x, y)$ be differentiable at (x_0, y_0), with $f(x_0, y_0) = u_0$, $g(x_0, y_0) = v_0$. Then

$$z = G(x, y) = F[f(x, y), g(x, y)]$$

is differentiable at (x_0, y_0) and

$$\frac{\partial z}{\partial x} = \frac{\partial G}{\partial x} = \frac{\partial z}{\partial u}\frac{\partial u}{\partial x} + \frac{\partial z}{\partial v}\frac{\partial v}{\partial x} \tag{12-103}$$

$$\frac{\partial z}{\partial y} = \frac{\partial G}{\partial y} = \frac{\partial z}{\partial u}\frac{\partial u}{\partial y} + \frac{\partial z}{\partial v}\frac{\partial v}{\partial y} \tag{12-104}$$

$$dz = dG = \frac{\partial z}{\partial u} du + \frac{\partial z}{\partial v} dv \tag{12-105}$$

where the partial derivatives with respect to u, v are evaluated at (u_0, v_0), those with respect to (x, y) at (x_0, y_0) and du, dv are the differentials of f and g at (x_0, y_0).

PROOF. If we fix y at y_0, then u and v become functions of x alone: $u = f(x, y_0)$, $v = g(x, y_0)$, and these functions of x are differentiable at x_0, with derivatives $f_x(x_0, y_0)$, $g_x(x_0, y_0)$ at x_0. Hence

$$G(x, y_0) = F[f(x, y_0), g(x, y_0)]$$

is composed as is the function g in Theorem 6. Therefore, by that theorem

$$\frac{\partial G}{\partial x} = \frac{\partial F}{\partial u}\frac{df(x, y_0)}{dx} + \frac{\partial F}{\partial v}\frac{df(x_0, y)}{dy} = \frac{\partial z}{\partial u}\frac{\partial u}{\partial x} + \frac{\partial z}{\partial v}\frac{\partial v}{\partial x}$$

Thus (12-103) is proved, and (12-104) is proved in the same way. Assuming that G has a differential at (x_0, y_0), we have at that point

$$\begin{aligned} dG &= \frac{\partial G}{\partial x}\Delta x + \frac{\partial G}{\partial y}\Delta y \\ &= \left(\frac{\partial z}{\partial u}\frac{\partial u}{\partial x} + \frac{\partial z}{\partial v}\frac{\partial v}{\partial x}\right)\Delta x + \left(\frac{\partial z}{\partial u}\frac{\partial u}{\partial y} + \frac{\partial z}{\partial v}\frac{\partial v}{\partial y}\right)\Delta y \\ &= \frac{\partial z}{\partial u}\left(\frac{\partial u}{\partial x}\Delta x + \frac{\partial u}{\partial y}\Delta y\right) + \frac{\partial z}{\partial v}\left(\frac{\partial v}{\partial x}\Delta x + \frac{\partial v}{\partial y}\Delta y\right) \\ &= \frac{\partial z}{\partial u} du + \frac{\partial z}{\partial v} dv \end{aligned}$$

Thus (12-105) follows. The proof that G has a differential is postponed to Section 12-13 (Theorem 9).

Note. In (12-103) $\partial z/\partial x$ is $(\partial z/\partial x)_y$, $\partial z/\partial u$ is $(\partial z/\partial u)_v$, $\partial u/\partial x$ is $(\partial u/\partial x)_y$ and so on.

The two chain rules given can be generalized to composition of functions of two and three variables, three and three, four and two, and so on. The general case is most easily described with the aid of matrices; see Section 12-13 below. For $z = F(u, v, w, \ldots)$, where u, v, $\ldots$ are functions of x,

$y, \ldots$ we find that

$$\frac{\partial z}{\partial x} = F_u \frac{\partial u}{\partial x} + F_v \frac{\partial v}{\partial x} + \cdots, \qquad \frac{\partial z}{\partial y} = F_u \frac{\partial u}{\partial y} + F_v \frac{\partial v}{\partial y} + \cdots, \quad \cdots$$

EXAMPLE 4. Let $z = u^2 + v^2$, where u and v are functions of x and y. Then

$$\frac{\partial z}{\partial x} = 2u \frac{\partial u}{\partial x} + 2v \frac{\partial v}{\partial x}, \qquad \frac{\partial z}{\partial y} = 2u \frac{\partial u}{\partial y} + 2v \frac{\partial v}{\partial y}$$

$$dz = 2u\,du + 2v\,dv$$

(where the differentials can be interpreted as referring to z, u and v as functions of x and y).

EXAMPLE 5. Let $z = (e^x + y)^5 + \cos v$, where v is a function of x and y. We can write $z = u^5 + \cos v$, where $u = e^x + y$ and u, v are functions of x and y [the function $v(x, y)$ being unspecified]. Hence

$$\frac{\partial z}{\partial x} = 5u^4 \frac{\partial u}{\partial x} - \sin v \frac{\partial v}{\partial x} = 5(e^x + y)^4 e^x - \sin v \frac{\partial v}{\partial x}$$

$$\frac{\partial z}{\partial y} = 5u^4 \frac{\partial u}{\partial y} - \sin v \frac{\partial v}{\partial y} = 5(e^x + y)^4 - \sin v \frac{\partial v}{\partial y}$$

EXAMPLE 6. Let $z = f(u, v)$, where $u = x^2 - y^2$, $v = 2xy$. Then

$$\frac{\partial z}{\partial x} = f_u \frac{\partial u}{\partial x} + f_v \frac{\partial v}{\partial x} = 2xf_u + 2yf_v$$

$$\frac{\partial z}{\partial y} = f_u \frac{\partial u}{\partial y} + f_v \frac{\partial v}{\partial y} = -2yf_u + 2xf_v$$

EXAMPLE 7. Let $s = F(u, v, w) = u^2 + v^2 + w^2$, where u, v, and w are functions of x and y. We find, for example, that

$$\frac{\partial s}{\partial x} = F_u \frac{\partial u}{\partial x} + F_v \frac{\partial v}{\partial x} + F_w \frac{\partial w}{\partial x} = 2u \frac{\partial u}{\partial x} + 2v \frac{\partial v}{\partial x} + 2w \frac{\partial w}{\partial x}$$

EXAMPLE 8. Let $w = F(x, y, z, t)$, where x, y, z are functions of t. Find dw/dt. Here, we can write $w = F(x, y, z, u)$, where x, y, z and u are all functions of t, and we happen to know $u(t) = t$. Hence

$$\frac{dw}{dt} = F_x \frac{dx}{dt} + F_y \frac{dy}{dt} + F_z \frac{dz}{dt} + F_u \frac{du}{dt} = F_x \frac{dx}{dt} + F_y \frac{dy}{dt} + F_z \frac{dz}{dt} + F_t$$

THEOREM 8. *Let f and g be functions of two variables having differentials at (x_0, y_0). Then $f + g$, $f - g$, $f \cdot g$, f/g all have differentials at (x_0, y_0) and*

$$d(f + g) = df + dg, \qquad d(f - g) = df - dg$$

$$d(fg) = f\,dg + g\,df, \qquad d\left(\frac{f}{g}\right) = \frac{g\,df - f\,dg}{g^2}$$

where on the right f and g are evaluated at (x_0, y_0) and for $d(f/g)$ it is assumed that $g(x_0, y_0) \neq 0$.

PROOF. Let $z = F(u, v) = u + v$, $u = f(x, y)$, $v = g(x, y)$. Then Theorem 7 is applicable at (x_0, y_0). Since $\partial z/\partial u = 1$, $\partial z/\partial v = 1$, we have by (12-105)

$$dz = d[f(x, y) + g(x, y)] = 1 \cdot du + 1 \cdot dv = df + dg$$

The other rules follow in the same way with $F = u - v$, $u \cdot v$, u/v ($v \neq 0$), respectively.

This result also extends to functions of any number of variables.

PROBLEMS

1. Let $x(t)$, $y(t)$, $u(t)$, $v(t)$, be differentiable for $a \leq t \leq b$. Apply a chain rule to find dz/dt for each of the following cases:

   (a) $z = 3x^2 - y^2$ (b) $z = x^2y^3$ (c) $z = xyu^2$

   (d) $z = xu - yv$ (e) $z = \sin(xu + ty)$ (f) $z = e^{xt}\cos(u + v)$

2. Let $z = f(x, y)$ and $x = x(t)$, $y = y(t)$ as given. Evaluate dz/dt under appropriate hypotheses:

   (a) $x = t^2 - 1$, $y = t^3 + t$ (b) $x = e^t + e^{-2t}$, $y = 2e^t - e^{-2t}$

   (c) $x^2 + xt + t^2 = 1$, $y^2 - yt + t^2 = 1$ (d) $t = x + e^x$, $t = y + \sin y$

3. For the following exercises make appropriate differentiability assumptions.

   (a) Let $z = f(u, v)$, $u = 3x^2y - y^3$, $v = x^3 - 3x^2y$. Find $\partial z/\partial x$, $\partial z/\partial y$.

   (b) Let $u = \phi(x, y)$, $x = z^2 + w^2$, $y = zw$. Find $\partial u/\partial z$, $\partial u/\partial w$.

   (c) Let $z = f(u, v, w)$, $u = xy^2$, $v = x^2y$, $w = x^2y^2$. Find $\partial z/\partial x$, $\partial z/\partial y$.

   (d) Let $w = f(x, y, z)$, $x = 2uv - t^2$, $y = tu + v^2$, $z = uvt$. Find $\partial w/\partial u$, $\partial w/\partial v$, $\partial w/\partial t$.

4. Make appropriate differentiability assumptions.

   (a) If $z = f(x - y)$ [that is, $z = f(u)$, $u = x - y$], show that $\partial z/\partial x + \partial z/\partial y = 0$.

   (b) If $z = f(2x - 3y)$, show that $3(\partial z/\partial x) + 2(\partial z/\partial y) = 0$.

   (c) If $z = f(x^2 - y^2)$, show that $2y(\partial z/\partial x) + 2x(\partial z/\partial y) = 0$.

   (d) If $z = f(xy)$, show that $xz_x - yz_y = 0$.

5. Let $z = F(u, v)$, $u = 2x - 3y$, $v = x + 2y$, so that z can also be expressed in x and y, and let F be differentiable for all u, v.

   (a) If $z_x + z_y \equiv 0$, show that $z_u - 3z_v \equiv 0$.

   (b) If $z_u - z_v \equiv 0$, show that $z_x + 3z_y \equiv 0$.

6. A function $z = f(x, y)$ becomes a function of r, θ when one expresses x and y in terms of polar coordinates r, θ. Assuming needed differentiability show that

$$z_r = z_x \cos\theta + z_y \sin\theta, \qquad z_\theta = -z_x\, r\sin\theta + z_y\, r\cos\theta$$

$$z_x = z_r\cos\theta - \frac{1}{r} z_\theta \sin\theta, \qquad z_y = z_r \sin\theta + \frac{1}{r} z_\theta \cos\theta$$

7. Let the hypotheses of Theorem 7 be satisfied and let the partial derivatives all be evaluated at (u_0, v_0) or (x_0, y_0) as described.

(a) If $z_u = 1$, $z_v = 3$, $u_x = 5$, $u_y = -1$, $v_x = 2$, $v_y = 7$, find z_x and z_y. Also find approximately $G(x_0 + 0.3, y_0 + 0.1) - G(x_0, y_0)$.

(b) If $z_x = 3$, $z_y = -2$, $u_x = 2$, $u_y = 3$, $v_x = -1$, $v_y = 4$, find z_u and z_v. Also find approximately $F(u_0 + 0.1, v_0 - 0.2) - F(u_0, v_0)$.

8. Let $z = f(x, y)$ be differentiable in an open region D including the circle $x^2 + y^2 = 1$. Show that, if f has its maximum value on the circle at the point (x_0, y_0), then $yf_x - xf_y = 0$ at this point. [*Hint.* Use the parametric representation $x = \cos t$, $y = \sin t$.]

9. Find the maximum value of f on the circle $x^2 + y^2 = 1$ (see Problem 8):

(a) $f(x, y) = x + y$ (b) $f(x, y) = x^3y^2$

10. Let $z = z(x, y)$ be a differentiable function in a given open region D.

(a) If $xyz + z^3 = 1$, find expressions for z_x and z_y in terms of x, y, and z.

(b) If $yz^3 - xz + yz = 1$, find expressions for z_x and z_y in terms of x, y, and z.

11. Let $u(x, y)$ and $v(x, y)$ be differentiable functions in an open region D.

(a) If $e^u + u^2 + uv \equiv 0$, show that $e^u u_x + 2uu_x + uv_x + vu_x \equiv 0$ and $e^u u_y + 2uu_y + uv_y + vu_y \equiv 0$.

(b) If $u^2 + v^2 - \cos(u + v) \equiv 0$, find two equations relating u_x, u_y, v_x and v_y.

(c) If $xu^3 + uvy + xv^3 \equiv 1$ and $xu^2v + yuv^2 \equiv 1$, find expressions for u_x and v_x in terms of x, y, u, and v.

(d) If $x^3u + xyv \ln u \equiv 1$ and $y^3uv + xyu \ln v \equiv 1$, find expressions for u_y and v_y in terms of x, y, u, and v.

12. Prove the parts of Theorem 8 relating to $d(fg)$ and $d(f/g)$.

12-11 THE DIRECTIONAL DERIVATIVE

Let F have a differential at (x_0, y_0), as above. We then consider a directed line through (x_0, y_0), as in Figure 12-27. Let α be the angle from the positive

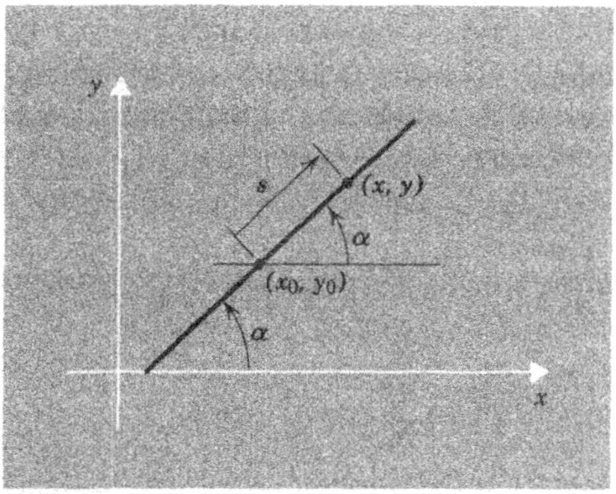

Figure 12-27 Definition of directional derivative.

x-axis to the direction chosen. It follows from geometry that the line has parametric equations

$$x = x_0 + s\cos\alpha = x(s), \qquad y = y_0 + s\sin\alpha = y(s) \qquad (12\text{-}110)$$

with s increasing in the chosen direction and $s = 0$ at (x_0, y_0). Since F is defined in a neighborhood of (x_0, y_0), F is defined along the line for s sufficiently close to 0. We evaluate F at the point $(x(s), y(s))$, obtaining a function:

$$g(s) = F(x(s), y(s))$$

By Theorem 6 (with t replaced by s), we know that $g'(s)$ exists and

$$g'(s) = F_x \frac{dx}{ds} + F_y \frac{dy}{ds} = F_x \cos\alpha + F_y \sin\alpha$$

where $F_x = \partial z/\partial x$ and $F_y = \partial z/\partial y$ are evaluated at $(x(s), y(s))$. The value of the derivative $g'(s)$ at $s = 0$ is termed the *directional derivative of F at* (x_0, y_0) *in direction* α. Thus at (x_0, y_0)

$$\text{directional derivative of } F \text{ in direction } \alpha = F_x(x_0, y_0)\cos\alpha + F_y(x_0, y_0)\sin\alpha \tag{12-111}$$

We note that when the chosen direction is that of the x-axis ($\alpha = 0$), we are simply finding the partial derivative $F_x(x_0, y_0)$, and that is just what (12-111) gives for $\alpha = 0$; similarly, for $\alpha = \pi/2$, we obtain $F_y(x_0, y_0)$. [Also, for $\alpha = \pi$, we obtain $-F_x(x_0, y_0)$.] Thus the directional derivative is a generalization of the partial derivative. It gives the rate of change of F (with respect to distance) in direction α. This idea occurs in many practical situations. In wading into a pond, one observes that the water becomes deeper more quickly in some directions than others; here, F is the depth and one is observing the directional derivative of F at various points in various directions. In driving across a city one observes that the traffic is getting worse in a certain direction and, hence, may try another direction; here F is the "traffic density."

The directional derivative is sometimes denoted by dF/ds, but this notation does not show the direction. A better notation is suggested by a vector interpretation of (12-111):

$$\frac{dF}{ds} = F_x \cos\alpha + F_y \sin\alpha = (F_x\mathbf{i} + F_y\mathbf{j}) \cdot (\cos\alpha\,\mathbf{i} + \sin\alpha\,\mathbf{j}) = \nabla F \cdot \mathbf{u}$$

where $\mathbf{u} = \cos\alpha\mathbf{i} + \sin\alpha\mathbf{j}$. Accordingly, the directional derivative is simply the *component of* ∇F *in the direction of the unit vector* $\mathbf{u}$ *in the chosen direction*. We, therefore, can denote the directional derivative by

$$\nabla_{\mathbf{u}} F$$

Other notations are $\nabla_\alpha F$ and $\nabla_{\mathbf{v}} F$ where $\mathbf{v}$ is any nonzero vector in the direction of $\mathbf{u}$, so that $\mathbf{u} = \mathbf{v}/\|\mathbf{v}\|$ and

$$\nabla_{\mathbf{v}} F = \nabla_{\mathbf{u}} F = \nabla F \cdot \frac{\mathbf{v}}{\|\mathbf{v}\|}$$

EXAMPLE 1. Let $F = x - 2y$. Then ∇F is the constant vector $\mathbf{i} - 2\mathbf{j}$. Hence, the directional derivative in the direction of $\mathbf{u} = \cos\alpha\,\mathbf{i} + \sin\alpha\,\mathbf{j}$ is

$$(\mathbf{i} - 2\mathbf{j}) \cdot (\cos\alpha\,\mathbf{i} + \sin\alpha\,\mathbf{j}) = \cos\alpha - 2\sin\alpha$$

If, for example, $\mathbf{u}$ is the unit vector $(\mathbf{i} - 2\mathbf{j})/\sqrt{5}$, then

$$\nabla_{\mathbf{u}} F = (\mathbf{i} - 2\mathbf{j}) \cdot \frac{(\mathbf{i} - 2\mathbf{j})}{\sqrt{5}} = \sqrt{5}$$

If $\mathbf{u} = (2\mathbf{i} + \mathbf{j})/\sqrt{5}$ (perpendicular to the previous choice), then

$$\nabla_{\mathbf{u}} F = (\mathbf{i} - 2\mathbf{j}) \cdot \frac{2\mathbf{i} + \mathbf{j}}{\sqrt{5}} = 0$$

The level curves of F are straight lines, as shown in Figure 12-28, and the gradient vector ∇F is perpendicular to the level curve at each point. This is the reason why ∇F has component 0 in the direction of a vector along the line.

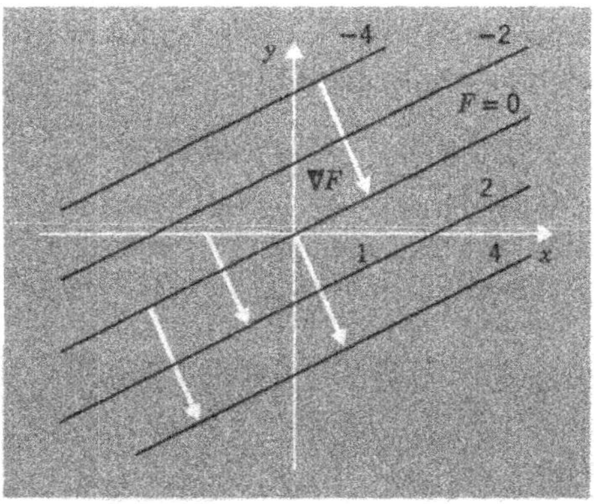

Figure 12-28 Level curves and gradient vector for $F(x, y) = x - 2y$.

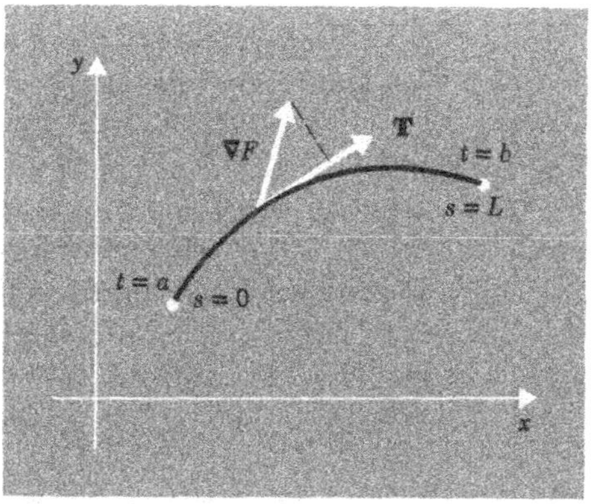

Figure 12-29 Directional derivative along a curve.

Directional Derivative Along a Curve. Let F be differentiable in the open region D and let a smooth path C: $x = f(t)$, $y = g(t)$, $a \leq t \leq b$, be given in D (Figure 12-29). As in Section 4-29, we can introduce arc length as parameter along C. We assume that $[f'(t)]^2 + [g'(t)]^2 > 0$ for $a \leq t \leq b$, so that s can be chosen to increase strictly with t. The path is now represented by equations $x = x(s)$, $y = y(s)$, and the function $G(s) = F(x(s), y(s))$ gives the value of the function F at the point $(x(s), y(s))$ on the curve. The derivative of $F(x(s), y(s))$ with respect to s is called the *directional derivative of F along C*, in the direction of increasing t (or s). Now, as in the case of the line above,

$$\frac{d}{ds} F(x(s), y(s)) = F_x \frac{dx}{ds} + F_y \frac{dy}{ds} = \nabla F \cdot \left(\frac{dx}{ds}\mathbf{i} + \frac{dy}{ds}\mathbf{j}\right)$$

Now

$$\frac{dx}{ds}\mathbf{i} + \frac{dy}{ds}\mathbf{j} = \mathbf{T} = \cos\alpha\,\mathbf{i} + \sin\alpha\,\mathbf{j}$$

where $\mathbf{T}$ is a unit tangent vector to the path in the chosen direction (Section 6-6). Thus

$$\frac{dF}{ds} = \nabla F \cdot \mathbf{T} = \nabla_{\mathbf{T}} F$$

At each point on C, the directional derivative along C is the same as the directional derivative of F in the direction of the tangent vector to C (in the direction of increasing s). As above, $\mathbf{T}$ can be replaced by any nonzero vector having the same direction, in particular, by the "velocity vector" $\mathbf{v} = f'(t)\mathbf{i} + g'(t)\mathbf{j}$:

$$\frac{dF}{ds} = \nabla_{\mathbf{v}} F = \nabla F \cdot \frac{\mathbf{v}}{\|\mathbf{v}\|} = \nabla F \cdot \mathbf{T}$$

If C happens to be a level curve of F, then F is constant along C and $dF/ds \equiv 0$. Hence, at each point $\nabla F \cdot \mathbf{T}$ must be 0 and ∇F is orthogonal to $\mathbf{T}$. *The gradient vector is normal to the level curve at each point.* This is illustrated in Figure 12-28.

Geometric Meaning of Gradient. At each point P_0: (x_0, y_0) in D, F has a directional derivative in each direction α, given by

$$\nabla F \cdot \mathbf{u} = \nabla_{\mathbf{u}} F$$

where $\mathbf{u}$ is the unit vector $\cos\alpha\,\mathbf{i} + \sin\alpha\,\mathbf{j}$. Now let us let α vary from 0 to 2π, so that, when $\mathbf{u}$ is represented by a line segment P_0Q, the end point Q of $\mathbf{u}$ traces a circle with center P_0 (Figure 12-30). The directional derivative

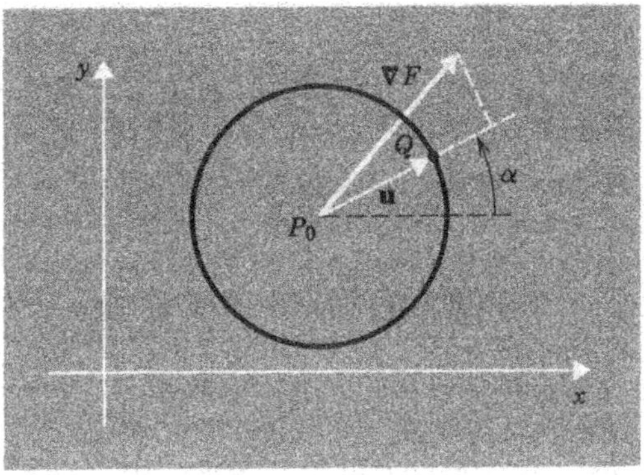

Figure 12-30 Geometric meaning of ∇F.

is the component of ∇F in the direction of $\mathbf{u}$. Hence, as α varies, the directional derivatives also varies, reaching a maximum when $\mathbf{u}$ has the same direction as ∇F and a minimum when $\mathbf{u}$ has the direction opposite to ∇F. The maximum value is thus

$$\nabla F \cdot \frac{\nabla F}{\|\nabla F\|} = \|\nabla F\|$$

The minimum is $-\|\nabla F\|$. When $\mathbf{u}$ is orthogonal to ∇F, $\nabla_{\mathbf{u}} F = 0$. Hence, we can say: *the gradient vector points in the direction in which F is increasing most rapidly; its magnitude $\|\nabla F\|$ is the directional derivative in that direction.*

This result gives a geometric meaning to the gradient vector, independent of position of coordinate axes. For example, in a room a person who is chilly will instinctively move in the direction of the gradient vector of the temperature; the benefit he receives in moving 1 ft depends on the size of that vector.

We note that if $\nabla F = \mathbf{0}$ at P_0, then $\nabla_{\mathbf{u}} F = 0$ for every $\mathbf{u}$. (This happens, in our example, when the person has found the warmest spot in the room.)

Extension to Functions of Several Variables. For a function $F(x_1, \ldots, x_n)$, the directional derivative $\nabla_{\mathbf{u}} F$ at $(x_1^0, \ldots, x_n^0)$ is defined as the derivative $(d/ds)F(x_1(s), \ldots, x_n(s))$ where

$$x_1(s) = x_1^0 + u_1 s, \qquad \ldots, \qquad x_n(s) = x_n^0 + u_n s$$

are the equations of a line through $(x_1^0, \ldots, x_n^0)$ in the direction of the unit vector $\mathbf{u} = (u_1, \ldots, u_n)$. One verifies, as above, that

$$\nabla_{\mathbf{u}} F = \nabla F \cdot \mathbf{u}$$

The gradient vector points in the direction of largest rate of increase of F, and its length gives the directional derivative in that direction.

PROBLEMS

1. Find the directional derivative of the function $z = F(x, y)$ at the point and in the direction given.
   (a) $z = x^3 + 5x^2y$ at $(2, 1)$ in a direction making angle $\pi/4$ with the positive x-axis.
   (b) $z = 3x^2 - y^4$ at $(2, 3)$ in a direction making angle π with the positive x-axis.
   (c) $z = ye^{xy}$ at $(0, 0)$ in the direction of $4\mathbf{i} + 3\mathbf{j}$.
   (d) $z = y\cos^2 x$ at $(\pi/2, 1)$ in the direction of $\mathbf{i} - \mathbf{j}$.
   (e) $z = x^2 - y^2$ at $(2, 3)$, direction of tangent vector of curve $x^2 + y^2 = 13$, with positive y-component at the point.
   (f) $z = xy$ at $(5, -1)$ in a direction of a tangent vector to the curve $2x + 5y^2 = 15$ at the point.
2. Let $F(x, y) = x^2 - 2y^2$. Find dF/ds along the curve given at the point specified, if s increases with t.
   (a) At $(1, 3)$ on curve $x = e^{2t}$, $y = 3e^t$.
   (b) At $(0, 0)$ on curve $x = t\cos t$, $y = t\sin t$.
3. Let $F(x, y)$ be differentiable at each point on the circle $x^2 + y^2 = 1$. Show that the directional derivative of F at (x, y) along the circle in the counterclockwise direction is $-yF_x(x, y) + xF_y(x, y)$.
4. With the aid of the result of problem 3, find the maximum and minimum values of each of the following functions on the circle $x^2 + y^2 = 1$:
   (a) $2x^2 + y^2$ (b) $x^2 - y^2$ (c) $x^2 + xy + y^2$ (d) $3x^2 + 2xy + 3y^2$
5. Let V be the vector space of all differentiable functions in a given open region D in the xy-plane. Let W be the vector space of all vector functions (vector fields) $\mathbf{u} = \mathbf{u}(x, y) = f(x, y)\mathbf{i} + g(x, y)\mathbf{j}$ in D.
   (a) Show that the equation $\mathbf{u} = \nabla F$ defines a linear mapping of V into W.
   (b) Show that ∇ obeys the laws:

$$\nabla(FG) = F\nabla G + G\nabla F,$$
$$\nabla(F^n) = nF^{n-1}\nabla F, \qquad n = 1, 2, \ldots$$

(c) Show that the kernel of the mapping ∇ consists of all constant functions in D.

6. *Temperature variation.* The temperature in a certain rectangular room is found to be given by

$$T = 60\frac{100 + 2d_1}{100 - 7d_2}$$

where d_1 is the distance from an outer wall, d_2 that from an adjacent inner wall; here T is in degrees Fahrenheit, d_1, d_2 are in feet. A person is standing 2 ft from each of these walls. In what direction should he start to move to get warmer?

12-12 DIFFERENTIAL OF A VECTOR FUNCTION, THE JACOBIAN MATRIX

Let a vector function $\mathbf{y} = \mathbf{F}(\mathbf{x}) = (F_1(\mathbf{x}), \ldots, F_m(\mathbf{x}))$ be defined in the open region D of V_n, with values in V_m.

We say that $\mathbf{F}$ is *differentiable* at $\mathbf{x}^0$, or that $\mathbf{F}$ has a differential at $\mathbf{x}^0$, if $\mathbf{x}^0$ is in D and

$$\Delta\mathbf{F} \equiv \mathbf{F}(\mathbf{x}^0 + \mathbf{h}) - \mathbf{F}(\mathbf{x}^0) = A\mathbf{h} + P(\mathbf{h})\mathbf{h} \tag{12-120}$$

where A is a constant m by n matrix and $P(\mathbf{h})$ is a matrix function of $\mathbf{h}$, defined in a neighborhood of $\mathbf{h} = \mathbf{0}$ with $P(\mathbf{0}) = O$ (the zero matrix) and P continuous at $\mathbf{0}$. The differential of $\mathbf{F}$ at $\mathbf{x}_0$ is then defined to be the linear function $A\mathbf{h}$:

$$d\mathbf{F} = A\mathbf{h} \tag{12-121}$$

One can also write $d\mathbf{F} = A\,\Delta\mathbf{x}$, $d\mathbf{F} = A\,d\mathbf{x}$, $d\mathbf{y} = A\,\Delta\mathbf{x}$ or $d\mathbf{y} = A\,d\mathbf{x}$.

From (12-120) we see that, when $\mathbf{F}$ has a differential at $\mathbf{x}^0$, the given mapping is closely approximated near $\mathbf{x}^0$ by a linear mapping:

$$\Delta\mathbf{F} \sim A\,\Delta\mathbf{x}$$

or

$$\mathbf{F}(\mathbf{x}) - \mathbf{F}(\mathbf{x}^0) \sim A(\mathbf{x} - \mathbf{x}^0)$$

This last relation is of the form:

$$\mathbf{y} = \mathbf{F}(\mathbf{x}) \sim A\mathbf{x} + \mathbf{b}$$

If we change to the geometric point of view and we regard $(x_1, \ldots, x_n)$, $(y_1, \ldots, y_m)$ as coordinates of points in R^n, R^m, respectively, then the equation says that (approximately) the mapping $\mathbf{F}$ is a linear mapping of R^n into R^m (Section 11-18). We can take new origins at $\mathbf{x}^0$ in R^n and at $\mathbf{y}^0$ in R^m, and introduce new coordinates:

$$\mathbf{x}^* = \mathbf{x} - \mathbf{x}^0, \qquad \mathbf{y}^* = \mathbf{y} - \mathbf{y}^0$$

Then relative to these origins our linear mapping is given by

$$\mathbf{y}^* = A\mathbf{x}^*$$

Actually, $\mathbf{x}^*$ is the same as $d\mathbf{x}$ and $\mathbf{y}^*$ can be interpreted as $d\mathbf{y}$, so that the last equation is the same as the equation $d\mathbf{y} = A\,d\mathbf{x}$ (see Figure 12-31).

The relationship is illustrated in Figure 12-31 for the case $m = n = 2$. The mapping **F** assigns vectors **y** to vectors **x**—that is, points (y_1, y_2) to points (x_1, x_2). In particular, $\mathbf{F}(\mathbf{x}^0) = \mathbf{y}^0$. The new coordinate axes in the x-space, with origin at $\mathbf{x}^0$, and those in the y-space, with origin $\mathbf{y}^0$, are also shown. Under the given mapping **F** the new axes in the x-space have as images, curves passing through $\mathbf{y}^0$. Under the linearized mapping $d\mathbf{y} = A\,d\mathbf{x}$, the curves are "straightened out" and become straight lines through the new origin; it can be shown that these lines are tangent at $\mathbf{y}^0$ to the previously obtained curved images of the new axes dx_1, dx_2 in the x-space.

In general, Equation (12-120) is equivalent to m scalar equations, of which the first is

$$\begin{aligned} F_1(x_1{}^0 + h_1, \ldots, x_n{}^0 + h_n) &- F_1(x_1{}^0, \ldots, x_n{}^0) \\ &= a_{11}h_1 + \cdots + a_{1n}h_n + p_{11}(\mathbf{h})h_1 + \cdots + p_{1n}(\mathbf{h})h_n \end{aligned} \tag{12-122}$$

From the definition of continuity of a matrix function we now conclude: the vector function **F** has a differential at $\mathbf{x}^0$ if, and only if, each of the scalar functions $F_1(x_1, \ldots, x_n), \ldots, F_m(x_1, \ldots, x_n)$ has a differential at $\mathbf{x}^0$. From (12-122) we deduce as in Theorems 3 and 2: *if* **F** *has a differential* $A\mathbf{h}$ *at* $\mathbf{x}^0$, *then* **F** *is continuous at* $\mathbf{x}^0$, *all partial derivatives* $\partial F_i/\partial x_j$ *exist at* $\mathbf{x}^0$ *and*

$$a_{11} = \frac{\partial F_1}{\partial x_1}(x_1{}^0, \ldots, x_n{}^0), \quad a_{12} = \frac{\partial F_1}{\partial x_2}(x_1{}^0, \ldots, x_n{}^0), \quad \ldots$$

and in general

$$a_{ij} = \frac{\partial F_i}{\partial x_j}(x_1{}^0, \ldots, x_n{}^0) = \frac{\partial F_i}{\partial x_j}(\mathbf{x}^0)$$

This shows that *the differential* $A\mathbf{h}$, *if it exists, is unique.*

The matrix

$$\begin{pmatrix} \dfrac{\partial F_1}{\partial x_1}(\mathbf{x}) & \cdots & \dfrac{\partial F_1}{\partial x_n}(\mathbf{x}) \\ \vdots & & \vdots \\ \dfrac{\partial F_m}{\partial x_1}(\mathbf{x}) & \cdots & \dfrac{\partial F_m}{\partial x_n}(\mathbf{x}) \end{pmatrix} = \left(\frac{\partial F_i}{\partial x_j}(\mathbf{x})\right) \tag{12-123}$$

is called the *Jacobian matrix* of the function or mapping **F**. Thus, when **F** has a differential at $\mathbf{x}^0$, that differential is $d\mathbf{F} = A\,d\mathbf{x}$, where A is the Jacobian matrix evaluated at $\mathbf{x}^0$:

$$d\mathbf{F} = \left(\frac{\partial F_i}{\partial x_j}\right)\bigg|_{\mathbf{x}^0} d\mathbf{x}$$

One also writes simply $\mathbf{F_x}$ or $\mathbf{y_x}$ for the Jacobian matrix, so that

$$d\mathbf{F} = \mathbf{F_x}(\mathbf{x}^0)\,d\mathbf{x} \qquad \text{or} \qquad d\mathbf{y} = \mathbf{y_x}(\mathbf{x}^0)\,d\mathbf{x}$$

As the notation reminds us, $d\mathbf{F}$ depends on the reference point $\mathbf{x}^0$ and on $d\mathbf{x}$, and, for each fixed $\mathbf{x}^0$, $d\mathbf{F}$ is linear in $d\mathbf{x}$.

From the preceding equation it follows that $d\mathbf{F}$ is an m by n matrix times

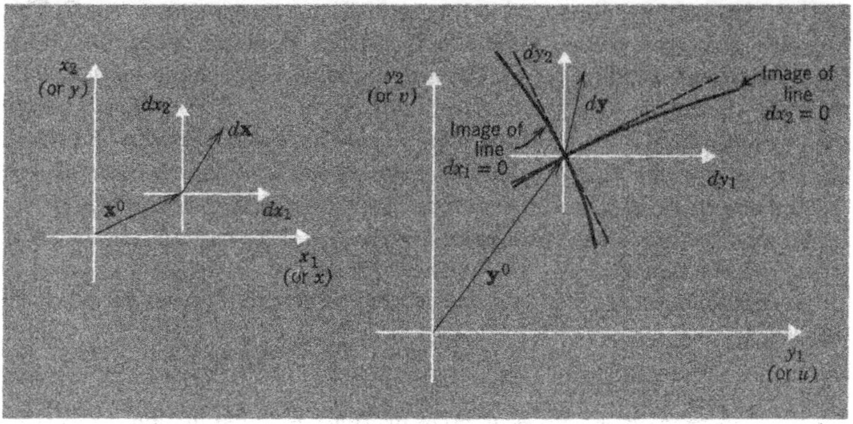

Figure 12-31 The differential as a linear mapping approximating a given mapping.

an n by 1 matrix; hence, $d\mathbf{F}$ is m by 1, a column vector. The ith element of the column is

$$\frac{\partial F_i}{\partial x_1} dx_1 + \cdots + \frac{\partial F_i}{\partial x_n} dx_n = dF_i$$

Hence, we can write

$$d\mathbf{F} = \begin{pmatrix} dF_1 \\ \vdots \\ dF_m \end{pmatrix} = \begin{pmatrix} dy_1 \\ \vdots \\ dy_m \end{pmatrix} = d\mathbf{y}$$

The Jacobian matrix and the associated linear mapping are the counterparts for vector functions of the derivative and differential, respectively, for a function of one variable. It will be seen that the Jacobian matrix measures "rate of change" in the sense of ratios of n-dimensional volumes. The counterpart of the condition $f'(x) > 0$ or $f'(x) < 0$, which is so important for functions of one variable, is that the Jacobian matrix have *maximum rank* (that is, rank equal to the smaller of the two numbers m, n). For example, if this condition holds true and $m = n$, then the mapping $\mathbf{y} = \mathbf{F}(\mathbf{x})$ has an inverse, at least, in a suitably restricted neighborhood.

EXAMPLE 1. Let a vector function from V_2 to V_2 be defined by the equation

$$(u, v) = \mathbf{F}(x, y) = (x^2 - y^2, 2xy)$$

The corresponding scalar equations are

$$u = x^2 - y^2, \quad v = 2xy$$

The Jacobian matrix is

$$A = \begin{pmatrix} 2x & -2y \\ 2y & 2x \end{pmatrix}$$

Now $\mathbf{F}(2, 1) = (3, 4)$ and at $(2, 1)$ $A = \begin{pmatrix} 4 & -2 \\ 2 & 4 \end{pmatrix}$, so that A has maximum rank, 2, and is nonsingular. Here

$$d\mathbf{F} = \begin{pmatrix} du \\ dv \end{pmatrix} = \begin{pmatrix} 4 & -2 \\ 2 & 4 \end{pmatrix} \begin{pmatrix} dx \\ dy \end{pmatrix}$$

or, in components,

$$du = 4dx - 2dy, \qquad dv = 2dx + 4dy \tag{12-124}$$

For $dy = 0$, $du = 4dx$ and $dv = 2dx$, so that (du, dv) follows a line of slope $\frac{1}{2}$ in the uv-plane; for $dx = 0$, $du = -2dy$, $dv = 4dy$, so that (du, dv) follows a line of slope -2 in the uv-plane (see Figure 12-31 p. 920). Both lines contain $(3, 4)$. The line $dy = 0$ is the line $y = 1$, and its image under the given mapping $\mathbf{F}$ is the curve $u = x^2 - 1$, $v = 2x$; these equations are *parametric equations,* with x as parameter, for a parabola: $u = (v^2/4) - 1$. The line $dx = 0$ (see Figure 12-31) is the line $x = 2$, and its image is the curve $u = 4 - y^2$, $v = 4y$. These equations are parametric equations, with y as parameter, for the parabola $u = 4 - (v^2/16)$. We note that, as in Figure 12-31, the images of the lines $dy = 0$, $dx = 0$ under the linearized mapping (12-124), are lines tangent to the images of those lines under the original nonlinear mapping $\mathbf{F}$.

EXAMPLE 2. Let $(y_1, y_2) = \mathbf{F}(x_1, x_2, x_3) = (x_1^2 + x_2^2 - x_3^2, x_1^2 - 3x_2^2 + x_3^2)$, so that

$$y_1 = x_1{}^2 + x_2{}^2 - x_3{}^2 = F_1(x_1, x_2, x_3)$$
$$y_2 = x_1{}^2 - 3x_2{}^2 + x_3{}^2 = F_2(x_1, x_2, x_3)$$

Then

$$\frac{\partial F_1}{\partial x_1} = 2x_1, \qquad \frac{\partial F_1}{\partial x_2} = 2x_2, \qquad \frac{\partial F_1}{\partial x_3} = -2x_3$$
$$\frac{\partial F_2}{\partial x_1} = 2x_1, \qquad \frac{\partial F_2}{\partial x_2} = -6x_2, \qquad \frac{\partial F_2}{\partial x_3} = 2x_3$$

Thus the Jacobian matrix is

$$A = \left(\frac{\partial F_i}{\partial x_j}\right) = \begin{pmatrix} 2x_1 & 2x_2 & -2x_3 \\ 2x_1 & -6x_2 & 2x_3 \end{pmatrix}$$

At $(x_1, x_2, x_3) = (1, 3, 1)$,

$$A = \begin{pmatrix} 2 & 6 & -2 \\ 2 & -18 & 2 \end{pmatrix}$$

and

$$d\mathbf{F} = \begin{pmatrix} dF_1 \\ dF_2 \end{pmatrix} = \begin{pmatrix} 2 & 6 & -2 \\ 2 & -18 & 2 \end{pmatrix} d\mathbf{x} = \begin{pmatrix} 2 & 6 & -2 \\ 2 & -18 & 2 \end{pmatrix} \begin{pmatrix} dx_1 \\ dx_2 \\ dx_3 \end{pmatrix}$$

Hence

$$dF_1 = 2dx_1 + 6dx_2 - 2dx_3, \qquad dF_2 = 2dx_1 - 18dx_2 + 2dx_3$$

At the point chosen, $F_1 = 9$, $F_2 = -25$. The preceding equations give approximately the changes of F_1, F_2 from these values if x_1, x_2, x_3 change by dx_1, dx_2, dx_3 respectively.

Special Cases of the Jacobian Matrix. For $n = 2$, $m = 1$—that is, one function of two variables, say $z = f(x, y)$, the Jacobian matrix is the row vector (f_x, f_y). Hence, it is the *gradient vector*. Similarly, for general n and $m = 1$ (one function of n variables), say $y = F(x_1, \ldots, x_n)$, the Jacobian matrix is the row vector $(F_{x_1}, \ldots, F_{x_n})$, which is the gradient vector ∇F. For $n = 1$ and $m = 1$, the matrix reduces to the derivative of a function of one variable.

For $n = 1$ and $m = 2$—that is, two functions of one variable, say $x = f(t)$, $y = g(t)$, the Jacobian matrix is the column vector

$$\begin{pmatrix} f'(t) \\ g'(t) \end{pmatrix}$$

Hence, it is our *velocity vector* $f'(t)\mathbf{i} + g'(t)\mathbf{j}$. Similarly for $n = 1$ and general m (m functions of one variable), say $x_i' = f_i(t)$, $i = 1, \ldots, m$, the Jacobian is the *column vector* $(f_1'(t), \ldots, f_m'(t))$, which can be interpreted as the velocity vector for a path in m-dimensional space. As for $m = 2$ (see Section 3-11), we can show that this vector is a vector tangent to the path.

12-13 THE GENERAL CHAIN RULE

We can now formulate and prove a general chain rule for composite vector functions:

THEOREM 9 (*The general chain rule*). *Let* $\mathbf{y} = \mathbf{F}(\mathbf{x})$ *be defined for* $\mathbf{x}$ *in the open region* D *of* V_n, *with values in* V_m. *Let* $\mathbf{F}$ *be expressed as the composition of two functions* $\mathbf{y} = \mathbf{f}(\mathbf{u})$, $\mathbf{u} = \mathbf{g}(\mathbf{x})$, *so that* $\mathbf{F} = \mathbf{f} \circ \mathbf{g}$. *If, at a particular* $\mathbf{x}$, $\mathbf{g}$ *has a differential* $d\mathbf{u} = B\,\Delta\mathbf{x}$ *and if, at the corresponding value* $\mathbf{u} = \mathbf{g}(\mathbf{x})$, $\mathbf{f}$ *has a differential* $d\mathbf{y} = A\,\Delta\mathbf{u}$, *then* F *has a differential at* $\mathbf{x}$:

$$d\mathbf{F} = AB\,\Delta\mathbf{x} = \mathbf{f_u g_x}\,\Delta\mathbf{x} \tag{12-130}$$

so that, for $\mathbf{y} = \mathbf{F}(\mathbf{x})$, *the Jacobian matrix is given by*

$$\mathbf{F_x} \equiv \mathbf{y_x} = \mathbf{y_u u_x} \tag{12-131}$$

PROOF. At the particular $\mathbf{x}$ and corresponding $\mathbf{u}$ considered,

$$\Delta\mathbf{u} = [B + P_1(\Delta\mathbf{x})]\,\Delta\mathbf{x}, \qquad \Delta\mathbf{y} = [A + P_2(\Delta\mathbf{u})]\,\Delta\mathbf{u}$$

where P_1, P_2 are matrix functions continuous at $\mathbf{0}$ with value O at $\mathbf{0}$. If we substitute the expression for $\Delta\mathbf{u}$ into that for $\Delta\mathbf{y}$, we find that

$$\begin{aligned}\Delta\mathbf{y} &= [A + P_2\{[B + P_1(\Delta\mathbf{x})]\,\Delta\mathbf{x}\}][B + P_1(\Delta\mathbf{x})]\,\Delta\mathbf{x} \\ &= [AB + AP_1(\Delta\mathbf{x}) + P_2\{\quad\}B + P_2\{\quad\}P_1(\Delta\mathbf{x})]\,\Delta\mathbf{x}\end{aligned}$$

where $P_2\{\quad\} = P_2\{[B + P_1(\Delta\mathbf{x})]\,\Delta\mathbf{x}\}$. Since $P_1(\Delta\mathbf{x})$ is continuous at $\Delta\mathbf{x} = \mathbf{0}$, with value O at $\Delta\mathbf{x} = \mathbf{0}$, $P_2\{\quad\}$ is continuous in $\Delta\mathbf{x}$ and also equals O for $\Delta\mathbf{x} = \mathbf{0}$ (Theorem C, Section 12-7; see the discussion of matrix functions in that section). Hence, $P(\Delta\mathbf{x}) = AP_1(\Delta\mathbf{x}) + P_2\{\quad\}B + P_2\{\quad\}P_1(\Delta\mathbf{x})$ is continuous at $\mathbf{0}$ with value O at $\mathbf{0}$, and we can write

$$\Delta\mathbf{y} = [AB + P(\Delta\mathbf{x})]\,\Delta\mathbf{x} = AB\,\Delta\mathbf{x} + P(\Delta\mathbf{x})\,\Delta\mathbf{x}$$

as required, so that $d\mathbf{y} = d\mathbf{F} = AB\,\Delta\mathbf{x}$, and (12-130) and (12-131) follow.

Remarks. The proof, as given, strictly parallels the proof of the corresponding theorem for real functions of one variable (Theorem 16, Section 3-23) and shows how linear algebra permits us to vastly generalize one-variable calculus by simply replacing real variables by vector variables. Certain precautions are necessary, of course. For example, the order of matrix functions in a product cannot be altered, as can the order of real-valued functions.

In terms of the components of the vector functions, Equation (12-131) becomes

$$\frac{\partial F_i}{\partial x_j} = \sum_{k=1}^{p} \frac{\partial f_i}{\partial u_k}\frac{\partial g_k}{\partial x_j}, \qquad i = 1, \ldots, m;\ j = 1, \ldots, n \qquad \text{(12-131')}$$

EXAMPLE 1. Let $\mathbf{y} = (y_1, y_2) = \mathbf{F}(\mathbf{x})$, [where $\mathbf{x} = (x_1, x_2)$], be defined as a composite function $\mathbf{f} \circ \mathbf{g}$ by the equations:

$$(y_1, y_2) = \mathbf{g}(u_1, u_2) = (u_1^2u_2^4 - u_1^4u_2^2,\ u_1^3u_2^3 + 3u_1^5u_2)$$
$$(u_1, u_2) = \mathbf{f}(x_1, x_2) = (x_1 \sin x_2 + x_1^2 \cos x_2,\ x_1^2 \sin x_2 - x_1 \cos x_2)$$

Then $\mathbf{y_x} = \mathbf{y_u}\mathbf{u_x}$, where

$$\mathbf{y_u} = \begin{pmatrix} 2u_1u_2^4 - 4u_1^3u_2^2 & 4u_1^2u_2^3 - 2u_1^4u_2 \\ 3u_1^2u_2^3 + 15u_1^4u_2 & 3u_1^3u_2^2 + 3u_1^5 \end{pmatrix}$$

$$\mathbf{u_x} = \begin{pmatrix} \sin x_2 + 2x_1 \cos x_2 & x_1 \cos x_2 - x_1^2 \sin x_2 \\ 2x_1 \sin x_2 - \cos x_2 & x_1^2 \cos x_2 + x_1 \sin x_2 \end{pmatrix}$$

We could easily multiply the two matrices. But it is far simpler to leave the result, as above, in indicated form. For any desired numerical values of x_1, x_2, we can easily calculate $\mathbf{y_u}$ and $\mathbf{u_x}$ and, therefore, $\mathbf{y_x}$. For example, for $x_1 = 1$ and $x_2 = \pi/2$, we find that $u_1 = 1$, $u_2 = 1$, and

$$\mathbf{y_x} = \begin{pmatrix} -2 & 2 \\ 18 & 6 \end{pmatrix}\begin{pmatrix} 1 & -1 \\ 2 & 1 \end{pmatrix} = \begin{pmatrix} 2 & 4 \\ 30 & -12 \end{pmatrix}$$

In this example, we also could have expressed the given function $\mathbf{y} = \mathbf{f} \circ \mathbf{g}$ directly in terms of $\mathbf{x}$, by elimination of u_1 and u_2. This would have given us very cumbersome expressions. The procedure followed above, based on the chain rule, is far simpler.

EXAMPLE 2. A differentiable function $\mathbf{F} = \mathbf{f} \circ \mathbf{g}$ is given, and it is known that $\mathbf{g}(\mathbf{0}) = \mathbf{0}$, $\mathbf{f}(\mathbf{0}) = (1, 3, 5)$, and

$$\mathbf{g}_{\mathbf{x}}(\mathbf{0}) = \begin{pmatrix} 3 & 4 & 2 \\ 5 & -1 & 6 \end{pmatrix}, \qquad \mathbf{f}_{\mathbf{u}}(\mathbf{0}) = \begin{pmatrix} 1 & 0 \\ 7 & 2 \\ -4 & 2 \end{pmatrix}$$

Find $\mathbf{F}(\mathbf{0})$ and $\mathbf{F}_{\mathbf{x}}(\mathbf{0})$.

Solution. We observe that $\mathbf{F}(\mathbf{0}) = \mathbf{f}[\mathbf{g}(\mathbf{0})] = \mathbf{f}(\mathbf{0}) = (1, 3, 5)$. Also

$$\mathbf{F}_{\mathbf{x}}(\mathbf{0}) = \mathbf{f}_{\mathbf{u}}(\mathbf{0})\mathbf{g}_{\mathbf{x}}(\mathbf{0}) = \begin{pmatrix} 1 & 0 \\ 7 & 2 \\ -4 & 2 \end{pmatrix}\begin{pmatrix} 3 & 4 & 2 \\ 5 & -1 & 6 \end{pmatrix} = \begin{pmatrix} 3 & 4 & 2 \\ 31 & 26 & 26 \\ -2 & -18 & 4 \end{pmatrix}$$

This example illustrates how we can use the chain rule to find the values of derivatives (that is, Jacobian matrices) of composite functions, without knowing any more about the functions than a pair of corresponding values and the Jacobian matrices at these values. In many practical situations, Jacobian matrices can be found empirically by experiment. Then, as in the example, through the chain rule, one can deduce Jacobian matrices for more complex relationships.

As for one-variable calculus, one may also have longer chains—for example,

$$\mathbf{y}_{\mathbf{x}} = \mathbf{y}_{\mathbf{u}}\mathbf{u}_{\mathbf{v}}\mathbf{v}_{\mathbf{w}}\mathbf{w}_{\mathbf{x}}$$

Here, there is a "chain reaction" linking vector $\mathbf{x}$ to vector $\mathbf{w}$, $\mathbf{w}$ to $\mathbf{v}$, $\mathbf{v}$ to $\mathbf{u}$, $\mathbf{u}$ to $\mathbf{y}$ as in Figure 12-32. Many phenomena of the physical world can

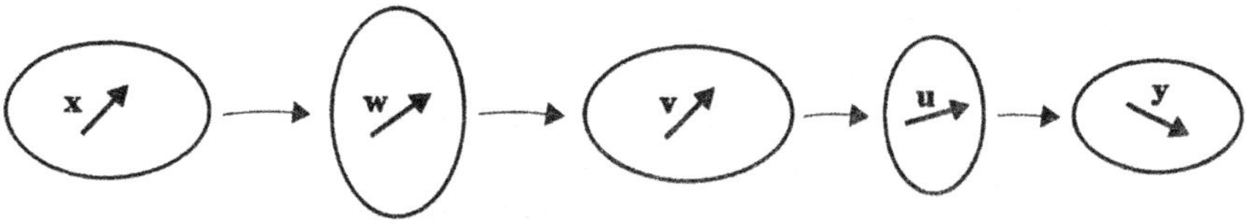

Figure 12-32 A chain of vector spaces.

be understood by such relationships. The vector $\mathbf{x}$ describes one set of physical variables. Once these variables are determined, $\mathbf{x}$ is known and another set of variables, forming vector $\mathbf{w}$, is also determined, and so on. An example would be the temperature, humidity, wind velocity, barometric pressure at 7 A.M. as the first set of variables $\mathbf{x}$; they, in turn, determine the amount of snow and ice on certain highways at 8 A.M.—another set of physical variables, forming vector $\mathbf{w}$; then $\mathbf{w}$ leads (in less predictable fashion) to the presence of so many snowplows on the highways at 8:30 A.M.—the vector $\mathbf{v}$. Similarly, $\mathbf{v}$ leads to $\mathbf{u}$, the amount of snow and ice on the highway at 9 A.M., and $\mathbf{u}$ leads to $\mathbf{y}$, the rate of occurrence of accidents on various roads between 9 and 10 A.M. Now we ask what happens if $\mathbf{x}$ changes slightly—that is, is replaced by $\mathbf{x} + d\mathbf{x}$. Then each vector of the chain also changes slightly. Each change is obtained from the previous one in the chain by multiplying by the appropriate Jacobian matrix:

$$d\mathbf{w} = D\,d\mathbf{x}, \qquad d\mathbf{v} = C\,d\mathbf{w}, \qquad d\mathbf{u} = B\,d\mathbf{v}, \qquad d\mathbf{y} = A\,d\mathbf{u}$$

so that the combined effect is described by multiplying the matrices:

$$d\mathbf{y} = ABCD\,d\mathbf{x}$$

The chain rules tells us that this naive theory is correct.

We remark that Theorem 9 includes all the previous chain rules (the ones of Sections 12-10 and 3-23) as special cases. The question of differentiability of the composite function, left open in the proof of Theorem 7 in Section 12-10, is also now settled.

PROBLEMS

1. Find the Jacobian matrix for each of the following mappings. State the dimensions concerned in each case.

(a) $(u_1, u_2) = (x_1 - 2x_2, 2x_1 + 3x_2)$

(b) $(u_1, u_2) = (x_1^2 + x_2^2, 2x_1 + 3x_1x_2)$

(c) $(u_1, u_2, u_3) = (x_1x_2x_3, x_1^2x_2 + x_2^2x_3, (x_1 + x_3)^2)$

(d) $(u_1, u_2, u_3) = (x_1 \cos x_2, x_1 \sin x_2, x_1 \cos x_2 \sin x_2)$

(e) $(u, v) = (x^4 + xy^3, x^2y^2 - 3y^4)$

(f) $(u, v) = (x \ln (x + y), (y - x) \ln (x + y))$

(g) $(u, v) = (x^2 + yz - z^2, xy - xz + 2z^2)$

(h) $(u, v) = (xe^y - yz, ye^y + xy - 2xz)$

2. (a) Find the linear equations approximating the mapping

$$u_1 = x_1 \cos x_2 - x_3^2, \qquad u_2 = x_1 \sin x_2 + x_1x_3, \qquad u_3 = x_1x_2x_3$$

near $x_1 = 2$, $x_2 = \pi/2$, $x_3 = -1$, and use these equations to find, approximately, the vector (u_1, u_2, u_3) for $x_1 = 2.1$, $x_2 = 1.6$, $x_3 = -0.9$.

(b) Proceed as in (a) for the mapping $(u_1, u_2, u_3) = (x_1x_2x_3, x_1^2x_2 - x_3^3, x_1x_2 - x_2x_3)$ near $\mathbf{x} = (1, 1, 1)$ and use the approximating mapping to evaluate $\mathbf{u}$ for $\mathbf{x} = (1.01, 1.03, 0.99)$ approximately.

3. For the mapping given find the Jacobian matrix and determine its rank. [*Remark.* The rank of a linear mapping gives the dimension of the range. Hence, in a region where the Jacobian matrix has rank r, one would expect the given nonlinear mapping to have an r-dimensional range: for $r = 0$, a point; for $r = 1$, a curve; for $r = 2$, a surface; and so on. This can be fully justified on the basis of the Implicit Function Theorem (Section 12-14).]

(a) $(u, v) = (\cos (x + 2y), \sin (x + 2y))$ (what is the range?)

(b) $(u, v) = (1 + x + y, (1 + x + y)^{-2})$, $x + y + 1 > 0$ (what is the range?)

(c) $(u, v, w) = (x + z, x - y, x^2 - xy - yz + xz)$

(d) $(u, v, w) = (e^{x+y+z}, \ln (x + y + z), (x + y + z)^2)$, $x + y + z > 0$

4. For each mapping find all points at which the Jacobian matrix exists and has less than maximum rank.

(a) $(u, v) = (x^3 - 3xy^2, 3x^2y - y^3)$ (b) $(u, v) = \left(\dfrac{x}{x^2 + y^2}, \dfrac{y}{x^2 + y^2}\right)$

(c) $z = x^2 + y^2$ (d) $z = xy$

(e) $y = x^2 - 5x$ (f) $y = x^3 - 3x^2 - 9x$

5. A differentiable mapping $\mathbf{y} = \mathbf{F}(\mathbf{x})$ is given as a composite of $\mathbf{y} = \mathbf{f}(\mathbf{u})$ and $\mathbf{u} = \mathbf{g}(\mathbf{x})$. At a certain $\mathbf{x}_0$, $\mathbf{g}_\mathbf{x}$ is known and, at $\mathbf{u}_0 = \mathbf{g}(\mathbf{x}_0)$, $\mathbf{f}_\mathbf{u}$ is known. Find $\mathbf{F}_\mathbf{x}(\mathbf{x}_0)$ from the given values of $\mathbf{g}_\mathbf{x}(\mathbf{x}_0)$ and $\mathbf{f}_\mathbf{u}(\mathbf{u}_0)$.

(a) $\mathbf{g}_\mathbf{x}(\mathbf{x}_0) = \begin{pmatrix} 1 & 0 & 2 \\ 3 & 1 & 4 \end{pmatrix}$, $\mathbf{f}_\mathbf{u}(\mathbf{u}_0) = \begin{pmatrix} 2 & -1 \\ 5 & 2 \end{pmatrix}$

(b) $\mathbf{g}_\mathbf{x}(\mathbf{x}_0) = \begin{pmatrix} 1 \\ 5 \end{pmatrix}$, $\mathbf{f}_\mathbf{u}(\mathbf{u}_0) = \begin{pmatrix} 3 & 5 \\ 0 & 1 \end{pmatrix}$

(c) $\mathbf{g}_\mathbf{x}(\mathbf{x}_0) = \begin{pmatrix} 2 & 5 \\ 1 & 7 \\ 3 & 2 \end{pmatrix}$, $\mathbf{f}_\mathbf{u}(\mathbf{u}_0) = \begin{pmatrix} 1 & 7 & 6 \\ 3 & -1 & 2 \end{pmatrix}$

(d) $\mathbf{g}_\mathbf{x}(\mathbf{x}_0) = \begin{pmatrix} 1 & 0 & 3 \\ 7 & 1 & -1 \\ 5 & 2 & 0 \end{pmatrix}$, $\mathbf{f}_\mathbf{u}(\mathbf{u}_0) = \begin{pmatrix} 1 & 0 & 1 \\ 2 & 1 & 0 \\ 3 & 0 & -2 \end{pmatrix}$

6. In each case a mapping $\mathbf{y} = \mathbf{F}(\mathbf{x})$ is defined by composition. Find $\mathbf{F}_\mathbf{x}$ at the point requested.

(a) $\mathbf{y} = (y_1, y_2) = (u_1^2 - u_1u_2, u_1u_2 + u_2^2)$, $\mathbf{u} = (x_1x_2, x_1^2 - x_2^2)$ at $\mathbf{x} = (3, 2)$

(b) $\mathbf{y} = (u_1 \sin u_2, u_1 \cos u_2)$, $\mathbf{u} = (x_1e^{x_1}, (x_1 - x_2)e^{x_1})$ at $(0, 0)$

(c) $\mathbf{y} = (u_1^2, u_1u_2, u_2^2)$, $\mathbf{u} = (x_1 - x_2, x_2 + 5x_3)$ at $(1, 2, 0)$

(d) $\mathbf{y} = (u_1/u_2, u_2/u_3, u_3/u_1)$, $\mathbf{u} = (x_1^3 + x_2^2 - x_3, x_1x_2x_3, x_1^2x_2 + x_2^2x_3)$ at $(1, 1, 1)$

7. (a) For the mapping $u = 2x + x^2 - y$, $v = x + 4y + y^2$ find the approximating linear mapping at $(x, y) = (0, 0)$ and compare graphically the images of the x- and y-axes under the given mapping and the approximating linear one.

(b) Proceed as in (a) for the mapping $u = e^x \cos y$, $v = e^x \sin y$.

12-14 IMPLICIT FUNCTIONS

In our study of functions of one variable, we were led to implicit functions: that is, functions defined by an equation such as

$$x^3 + xy^2 + y^3 - 1 = 0$$

In general, we said that a function $y = f(x)$, $a \le x \le b$, was defined implicitly by an equation

$$F(x, y) = 0$$

if

$$F(x, f(x)) \equiv 0, \qquad a \le x \le b$$

Similarly, we say that a function $z = f(x, y)$, (x, y) in region D, is defined implicitly by the equation

$$F(x, y, z) = 0 \tag{12-140}$$

if one has

$$F(x, y, f(x, y)) \equiv 0, \qquad (x, y) \text{ in } D$$

For example, the function

$$z = \sqrt{1 - x^2 - y^2}, \qquad x^2 + y^2 < 1$$

is defined implicitly by the equation

$$x^2 + y^2 + z^2 = 1 \tag{12-141}$$

We note that the function

$$z = -\sqrt{1 - x^2 - y^2}, \qquad x^2 + y^2 < 1$$

is defined implicitly by the same equation.

The concept clearly extends to functions of any number of variables. In practice, one is often given only an implicit equation, such as (12-140), and is unable to solve explicitly for a function of x and y. However, as for functions of one variable, one can nevertheless obtain an expression for derivatives. For example, from (12-141) one can reason that if $z = f(x, y)$ is a differentiable function defined by the equation, then we can differentiate partially with respect to x and with respect to y:

$$2x + 2z\frac{\partial z}{\partial x} = 0 \quad (y = \text{const}), \qquad 2y + 2z\frac{\partial z}{\partial y} = 0 \quad (x = \text{const})$$

Hence

$$\frac{\partial z}{\partial x} = -\frac{x}{z}, \qquad \frac{\partial z}{\partial y} = -\frac{y}{z}$$

Thus we have obtained the partial derivatives in terms of x, y, and z, where z stands for $f(x, y)$. Even though we do not know $f(x, y)$ explicitly, the expressions for the partial derivatives are useful.

THEOREM 10. *Let the function $F = F(x, y, z)$ be defined and differentiable in an open region D of R^3. Let the function $z = f(x, y)$ be defined and differentiable in an open region D_1 of R^2. Let the function f be such that $F(x, y, f(x, y)) \equiv 0$ in D_1, so that f is a function defined implicitly by the equation $F(x, y, z) = 0$. Then for (x, y) in D_1*

$$\frac{\partial f}{\partial x}(x, y) = -\frac{F_x}{F_z}, \qquad \frac{\partial f}{\partial y}(x, y) = -\frac{F_y}{F_z} \tag{12-142}$$

where on the right the partial derivatives F_x, F_y, F_z are evaluated at (x, y, z), with $z = f(x, y)$, and it is assumed that $F_z \neq 0$ at each such point.

PROOF. We are given that

$$F(x, y, f(x, y)) = 0$$

for every (x, y) in D_1. Hence, the derivative of the left side of this equation with respect to x is 0. By the chain rule, we conclude that

$$F_x + F_z \frac{\partial f}{\partial x} = 0$$

where $\partial f/\partial x$ is evaluated at (x, y) and F_x, F_z are evaluated at (x, y, z) with $z = f(x, y)$. Since $F_z \neq 0$ for these values, we can divide by F_z and obtain the first of (12-142). The other is obtained in the same way.

Remark 1. Instead of taking partial derivatives, we can take the differential of both sides:

$$F_x\,dx + F_y\,dy + F_z\,dz = 0$$

where $dz = df(x, y)$. Hence

$$df = dz = -\frac{F_x}{F_z}\,dx - \frac{F_y}{F_z}\,dy$$

from which we read off the partial derivatives as given by Equation (12-142).

Remark 2. Theorem 10 states that *if* a differentiable function f is defined by an implicit equation, then that function has partial derivatives as given. The theorem does not assert that there *is* such a function f. The Implicit Function Theorem (see the next section) does provide sufficient conditions for existence of a solution.

The reasoning extends at once to an equation

$$F(x_1, \ldots, x_n) = 0$$

which may define a function $x_n = f(x_1, \ldots, x_{n-1})$ implicitly. Under the analogous hypotheses, one obtains formulas for derivatives as follows:

$$\frac{\partial f}{\partial x_1}(x_1, \ldots, x_{n-1}) = -\frac{F_{x_1}}{F_{x_n}}, \ldots, \frac{\partial f}{\partial x_{n-1}}(x_1, \ldots, x_{n-1}) = -\frac{F_{x_{n-1}}}{F_{x_n}} \qquad (12\text{-}143)$$

Here the F_{x_i} are evaluated at $(x_1, \ldots, x_n)$ with $x_n = f(x_1, \ldots, x_{n-1})$.

When $n = 2$ we are considering an equation $F(x, y) = 0$, and the formula becomes

$$\frac{dy}{dx} = -\frac{F_x}{F_y}$$

EXAMPLE 1. Given the implicit equation

$$x^3 + xy^2 + xu^3 + yu + u^2v - v^3 - 1 = 0$$

we assume that $v = f(x, y, u)$ is a differentiable function satisfying the equation and conclude from (12-143) that

$$\frac{\partial v}{\partial x} = -\frac{3x^2 + y^2 + u^3}{u^2 - 3v^2}, \quad \frac{\partial v}{\partial y} = -\frac{2xy + u}{u^2 - 3v^2}, \quad \frac{\partial v}{\partial u} = -\frac{3xu^2 + y + 2uv}{u^2 - 3v^2}$$

where on the right v is $f(x, y, u)$ and the denominators are assumed to be nonzero. We could also take differentials:

$$(3x^2 + y^2 + u^3)\,dx + (2xy + u)\,dy + (3xu^2 + y + 2uv)\,du + (u^2 - 3v^2)\,dv = 0$$

Hence

$$dv = -\frac{3x^2 + y^2 + u^3}{u^2 - 3v^2}\,dx - \frac{2xy + u}{u^2 - 3v^2}\,dy - \frac{3xu^2 + y + 2uv}{u^2 - 3v^2}\,du$$

and we can again obtain the partial derivatives.

In some cases, one is led to simultaneous equations: for example,

$$F(x, y, u, v) = 0, \qquad G(x, y, u, v) = 0 \tag{12-144}$$

Here we hope to be able to solve for two unknowns in terms of the other two: say, $x = f(u, v)$, $y = g(u, v)$. If we *assume* that such functions exist and that appropriate differentiability conditions hold true, then we can differentiate (12-144) implicitly with respect to u, with v held constant, to obtain

$$F_x \frac{\partial x}{\partial u} + F_y \frac{\partial y}{\partial u} + F_u = 0, \qquad G_x \frac{\partial x}{\partial u} + G_y \frac{\partial y}{\partial u} + G_u = 0$$

These are simultaneous linear equations for $\partial x/\partial u$, $\partial y/\partial u$. If we solve them by Cramer's rule, we obtain

$$\frac{\partial x}{\partial u} = -\frac{\begin{vmatrix} F_u & F_y \\ G_u & G_y \end{vmatrix}}{\begin{vmatrix} F_x & F_y \\ G_x & G_y \end{vmatrix}}, \qquad \frac{\partial y}{\partial u} = -\frac{\begin{vmatrix} F_x & F_u \\ G_x & G_u \end{vmatrix}}{\begin{vmatrix} F_x & F_y \\ G_x & G_y \end{vmatrix}} \tag{12-145}$$

Here it is essential that the denominators be nonzero. The determinants that appear here are known as *Jacobian determinants* or, simply, as *Jacobians*. One writes, for example,

$$\frac{\partial(F, G)}{\partial(x, y)} = \begin{vmatrix} F_x & F_y \\ G_x & G_y \end{vmatrix}$$

With this notation, (12-145) becomes

$$\frac{\partial x}{\partial u} = -\frac{\dfrac{\partial(F, G)}{\partial(u, y)}}{\dfrac{\partial(F, G)}{\partial(x, y)}}, \qquad \frac{\partial y}{\partial u} = -\frac{\dfrac{\partial(F, G)}{\partial(x, u)}}{\dfrac{\partial(F, G)}{\partial(x, y)}} \tag{12-145'}$$

where we assume that $\partial(F, G)/\partial(x, y) \neq 0$. These formulas can be remembered by observing that both denominators are the Jacobian of F, G with respect to x and y. In going from the denominator to the numerator, one replaces the dependent variable whose derivative we are finding by the independent variable in question; in the first equation, x is replaced by u; in the second, y is replaced by u. There are formulas analogous to (12-145′) for $\partial x/\partial v$ and $\partial y/\partial v$. In fact, one can simply replace u by v throughout.

We remark that the Jacobian determinants appearing here are, except for sign, second-order minors of the Jacobian matrix of the vector function (F, G):

$$\begin{pmatrix} F_x & F_y & F_u & F_v \\ G_x & G_y & G_u & G_v \end{pmatrix}$$

EXAMPLE 2. $x^2 + xu - xy + v^2 = 0,\ xy + y^2 - uv = 0$
We find that

$$\frac{\partial(F, G)}{\partial(x, y)} = \begin{vmatrix} 2x + u - y & -x \\ y & x + 2y \end{vmatrix} = 2x^2 + 4xy - 2y^2 + xu + 2yu$$

$$\frac{\partial(F, G)}{\partial(u, y)} = \begin{vmatrix} x & -x \\ -v & x + 2y \end{vmatrix} = x^2 + 2xy - xv$$

and, hence, that

$$\frac{\partial x}{\partial u} = -\frac{x^2 + 2xy - xv}{2x^2 + 4xy - 2y^2 + xu + 2yu}$$

The other derivatives are found in the same way.

The formal discussion extends at once to the case of n equations in $n + m$ unknowns:

$$\begin{aligned} &F_1(x_1, \ldots, x_n, u_1, \ldots, u_m) = 0, \\ &\vdots \\ &F_n(x_1, \ldots, x_n, u_1, \ldots, u_m) = 0 \end{aligned} \tag{12-146}$$

(When the number of equations equals or exceeds the numbers of unknowns, usually no functions are defined implicitly.) From (12-146) one may hope to solve for $x_1, x_2, \ldots, x_n$ as functions of $u_1, \ldots, u_m$:

$$x_1 = f_1(u_1, \ldots, u_n), \quad \ldots, \quad x_n = f_n(u_1, \ldots, u_m)$$

If one proceeds as for Equations (12-144), one obtains simultaneous equations for the partial derivatives $\partial x_i/\partial u_j$, for fixed j. The solution by Cramer's rule leads to formulas of which the following is typical:

$$\frac{\partial f_2}{\partial u_1} = -\frac{\dfrac{\partial(F_1, \ldots, F_n)}{\partial(x_1, u_1, x_3, \ldots, x_n)}}{\dfrac{\partial(F_1, \ldots, F_n)}{\partial(x_1, \ldots, x_n)}} \tag{12-147}$$

Here, for example, the denominator is the Jacobian determinant

$$\begin{vmatrix} \partial F_1/\partial x_1 & \cdots & \partial F_1/\partial x_n \\ \vdots & & \vdots \\ \partial F_n/\partial x_1 & \cdots & \partial F_n/\partial x_n \end{vmatrix} \tag{12-148}$$

and the numerator is obtained from this by replacing the second column by $\partial F_1/\partial u_1, \ldots, \partial F_n/\partial u_1$.

These formulas can also be obtained by matrices. If we differentiate (12-146) with respect to $u_1, \ldots, u_m$, one obtains a set of equations that can be written

$$\mathbf{F_x f_u} + \mathbf{F_u} = O \tag{12-149}$$

Here $\mathbf{F_x}$ is the matrix whose determinant is (12-148). It is a Jacobian matrix of $\mathbf{F}$ with respect to $x_1, \ldots, x_n$, treating $u_1, \ldots, u_m$ as constants; $\mathbf{F_u}$ is defined similarly. If (12-148) is not 0, then the matrix $\mathbf{F_x}$ is nonsingular, and we solve (12-149) by multiplying both sides by the inverse of $\mathbf{F_x}$ to obtain

$$\mathbf{f_u} = -\mathbf{F_x}^{-1}\mathbf{F_u} \tag{12-149'}$$

From the formulas for the inverse of a matrix (Theorem 18 in Section 10-13), we again obtain the equations such as (12-147). We remark that (12-149′) is an exceptionally concise form for all the formulas.

For Example 2 above, we regard x as x_1, y as x_2, u as u_1, v as u_2. Then

$$\mathbf{F_x} = \begin{pmatrix} 2x+u-y & -x \\ y & x+2y \end{pmatrix}, \qquad \mathbf{F_u} = \begin{pmatrix} x & 2v \\ -v & -u \end{pmatrix}$$

and, hence,

$$\begin{pmatrix} \partial x/\partial u & \partial x/\partial v \\ \partial y/\partial u & \partial y/\partial v \end{pmatrix} = -\begin{pmatrix} 2x+u-y & -x \\ y & x+2y \end{pmatrix}^{-1} \begin{pmatrix} x & 2v \\ -v & -u \end{pmatrix}$$

$$= \frac{-1}{(2x^2+4xy-2y^2+xu+2yu)} \begin{pmatrix} x+2y & x \\ -y & 2x+u-y \end{pmatrix} \begin{pmatrix} x & 2v \\ -v & -u \end{pmatrix}$$

$$= \frac{-1}{(2x^2+\cdots+2yu)} \begin{pmatrix} x^2+2xy-xv & 2xv+4yv-xu \\ -xy-2xv-uv+yv & -2yv-2xu-u^2+uy \end{pmatrix}$$

We can now read off $\partial x/\partial u$ (getting the same expression as above) as well as $\partial x/\partial v$, $\partial y/\partial u$, $\partial y/\partial v$.

EXAMPLE 3. $F_1 = 3x_1 - 5x_2 - u_1 + 2u_2 - 5u_3 = 0$

$F_2 = 2x_1 - 3x_2 + u_1 + 4u_2 + 2u_3 = 0$

Here $n = 2$, $m = 3$. Also

$$\mathbf{F_x} = \begin{pmatrix} 3 & -5 \\ 2 & -3 \end{pmatrix}$$

$$\det \mathbf{F_x} = \frac{\partial(F_1, F_2)}{\partial(x_1, x_2)} = \begin{vmatrix} 3 & -5 \\ 2 & -3 \end{vmatrix} = 1$$

so that the formulas are applicable. If $x_1 = f_1(u_1, u_2, u_3)$, $x_2 = f_2(u_1, u_2, u_3)$, then

$$\frac{\partial f_1}{\partial u_1} = -\frac{\dfrac{\partial(F_1, F_2)}{\partial(u_1, x_2)}}{\dfrac{\partial(F_1, F_2)}{\partial(x_1, x_2)}} = -\frac{\begin{vmatrix} -1 & 5 \\ 1 & -3 \end{vmatrix}}{\begin{vmatrix} 3 & -5 \\ 2 & -3 \end{vmatrix}} = 2$$

and so on. Also, we verify that $\mathbf{F_x}$ has inverse $\begin{pmatrix} -3 & 5 \\ -2 & 3 \end{pmatrix}$ and that $\mathbf{F_u}$ is $\begin{pmatrix} -1 & 2 & -5 \\ 1 & 4 & 2 \end{pmatrix}$. Hence

$$\mathbf{f_u} = -\begin{pmatrix} -3 & 5 \\ -2 & 3 \end{pmatrix}\begin{pmatrix} -1 & 2 & -5 \\ 1 & 4 & 2 \end{pmatrix} = \begin{pmatrix} -8 & -14 & -25 \\ -5 & -8 & -16 \end{pmatrix}$$

from which we read off

$$\frac{\partial f_1}{\partial u_1} = -8, \quad \frac{\partial f_1}{\partial u_2} = -14, \quad \frac{\partial f_1}{\partial u_3} = -25, \quad \frac{\partial f_2}{\partial u_1} = -5, \ldots.$$

In this example, the given equations happen to be linear and can be written in the form

$$A\mathbf{x} + B\mathbf{u} = \mathbf{0}, \qquad A = \begin{pmatrix} 3 & -5 \\ 2 & -3 \end{pmatrix}, \qquad B = \begin{pmatrix} -1 & 2 & -5 \\ 1 & 4 & 2 \end{pmatrix}$$

Since A is nonsingular, we can solve for $\mathbf{x}$ by multiplying by $A^{-1} = \begin{pmatrix} -3 & 5 \\ -2 & 3 \end{pmatrix}$:

$$\mathbf{x} = -A^{-1}B\mathbf{u} = \begin{pmatrix} -8 & -14 & -25 \\ -5 & -8 & -16 \end{pmatrix}\mathbf{u}$$

That is

$$x_1 = f_1(u_1, u_2, u_3) = -8u_1 - 14u_2 - 25u_3$$
$$x_2 = f_2(u_1, u_2, u_3) = -5u_1 - 8u_2 - 16u_3$$

From these equations we read off the matrix $\mathbf{f_u}$ and the partial derivatives as above.

When the equations are not linear as in the example, it is normally very difficult to solve explicitly for the functions f_i. However, by taking differentials, we in effect replace given nonlinear equations by linear equations in the differentials. These linear equations are treated exactly as in the example; that is, they are solved for the differentials dx_i by multiplying by the inverse of the matrix $\mathbf{F_x} = (\partial F_i/\partial x_j)$. (If we take differentials in the example we simply replace x_i by dx_i, u_j by du_j; the equations are already linearized.) Once we have the dx_i in terms of the du_j, we can read off partial derivatives. Thus the crucial idea is that of *linearizing the equations*. Since the differentials give linear approximations, we can also say we are approximating given nonlinear equations by linear equations; the approximations are such that, at each point concerned, the derivatives can be obtained exactly from the linear equations. Geometrically, the procedure is analogous to finding the derivative of $y = f(x)$ at a point as the slope of the tangent line (graph of linear approximation) at the point.

Remark 3. When nonlinear implicit equations are difficult to solve, one can use the derivative formulas obtained above as an aid to finding the solutions. We illustrate this for an equation of form $F(x, y) = 0$, selecting one which we, in fact, can solve explicitly:

EXAMPLE 4 $\quad F(x, y) = x^2 + 2xy + 3y^2 - 6 = 0$

We observe that the graph passes through the point (1, 1) in the xy-plane and try to obtain a solution through this point. By implicit differentiation,

$$2x\,dx + 2y\,dx + 2x\,dy + 6y\,dy = 0$$

$$\frac{dy}{dx} = -\frac{2x + 2y}{2x + 6y} = -\frac{F_x}{F_y}$$

Hence

$$\frac{dy}{dx} = y' = -\frac{x + y}{x + 3y}$$

However, this is the derivative of an unknown function $y = f(x)$, concerning which we know only that $f(1) = 1$. Our equation gives

$$f'(1) = -\frac{1 + 1}{1 + 3} = -\frac{1}{2}$$

We now use the idea that the tangent line approximates the curve closely, and we follow this line (of slope $-\frac{1}{2}$) as x increases from 1 to 1.5. Thus y changes by $-\frac{1}{4}$, and we reach the point (1.5, 0.75) (see Figure 12-33). At this

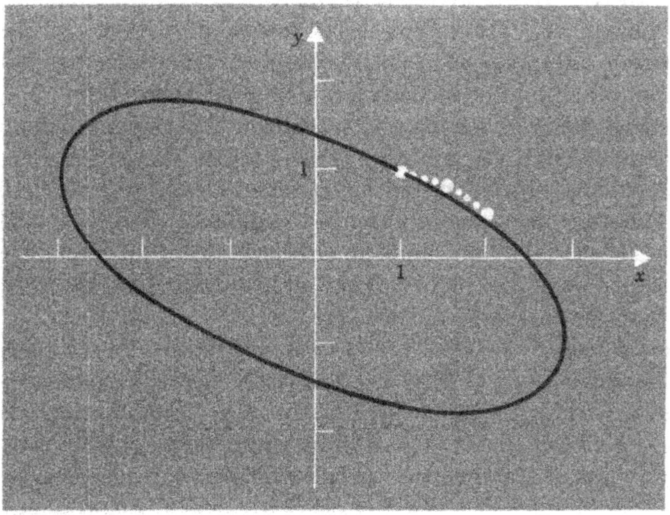

Figure 12-33 The approximate solution of equation $x^2 + 2xy + 3y^2 - 6 = 0$.

point our slope is

$$y' = -\frac{1.5 + 0.75}{1.5 + 3(0.75)} = -0.60$$

Hence, we follow the line with slope -0.60 to $x = 2$, with y decreasing by -0.3, and we reach the point $(2, 0.45)$. The procedure can be continued indefinitely. Also, we can get better approximations by using shorter intervals.

The exact solution is given by

$$y = \frac{-x \pm \sqrt{18 - 2x^2}}{3}$$

We take the plus sign and obtain a differentiable function f with $f(1) = 1$:

$$y = f(x) = \frac{-x + \sqrt{18 - 2x^2}}{3}, \qquad -3 < x < 3$$

This function is graphed in Figure 12-33, along with the approximate solution. (The graph is part of an ellipse; see Section 6-5.)

The procedure described can be applied generally to an equation $F(x, y) = 0$ for which we know a starting point (x_0, y_0) on the graph. We obtain a formula for y':

$$y' = -\frac{F_x}{F_y} = g(x, y) \qquad (12\text{-}143')$$

and can then move short distances along straight lines to obtain a broken line approximating the solution. Equation (12-143′) is a first-order differential equation, and our procedure is simply a standard numerical method for solving such equations (Chapter 14). The method has been highly refined to one yielding high accuracy and adapted to digital computers. The procedure can be extended to general simultaneous equations in several unknowns.

For all the formulas derived, we have had to assume that the appropriate denominator was not 0. That denominator is the Jacobian of the functions $F, G, \ldots$ with respect to the chosen dependent variables. It can be shown that, when this determinant is not 0 and appropriate continuity conditions hold true, one, in fact, is sure that the implicit equations define differentiable functions as sought. This is the essence of the Implicit Function Theorem, discussed in the next section.

If the Jacobian in question is 0 at the point considered, then normally no differentiable function is defined. This is illustrated by the equation $x^2 + y^2 = 1$. The formulas lead to $dy/dx = -(2x/2y) = -(x/y)$, and the denominator is 0 where $y = 0$. But $y = \pm\sqrt{1 - x^2}$ and, with either sign, we have trouble where $y = 0$, that is, where $x = \pm 1$; for the derivative dy/dx does not exist (the slope of the tangent to the circle is ∞).

‡12-15 IMPLICIT FUNCTION THEOREM

The discussion of implicit equations in the preceding section is purely formal; it is assumed that we can solve them, and then it gives formulas for derivatives of the solutions. It is important to know when we can solve. In practice, the typical situation is that one initially knows just one point on the graph of the solution sought and then one tries to find a continuous solution through this point. The Implicit Function Theorem, in its various forms, insures that, *if the relevant Jacobian is not zero at the point, then this is possible at least in a sufficiently small neighborhood of the initial point.* The basis of the theorem is the fact that, near the point, our equation can be approximated by linear equations (by taking differentials), so that, provided that the relevant matrix is nonsingular, one can always solve.

We here consider only the simplest case of one equation in two unknowns, which we write as

$$F(x, y) = 0 \tag{12-150}$$

For a discussion of the general case, we refer the reader to Chapter 9 of the book *Principles of Mathematical Analysis* by Walter Rudin (McGraw-Hill, New York, 1964).

THEOREM 11. *Let $F(x, y)$ be defined in an open region D of the xy-plane, let (x_0, y_0) be in D and let $F(x_0, y_0) = 0$; let the partial derivatives F_x, F_y be continuous in D and suppose that $F_y(x_0, y_0) \neq 0$. Then there is a function*

$$y = f(x), \text{ with domain } |x - x_0| < \delta, \qquad \delta > 0 \tag{12-151}$$

whose graph is in D, with $f(x_0) = y_0$, and which satisfies Equation (12-150). *Furthermore, a positive number η can be chosen so that the graph of* (12-151) *lies in the set*

$$|x - x_0| < \delta, \qquad |y - y_0| < \eta$$

and provides all solutions of (12-150) *in this set. The function f is differentiable and*

$$f'(x) = -\frac{F_x(x, f(x))}{F_y(x, f(x))}$$

PROOF. We let

$$g(x, y) = -\frac{F_x(x, y)}{F_y(x, y)}$$

wherever $F_y \neq 0$, so that

$$F_x = -F_y g$$

Now $F_y(x_0, y_0) \neq 0$. Let us suppose, to be specific, that $F_y(x_0, y_0) > 0$. Then by continuity $F_y(x, y) > 0$ in a neighborhood of (x_0, y_0). Therefore, we can choose δ, η so small and positive that the closed rectangular region

$$E: |x - x_0| \leq \delta, \quad |y - y_0| \leq \eta$$

is in D, and $F_y > 0$ in E. The function g is also continuous in E and, hence, has absolute minimum m and maximum M in E (Theorem F, Section 12-7 and Section 12-25 below):

$$m \leq g(x, y) \leq M \qquad \text{in } E$$

We replace δ by a smaller number, if necessary, to insure that $|m|\,\delta < \eta$ and $|M|\,\delta < \eta$. This is done to insure that the graphs of the linear functions

$$y - y_0 = M(x - x_0), \qquad y - y_0 = m(x - x_0), \qquad |x - x_0| \leq \delta \quad (12\text{-}152)$$

lie in E (see Figure 12-34). We assume δ so chosen.

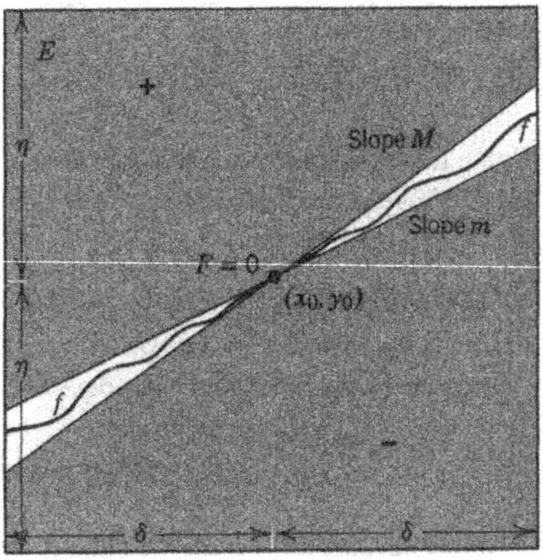

Figure 12-34 The proof of Implicit Function Theorem.

Now, since $F_y > 0$ in E and $F(x_0, y_0) = 0$, F is monotone strictly increasing along the line $x = x_0$ in E and, hence, is positive for $y > y_0$, negative for $y < y_0$. Along a line

$$y - y_0 = \lambda(x - x_0)$$

F becomes a function of x and

$$\frac{dF}{dx} = F_x + F_y \frac{dy}{dx} = F_x + \lambda F_y = -F_y g + \lambda F_y = F_y(\lambda - g)$$

Hence, if λ is greater than M, the maximum of g, dF/dx is positive along the line; thus F itself is positive for $x > x_0$, negative for $x < x_0$. There is similar reasoning, with reversal of signs, for $\lambda < m$. Thus the sign of F is as in Figure 12-34. Since $F_y > 0$, F is monotone strictly increasing in y on each line $x = \text{const}$ in E and, hence, must go from negative to positive. Therefore, by the Intermediate Value Theorem (Section 2-7), $F(x, y) = 0$ for exactly one y, for each x. This value of y we denote by $f(x)$. Thus

$$F(x, f(x)) = 0$$

and $y = f(x)$ provides all solutions of the implicit equation in E.

The graph of f is squeezed between the two lines (12-152). Hence, f is continuous at x_0, with $f(x) \to y_0 = f(x_0)$ as $x \to x_0$. But the same argument applies to each point (x, y) on the graph of f; we can find a rectangle E centered at the point and so on, as for (x_0, y_0). Therefore, f is continuous for all x in the interval $|x - x_0| < \delta$. (We can also define f at the end points of the interval, $x_0 \pm \delta$, but ignore these values in the subsequent discussion.)

Finally, we seek the derivative of f at x_0. Along a line $y - y_0 = \lambda(x - x_0)$ with $\lambda = g(x_0, y_0) + \epsilon$, $\epsilon > 0$, we have, as above,

$$\frac{dF}{dx} = F_y \cdot (\lambda - g) = F_y \cdot [g(x_0, y_0) - g(x, y) + \epsilon]$$

Since g is continuous at (x_0, y_0), $|g(x, y) - g(x_0, y_0)| < \epsilon$ for (x, y) sufficiently close to (x_0, y_0): thus the quantity in brackets is positive and $dF/dx > 0$. Therefore, F itself is positive along the line, for $x > x_0$ and x sufficiently close to x_0. But this means that the graph of f must be below the line; that is

$$f(x) < y_0 + [g(x_0, y_0) + \epsilon](x - x_0)$$

or

$$\frac{f(x) - f(x_0)}{x - x_0} < g(x_0, y_0) + \epsilon$$

for $x > x_0$ as above. Similarly,

$$\frac{f(x) - f(x_0)}{x - x_0} > g(x_0, y_0) - \epsilon$$

for $x > x_0$ and x sufficiently close to x_0. Thus

$$\lim_{x \to x_0+} \frac{f(x) - f(x_0)}{x - x_0} = g(x_0, y_0)$$

In the same way, the same limit is found as $x \to x_0-$. Therefore

$$f'(x_0) = g(x_0, y_0)$$

Again, the argument applies to every point (x, y) on the graph of f, and we conclude that at each such point

$$f'(x) = g(x, y) = -\frac{F_x(x, y)}{F_y(x, y)}$$

Thus the theorem is proved.

Remark. We note that the proof gives some information as to the size of the x-interval in which a solution $y = f(x)$ can be found. Specifically, one must choose a rectangle E: $|x - x_0| \leq \delta$, $|y - y_0| \leq \eta$ in which $F_y > 0$ (or $F_y < 0$) and in which δ has been restricted so that $\delta K < \eta$, where $K = \text{Max}\,|g(x, y)|$ in E. As pointed out in Section 3-8, one sometimes has more information about F that permits one to give a better estimate of the size of the interval. For example, if $F_y > 0$ in E and F is positive for $y = y_0 + \eta$, negative for $y = y_0 - \eta$, then the solution is defined and unique for $|x - x_0| < \delta$. The restriction $|y - y_0| \leq \eta$, in general, is needed to insure uniqueness of the solution. For example, the equation

$$y^2 - y \sin x - e^x y + e^x \sin x = 0$$

satisfies the hypotheses of the theorem with $x_0 = 0$, $y_0 = 0$, and a solution is given by $y = \sin x$. However, another solution is $y = e^x$; we exclude this solution by restricting to a sufficiently small rectangle E about $(0, 0)$.

PROBLEMS

1. Find the indicated partial derivatives, assuming that the implicit function theorem is applicable in each case:

(a) $\left(\frac{\partial z}{\partial x}\right)_y$ and $\left(\frac{\partial z}{\partial y}\right)_x$, if $x^2 e^z + yz^2 + xy - 1 = 0$.

(b) $\left(\frac{\partial u}{\partial v}\right)_w$ and $\left(\frac{\partial u}{\partial w}\right)_v$, if $u^3 v - uv^3 + uvw - 2w = 0$.

(c) $\left(\frac{\partial z}{\partial x}\right)_{ty}$, if $xyzt - x^2 - y^2 - z^2 - t^2 + 1 = 0$.

(d) $\left(\frac{\partial x_1}{\partial x_3}\right)_{x_2 x_4}$, if $x_1 \ln (x_2^2 + x_3^2) + x_3 \sin (x_1 - x_4) + \cos (x_1 + x_2) = 0$.

(e) dx/dt and dy/dt, if $\begin{cases} x^2 - xt - y^2 t^2 - y + 2 = 0 \\ y^2 - 2yt + xt^2 - y - t - x = 0 \end{cases}$

(f) du/dv, if $\begin{cases} u^2 - v^2 + w^2 + uv - vw = 0 \\ u^2 + 2v^2 - w^2 + uv - v - 1 = 0 \end{cases}$

(g) $\left(\frac{\partial x}{\partial u}\right)_v$ and $\left(\frac{\partial x}{\partial v}\right)_u$, if $\begin{cases} \cos x + x \sin u - y \cos v - \sin y = 0 \\ \sin x - y \sin u + x \cos v + \cos y = 0 \end{cases}$

(h) $\left(\frac{\partial u}{\partial x}\right)_y$ and $\left(\frac{\partial u}{\partial y}\right)_x$, for the equations of part (g).

(i) $\left(\frac{\partial z}{\partial x}\right)_y$ if $\begin{cases} x^2 - ty + xtz - uy^2 + z^2 + 1 = 0 \\ x^2 + xt - ytz^2 + ux^2 - yz - 2 = 0 \\ t^2 - xy + xz + tz - x^2 - u = 0 \end{cases}$

(j) $\left(\frac{\partial u}{\partial x}\right)_{yz}$ if $\begin{cases} ue^u + \sin v - vw + x^2 - xyz + 1 = 0 \\ ue^v + \cos u - uv + x^2 + 2xyz - y = 0 \\ we^v + \sin u - vw + \cos w - xy = 0 \end{cases}$

2. In each of the following cases, one has equations $F_i(x_1, \ldots, x_n) = 0$, $i = 1, \ldots, m$. The Jacobian matrix $(\partial F_i/\partial x_j)$ is given, evaluated at a certain point $(x_1^0, \ldots, x_n^0)$ at which all equations are satisfied. From this information obtain the Jacobian matrix requested, in indicated form as a product of matrices; for example, in (a), one is solving for x_1 and x_2 in terms of x_3 and x_4.

(a) $\begin{pmatrix} 2 & 3 & 1 & 0 \\ 3 & 5 & 2 & 5 \end{pmatrix}$, find $\left(\frac{\partial x_i}{\partial x_j}\right)$, for $i = 1, 2, j = 3, 4$

(b) $\begin{pmatrix} 5 & 1 & 0 & 5 \\ -1 & 1 & 3 & 6 \end{pmatrix}$, find $\left(\dfrac{\partial x_i}{\partial x_j}\right)$, for $i = 3, 4, j = 1, 2$

(c) $\begin{pmatrix} 6 & 2 & 3 & 4 & 5 \\ 1 & 0 & 1 & 1 & 2 \\ 5 & 1 & 3 & 0 & 2 \end{pmatrix}$, find $\left(\dfrac{\partial x_i}{\partial x_j}\right)$, for $i = 1, 2, 3, j = 4, 5$

(d) $\begin{pmatrix} 4 & 2 & 1 & 2 & 1 & 7 \\ 1 & 0 & 2 & -1 & 0 & 5 \\ 6 & 1 & 1 & 5 & 2 & 3 \end{pmatrix}$, find $\left(\dfrac{\partial x_i}{\partial x_j}\right)$, for $i = 2, 3, 5, j = 1, 4, 6$

3. (a) For the system of Problem 2(a), assume that $(x_1{}^0, x_2{}^0, x_3{}^0, x_4{}^0)$ is $(0, 0, 0, 0)$ and find x_1, x_2 approximately for $x_3 = 0.5$, $x_4 = 0.2$.

(b) For the system of Problem 2(b), assume that $(x_1{}^0, x_2{}^0, x_3{}^0, x_4{}^0)$ is $(2, 1, 3, 5)$ and find x_3, x_4 approximately for $x_1 = 2.2$, $x_2 = 0.9$.

‡4. Show that the equation $(1 + x^2 + y^2)z + z^3 + 1 = 0$ can be solved for $z = f(x, y)$, where f is defined and differentiable for all (x, y).

5. *Algebraic equations.* Let $F(x, y)$ be a polynomial in x, y without constant term, so that $F(0, 0) = 0$. The Implicit Function Theorem may then be applicable to show that, near $(0, 0)$, the graph of the equation $F(x, y) = 0$ can be represented in the form $y = f(x)$ or $x = g(y)$.

(a) Show that, if $F(x, y) = ax + by +$ terms of higher degree and $a^2 + b^2 \neq 0$, then $ax + by = 0$ is the equation of the tangent line to the graph of $F(x, y) = 0$ at $(0, 0)$.

(b) Find the tangent line to the graph at $(0, 0)$ (see part (a)):

(i) $2x - 3y + x^2 = 0$ (ii) $x + y^2 + xy^3 = 0$ (iii) $2y - x^2y = 0$

‡(c) For each of the following equations, show that the Implicit Function Theorem is not applicable at $(0, 0)$ and describe the nature of the graph near $(0, 0)$:

(i) $x^2 - y^3 = 0$, (ii) $xy = 0$, (iii) $x^2 - y^4 = 0$, (iv) $x^3 - x^2 + y^2 = 0$

12-16 INVERSE FUNCTIONS

Equations of the form

$$u = f(x, y), \qquad v = g(x, y) \tag{12-160}$$

can be interpreted as a mapping from a set in the xy-plane into the uv-plane, as in Figure 12-35. If this mapping happens to be one-to-one, one can form an inverse mapping, of the form

$$x = \varphi(u, v), \qquad y = \psi(u, v) \tag{12-161}$$

Finding the inverse mapping is equivalent to solving equations (12-160) for x and y in terms of u and v. If we write (12-160) in the form

$$f(x, y) - u = 0, \qquad g(x, y) - v = 0 \tag{12-160$'$}$$

we see that we are really dealing with two implicit equations in four unknowns and that we are trying to solve for two of the unknowns—namely, x and y. Hence, the Implicit Function Theorem for this case is relevant. To apply the

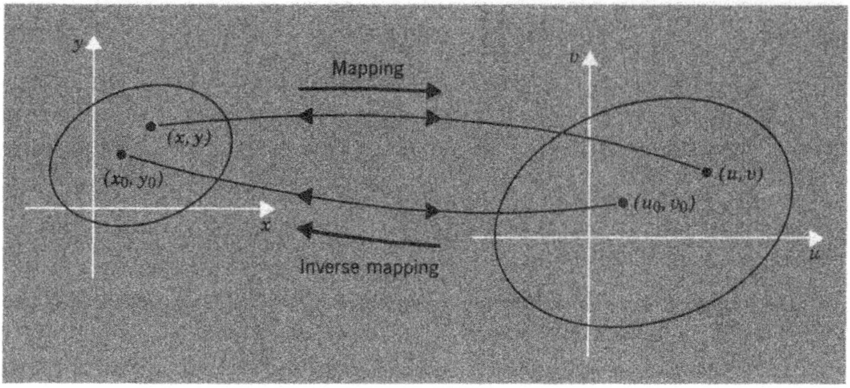

Figure 12-35 Inverse mapping.

theorem, we select a point (x_0, y_0) at which f and g are defined, and let

$$u_0 = f(x_0, y_0), \qquad v_0 = g(x_0, y_0)$$

as suggested in Figure 12-35. Thus (x_0, y_0, u_0, v_0) is a set of values satisfying both equations (12-160′). We assume that f, g and their partial derivatives satisfy usual continuity conditions in a neighborhood of (x_0, y_0). Then, as in Section 12-14, the crucial condition is that the Jacobian of the system (12-160′) have a value different from zero; we set $F_1 = F_1(x, y, u, v) = f(x, y) - u$, $F_2 = g(x, y) - v$, and the condition becomes

$$\frac{\partial(F_1, F_2)}{\partial(x, y)} \neq 0$$

at (x_0, y_0, u_0, v_0). But

$$\frac{\partial(F_1, F_2)}{\partial(x, y)} = \begin{vmatrix} f_x & f_y \\ g_x & g_y \end{vmatrix} = \frac{\partial(f, g)}{\partial(x, y)}$$

Hence, if

$$\frac{\partial(f, g)}{\partial(x, y)} \neq 0 \qquad \text{at } (x_0, y_0) \tag{12-162}$$

we get a solution (the inverse mapping (12-161)) in a neighborhood of (u_0, v_0). The solution functions are continuous and have continuous partial derivatives:

$$\frac{\partial \varphi}{\partial u} = -\frac{\dfrac{\partial(F_1, F_2)}{\partial(u, y)}}{\dfrac{\partial(F_1, F_2)}{\partial(x, y)}} = -\frac{\begin{vmatrix} -1 & f_y \\ 0 & g_y \end{vmatrix}}{\dfrac{\partial(f, g)}{\partial(x, y)}} = \frac{g_y}{\dfrac{\partial(f, g)}{\partial(x, y)}}$$

and so on. We remark that (u_0, v_0) do not appear in (12-162).

The reasoning extends to a system of n equations:

$$u_i = g_i(x_1, \ldots, x_n), \qquad i = 1, \ldots, n \tag{12-163}$$

We write these as implicit equations

$$g_i(x_1, \ldots, x_n) - u_i = 0, \qquad i = 1, \ldots, n$$

and want to solve for $x_1, \ldots, x_n$ in terms of $u_1, \ldots, u_n$ (inverse mapping). The Jacobian condition becomes

$$\frac{\partial(g_1, \ldots, g_n)}{\partial(x_1, \ldots, x_n)} \neq 0 \qquad \text{at } (x_1^0, \ldots, x_n^0). \tag{12-164}$$

We can interpret this result in matrix form: we are given a differentiable vector function

$$\mathbf{u} = \mathbf{g}(\mathbf{x})$$

for $\mathbf{x}$ in an open region D_0 of V_n, with values also in V_n. We have

$$d\mathbf{u} = \mathbf{g_x}\, d\mathbf{x}$$

and wish to solve for $\mathbf{x}$ near $\mathbf{x}^0$; that is, we wish to express $d\mathbf{x}$ in terms of $d\mathbf{u}$. But the last equation is a matrix equation:

$$\begin{pmatrix} du_1 \\ \vdots \\ du_n \end{pmatrix} = \begin{pmatrix} \dfrac{\partial g_1}{\partial x_1} & \cdots & \dfrac{\partial g_1}{\partial x_n} \\ \vdots & & \vdots \\ \dfrac{\partial g_n}{\partial x_1} & \cdots & \dfrac{\partial g_n}{\partial x_n} \end{pmatrix} \begin{pmatrix} dx_1 \\ \vdots \\ dx_n \end{pmatrix}$$

and can be solved uniquely, provided that the n by n matrix $\mathbf{g_x}$ is nonsingular at the point considered [condition (12-164)]. The solution is

$$d\mathbf{x} = \mathbf{g_x}^{-1}\, d\mathbf{u} \tag{12-165}$$

From this equation we read off the partial derivatives $\partial x_i/\partial u_j$. They are the entries of the matrix $\mathbf{x_u}$, where $d\mathbf{x} = \mathbf{x_u}\, d\mathbf{u} = \mathbf{g_x}^{-1}\, d\mathbf{u}$, so that

$$\mathbf{x_u} = \mathbf{g_x}^{-1} = \mathbf{u_x}^{-1} \tag{12-166}$$

Thus *the Jacobian matrix of the inverse mapping is the inverse of the Jacobian matrix of the mapping.*

Remark. For $n = 1$, we have one equation

$$u = g(x)$$

and (12-166) becomes the rule of elementary calculus:

$$\frac{dx}{du} = \frac{1}{(du/dx)}$$

If the equations (12-163) themselves happen to have linear form

$$u_i = \sum_{j=1}^{n} b_{ij} x_j, \qquad i = 1, \ldots, n$$

we can take $(0, \ldots, 0)$ as the given point $(x_1^0, \ldots, x_n^0)$ and identify dx_i with

x_i, du_i with u_i for all i. Hence, we have a matrix equation

$$\mathbf{u} = B\mathbf{x}$$

representing a linear mapping from V_n to V_n. We know (Section 10-12) that this mapping is one-to-one and has an inverse precisely when B is nonsingular (that is, det $B \neq 0$). The inverse is linear and is given by

$$\mathbf{x} = B^{-1}\mathbf{u}$$

The calculus in effect permits us to reduce the general nonlinear problem to the linear one.

EXAMPLE 1. $u = 2x - 5y$, $v = x + y$. Here the Jacobian determinant is

$$\frac{\partial(f, g)}{\partial(x, y)} = \frac{\partial(u, v)}{\partial(x, y)} = \begin{vmatrix} 2 & -5 \\ 1 & 1 \end{vmatrix} = 7$$

The mapping is itself linear, with matrix $B = \begin{pmatrix} 2 & -5 \\ 1 & 1 \end{pmatrix}$; the Jacobian determinant is det B. Since det $B \neq 0$, B is nonsingular, and the mapping is one-to-one. We solve for x, y to obtain the inverse mapping:

$$x = \frac{1}{7}(u + 5v), \qquad y = \frac{1}{7}(-u + 2v)$$

Thus

$$B^{-1} = \frac{1}{7}\begin{pmatrix} 1 & 5 \\ -1 & 2 \end{pmatrix}$$

The mapping can be studied in detail by considering the images of lines parallel to the axes. For example, lines $x = \text{const} = k$ correspond to parallel lines

$$u + 5v = 7k$$

while lines $y = \text{const} = k$ correspond to lines

$$-u + 2v = 7k$$

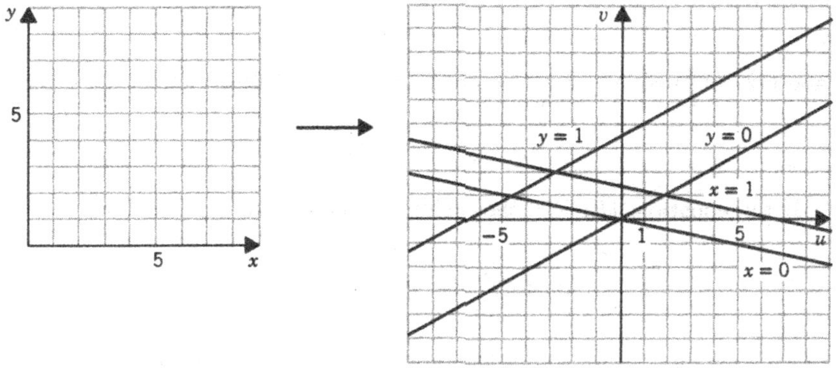

Figure 12-36 The mapping $u = 2x - 5y$, $v = x + y$.

We note that these systems of lines provide oblique coordinates in the uv-plane (Figure 12-36).

Remark. In the example a square with sides parallel to the axes in the xy-plane corresponds to a parallelogram in the uv-plane, and one can verify that the area of the parallelogram is 7 times the area of the square (Problem 4 below). In fact, here all areas are multiplied by 7 in going from a set in the xy-plane to its image in the uv-plane. There is an analogous statement for the general linear mapping from R^n to R^n with matrix B. The absolute value of det B gives the ratio of the n-volume of the image of set E to the n-volume of the set E (see Section 11-16). For a nonlinear mapping $\mathbf{u} = \mathbf{g}(\mathbf{x})$ we have an approximating linear mapping near a particular point; its matrix is $\mathbf{g_x}$ and its determinant is the corresponding Jacobian determinant. Thus the absolute value of the Jacobian determinant measures, approximately, the ratio of n-volume of image to n-volume of set, near a given point; roughly, it is the expansion factor. When the determinant is 0, something has gone wrong.

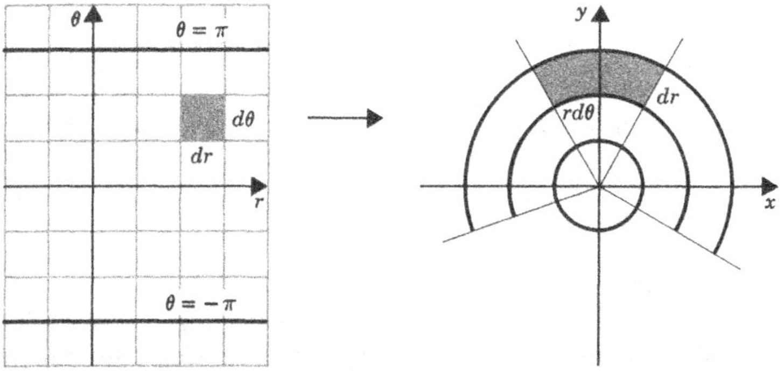

Figure 12-37 Mapping associated with polar coordinates.

EXAMPLE 2. $x = r\cos\theta$, $y = r\sin\theta$. These are the equations relating rectangular and polar coordinates. However, we can consider them as representing a mapping from the $r\theta$-plane to the xy-plane. The Jacobian determinant is

$$\frac{\partial(x, y)}{\partial(r, \theta)} = \begin{vmatrix} \cos\theta & -r\sin\theta \\ \sin\theta & r\cos\theta \end{vmatrix} = r$$

Hence, the determinant is 0 only when $r = 0$, so that $x = 0$, $y = 0$. For $r \neq 0$, the Implicit Function Theorem applies and we get inverse functions. We can in fact solve:

$$r = \pm\sqrt{x^2 + y^2}, \qquad \theta = \cos^{-1}\frac{x}{r} = \sin^{-1}\frac{y}{r}$$

However, we get many values of r and θ for each x and y. To obtain single-valued expressions, we must restrict r and θ (and x, y correspondingly), for example, by requiring $r > 0$ and $-\pi < \theta < \pi$. (The restriction is in accordance with the Implicit Function Theorem, which provides a unique solution only in a *neighborhood* of a given point.) We remark finally that the lines $\theta =$ constant, $r =$ constant correspond to lines and concentric circles in the xy-plane, as in Figure 12-37.

Polar coordinates are an example of curvilinear coordinates in the plane. Other curvilinear coordinates are obtained through equations $x = f(u, v)$, $y = g(u, v)$, or, in R^n, by $\mathbf{x} = \mathbf{f}(\mathbf{u})$; see Problem 3 below.

We note that for the mapping associated with polar coordinates the Jacobian determinant is r and that a small rectangle of sides dr, $d\theta$ in the $r\theta$-plane corresponds to an approximately rectangular figure in the xy-plane. The area of the image should be approximately $r\,dr\,d\theta$, and we can check by geometry that this, indeed, is a very good approximation to the area (Problem 5 below).

PROBLEMS

1. For each of the following mappings, find the Jacobian determinant of the mapping and of the inverse mapping.

   (a) $u = 2x - y,\ v = x + 4y$ (b) $u = 3x + 2y,\ v = x - y$

   (c) $u = x - y - z,\ v = 2x + 2y + 5z,\ w = x + 3z$

   (d) $u = 2x + y,\ v = 2y - z,\ w = 3x$ (e) $u = e^x + y^3,\ v = 3e^x - 2y^3$

   (f) $u = \dfrac{x}{x^2 + y^2},\ v = -\dfrac{y}{x^2 + y^2}$

   (g) $u = x^3 + xyz,\ v = x^2y - xy^2,\ w = x^3 - xz^2 + z^3$

   (h) $u = x\cos y - z,\ v = x\sin y + 2z,\ w = x^2 + z^2$

2. For each of the following mappings find linear equations approximating the mapping and linear equations approximating the inverse mapping near the given point:

   (a) as in Problem 1(e) near $x = 0,\ y = 1$,

   (b) as in Problem 1(f) near $x = 1,\ y = 0$,

   (c) as in Problem 1(g) at $x = 1,\ y = 1,\ z = 1$,

   (d) as in Problem 1(h) at $x = 1,\ y = 0,\ z = 1$.

3. (a) Let the equations of Problem 1(e) be used to define curvilinear coordinates x, y in the uv-plane. Sketch the lines $x =$ const, $y =$ const and their images for several choices of the constants.

   (b) Proceed as in part (a) with the equations of Problem 1(f).

4. Let a, b, c, d be constants and let the equations

$$u = ax + by, \qquad v = cx + dy$$

   define a one-to-one mapping of the xy-plane onto the uv-plane.

   (a) Show that the square of vertices $(0, 0)$, $(1, 0)$, $(1, 1)$, $(0, 1)$ in the xy-plane has as image a parallelogram in the uv-plane whose area is $\pm\begin{vmatrix} a & b \\ c & d \end{vmatrix}$ (see Section 1-12).

(b) Show generally that every triangle of area A in the xy-plane has as image a triangle in the uv-plane whose area is $\pm\begin{vmatrix} a & b \\ c & d \end{vmatrix} A$.

5. Let r, θ be polar coordinates in the xy-plane. Show that the area of the portion of the xy-plane defined by the inequalities

$$r_0 \leq r \leq r_0 + dr, \qquad \theta_0 \leq \theta \leq \theta_0 + d\theta$$

where $r_0 > 0$, $dr > 0$, $0 < d\theta < 2\pi$, is $r^* \, dr \, d\theta$, where r^* is an appropriate value between r_0 and $r_0 + dr$. Hence for small dr, the area is approximately $r_0 \, dr \, d\theta$.

6. Show on the basis of the geometric meaning of the Jacobian determinant that, in cylindrical coordinates r, θ, z, the volume of the portion of space for which

$$r_0 \leq r \leq r_0 + dr, \qquad \theta_0 \leq \theta \leq \theta_0 + d\theta, \qquad z_0 \leq z \leq z_0 + dz$$

where $r_0 > 0$, $dr > 0$, $0 < d\theta < 2\pi$, $dz > 0$, is given approximately (for small dr, $d\theta$, dz) by $r_0 \, dr \, d\theta \, dz$ and justify by a sketch (see Problem 5).

‡7. Find a result analogous to that of Problem 6 for spherical coordinates (Section 11-20).

12-17 CURVES IN SPACE

As in the plane, it is most convenient to represent a curve in space by parametric equations:

$$x = f(t), \qquad y = g(t), \qquad z = h(t) \tag{12-170}$$

where all functions are defined and continuous in a given interval. These equations then define a path in space (Figure 12-38). One can think of t as

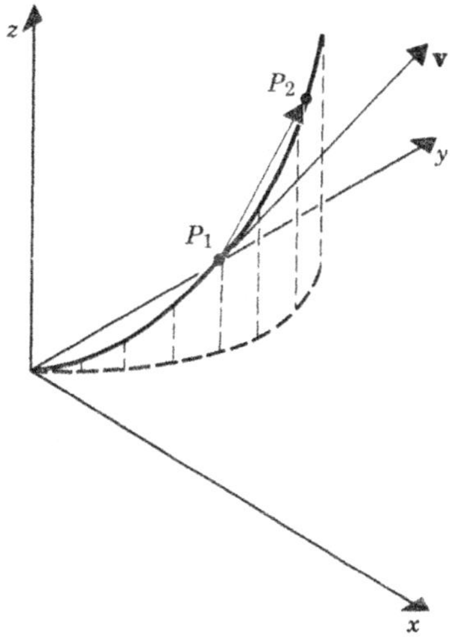

Figure 12-38 Path in space and tangent line.

time and (x, y, z) as the position of a point P moving in space. If we write

$$\mathbf{r} = \overrightarrow{OP} = x\mathbf{i} + y\mathbf{j} + z\mathbf{k}$$

then Equation (12-170) can be replaced by one vector equation

$$\mathbf{r} = \mathbf{F}(t) \tag{12-170'}$$

In particular, if $\mathbf{a}$ and $\mathbf{b}$ are constant vectors,

$$\mathbf{r} = \mathbf{a} + t\mathbf{b}, \qquad -\infty < t < \infty$$

represents a straight line in space, as in Section 11-4.

The derivative

$$\frac{d\mathbf{r}}{dt} = \mathbf{F}'(t) = f'(t)\mathbf{i} + g'(t)\mathbf{j} + h'(t)\mathbf{k}$$

can be interpreted as the velocity $\mathbf{v}$ of the moving point P. Let $\overrightarrow{OP_1} = \mathbf{F}(t_1)$ and let $\mathbf{F}'(t_1)$ exist and be different from $\mathbf{0}$. Then the vector $\mathbf{v} = \mathbf{F}'(t_1)$ gives the direction of a straight line through P_1; this direction is the limiting direction of a chord P_1P_2, as in Figure 12-38. Hence, just as in the plane, we can interpret this line as the *tangent* to the path at P_1. If Q is a general point on the tangent line, the line has the equation

$$\overrightarrow{OQ} = \overrightarrow{OP_1} + \tau\mathbf{v} = \overrightarrow{OP_1} + \tau\mathbf{F}'(t_1) \tag{12-171}$$

in terms of the parameter τ. If P_1 is (x_1, y_1, z_1) and Q is (x, y, z), (12-171) becomes

$$x = x_1 + a\tau, \qquad y = y_1 + b\tau, \qquad z = z_1 + c\tau \tag{12-171'}$$

where $\mathbf{F}'(t_1) = a\mathbf{i} + b\mathbf{j} + c\mathbf{k}$.

EXAMPLE 1. $x = t$, $y = t$, $z = t^2$. Here $\mathbf{F}(t) = t\mathbf{i} + t\mathbf{j} + t^2\mathbf{k}$, $\mathbf{F}'(t) = \mathbf{i} + \mathbf{j} + 2t\mathbf{k}$. For $t = t_1 = 1$, we find the tangent line as follows: $\mathbf{F}(1) = \mathbf{i} + \mathbf{j} + \mathbf{k}$, so that P_1 is $(1, 1, 1)$, and $\mathbf{F}'(1) = \mathbf{i} + \mathbf{j} + 2\mathbf{k}$; the vector equation of the tangent line is

$$\overrightarrow{OQ} = \mathbf{i} + \mathbf{j} + \mathbf{k} + \tau(\mathbf{i} + \mathbf{j} + 2\mathbf{k})$$

The parametric equations of the tangent line are

$$x = 1 + \tau, \qquad y = 1 + \tau, \qquad z = 1 + 2\tau, \qquad -\infty < \tau < \infty$$

Such a path and tangent line are shown in Figure 12-38.

If in (12-171') we eliminate τ, we obtain two linear equations in x, y, z representing two planes intersecting in the tangent line (see Section 11-12). For the line found in Example 1, we obtain

$$x - y = 0, \qquad 2y - z = 1$$

These equations also represent the tangent line. As shown in Section 11-12, all such representations are obtained by choosing two linearly independent vectors $\mathbf{u}$, $\mathbf{w}$ orthogonal to $\mathbf{v}$ and forming the equations:

$$\overrightarrow{P_1Q} \cdot \mathbf{u} = 0, \qquad \overrightarrow{P_1Q} \cdot \mathbf{w} = 0$$

In general, a plane through the tangent line is said to be a *tangent plane* of the curve at the point P_1. Thus a curve has infinitely many tangent planes at each point.

One can define arc length s for a path in space in exactly the same way as for a path in the plane (Section 4-27). By taking the limit of inscribed polygons, we obtain

$$L = \int_a^b \sqrt{[f'(t)]^2 + [g'(t)]^2 + [h'(t)]^2}\, dt \qquad (12\text{-}172)$$

where L is the length of the path (12-170) between $t = a$ and $t = b$; we assume that f, g, and h have continuous derivatives in this interval. We can write (12-172) in other forms:

$$L = \int_a^b \|\mathbf{F}'(t)\|\, dt \qquad \text{or} \qquad L = \int_a^b \|\mathbf{v}\|\, dt \qquad (12\text{-}172')$$

Thus $\|\mathbf{v}\|$, the magnitude of the velocity vector, can be interpreted as speed:

$$\|\mathbf{v}\| = \frac{ds}{dt}$$

where s is the arc length from a fixed initial time t_0 to a general t:

$$s = \int_{t_0}^{t} \sqrt{[f'(u)]^2 + [g'(u)]^2 + [h'(u)]^2}\, du \qquad (12\text{-}173)$$

As in Section 4-29, we can consider different parametrizations and define two paths to be equivalent if one is obtained from the other by an equation $t = \varphi(\tau)$ relating the two parameters, where φ' is continuous and positive (or negative) over the corresponding interval. When $\mathbf{F}'(t)$ is itself continuous and not $\mathbf{0}$ on the path (12-170), we can use Equation (12-173) to introduce s as parameter along the path; s then increases with t. In terms of s as parameter, the tangent vector has components dx/ds, dy/ds, dz/ds:

$$\frac{dx}{ds} = \frac{dx}{dt}\bigg/\frac{ds}{dt} = \frac{1}{\|\mathbf{v}\|}\frac{dx}{dt}, \qquad \frac{dy}{ds} = \frac{1}{\|\mathbf{v}\|}\frac{dy}{dt}, \qquad \frac{dz}{ds} = \frac{1}{\|\mathbf{v}\|}\frac{dz}{dt}$$

Accordingly, the new tangent vector is

$$\mathbf{T} = \frac{dx}{ds}\mathbf{i} + \frac{dy}{ds}\mathbf{j} + \frac{dz}{ds}\mathbf{k} = \frac{1}{\|\mathbf{v}\|}\mathbf{v} \qquad (12\text{-}174)$$

Thus $\mathbf{T}$ is simply the unit vector in the direction of $\mathbf{v}$ (see Section 6-6).

Normal Plane to a Curve. For our path (12-170), the plane through P_1 orthogonal to the tangent line at that point is called the *normal plane* to the path at the point. Since $\mathbf{v} = \mathbf{F}'(t_1)$ is a normal vector of the plane, the normal plane has the equation:

$$\overrightarrow{P_1Q} \cdot \mathbf{F}'(t_1) = 0 \qquad (12\text{-}175)$$

At the point P_1: (1, 1, 1) of Example 1, $\mathbf{F}'(t_1) = \mathbf{i} + \mathbf{j} + 2\mathbf{k}$, so that the normal plane is

$$(x-1)+(y-1)+2(z-1)=0 \qquad \text{or} \qquad x+y+2z=4$$

Each line through P_1 in the normal plane is called a *normal line* of the path at the point; equivalently, we can define a normal line as a line through P_1 perpendicular to the tangent line.

12-18 SURFACES IN SPACE

Let us first consider a surface given by an equation

$$z=f(x, y) \tag{12-180}$$

where f is defined and differentiable in an open region D and, hence, $\partial z/\partial x$ and $\partial z/\partial y$ exist in D. We have seen in Section 12-8 that $\partial z/\partial x$ at (x_1, y_1) can be interpreted as the slope of a line tangent to the curve $z=f(x, y_1)$ in the plane $y=y_1$; $\partial z/\partial y$ can be interpreted similarly (Figure 12-22).

We also showed in Section 12-8 that the plane containing the two lines had the equation

$$z-z_1=f_x(x_1, y_1)(x-x_1)+f_y(x_1, y_1)(y-y_1) \tag{12-181}$$

and that it was reasonable to call this plane the *tangent plane* to the surface at the point (x_1, y_1, z_1), where $z_1=f(x_1, y_1)$.

A surface in space may also be defined by an equation

$$F(x, y, z)=0 \tag{12-182}$$

as we now show. Let P_1: (x_1, y_1, z_1) be a point satisfying the equation, let F have a differential in a neighborhood of P_1 and let ∇F not be $\mathbf{0}$ at P_1. If, for example, $F_z(x_1, y_1, z_1) \neq 0$, then the Implicit Function Theorem (Sections 12-14 and 12-15) implies that, at least in a neighborhood of P_1, the graph of (12-182) coincides with the graph of a differentiable function $z=f(x, y)$, so that we do obtain a surface (12-180). Now let a differentiable path $\mathbf{r}=\varphi(t)\mathbf{i}+\psi(t)\mathbf{j}+\chi(t)\mathbf{k}$ lie in the surface (12-182) and pass through P_1 for $t=t_1$. Let the velocity vector

$$\mathbf{v}=\varphi'(t_1)\mathbf{i}+\psi'(t_1)\mathbf{j}+\chi'(t_1)\mathbf{k}$$

exist and be different from $\mathbf{0}$. Since the path lies on the surface,

$$F[\varphi(t), \psi(t), \chi(t)]=0$$

for all t in the given interval. Hence, by the chain rule,

$$F_x\varphi'(t)+F_y\psi'(t)+F_z\chi'(t)=0 \tag{12-183}$$

When $t=t_1$, $F_x=F_x(x_1, y_1, z_1)$ and so on; that is, at P_1 we have

$$\mathbf{v}\cdot\nabla F=0$$

Now ∇F is a fixed nonzero vector. We conclude that the tangent vectors at P_1 to *all* paths in the surface passing through P_1 are orthogonal to ∇F or, equivalently, that the tangent lines at P_1 to these paths all lie in the plane through P_1 with normal vector ∇F. Therefore, the plane in question must be the *tangent plane* to the surface (12-182) at P_1 (see Figure 12-39). The equation for the tangent plane is

$$\nabla F \cdot \overrightarrow{P_1P} = 0$$

or, in scalar form,

$$F_x(x_1, y_1, z_1)(x - x_1) + F_y(x_1, y_1, z_1)(y - y_1) + F_z(x_1, y_1, z_1)(z - z_1) = 0 \quad (12\text{-}184)$$

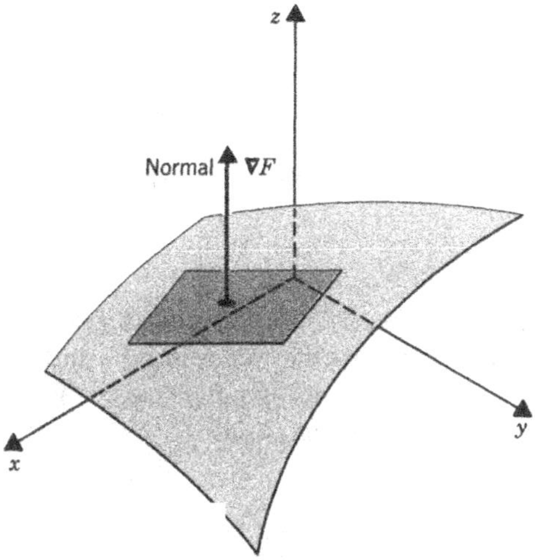

Figure 12-39 Tangent plane and normal line to surface $F(x, y, z) = 0$.

As a check, we consider the surface (12-180). Here we can take $F(x, y, z) = z - f(x, y)$ and $F_x = -f_x$, $F_y = -f_y$, $F_z = 1$ and (12-184) reduces to (12-181).

The vector ∇F (or any nonzero scalar multiple of this vector) is called a *normal vector* of the surface at P_1, and the line $\nabla F \times \overrightarrow{P_1P} = \mathbf{0}$ is called the *normal line* to the surface at P_1 (see Figure 12-39).

Since, near P_1, the surface is representable in a form such as $z = f(x, y)$, we can reason as in Section 12-9 that near P_1 the surface is closely approximated by the tangent plane at P_1.

EXAMPLE 1. The equation $x^2 + y^2 + z^2 = 9$ defines a surface, a sphere. We can write the equation as in (12-182), with $F(x, y, z) = x^2 + y^2 + z^2 - 9$. We seek the tangent plane at P_1: (2, 2, 1). We find that $F_x = 2x = 4$, $F_y = 2y = 4$, $F_z = 2z = 2$ at P_1. Hence, (12-184) becomes

$$4(x - 2) + 4(y - 2) + 2(z - 1) = 0 \qquad \text{or} \qquad 2x + 2y + z = 9$$

We could also solve for z:

$$z = \pm\sqrt{9 - x^2 - y^2}$$

and must use the + sign for the point (2, 2, 1). At this point we find that

$$\frac{\partial z}{\partial x} = \frac{-x}{\sqrt{9 - x^2 - y^2}} = -2, \qquad \frac{\partial z}{\partial y} = \frac{-y}{\sqrt{9 - x^2 - y^2}} = -2$$

and, hence, by (12-181)

$$z - 1 = -2(x - 2) - 2(y - 2) \qquad \text{or} \qquad 2x + 2y + z = 9$$

as before.

At the point P_2: $(3, 0, 0)$, F_z is found to be 0, and we cannot solve for z as a differentiable function of x and y. At this point we could solve for x: $x = \sqrt{9 - y^2 - z^2}$ and proceed as before with the aid of partial derivatives $\partial x/\partial y$, $\partial x/\partial z$. By treating the equation implicitly (not solving for one letter), we avoid this difficulty and obtain the tangent plane at a general point of the surface; however, (12-184) shows (as does the Implicit Function Theorem, see Section 12-14) that at least one partial derivative of F must differ from 0 in order that we may be sure that a plane is determined.

Remark. For a function F, having a continuous gradient vector, each *level surface* of F has an equation

$$F(x, y, z) = c$$

If we write this in the form $F(x, y, z) - c = 0$, we see that the preceding theory is applicable and that, wherever $\nabla F \neq \mathbf{0}$, the equation does define a surface with ∇F as normal vector. At each point (x, y, z) the *gradient vector* ∇F *is normal to the level surface of* F *passing through that point.*

By similar reasoning, we see that an equation

$$F(x, y) = c$$

which defines the *level curves of* $F(x, y)$ in the xy-plane, also defines a true curve through each point where $\nabla F \neq \mathbf{0}$ and that ∇F is a normal vector to the level curve at such a point (see Section 12-11).

A Curve as Intersection of Two Surfaces. Let two surfaces

$$F(x, y, z) = 0, \qquad G(x, y, z) = 0 \tag{12-185}$$

contain the point P_1: (x_1, y_1, z_1) and let both have tangent planes at P_1 as above. Then both surfaces are closely approximated, near P_1, by their tangent planes at P_1 and, if these planes are not coincident, they must intersect in a line L. Therefore, we then expect the surfaces to intersect in a curve, and we expect the line L to be the tangent to the curve at P_1. Indeed, if there is a curve as described, its tangent line must lie in both tangent planes, as above and, hence, must be the line L. The condition that the tangent planes not coincide is that the normal vectors ∇F, ∇G be linearly independent at P_1, that is, that the matrix

$$\begin{pmatrix} F_x & F_y & F_z \\ G_x & G_y & G_z \end{pmatrix} \tag{12-186}$$

have rank 2 at (x_1, y_1, z_1), and so, at least, one second-order minor be nonzero.

If, for example,

$$\begin{vmatrix} F_x & F_y \\ G_x & G_y \end{vmatrix} \neq 0 \quad \text{or} \quad \frac{\partial(F, G)}{\partial(x, y)} \neq 0$$

then, as in Section 12-14, we can solve (12-185) for x and y in terms of z, near z_1: say, $x = f(z)$, $y = g(z)$. Then the equations

$$x = f(t),\ y = g(t),\ z = t, \qquad z_1 - \delta < t < z_1 + \delta$$

are parametric equations for the curve of intersection, near P_1. Also, as in Section 11-12, $\nabla F \times \nabla G$ is a vector along the tangent line to the path.

EXAMPLE 2. The equations

$$x + y - z = 0, \qquad x^2 + y^2 = 1$$

describe a plane and a cylinder in space (Figure 12-40, see Section 11-19).

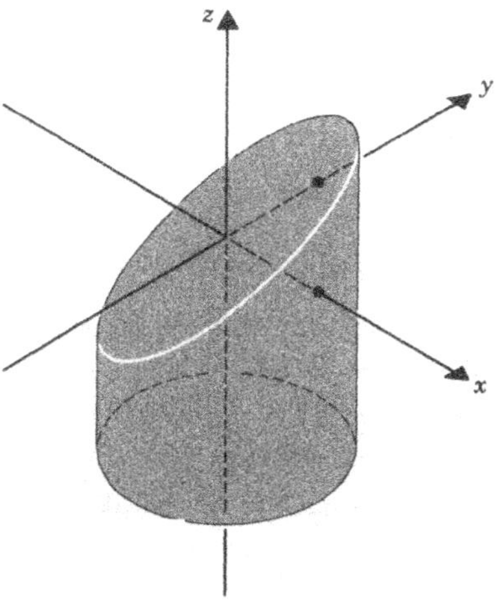

Figure 12-40 The curve of intersection of plane and cylinder.

For $x = 1$, $y = 0$ and $z = 1$ both equations are satisfied; let P_1 be the point $(1, 0, 1)$. At P_1, the Jacobian matrix (12-186) is

$$\begin{pmatrix} 1 & 1 & -1 \\ 2 & 0 & 0 \end{pmatrix}$$

and this has rank 2. Hence, there is a curve of intersection. Also, at P_1

$$\nabla F \times \nabla G = \begin{vmatrix} \mathbf{i} & \mathbf{j} & \mathbf{k} \\ 1 & 1 & -1 \\ 2 & 0 & 0 \end{vmatrix} = -2\mathbf{j} - 2\mathbf{k}$$

Hence the tangent line has the equation $\overrightarrow{P_1P} \times (-2\mathbf{j} - 2\mathbf{k}) = \mathbf{0}$ or $\overrightarrow{P_1P} = t(\mathbf{j} + \mathbf{k})$.

Parametric Representation of Surfaces. In Section 11-10 we saw that a plane could be represented parametrically by equations of form

$$x = a_1 + b_1u + c_1v, \qquad y = a_2 + b_2u + c_2v, \qquad z = a_3 + b_3u + c_3v,$$

where $-\infty < u < \infty$, $-\infty < v < \infty$. An analogous representation is available for surfaces in space:

$$x = f(u, v), \qquad y = g(u, v), \qquad z = h(u, v) \tag{12-187}$$

For example, $x = \cos u$, $y = \sin u$, $z = v$ are parametric equations for a circular cylinder. The discussion of such representations is left to the exercises (Problems 8-13 below).

PROBLEMS

1. For each of the following paths in space, find the tangent line and normal plane at the given point and graph:
   (a) $x = \cos t$, $y = \sin t$, $z = t$ (helix) for $t = \pi/4$
   (b) $x = t$, $y = t^2$, $z = t + t^2$ for $t = 1$
   (c) $x = \cos t$, $y = \sin t$, $z = \cos t$ for $t = \pi/4$
   (d) $x = t$, $y = t$, $z = e^{-t}$ for $t = 1$
   (e) $x = t^2 + 1$, $y = 3 - t^2$, $z = 2 + t$ at $(1, 3, 2)$
   (f) $x = \operatorname{Sin}^{-1} t$, $y = \operatorname{Cos}^{-1} t$, $z = 3$, $-1 < t < 1$, at $(0, \pi/2, 3)$
2. For each of the following surfaces in space, find the tangent plane and normal line at the given point. Also graph the surface near this point.
   (a) $z = x^2 + y^2$ for $x = 1$, $y = 2$ (b) $z = e^x \cos y$ for $x = 0$, $y = 0$
   (c) $z = e^{-x^2-y^2}$ for $x = 1$, $y = 2$ (d) $z = x^y$ for $x = 1$, $y = 1$
   (e) $x^2 + y^2 + z^2 = 11$ at $(3, 1, 1)$ (f) $2x^2 + y^2 + 2z^2 = 5$ at $(1, 1, 1)$
   (g) $x^2 - y^2 - z^2 = 0$ at $(5, 3, 4)$ (h) $x^2 + y^2 = 13$ at $(3, 2, 1)$
3. Show that, near the point given, the surfaces intersect in a curve, find the equation of the tangent line to the curve and graph.
   (a) $2x + y + z = 4$, $x^2 + y^2 + z^2 = 3$ at $(1, 1, 1)$
   (b) $x + y + z = 1$, $x^2 + y^2 - z^2 = 0$ at $(1, 0, 1)$
   (c) $x^2 + y^2 + z^2 = 3$, $x^2 - y^2 + 2z^2 = 2$ at $(1, 1, 1)$
   (d) $x^2 - y = 0$, $y^2 - z = 0$ at $(1, 1, 1)$
4. For a path $\mathbf{r} = \mathbf{f}(t)$, $a \leq t \leq b$, in space, let the velocity $\mathbf{v} = \mathbf{f}'(t)$ and acceleration $\mathbf{a} = \mathbf{f}''(t)$ exist for $a \leq t \leq b$. If for a particular t these vectors are linearly independent, then the plane through the corresponding point with base space Span $(\mathbf{v}, \mathbf{a})$ is called the *osculating plane* of the curve at the point. This plane is a tangent plane of the curve and can be regarded as the tangent plane that is closest to the curve near the point.
   (a) Show that the osculating plane has the equation $\overrightarrow{P_1P} \cdot \mathbf{v} \times \mathbf{a} = 0$.
   (b) Find $\mathbf{v}$, $\mathbf{a}$ and the osculating plane for the path of Problem 1(a) at the given point.

(c) Proceed as in part (b) using the path of Problem 1(c).

5. Let $\mathbf{f}(t)$ and $\mathbf{g}(t)$ define paths in space and let $\mathbf{f}$ and $\mathbf{g}$ have derivatives at t. Prove that

$$[\mathbf{f}(t) \times \mathbf{g}(t)]' = \mathbf{f}(t) \times \mathbf{g}'(t) + \mathbf{f}'(t) \times \mathbf{g}(t)$$

‡6. *Curvature of a path in space.* Let $\mathbf{f}(t)$ have derivatives through the third order for $a \leq t \leq b$ and let $\mathbf{v} = \mathbf{f}'(t) \neq \mathbf{0}$ for $a \leq t \leq b$. Let $\mathbf{v} = \|\mathbf{v}\|\mathbf{T} = (ds/dt)\mathbf{T}$ and let $d\mathbf{T}/ds \neq \mathbf{0}$ for $a \leq t \leq b$.

(a) Show that $\mathbf{T} \cdot d\mathbf{T}/ds = \mathbf{0}$ and, hence, one can write

$$\frac{d\mathbf{T}}{ds} = \kappa\mathbf{N}, \qquad \kappa > 0$$

where $\mathbf{N}$ is a unit normal vector; κ is called the *curvature*, $\rho = 1/\kappa$ the *radius of curvature*, $\mathbf{N}$ the unit *principal normal vector.*

(b) The vector $\mathbf{T} \times \mathbf{N} = \mathbf{B}$ is called the *binormal vector.* Prove that $\mathbf{B}$ is a unit vector, normal to the curve, and prove the *Frenet formulas*

$$\frac{d\mathbf{T}}{ds} = \kappa\mathbf{N}, \qquad \frac{d\mathbf{N}}{ds} = -\kappa\mathbf{T} + \tau\mathbf{B}, \qquad \frac{d\mathbf{B}}{ds} = -\tau\mathbf{N}$$

where τ is a scalar, called the *torsion.* (*Hint.* Use the relations $\mathbf{N} \cdot \mathbf{N} = 1$, $\mathbf{N} \cdot \mathbf{T} = \mathbf{N} \cdot \mathbf{B} = \mathbf{B} \cdot \mathbf{T} = 0$, $\mathbf{B} \cdot \mathbf{B} = 1$, and differentiate with respect to s.)

(c) Show that $\mathbf{a} = \dfrac{d\mathbf{v}}{dt} = \dfrac{d^2s}{dt^2}\mathbf{T} + \dfrac{\|\mathbf{v}\|^2}{\rho}\mathbf{N}$ (see Section 6-7).

7. *Motion under a central force.* Let a particle P of mass m move in space under the influence of a force $\mathbf{F}$ toward or away from a fixed center O, distinct from P, so that $\mathbf{F} = \varphi(t)\overrightarrow{OP}$. Let $\mathbf{r} = \overrightarrow{OP}$, $\mathbf{v} = d\mathbf{r}/dt$. Assume all needed differentiability. Also note the result of Problem 5.

(a) Show that $\dfrac{d}{dt}(\mathbf{r} \times \mathbf{v}) = \mathbf{0}$, so that $\mathbf{r} \times \mathbf{v} = \mathbf{h}$, where $\mathbf{h}$ is a constant vector.

(b) Show from (a) that, if $\mathbf{h} \neq \mathbf{0}$, P moves in a fixed plane through O.

(c) Show from (a) that if $\mathbf{h} = \mathbf{0}$, then P moves on a fixed line through O. (*Hint.* write $\mathbf{r} = r\mathbf{R}$, where $r = \|\mathbf{r}\|$, and show that $d\mathbf{R}/dt = \mathbf{0}$.)

8. Let the functions (12-187) be defined and differentiable in an open region D of the uv-plane. At (u_1, v_1) let the Jacobian matrix $\begin{pmatrix} f_u & g_u & h_u \\ f_v & g_v & h_v \end{pmatrix}'$ (note the transpose) have rank 2. Regard the equations as defining a mapping from D into R^3. Let $P_1\colon(x_1, y_1, z_1)$ be the image of (u_1, v_1), so that $x_1 = f(u_1, v_1), \ldots$

(a) Show that the line $v = v_1$ in D has as image a path in R^3, with u as parameter, and that for $u = u_1$ the vector $\mathbf{w}_1 = f_u(u_1, v_1)\mathbf{i} + g_u(u_1, v_1)\mathbf{j} + h_u(u_1, v_1)\mathbf{k}$ is tangent to this path at P_1.

(b) Obtain a result similar to (a) for the image of the line $u = u_1$; let $\mathbf{w}_2$ be the tangent vector found.

(c) Conclude from the results of (a) and (b) that, if the range of the mapping is a surface in R^3, then $\mathbf{w}_1 \times \mathbf{w}_2$ is a nonzero normal vector to the surface at P_1, so that $\mathbf{w}_1 \times \mathbf{w}_2 \cdot \overrightarrow{P_1P} = 0$ is the equation of the tangent plane at P_1.

9. The equations $x = \cos u$, $y = \sin u$, $z = v$ are parametric equations for a right circular *cylinder* in space.

   (a) Graph the curves on the surface corresponding to each of the following choices of $u =$ const. or $v =$ const., as in Problem 8.

   (i) $u = 0$ (ii) $u = \pi/4$ (iii) $u = \pi/2$ (iv) $v = 0$ (v) $v = 1$ (vi) $v = 2$

   (b) Find the tangent plane and normal line to the surface for $u = \pi/4$, $v = 1$.

10. The equations $x = v \cos u$, $y = v \sin u$, $z = v$ are parametric equations for a right circular *cone* in space.

   (a) Proceed as in Problem 9(a) using

   (i) $v = 1$ (ii) $v = 2$ (iii) $v = 0$ (iv) $u = 0$ (v) $u = \pi/4$

   (b) Proceed, as in Problem 9(b), using $u = \pi/4$, $v = 1$.

11. The equations $x = (2 + \cos u) \cos v$, $y = (2 + \cos u) \sin v$, $z = \sin u$ are parametric equations for the surface of a *torus* in space.

   (a) Proceed as in Problem 9(a) using

   (i) $v = 0$ (ii) $v = \pi/2$ (iii) $v = \pi/4$

   (iv) $u = 0$ (v) $u = \pi$ (vi) $u = \pi/2$

   (b) Proceed as in Problem 9(b) using $u = \pi/4$, $v = \pi/4$.

12. The equations $x = 3 \cos u \sin v$, $y = 2 \sin u \sin v$, $z = \cos v$ are parametric equations for the surface of an *ellipsoid* in space.

   (a) Proceed as in Problem 9(a) using

   (i) $u = 0$ (ii) $u = \pi/3$ (iii) $v = 0$ (iv) $v = \pi/6$

   (b) Proceed as in Problem 9(b) using $u = \pi/3$, $v = \pi/6$.

13. Let a surface be given in parametric form by Equations (12-187). A path in the surface can then be specified by giving u and v as functions of t, so that ultimately x, y, and z become functions of t. Show, assuming needed differentiability, that the tangent line to such a path lies in the corresponding tangent plane of Problem 8(c).

12-19 PARTIAL DERIVATIVES OF HIGHER ORDER

Let the function $z = f(x, y)$ be defined in the open region D of the xy-plane and let the partial derivatives f_x and f_y exist in D. Then each of these partial derivatives is a function with domain D, and we can seek its partial derivatives:

$$\begin{aligned} f_{xx} &= \frac{\partial}{\partial x}(f_x), & f_{xy} &= \frac{\partial}{\partial y}(f_x) \\ f_{yx} &= \frac{\partial}{\partial x}(f_y), & f_{yy} &= \frac{\partial}{\partial y}(f_y) \end{aligned} \tag{12-190}$$

The new derivatives, when they exist, are called *partial derivatives of f of second order.* It appears from (12-190) that there are 4 such second-order derivatives. However, it will be seen that, if f, f_x, f_y, f_{xy} *and* f_{yx} *are continuous in D, then the mixed derivatives are equal:*

$$f_{xy} = f_{yx} \tag{12-191}$$

Hence, there are, in effect, only 3 second-order partial derivatives: f_{xx}, f_{xy}, and f_{yy}. A proof of (12-191) is given in the next section.

EXAMPLE 1. Let $z = x^2y^3 + x\cos(xy) = f(x, y)$. Then we write

$$\frac{\partial z}{\partial x} = f_x(x, y) = 2xy^3 + \cos(xy) - xy\sin(xy)$$

$$\frac{\partial z}{\partial y} = f_y(x, y) = 3x^2y^2 - x^2\sin(xy)$$

$$\frac{\partial^2 z}{\partial x^2} = f_{xx}(x, y) = 2y^3 - 2y\sin(xy) - xy^2\cos(xy)$$

$$\frac{\partial^2 z}{\partial y\,\partial x} = f_{xy}(x, y) = \frac{\partial}{\partial y}(f_x(x, y)) = 6xy^2 - 2x\sin(xy) - x^2y\cos(xy)$$

$$\frac{\partial^2 z}{\partial x\,\partial y} = f_{yx}(x, y) = \frac{\partial}{\partial x}(f_y(x, y)) = 6xy^2 - 2x\sin(xy) - x^2y\cos(xy)$$

$$\frac{\partial^2 z}{\partial y^2} = f_{yy}(x, y) = \frac{\partial}{\partial y}(f_y(x, y)) = 6x^2y - x^3\cos(xy)$$

We note that $f_{xy} = f_{yx}$, in accordance with (12-191). The notations $\partial^2 z/\partial x^2$, $\partial^2 z/\partial y \partial x, \ldots$ are also introduced here. The difference between $\partial^2 z/\partial y\partial x$ and $\partial^2 z/\partial x\partial y$ should be noted; however, by (12-191) these are equal (under the conditions stated).

If the second-order partial derivatives exist in D, one can form partial derivatives of third order:

$$\frac{\partial^3 z}{\partial x^3} = f_{xxx}(x, y) = \frac{\partial}{\partial x}(f_{xx}(x, y)), \qquad \frac{\partial^3 z}{\partial y\,\partial x^2} = f_{xxy} = \frac{\partial}{\partial y}(f_{xx}(x, y))$$

and so on. By virtue of (12-191), the order of differentiation does not matter for smooth functions; that is, for example,

$$\frac{\partial^3 z}{\partial y\,\partial x^2} = \frac{\partial^3 z}{\partial x\,\partial y\,\partial x} = \frac{\partial^3 z}{\partial x^2\,\partial y}$$

Hence, one obtains in all (under the appropriate continuity conditions) four third-order partial derivatives:

$$f_{xxx},\ f_{xxy},\ f_{xyy},\ f_{yyy}$$

Similarly, one obtains in general $n + 1$ partial derivatives of order n:

$$\frac{\partial^n z}{\partial x^n},\quad \frac{\partial^n z}{\partial x^{n-1}\,\partial y},\quad \ldots,\quad \frac{\partial^n z}{\partial x^n}$$

The definitions extend naturally to functions of three or more variables. For example, a function $f(x, y, z)$ has (under appropriate continuity conditions) 3 first partial derivatives, 6 second partial derivatives:

$$f_{xx},\ f_{yy},\ f_{zz},\ f_{xy},\ f_{yz},\ f_{xz}$$

and 10 third partial derivatives:

$$f_{xxx},\ f_{yyy},\ f_{zzz},\ f_{xxy},\ f_{xxz},\ f_{xyy},\ f_{xzz},\ f_{yzz},\ f_{yyz},\ f_{xyz}$$

It should be remarked that the rule (12-191) needs to be proved only for functions of two variables. For once that has been done, it follows that

$$f_{xy}(x, y, z, \ldots) = f_{yx}(x, y, z, \ldots)$$

since on both sides $z, \ldots$ are treated as constants. Hence, the one rule (12-191) establishes that (under the relevant continuity conditions) *for all partial derivatives the order of differentiation does not matter.*

Partial derivatives of higher order arise in many physical problems: for example, in the theories of electromagnetism, heat conduction, vibration of solid bodies, fluid motion, and thermodynamics. In all these cases, basic physical laws are expressed as equations relating the partial derivatives of appropriate functions. For example, the *heat equation*

$$\frac{\partial u}{\partial t} = k^2\left(\frac{\partial^2 u}{\partial x^2} + \frac{\partial^2 u}{\partial y^2} + \frac{\partial^2 u}{\partial z^2}\right) \tag{12-192}$$

governs the variation of temperature $u = f(t, x, y, z)$ with time t and position (x, y, z) within a homogeneous solid body subject to certain temperature variations in the surrounding medium. Other examples are given in Problems 5 to 7 below. Equation (12-192) is an example of a *partial differential equation.* A function satisfying such an equation in an open region is called a *solution* of the partial differential equation.

Other Notations. One writes, for example,

$$f_{xx}(x, y) = f_{11}(x, y) = \frac{\partial^2 f}{\partial x^2}(x, y), \quad f_{12} = f_{xy}, \quad f_{22} = f_{yy}$$

To emphasize the variables that are held constant, one writes, for example,

$$\left(\frac{\partial^2 w}{\partial x^2}\right)_{yz} \qquad \text{for} \qquad f_{xx}(x, y, z), \qquad \text{where } w = f(x, y, z)$$

One also uses an operator symbol $\nabla_x, \nabla_y, \ldots$:

$$\nabla_x f = f_x, \qquad \nabla_x^2 f = f_{xx}, \qquad \nabla_x \nabla_y f = f_{xy}, \ldots$$

Here one can think of ∇_x, ∇_y as components of a gradient operator ∇; in terms of x, y, and z

$$\nabla = \frac{\partial}{\partial x}\mathbf{i} + \frac{\partial}{\partial y}\mathbf{j} + \frac{\partial}{\partial z}\mathbf{k} = \nabla_x\mathbf{i} + \nabla_y\mathbf{j} + \nabla_z\mathbf{k},$$

so that

$$\nabla f = f_x\mathbf{i} + f_y\mathbf{j} + f_z\mathbf{k}$$

One also writes

$$\nabla \cdot \nabla = \nabla^2 = \frac{\partial^2}{\partial x^2} + \frac{\partial^2}{\partial y^2} + \frac{\partial^2}{\partial z^2}$$

so that

$$\nabla^2 f = \frac{\partial^2 f}{\partial x^2} + \frac{\partial^2 f}{\partial y^2} + \frac{\partial^2 f}{\partial z^2}$$

One calls this expression the *Laplacian of f*. The same symbols are used for functions of two variables x and y, the terms relating to z being dropped. The equation

$$\frac{\partial^2 f}{\partial x^2} + \frac{\partial^2 f}{\partial y^2} + \frac{\partial^2 f}{\partial z^2} = 0$$

is called the *Laplace equation*. A function $f(x, y, z)$ [or $f(x, y)$] satisfying this equation in an open region is said to be *harmonic*.

‡12-20 PROOF OF THEOREM ON MIXED PARTIAL DERIVATIVES

We first formulate a new mean value theorem.

THEOREM 12. *Let $f(x, y)$ be defined in an open region D that includes the rectangular region R: $x_0 \leq x \leq x_1$, $y_0 \leq y \leq y_1$. Let f, f_x and f_{xy} be continuous in D. Then there exists a point (ξ, η) in R such that*

$$f(x_1, y_1) - f(x_1, y_0) - f(x_0, y_1) + f(x_0, y_0) = f_{xy}(\xi, \eta)(x_1 - x_0)(y_1 - y_0) \tag{12-200}$$

PROOF. Since f_{xy} is continuous in R, it has an absolute minimum m and an absolute maximum M in R (see Sections 12-7 and 12-22). Then $f_{xy} \leq M$ implies

$$\int_{y_0}^{y_1} f_{xy}(x, y)\, dy \leq \int_{y_0}^{y_1} M\, dy \qquad \text{or} \qquad f_x(x, y_1) - f_x(x, y_0) \leq M(y - y_0)$$

and, hence, that

$$\int_{x_0}^{x_1} [f_x(x, y_1) - f_x(x, y_0)]\, dx \leq \int_{x_0}^{x_1} M(y_1 - y_0)\, dx$$

or

$$f(x_1, y_1) - f(x_1, y_0) - f(x_0, y_1) + f(x_0, y_0) \leq M(x_1 - x_0)(y_1 - y_0)$$

Similarly, we show that

$$f(x_1, y_1) - f(x_1, y_0) - f(x_0, y_1) + f(x_0, y_0) \geq m(x_1 - x_0)(y_1 - y_0)$$

Since $f_{xy}(x, y)$ is continuous in R, it takes on every value between m and M in R. Hence, for some (ξ, η), Equation (12-200) holds true.

Now we can deduce the rule on mixed derivatives:

THEOREM 13. *Let the function f be defined in the open region D of the xy-plane. Let f, f_x, f_y, f_{xy}, and f_{yx} be continuous in D. Then*

$$f_{xy}(x, y) = f_{yx}(x, y) \text{ in } D$$

PROOF. Let (x_0, y_0) be a fixed point of D, let $x_1 = x_0 + h$, $y_1 = y_0 + h$, and let h be so small and positive that the rectangular region R of Theorem

12 lies in D. Then Theorem 12 gives

$$f(x_0 + h, y_0 + h) - f(x_0 + h, y_0) - f(x_0, y_0 + h) + f(x_0, y_0) = f_{xy}(\xi, \eta)h^2 \tag{12-201}$$

where $x_0 \leq \xi \leq x_0 + h$, $y_0 \leq \eta \leq y_0 + h$. If we interchange the roles of x and y, we obtain a similar relation involving f_{yx}:

$$f(x_0 + h, y_0 + h) - f(x_0, y_0 + h) - f(x_0 + h, y_0) + f(x_0, y_0) = f_{yx}(\xi', \eta')h^2 \tag{12-201'}$$

However, the left side of (12-201) and the left side of (12-201′) are the same! Hence, if we subtract the second equation from the first, we obtain

$$0 = [f_{xy}(\xi, \eta) - f_{yx}(\xi', \eta')]h^2$$

or

$$f_{xy}(\xi, \eta) = f_{yx}(\xi', \eta') \tag{12-202}$$

If now we let $h \to 0+$, then $(\xi, \eta) \to (x_0, y_0)$ and $(\xi', \eta') \to (x_0, y_0)$. Since f_{xy} and f_{yx} are continuous, we conclude from (12-202) that

$$f_{xy}(x_0, y_0) = f_{yx}(x_0, y_0)$$

as was to be proved.

Remark. We can interpret the left side of (12-201) or (12-201′) as a *second difference* of f. We can write

$$\Delta_x f = f(x + h, y) - f(x), \qquad \Delta_y f = f(x, y + h) - f(x, y)$$

These are called *first differences* of f. From them we obtain the second difference

$$\begin{aligned} \Delta_{xy} f &= \Delta_y[\Delta_x f] = \Delta_y[f(x + h, y) - f(x, y)] \\ &= [f(x + h, y + h) - f(x, y + h)] - [f(x + h, y) - f(x, y)] \\ &= f(x + h, y + h) - f(x, y + h) - f(x + h, y) + f(x, y) \end{aligned}$$

Thus $\Delta_{xy} f$ is the left side of (12-201) evaluated at (x_0, y_0). The symmetry of this expression shows that

$$\Delta_{xy} f = \Delta_{yx} f$$

It is the symmetry of the second difference $\Delta_{xy} f$ which leads to the equality of the mixed derivatives f_{xy} and f_{yx}. From (12-201) we can therefore now write

$$f_{xy}(x_0, y_0) = \lim_{h \to 0} f_{xy}(\xi, \eta) = \lim_{h \to 0} \frac{\Delta_{xy} f}{h^2}$$

and, similarly, from (12-201′)

$$f_{yx}(x_0, y_0) = \lim_{h \to 0} \frac{\Delta_{yx} f}{h^2}$$

Since $\Delta_{xy} f = \Delta_{yx} f$, it follows that $f_{xy}(x_0, y_0) = f_{yx}(x_0, y_0)$.

PROBLEMS

1. Find the derivatives requested:
   (a) f_{xx}, f_{xy}, and f_{yy} for $f(x, y) = 2x^3y^2 - 3xy^2 + x - 2y$
   (b) f_{xx}, f_{yx}, and f_{yy} for $f(x, y) = x^2/(y^2 + 1)$
   (c) f_{xy}, f_{yz}, f_{xz} for $f(x, y, z) = x^2e^{yz}$
   (d) f_{xyz}, f_{xxz}, and f_{xzz} for $f(x, y, z) = e^x \cos(2y - 3z)$
2. Verify that $f_{xy} = f_{yx}$ in each of the following cases:
   (a) $f(x, y) = x^7y^5$ (b) $f(x, y) = \dfrac{x^9}{y^4}$
   (c) $f(x, y) = x^y$ (d) $f(x, y) = x \ln(x^2 - y^2)$
   (e) $f(x, y, z) = x^2z^4 - y^3z^2 + x^3y^2$ (f) $f(x, y, z) = x/(y + z)$
3. Verify that f is harmonic in each of the following cases:
   (a) $f(x, y) = x^2 - y^2$ (b) $f(x, y) = xy$
   (c) $f(x, y) = x^3 - 3xy^2$ (d) $f(x, y) = x^4 - 6x^2y^2 + y^4$
   (e) $f(x, y, z) = x^2 + y^2 - 2z^2$ (f) $f(x, y, z) = (x^2 + y^2 + z^2)^{-1/2}$
4. Show that each of the following functions is a solution of the heat equation (12-192); throughout, a and k are nonzero constants.
   (a) $e^{-k^2a^2t} \sin ax$ (b) $e^{-2k^2a^2t} \cos a(x + y)$ (c) $\dfrac{e^{-ax^2/t}}{\sqrt{t}}$, $a = (4k^2)^{-1}$
5. The *biharmonic equation*

$$\nabla^4 f \equiv \nabla^2(\nabla^2 f) = 0$$

   is important in the theory of elasticity.
   (a) Show that, in two dimensions, the equation is

$$\frac{\partial^4 f}{\partial x^4} + 2\frac{\partial^4 f}{\partial x^2\,\partial y^2} + \frac{\partial^4 f}{\partial y^4} = 0$$

   (b) Show that every harmonic function is biharmonic.
   (c) For what values of a is $ax^4 + 2x^2y^2 + y^4$ biharmonic?
6. The *wave equation*

$$\frac{\partial^2 u}{\partial t^2} = c^2\left(\frac{\partial^2 u}{\partial x^2} + \frac{\partial^2 u}{\partial y^2} + \frac{\partial^2 u}{\partial z^2}\right) \qquad (c = \text{const} > 0)$$

   is important in the theory of electromagnetism.
   (a) Show that $u = \sin(x - ct)$ is a solution (see Figure 12-11 in Section 12-2).
   (b) Show generally that every function of form $u = f(x - ct)$ or of form $u = f(x + ct)$ is a solution, provided that f is appropriately differentiable.
7. The *Poisson equation*

$$\frac{\partial^2 u}{\partial x^2} + \frac{\partial^2 u}{\partial y^2} + \frac{\partial^2 u}{\partial z^2} = g(x, y, z)$$

   is important in the theory of gravitation. Find a solution for each of the following cases:

(a) $g(x, y, z) \equiv 1$ (b) $g(x, y, z) \equiv e^x$ (c) $g(x, y, z) \equiv x - y + 2z$

8. Show that the solutions of the following equations in a given open region form a vector space:

   (a) Laplace's equation in 2 dimensions (b) The heat equation for fixed k

   (c) The biharmonic equation in 2 dimensions (Problem 5)

   (d) The wave equation (Problem 6) for fixed c

9. Show that, for given g, the solutions of the Poisson equation (Problem 7) in a given open region D, if there are any, form a linear variety in the vector space of all functions defined in D.

10. Let f have continuous derivatives through the second order in an open region including the square of vertices $(0, 0)$, $(1, 0)$, $(0, 1)$, $(1, 1)$ and let f have a zero at each vertex. Show that f_{xy} has a zero inside the square.

11. Prove: if f has continuous derivatives through the second order for all (x, y) and if $f_{xy} \equiv 0$, then f can be expressed as the sum of a function of x and a function of y. (*Hint*. Apply Theorem 12 with $x_0 = 0$, $y_0 = 0$, $x_1 = x$, $y_1 = y$.)

12. To prove Theorem 12, let

$$F(x, y) = (x_1 - x_0)(y_1 - y_0)f(x, y) \\ - (x - x_0)(y - y_0)[f(x_1, y_1) - f(x_1, y_0) - f(x_0, y_1) + f(x_0, y_0)]$$

Now apply Rolle's theorem to $F(x, y_1) - F(x, y_0)$ to show that $F_x(\xi, y_1) - F_x(\xi, y_0) = 0$ for some ξ, $x_0 < \xi < x_1$. Then apply Rolle's theorem to $F_x(\xi, y)$ to show that $F_{xy}(\xi, \eta) = 0$ for some η, $y_0 < \eta < y_1$, and verify that this last equation gives the desired result.

13. *Chain rules*. Let $z = F(u, v)$, $u = f(x, y)$, $v = g(x, y)$, so that $z = F[f(x, y), g(x, y)]$. Prove, under appropriate hypotheses:

$$\text{(a)}\ \frac{\partial^2 z}{\partial x^2} = F_{uu}\left(\frac{\partial u}{\partial x}\right)^2 + 2F_{uv}\frac{\partial u}{\partial x}\frac{\partial v}{\partial x} + F_{vv}\frac{\partial^2 v}{\partial x^2} + F_u\frac{\partial^2 u}{\partial x^2} + F_v\frac{\partial^2 v}{\partial x^2}$$

$$\text{(b)}\ \frac{\partial^2 z}{\partial x\,\partial y} = F_{uu}\frac{\partial u}{\partial x}\frac{\partial u}{\partial y} + F_{uv}\left(\frac{\partial u}{\partial x}\frac{\partial v}{\partial y} + \frac{\partial u}{\partial y}\frac{\partial v}{\partial x}\right) + F_{vv}\frac{\partial v}{\partial x}\frac{\partial v}{\partial y} \\ + F_u\frac{\partial^2 u}{\partial x\,\partial y} + F_v\frac{\partial^2 v}{\partial x\,\partial y}$$

12-21 TAYLOR'S FORMULA

For a function f of one variable, Taylor's formula can be written as follows:

$$f(x) = f(a) + (x - a)f'(a) + (x - a)^2\frac{f''(a)}{2!} + \cdots + (x - a)^n\frac{f^{(n)}(a)}{n!} + R_n \tag{12-210}$$

with

$$R_n = \frac{(x - a)^{n+1}f^{(n+1)}(a + \mu(x - a))}{(n + 1)!}, \qquad 0 < \mu < 1$$

Here we have given the remainder R_n in Lagrange's form (see Sections 6-12 and 8-19). The formula has the effect of replacing f by a *polynomial* plus a remainder term. Under appropriate conditions, the remainder term is small, and the formula provides an approximation of f by a polynomial. Also in many cases $R_n \to 0$ as $n \to \infty$, so that

$$f(x) = \lim_{n\to\infty}\left[f(a) + \cdots + (x-a)^n \frac{f^{(n)}(a)}{n!}\right] = \sum_{n=0}^{\infty} \frac{f^{(n)}(a)}{n!}(x-a)^n$$

This is the representation of f by its Taylor series (Section 8-20).

For a function $f(x, y)$ of two variables, a power series with center (a, b) has the form:

$$c_{00} + c_{10}(x-a) + c_{01}(y-b) + c_{20}(x-a)^2 + c_{11}(x-a)(y-b) + c_{02}(y-b)^2 + \cdots$$

For simplicity we concentrate on the case $a = 0$, $b = 0$; the general case is obtainable from the special case by a simple substitution (translation). Thus we consider the equation

$$f(x, y) = c_{00} + c_{10}x + c_{01}y + c_{20}x^2 + \cdots + c_{mn}x^m y^n + \cdots \qquad (12\text{-}211)$$

and proceed *formally*, assuming all steps to be justified. From (12-211) we see that $c_{00} = f(0, 0)$. Also

$$\begin{aligned}
f_x &= c_{10} + 2c_{20}x \cdots + mc_{mn}x^{m-1}y^n + \cdots \\
f_y &= c_{01} + c_{11}x + \cdots + nc_{mn}x^m y^{n-1} + \cdots \\
f_{xx} &= 2c_{20} + 6c_{30}x + \cdots + m(m-1)c_{mn}x^{m-2}y^n + \cdots \\
f_{xy} &= c_{11} + 2c_{21}x + \cdots + mnc_{mn}x^{m-1}y^{n-1} + \cdots
\end{aligned}$$

Accordingly, $f_x(0, 0) = c_{10}$, $f_y(0, 0) = c_{01}$, $f_{xx}(0, 0) = 2c_{20}$, $f_{xy}(0, 0) = c_{11}$, and so on. In general,

$$\frac{\partial^{r+s} f}{\partial x^r\, \partial y^s}(0, 0) = r!s!c_{rs}$$

so that

$$c_{rs} = \frac{1}{r!s!}\frac{\partial^{r+s} f}{\partial x^r\, \partial y^s}(0, 0) = \frac{1}{n!}\binom{n}{r}\frac{\partial^n f}{\partial x^r\, \partial y^{n-r}}(0, 0)$$

where $n = r + s$. Hence, from (12-211),

$$\begin{aligned}
f(x, y) = f(0, 0) &+ xf_x(0, 0) + yf_y(0, 0) \\
&+ \frac{1}{2!}[x^2 f_{xx}(0, 0) + 2xyf_{xy}(0, 0) + y^2 f_{yy}(0, 0)] \\
&+ \cdots + \frac{1}{n!}\left[x^n \frac{\partial^n f}{\partial x^n} + \binom{n}{1} x^{n-1} y \frac{\partial^n f}{\partial x^{n-1}\, \partial y} + \cdots + y^n \frac{\partial^n f}{\partial y^n}\right] + \cdots
\end{aligned}$$

where all partial derivatives are evaluated at $(0, 0)$. This is the *Taylor series of* $f(x, y)$ *with center* $(0, 0)$. The representation can be shown to be valid for many common functions in appropriate regions. We do not pursue this topic further here but, rather, consider the polynomial approximation that it suggests. We write

$$p_n(x, y) = \frac{1}{n!}\sum_{r=0}^{n}\binom{n}{r}\frac{\partial^n f}{\partial x^r\,\partial y^{n-r}}(0, 0)\,x^r y^{n-r} \qquad (12\text{-}212)$$

Thus the Taylor series of f is simply the series $\Sigma p_n(x, y)$. We now write

$$f(x, y) = p_0(x, y) + \cdots + p_n(x, y) + R_n \qquad (12\text{-}213)$$

This will be our *Taylor's formula for functions of two variables,* provided that we give a representation for the remainder R_n.

THEOREM 14. *Let $f(x, y)$ be defined and have continuous derivatives through the $(n + 1)$-st order in an open region D which includes the origin $(0, 0)$. Let (x, y) be a point of D such that the line segment from $(0, 0)$ to (x, y) lies in D. Then Taylor's formula (12-213) is valid, where*

$$R_n = \frac{1}{(n+1)!}\sum_{r=0}^{n+1}\binom{n+1}{r}\frac{\partial^{n+1} f}{\partial x^r\,\partial y^{n+1-r}}(\mu x, \mu y)\,x^r y^{n+1-r} \qquad (12\text{-}214)$$

and μ is a number, depending on (x, y), between 0 and 1.

Remarks. In the formula for R_n the partial derivatives of f are evaluated at a point $(\mu x, \mu y)$ between (x, y) and $(0, 0)$, as in Figure 12-41. This is in direct

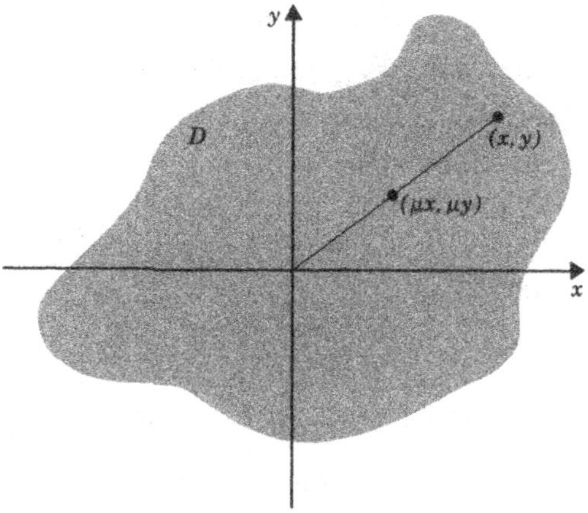

Figure 12-41 Proof of Taylor's formula.

analogy with the formula (12-210) for functions of one variable. We note two special cases of Theorem 14:

$n = 0$: $f(x, y) = f(0, 0) + xf_x(\mu x, \mu y) + yf_y(\mu x, \mu y)$;

$n = 1$: $f(x, y) = f(0, 0) + xf_x(0, 0) + yf_y(0, 0)$

$$+ \frac{1}{2!}[x^2 f_{xx}(\mu x, \mu y) + 2xyf_{xy}(\mu x, \mu y) + y^2 f_{yy}(\mu x, \mu y)]$$

The case $n = 0$ is called the Mean Value Theorem for functions of two variables.

PROOF OF THEOREM 14. Let (x, y) be as stated in the Theorem and let

$$g(t) = f(tx, ty), \qquad 0 \le t \le 1$$

As t varies from 0 to 1, (tx, ty) moves from $(0, 0)$ to (x, y) on the line segment joining these points; hence, f is defined at all these points and g is defined for $0 \le t \le 1$. By the Chain Rule

$$g'(t) = xf_x(tx, ty) + yf_y(tx, ty)$$
$$g''(t) = x^2 f_{xx}(tx, ty) + 2xyf_{xy}(tx, ty) + y^2 f_{yy}(tx, ty)$$

and by induction (Problem 5 below)

$$g^{(k)}(t) = \sum_{r=0}^{k} \binom{k}{r} x^r y^{k-r} \frac{\partial^k f}{\partial x^r\, \partial y^{k-r}}(tx, ty), \quad k = 1, \ldots, n+1 \qquad (12\text{-}215)$$

Therefore,

$$g^{(k)}(0) = \sum_{r=0}^{k} \binom{k}{r} x^r y^{k-r} \frac{\partial^k f}{\partial x^r\, \partial y^{k-r}}(0, 0) = k!p_k(x, y)$$

Now on putting $t = 1$ in Taylor's formula for $g(t)$ on $[0, 1]$, we obtain

$$g(1) = g(0) + g'(0) + \frac{g''(0)}{2!} + \cdots + \frac{g^{(n)}(0)}{n!} + \frac{g^{(n+1)}(\mu)}{(n+1)!}$$

But $g(1) = f(x, y)$, $g(0) = f(0, 0)$, $g^{(k)}(0) = k!p_k(x, y)$ for $k = 1, \ldots, n$ as above and by (12-215) $g^{(n+1)}(\mu)/(n+1)!$ equals R_n, as given in (12-214). Hence, (12-213) follows.

EXAMPLE 1. Let $f(x, y) = x + y + e^y \cos x$. Then

$$f_x = 1 - e^y \sin x, \qquad f_y = 1 + e^y \cos x,$$
$$f_{xx} = -e^y \cos x = -f_{yy}, \qquad f_{xy} = -e^y \sin x$$

Hence

$$f(x, y) = 1 + x + 2y + \frac{1}{2!}(-x^2 e^{\mu y} \cos \mu x - 2xye^{\mu y} \sin \mu x + y^2 e^{\mu y} \cos \mu x)$$

If (x, y) is very close to $(0, 0)$ the first three terms on the right give a good approximation of f; this is in effect the approximation of the graph of $z = f(x, y)$ by the graph of the tangent plane at $(0, 0, 1)$, or the approximation of $\Delta f = f(x, y) - f(0, 0)$ by df.

Discussion. For $n = 1$ the remainder term can be written as

$$R_1 = x^2 q_1(x, y) + xyq_2(x, y) + y^2 q_3(x, y) \qquad (12\text{-}216)$$

Here, for each (x, y) other than $(0, 0)$

$$q_1(x, y) = \frac{1}{2!} f_{xx}(\mu x, \mu y)$$

for a proper choice of μ, $0 < \mu < 1$. We also set $q_1(0, 0) = (1/2!)f_{xx}(0, 0)$ and, by continuity of f_{xx}, we conclude that $q_1(x, y)$ is continuous at $(0, 0)$. A similar reasoning applies to q_2, q_3. Hence, R_1 can be written as in (12-216) where q_1, q_2, q_3 are continuous at $(0, 0)$. It follows that all three functions are bounded in some neighborhood of $(0, 0)$, say $|q_i| < K/3$, $i = 1, 2, 3$, for an appropriate constant K. If we write $x = r\cos\theta$, $y = r\sin\theta$, then in this neighborhood

$$|R_1| = |r^2(q_1\cos^2\theta + q_2\cos\theta\sin\theta + q_3\sin^2\theta)| \leq r^2(|q_1| + |q_2| + |q_3|) \leq Kr^2$$

Hence, as $r \to 0$, the remainder term approaches 0 at least as fast as a constant times r^2.

For general n we reason similarly and conclude that in a neighborhood of $(0, 0)$,

$$\begin{aligned} f(x, y) &= p_0(x, y) + \cdots + p_n(x, y) + R_n \\ R_n &= x^{n+1}q_1(x, y) + x^n y q_2(x, y) + \cdots + y^{n+1}q_{n+1}(x, y) \\ |R_n| &\leq Kr^{n+1} \end{aligned}$$

Here the q_i are all continuous at $(0, 0)$. Thus f is approximated by a polynomial of degree n plus a remainder term which approaches 0, as $(x, y) \to (0, 0)$, faster than a constant times r^{n+1}.

We remark that *our function $f(x, y)$ can be represented in only one way in the form just described:* as the sum of a polynomial of degree at most n and a remainder whose absolute value is less than or equal to a constant times r^{n+1}. The proof is left as an exercise (Problem 4 below). This result is very useful, for it tells us that if we have, *by any means,* found a representation of f as such a sum, then *the polynomial part must be* $p_0(x, y) + p_1(x, y) + \cdots + p_n(x, y)$.

EXAMPLE 2. We know from one-variable calculus that, on each interval $|u| \leq c$,

$$e^u = 1 + u + \frac{u^2}{2!} + g(u)$$

where the remainder $g(u)$ satisfies $|g(u)| \leq k|u|^3$, for some constant k [see the last paragraph of Section 6-12]. We can now write

$$e^{2x+3y} = 1 + 2x + 3y + \frac{(2x + 3y)^2}{2!} + g(2x + 3y)$$

If we set $g(2x + 3y) = R_2(x, y)$, then we know that, for $|2x + 3y| \leq c$ [hence, in a neighborhood of $(0, 0)$],

$$|R_2(x, y)| \leq k|2x + 3y|^3 = kr^3|2\cos\theta + 3\sin\theta|^3 \leq 125kr^3$$

since $|2\cos\theta + 3\sin\theta|$ can surely not exceed 5. Hence, $R_2(x, y)$ satisfies the proper inequality, and we, in fact, have found the Taylor's formula for $n = 2$ for the function e^{2x+3y}:

$$e^{2x+3y} = 1 + 2x + 3y + \frac{(2x + 3y)^2}{2!} + R_2(x, y)$$

This result was found without calculating any partial derivatives of our function.

EXAMPLE 3. Since $\sin u = u - (u^3/6) + g(u)$, where $|g(u)| \le k|u|^4$ for $|u| \le c$, we find that

$$\sin(x + y) = x + y - [(x + y)^3/6] + R_3(x, y)$$

The reasoning is the same as for Example 2.

In this way we can find many Taylor's formula representations without finding partial derivatives. We are not obtaining the detailed representation of the remainder, but for many purposes this is not needed.

Theorem 14 can be generalized to functions of more than two variables. The formulas are considered in Problem 3 below. In general the formula provides an approximation of $f(\mathbf{x}) = f(x_1, \ldots, x_k)$ by a polynomial P_n in $x_1, \ldots, x_k$ of degree n plus a remainder $R_n(\mathbf{x})$, where $|R_n(\mathbf{x})| \le K\|\mathbf{x}\|^{n+1}$ for $\|\mathbf{x}\| < \delta$. The polynomial P_n can be described as the unique polynomial of degree n in $x_1, \ldots, x_k$ such that $f(\mathbf{x}) = P_n(\mathbf{x}) + R_n(\mathbf{x})$, and $|R_n(\mathbf{x})| \le K\|\mathbf{x}\|^{n+1}$.

PROBLEMS

1. Follow the method of Example 2 to obtain a Taylor's formula for the given function at (0, 0) with the given value for n (no detailed expression for the remainder is required).

   (a) $\cos(x + y)$, $n = 3$ (b) $\tan(2x - y)$, $n = 2$
   (c) $\ln(1 + 2x + y)$, $n = 2$ (d) $\sqrt{1 + x + y}$, $n = 2$
   (e) e^{xy}, $n = 4$ (f) $(2x + y)\sin(2x + y)$, $n = 5$
   (g) $\sin(1 + x + y)$, $n = 2$ [*Hint.* First expand $\sin(1 + u)$ at $u = 0$.]
   (h) $\cos(1 + 2x + y)$, $n = 3$

2. Write out Taylor's formula for the given function and given value of n, giving the remainder as in Example 1.

   (a) $f(x, y) = x\cos xy$, $n = 0$ (b) $f(x, y) = x\sin xy$, $n = 0$
   (c) $f(x, y) = xe^y$, $n = 1$ (d) $f(x, y) = \dfrac{x}{1 + y}$, $n = 1$
   (e) $f(x, y) = \cos xy$, $n = 2$ (f) $f(x, y) = \ln(1 + x^2 + y^2)$, $n = 2$

3. (a) Let $P(x, y, z) = \Sigma c_{ijk} x^i y^j z^k$ be a polynomial in x, y, z. Show that

$$c_{ijk} = \frac{1}{i!j!k!} \frac{\partial^{i+j+k} P}{\partial x^i\, \partial y^j\, \partial z^k}(0, 0, 0)$$

   (b) On the basis of the result of (a) write out Taylor's formula for $f(x, y, z)$ for $n = 1$.

   (c) Extend the results of parts (a) and (b) to $f(x_1, \ldots, x_k)$ and general $n \ge 0$.

‡4. Let $f(x, y)$ satisfy the hypotheses of Theorem 14 and, in a neighborhood of (0, 0), let $f(x, y) = P_1(x, y) + Q_1(x, y)$, $f(x, y) = P_2(x, y) + Q_2(x, y)$, where $P_1(x, y)$ and $P_2(x, y)$ are polynomials of degree at most n and $|Q_1(x, y)| \le K_1 r^{n+1}$,

$|Q_2(x, y)| \leq K_2 r^{n+1}$ for certain constants K_1 and K_2. Prove that $P_1(x, y) \equiv P_2(x, y)$ and $Q_1(x, y) \equiv Q_2(x, y)$. [*Hint.* Let $P(x, y) = P_1(x, y) - P_2(x, y)$ and show that P is a polynomial of degree at most n, satisfying an inequality $|P(x, y)| \leq Kr^{n+1}$ for some K. Show that, for each fixed t, $P(x, tx)$ is a polynomial in x with coefficients which are polynomials in t and that $|P(x, tx)| \leq h(t)|x|^{n+1}$. Conclude from this that each coefficient of a power of x in $P(x, tx)$ must be identically 0, so that $P(x, y) \equiv 0$.]

5. Prove (12-215) by induction. $\left[\textit{Hint.} \text{ Use the rule: } \binom{k}{r} + \binom{k}{r-1} = \binom{k+1}{r}.\right]$

12-22 MAXIMA AND MINIMA OF FUNCTIONS OF TWO VARIABLES

For the theory up to this point we have been mainly concerned with *open regions*, the analogues of intervals $a < x < b$ on the line. For problems about maxima and minima, we must allow for *bounded closed regions*—the analogues of the closed intervals on the line. We recall that a function $f(x)$ continuous on a closed interval must have an absolute maximum M and an absolute minimum m in that interval, and that there is no such assertion for a function $f(x)$ defined and continuous on an open interval. The difficulty arises because the maximum or minimum may occur at the end of the interval (or the function may have an infinite limit as one approaches the ends of the interval). For a function $f(x, y)$ of two variables, we have an analogous statement: *if f is defined and continuous in a bounded closed region, then f has an absolute maximum M and an absolute minimum m; for a function defined in an open region, there may be no absolute maximum or minimum.* For example, if $f(x, y) = x + y$ in the circular region $x^2 + y^2 \leq 1$ (see Figure 12-42), then f has an absolute maximum of $\sqrt{2}$ at the point $(\sqrt{2}/2, \sqrt{2}/2)$, and an absolute minimum of $-\sqrt{2}$ at the point $(-\sqrt{2}/2, -\sqrt{2}/2)$ (this will be proved below). The function $f(x, y) = 1/(x^2 + y^2)$ for $0 < x^2 + y^2 < 1$ has

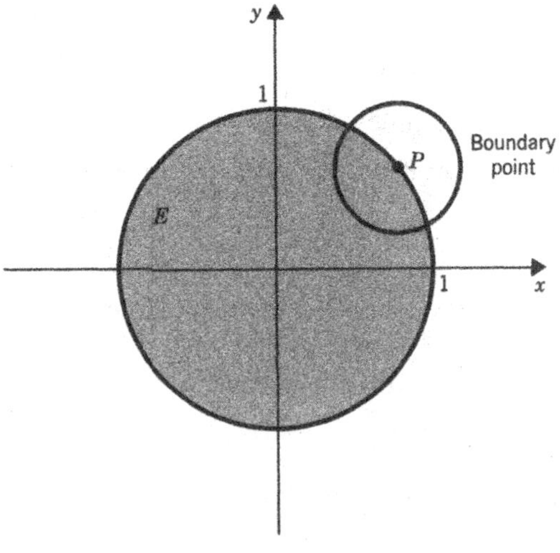

Figure 12-42 Boundary point.

no absolute maximum or minimum; it becomes arbitrarily large positive (limit ∞) as (x, y) approaches $(0, 0)$; it is always greater than 1 but can be made as close to 1 as desired by choosing (x, y) sufficiently close to the boundary circle $x^2 + y^2 = 1$.

In order to state our results precisely, we need several definitions concerning sets, extending those of Section 12-1. We state these for sets in the plane, but the definitions extend at once to R^n, as does the basic Theorem F on maxima and minima.

By a *boundary* point of a set E in the plane we mean a point P such that every neighborhood of P contains at least a point of E and a point not in E. For example, if E is the set of all (x, y) such that $x^2 + y^2 \leq 1$ (Figure 12-42), then each point of the circumference is a boundary point of E.

A set E is said to be *closed* if every boundary point of E belongs to E. For technical reasons, the empty set also is considered to be closed.

If to any set E we adjoin all boundary points not already in E, then we obtain a set E_1 *which is closed.*

‡To prove this assertion, we let P_1 be a boundary point of E_1 and choose a neighborhood U_1 of P_1, of radius ρ_1. Then this neighborhood must contain a point of E_1, say Q, and a point not in E_1 and so not in E. Next we choose a neighborhood U_2 of Q, with radius ρ_2 so small that the neighborhood U_2 is contained in the previously chosen neighborhood of P_1 (see Figure 12-43).

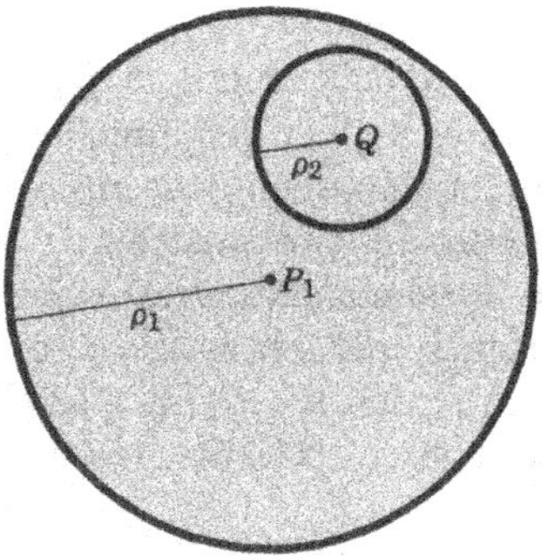

Figure 12-43 Proof that set plus boundary is a closed set.

Then, since Q is in E_1, the neighborhood U_2 must contain points in E. Therefore, the neighborhood U_1 of P_1 contains both a point in E and a point not in E. This holds true for *every* neighborhood U_1 of P_1. Therefore, P_1 is a boundary point of E, and P_1 belongs to E_1. Therefore, every boundary point of E_1 belongs to E_1, and E_1 is a closed set.‡

We defined open regions in Section 12-1: they are sets that are nonempty, open, and pathwise connected. We now define a *closed region* as an open region to which all boundary points are adjoined. Therefore, every closed region is a closed set.

A set E is said to be *bounded* if there is a number $K > 0$ such that $\|OP\| < K$ for all points P in E. Thus the circular region of Figure 12-42 is bounded, as are all neighborhoods. But a half-plane: $x > 0$ is not bounded (we say it is *unbounded*).

We can now state the main theorem on maxima and minima.

THEOREM F. *Let the function $f(x, y)$ be defined and continuous on the nonempty bounded closed set E in the xy-plane. Then f has an absolute maximum M and an absolute minimum m; that is, there are numbers m and M such that*

$$m \leq f(x, y) \leq M \quad \text{in } E$$

and $f(x_1, y_1) = m$, $f(x_2, y_2) = M$ for certain points (x_1, y_1), (x_2, y_2) in E.

This theorem is proved in Section 12-25 below. We remark that each of the following is a bounded closed set: a line segment, including its end points; the circumference of a circle; a rectangle, including all points inside and all points on the edges; a circular disk, including all points inside and on the circumference.

For a function continuous on a bounded closed region E, such as the set E of Figure 12-42, we know that f has both an absolute maximum and an absolute minimum in E. To seek the absolute maximum, for example, we first try to determine whether it occurs at an interior point of E. If not, then we examine the boundary of E and try to locate the maximum there. If the absolute maximum occurs at an interior point P_0, then f has also a *local* maximum at P_0 [that is, in geometric notation, $f(P) \leq f(P_0)$ in some neighborhood of P_0] and, by Theorem 1 of Section 12-8, if f is differentiable at P_0, then $\nabla f = \mathbf{0}$ at P_0. Hence, to seek interior maximum points, we first seek the points where $\nabla f = \mathbf{0}$ (that is, where both f_x and f_y are 0). We call these points the *critical points* of f. As for functions of one variable, a critical point may provide a local maximum, a local minimum, or neither. For functions of one variable, we had useful tests (see Section 6-1) for classifying the critical points. We shall here develop such tests for functions of two variables.

In order to prepare the way for these tests, we first consider several examples. We might suspect that, if (x_0, y_0) is a point such that, for fixed $y = y_0$, $f(x, y_0)$ has a local maximum (as a function of x) at x_0, and $f(x_0, y)$ has a local maximum at y_0, then f must have a local maximum at (x_0, y_0). Our first example shows that this is *not* the case.

EXAMPLE 1. The function

$$z = -x^2 + 3xy - y^2 = f(x, y)$$

is such that

$$f(x, 0) = -x^2, \qquad f(0, y) = -y^2$$

Thus in both cases there is a local maximum at $(0, 0)$ (Figure 12-44). However, for $x = y$, $z = x^2$, so that along the line $x = y$, f has a local *minimum* at $(0, 0)$. The level curves of f are shown in Figure 12-45. They form a *saddle point* at $(0, 0)$. From the level curves (hyperbolas) it is clear that along the x-axis (as function of x) f has a local maximum at $x = 0$, along the y-axis (as function of y) f has a local maximum at $y = 0$; however, $f(x, y)$ does not have a local maximum at $(0, 0)$, since in every neighborhood of $(0, 0)$ f takes on values both greater than and less than $f(0, 0) = 0$.

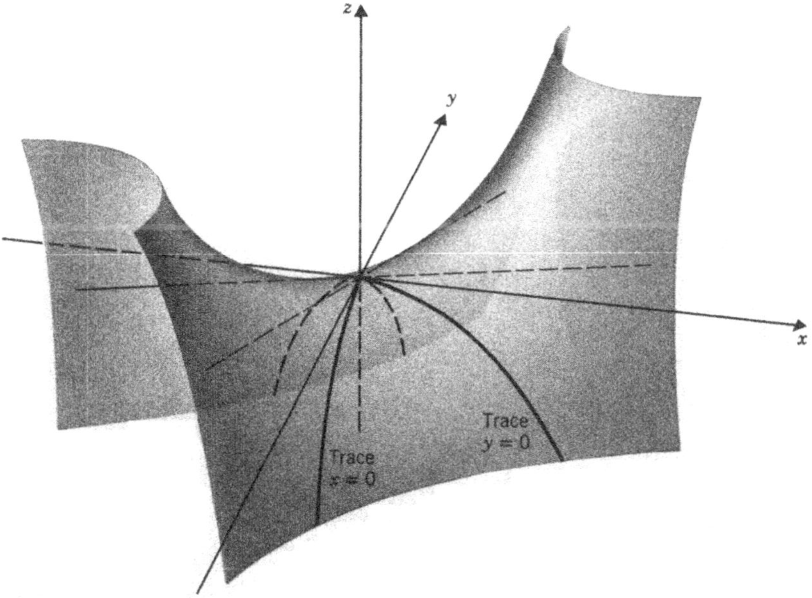

Figure 12-44 The function $z = -x^2 + 3xy - y^2$.

EXAMPLE 2. $f(x, y) = 2x^2 + 3y^2$. Since $f_x = 4x$, $f_y = 6y$, there is a critical point at $(0, 0)$. Clearly $f(x, y) > f(0, 0) = 0$ for $(x, y) \neq (0, 0)$. Hence, there is a local (in fact, absolute) minimum at $(0, 0)$. The function and its level curves are graphed in Figures 12-46 and 12-47. The level curves are ellipses.

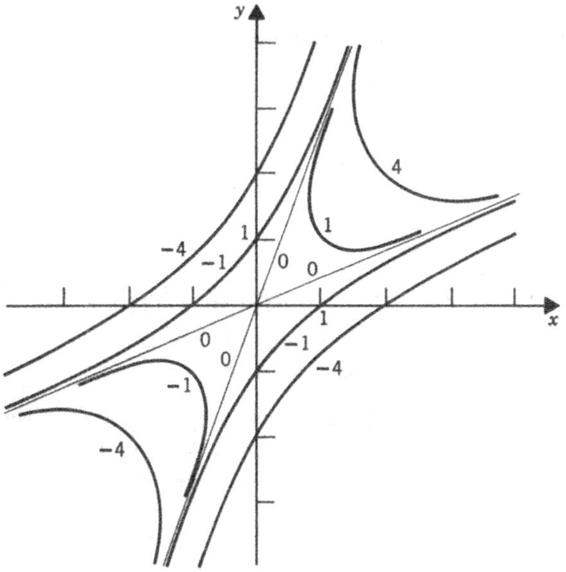

Figure 12-45 Level curves for function of Figure 12-44.

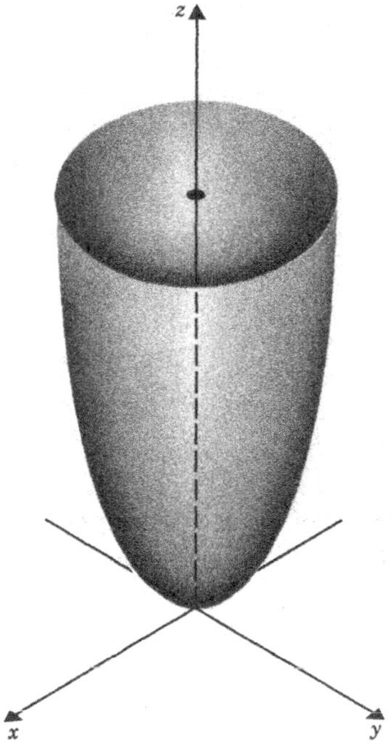

Figure 12-46 $z = 2x^2 + 3y^2$.

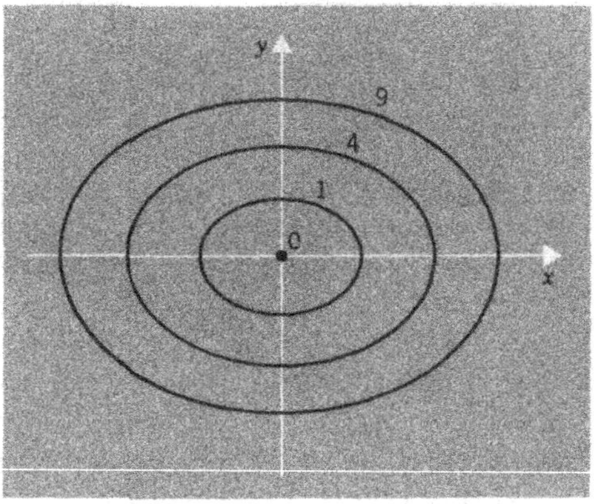

Figure 12-47 Level curves of function of Figure 12-46.

EXAMPLE 3. $f(x, y) = 3x^2 - 2xy + 3y^2$. Here, $f_x = 6x - 2y$, $f_y = -2x + 6y$. Hence, the critical points are the solutions of the equations

$$6x - 2y = 0, \qquad -2x + 6y = 0$$

We verify that (0, 0) is the only solution. The level curves of f are the curves

$$3x^2 - 2xy + 3y^2 = c$$

for different values of c. By the rules of Section 6-5, they are ellipses, possibly degenerate. One can, in fact, rotate axes by 45°, as in Section 6-5, to obtain the new equation

$$2x'^2 + 4y'^2 = c$$

This equation shows at once that the curves are ellipses. Also, in terms of the new coordinates, our function is given by

$$z = 2x'^2 + 4y'^2$$

Thus, as in Example 2, there is a local (in fact, absolute) minimum at the origin. The function and its level curves are graphed in Figures 12-48 and 12-49. The figure also shows the new coordinate axes, x' and y'.

EXAMPLE 4. $f(x, y) = -3x^2 + 2xy - 3y^2$. This is the function of Example 3 reversed in sign. Hence, the level curves are ellipses, as in Figure 12-49 (with the values of f multiplied by -1) and f has a relative maximum at (0, 0).

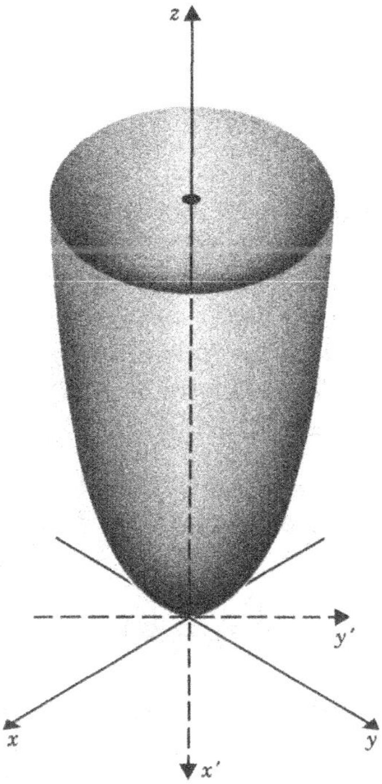

Figure 12-48 $z = 3x^2 - 2xy + 3y^2$.

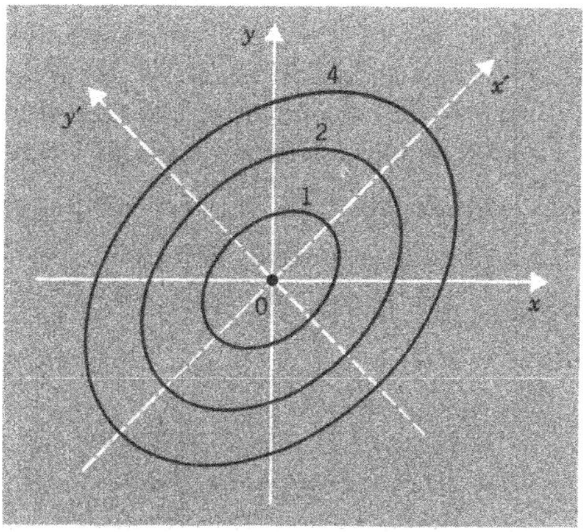

Figure 12-49 Level curves for function of Figure 12-48.

From these examples it is clear that for a general quadratic function

$$f(x, y) = Ax^2 + Bxy + Cy^2$$

for which the corresponding level curves are ellipses, the function f has a local (in fact, absolute) maximum or minimum at $(0, 0)$. To tell whether one has a maximum or a minimum, one need only consider the function along the x or y axes: $f(x, 0) = Ax^2$ or $f(0, y) = Cy^2$. If A and C are positive, there is a local minimum; if A and C are negative, there is a local maximum. We note that for ellipses one has

$$B^2 - 4AC < 0$$

so that A and C must have the same sign. (The expression $B^2 - 4AC$ is called the *discriminant* of the quadratic function.)

When $B^2 - 4AC > 0$, the level curves are hyperbolas as in Example 1 and one has a saddle point, with no local maximum or minimum. When $B^2 - 4AC = 0$, the level curves are (possibly degenerate) parabolas. This case is left to Problem 5 below.

We have considered the quadratic function at length because it is the key to the general case. Let a function f be given with a critical point at (x_0, y_0). By translating axes we can change to coordinates with origin at (x_0, y_0) and, hence, reduce our discussion to the case when the critical point is $(0, 0)$. Thus

$f_x(0, 0) = 0$, $f_y(0, 0) = 0$. Under the usual hypotheses, Taylor's formula gives

$$f(x, y) = f(0, 0) + \frac{1}{2!}(x^2 f_{xx}(0, 0) + 2xy f_{xy}(0, 0) + y^2 f_{yy}(0, 0)) + \cdots$$
$$= f(0, 0) + Ax^2 + Bxy + Cy^2 + \cdots$$

The remaining terms can be considered to be negligible for (x, y) sufficiently close to $(0, 0)$. Hence, f should behave like the quadratic function $Ax^2 + Bxy + Cy^2$. In particular, if $B^2 - 4AC < 0$, we expect f to have a local maximum (if $A < 0$) or minimum (if $A > 0$) at $(0, 0)$; if $B^2 - 4AC > 0$, we expect a saddle point; if $B^2 - 4AC = 0$, we are in a borderline case (and, in this case, the terms of higher degree in Taylor's formula must normally be considered). We thus are led to formulate a theorem:

THEOREM 15. *Let $f(x, y)$ be defined in the open region D and have continuous partial derivatives through the second order in D. Let f have a critical point at (x_0, y_0) and let*

$$A = \frac{1}{2} f_{xx}(x_0, y_0), \qquad B = f_{xy}(x_0, y_0), \qquad C = \frac{1}{2} f_{yy}(x_0, y_0)$$

I. *If $B^2 - 4AC < 0$, then f has a local maximum at (x_0, y_0) if $A < 0$ and a local minimum at (x_0, y_0) if $A > 0$.*

II. *If $B^2 - 4AC > 0$, then f has neither maximum nor minimum at (x_0, y_0).*

The proof is given in the next section.

EXAMPLE 5. Let $f(x, y) = x^3 - 2x^2y - x^2 - 2y^2 - 3x$. The critical points are given by the equations

$$f_x \equiv 3x^2 - 4xy - 2x - 3 = 0, \qquad f_y \equiv -2x^2 - 4y = 0$$

Eliminating y, we are led to a cubic equation for x:

$$2x^3 + 3x^2 - 2x - 3 = 0$$

We find the roots to be $x = 1$, $x = -1$, $x = -3/2$; the corresponding values of y are $-1/2$, $-1/2$, and $-9/8$. Now

$$f_{xx} = 6x - 4y - 2, \qquad f_{xy} = -4x, \qquad f_{yy} = -4$$

Thus at $(1, -1/2)$ we find that $A = 3$, $B = -4$, $C = -2$. Therefore, $B^2 - 4AC = 40 > 0$, and there is neither maximum nor minimum at $(1, -1/2)$. At $(-1, -1/2)$, $A = -3$, $B = 4$, $C = -2$, so that $B^2 - 4AC = -8 < 0$; thus f has a local maximum at $(-1, -1/2)$. Finally, at $(-3/2, -9/8)$, $A = -13/4$, $B = 6$, $C = -2$, so that $B^2 - 4AC = 10 > 0$; there is no local maximum or minimum at this point.

EXAMPLE 6. Let $f(x, y) = x + y$ for $x^2 + y^2 \leq 1$. We seek the *absolute* maximum and minimum of f. Here, $f_x \equiv 1$, $f_y \equiv 1$, so that there are *no* critical points. Hence, the absolute maximum and minimum must occur on the boundary. The boundary is the circle $x^2 + y^2 = 1$. We represent this parametrically as $x = \cos t$, $y = \sin t$, $0 \leq t \leq 2\pi$. Then on the boundary

$$f(x, y) = \cos t + \sin t = g(t), \qquad 0 \leq t \leq 2\pi$$

The function $g(t)$ has derivative $-\sin t + \cos t$ and, hence, g has critical points at $t = \pi/4$ and $t = 5\pi/4$. At these points, g has values $\sqrt{2}$ and $-\sqrt{2}$, respectively. Also $g''(t) = -\sin t - \cos t$ and $g''(\pi/4) = -\sqrt{2}$, $g''(5\pi/4) = \sqrt{2}$. Hence, g has a local maximum at $\pi/4$ and a local minimum at $5\pi/4$. This information is enough (as shown in Section 6-1) to ensure us that we have found the absolute maximum and minimum of g, hence of f: f has its absolute minimum of $-\sqrt{2}$ at $(-\sqrt{2}/2, -\sqrt{2}/2)$, its absolute maximum of $\sqrt{2}$ at $(\sqrt{2}/2, \sqrt{2}/2)$. The function and its level curves are graphed in Figures 12-50 and 12-51.

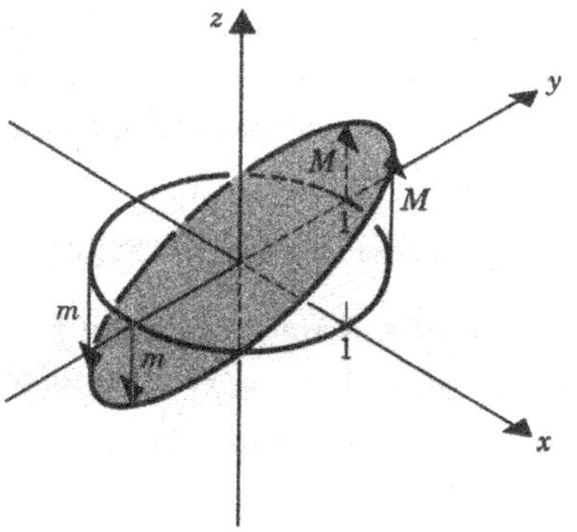

Figure 12-50 $f(x, y) = x + y$, $x^2 + y^2 \leq 1$.

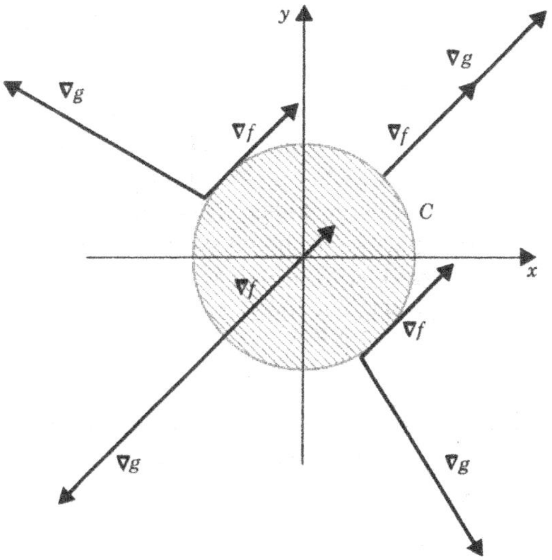

Figure 12-51 Level curves for function of Figure 12-50.

EXAMPLE 7. $f(x, y) = x^2 + y^2$ for $|x| \leq 1$, $|y| \leq 1$. Again, we seek the absolute maximum and minimum. Here, $f_x = 2x$ and $f_y = 2y$ and there is a critical point at $(0, 0)$, where f has value 0. This is clearly the absolute minimum of f. The absolute maximum must occur on the boundary. The boundary is formed of four segments: $x = 1$, $-1 \leq y \leq 1$; $y = 1$, $-1 \leq x \leq 1$; $x = -1$, $-1 \leq y \leq 1$; $y = -1$, $-1 \leq x \leq 1$. On the first segment, $f(x, y) = f(1, y) = 1 + y^2$, so that f has a maximum of 2 at $(1, 1)$ and at $(1, -1)$; on the second segment, $f(x, y) = f(x, 1) = x^2 + 1$, and f has a maximum of 2 at $(1, 1)$ and $(-1, 1)$. Continuing in this way, we find that the absolute maximum of f is 2 and that this occurs at all four points $(\pm 1, \pm 1)$.

‡12-23 LAGRANGE MULTIPLIERS

This topic is treated briefly in Section 6-2. We now discuss it again, with more powerful tools at our disposal.

We consider the problem of maximizing $z = f(x, y)$, where x and y are related by an equation $g(x, y) = 0$, called a *side condition*. We assume that the equation $g(x, y) = 0$ has as graph a curve C in the xy-plane and that ∇f and ∇g are continuous in an open region D containing the whole curve C. We also assume that $\nabla g \neq \mathbf{0}$ along C. By the Implicit Function Theorem (Sections 12-14 and 12-15), this means that each point on the curve C has a neighborhood in which the curve can be represented in one of the two forms, $y = F(x)$ or $x = G(y)$, where F or G is differentiable, and the corresponding derivative is given by the equation $g_x\,dx + g_y\,dy = 0$. It follows that the vector $\nabla g = (g_x, g_y)$ is a nonzero normal vector to C.

Now in a neighborhood of each point (x_0, y_0) on C, $f(x, y)$ can be expressed in one of the two forms $f(x, F(x)) = \phi(x)$, or $f(G(y), y) = \psi(y)$, and we ask whether either of these functions has a local maximum or minimum at the value x_0 or y_0, respectively (see Examples 6 and 7 of the previous section). Equivalently, we can introduce arc length s as parameter along C and ask whether $f(x(s), y(s)) = H(s)$ has a local minimum or maximum at the corresponding value s_0. But $H'(s)$ is just the *directional derivative* of f along the curve C and, hence, $H'(s)$ can be 0 only if ∇f has a zero component along the tangent to the curve (Section 12-11). But this last condition is equivalent to the condition that ∇f be a scalar multiple of the normal vector ∇g at the point. That is, there must be a number λ such that

$$\nabla f + \lambda \nabla g = \mathbf{0} \tag{12-230}$$

Therefore, *to find the local maxima and minima of f along C, we must find all (x, y) on C such that* (12-230) *is satisfied for some* λ: that is, we must solve the three equations

$$\frac{\partial f}{\partial x} + \lambda \frac{\partial g}{\partial x} = 0, \qquad \frac{\partial f}{\partial y} + \lambda \frac{\partial g}{\partial y} = 0, \qquad g(x, y) = 0 \tag{12-231}$$

simultaneously for x, y and λ. This is the method of Lagrange multipliers in its simplest form; λ is the Lagrange multiplier. If we have found all local maxima and minima of f on C and C is a bounded closed set, then we can find the absolute maximum of f on C by simply seeing which local maximum gives the largest value to f.

EXAMPLE 1. $f(x, y) = x + y$, where $x^2 + y^2 = 1$ (see Example 6 of the preceding section). Here (12-231) becomes

$$1 + 2\lambda x = 0, \qquad 1 + 2\lambda y = 0, \qquad x^2 + y^2 = 1$$

(g being $x^2 + y^2 - 1$). We solve the first two equations for x and y and substitute in the third and find $\lambda^2 = 1/2$. For $\lambda = \sqrt{2}/2$, $x = -\sqrt{2}/2$, $y = -\sqrt{2}/2$; for $\lambda = -\sqrt{2}/2$, $x = y = \sqrt{2}/2$. Thus the local maxima and minima occur only at the points $(-\sqrt{2}/2, -\sqrt{2}/2)$ and $(\sqrt{2}/2, \sqrt{2}/2)$.

Since $f = -\sqrt{2}$ at the first point and $f = \sqrt{2}$ at the second, we have a local and absolute minimum at the first point, and we have a local and absolute maximum at the second point. This agrees with the results in Example 6 in the preceding section. The vectors ∇f and ∇g are shown in Figure 12-51. It is clear that the vectors are linearly dependent only at the two points shown.

The method extends to functions of several variables. For example, for $f(x, y, z)$ with a side condition $g(x, y, z) = 0$, describing an appropriate surface S in space, we reason that at a local maximum P_0, for example, f must have zero directional derivative at P_0 along each curve in the surface S through P_0 and, hence, ∇f must again be a scalar multiple of the normal vector ∇g. Thus we obtain (12-230) again and must solve the simultaneous equations

$$\frac{\partial f}{\partial x} + \lambda \frac{\partial g}{\partial x} = 0, \qquad \frac{\partial f}{\partial y} + \lambda \frac{\partial g}{\partial y} = 0, \qquad \frac{\partial f}{\partial z} + \lambda \frac{\partial g}{\partial z} = 0, \qquad g(x, y, z) = 0 \tag{12-232}$$

EXAMPLE 2. $f(x, y, z) = 3x + y + z$, where $2x^2 + y^2 + z^2 = 1$ (the surface S is an ellipsoid). Here (12-232) becomes

$$3 + 4\lambda x = 0, \qquad 1 + 2\lambda y = 0, \qquad 1 + 2\lambda z = 0, \qquad 2x^2 + y^2 + z^2 = 1$$

We find easily that there are two points at which f can have a local maximum or minimum, namely, $(\pm 3/\sqrt{26}, \pm 2/\sqrt{26}, \pm 2/\sqrt{26})$. Therefore, f has an absolute maximum of $13/\sqrt{26}$ and an absolute minimum of $-13/\sqrt{26}$.

The method also extends to the case of several side conditions: say to $f(x, y, z)$ with side conditions

$$g(x, y, z) = 0 \qquad \text{and} \qquad h(x, y, z) = 0 \tag{12-233}$$

We assume that the surfaces $g(x, y, z) = 0$ and $h(x, y, z) = 0$ intersect in a curve C in space, as in Section 12-18, and that ∇g and ∇h are continuous and linearly independent at each point of C. Hence, as in Section 12-18, the tangent vector to C is normal to both ∇g and ∇h. To require that ∇f have zero component along the tangent vector is the same as to require that ∇f be in Span $(\nabla g, \nabla h)$; that is, that there exist scalars λ_1, λ_2 such that

$$\nabla f + \lambda_1 \nabla g + \lambda_2 \nabla h = \mathbf{0} \tag{12-234}$$

Equations (12-233) and (12-234) provide five simultaneous equations for $\lambda_1, \lambda_2, x, y$, and z.

By similar reasoning, we obtain equations for maximizing $f(x_1, \ldots, x_n)$ with side conditions

$$g_1(x_1, \ldots, x_n) = 0, \ \ldots, \ g_k(x_1, \ldots, x_n) = 0 \tag{12-235}$$

Here k must be less than n. (Why?) The condition found is

$$\nabla f + \lambda_1 \nabla g_1 + \cdots + \lambda_k \nabla g_k = \mathbf{0} \tag{12-236}$$

Equations (12-235) and (12-236) provide $n + k$ simultaneous equations for $\lambda_1, \ldots, \lambda_k, x_1, \ldots, x_n$.

‡12-24 PROOF OF THEOREM ON LOCAL MAXIMA AND MINIMA

We proceed to prove Theorem 15. As above, we can without loss of generality take (x_0, y_0) to be $(0, 0)$.

Case I. $B^2 - 4AC < 0$. Here $A \neq 0$. First suppose $A > 0$. We then assert that there is a number $m > 0$ such that

$$Ax^2 + Bxy + Cy^2 \geq m(x^2 + y^2), \qquad (x, y) \neq (0, 0) \qquad (12\text{-}240)$$

To show this we let

$$u = g(x, y) = \frac{Ax^2 + Bxy + Cy^2}{x^2 + y^2}, \qquad (x, y) \neq (0, 0)$$

If we write $x = r\cos\theta$, $y = r\sin\theta$, $0 \leq \theta \leq 2\pi$, then

$$u = \frac{Ar^2\cos^2\theta + \cdots}{r^2} = A\cos^2\theta + B\sin\theta\cos\theta + C\sin^2\theta = \varphi(\theta)$$

Here φ is continuous for $0 \leq \theta \leq 2\pi$. Also, by completing the square,

$$\varphi(\theta) = A\left(\cos\theta + \frac{B}{2A}\sin\theta\right)^2 - \sin^2\theta\,\frac{B^2 - 4AC}{4A} \qquad (12\text{-}241)$$

Since $B^2 - 4AC < 0$ and $A > 0$, both terms are nonnegative. The second term is 0 only for $\sin\theta = 0$, in which case the first term equals $A > 0$. Hence, $\varphi(\theta) > 0$ for $0 \leq \theta \leq 2\pi$. Therefore, $\varphi(\theta)$ has a positive minimum m, and $u = \varphi(\theta) \geq m > 0$ for all θ or, in rectangular coordinates,

$$\frac{Ax^2 + Bxy + Cy^2}{x^2 + y^2} \geq m > 0$$

which gives (12-240).

Now by Taylor's formula (Section 12-21, Equation (12-216))

$$f(x, y) = f(0, 0) + q_1x^2 + q_2xy + q_3y^2$$

where $q_i = q_i(x, y)$ is continuous at $(0, 0)$ $(i = 1, 2, 3)$ and $q_1(0, 0) = A$, $q_2(0, 0) = B$, $q_3(0, 0) = C$. Hence, we can choose δ so small and positive that

$$|q_1 - A| < \frac{m}{6}, \qquad |q_2 - B| < \frac{m}{6}, \qquad |q_3 - C| < \frac{m}{6} \qquad \text{for } x^2 + y^2 < \delta^2$$

For (x, y) so restricted,

$$|(q_1 - A)x^2 + (q_2 - B)xy + (q_3 - C)y^2|$$
$$\leq |q_1 - A|r^2 + |q_2 - B|r^2 + |q_3 - C|r^2 \leq \frac{m}{6}\cdot 3r^2 = \frac{m}{2}(x^2 + y^2)$$

From this inequality and (12-240) we conclude that

$$f(x, y) = f(0, 0) + Ax^2 + Bxy + Cy^2 + (q_1 - A)x^2 + (q_2 - B)xy + (q_3 - C)y^2$$
$$\geq f(0, 0) + m(x^2 + y^2) - \frac{m}{2}(x^2 + y^2) = f(0, 0) + \frac{m}{2}(x^2 + y^2)$$

for $0 < x^2 + y^2 < \delta^2$. Accordingly, $f(x, y) > f(0, 0)$ for $0 < x^2 + y^2 < \delta^2$, and f has a local minimum at $(0, 0)$.

For $A < 0$ the proof that f has a local maximum is similar.

Case II. $B^2 - 4AC > 0$. We introduce $u = g(x, y)$ as in Case I and write u as $\varphi(\theta)$. We assert that $\varphi(\theta)$ changes sign in the interval $[0, 2\pi]$. If A and C are both 0, then $B \neq 0$ and $\varphi(\theta) = B \sin\theta \cos\theta$, so that the assertion follows. If, for example, $A \neq 0$, then we can write $\varphi(\theta)$ as in (12-241); then $\varphi(0) = A$ and when $\theta = \mathrm{Cot}^{-1}(-B/2A)$, $\varphi(\theta)$ has the sign of $-A$. Hence, $\varphi(\theta)$ changes sign. Let $\varphi(\theta_1) = m_1 > 0$, $\varphi(\theta_2) = -m_2 < 0$. Hence, in rectangular coordinates,

$$Ax^2 + Bxy + Cy^2 = m_1(x^2 + y^2) \qquad \text{for } x = r\cos\theta_1,\ y = r\sin\theta_1$$

The same reasoning as for Case I allows us to conclude that

$$f(x, y) \geq f(0, 0) + \frac{m_1}{2}(x^2 + y^2) \qquad \text{for } x = r\cos\theta_1,\ y = r\sin\theta_1$$

provided that $0 < x^2 + y^2 < \delta_1^2$. Similarly

$$f(x, y) \leq f(0, 0) - \frac{m_2}{2}(x^2 + y^2) \qquad \text{for } x = r\cos\theta_2,\ y = r\sin\theta_2$$

provided that $0 < x^2 + y^2 < \delta_2^2$. Hence, f can have neither a local maximum nor a local minimum at $(0, 0)$.

PROBLEMS

1. For each of the following sets determine if the set is closed or not, and in any case determine the boundary of the set. For (a), . . . , (d) consider the sets as contained in R^1.
   (a) The set of rational numbers of the form $1/n$, n a positive integer.
   (b) The set of rational numbers consisting of 0 and of the numbers of the form $1/n$, n an integer.
   (c) The set of all rational numbers with absolute values at most 1.
   (d) The set of all irrational numbers with absolute value at most 1.
   (e) The set of points in the plane with both x and y coordinates rational numbers.
   (f) The set of points in the plane satisfying the inequalities: $-1 \leq y \leq \sin x$, $0 \leq x \leq 2$.
   (g) The set of all points in the plane satisfying the two inequalities: $-1 \leq y(1 + x^2)^{-1} < 2, 0 \leq x \leq 9$.
   ‡(h) The set of points in the plane satisfying the inequalities: $-1 \leq y \leq \sin(1/x)$, $0 < x \leq 1/\pi$.

‡2. (a) By the *complement of a set E* in R^n we mean the set of all points of R^n not in E. Show: if E is open, then the complement of E is closed; if E is closed, then the complement of E is open.
   (b) Show that the complement of the union of two sets is the intersection of their complements.

(c) Show that the intersection of two closed sets is closed.

(d) Can a set E in R^n be both open and closed?

3. Locate all critical points and determine whether a local maximum or minimum occurs at each:

(a) $z = xy$ (b) $z = 1 - x^2 - 2y^2$ (c) $z = x^2 + 2x + y^2$

(d) $z = 2x^2 + 4x + y^2 - 6y$ (e) $z = x^2 - xy + y^2$

(f) $z = 2x^2 + 5xy + y^2$ (g) $z = x^2 + 2xy + 3y^2 + 4x + 6y$

(h) $z = 2x^2 - 4xy + y^2 - x + 2y$ (i) $z = e^{-x^2-y^2}$

(j) $z = x^2 - 2x(\sin y + \cos y) + 1$ (k) $z = xy^2 + x^2y - xy$

(l) $z = x^3 + y^3$ (m) $z = x^2 - 2xy + y^2 - 2x + 2y$

(n) $z = x^4 + 3x^2y^2 + y^4$

(o) $z = [x^2 + (y + 1)^2][x^2 + (y - 1)^2]$ (interpret geometrically)

4. Find the absolute maximum and minimum of $z = f(x, y)$ in the given region.

(a) $z = 3x + 4y,\ x^2 + y^2 \leq 1$

(b) $z = 5x + 12y,\ x^2 + y^2 \leq 1$

(c) $z = 3x^2 + 2y^2,\ x^2 + y^2 \leq 1$

(d) $z = x^2 - y^2,\ x^2 + y^2 \leq 1$

(e) $z = xy,\ 2x^2 + y^2 \leq 1$

(f) $z = xe^{-x} \cos y,\ -1 \leq x \leq 1,\ -\pi \leq y \leq \pi$

(g) $z = x^2 + 2xy + 3y^2 + 4x + 6y,\ -2 \leq x \leq 0,\ -1 \leq y \leq 0$ [see Problem 3(g)]

(h) $z = x^2 - 2xy + y^2 - 2x + 2y,\ 0 \leq x \leq 2,\ 0 \leq y \leq 2$ [see Problem 3(*m*)]

5. Show that if $B^2 - 4AC = 0$ but $A \neq 0$, then the function $z = Ax^2 + Bxy + Cy^2$ has a local maximum or minimum at $(0, 0)$, but also has other critical points. Discuss the nature of the level curves.

6. Find the common perpendicular to the two skew lines

$$L_1\colon \overrightarrow{OP} = (3, 4, 0) + t(2, -2, 1), \qquad L_2\colon \overrightarrow{OP} = (4, 6, 2) + \tau(2, 1, -2)$$

by choosing t, τ to minimize the square of the distance between the corresponding points on the lines.

7. *Method of least squares.* To fit given data (x_i, y_i), $i = 1, \ldots, n$ by this method, one selects a formula $y = \varphi(x, \alpha_1, \ldots, \alpha_k)$, depending on k parameters $\alpha_1, \ldots, \alpha_k$, and tries to choose $\alpha_1, \ldots, \alpha_k$ so that the total square error

$$z = \sum_{i=1}^{n} [y_i - \varphi(x_i, \alpha_1, \ldots, \alpha_k)]^2 = F(\alpha_1, \ldots, \alpha_k)$$

is minimized (see Problem 20 following Section 6-1).

(a) Find the best fitting straight line $y = \alpha_1 x + \alpha_2$.

(b) Find the best fitting parabola $y = \alpha_1 x^2 + \alpha_2$.

(c) Find the best fitting sinusoidal function $y = \alpha_1 \cos x + \alpha_2 \sin x$.

(d) Find the best fitting parabola $y = \alpha_1 + \alpha_2 x + \alpha_3 x^2$.

[*Note.* In each case one is led to simultaneous equations for the α's. Assume the determinant of coefficients is not 0; this can be shown to be the case under reasonable assumptions on the values $x_1, \ldots, x_n$—for example, that they are not all the same for parts (a) and (b).]

8. In mechanics a particle P is said to move in a conservative force field, if the force $\mathbf{F}$ acting on P depends only on position and

$$\mathbf{F} = -\nabla U$$

where U is the *potential energy* of P. At each critical point of U, the force is **0**, and one has an *equilibrium position* for the particle. It can be shown that the equilibrium is stable if U has a relative minimum at the point, is unstable otherwise.

(a) Let P move in the xy-plane with potential energy $k(x^2 + y^2)$. Determine the equilibrium position and whether it is stable. Assume $k > 0$.

(b) Let P move in the xy-plane with potential energy $U = k(r_1{}^2 + r_2{}^2 + r_3{}^2)$, where k is a positive constant, $r_i = \|\overrightarrow{PP_i}\|$ and P_1 is $(0, 2)$, P_2 is $(\sqrt{3}, -1)$, P_3 is $(-\sqrt{3}, -1)$. Show that the only equilibrium position is $(0, 0)$ and determine whether it is stable. (Here P can be considered to be attracted to P_1, P_2, and P_3 by three springs, in accordance with Hooke's law.)

(c) Let P move in the xy-plane with potential energy $U = -k[(1/r_1) + (1/r_2)]$, where r_1 is the distance from P to $(1, 0)$, r_2 is the distance from P to $(-1, 0)$, and k is a positive constant. Show that $(0, 0)$ is an equilibrium position and determine whether it is stable. (This problem corresponds to the motion of a particle in a plane subject to the gravitational attraction of two equal fixed masses.)

(d) Consider the effect on part (c) of replacing k by a negative constant h. [The forces become repulsive, as would be the case if there were electric charges of same sign at $(1, 0)$, $(-1, 0)$ and P.]

‡**9.** Use the method of Lagrange multipliers to locate all local maxima and minima and also to find the absolute maximum and minimum.

(a) $f(x, y) = 3x + 4y$, where $x^2 + y^2 = 1$

(b) $f(x, y) = 5x + 12y$, where $x^2 + y^2 = 1$

(c) $f(x, y) = x^2 + y^2$, where $x^4 + y^4 = 1$

(d) $f(x, y) = xy$, where $2x^2 + y^2 = 1$

(e) $f(x, y) = x + y$, where $(1/x) + (1/y) = 1$, $x > 0$, $y > 0$

(f) $f(x, y) = x^2 + y^2$, where $x^{2/3} + y^{2/3} = 1$, $x > 0$, $y > 0$

(g) $f(x, y, z) = x + y + z$, where $x^2 + y^2 + z^2 = 1$

(h) $f(x, y, z) = 2x - y + 3z$, where $x^2 + 2y^2 + 3z^2 = 1$

(i) $f(x, y, z) = xy + yz + xz$, where $x^2 + y^2 + z^2 = 1$ (see Problem 12)

(j) $f(x, y, z) = x^2 - 2y^2 - 3z^2$, where $x^2 + y^2 + z^2 = 1$

(k) $f(x, y, z) = x + y + z$, where $x^2 + y^2 = 1$ and $y^2 + z^2 = 1$

(l) $f(x, y, z) = x^2 + y^2 + z^2$, where $x + y + z = 1$ and $x^2 + y^2 - z^2 = 0$

‡**10.** Find the foot of the perpendicular from P: $(6, 2, 3)$ to the plane $z = 5x - y + 2$ by minimizing the square of the distance from P to (x, y, z), where (x, y, z) is in the plane.

‡11. Find the local maxima and minima of $z = x^2 + y^2$ subject to the side condition stated (see Section 6-2 for discussion of the geometric interpretation; one should note that the λ of Section 6-2 corresponds to $-1/\lambda$ as used here).

(a) $3x^2 + xy + 3y^2 = 1$ (b) $2x^2 + xy + 2y^2 = 1$

(c) $xy = 1$ (d) $x^2 + 4xy + y^2 = 1$

12. Show that maximizing $f(x_1, \ldots, x_n) = \sum_{i,j=1}^{n} a_{ij}x_ix_j$, subject to the side condition $\sum_{i=1}^{n} x_i^2 = 1$, leads to the equations

$$(A - \lambda I)\mathbf{x} = \mathbf{0}, \qquad \|\mathbf{x}\|^2 = 1$$

where $A = (a_{ij})$. Hence, the solutions are given by values λ and vectors $\mathbf{x}$, where λ is an eigenvalue of A and $\mathbf{x}$ is an associated eigenvector, chosen as a unit vector.

‡12-25 SOME DEEPER RESULTS ON CONTINUITY

For some important properties of continuous functions one needs a more profound analysis of sets in R^n. We carry out such an analysis here and establish consequences for continuous functions. For convenience, we work in R^2, the xy-plane, since the extension to the general case is immediate.

Definition. Let $\{P_n\}$ $(n = 1, 2, \ldots)$ be a sequence of points in R^m space. By a *subsequence* of $\{P_n\}$ we mean a sequence $\{P_{nk}\}$, $k = 1, 2, \ldots$, where n_k is strictly increasing in k.

For example, a subsequence of $\{P_n\}$ is the sequence $P_2, P_4, P_6, \ldots, P_{2k}, \ldots$. Here $n_k = 2k$. Another subsequence is $P_4, P_7, P_{10}, \ldots, P_{3k+1}, \ldots$. Here $n_k = 3k + 1$. In general, to obtain a subsequence of $\{P_n\}$, we simply drop some members of the sequence, leaving an infinite number, and renumber the rest in the order in which they occur. Thus $\{2^{-n}\}$ is a subsequence of $\{1/n\}$. Here we omit all those members $1/n$ for which n is not a power of 2.

THEOREM J (Weierstrass-Bolzano Theorem). *Let E be a bounded closed set in the xy-plane, let $\{P_n\}$ be an infinite sequence of points of E. Then $\{P_n\}$ contains a convergent subsequence. More precisely, there is a subsequence $\{P_{n_k}\}$ for which*

$$\lim_{k\to\infty} P_{n_k} = P_0 \tag{12-250}$$

and P_0 is in E.

Remark. Condition (12-250) is interpreted as in all limit definitions: for each $\epsilon > 0$ there is a K such that $d(P_0, P_{n_k}) = \|\overrightarrow{P_0P_{n_k}}\| < \epsilon$ for $k > K$.

PROOF OF THEOREM J. Since E is a bounded closed set, we can enclose E in a circle and, hence, also in a rectangle E_0: $a \leq x \leq b$, $c \leq y \leq d$. We now divide E_0 into four rectangles by joining the midpoints of opposite sides, as in Figure 12-52. At least one of these four rectangles must contain P_n for infinitely many values of n. We choose such a rectangle: $a_1 \leq x \leq b_1$,

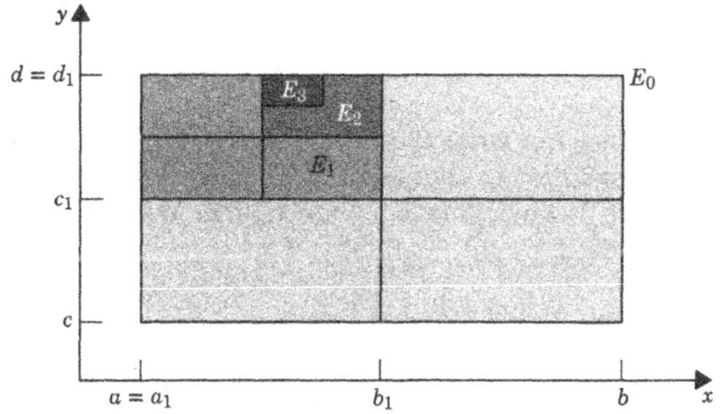

Figure 12-52 Proof of Weierstrass-Bolzano theorem.

$c_1 \le x \le d_1$, divide it into four rectangles in the same way, and again select one of the four containing P_n for infinitely many n, and so on. In this way we obtain an infinite sequence of rectangles

$$E_m\colon\ a_m \le x \le b_m,\ c_m \le y \le d_m$$

each containing P_n for infinitely many n. By the construction, $\{a_m\}$ and $\{c_m\}$ are bounded monotone nondecreasing sequences, $\{b_m\}$ and $\{d_m\}$ are bounded monotone nonincreasing sequences, and $b_m - a_m = 2^{-m}(b - a)$, $d_m - c_m = 2^{-m}(d - c)$. By Theorem H in Section 2-12, it follows that all four sequences converge and

$$\lim_{m\to\infty} a_m = \lim_{m\to\infty} b_m = x_0, \qquad \lim_{m\to\infty} c_m = \lim_{m\to\infty} d_m = y_0$$

Also, every circular neighborhood of P_0: (x_0, y_0) contains the rectangle E_m for m sufficiently large and, hence, contains P_n for infinitely many n. Thus P_0 is either in E or is a boundary point of E. Since E is closed, P_0 must lie in E. Now we let P_{n_1} be the P_n of lowest index lying in the neighborhood of P_0 of radius 1, we let P_{n_2} be the P_n of lowest index after n_1 lying in the neighborhood of P_0 of radius $\frac{1}{2}$; we proceed by inductive definition (Section 0-20) to obtain a subsequence $\{P_{n_k}\}$ such that $\|\overrightarrow{P_0P_{n_k}}\| < 2^{-k}$. Accordingly, (12-250) holds true and we are done.

PROOF OF THEOREM F. We are given a set E as in Theorem J (nonempty, bounded, and closed) and a function f defined and continuous on E. We show that f has an absolute maximum on E, the proof for a minimum being analogous.

The range of our function f forms a set B of real numbers, which we can consider to lie on the z-axis. If B has a largest number, then that is our absolute maximum M. Suppose that B is bounded above; then B has a least upper bound z^*. If z^* is not in B, we can choose numbers in B as close as we wish to z^* (see Remark 1 in Section 2-14); we choose z_n in B so that $z^* - 2^{-n} < z_n < z^*$. Then $\{z_n\}$ is a sequence in B, converging to z^*. If z^* is in B, such a sequence also exists, for we can set $z_1 = z_2 = \cdots = z^*$. If B is not bounded above, we can find arbitrarily large numbers in B and, hence, can find a sequence

$\{z_n\}$ in B, such that $\lim z_n = \infty$. Thus we always have a sequence $\{z_n\}$ in B such that $\{z_n\}$ converges to z^* if B is bounded above and $\lim z_n = \infty$ otherwise.

Since B is the range of f, we can write $z_n = f(P_n)$, where P_n is a point of E. The sequence $\{P_n\}$ now satisfies the conditions of Theorem J. Hence, we can choose a subsequence $\{P_{n_k}\}$ converging to a point P_0 in E. We assert that f has its maximum at P_0. Indeed, by the continuity of f,

$$\lim_{k\to\infty} z_{n_k} = \lim_{k\to\infty} f(P_{n_k}) = f(P_0)$$

Since $f(P_0)$ is defined, it is a finite real number; it follows that B must be bounded above and, hence, that z_{n_k} converges to z^*. Thus $f(P_0) = z^*$ and z^* must be the absolute maximum of f.

THEOREM K (*Uniform Continuity Theorem*). *Let E be a bounded closed set in the xy-plane and let f be continuous on E. Then for every* $\epsilon > 0$, *there is a* $\delta > 0$ *such that whenever P, Q are in E and* $d(P, Q) = \|\overrightarrow{PQ}\| < \delta$ *we have* $|f(P) - f(Q)| < \epsilon$.

Remark. The conclusion looks just like the definition of continuity. However, previously we fixed a point Q and then, for each ϵ, could choose δ so that $\|\overrightarrow{PQ}\| < \delta$ implies $|f(P) - f(Q)| < \epsilon$. As we change Q, we generally have to find a new δ for each ϵ and Q; thus, we can think of δ as a function of Q and ϵ: $\delta = \delta(Q, \epsilon)$. Theorem K asserts that, if our domain is a bounded closed set E, then for each ϵ, one δ works for all Q in E. The conclusion of the theorem is usually referred to as the "uniform continuity" of f on the set E.

PROOF OF THEOREM K. We shall use a proof by contradiction. Let us suppose that the conclusion fails. Then for some $\epsilon > 0$ there is no δ with the asserted property. In particular for $\delta = \delta_n = 2^{-n}$, there must be points P_n, Q_n in E such that $\|\overrightarrow{P_nQ_n}\| < 2^{-n}$ but $|f(P_n) - f(Q_n)| \geq \epsilon$. By Theorem J, the sequence $\{P_n\}$ has a convergent subsequence $\{P_{n_k}\}$, with limit P_0 in E. By the continuity of f we can choose $\delta_0 = \delta_0(Q, \epsilon)$ such that $\|\overrightarrow{P_0Q}\| < \delta_0$ implies $|f(P_0) - f(Q)| < \epsilon/2$. Now choose n_k so large that $2^{-n_k} < \delta_0/2$ and $\|\overrightarrow{P_{n_k}P_0}\| < \delta_0/2$. Then $\|\overrightarrow{P_{n_k}Q_{n_k}}\| < 2^{-n_k} < \delta_0/2$, so that (see Figure 12-53)

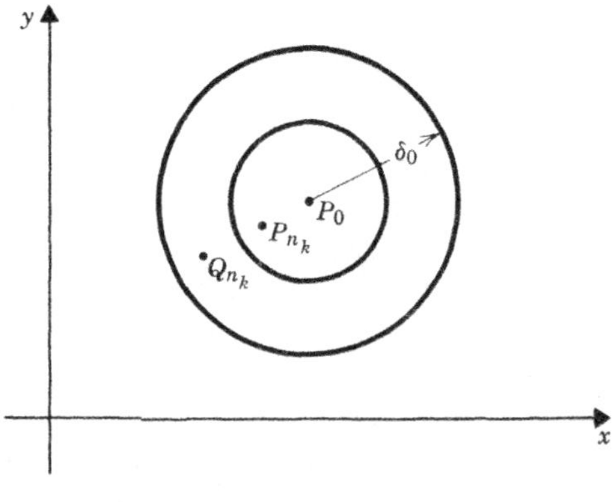

Figure 12-53 Proof of Uniform Continuity Theorem.

$$\|\overrightarrow{Q_{n_k}P_0}\| \leq \|\overrightarrow{Q_{n_k}P_{n_k}}\| + \|\overrightarrow{P_{n_k}P_0}\| < \frac{\delta_0}{2} + \frac{\delta_0}{2} = \delta_0$$

Accordingly, by the choice of δ_0,

$$|f(Q_{n_k}) - f(P_0)| < \frac{\epsilon}{2}$$

Also $\|\overrightarrow{P_{n_k}P_0}\| < \delta_0/2 < \delta_0$, so that in the same way

$$|f(P_{n_k}) - f(P_0)| < \frac{\epsilon}{2}$$

Accordingly,

$$|f(P_{n_k}) - f(Q_{n_k})| = |[f(P_{n_k}) - f(P_0)] + [f(P_0) - f(Q_{n_k})]|$$

$$\leq |f(P_{n_k}) - f(P_0)| + |f(P_0) - f(Q_{n_k})| < \frac{\epsilon}{2} + \frac{\epsilon}{2} = \epsilon$$

This contradicts our choice of the sequences $\{P_n\}$, $\{Q_n\}$. Therefore, the conclusion of the theorem must be true: for every $\epsilon > 0$, there is a $\delta > 0$ such that P, Q in E, and $\|\overrightarrow{PQ}\| < \delta$ imply $|f(P) - f(Q)| < \epsilon$.

Existence of the Riemann Integral. Theorem K can be used to simplify greatly the proof (Section 4-26) of existence of the Riemann integral of a continuous function. We need the one-dimensional form of Theorem K: if f is continuous on a bounded closed set E on the x-axis, then f is uniformly continuous: for each $\epsilon > 0$, there is a $\delta > 0$ such that x_1, x_2 in E, and $|x_1 - x_2| < \delta$ imply $|f(x_1) - f(x_2)| < \epsilon$. We defined the definite integral of a continuous function f on $[a, b]$ as the glb of all sums

$$\sum_{i=1}^{n} M_i \, \Delta_i x$$

where the interval $[a, b]$ has been subdivided as usual, and M_i is the maximum of f in $[x_{i-1}, x_i]$. We want to show that the integral also equals the Riemann integral

$$\lim_{\text{mesh}\to 0} \sum_{i=1}^{n} f(\xi_i) \, \Delta_i x,$$

where $x_{i-1} \leq \xi_i \leq x_i$.

Let a positive ϵ be given and let $\epsilon_1 = \epsilon/(b - a)$. Since f is continuous on $[a, b]$ and $[a, b]$ is a bounded closed set, we can choose δ so that ξ, η on $[a, b]$ and $|\xi - \eta| < \delta$ imply $|f(\xi) - f(\eta)| < \epsilon_1$. We now choose a subdivision of mesh less than δ. Then in each subinterval $[x_{i-1}, x_i]$, f has its maximum M_i at some ξ_i^* and its minimum m_i at some η_i^*. Since our mesh is less than δ, $|\xi_i^* - \eta_i^*| < \delta$, and, hence, $|M_i - m_i| < \epsilon_1$. Since $m_i \leq M_i$, we can write $M_i - m_i < \epsilon_1$. Hence

$$\sum_{i=1}^{n} M_i \, \Delta_i x - \sum_{i=1}^{n} m_i \, \Delta_i x = \sum_{i=1}^{n} (M_i - m_i) \, \Delta_i x < \epsilon_1 \cdot \sum_{i=1}^{n} \Delta_i x = \epsilon_1(b - a) = \epsilon$$

But we know that

$$\sum_{i=1}^{n} m_i \, \Delta_i x \leq \int_a^b f(x)\, dx \leq \sum_{i=1}^{n} M_i \, \Delta_i x$$

and that

$$\sum_{i=1}^{n} m_i \, \Delta_i x \leq \sum_{i=1}^{n} f(\xi_i)\, \Delta_i x \leq \sum_{i=1}^{n} M_i \, \Delta_i x$$

It follows that, for all subdivisions of mesh less than δ, we have

$$\left| \sum_{i=1}^{n} f(\xi_i)\, \Delta_i x - \int_a^b f(x)\, dx \right| < \epsilon$$

or

$$\lim_{\text{mesh}\to 0} \sum_{i=1}^{n} f(\xi_i)\, \Delta_i x = \int_a^b f(x)\, dx \tag{12-251}$$

as was to be proved.

Riemann Integral of Composite Functions. At several points in Chapters 4 and 7 we had to show that a Riemann integral involving two functions $f(x)$, $g(x)$ could be obtained as a limit of sums in which f and g are evaluated at different points ξ_i, η_i in the subdivision interval (see, for example, Problem 5 following Section 4-26 and the discussion of arc length in Section 4-27). We take care of all these cases by proving the following theorem.

THEOREM L. *Let f and g be continuous in $[a, b]$, with $c \leq f(x) \leq d$ and $h \leq g(x) \leq k$ for $a \leq x \leq b$. Let $F(u, v)$ be continuous in the rectangle E: $c \leq u \leq d$, $h \leq v \leq k$. Then*

$$\int_a^b F(f(x), g(x))\, dx = \lim_{\text{mesh}\to 0} \sum_{i=1}^{n} F(f(\xi_i), g(\eta_i))\, \Delta_i x \tag{12-252}$$

where on the right, subdivisions of the interval $[a, b]$ as usual are assumed, and ξ_i, η_i are chosen in $[x_{i-1}, x_i]$.

PROOF. Let $\epsilon > 0$ be given and let $\epsilon_1 = \epsilon/(b - a)$. Since F is continuous on the bounded closed set E, we can choose $\delta_1 > 0$ so that P, Q in E, and $\|\overrightarrow{PQ}\| < \delta_1$ imply $|F(P) - F(Q)| < \epsilon_1/2$. Since g is continuous on the closed interval $[a, b]$, we can choose $\delta_2 > 0$ so that ξ, η in $[a, b]$ and $|\xi - \eta| < \delta_2$ imply $|g(\xi) - g(\eta)| < \delta_1$. Since $F(f(x), g(x))$ is continuous on $[a, b]$, we can choose $\delta_3 > 0$ so that for every subdivision of mesh less than δ_3

$$\left| \sum_{i=1}^{n} F(f(\xi_i), g(\xi_i))\, \Delta_i x - \int_a^b F(f(x), g(x))\, dx \right| < \frac{\epsilon}{2} \tag{12-253}$$

We let δ be the smaller of δ_2, δ_3 and consider a subdivision of mesh less than

δ. If we now choose the ξ_i, η_i in $[x_{i-1}, x_i]$, then $|\xi_i - \eta_i| < \delta_2$, so that $|g(\xi_i) - g(\eta_i)| < \delta_1$. Hence, the two points

$$(f(\xi_i), g(\xi_i)), \qquad (f(\xi_i), g(\eta_i))$$

in E are at most distance δ_1 apart. Therefore, by the choice of δ_1,

$$|F(f(\xi_i), g(\xi_i)) - F(f(\xi_i), g(\eta_i))| < \frac{\epsilon_1}{2}$$

and

$$\left|\sum_{i=1}^{n} F(f(\xi_i), g(\xi_i))\,\Delta_i x - \sum_{i=1}^{n} F(f(\xi_i), g(\eta_i))\,\Delta_i x\right|$$

$$\leq \sum_{i=1}^{n} |F(f(\xi_i), g(\xi_i)) - F(f(\xi_i), g(\eta_i))|\,\Delta_i x$$

$$< \frac{\epsilon_1}{2}\sum_{i=1}^{n} \Delta_i x = \frac{\epsilon_1}{2}(b-a) = \frac{\epsilon}{2}$$

Accordingly, by (12-253),

$$\left|\sum_{i=1}^{n} F(f(\xi_i), g(\eta_i))\,\Delta_i x - \int_a^b F(f(x), g(x))\,dx\right|$$

$$\leq \left|\sum_{i=1}^{n} F(f(\xi_i), g(\eta_i))\,\Delta_i x - \sum_{i=1}^{n} F(f(\xi_i), g(\xi_i))\,\Delta_i x\right|$$

$$+ \left|\sum_{i=1}^{n} F(f(\xi_i), g(\xi_i))\,\Delta_i x - \int_a^b F(f(x), g(x))\,dx\right|$$

$$< \frac{\epsilon}{2} + \frac{\epsilon}{2} = \epsilon$$

and (12-252) is proved.

Continuity of the Integral. As one other application of uniform continuity we consider an integral

$$\int_a^b F(x, y)\,dx$$

For each fixed y, we have a usual definite integral which, under a continuity assumption, has a value. This value depends on the y chosen. For example,

$$\int_0^1 e^{xy}\,dx = \left.\frac{e^{xy}}{y}\right|_0^1 = \frac{e^y - 1}{y}$$

for $y \neq 0$. For $y = 0$, the value is 1. We should like to show that, when $F(x, y)$ is continuous in (x, y), the value of the integral $\int_a^b F(x, y)\,dx$ is a continuous function of y.

THEOREM M. *Let F be continuous in the rectangle E: $a \leq x \leq b$, $c \leq y \leq d$. Then*

$$g(y) = \int_a^b F(x, y)\, dx$$

defines a function continuous for $c \leq y \leq d$.

PROOF. As remarked at the end of Section 12-7, for each fixed y, $F(x, y)$ is continuous in x—here for $a \leq x \leq b$. Thus $g(y)$ is defined for $c \leq y \leq d$. By uniform continuity, given $\epsilon > 0$, we can choose $\delta > 0$, so that

$$|F(x_1, y_1) - F(x_2, y_2)| < \epsilon_1 = \epsilon/(b - a)$$

for (x_1, y_1), (x_2, y_2) in E and less than distance δ apart. Therefore, for $|y_1 - y_2| < \delta$, we have

$$\begin{aligned} |g(y_1) - g(y_2)| &= \left| \int_a^b F(x, y_1)\, dx - \int_a^b F(x, y_2)\, dx \right| \\ &= \left| \int_a^b [F(x, y_1) - F(x, y_2)]\, dx \right| \\ &\leq \int_a^b |F(x, y_1) - F(x, y_2)|\, dx < \int_a^b \epsilon_1\, dx = \epsilon_1(b - a) = \epsilon \end{aligned}$$

Accordingly, g is continuous in $[c, d]$.

For our example with $F(x, y) = e^{xy}$, $g(y)$ must be continuous at $y = 0$. Hence, we conclude:

$$\lim_{y \to 0} \frac{e^y - 1}{y} = 1$$

PROBLEMS

1. Choose convergent subsequences of the following sequences:

   (a) $\{(-1)^n\}$ (b) $\left\{\sin \frac{n\pi}{6}\right\}$

   (c) $\left\{\left(\cos \frac{n\pi}{3}, \sin \frac{n\pi}{5}\right)\right\}$ (d) $\left\{\left(n \sin \frac{n\pi}{4}, n \cos \frac{n\pi}{8}\right)\right\}$

2. Give a proof of the Weierstrass-Bolzano theorem for a bounded closed set on the x-axis.
3. Use the Weierstrass-Bolzano theorem to prove the existence of a maximum for a function f continuous on $[a, b]$.
4. Let E_1, E_2 be two bounded closed sets in the plane with no points in common. Prove that there are two closest points of the sets. That is, prove there are points P_1 in E_1, P_2 in E_2 such that $\|\overrightarrow{Q_1Q_2}\| \geq \|\overrightarrow{P_1P_2}\|$ for all Q_1 in E_1, Q_2 in E_2.

5. Find $\delta(\epsilon)$ so that $|f(x_1) - f(x_2)| < \epsilon$ for $|x_1 - x_2| < \delta$ if

   (a) $f(x) = 3x - 1,\ 0 \le x \le 1$ (b) $f(x) = x^2,\ 0 \le x \le 1$

6. Prove: if $|f'(x)| < K$ for $a < x < b$, then $|f(x_1) - f(x_2)| < \epsilon$ for $|x_1 - x_2| < \epsilon/K$.

7. Prove: if $f(x, y)$ is defined for $a \le x \le b,\ c \le y \le d$ and

$$|f(x_1, y_1) - f(x_2, y_2)| \le K_1|x_1 - x_2| + K_2|y_1 - y_2|$$

   for all (x_1, y_1), (x_2, y_2) in the domain, where K_1, K_2 are constants, then f is continuous and find a $\delta(\epsilon)$ such that

$$|f(x_1, y_1) - f(x_2, y_2)| < \epsilon$$

   for (x_1, y_1), (x_2, y_2) at most distance $\delta(\epsilon)$ apart.

8. Generalize Theorem L and its proof to $F(f(x), g(x), h(x))$.

9. Evaluate the integrals and verify continuity as in Theorem M:

   (a) $\int_0^1 \sin xy\, dx$ (b) $\int_0^1 \frac{1}{x + y}\, dx$ (c) $\int_{0.5}^1 \sqrt{x^2 + y}\, dx$

10. Prove: if $\varphi(y)$ is continuous for $c \le y \le d$, $\varphi(y) > a$ for $c \le y \le d$, and $F(x, y)$ is continuous for $a \le x \le \varphi(y)$, $c \le y \le d$, then

$$g(y) = \int_a^{\varphi(y)} F(x, y)\, dx$$

   is continuous for $c \le y \le d$. [*Hint.* set $x = a + u(\varphi(y) - a)$ and apply Theorem M.]

13

INTEGRAL CALCULUS OF FUNCTIONS OF SEVERAL VARIABLES

In this chapter we present the theory of double integrals and triple integrals, and related concepts. The ideas are natural extensions of the ones of Chapter 4, but several novel processes are introduced.

13-1 THE DOUBLE INTEGRAL

We begin by an intuitive discussion, in order to bring out the main ideas. In Section 13-2 we make the theory precise.

One is led naturally to the double integral in seeking the volume of a region "beneath a surface"—just as we are led to the definite integral in seeking the area beneath a curve. Let the surface be given by $z = f(x, y)$, for (x, y) in a region R of the xy-plane, as in Figure 13-1. Let $f(x, y)$ be continuous and

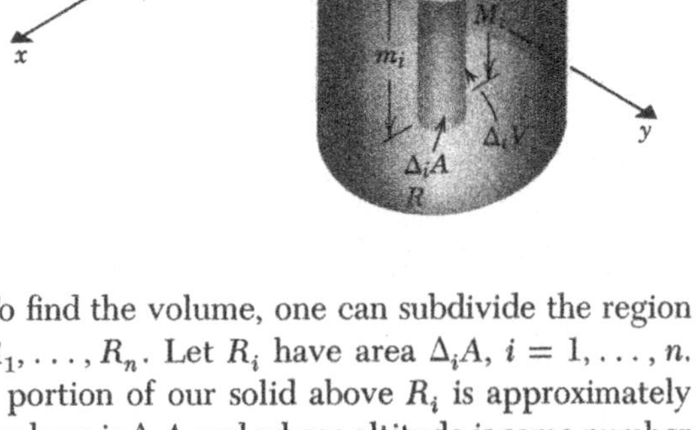

Figure 13-1 The volume represented by $\iint_R f(x, y)\, dA$.

nonnegative for (x, y) in R. To find the volume, one can subdivide the region R into many small regions $R_1, \ldots, R_n$. Let R_i have area $\Delta_i A$, $i = 1, \ldots, n$. Then the volume $\Delta_i V$ of the portion of our solid above R_i is approximately the volume of a cylinder whose base is $\Delta_i A$ and whose altitude is some number between the minimum m_i and maximum M_i of f in R_i. Hence

$$m_i \, \Delta_i A \leq \Delta_i V \leq M_i \, \Delta_i A$$

and for the total volume $V = \sum_{i=1}^{n} \Delta_i V$ we can write

$$\sum_{i=1}^{n} m_i \, \Delta_i A \leq V \leq \sum_{i=1}^{n} M_i \, \Delta_i A$$

It is plausible that if the regions R_i are made small enough, then both sums can be made as close to V as desired; that is, we expect

$$V = \text{glb} \sum_{i=1}^{n} M_i \, \Delta_i A \qquad \text{and} \qquad V = \text{lub} \sum_{i=1}^{n} m_i \, \Delta_i A$$

Here the glb and lub refer to all values of the sums obtained from all possible subdivisions of R. We shall show that, for a continuous function $z = f(x, y)$ and a region R satisfying reasonable hypotheses, the glb and lub exist and are equal. The common value is the *double integral* of f over R, and we write

$$\iint_R f(x, y) \, dA = \text{glb} \left\{ \sum M_i \, \Delta_i A \right\} = \text{lub} \left\{ \sum m_i \, \Delta_i A \right\} \tag{13-10}$$

The double integral is also denoted by

$$\iint_R f(x, y) \, dx \, dy$$

but it is important to distinguish this from the iterated integral, which we discuss next. One also writes simply $\int_R f(x, y) \, dA$, with just one integral sign, since the dA indicates that one is dealing with a 2-dimensional integral. For $f \geq 0$, the double integral can always be interpreted as the volume of the corresponding solid region. However, the double integral has other applications, including many for which f has variable sign.

Evaluation of Double Integral by Means of an Iterated Integral. We have given a definition of the double integral. We now ask: How can we evaluate one?

Suppose that R is the region between the curves $y = 1 + \cos x$ and $y = \sin x$ for $0 \leq x \leq \pi/2$ (Figure 13-2). For such a region R we saw in Section 7-7 that it was reasonable to assign volume to the solid region beneath a surface $z = f(x, y)$, for (x, y) in R, by the formula

$$V = \int_0^{\pi/2} A(x) \, dx \tag{13-11}$$

where $A(x)$ is the area of the cross section of the solid by a plane perpendicular to the x-axis. Here each such cross section can be regarded as a planar region

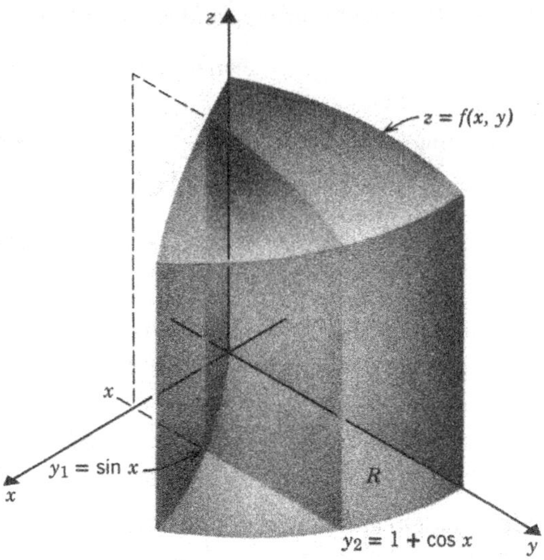

Figure 13-2 The volume as integral of area of cross section.

in the yz-plane (since x is fixed): namely, the region beneath the curve $z = f(x, y)$ for $y_1 \leq y \leq y_2$, where $y_1 = \sin x$ and $y_2 = 1 + \cos x$. Thus

$$A(x) = \int_{y_1}^{y_2} f(x, y)\, dy = \int_{\sin x}^{1+\cos x} f(x, y)\, dy \tag{13-12}$$

In these two integrals x is treated as constant; the integration is with respect to y. From (13-12) and (13-11) we obtain

$$V = \int_0^{\pi/2} \left[\int_{\sin x}^{1+\cos x} f(x, y)\, dy \right] dx$$

It is customary to omit the brackets and to write

$$V = \int_0^{\pi/2} \int_{\sin x}^{1+\cos x} f(x, y)\, dy\, dx \tag{13-13}$$

In such expressions we are first to integrate with respect to the *inside* variable—here that variable is y, and then with respect to the *outside* variable—here, x. The righthand side of (13-13) is an example of an *iterated integral.*

We saw above that the volume is also given by a double integral:

$$V = \iint_R f(x, y)\, dA \tag{13-14}$$

where R is the region of Figure 13-2. From (13-13) and (13-14) we now conclude:

$$\iint_R f(x, y)\, dA = \int_0^{\pi/2} \int_{\sin x}^{1+\cos x} f(x, y)\, dy\, dx \tag{13-15}$$

Thus for the case considered *the double integral can be replaced by an iterated integral.*

The procedure illustrated will be shown to be valid whenever R is a region between two curves:

$$\varphi_1(x) \leq y \leq \varphi_2(x) \qquad \text{for } a \leq x \leq b \tag{13-16}$$

where φ_1 and φ_2 are continuous in $[a, b]$ (see Figure 13-3*a*). For such a region R we have the iterated integral

$$\int_a^b \int_{\varphi_1(x)}^{\varphi_2(x)} f(x, y)\, dy\, dx$$

If f is continuous on R, then the iterated integral exists. In many cases the methods of Chapter 4 (and integral tables) permit us to evaluate the integral exactly.

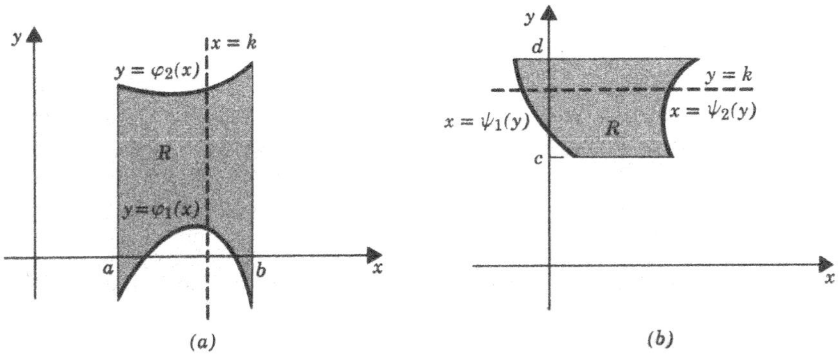

Figure 13-3 Regions for iterated integrals. (*a*) Region for equation (13-16); (*b*) region for equation (13-16′).

For a region as in Figure 13-3*a* and a nonnegative function f, the iterated integral and double integral both give the volume of the solid region beneath the surface $z = f(x, y)$, for (x, y) in R. Hence, the iterated integral and double integral are equal. We shall see that the equality holds true generally:

$$\iint_R f(x, y)\, dA = \int_a^b \int_{\varphi_1(x)}^{\varphi_2(x)} f(x, y)\, dy\, dx \tag{13-17}$$

with no assumption about the sign of f.

A similar discussion applies to the case of a region R as follows,

$$\psi_1(y) \leq x \leq \psi_2(y) \quad \text{for } c \leq y \leq d \tag{13-16′}$$

where ψ_1 and ψ_2 are continuous in $[c, d]$. Figure 13-3*b* shows such a region. A double integral of a function f over R can be evaluated as an iterated integral:

$$\iint_R f(x, y)\, dA = \int_c^d \int_{\psi_1(y)}^{\psi_2(y)} f(x, y)\, dx\, dy \tag{13-17′}$$

We note that the region R of Figure 13-1 could be described either by inequalities of form (13-16) or by inequalities of form (13-16′). Hence, for this region we can use either (13-17) or (13-17′) to evaluate the double integral. We can use both ways in order to check a calculation.

EXAMPLE 1. Let $f(x, y) = x^2 + y^2$ for $0 \le x \le 1$, $0 \le y \le 2$, as in Figure 13-4. Then the volume V beneath the surface $z = f(x, y)$ is given by

$$V = \iint_R f(x, y)\, dA$$

where R is the rectangular region $0 \le x \le 1$, $0 \le y \le 2$. By (13-17) we can write

$$V = \iint_R f(x, y)\, dA = \int_0^1 \int_0^2 f(x, y)\, dy\, dx = \int_0^1 \int_0^2 (x^2 + y^2)\, dy\, dx$$
$$= \int_0^1 \left(x^2 y + \frac{y^3}{3}\right)\Bigg|_{y=0}^{y=2} dx = \int_0^1 \left\{2x^2 + \left(\frac{8}{3}\right)\right\} dx = \frac{10}{3}$$

Similarly by (13-17′)

$$V = \int_0^2 \int_0^1 (x^2 + y^2)\, dx\, dy = \int_0^2 \left(\frac{x^3}{3} + xy^2\right)\Bigg|_{x=0}^{x=1} dy$$
$$= \int_0^2 \left(\frac{1}{3} + y^2\right) dy = \frac{10}{3}$$

EXAMPLE 2. Let $f(x, y) = 1$ for $(x - 1)^2 + (y - 2)^2 = 1$, as in Figure 13-5.

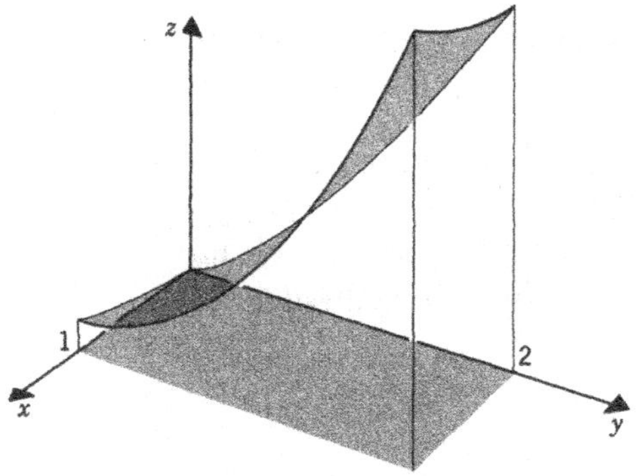

Figure 13-4 Example 1.

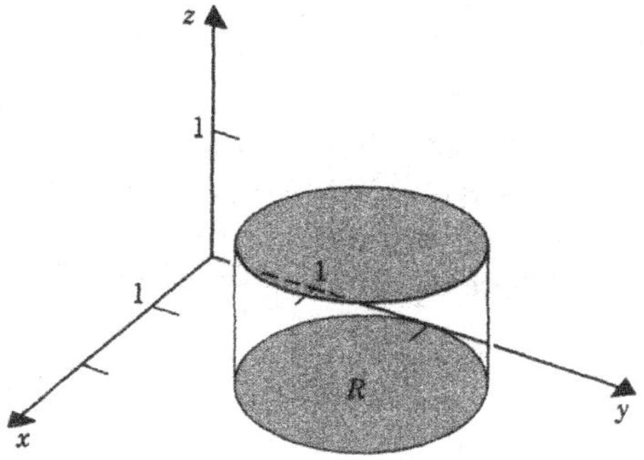

Figure 13-5 Example 2.

The region R is circular and the solid is a circular cylinder of altitude $h = 1$ and radius $a = 1$, and so has volume $\pi a^2 h = \pi$. We can now also write

$$V = \iint_R f(x, y)\, dA = \int_0^2 \int_{2-\sqrt{1-(x-1)^2}}^{2+\sqrt{1-(x-1)^2}} dy\, dx = \int_0^2 y \Bigg|_{y=2-\sqrt{\cdots}}^{y=2+\sqrt{\cdots}} dx$$
$$= \int_0^2 2\sqrt{1 - (x - 1)^2}\, dx = [(x - 1)\sqrt{1 - (x - 1)^2} + \mathrm{Sin}^{-1}(x - 1)]\Bigg|_0^2 = \pi$$

Similarly, we can integrate in the other order:

$$V = \int_1^3 \int_{1-\sqrt{1-(y-2)^2}}^{1+\sqrt{1-(y-2)^2}} dx\, dy.$$

EXAMPLE 3. Let $f(x, y) = 1 + x$ for $0 \leq x \leq 3$, $0 \leq y \leq 2x^3 - 9x^2 + 12x$.

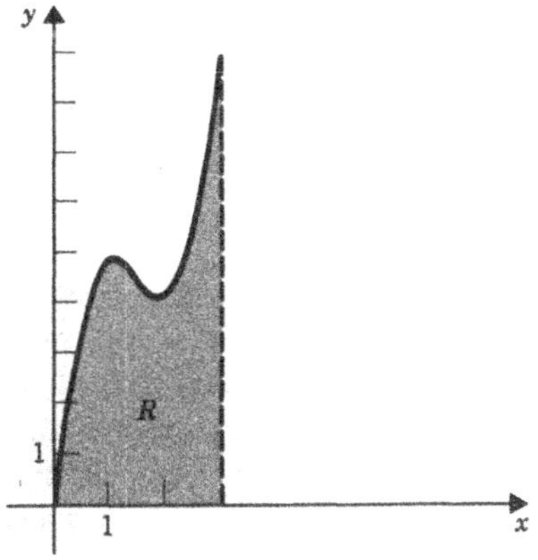

Figure 13-6 Example 3.

The inequalities describe a region R as in Figure 13-6. By (13-17)

$$\iint_R f(x, y)\, dA = \int_0^3 \int_0^{2x^3-9x^2+12x} (1 + x)\, dy\, dx$$

$$= \int_0^3 (y + xy)\Big|_{y=0}^{y=2x^3-9x^2+12x} dx$$

$$= \int_0^3 (2x^4 - 7x^3 + 3x^2 + 12x)\, dx = 36\frac{9}{20}$$

For this region R, the formula (13-17′) cannot be applied, since R cannot be described by inequalities of the form (13-16′); for fixed y, x is not confined to a single interval. For such cases, one can try to decompose R into several pieces, to each of which a formula of form (13-16′) is applicable, and then can add the results.

In general, for (13-17) we need a region R as in Figure 13-3a, describable by (13-16); here each line $x = \text{const} = k$, $a < k < b$, meets R in a single line segment. For (13-17′), we need a region R as in Figure 13-3b, describable by (13-16′); here each line $y = \text{const} = k$, $c < k < d$ meets R in a single line segment.

For a region R as in Figure 13-7, neither (13-17) nor (13-17′) is applicable. We could with some effort decompose R into a number of pieces, to each of which one of the two formulas (13-17), (13-17′) is applicable. Alternatively, one can seek a direct numerical method for calculating the double integral of f over R. To this end we reason that, for a fine enough subdivision of R, the double integral should be approximately equal to

$$\sum_{i=1}^{n} f(\xi_i, \eta_i)\, \Delta_i A \qquad [(\xi_i, \eta_i) \text{ in } R_i]$$

since for each i, $m_i\, \Delta_i A \leq f(\xi_i, \eta_i)\, \Delta_i A \leq M_i\, \Delta_i A$. More precisely, we expect that

$$\iint_R f(x, y)\, dA = \lim_{\text{mesh}\to 0} \sum_{i=1}^{n} f(\xi_i, \eta_i)\, \Delta_i A \tag{13-18}$$

the limit being as for the Riemann integral $\int_a^b f(x)\, dx$ (Section 4-25). The mesh should be the maximum "width" of the subregions $R_1, \ldots, R_n$, in some reasonable sense. To evaluate the integral numerically, we can simply subdivide R as in Figure 13-7, with the aid of parallels to the axes. We then choose a point (ξ_i, η_i) in each little region R_i, multiply $f(\xi_i, \eta_i)$ by $\Delta_i A$ and add. For the subdivision as in Figure 13-7 the areas $\Delta_i A$ are easy to find for all the square regions R_i—all are equal to $(0.2)^2 = 0.04$; for the irregular ones, one can estimate the area graphically. In fact, for a fine enough subdivision, the contribution to the sum in (13-18) from the irregular regions can be shown to be negligible; that is, for a reasonable region R, (13-18) remains correct if one simply drops the terms on the right coming from subregions R_i meeting the boundary of R.

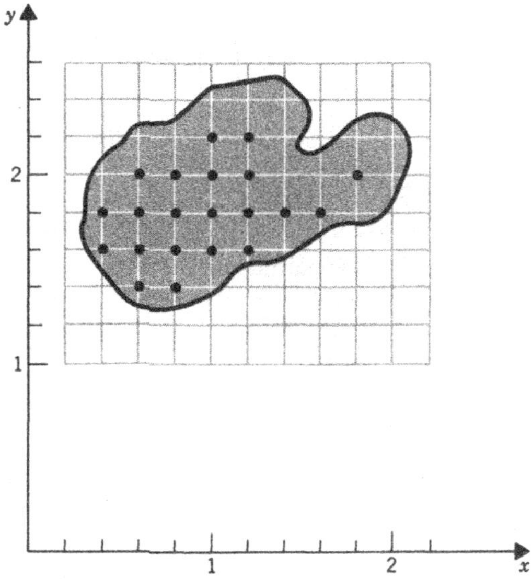

Figure 13-7 Example 4.

EXAMPLE 4. $\iint_R (x + y)\, dA$ over the region R of Figure 13-7. We then choose the squares R_i having lower left vertex at a marked point and also use this point as (ξ_i, η_i). The numerical work is indicated in Table 13-1. The sum

Table 13-1

x	y	$f(x, y)$	x	y	$f(x, y)$	x	y	$f(x, y)$
.4	1.6	2.0	.8	1.6	2.4	1.2	1.6	2.8
	1.8	2.2		1.8	2.6		1.8	3.0
.6	1.4	2.0		2.0	2.8		2.0	3.2
	1.6	2.2	1.0	1.6	2.6		2.2	3.4
	1.8	2.4		1.8	2.8	1.4	1.8	3.2
	2.0	2.6		2.0	3.0	1.6	1.8	3.4
.8	1.4	2.2		2.2	3.2	1.8	2.0	3.8
sum		15.6			19.4			22.8

$\Sigma f(x_i, y_i)\, \Delta_i A$ equals

$$0.04 \sum f(x_i, y_i) = 0.04(57.8) = 2.312$$

Hence, the value of the integral is approximately 2.31. We could improve the accuracy by making some estimate of the contribution from the partial squares near the boundary, but the correction will not be very large. The accuracy can also be improved by using smaller squares.

EXAMPLE 5. $\iint\limits_R e^{xy} \sin (x^2 + y^2)\, dA$, where R is as in Example 3. We could use an iterated integral, with the same limits of integration as for Example 3. However, the integrand has no familiar indefinite integral with respect to y. Hence, we cannot evaluate the iterated integral as in Examples 1 to 3, and some numerical procedure is called for. The simplest is to use (13-18) and to replace the integral by a sum $\Sigma f(\xi_i, \eta_i)\, \Delta A_i$, as for Example 4. The work is just like that for that example.

Hence, we see that Equation (13-18) is of much more than theoretical interest. It is an important tool for numerical evaluation of the integral when the region R is awkward or when the integrand f is awkward.

PROBLEMS

1. Find the volume of the solid region below the given surface $z = f(x, y)$ for (x, y) in the region R defined by the given inequalities.
   (a) $z = e^x \cos y$, $0 \le x \le 1$, $0 \le y \le \pi/2$
   (b) $z = x^2 e^{-x-y}$, $0 \le x \le 1$, $0 \le y \le 2$
   (c) $z = x^2 y$, $0 \le x \le 1$, $x + 1 \le y \le x + 2$
   (d) $z = x + y$, $0 \le y \le 1$, $0 \le x \le 1 + y^2$
   (e) $z = \sin (x + y)$, $0 \le y \le 1$, $1 - y \le x \le 1$
   (f) $z = (x + y)^2$, $0 \le x \le 1$, $x^5 \le y \le x^4$
   (g) $z = 1 - x^2 - 2y^2$, $0 \le x \le 1 - 2y^2$
   (h) $z = \sqrt{1 - x^2 - y^2}$, $x^2 + y^2 \le 1$ (hemisphere)
   (i) $z = \sqrt{x^2 - y^2}$, $x^2 - y^2 \ge 0$, $0 \le x \le 1$

(j) $z = 1$, $x^2 + 2y^2 \leq 1$ (elliptic cylinder)

(k) $z = 1$, $0 \leq y$, $0 \leq x \leq 1 - y$ (triangular prism)

(l) $z = 1 - x - y$, $0 \leq y$, $0 \leq x \leq 1 - y$ (tetrahedron)

(m) $z = \sqrt{1 - 2x^2 - 3y^2}$, $2x^2 + 3y^2 \leq 1$ (half of ellipsoid)

(n) $z = \sqrt{1 - x^2}$, $0 \leq x$, $0 \leq y \leq 1 - x$

(o) $z = 3 - x - y$, $x^2 - 1 \leq y \leq 1 - x^2$

(p) $z = xy$, $0 \leq y \leq 4$, $4 - y \leq x \leq 8 - (y - 2)^3$

2. For each of the following choices of R, represent $\iint_R f(x, y)\, dA$ as an iterated integral of both forms (13-17) and (13-17′), wherever possible:

(a) $1 \leq x \leq 2$, $1 - x \leq y \leq 1 + x$ (b) $-1 \leq x \leq 2$, $-x^2 \leq y \leq x^2 + 1$

(c) region bounded by lines $y = x$, $x + y + 1 = 0$, $x - 2y + 3 = 0$

(d) region bounded by $y = 1 - x^2$ and $y = x^2 - 1$

(e) $y^2 + x(x - 1) \leq 0$ (f) $x^{2/3} + y^{2/3} \leq 1$

(g) $x + y - 1 \geq 0$, $x^2 + y^2 \leq 1$

(h) $xy \leq 2$, $x - y + 1 \geq 0$, $x - y - 1 \leq 0$

(i) $x + y - 1 \geq 0$, $-x + y + 1 \geq 0$, $y - 1 \leq 0$

(j) $x - 2y - 3 \leq 0$, $x + 3y + 2 \geq 0$, $x + 13y - 18 \leq 0$

3. For each of the following iterated integrals, find the region R and write the integral in the other form (interchanging the order of integration):

(a) $\displaystyle\int_{1/2}^{1}\int_{0}^{1-x} f(x, y)\, dy\, dx$ (b) $\displaystyle\int_{0}^{1}\int_{0}^{y} f(x, y)\, dx\, dy$

(c) $\displaystyle\int_{0}^{1}\int_{0}^{\sqrt{1-x^2}} f(x, y)\, dy\, dx$ (d) $\displaystyle\int_{0}^{1}\int_{-\sqrt{1-x^2}}^{\sqrt{1-x^2}} f(x, y)\, dy\, dx$

(e) $\displaystyle\int_{0}^{1}\int_{y-1}^{0} f(x, y)\, dx\, dy$ (f) $\displaystyle\int_{0}^{1}\int_{1-x}^{1-x^2} f(x, y)\, dy\, dx$

(g) $\displaystyle\int_{0}^{1}\int_{1-x}^{1+x} f(x, y)\, dy\, dx$ (h) $\displaystyle\int_{0}^{1}\int_{\mathrm{Sin}^{-1}x}^{\pi x/2} f(x, y)\, dy\, dx$

4. Evaluate by interchanging the order of integration:

(a) $\displaystyle\int_{0}^{c}\int_{x}^{c} \frac{x\, dy}{\sqrt{x^2 + y^2}}\, dx$, $c > 0$

(b) $\displaystyle\int_{0}^{\pi/2}\int_{0}^{y} \cos 2y \sqrt{1 - k^2 \sin^2 x}\; dx\, dy$, $0 < k^2 < 1$

5. (a) Evaluate $\iint_R e^{(x+2y)/10}\, dA$, where R is the square $0 \leq x \leq 2$, $0 \leq y \leq 2$, by subdividing R into 4 equal squares and by evaluating the corresponding sum $\Sigma f(\xi_i, \eta_i)\, \Delta_i A$ for appropriate choices of the (ξ_i, η_i).

(b) Proceed as in part (a), using 16 equal squares instead of 4.

6. Evaluate $\iint_R (x^2 - y)\, dx\, dy$, where R is the region outside the square of vertices $(1, 0)$, $(1, 2)$, $(-1, 2)$, $(-1, 0)$ and inside the square of vertices $(3, 0)$, $(\pm 3, 6)$, $(-3, 0)$, by subdividing R into squares of side 1 and evaluating the corresponding sum $\Sigma f(\xi_i, \eta_i)\, \Delta_i A$ for appropriate choices of (ξ_i, η_i).

7. Let R be the region of Figure 13-7. Use the subdivision in that figure to evaluate numerically:

(a) $\iint\limits_R xy\,dA$ (b) $\iint\limits_R e^{xy}\,dA$ (c) $\iint\limits_R y^2\,dA$ (d) $\iint\limits_R e^x\,dA$

13-2 THEORY OF THE DOUBLE INTEGRAL

We now develop the concept of double integral in detail. Let C be a piecewise smooth simple closed path in the xy-plane; thus, C is the path $\overrightarrow{OP} = \mathbf{r}(t)$, $a \leq t \leq b$, where $\mathbf{r}$ is continuous in $[a, b]$ and (except for finitely many values of t) $\mathbf{r}'(t)$ is continuous in $[a, b]$ and $\mathbf{r}(s) = \mathbf{r}(t)$ with $a \leq s < t \leq b$ if, and only if, $s = a$, $t = b$ (see Figure 13-8). Let R be the bounded closed region enclosed by C. Thus R consists of C plus interior. By a *subdivision* of R we mean a decomposition of R into a finite number of regions $R_1, \ldots, R_n$, of the same type as R—that is, each enclosed by a piecewise smooth simple closed path. We denote by $\Delta_i A$ the area of R_i $(i = 1, \ldots, n)$. The area of R is the sum of the $\Delta_i A$.

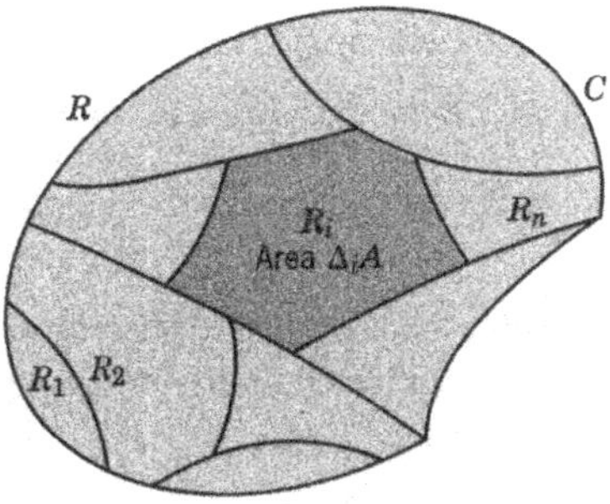

Figure 13-8 Subdivision of region R.

Let f be a function defined and continuous in R and having absolute minimum m and maximum M in R (Theorem F, Section 12-7). Let m_i and M_i be the absolute minimum and maximum, respectively, of f in R_i. We call the number

$$\sum_{i=1}^{n} M_i \,\Delta_i A$$

an *upper sum* for f and the number

$$\sum_{i=1}^{n} m_i \,\Delta_i A$$

a *lower sum* for f. It follows that every upper sum is greater than or equal to

$$\sum_{i=1}^{n} m \,\Delta_i A = m \sum_{i=1}^{n} \Delta_i A = mA$$

where A is the area of R. Similarly, every lower sum is less than or equal to MA.

We now define our double integral:

$$\iint_R f(x, y)\, dA = \text{glb} \sum_{i=1}^{n} M_i\, \Delta_i A \tag{13-20}$$

where the glb refers to all subdivisions of R as above.

In making our definition, we have tacitly assumed that R and the smaller regions R_i have area. A proof of this in general can be given on the basis of the theory in Section 7-4. However, we shall be mainly concerned with very simple regions of the type of Figure 13-3 above, or obtainable from such regions by piecing together or removing pieces, and for all such regions the area can be found as usual by definite integrals. Also for such regions, subdivisions can be obtained easily as in Figure 13-10 on page 1001.

For the definite integral we saw in Chapter 4 that every lower sum is at most equal to the integral, and hence every lower sum is less than or equal to every upper sum. For the double integral there is a similar statement, which we formulate as a lemma:

LEMMA. *For a function f continuous in R as above, every lower sum is less than or equal to every upper sum.*

A proof of the lemma is given in Section 13-3.

THEOREM 1. *Let R be a bounded closed region enclosed by a piecewise smooth simple closed path C. Let f be continuous in R, with absolute minimum m and absolute maximum M in R. Then*

$$\iint_R f(x, y)\, dA$$

exists. Furthermore,

$$mA \le \iint_R f(x, y)\, dA \le MA \tag{13-21}$$

and, more generally, if $K \le f(x, y) \le L$ in R, then

$$KA \le \iint_R f(x, y)\, dA \le LA \tag{13-22}$$

PROOF. The existence of the double integral follows from the fact that the upper sums $\Sigma M_i\, \Delta_i A$ are all bounded below—for example, by mA. Hence, as in Section 2-13, the glb of the set of all upper sums exists. Thus the double integral is defined precisely by (13-20). If $K \le f(x, y) \le L$ in R, then

$K \le M_i \le L$ for all i, so that

$$\sum_{i=1}^{n} K\,\Delta_i A = KA \le \sum_{i=1}^{n} M_i\,\Delta_i A \le LA = \sum_{i=1}^{n} L\,\Delta_i A$$

Hence, (13-22) follows, and (13-21) is a special case of (13-22).

A set E in the plane is said to be *bounded* (Section 12-22) if it can be enclosed in a circular region (of finite radius). If E is bounded, the set of all distances $\|\overrightarrow{PQ}\|$ between points P, Q in E is bounded above. The lub of all these distances is called the *diameter* of E. This concept is illustrated in Figure 13-9.

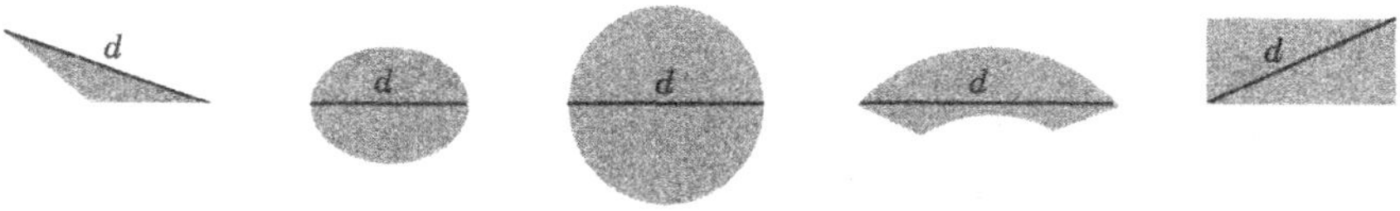

Figure 13-9 Diameter d of a set.

Let R be a region as in Theorem 1. By the *mesh* of a subdivision of R, we mean the largest diameter of the subregions $R_1, \ldots, R_n$. With the aid of parallels to the coordinate axes as in Figures 13-7 and 13-10, we can obtain a subdivision whose mesh is as small as desired. A proof of this assertion in general is difficult; a proof for the case of a region R as in Figure 13-3 is given in the next section.

THEOREM 2. *Under the hypotheses of Theorem 1,*

$$\iint_R f(x, y)\,dA = \lim_{\text{mesh}\to 0} \sum_{i=1}^{n} f(\xi_i, \eta_i)\,\Delta_i A \tag{13-23}$$

That is, if I denotes the value of the double integral, then for every $\epsilon > 0$ there is a $\delta > 0$ such that

$$\left| I - \sum_{i=1}^{n} f(\xi_i, \eta_i)\,\Delta_i A \right| < \epsilon$$

for every subdivision of R of mesh less than δ and every choice of the (ξ_i, η_i) in R_i $(i = 1, \ldots, n)$.

A proof is given in the next section. The theorem shows that for a continuous function f the double integral can be defined in the same way as the Riemann integral (Section 4-25).

THEOREM 3. *Under the hypotheses of Theorem 1,*

$$\iint_R f(x, y)\,dA = \text{lub} \sum_{i=1}^{n} m_i\,\Delta_i A \tag{13-24}$$

where the lub is taken over all subdivisions of R, and m_i is the absolute minimum of f in R_i.

PROOF. Since every lower sum of f is less than or equal to every upper sum, it follows that every lower sum is less than or equal to the double integral of f:

$$\sum_{i=1}^{n} m_i \, \Delta_i A \leq \iint_R f(x, y)\, dA = I \tag{13-25}$$

Now given $\epsilon > 0$, we choose δ as in Theorem 2 and select a subdivision of mesh less than δ. In each R_i we can choose (ξ_i, η_i) at a point where f attains its absolute minimum in R_i; that is, so that $f(\xi_i, \eta_i) = m_i$. Hence, by Theorem 2,

$$\left| \sum_{i=1}^{n} m_i \, \Delta_i A - I \right| < \epsilon$$

and, hence, for this subdivision,

$$\sum_{i=1}^{n} m_i \, \Delta_i A > I - \epsilon$$

Therefore, lub $\{\Sigma m_i \, \Delta_i A\}$ is at least equal to I. By (13-25) the lub is at most equal to I. Hence, it must be equal to I.

We can now conclude, as for the definite integral, that $\iint_R f(x, y)\, dA$ is the one and only number I such that

$$\sum_{i=1}^{n} m_i \, \Delta_i A \leq I \leq \sum_{i=1}^{n} M_i \, \Delta_i A$$

for all subdivisions of R.

THEOREM 4. *Let R be given by the inequalities*

$$a \leq x \leq b, \qquad \varphi_1(x) \leq y \leq \varphi_2(x)$$

where $\varphi_1(x)$ and $\varphi_2(x)$ are continuous in $[a, b]$ and let f be continuous in R. Then

$$\iint_R f(x, y)\, dA = \int_a^b \int_{\varphi_1(x)}^{\varphi_2(x)} f(x, y)\, dy\, dx \tag{13-26}$$

‡PROOF. Under the hypotheses stated, the integral

$$\int_{\varphi_1(x)}^{\varphi_2(x)} f(x, y)\, dy$$

is a continuous function of x in $[a, b]$. This follows from Theorem M in Section 12-25; see also Problem 10 following that section. Hence, the right hand side of (13-26) has meaning as do the other iterated integrals occurring in our proof.

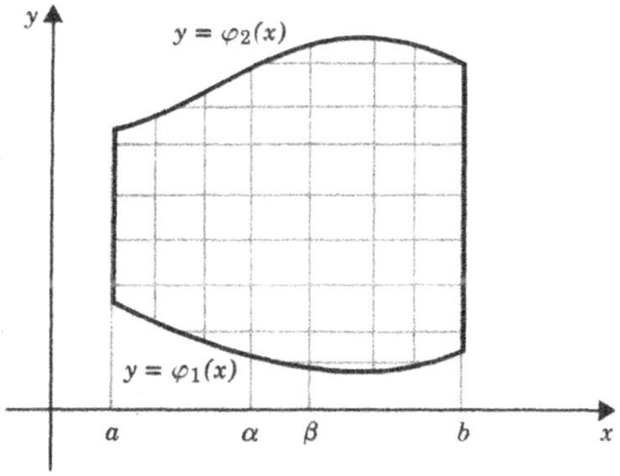

Figure 13-10 Subdivision of region $a \le x \le b$, $\varphi_1(x) \le y \le \varphi_2(x)$.

Now, let ϵ be a given positive number and choose δ as in Theorem 2. We select a subdivision of R of mesh less than δ as suggested in Figure 13-10. Here each subregion is of the form $\alpha_i \le x \le \beta_i$, $p_i(x) \le y \le q_i(x)$, where $p_i(x)$ and $q_i(x)$ are continuous; when p_i and q_i are constant functions the subregion is a rectangle. A proof that such a subdivision can be chosen with mesh less than δ is given in the next section. From the form of R_i, we can evaluate the corresponding iterated integral:

$$\int_{\alpha_i}^{\beta_i} \int_{p_i(x)}^{q_i(x)} f(x, y)\, dy\, dx = J_i$$

If one adds all those iterated integrals having the same x-interval, say $\alpha \le x \le \beta$ (see Figure 13-10) and repeatedly applies the rule $\int_a^b + \int_b^c = \int_a^c$, one obtains the iterated integral

$$\int_{\alpha}^{\beta} \int_{\varphi_1(x)}^{\varphi_2(x)} f(x, y)\, dy\, dx$$

If one adds all these integrals, for all x subintervals, one obtains the iterated integral on the right of (13-26). Thus

$$\int_a^b \int_{\varphi_1(x)}^{\varphi_2(x)} f(x, y)\, dy\, dx = J_1 + \cdots + J_n$$

Now in R_i, $m_i \le f(x, y) \le M_i$. Hence, for $\alpha_i \le x \le \beta_i$,

$$m_i[q_i(x) - p_i(x)] \le \int_{p_i(x)}^{q_i(x)} f(x, y)\, dy \le M_i[q_i(x) - p_i(x)]$$

Therefore,

$$m_i \int_{\alpha_i}^{\beta_i} [q_i(x) - p_i(x)]\, dx \le \int_{\alpha_i}^{\beta_i} \int_{p_i(x)}^{q_i(x)} f(x, y)\, dy\, dx \le M_i \int_{\alpha_i}^{\beta_i} [q_i(x) - p_i(x)]\, dx$$

The integrals to the left and right are equal to $\Delta_i A$, the area of R_i. Therefore

$$m_i \, \Delta_i A \leq J_i \leq M_i \, \Delta_i A$$

and, accordingly, by Theorem D of Section 12-7,†

$$J_i = f(\xi_i, \eta_i) \, \Delta_i A$$

for some (ξ_i, η_i) in R_i. Therefore

$$\int_a^b \int_{\varphi_1(x)}^{\varphi_2(x)} f(x, y) \, dy \, dx = J_1 + \cdots + J_n = \sum_{i=1}^{n} f(\xi_i, \eta_i) \, \Delta_i A$$

Since our mesh is less than δ, the last sum differs from the double integral by at most ϵ. Therefore, this holds true for the iterated integral. But the iterated integral is a fixed number, not depending on ϵ. Therefore, it equals the double integral, and (13-26) is proved.‡

In the same way, one proves that for an appropriate region R, as in Figure 13-6,

$$\iint_R f(x, y) \, dA = \int_a^b \int_{\psi_1(y)}^{\psi_2(y)} f(x, y) \, dx \, dy$$

THEOREM 5. *Let R be as in Theorem 1. Let f_1 and f_2 be continuous in R and let c_1 and c_2 be constants. Then*

$$\iint_R (c_1 f_1 + c_2 f_2) \, dA = c_1 \iint_R f_1 \, dA + c_2 \iint_R f_2 \, dA \tag{13-27}$$

PROOF. For each subdivision of R and choice of the (ξ_i, η_i) in R_i, we have

$$\sum_{i=1}^{n} [c_1 f_1(\xi_i, \eta_i) + c_2 f_2(\xi_i, \eta_i)] \, \Delta_i A = c_1 \sum_{i=1}^{n} f_1(\xi_i, \eta_i) \, \Delta_i A + c_2 \sum_{i=1}^{n} f_2(\xi_i, \eta_i) \, \Delta_i A$$

But if the subdivision has mesh less than δ and δ is sufficiently small, then each of the three sums differs from the corresponding double integral in (13-27) by less than ϵ. Hence, the double integral on the left of (13-27) differs from the sum of the terms on the right of (13-27) by less than $\epsilon + |c_1|\epsilon + |c_2|\epsilon = \epsilon(1 + |c_1| + |c_2|)$. This last number can be made as small as desired by the proper choice of ϵ. Hence, both sides of (13-27) must be equal.

THEOREM 6. *Let R be as in Theorem 1 and let R be subdivided into two regions R', R''. Let f be continuous in R. Then*

$$\iint_R f(x, y) \, dA = \iint_{R'} f(x, y) \, dA + \iint_{R''} f(x, y) \, dA$$

PROOF. Let R' be subdivided into regions $R_1, \ldots, R_m$; let R'' be subdivided

† Theorem D is stated for open regions, but it can be proved in similar manner for a closed region.

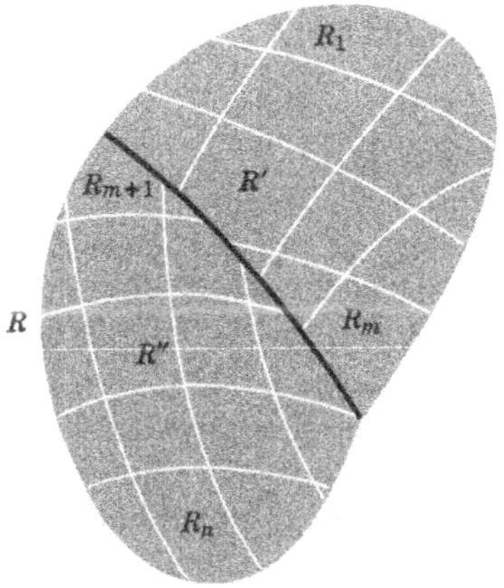

Figure 13-11 Proof of Theorem 6.

into regions $R_{m+1}, \ldots, R_n$. Then R is subdivided into $R_1, \ldots, R_n$ (see Figure 13-11). Let (ξ_i, η_i) be chosen in R_i for $i = 1, \ldots, n$. Then

$$\sum_{i=1}^{n} f(\xi_i, \eta_i)\, \Delta_i A = \sum_{i=1}^{m} f(\xi_i, \eta_i)\, \Delta_i A + \sum_{i=m+1}^{n} f(\xi_i, \eta_i)\, \Delta_i A$$

The rest of the proof is essentially the same as for Theorem 5 (Problem 9 below).

Ignoring Subregions Meeting the Boundary. We have suggested at several points that in the limit process (13-23) we can ignore some or all terms coming from subregions meeting the boundary, since the total area of such subregions approaches 0 as the mesh approaches 0. We formulate this result and an application as a supplement to Theorem 2:

THEOREM 2′. *Under the hypotheses of Theorem 1, Equation (13-23) remains valid if, in each sum, some or all terms coming from subregions meeting the boundary of R are replaced by zeros. In particular, for R as in Theorem 4,*

$$\iint_R f(x, y)\, dA = \lim_{\text{mesh}\to 0} \sum_{i=1}^{n} f(\xi_i, \eta_i)\, \Delta_i x\, \Delta_i y \tag{13-28}$$

where each subdivision of R is formed of rectangles R_i, of sides $\Delta_i x$, $\Delta_i y$ $(i = 1, \ldots, n)$ as in Figure 13-10, and other subregions (ignored in the sum) meeting the boundary curves $y = \varphi_1(x)$, $y = \varphi_2(x)$, $a \le x \le b$.

A proof for the case of Figure 13-10—that is, of (13-28)—is given in the next section.

Integration of Piecewise Continuous Functions. Let R be as in Theorem 1. We say that a function f, defined in R, is *piecewise continuous* in R if there is a subdivision of R into regions $R_1, \ldots, R_k$ and for each i there is a function f_i continuous on the *closed* region R_i such that $f(x, y) = f_i(x, y)$

inside R_i. For example, the function f defined as follows:

$$f(x, y) = \begin{cases} e^{xy}, & \text{for } 0 \le x \le y \le 1 \\ x - y, & \text{for } 0 \le y < x \le 1 \end{cases}$$

is piecewise continuous on the square R: $0 \le x \le 1$, $0 \le y \le 1$. Here R is subdivided into two regions R_1, R_2 by the diagonal $y = x$; let R_1 be above the diagonal, R_2 below it. In R_1, f coincides with $f_1(x, y) = e^{xy}$; *inside* R_2, f coincides with $f_2(x, y) = x - y$. In general, f is continuous inside each R_i and has a limit at each boundary point of R_i as the point is approached from within R_i; at each boundary point common to two subregions one in general obtains different limits from the two regions. For our example, f has limit $e^{1/4}$ as $(\frac{1}{2}, \frac{1}{2})$ is approached from inside R_1, has limit 0 as $(\frac{1}{2}, \frac{1}{2})$ is approached from inside R_2.

For a piecewise continuous function f, one can define the double integral of f over R by the equation:

$$\iint\limits_{R} f\, dA = \iint\limits_{R_1} f_1\, dA + \cdots + \iint\limits_{R_k} f_k\, dA$$

One can verify that the double integral can even be obtained as a limit as in Theorem 2. The discussion is analogous to that of Section 4-30.

More General Regions of Integration. We have developed the theory thus far only for a region bounded by a single simple closed curve C. With very minor changes, everything we have done can be extended to a region R bounded by a finite number of piecewise smooth simple closed paths, as in Figure 13-12. The double integral over such region can also be represented as a sum of a finite number of integrals over regions $S_1, \ldots, S_k$, each of which is bounded by one piecewise smooth simple closed path. The process of breaking R up into such regions is suggested in Figure 13-12.

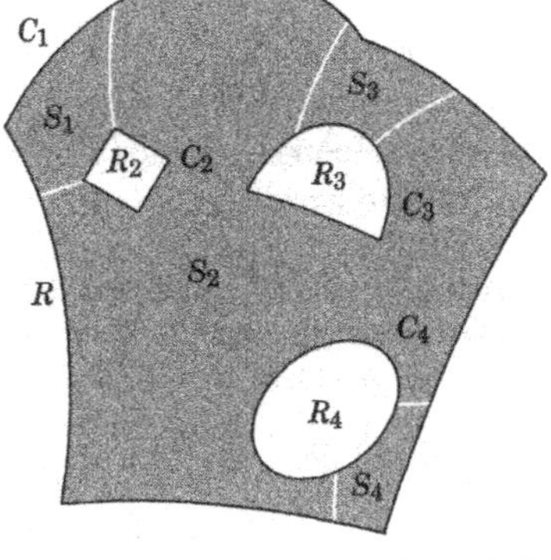

Figure 13-12 Region R bounded by several curves.

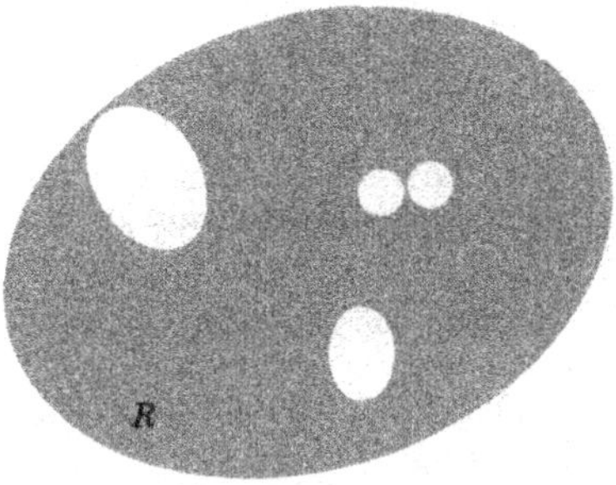

Figure 13-13 Region for double integral.

If the integrand f for the region R of Figure 13-12 is continuous over the whole region R_1 enclosed by C_1, then as in Theorem 6

$$\iint_R f(x, y)\, dA = \iint_{R_1} f(x, y)\, dA - \iint_{R_2} f(x, y)\, dA - \iint_{R_3} f(x, y)\, dA - \iint_{R_4} f(x, y)\, dA$$

where R_2, R_3, R_4 are the regions enclosed by C_2, C_3, C_4, respectively. The theory also extends to more complicated types of regions, as in Figure 13-13.

‡13-3 PROOF THAT DOUBLE INTEGRAL CAN BE REPRESENTED AS A LIMIT

In this section our goal is to prove Theorem 2: that the double integral can be represented as $\lim \Sigma f(\xi_i, \eta_i)\, \Delta_i A$, as the mesh approaches 0.

We are given a region R as in Theorem 1. Then R has a subdivision of arbitrarily small mesh; see below. Also R has area (see Section 7-4).

We now imitate the proof of the existence of the Riemann integral of a continuous function of one variable in Section 12-25. Given $\epsilon > 0$, we let $\epsilon_1 = \epsilon/A$, where A is the area of R. Since $f(x, y)$ is continuous in R and R is a bounded closed set, f is also uniformly continuous on R (Theorem K, Section 12-25). Therefore, we can choose $\delta > 0$ so that $|f(P) - f(Q)| < \epsilon_1$ for each two points P, Q in R for which $\|\overrightarrow{PQ}\| < \delta$. We now choose a subdivision of R of mesh less than δ. Then in each subregion R_i the function f takes its absolute maximum M_i on R_i at a point P_i, its absolute minimum m_i at a point Q_i. Hence

$$0 \le M_i - m_i = f(P_i) - f(Q_i) < \epsilon_1$$

Therefore

$$\sum_{i=1}^{n} M_i\, \Delta_i A - \sum_{i=1}^{n} m_i\, \Delta_i A = \sum_{i=1}^{n} (M_i - m_i)\, \Delta_i A < \epsilon_1 \sum_{i=1}^{n} \Delta_i A = \epsilon_1 A = \epsilon$$

But we know that

$$\sum_{i=1}^{n} m_i\, \Delta_i A \le \iint_R f(x, y)\, dA \le \sum_{i=1}^{n} M_i\, \Delta_i A$$

and that

$$\sum_{i=1}^{n} m_i\, \Delta_i A \le \sum_{i=1}^{n} f(\xi_i, \eta_i)\, \Delta_i A \le \sum_{i=1}^{n} M_i\, \Delta_i A$$

It follows that for every subdivision of mesh less than δ we have

$$\left| \sum_{i=1}^{n} f(\xi_i, \eta_i)\, \Delta_i A - \iint_R f(x, y)\, dA \right| < \epsilon$$

or

$$\lim_{\text{mesh}\to 0} \sum_{i=1}^{n} f(\xi_i, \eta_i)\, \Delta_i A = \iint_R f(x, y)\, dA$$

as was to be proved.

Proof of Existence of a Subdivision of Arbitrarily Small Mesh. To prove this for the most general region R is quite awkward, and we therefore restrict ourselves to a region $a \le x \le b$, $\varphi_1(x) \le y \le \varphi_2(x)$, as in Figure 13-10; here φ_1 and φ_2 are continuous and $\varphi_1(x) < \varphi_2(x)$, except possibly at a and b where equality may hold true. Let a positive number ϵ be given. Now by Theorem K, Section 12-25, $\varphi_1(x)$ is uniformly continuous; hence, we can choose $\delta_1 > 0$ so that $|\varphi_1(x') - \varphi_1(x'')| < \epsilon/4$ for all x', x'' in $[a, b]$ for which $|x' - x''| < \delta_1$. We choose δ_2 similarly with respect to φ_2 and let δ be the smallest of δ_1, δ_2, $\epsilon/4$. We then subdivide the interval $[a, b]$ into equal parts of width less than δ, by points $x_0 = a, x_1, \ldots, x_m = b$. The vertical lines $x = x_0$, $x = x_1, \ldots, x = x_m$ then subdivide R into m subregions. We subdivide each of these further with the aid of the lines $y = k\delta/2$, $k = 0, \pm 1, \pm 2, \ldots$, for those values of k for which

$$\varphi_1(x) < k\delta/2 < \varphi_2(x), \qquad x_{j-1} \le x \le x_j$$

where $[x_{j-1}, x_j]$ is the x-interval for the subregion in question (see Figure 13-14); it may happen that no segments parallel to the x-axis are allowed.

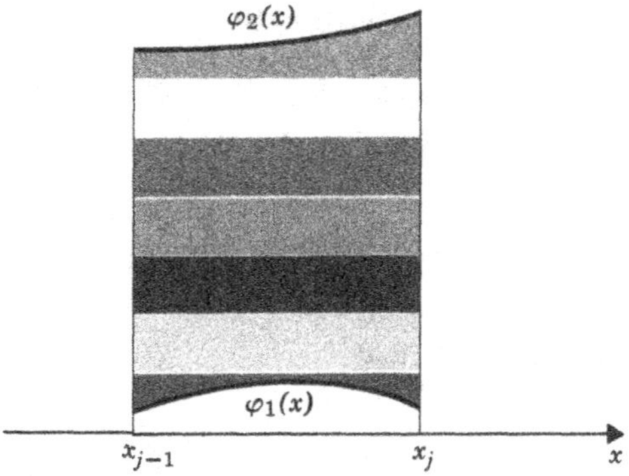

Figure 13-14 Subdivision process.

After we have carried out this process for all m intervals $[x_0, x_1], \ldots, [x_{m-1}, x_m]$, we have a subdivision of R into regions R_i, $i = 1, \ldots, n$. Each R_i is either (1) a rectangle, or (2) a region whose upper boundary is part of the graph of $y = \varphi_2(x)$ and whose lower boundary is a segment $y = k\,\delta/2$ as above, or (3) a region whose upper boundary is such a segment and whose lower boundary is part of the graph of $y = \varphi_1(x)$, or (4) a region $x_{j-1} \le x \le x_j$, $\varphi_1(x) \le y \le \varphi_2(x)$. Now let m_j', M_j' be the absolute minimum and maximum of $\varphi_2(x)$ for $x_{j-1} \le x \le x_j$. By the choice of δ, $M_j' - m_j' < \epsilon/4$. Hence, each R_i of type (2) lies in a rectangle of base less than δ and altitude less than $(\delta/2) + (\epsilon/4)$. The diameter of R_i is less than or equal to that of the rectangle and, hence, is less than the sum of base and altitude. Therefore, the diameter of each R_i of type (2) is less than

$$\delta + \left(\frac{\delta}{2}\right) + \left(\frac{\epsilon}{4}\right) < \left(\frac{\epsilon}{4}\right) + \left(\frac{\epsilon}{8}\right) + \left(\frac{\epsilon}{4}\right) < \epsilon$$

The same reasoning applies to each R_i of type (3). Each R_i of type (1) is a rectangle of diameter less than $\delta + (\delta/2)$, which is less than $3\epsilon/8$. Each R_i of type (4) is contained in a rectangle of base less than δ and altitude less than $(\delta/2) + (\epsilon/4) + (\epsilon/4)$; hence, its diameter is less than

$$\delta + \frac{\delta}{2} + \frac{\epsilon}{4} + \frac{\epsilon}{4} < \frac{3\epsilon}{8} + \frac{\epsilon}{2} = \frac{7}{8}\epsilon < \epsilon$$

Therefore, every R_i has diameter less than ϵ, as required.

PROOF OF A SPECIAL CASE OF THEOREM 2′. We note that the total area of the regions R_i of type (2) is less than

$$(x_1 - x_0)\left(\frac{\delta}{2} + \frac{\epsilon}{4}\right) + (x_2 - x_1)\left(\frac{\delta}{2} + \frac{\epsilon}{4}\right) + \cdots + (x_m - x_{m-1})\left(\frac{\delta}{2} + \frac{\epsilon}{4}\right)$$
$$= (b - a)\left(\frac{\delta}{2} + \frac{\epsilon}{4}\right) < (b - a)\frac{3\epsilon}{8}$$

There are similar estimates for the total area of the regions of types (3) and (4); in each case, the total area is less than $(b - a)\epsilon$. This explains why, in computing the double integral numerically, we can ignore the contributions from the subregions meeting the boundary curves $y = \varphi_1(x)$, $y = \varphi_2(x)$. For let K be the maximum of $|f(x, y)|$ in R. Then in the sum $\Sigma f(\xi_i, \eta_i)\, \Delta_i A$ the terms arising from these subregions have total absolute value less than K times their total area; hence, in all, less than $3K\epsilon(b - a)$. For ϵ sufficiently small, this number is negligible in a computation seeking a given accuracy. Therefore, Equation (13-28) is established. A similar remark applies to the subregions meeting the boundary segments on the lines $x = a$, $x = b$. For the general region R bounded by a piecewise smooth simple closed path C, the contribution to the sum from subregions touching the boundary can be ignored simply because the boundary can be enclosed in a finite number of rectangles whose total area is as small as desired; that is, the boundary curve has zero area, as in Section 4-18.

PROOF OF THE LEMMA OF SECTION 13-2. To prove the lemma, we let a lower sum $\Sigma m_i'\, \Delta_i' A$ and an upper sum $\Sigma M_j''\, \Delta_j'' A$ be given. Here in general the two sums refer to two different subdivisions of R: one of regions $R_1', \ldots, R_p'$, one of regions $R_1'', \ldots, R_q''$.

Now let us assume that we can find a new common subdivision of R: that is, a subdivision of R by regions $R_1, \ldots, R_n$, such that each R_i' is subdivided by a finite number of the R_i and each R_j'' is subdivided by a finite number of the R_i. Let $\Sigma m_i\, \Delta_i A$ be the lower sum, $\Sigma M_i\, \Delta_i A$ the upper sum, for the subdivision by the R_i. Let, for example, R_1' be the union of R_1, R_2 and R_3. Since R_1, R_2 and R_3 are contained in R_1', m_1' cannot exceed m_1, m_2, or m_3

and hence

$$m_1' \Delta_1' A = m_1'(\Delta_1 A + \Delta_2 A + \Delta_3 A) = m_1' \Delta_1 A + m_1' \Delta_2 A + m_1' \Delta_3 A$$
$$\leq m_1 \Delta_1 A + m_2 \Delta_2 A + m_3 \Delta_3 A$$

Reasoning similarly for all R_i' and for the upper sums, we conclude that

$$\sum m_i' \Delta_i' A \leq \sum m_i \Delta_i A \leq \sum M_i \Delta_i A \leq \sum M_j'' \Delta_j'' A$$

and our assertion follows.

In general, one cannot find a common subdivision $R_1, \ldots, R_n$. However, one can subdivide R by lines parallel to the axes as in Figure 13-7, thereby also subdividing each R_i' and each R_j'' into little regions which are either squares or else partial squares, meeting the boundary curves of the R_i' and R_j''. If we ignore the partial squares, considering only those squares R_1, $R_2, \ldots$ which are wholly contained in the R_i' and in the R_j'', then one has in effect a common subdivision $R_1, \ldots, R_n$ and can reason as before. For a fine enough subdivision, the error committed by ignoring the partial squares can be made as small as desired and one concludes that the given lower sum cannot exceed the given upper sum.

PROBLEMS

1. Justify the inequalities:

   (a) $0 \leq \iint\limits_R (x^2 + y^2)\, dA \leq 30$, where R is the rectangle $0 \leq x \leq 1$, $0 \leq y \leq 3$.

   (b) $\frac{4}{3} \leq \iint\limits_R e^{xy}\, dA \leq \frac{4e^2}{3}$, where R is the region $0 \leq x \leq 1$, $0 \leq y \leq 1 + x^2$.

   (c) $1 + e \leq \iint\limits_R x^2 e^{y^2}\, dA \leq 5e + 5e^4$, where R is the square: $0 \leq x \leq 2$, $0 \leq y \leq 2$. (*Hint*. Subdivide R into four squares.)

   (d) $\frac{\pi - 2}{6} \leq \iint\limits_R \frac{x^2}{1 + y^2}\, dA \leq 1 + \frac{2\pi - 4}{3}$, where R is the circular region $x^2 + y^2 \leq 1$. (*Hint*. Subdivide R by the lines $x = \pm\sqrt{2}/2$, $y = \pm\sqrt{2}/2$.)

2. Show that if k is a constant, then $\iint\limits_R k\, dA = kA$, where A is the area of R.

3. (a) Evaluate $\iint\limits_R (x + y)\, dA$, where R is the region bounded by the square of vertices $(1, 1)$, $(1, -1)$, $(-1, -1)$, $(-1, 1)$ and the square of vertices $(2, 2)$, $(2, -2)$, $(-2, -2)$, $(-2, 2)$.

   (b) Evaluate $\iint\limits_R (x^2 + y^2)\, dA$, where R is the annulus bounded by the circles $x^2 + y^2 = 1$, $x^2 + y^2 = 4$.

4. Let R be the region between the circle $x^2 + y^2 = 1$ and the ellipse $4x^2 + y^2 = 16$. Show that

$$\iint_R f(x, y)\, dA = \int_{-2}^{2}\int_{-\sqrt{16-4x^2}}^{\sqrt{16-4x^2}} f_1(x, y)\, dy\, dx$$

where $f_1(x, y) = f(x, y)$ in R, $f_1(x, y) = 0$ inside the circle $x^2 + y^2 = 1$. (Thus for fixed x between -1 and 1, f_1 is piecewise continuous in y.)

5. Let f be continuous in R_1: $x^2 + y^2 \leq 25$. Let $R_2, \ldots, R_5$ be defined as follows:

$$R_2\colon x^2 + y^2 \leq 4, \qquad R_3\colon (x-3)^2 + y^2 \leq 1,$$
$$R_4\colon x^2 + y^2 \leq 1, \qquad R_5\colon (x-1)^2 + y^2 \leq 1$$

Let C_i be the boundary of R_i for $i = 1, \ldots, 5$. Let I_i be the double integral of f over R_i and let

$$I_1 = 10, \qquad I_2 = 3, \qquad I_3 = 1, \qquad I_4 = 6, \qquad I_5 = 2$$

Find the double integral of f over the region bounded by the curves named:

(a) C_1 and C_2 (b) C_1, C_3, and C_4 (c) C_2 and C_4
(d) C_1, C_2, and C_3 (e) C_2 and C_5 (f) C_1, C_3, and C_5

‡6. Let $f(x)$ and $g(x)$ be continuous for $0 \leq x < \infty$, let $F_1(x) = \int_0^x f(t)\, dt$, $F_2(x) = \int_0^x F_1(t)\, dt$, $\ldots$, $F_n(x) = \int_0^x F_{n-1}(t)\, dt$

(a) Show that for $x \geq 0$

$$\int_0^x\int_0^v f(t)g(v-t)\, dt\, dv = \int_0^x\int_0^{x-t} f(t)g(u)\, du\, dt$$

(b) Show that $F_2(x) = \int_0^x f(t)(x-t)\, dt$

(c) Show that $F_n(x) = \int_0^x f(t)\dfrac{(x-t)^{n-1}}{(n-1)!}\, dt$

7. Let R be a region of area A, as in Theorem 1, and let R be subdivided into convex regions R_i, with mesh less than δ. Let f have continuous first partial derivatives in an open region containing R and let $\|\nabla f\| < K$ in R. Show that if $\delta < \epsilon/(KA)$, then

$$\left| \iint_R f(x, y)\, dA - \sum_{i=1}^{n} f(\xi_i, \eta_i)\, \Delta_i A \right| < \epsilon$$

(cf. Theorem 24, Section 4-25). *Note:* A set E in R^n is said to be *convex* if for every pair of points A, B in E the line segment AB is in E.

8. Describe in detail a subdivision of R: $0 \leq x \leq \pi$, $\sin x \leq y \leq 4 - 2\sin x$, of mesh less than $\frac{1}{2}$.

9. Complete the proof of Theorem 6.

13-4 DOUBLE INTEGRALS IN POLAR COORDINATES

We first consider again the expression for the double integral of a function $g(x, y)$ over a region R: $a \leq x \leq b$, $\varphi_1(x) \leq y \leq \varphi_2(x)$, as in Figure 13-10. If we subdivide as in that figure and exclude those subregions that meet the upper and lower boundaries (as we are allowed to do), then we are dealing only with rectangles and $\Delta_i A = \Delta_i y\, \Delta_i x$ for each subregion. Hence, as in Theorem 2′,

$$\begin{aligned}\iint_R g(x, y)\, dA &= \int_a^b \int_{\varphi_1(x)}^{\varphi_2(x)} g(x, y)\, dy\, dx \\ &= \lim_{\text{mesh}\to 0} \sum_{i=1}^{n} g(\xi_i, \eta_i)\, \Delta_i y\, \Delta_i x\end{aligned} \tag{13-40}$$

Now let a region R be given in terms of polar coordinates r, θ by the inequalities

$$\alpha \leq \theta \leq \beta, \qquad p_1(\theta) \leq r \leq p_2(\theta) \tag{13-41}$$

where $p_1(\theta)$ and $p_2(\theta)$ are continuous for $\alpha \leq \theta \leq \beta$, $\beta - \alpha \leq 2\pi$, and $0 \leq p_1(\theta) < p_2(\theta)$, except possibly for $p_1(\alpha) = p_2(\alpha)$, $p_1(\beta) = p_2(\beta)$ (see Figure 13-15). Then by analogy with Figure 13-10, we subdivide the region R into regions R_i with the aid of rays and circular arcs, as in Figure 13-15. If we disregard those subregions meeting the boundaries $r = p_1(\theta)$, $r = p_2(\theta)$, then each R_i has area

$$\Delta_i A = r_i^* \,\Delta_i r\, \Delta_i \theta$$

where r_i^* is the average of the inner and outer radii of R_i, as in Figure 13-16 (Problem 3 below).

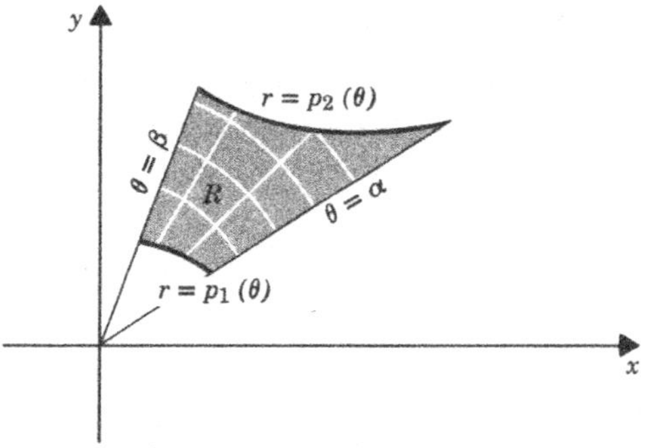

Figure 13-15 Region for double integral in polar coordinates.

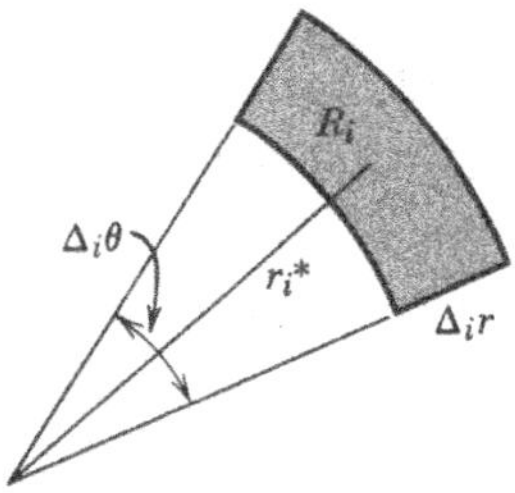

Figure 13-16 Basic subregion.

Now let the function f be defined and continuous in R. Then by the equations $x = r\cos\theta$, $y = r\sin\theta$, f can be expressed in terms of polar coordinates and becomes a continuous function of r and θ; in fact $f(x, y) =$

$f(r\cos\theta, r\sin\theta) = F(r,\theta)$. Ignoring the subregions meeting the boundaries $r = p_1(\theta)$, $r = p_2(\theta)$, the sum for the double integral of f over R becomes

$$\sum_{i=1}^{n} F(r_i', \theta_i')\,\Delta_i A = \sum_{i=1}^{n} F(r_i', \theta_i') r_i^* \,\Delta_i r\,\Delta_i\theta$$

Here (r_i', θ_i') are polar coordinates of an arbitrary point in R_i. We may in particular take $r_i' = r_i^*$. Then our sum becomes

$$\sum_{i=1}^{n} F(r_i^*, \theta_i') r_i^*\,\Delta_i r\,\Delta_i\theta$$

This is of the same form as the sum in (13-40) with $\Delta_i y$ replaced by $\Delta_i r$, $\Delta_i x$ by $\Delta_i\theta$, the function $g(x, y)$ by the function $rF(r,\theta)$. The condition "mesh $\to 0$" in (13-40) is equivalent to "maximum value of $|\Delta_i x| + |\Delta_i y| \to 0$" and, therefore, becomes "maximum value of $|\Delta_i r| + |\Delta_i\theta| \to 0$", which in turn implies "mesh $\to 0$" for the subdivisions of R in Figure 13-15. Hence, (13-40) allows us to write

$$\iint_R f\,dA = \int_\alpha^\beta \int_{p_1(\theta)}^{p_2(\theta)} F(r,\theta) r\,dr\,d\theta = \lim_{\text{mesh}\to 0} \sum_{i=1}^{n} F(r_i^*, \theta_i') r_i^*\,\Delta_i r\,\Delta_i\theta \qquad (13\text{-}42)$$

This equation allows us to evaluate double integrals over regions more easily described in terms of polar coordinates, especially circular regions $0 \le \theta \le 2\pi$, $0 \le r \le a$. The formula is most easily remembered in the form

$$dA = r\,dr\,d\theta \qquad (13\text{-}43)$$

which describes the fact that our basic subregion R_i is approximately a rectangle with sides dr and $r\,d\theta$, as in Figure 13-17.

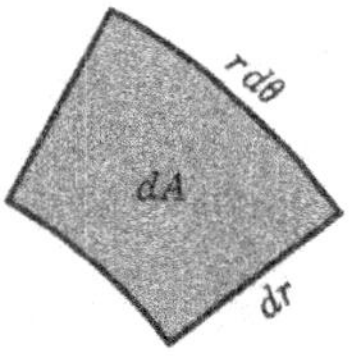

Figure 13-17 dA in polar coordinates.

The formula (13-42) also follows from a general formula for double integrals in curvilinear coordinates:

$$\iint_{R_{xy}} f(x, y)\,dx\,dy = \iint_{R_{uv}} f[x(u, v), y(u, v)] \left|\frac{\partial(x, y)}{\partial(u, v)}\right| du\,dv$$

This is discussed in the next section. For polar coordinates we take $u = r$, $v = \theta$, and the Jacobian factor (note the *absolute value*) becomes

$$\left|\frac{\partial(x, y)}{\partial(r, \theta)}\right| = \left|\,\begin{vmatrix} \partial x/\partial r & \partial x/\partial\theta \\ \partial y/\partial r & \partial y/\partial\theta \end{vmatrix}\,\right| = \left|\,\begin{vmatrix} \cos\theta & -r\sin\theta \\ \sin\theta & r\cos\theta \end{vmatrix}\,\right|$$

$$= r\cos^2\theta + r\sin^2\theta = r$$

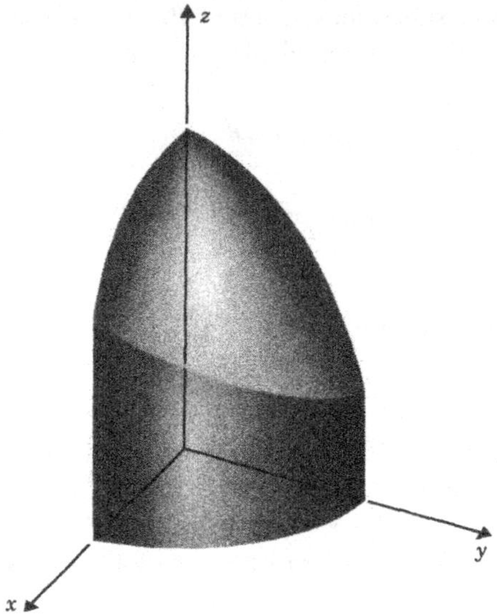

Figure 13-18 Example 1.

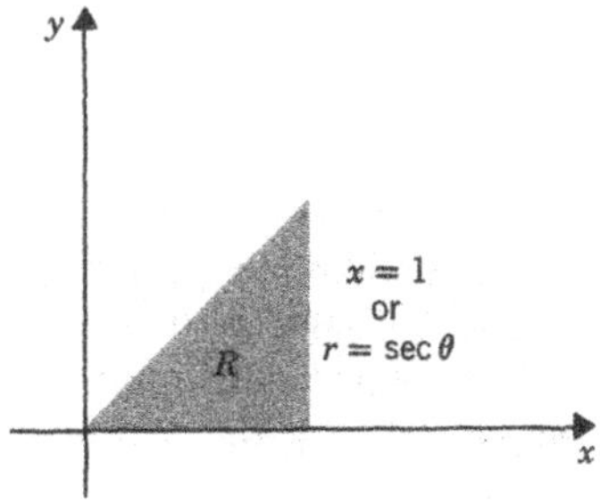

Figure 13-19 Example 2.

EXAMPLE 1. Find the volume of the solid region beneath the surface $z = 3 - x^2 - 2y^2$ for $x^2 + y^2 \leq 1$ (see Figure 13-18). Here R is given by the inequalities $0 \leq \theta \leq 2\pi$, $0 \leq r \leq 1$ and f is given in polar coordinates by $3 - r^2(\cos^2\theta + 2\sin^2\theta)$. Hence

$$V = \iint_R f\,dA = \int_0^{2\pi}\int_0^1 (3 - r^2(\cos^2\theta + 2\sin^2\theta))r\,dr\,d\theta$$

$$= \int_0^{2\pi}\left(\frac{3}{2} - \frac{\cos^2\theta}{4} - \frac{2\sin^2\theta}{4}\right)d\theta = \frac{9\pi}{4}$$

EXAMPLE 2. Evaluate the double integral of $z = f(x, y) = \sqrt{x^2 + y^2}$ over the triangular region of vertices (0, 0), (1, 0), (1, 1). Here $f = r$ in polar coordinates, R is given by $0 \leq \theta \leq \pi/4$, $0 \leq r \leq \sec\theta$ (see Figure 13-19). Hence, the integral is

$$\int_0^{\pi/4}\int_0^{\sec\theta} r^2\,dr\,d\theta = \frac{1}{3}\int_0^{\pi/4} \sec^3\theta\,d\theta$$

$$= \frac{1}{6}[\sec\theta\tan\theta + \ln(\sec\theta + \tan\theta)]\Big|_0^{\pi/4}$$

$$= \frac{1}{6}[\sqrt{2} + \ln(1 + \sqrt{2})]$$

Remark. Once we have an iterated integral, as in the middle expression in (13-42), we can forget that r, θ are polar coordinates and can treat the integral just like one in x and y. The region of integration can be represented as in Figure 13-20. When this region is of the appropriate type, one can also integrate in the other order.

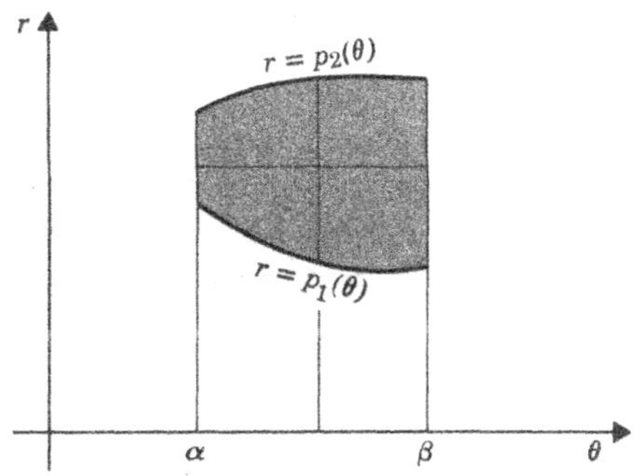

Figure 13-20 Region of integration for (13-42), with r, θ represented as rectangular coordinates.

‡13-5 OTHER CURVILINEAR COORDINATES

Polar coordinates are one example of "curvilinear coordinates" in the plane. Other curvilinear coordinates are obtained through equations

$$x = H(u, v), \qquad y = G(u, v) \tag{13-50}$$

for (u, v) in a region R_{uv} in the uv-plane. Under appropriate conditions on H and G, these equations describe a one-to-one continuous mapping of R_{uv} onto a region R_{xy} in the xy-plane, as in Figure 13-21 (see Section 12-16). The lines $u = \text{const}$, $v = \text{const}$ in R_{uv} become curves in R_{xy} with whose aid we assign curvilinear coordinates (u, v) to the points of R_{xy}. In the case of polar coordinates, the region R_{uv} is typically as in Figure 13-20, and R_{xy} as in Figure 13-15; the lines $u = \text{const}$, $v = \text{const}$ are the lines $r = \text{const}$, $\theta = \text{const}$, which become the rays and circles in Figure 13-15; the mapping equations (13-50) are the equations $x = r\cos\theta$, $y = r\sin\theta$.

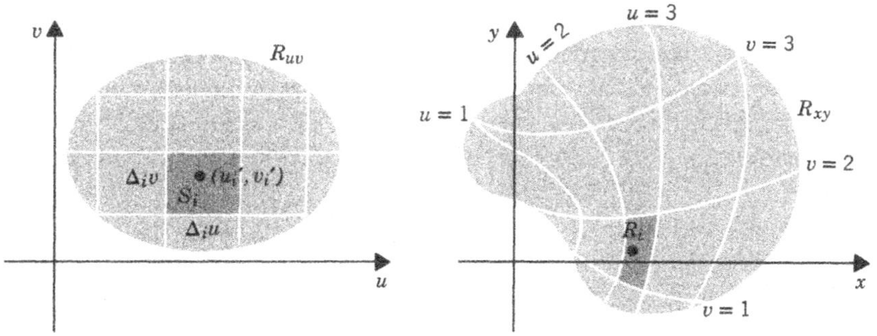

Figure 13-21 Curvilinear coordinates introduced by a mapping.

If we subdivide R_{uv} by parallels to the u- and v-axes and ignore subregions meeting the boundary, we obtain rectangles of sides $\Delta_i u$, $\Delta_i v$. Such a rectangle S_i corresponds to a curvilinear rectangle R_i in the xy-plane; the area $\Delta_i A$ of R_i can be found from the area of the corresponding rectangle in R_{uv} and the mapping functions (13-50). When F and G have continuous first partial derivatives, we have the formula

$$\Delta_i A = \left|\frac{\partial(x, y)}{\partial(u, v)}\right|\bigg|_{\substack{u=u' \\ v=v'}} \Delta_i u \, \Delta_i v \tag{13-51}$$

where (u', v') is an appropriately chosen point in S_i. From (13-51) one deduces that the Jacobian determinant is, apart from sign, the limit of ratios of areas:

$$\frac{\partial(x, y)}{\partial(u, v)}\bigg|_{\substack{u=u_0 \\ v=v_0}} = \pm \lim_{\text{diam of } S_i \to 0} \frac{\Delta_i A}{\Delta_i u \, \Delta_i v}$$

Thus the Jacobian determinant is in a sense a generalization of the derivative to two dimensions. From (13-51) one deduces, by reasoning as in Section 13-4, that

$$\iint\limits_{R_{xy}} f \, dA_{xy} = \iint\limits_{R_{uv}} F(u, v) \left|\frac{\partial(x, y)}{\partial(u, v)}\right| dA_{uv} \tag{13-52}$$

where f is continuous in R_{xy} and, on the right, $F(u, v) = f[H(u, v), G(u, v)]$. In the case of polar coordinates, the Jacobian determinant is $|\partial(x, y)/\partial(r, \theta)| = r$, as shown in Section 13-4, and (13-52) becomes, for a region as in Figure 13-20,

$$\iint\limits_{R_{xy}} f \, dA_{xy} = \iint\limits_{R_{r\theta}} F r \, dA_{r\theta} = \int_\alpha^\beta \int_{p_1(\theta)}^{p_2(\theta)} F(r, \theta) r \, dr \, d\theta$$

In Section 11-17 we gave a proof of (13-51) for the case where the mapping (13-50) is linear, and the case of polar coordinates was treated in Problem 5 following Section 12-16.

For a proof of (13-51) or (13-52) one is referred to texts on advanced calculus, for example, Section 5-14, of the text by W. Kaplan, *Advanced Calculus*, Addison-Wesley, Reading, Mass., 1955. We can derive (13-51) intuitively as follows. Near a particular point (u_0, v_0), H and G can be approximated by linear functions, so that our mapping is given approximately by

$$\begin{aligned} x &= x_0 + a(u - u_0) + b(v - v_0) \\ y &= y_0 + c(u - u_0) + d(v - v_0) \end{aligned} \tag{13-53}$$

where

$$a = H_u(u_0, v_0), \quad b = H_v(u_0, v_0), \quad c = G_u(u_0, v_0), \quad d = G_v(u_0, v_0)$$

We obtain this linear approximation of the mapping either by Taylor's formula or, equivalently, by writing $dx = H_u du + H_v dv$, $dy = G_u du + G_v dv$ and approximating $\Delta x = x - x_0$ by dx, $\Delta y = y - y_0$ by dy. From (13-53) we find that the rectangle $u_0 \leq u \leq u_0 + \Delta u$, $v_0 \leq v \leq v_0 + \Delta v$ corresponds to a parallelogram in the xy-plane, as in Figure 13-22. Two directed edges of this parallelogram represent the vectors $\Delta u(a\mathbf{i} + c\mathbf{j})$, $\Delta v(b\mathbf{i} + d\mathbf{j})$. Hence, (see Section 1-12) the area of the parallelogram is

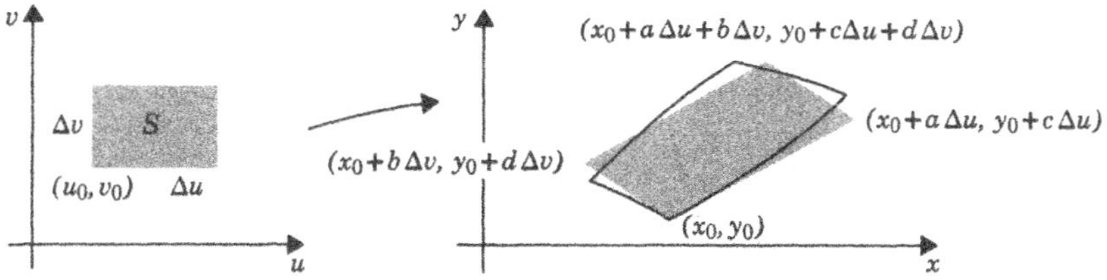

Figure 13-22 The linear approximation to a mapping.

$$\pm \begin{vmatrix} a\,\Delta u & c\,\Delta u \\ b\,\Delta v & d\,\Delta v \end{vmatrix} = \pm\,\Delta u\,\Delta v \begin{vmatrix} a & b \\ c & d \end{vmatrix} = \pm\,\Delta u\,\Delta v \begin{vmatrix} H_u & H_v \\ G_u & G_v \end{vmatrix}$$

where the sign is chosen to give a positive result and $H_u, \ldots, G_v$ are evaluated at (u_0, v_0). The parallelogram approximates a subregion R_i and, hence, its area approximates $\Delta_i A$. Therefore, it is reasonable that $\Delta_i A$ be given as in (13-51).

EXAMPLE. $x = u^2 - v^2$, $y = 2uv$ in R_{uv}: $1 \le u \le 2$, $-1 \le v \le 1$. Here we find that the region R_{xy} is as in Figure 13-23, bounded by four parabolic

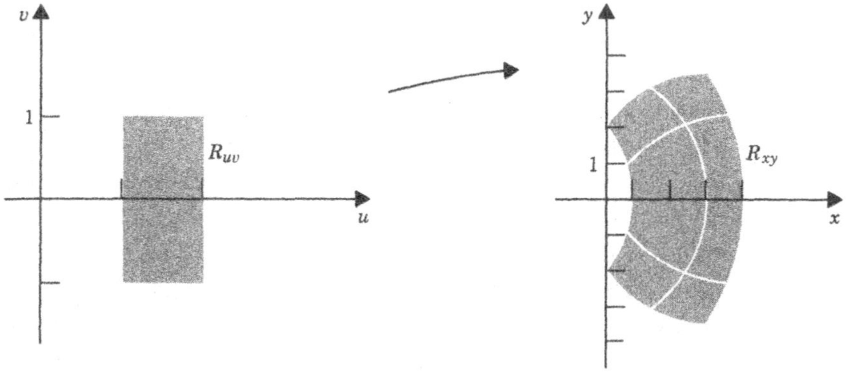

Figure 13-23 Mapping $x = u^2 - v^2$, $y = 2uv$.

arcs (Problem 4 below). The lines $u =$ const, $v =$ const are also parabolic arcs. To evaluate the double integral of $f(x, y) = x$ over R_{xy}, we write

$$\iint_{R_{xy}} x\, dA_{xy} = \int_1^2\int_{-1}^1 (u^2 - v^2) \left| \begin{vmatrix} 2u & -2v \\ 2v & 2u \end{vmatrix} \right| dv\, du$$

$$= \int_1^2\int_{-1}^1 (u^2 - v^2)(4u^2 + 4v^2)\, dv\, du$$

$$= 4\int_1^2\int_{-1}^1 (u^4 - v^4)\, dv\, du = 48$$

Remark. For more discussion of the interpretation of a determinant as ratio of areas or, in n-dimensional space, as ratio of n-dimensional volumes for linear mappings, see Sections 11-18 and 12-16.

PROBLEMS

1. Evaluate with the aid of polar coordinates:

(a) $\iint\limits_R \sqrt{x^2 + y^2}\, dA$, R: $x^2 + y^2 \le 4$

(b) $\iint\limits_R (x^2 + y)\, dA$, R: $x^2 + y^2 \le 1$

(c) $\iint\limits_R xy\, dA$, R: $0 \le y \le \sqrt{2}/2$, $y \le x \le \sqrt{1 - y^2}$ (circular sector)

(d) $\iint\limits_R xy^2\, dA$, R: $0 \le x \le 2$, $0 \le y < \sqrt{4 - x^2}$

(e) $\iint\limits_R f\, dA$, $f(r, \theta) = r^2 \cos\theta$, R: $0 \le \theta \le \pi/2$, $1 \le r \le 1 + \sin\theta$

(f) $\iint\limits_R f\, dA$, $f(r, \theta) = \cos\theta$, R: $0 \le \theta \le \pi$, $0 \le r \le 2 - \sin\theta$

(g) $\iint\limits_R f\, dA$, $f(r, \theta) = r$, R: $0 \le \theta \le \pi/2$, $0 \le r \le \theta$

(h) $\iint\limits_R f\, dA$, $f(r, \theta) = \theta^2$, R: $0 \le \theta \le 3\pi$, $e^{\theta}/2 \le r \le e^{\theta}$ (graph the region R!)

2. Find the volume of the solid with the aid of cylindrical coordinates r, θ, z.

(a) The solid cylinder: $0 \le r \le a$, $0 \le \theta \le 2\pi$, $0 \le z \le h$.

(b) The solid cylinder of altitude h whose base is the region enclosed by the cardioid $r = a(1 + \cos\theta)$.

(c) The volume of the region below the plane $z = 2 + x + y$ for $x^2 + y^2 \le 1$.

(d) The volume below the plane $z = 2 - x - y$ for $x^2 + y^2 \le 1$.

(e) The volume of the solid cone above the surface $z^2/h^2 = (x^2 + y^2)/a^2$ and below the plane $z = h$, for $x^2 + y^2 \le a^2$.

(f) The volume of the torus obtained by rotating about the z-axis the circular region in the xz-plane with center at $(b, 0, 0)$ and radius a, where $0 < a < b$.

3. Show by geometry that the region R_i of Figure 13-16 has area $r_i^* \Delta_i r\, \Delta_i \theta$, where r_i^* is the average of the radii of the two circular arcs.

4. Show that the Example of Section 13-5 leads to the graph of Figure 13-23.

5. Let the mapping $x = 3u - v$, $y = u + 2v$ be given in R_{uv}: $0 \le u \le 1, 0 \le v \le 1$. Show that the corresponding region R_{xy} in the xy-plane is a parallelogram and evaluate

$$\iint\limits_{R_{xy}} 5xy\, dA_{xy}$$

by means of (13-52).

6. Let the mapping $x = e^u \cos v$, $y = e^u \sin v$ be given in R_{uv}: $0 \le u \le 1$, $0 \le v \le \pi/2$. Find the corresponding region R_{xy} in the xy-plane and evaluate

$$\iint\limits_{R_{xy}} x^2 y\, dA_{xy}$$

13-6 TRIPLE INTEGRALS

The double integral has a natural extension to 3-dimensional space: the *triple integral*

$$\iiint_R f(x, y, z)\, dV \qquad \text{or} \qquad \iiint_R f(x, y, z)\, dx\, dy\, dz \tag{13-60}$$

and to n-dimensional space: the *n-fold multiple integral*

$$\int_R\cdots\int f(x_1, \ldots, x_n)\, dV_n \qquad \text{or} \qquad \int_R\cdots\int f(x_1, \ldots, x_n)\, dx_1 \cdots dx_n$$

These integrals (for $n \geq 2$) are all called *multiple integrals*. In this section we develop the triple integral. The n-fold multiple integral can be developed in the same way, with n-volume replacing ordinary volume (see Section 11-16; see also Problem 10 following Section 13-7).

For the triple integral (13-60) we consider a function $w = f(x, y, z)$ defined in a region R of xyz-space. By analogy with the double integral, when $f \geq 0$, we can then interpret the integral as the (4-dimensional) volume below the "surface" $w = f(x, y, z)$ in $xyzw$-space. Thus we are forced to go up to 4-dimensional space for this interpretation. A model easier to grasp is that of mass and density. We think of f as density, in units of mass per unit volume. Then (13-60) gives the total mass of the solid filling the region R.

A careful discussion of regions R appropriate for triple integrals leads us into an area of some complexity. We do not attempt such a discussion here and proceed intuitively. We think of the typical solid object we find about us: a table, a chair, a building, a bridge, the whole earth. In each case we are dealing with a bounded closed region, whose boundary appears to be formed of a finite number of surfaces; each bounding surface is smooth, except for a finite number of edges and corners, where smooth portions come together. Typical precise mathematical examples are a solid sphere or ball, a solid cylinder (bounded by three smooth surfaces, fitting together along two circles), a solid cube (bounded by six smooth surfaces, fitting together along 12 edges), a solid torus (bounded by one smooth surface), a spherical shell, between two concentric spheres (bounded by two smooth surfaces). We shall refer to each such solid region as a *standard solid region*. A very important class of such regions consists of those regions R which, for appropriate choice of coordinates in space, are defined by inequalities

$$a \leq x \leq b, \qquad \varphi_1(x) \leq y \leq \varphi_2(x), \qquad \psi_1(x, y) \leq z \leq \psi_2(x, y) \tag{13-61}$$

where φ_1 and φ_2 are continuous in $[a, b]$, and ψ_1 and ψ_2 are continuous for $a \leq x \leq b$, $\varphi_1(x) \leq y \leq \varphi_2(x)$ (Figure 13-24). Most regions occurring in applications are of this type, or can be subdivided into a finite number of regions of this type.

The *diameter* of a bounded closed solid region R is defined as in two dimensions: it is the least upper bound of the set of all distances $\|\overrightarrow{AB}\|$ between pairs of points of R. By a *subdivision* of a standard region R we mean a representation of R as a union of standard regions $R_1, \ldots, R_k$, each two of

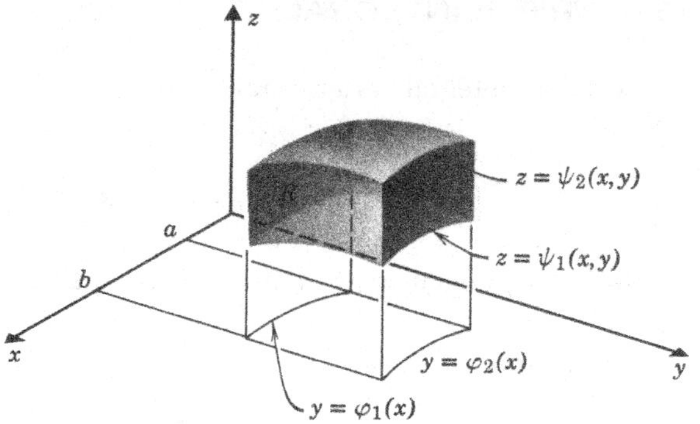

Figure 13-24 Typical region for triple integral.

which intersect only along boundary surfaces. By the *mesh* of the subdivision we mean the maximum diameter of $R_1, \ldots, R_k$.

We now assume that every standard solid region R has volume V. A proof of this, based on a precise definition of standard region and a mathematical theory of volume, requires more advanced tools; the analogous problem for area is discussed in Sections 4-18 and 7-4. For a region as in Figure 13-24 the reasoning of Section 13-1 leads us to an expression for V as a double integral

$$V = \iint_{R_{xy}} [\psi_2(x, y) - \psi_1(x, y)]\, dA$$

where R_{xy} is the region $a \le x \le b$, $\varphi_1(x) \le y \le \varphi_2(x)$, and ψ_1, ψ_2 are as in (13-161).

We also assume, as was done for double integrals, that every standard solid region R has a subdivision of arbitrarily small mesh. This can be proved as in Section 13-3 for a region as in Figure 13-24. One can also prove, as in that section, that the total volume of the subregions meeting the boundary of R approaches 0 as the mesh approaches 0; we shall assume that this has been proved for all standard regions. It then follows, as for double integrals, that in obtaining the triple integral as limit of a sum, one can include or ignore terms corresponding to subregions meeting the boundary.

Definition. Let R be a standard solid region in xyz-space and let the function f be defined and continuous in R. Then

$$\iiint_R f(x, y, z)\, dV = \text{glb}\left\{\sum_{i=1}^{n} M_i\, \Delta_i V\right\}$$

where the greatest lower bound is taken over all subdivisions of R into subregions $R_1, \ldots, R_n$, $\Delta_i V$ is the volume of R_i and M_i is the absolute maximum of f in R_i.

From the definition one deduces, as for the double integral, that

$$\iiint_R f(x, y, z)\, dV = \text{lub}\left\{\sum_{i=1}^{n} m_i\, \Delta_i V\right\}$$

where m_i is now the absolute minimum of f in R_i. It follows that the value I of the triple integral is the one and only number such that

$$\sum_{i=1}^{n} m_i\, \Delta_i V \leq I \leq \sum_{i=1}^{n} M_i\, \Delta_i V$$

for all subdivisions of R.

Furthermore, the triple integral can be obtained by a limit process like that for the Riemann integral:

$$\iiint\limits_R f(x, y, z)\, dV = \lim_{\text{mesh}\to 0} \sum_{i=1}^{n} f(\xi_i, \eta_i, \zeta_i)\, \Delta_i V \qquad (13\text{-}62)$$

where (ξ_i, η_i, ζ_i) is now an arbitrary point in R_i. In the sum on the right one can omit some or all of the terms arising from subregions R_i meeting the boundary of R.

Iterated Integral. We are led to an iterated integral by the following intuitive reasoning. We consider f as density and assume R, as in Figure 13-24, is described by (13-61). We can compute the total mass by slicing R into thin slabs by planes perpendicular to the x-axis. Within each slab x is effectively constant, so that the density f effectively depends only on y and z. Hence, the total mass of the slab is, approximately, given by a double integral with respect to y and z. One can write

$$\Delta m = f(x, \eta, \zeta)\, \Delta A_{yz}\, \Delta x$$

where Δm is the mass of the portion of the slab for which (y, z) is in a tiny region R_{yz}, ΔA_{yz} is the area of R_{yz}, Δx is the thickness of the slab, (η, ζ) is in R_{yz}. Hence, the total mass of the slab is, approximately,

$$\Delta x \iint\limits_{R_{yz}} f(x, y, z)\, dz\, dy = \Delta x \int_{\varphi_1(x)}^{\varphi_2(x)} \int_{\psi_1(x, y)}^{\psi_2(x, y)} f(x, y, z)\, dz\, dy$$

where we integrate over the whole cross section at the given x. The sum of the masses of the slabs is a sum for a Riemann integral with respect to x. Passing to the limit we obtain:

$$\text{total mass} = \int_a^b \int_{\varphi_1(x)}^{\varphi_2(x)} \int_{\psi_1(x, y)}^{\psi_2(x, y)} f(x, y, z)\, dz\, dy\, dx$$

We are thus led to the basic rule: for a region defined by (13-61), as in Figure 13-24,

$$\iiint\limits_R f(x, y, z)\, dV = \int_a^b \int_{\varphi_1(x)}^{\varphi_2(x)} \int_{\psi_1(x, y)}^{\psi_2(x, y)} f(x, y, z)\, dz\, dy\, dx \qquad (13\text{-}63)$$

A strict proof can be given as in Section 13-2. In applying (13-63), one should note that a and b are the smallest and largest values of x in the solid, $\varphi_1(x)$ and $\varphi_2(x)$ are the smallest and largest values of y in the cross section of R by a plane $x = \text{const}$, $\psi_1(x, y)$ and $\psi_2(x, y)$ are the smallest and largest values of z in R, for fixed x and y. There are analogous formulas for the cases in which R is representable by formulas like (13-61), but with the roles of x, y, and z permuted, for example:

$$a \leq z \leq b, \qquad \varphi_1(z) \leq y \leq \varphi_2(z), \qquad \psi_1(y, z) \leq x \leq \psi_2(y, z)$$

There are 6 cases in all (corresponding to the 6 permutations xyz, zxy, yzx, xzy, zyx, yxz). For some regions two or more different representations may be available. Then one has one's choice of two or more different orders of integration.

For all cases one should work from the outside in. In (13-63) the outer variable is x, the next variable is y, the inner variable is z. To set up limits for a typical integral, one should decide which variable is to be outer, which next, which inner. Then one should follow these steps in order:

Step 1. *Find the smallest and largest values of the outer variable;* if these values are a and b, respectively, and the outer variable is x, then one can now write:

$$\int_a^b \int \int f(\ldots)\, d\ldots d\ldots dx$$

Step 2. *Consider the outer variable to be fixed at a typical value, determining a cross section of the solid region; determine the smallest and largest values of the next variable in that cross section.* If x is the outer variable and the next variable is y, then the smallest and largest values of y are $\varphi_1(x)$ and $\varphi_2(x)$, respectively, and one can now write

$$\int_a^b \int_{\varphi_1(x)}^{\varphi_2(x)} \int f(\ldots)\, dz\, dy\, dx$$

Step 3. *Finally fix both the outer variable and the next variable at typical values, thereby restricting attention to the points of the region on a straight line; determine the smallest and largest values of the inner variable on that line.* If x is the outer variable, y the next variable, then z is the inner variable, and the smallest and largest values of z on the line $x = \text{const}$, $y = \text{const}$ are $\psi_1(x, y)$ and $\psi_2(x, y)$, respectively, and one obtains (13-63).

EXAMPLE 1. Let R be the rectangular parallelepiped $0 \leq x \leq h$, $0 \leq y \leq k$, $0 \leq z \leq l$. Then

$$\begin{aligned}
\iiint_R \frac{x^2 e^y}{1+z^2}\, dV &= \int_0^h \int_0^k \int_0^l \frac{x^2 e^y}{1+z^2}\, dz\, dy\, dx \\
&= \int_0^h \int_0^k x^2 e^y \operatorname{Tan}^{-1} z \Big|_{z=0}^{z=l} dy\, dx \\
&= \int_0^h \int_0^k x^2 e^y \operatorname{Tan}^{-1} l\, dy\, dx = \int_0^h x^2 \operatorname{Tan}^{-1} l\, e^y \Big|_{y=0}^{y=k} dx \\
&= \int_0^h x^2 \operatorname{Tan}^{-1} l\,(e^k - 1)\, dx = \frac{h^3}{3} \operatorname{Tan}^{-1} l\,(e^k - 1)
\end{aligned}$$

Here all 6 orders could have been used. The example illustrates the fact that in an iterated integral one can often factor and write the integral as a product of integrals. Thus, when $a_1, \ldots, b_3$ are constants and the integrand factors: $f(x)g(y)h(z)$,

$$\int_{a_1}^{b_1}\int_{a_2}^{b_2}\int_{a_3}^{b_3} f(x)g(y)h(z)\,dz\,dy\,dx = \int_{a_1}^{b_1} f(x)\,dx \int_{a_2}^{b_2} g(y)\,dy \int_{a_3}^{b_3} h(z)\,dz$$

For the result of integrating with respect to z is a constant which can be factored out; the same applies to the integration with respect to y. Similarly, if ψ_1 and ψ_2 are constants, say a_3, b_3, and the integrand factors, then

$$\int_{a_1}^{b_1}\int_{\varphi_1(x)}^{\varphi_2(x)}\int_{a_3}^{b_3} f(x, y)h(z)\,dz\,dy\,dx = \int_{a_1}^{b_1}\int_{\varphi_1(x)}^{\varphi_2(x)} f(x, y)\,dy\,dx \int_{a_3}^{b_3} h(z)\,dz$$

by the same reasoning.

EXAMPLE 2. Let R be the tetrahedron of vertices $(0, 0, 0)$, $(1, 0, 0)$, $(0, 2, 0)$,

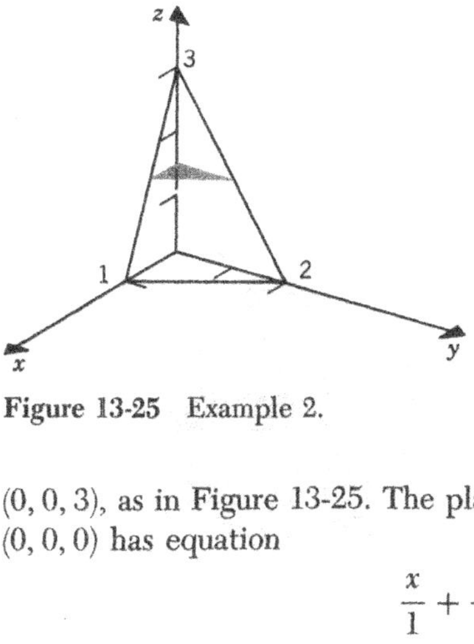

Figure 13-25 Example 2.

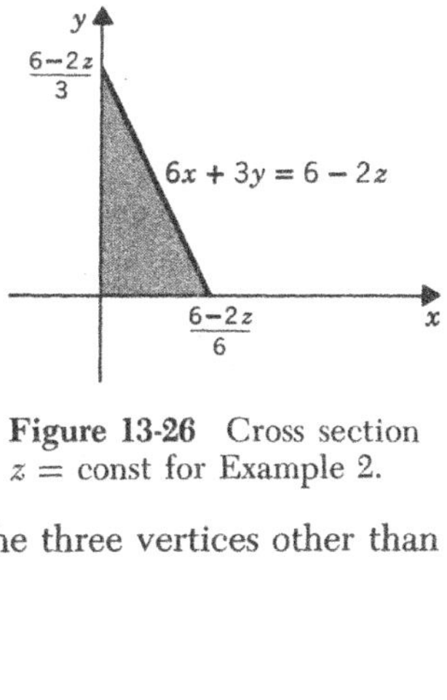

Figure 13-26 Cross section $z = \text{const}$ for Example 2.

$(0, 0, 3)$, as in Figure 13-25. The plane through the three vertices other than $(0, 0, 0)$ has equation

$$\frac{x}{1} + \frac{y}{2} + \frac{z}{3} = 1$$

and, hence, R is defined by the four inequalities

$$x \geq 0, \qquad y \geq 0, \qquad z \geq 0, \qquad 6x + 3y + 2z - 6 \leq 0$$

(see Section 11-15). We choose z to be the outer variable, y next, and x to be inner. *Step* 1: the smallest value of z in R is clearly 0, and the largest value of z is 3. *Step* 2: for fixed z, between 0 and 3, (x, y) varies over the triangle $x \geq 0$, $y \geq 0$, $6x + 3y \leq 6 - 2z$ (Figure 13-26). Hence, the smallest and largest values of y are 0 and $(6 - 2z)/3$, respectively. *Step* 3: for fixed z and y, x varies between 0 and $(6 - 2z - 3y)/6$. Therefore,

$$\iiint_R (x + z)\,dV = \int_0^3 \int_0^{\frac{6-2z}{3}} \int_0^{\frac{6-2z-3y}{6}} (x + z)\,dx\,dy\,dz$$

$$= \int_0^3 \int_0^{\frac{6-2z}{3}} \left(\frac{x^2}{2} + xz\right)\Bigg|_{x=0}^{x=\frac{6-2z-3y}{6}} dy\,dz = \cdots$$

In this example, all 6 orders can be used. Of course, the limits of integration change if we change the order of integration.

EXAMPLE 3. Let R be the hemisphere $x^2 + y^2 + z^2 \leq 1, y \geq 0$, as in Figure 13-27. Then

$$\iiint_R f(x, y, z)\, dV = \int_{-1}^{1} \int_{-\sqrt{1-x^2}}^{\sqrt{1-x^2}} \int_0^{\sqrt{1-x^2-z^2}} f(x, y, z)\, dy\, dz\, dx$$

$$= \int_{-1}^{1} \int_0^{\sqrt{1-z^2}} \int_{-\sqrt{1-y^2-z^2}}^{\sqrt{1-y^2-z^2}} f(x, y, z)\, dx\, dy\, dz$$

For the first iterated integral, Step 1 gives -1 and 1 as the smallest and largest values of x. Then Step 2 gives $-\sqrt{1-x^2}$, $\sqrt{1-x^2}$ as the smallest and largest values of z for fixed x; the cross section $x = \text{const}$ is a semicircular region (Figure 13-27). Finally, Step 3 gives 0 and $\sqrt{1-x^2-z^2}$ as the smallest and largest values of y for fixed x and z; here one considers a line on which x and z are constant, as in Figure 13-27. For the second iterated integral, the reasoning is similar.

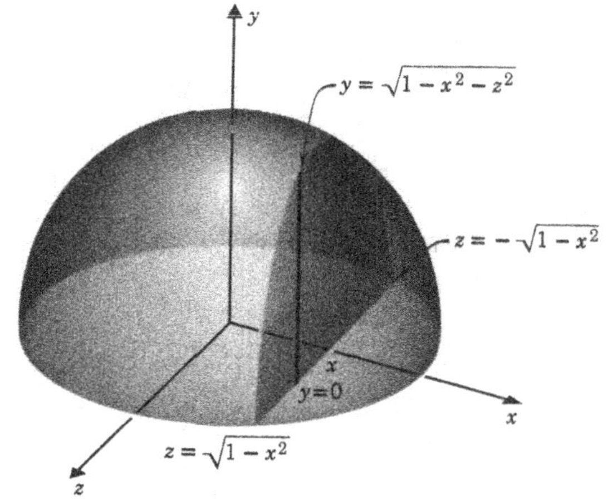

Figure 13-27 Example 3.

Further Properties of Triple Integrals. We here list properties analogous to those of double integrals. The proofs are the same as for double integrals. Throughout R is a standard solid region of volume V, and f and g are continuous in R.

1. *If* $K \leq f(x, y, z) \leq L$ *in* R, *then*

$$KV \leq \iiint_R f(x, y, z)\, dV \leq LV$$

2. *If* c_1 *and* c_2 *are constants, then*

$$\iiint_R (c_1 f + c_2 g)\, dV = c_1 \iiint_R f\, dV + c_2 \iiint_R g\, dV$$

3. *If R is subdivided into the standard regions R_1, R_2, then*

$$\iiint_R f\,dV = \iiint_{R_1} f\,dV + \iiint_{R_2} f\,dV$$

4. *For a region R as in* (13-61)

$$\iiint_R f(x, y, z)\,dV = \int_a^b \int_{\varphi_1(x)}^{\varphi_2(x)} \int_{\psi_1(x,y)}^{\psi_2(x,y)} f(x, y, z)\,dz\,dy\,dx$$

$$= \lim_{\text{mesh}\to 0} \sum_{i=1}^{n} f(\xi_i, \eta_i, \zeta_i)\,\Delta_i x\,\Delta_i y\,\Delta_i z$$

where for the last sum R is subdivided by planes perpendicular to the coordinate axes into rectangular parallelepipeds R_i ($i = 1, \dots, n$), of edges $\Delta_i x$, $\Delta_i y$, $\Delta_i z$ and other subregions which meet the boundary of R and which are ignored in the sum.

The following additional properties are discussed in Section 13-8 but, for convenience, we also list them here.

5. (*Mean Value Theorem for triple integrals*). *For some* (ξ, η, ζ) *in R*

$$\iiint_R f(x, y, z)\,dV = f(\xi, \eta, \zeta)V$$

6. *If* $f(x, y, z) \le g(x, y, z)$ *in R, then*

$$\iiint_R f(x, y, z)\,dV \leqq \iiint_R g(x, y, z)\,dV$$

7. *If R_1 is a standard region contained in R and* $f(x, y, z) \ge 0$ *in R, then*

$$\iiint_{R_1} f(x, y, z)\,dV \leqq \iiint_R f(x, y, z)\,dV$$

8. $$\lim_{\text{diam of } \Delta R\to 0} \frac{\displaystyle\iiint_{\Delta R} f\,dV}{\Delta V} = f(x_0, y_0, z_0)$$

where in the limit process one considers standard regions ΔR of volume ΔV, contained in R and containing (x_0, y_0, z_0).

9. *If* $f(x, y, z) \ge 0$ *in R and* $\iiint_R f(x, y, z)\,dV = 0$, *then* $f(x, y, z) \equiv 0$ *in R.*

13-7 TRIPLE INTEGRALS IN CYLINDRICAL AND SPHERICAL COORDINATES

Cylindrical coordinates r, θ, z (see Section 11-20) are most appropriate for a standard region R defined by inequalities

$$\alpha \le \theta \le \beta, \qquad \varphi_1(\theta) \le r \le \varphi_2(\theta), \qquad \psi_1(r, \theta) \le z \le \psi_2(r, \theta)$$

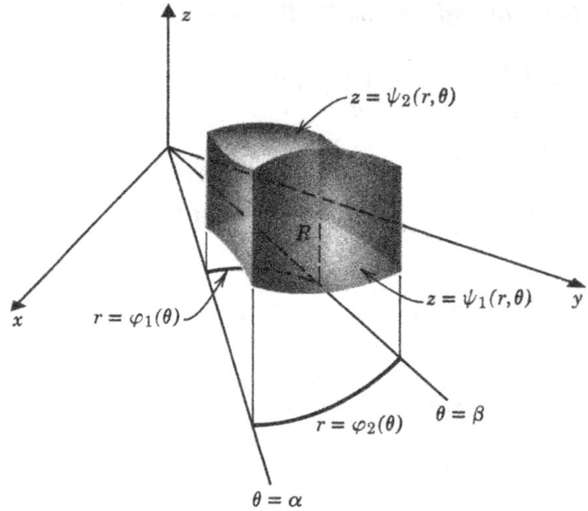

Figure 13-28 Region for triple integral in cylindrical coordinates.

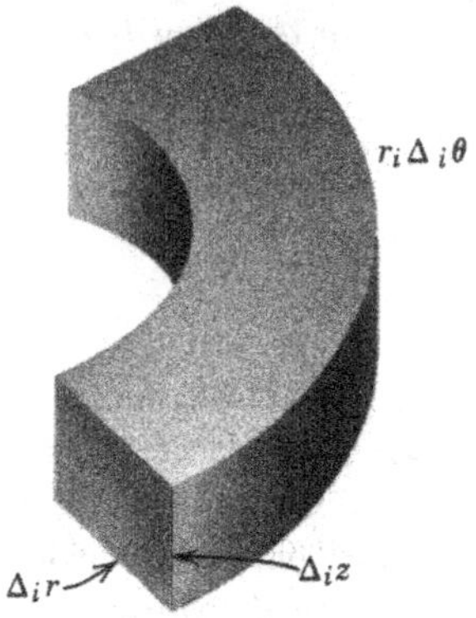

Figure 13-29 The element of volume for cylindrical coordinates.

as suggested in Figure 13-28. If one subdivides R by planes through the z-axis (planes $\theta = \text{const}$), planes perpendicular to the z-axis (planes $z = \text{const}$) and right circular cylinders $r = \text{const}$, and ignores subregions meeting the boundary of R, one obtains subregions as in Figure 13-29. For small mesh, each subregion R_i is approximately a rectangular parallelepiped of edges $\Delta_i r$, $\Delta_i z$ and $r_i' \Delta_i \theta$, where r_i' is an average value of r in R_i. In fact, as for $\Delta_i A$ in polar coordinates (Section 13-4),

$$\Delta_i V = r_i' \, \Delta_i z \, \Delta_i r \, \Delta_i \theta$$

where r_i' is the average of the inner and outer radii (Problem 6 below). To evaluate $\iiint\limits_R f\, dV$, where $f = f(x, y, z)$ is continuous in R, we write

$$f(x, y, z) = f(r\cos\theta, r\sin\theta, z) = F(r, \theta, z)$$

Then we can write:

$$\sum_{i=1}^{n} f(\xi_i, \eta_i, \zeta_i)\, \Delta_i V = \sum_{i=1}^{n} F(r_i'', \theta_i'', z_i'') r_i' \, \Delta_i z \, \Delta_i r \, \Delta_i \theta$$

Here we can assume (ξ_i, η_i, ζ_i) chosen so that $r_i'' = r_i'$. We can then apply Rule 4 of Section 13-6, exactly as in Section 13-4, and conclude:

$$\iiint\limits_R f\, dV = \int_\alpha^\beta \int_{\varphi_1(\theta)}^{\varphi_2(\theta)} \int_{\psi_1(r,\theta)}^{\psi_2(r,\theta)} F(r, \theta, z) r \, dz\, dr\, d\theta \qquad (13\text{-}70)$$

As for polar coordinates, the right-hand side can be interpreted as a triple integral over a region $R_{r\theta z}$ in $r\theta z$-space, with r, θ, z regarded as rectangular coordinates. We then write (13-70) as follows:

$$\iiint\limits_{R_{xyz}} f(x, y, z)\, dV_{xyz} = \iiint\limits_{R_{r\theta z}} f(r\cos\theta, r\sin\theta, z) r\, dV_{r\theta z} \qquad (13\text{-}71)$$

This formula is applicable, not only to a region as in Figure 13-28 but to an arbitrary standard region R_{xyz}.

For spherical coordinates ρ, φ, θ (Section 11-20)

$$x = \rho\sin\varphi\cos\theta, \qquad y = \rho\sin\varphi\sin\theta, \qquad z = \rho\cos\varphi$$

and we go through an analogous reasoning. We start with a standard region R as in Figure 13-30:

$$\alpha \le \theta \le \beta, \qquad g_1(\theta) \le \varphi \le g_2(\theta), \qquad h_1(\varphi, \theta) \le \rho \le h_2(\varphi, \theta)$$

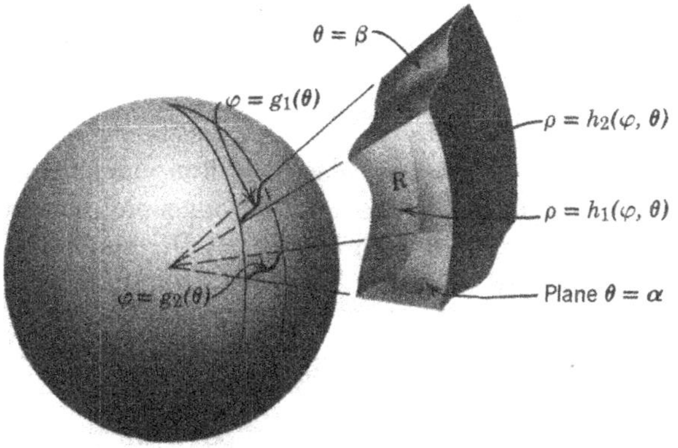

Figure 13-30 Region for triple integral in spherical coordinates.

We then subdivide R by *planes* θ = const, *cones* φ = const, and *spheres* ρ = const. We obtain a typical subregion as in Figure 13-31. Here we reason intuitively that if the subregion R_i has small mesh, then R_i is approximately

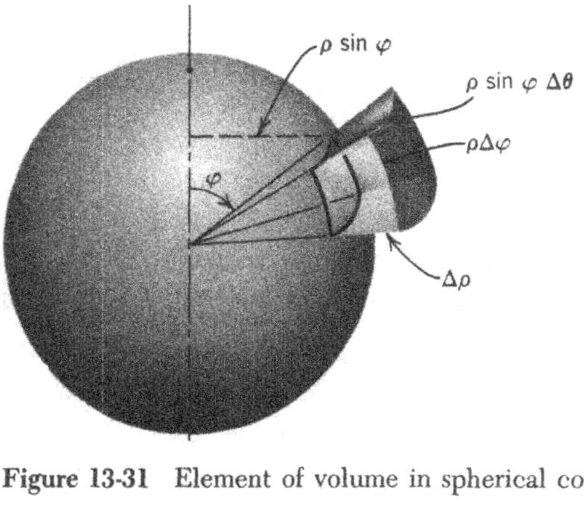

Figure 13-31 Element of volume in spherical coordinates.

a rectangular parallelepiped with edges $\Delta_i\rho$, $\rho_i' \sin\varphi_i' \Delta_i\theta$, $\rho_i'' \Delta_i\varphi$, where $\rho_i', \rho_i'', \varphi_i'$ are suitable average values of ρ and φ in R_i. Hence, we expect

$$\Delta_i V = \rho_i'\rho_i'' \sin\varphi_i' \Delta_i\rho\, \Delta_i\varphi\, \Delta_i\theta \tag{13-72}$$

and in the limit (as in Section 13-3)

$$\iiint\limits_R f(x, y, z)\, dV = \int_\alpha^\beta \int_{g_1(\theta)}^{g_2(\theta)} \int_{h_1(\varphi, \theta)}^{h_2(\varphi, \theta)} f(\rho \sin\varphi \cos\theta, \dots)\rho^2 \sin\varphi \, d\rho\, d\varphi\, d\theta \tag{13-73}$$

and more generally, for an arbitrary standard region R_{xyz},

$$\iiint\limits_{R_{xyz}} f(x, y, z)\, dV_{xyz} = \iiint\limits_{R_{\rho\varphi\theta}} f(\rho\sin\varphi\cos\theta, \dots)\rho^2 \sin\varphi\, dV_{\rho\varphi\theta} \tag{13-74}$$

The formula (13-72) can be justified by first obtaining an exact formula for the volume $\Delta_i V$ or from the general formula (13-75) (Problem 7 below). It should be noted that our intuitive derivation relies on the fact that the surfaces $\rho =$ const, $\varphi =$ const, $\theta =$ const meet *orthogonally:* that is, the normal lines to two surfaces meeting at a point are orthogonal.

Other Curvilinear Coordinates. The special geometric arguments used for cylindrical and spherical coordinates lead us to ask for a general method for computing triple integrals in curvilinear coordinates. By the methods of Section 13-5 one is led to the general formula

$$\iiint\limits_{R_{xyz}} f(x, y, z)\, dV_{xyz} = \iiint\limits_{R_{uvw}} F(u, v, w) \left|\frac{\partial(x, y, z)}{\partial(u, v, w)}\right| dV_{uvw} \tag{13-75}$$

where $x = p_1(u, v, w)$, $y = p_2(u, v, w)$, $z = p_3(u, v, w)$ and $F(u, v, w) = f(p_1(u, v, w), \dots)$. For a discussion of this formula one is referred to texts on advanced calculus, for example, Chapter 5 of the text by W. Kaplan, previously cited. For cylindrical coordinates we take $u = r$, $v = \theta$, $w = z$ and find that the Jacobian is r. For spherical coordinates, we take $u = \rho$, $v = \theta$, $w = \varphi$ and find that the Jacobian is $-\rho^2 \sin\varphi$ (Problem 8 below). We thus obtain (13-71) and (13-74).

EXAMPLE 1. Let $f(x, y, z) = 4xy$. Let R be the cylindrical region $x^2 + y^2 \le 1, 0 \le z \le 1$. Then R is given in cylindrical coordinates by inequalities: $0 \le \theta \le 2\pi$, $0 \le r \le 1$, $0 \le z \le 1$ and

$$\iiint\limits_R 4xy\, dV = \int_0^{2\pi}\int_0^1\int_0^1 4r^2 \sin\theta\cos\theta\, r\, dz\, dr\, d\theta$$

Here the limits of integration are constants, and the integrand is expressible as product of a function of r times a function of θ. Hence, we can write the integral as

$$\int_0^{2\pi} \sin\theta\cos\theta\, d\theta \int_0^1 dz \int_0^1 4r^3\, dr = \frac{\sin^2\theta}{2}\Bigg|_0^{2\pi} z\Bigg|_0^1 r^4\Bigg|_0^1 = 0$$

EXAMPLE 2. Let $f(x, y, z) = z^2$. Let R be the spherical region $x^2 + y^2 +$

$z^2 \leq 1$. Then R is given in spherical coordinates by the inequalities $0 \leq \theta \leq 2\pi$, $0 \leq \varphi \leq \pi$, $0 \leq \rho \leq 1$ and

$$\iiint_R z^2\,dV = \int_0^{2\pi}\int_0^{\pi}\int_0^1 \rho^2\cos^2\varphi\,\rho^2\sin\varphi\,d\rho\,d\varphi\,d\theta$$

$$= \int_0^{2\pi} d\theta \int_0^{\pi} \cos^2\varphi\sin\varphi\,d\varphi \int_0^1 \rho^4\,d\rho = \frac{4\pi}{15}$$

EXAMPLE 3. Let $f(x, y, z) = x^2 + y^2 + z^2$. Let R be the region $x^2 + y^2 \leq 1$, $0 \leq z \leq 4 - x^2 - y^2$ (Figure 13-32). We use cylindrical coordinates:

$$\iiint_R (x^2 + y^2 + z^2)\,dV = \int_0^{2\pi}\int_0^1\int_0^{4-r^2} (r^2 + z^2)r\,dz\,dr\,d\theta$$

$$= \int_0^{2\pi}\int_0^1 \left(r^3 z + \frac{rz^3}{3}\right)\bigg|_{z=0}^{z=4-r^2} dr\,d\theta$$

$$= \int_0^{2\pi}\int_0^1 \left[r^3(4 - r^2) + r\frac{(4 - r^2)^3}{3}\right] dr\,d\theta$$

$$= \int_0^{2\pi} \left(r^4 - \frac{r^6}{6} - \frac{(4 - r^2)^4}{24}\right)\bigg|_{r=0}^{r=1} d\theta = 65\pi/4$$

Here we followed the three steps of Section 13-7: *Step* 1: the smallest and largest values of θ can be taken as 0 and 2π. *Step* 2: for fixed θ, the smallest

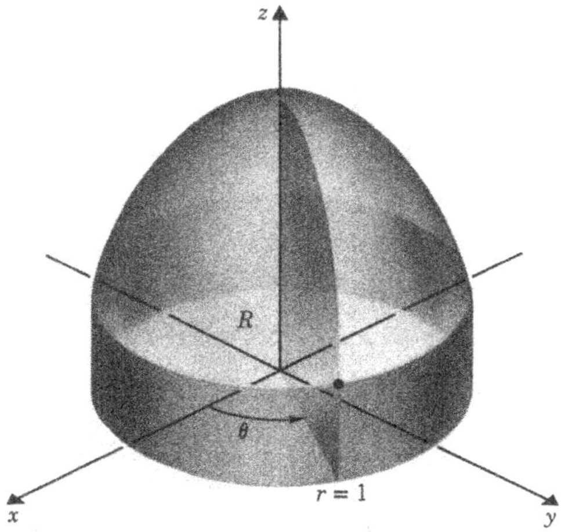

Figure 13-32 Example 3.

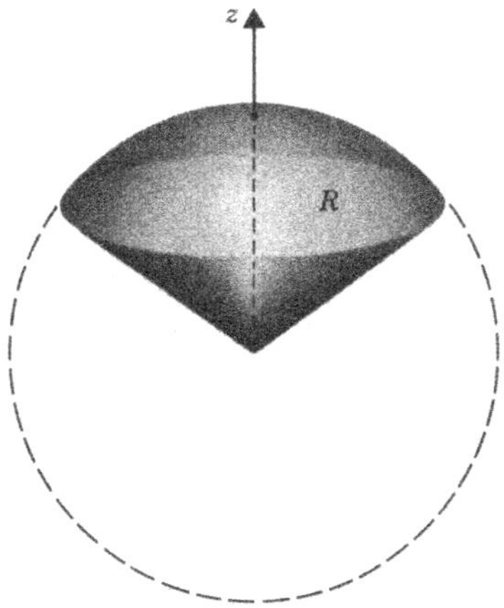

Figure 13-33 Example 4.

and largest values of r are 0 and 1. *Step* 3: for fixed θ and r, the smallest and largest values of z are 0 and $4 - r^2$ (see Figure 13-32). (Here the θ-integration could have been factored out and, hence, replaced by a factor of 2π).

EXAMPLE 4. Let $f(x, y, z) = x^2 + y^2$. Let R be the region $x^2 + y^2 + z^2 \leq 1$, $x^2 + y^2 \leq z^2$, $z \geq 0$, as in Figure 13-33. Here we can describe R in spherical coordinates by the inequalities $0 \le \theta \le 2\pi$, $0 \le \varphi \le \pi/4$, $0 \le \rho \le 1$ and

$$\iiint\limits_R (x^2 + y^2)\,dV$$

$$= \int_0^{2\pi}\int_0^{\pi/4}\int_0^1 (\rho^2 \sin^2\varphi \cos^2\theta + \rho^2 \sin^2\varphi \sin^2\theta)\rho^2 \sin\varphi\,d\rho\,d\varphi\,d\theta$$

$$= \int_0^{2\pi}\int_0^{\pi/4}\int_0^1 \rho^4 \sin^3\varphi\,d\rho\,d\varphi\,d\theta$$

$$= \int_0^{2\pi} d\theta \int_0^{\pi/4} \sin^3\varphi\,d\varphi \int_0^1 \rho^4\,d\rho = \frac{\pi(4\sqrt{2} - 5)}{15\sqrt{2}}$$

In this example, R could also be described in cylindrical coordinates by the inequalities:

$$0 \le \theta \le 2\pi, \qquad 0 \le r \le \sqrt{2}/2, \qquad r \le z \le \sqrt{1 - r^2}$$

Hence

$$\iiint\limits_R (x^2 + y^2)\,dV = \int_0^{2\pi}\int_0^{\sqrt{2}/2}\int_r^{\sqrt{1-r^2}} r^3\,dz\,dr\,d\theta$$

The integration can be carried out with a little more effort than in spherical coordinates.

EXAMPLE 5. The region between the two hyperbolas $4y^2 - z^2 = 4$, $4y^2 - z^2 = 16$ and between the rays $z = \pm y$ for $y \ge 0$ is rotated about the z-axis to obtain a solid region R, as in Figure 13-34. We seek the triple integral

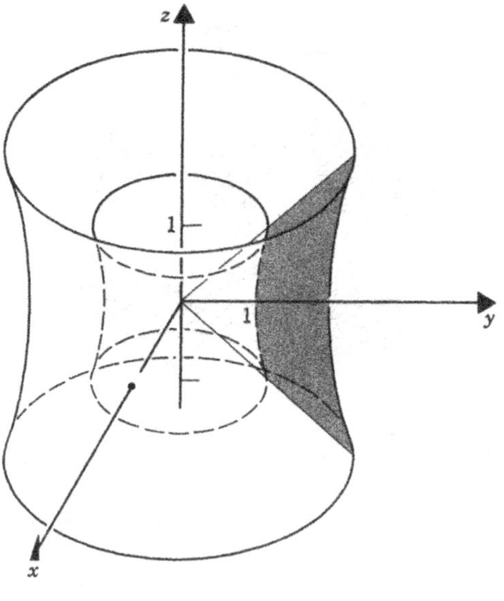

Figure 13-34 Example 5.

of $f(x, y, z) = x^2 + y^2 + z^2$ over R. We use spherical coordinates. In the yz-plane φ varies between $\pi/4$ and $3\pi/4$ and, for fixed φ, ρ varies between its values on the two hyperbolas. Now the hyperbolas have equations in spherical coordinates (θ having the value $\pi/2$ in the yz plane):

$$4\rho^2 \sin^2 \varphi - \rho^2 \cos^2 \varphi = 4, \qquad 4\rho^2 \sin^2 \varphi - \rho^2 \cos^2 \varphi = 16$$

Hence

$$g_1(\varphi) = \left(\frac{4}{4 \sin^2 \varphi - \cos^2 \varphi}\right)^{1/2} \leq \rho \leq \left(\frac{16}{4 \sin^2 \varphi - \cos^2 \varphi}\right)^{1/2} = g_2(\varphi)$$

in the region. Since R is a solid of revolution about the z-axis, the limits for ρ do not depend on θ and θ ranges from 0 to 2π. Hence

$$\iiint_R f\,dV = \int_0^{2\pi} \int_{\pi/4}^{3\pi/4} \int_{g_1(\varphi)}^{g_2(\varphi)} \rho^2 \rho^2 \sin \varphi \, d\rho \, d\varphi \, d\theta$$

The integration is somewhat lengthy, but it can be carried out.

PROBLEMS

1. Evaluate $\iiint_R f(x, y, z)\,dV$ for the following choices of f and R:
   (a) $f(x, y, z) = \sqrt{x + y + z}$, R: the cube of vertices $(0, 0, 0)$, $(1, 0, 0)$, $(0, 1, 0)$, $(0, 0, 1)$, $(1, 1, 0)$, $(1, 0, 1)$, $(0, 1, 1)$, $(1, 1, 1)$.
   (b) $f(x, y, z) = (2x - y - z)^5$, R: the rectangular parallelepiped $0 \leq x \leq 1$, $0 \leq y \leq 2$, $0 \leq z \leq 3$.
   (c) $f(x, y, z) = x^2yz$, R: the tetrahedron of vertices $(0, 0, 0)$, $(1, 0, 0)$, $(1, 1, 0)$, $(1, 1, 1)$.
   (d) $f(x, y, z) = x^2 + z^2$, R: the pyramid of vertices $(\pm 1, \pm 1, 0)$ and $(0, 0, 1)$.
   (e) $f(x, y, z) = x + z$, R: $x^2 + y^2 + z^2 \leq 1$, $x \geq 0$, $y \geq 0$, $z \geq 0$.
   (f) $f(x, y, z) = 2yz$, R: the region $0 \leq z \leq 1 - x^2 - y^2$.
2. Write each of the following integrals in the form

$$\int_a^b \int_{\varphi_1(x)}^{\varphi_2(x)} \int_{\psi_1(x,y)}^{\psi_2(x,y)} f(x, y, z)\,dz\,dy\,dx$$

   (a) $\int_0^1 \int_2^3 \int_4^5 e^{xy}\,dx\,dy\,dz$ (b) $\int_1^2 \int_2^4 \int_0^1 (x^2 + y^2)^{3/2}\,dz\,dx\,dy$
   (c) $\int_0^1 \int_0^y \int_{\sqrt{x^2+y^2}}^1 \sin(xy)\,dz\,dx\,dy$ (d) $\int_0^1 \int_0^{1-x} \int_0^{\sqrt{(z-1)^2-x^2}} x^3\,dy\,dz\,dx$
3. Evaluate with the aid of cylindrical coordinates:
   (a) $\iiint_R xy\,dV$, R: $x^2 + y^2 \leq 1$, $x \geq 0$, $y \geq 0$, $0 \leq z \leq 1$
   (b) $\iiint_R \sin^2 \theta\,dV$, R: $0 \leq r \leq 1 - \cos 2\theta$, $0 \leq \theta \leq \pi/2$, $0 \leq z \leq 1$
   (c) $\iiint_R r^2\,dV$, R: $0 \leq r \leq e^\theta$, $0 \leq \theta \leq \pi$, $0 \leq z \leq r \sin \theta$
   (d) the integral of Problem 1(e)

4. Evaluate with the aid of spherical coordinates:

(a) $\iiint_R (x^2 + y^2 + z^2)^{3/2}\, dV$, R: $x^2 + y^2 + z^2 \leq 1$

(b) $\iiint_R \rho^3 \sin\varphi\, dV$, R: $x^2 + y^2 + z^2 \leq 1$

(c) $\iiint_R \rho \cos(\varphi + \theta)\, dV$, R: $0 \leq \rho \leq 1$, $0 \leq \theta \leq \pi/2$, $0 \leq \varphi \leq \theta$

(d) $\iiint_R e^{-\rho} \sin\theta\, dV$, R: $0 \leq \rho \leq 1$, $0 \leq \theta \leq 2\pi$, $0 \leq \varphi \leq \pi/4$

5. Justify each assertion:

(a) $84\pi e^{-16} \leq \iiint_R e^{-x^2-y^2-z^2}\, dV \leq 84\pi e^{-1}$, where R: $1 \leq x^2 + y^2 + z^2 \leq 16$.

(b) $\iiint_R \sqrt{x^2 + y^2 + z^2}\, dV \leq \iiint_R (x + y + z)\, dV = \frac{3}{2}$, where R is the cube of Problem 1(a).

6. In cylindrical coordinates, let R be the region $r_1 \leq r \leq r_2 = r_1 + \Delta r$, $\alpha \leq \theta \leq \alpha + \Delta\theta$, $h \leq z \leq h + \Delta z$. Show by geometry that R has volume $\Delta V = \frac{1}{2}(r_1 + r_2)\, \Delta r\, \Delta\theta\, \Delta z$.

7. (a) Justify the assertion: if a region in a plane passing through the z-axis is rotated about the z-axis, but only through angle α (instead of 2π), to form a "partial solid of revolution," then the volume of this solid is $\alpha/(2\pi)$ times the volume of the complete solid of revolution.

(b) Apply the result of part (a) to obtain the volume of the region (described by spherical coordinates)

$$\rho_0 \leq \rho \leq \rho_0 + \Delta\rho, \qquad \varphi_0 \leq \varphi \leq \varphi_0 + \Delta\varphi, \qquad \theta_0 \leq \theta \leq \theta_0 + \Delta\theta$$

and, hence, justify (13-172) and (13-173). [*Hint.* Here $\alpha = \Delta\theta$ and the complete solid of revolution is a solid region between two conical shells. By the result of Section 7-6, its volume is

$$2\pi[(\rho_0 + \Delta\rho)^3 - \rho_0{}^3]\, \frac{\cos\varphi_0 - \cos(\varphi_0 + \Delta\varphi)}{3}$$

Show that this can be written in the form

$$2\pi \cdot 3\rho'^2\, \Delta\rho \cdot \frac{1}{3} \sin\varphi'\, \Delta\varphi$$

where $\rho_0 < \rho' < \rho_0 + \Delta\rho$, $\varphi_0 < \varphi' < \varphi_0 + \Delta\varphi$.]

8. Prove: (a) for cylindrical coordinates, $\partial(x, y, z)/\partial(r, \theta, z) = r$.

(b) For spherical coordinates, $\partial(x, y, z)/\partial(\rho, \theta, \varphi) = -\rho^2 \sin\varphi$.

‡9. Let R_{xy} be a bounded closed region in the xy-plane enclosed by the simple closed piecewise smooth path C. Let $\psi_1(x, y)$ and $\psi_2(x, y)$ be continuous for all (x, y), let $f(x, y, z)$ be continuous for all (x, y, z). Let R be the region formed of all points

(x, y, z) for which (x, y) is in R_{xy} and $\psi_1(x, y) \leq z \leq \psi_2(x, y)$. Justify the relation:

$$\iiint\limits_R f(x, y, z)\, dV = \iint\limits_{R_{xy}} \left[\int_{\psi_1(x,y)}^{\psi_2(x,y)} f(x, y, z)\, dz\right] dA$$

10. To find the n-volume of a solid n-sphere B in R^n, one can assume coordinates chosen so that B is given by the inequality $x_1^2 + \cdots + x_n^2 \leq a^2$. We then imitate the reasoning of Section 13-1 to obtain the formula

$$V = \int_{-a}^{a} f(x_n)\, dx_n$$

where $f(x_n)$ is the $(n-1)$-volume of the cross section of B by a hyperplane $x_n = \text{const}$. But this cross section is a solid $(n-1)$-sphere of radius $\sqrt{a^2 - x_n{}^2}$. We now use induction and assume that we have shown that the $(n-1)$-volume of the solid $(n-1)$-sphere of radius b is $\beta_{n-1}b^{n-1}$, where β_{n-1} is a constant to be found. Then we conclude that

$$V = \int_{-a}^{a} \beta_{n-1}(a^2 - x_n{}^2)^{(n-1)/2}\, dx_n = 2\beta_{n-1}a^n \int_0^{\pi/2} \cos^n\theta\, d\theta$$

Hence, $V = \beta_n a^n$, with

$$\beta_n = 2\beta_{n-1} \int_0^{\pi/2} \cos^n\theta\, d\theta$$

Now use integral tables and the known value of β_2 (what is β_1?) to show that

$$\beta_n = \begin{cases} \pi^{n/2}\Big/\left(1 \cdot 2 \cdots \dfrac{n}{2}\right), & \text{if } n \text{ even} \\ 2(2\pi)^{(n-1)/2}/(1 \cdot 3 \cdots n), & \text{if } n \text{ odd} \end{cases}$$

11. Evaluate:

(a) $\iiiint\limits_R dx_1\, dx_2\, dx_3\, dx_4$, where R is the region $0 \leq x_1 \leq a$, $0 \leq x_2/b \leq 1 - x_1/a$, $0 \leq x_3/c \leq 1 - x_1/a - x_2/b$, $0 \leq x_4/d \leq 1 - x_1/a - x_2/b - x_3/c$, where a, b, c, d are positive constants. (Remark. R is a 4-simplex and the integral gives the 4-volume of R. This can also be found as in Section 11-16).

(b) $\iiiint\limits_R e^{x_1+x_2+x_3+x_4}\, dx_1\, dx_2\, dx_3\, dx_4$, where R is the region $0 \leq x_1 \leq 1$, $0 \leq x_2 \leq 2$, $0 \leq x_3 \leq 3$, $0 \leq x_4 \leq 4$.

13-8 FURTHER PROPERTIES OF MULTIPLE INTEGRALS

In this section we consider several additional properties of multiple integrals. The ideas involved have important interpretations in terms of mass and density. Therefore, we develop the theory along with its application. For simplicity, we consider only double integrals, but all the results have their counterparts for triple and n-fold multiple integrals.

We shall be thinking of mass spread out in a plane. A good physical model is a layer of paint on a flat surface. There are many other models, not necessarily related to mass: for example, distribution of electric charge on a plate, distribution of heat energy on a surface, and distribution of a population of any kind in a geographical area. Here we shall refer only to mass and density in terms of mass per unit area; however, the ideas have the broader interpretation suggested. In the case of mass, the corresponding integrals (total mass) and integrands (density) are nonnegative. However, our theory will include the more general case of variable sign, for example, as occurs for distributions of electric charge.

In Section 7-9 to 7-12 we considered mass distributions of various types; in particular, distributions in the xy-plane. However, we considered only those distributions in the plane for which the density is constant over whole regions $R_1, \ldots, R_k$. The total mass was then found to be

$$M_0 = \delta_1 A_1 + \delta_2 A_2 + \cdots + \delta_k A_k$$

where δ_i is the density in R_i and A_i is the area of R_i $(i = 1, \ldots, k)$. For a general nonhomogeneous continuous distribution of mass over a region R, the density δ becomes a continuous function of x and y, say $\delta = f(x, y)$. If R is subdivided into small regions $R_1, \ldots, R_n$, it is reasonable to approximate the total mass by the sum

$$f(\xi_1, \eta_1)\, \Delta_1 A + \cdots + f(\xi_n, \eta_n)\, \Delta_n A$$

where (ξ_i, η_i) is in R_i and R_i has area $\Delta_i A$ $(i = 1, \ldots, n)$. Here $f(\xi_i, \eta_i)$ can be regarded as an estimate of the average density over R_i. As we let the mesh of our subdivision approach 0, we expect the sum to approach the exact value of the total mass. Thus we write

$$M_0 = \lim_{\text{mesh}\to 0} \sum_{i=1}^{n} f(\xi_i, \eta_i)\, \Delta_i A = \iint_R f(x, y)\, dA \tag{13-80}$$

This formula, which we have only shown to be plausible, will be justified in several different ways in the following discussion.

Throughout this section we consider bounded regions, bounded by a finite number of piecewise smooth simple closed paths, as at the end of Section 13-2. For brevity, we refer to such a region as a *standard region*. As remarked at the end of Section 13-2, all the theorems of that section can be extended to standard regions.

THEOREM 7 (*Mean Value Theorem for Double Integrals*). *Let f be continuous in the standard region R, of area A. Then there is a point (ξ, η) in R such that*

$$\iint_R f(x, y)\, dA = f(\xi, \eta) A \tag{13-81}$$

PROOF. By Theorem 1, the value of the double integral is a number between mA and MA (inclusive); hence, the integral divided by A is a number k between m and M. By Theorem D of Section 12-7, $f(\xi, \eta) = k$ for some (ξ, η) in R. Hence, (13-81) follows.

We can write (13-81) as follows

$$f(\xi, \eta) = \frac{\displaystyle\iint_R f(x, y)\, dA}{A} \tag{13-81$'$}$$

and consider the right side as an average value or mean value of f in R. If we divide R into n parts of equal area A/n, then the right side is approximated by a corresponding sum, divided by A:

$$\frac{\displaystyle\sum_{i=1}^{n} f(\xi_i, \eta_i)(A/n)}{A} = \frac{f(\xi_1, \eta_1) + \cdots + f(\xi_n, \eta_n)}{n}$$

This is just the ordinary average of the values $f(\xi_1, \eta_1), \ldots, f(\xi_n, \eta_n)$. Thus it is natural to term the right side of (13-81′) the *average* or *mean value* of f in R. Theorem 7 asserts that f equals its mean value somewhere in R. In terms of mass and density, the theorem states that the total mass in R can always be written as the area of R times the average density, and the average density equals the actual density at some point (ξ, η) in R.

THEOREM 8. *Let R be a standard region, let f and g be continuous in R and let $f(x, y) \le g(x, y)$ in R. Then*

$$\iint_R f(x, y)\, dA \le \iint_R g(x, y)\, dA$$

In particular, if $f(x, y) \ge 0$ in R, then

$$\iint_R f(x, y)\, dA \ge 0$$

The theorem follows from Theorem 1. The details are left as an exercise (Problem 7 below). The theorem can be interpreted as follows: if we increase the density, then the total mass increases; also, if the density is nonnegative (as usually assumed), then the total mass is nonnegative.

THEOREM 9. *Let the standard region R be subdivided into two standard regions R_1, R_2; let f be continuous in R and let $f(x, y) \ge 0$ in R. Then*

$$\iint_{R_1} f(x, y)\, dA \le \iint_R f(x, y)\, dA \tag{13-82}$$

PROOF. By Theorem 6 for standard regions (as at the end of Section 13-2),

$$\iint_{R_1} f(x, y)\, dA + \iint_{R_2} f(x, y)\, dA = \iint_R f(x, y)\, dA$$

Since $f(x, y) \ge 0$ in R, the second term on the left is nonnegative, by Theorem 8, so that (13-82) follows.

Remark 1. By more advanced methods one can prove: if f and R are as stated in Theorem 9 and R_1 is any standard region contained in R, then (13-82) holds true. In terms of mass, (13-82) states simply that the larger the region the more mass it contains.

THEOREM 10. *Let R be a standard region, let f be continuous in R and let (x_0, y_0) be a point of R. Then for every $\epsilon > 0$, a number $d > 0$ can be found such that*

$$\left| \frac{\iint\limits_{R^*} f(x, y)\, dA}{A^*} - f(x_0, y_0) \right| < \epsilon \tag{13-83}$$

for every standard region R^* contained in R, containing (x_0, y_0), having area A^* and having diameter less than d.

PROOF. The hypotheses are illustrated in Figure 13-35. The conclusion

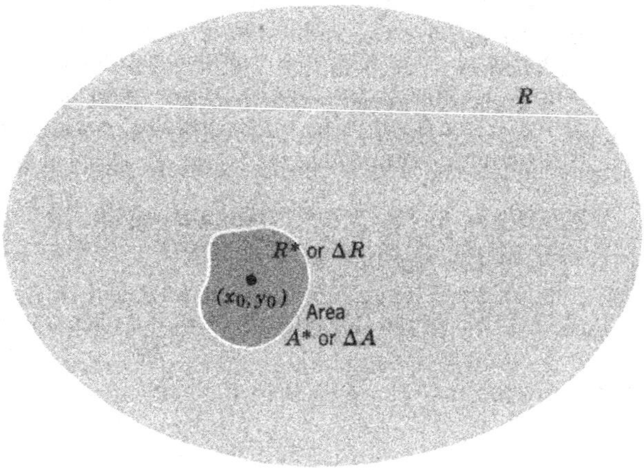

Figure 13-35 Density as "derivative," of mass.

of Theorem 10 can be written:

$$\lim_{\text{diam of } \Delta R \to 0} \frac{\iint\limits_{\Delta R} f\, dA}{\Delta A} = f(x_0, y_0) \tag{13-83'}$$

The ratio of integral to area in (13-83) or (13-83′) is just the mean value in the region R^* or ΔR. Thus the theorem asserts that the mean value of f, in a region R^* containing (x_0, y_0), approaches $f(x_0, y_0)$ as the diameter of R^* approaches 0. By Theorem 7, the mean value is equal to $f(\xi^*, \eta^*)$, where (ξ^*, η^*) is a point of R^*. As the diameter of R^* approaches 0, the distance from (ξ^*, η^*) to (x_0, y_0) approaches 0 and, hence, by the continuity of f, $f(\xi^*, \eta^*)$ approaches $f(x_0, y_0)$. Therefore, (13-83′) is proved.

COROLLARY. *Let f and g be continuous in the standard region R and* *let*

$$\iint_{R^*} f(x, y)\, dA = \iint_{R^*} g(x, y)\, dA \tag{13-84}$$

for every standard region R^* contained in R. Then $f(x, y) \equiv g(x, y)$ in R.

PROOF. By (13-84) and (13-83′), for every (x_0, y_0) in R

$$f(x_0, y_0) = \lim \frac{\iint_{\Delta R} f\, dA}{\Delta A} = \lim \frac{\iint_{\Delta R} g\, dA}{\Delta A} = g(x_0, y_0)$$

Remark 2. Theorem 10 is a 2-dimensional version of the fundamental theorem of calculus (Theorem 11, Section 4-17). For the latter theorem can be written

$$\lim_{\Delta x \to 0} \frac{\int_{x_0-\Delta x}^{x_0+\Delta x} f(x)\, dx}{2\Delta x} = f(x_0)$$

We, therefore, can consider the limit in (13-83) or (13-83′) as a kind of derivative: the derivative of the integral of f with respect to area. The theorem asserts that we can recover the integrand from the integral over variable regions (a sort of indefinite integral) by taking the derivative in this sense. Equation (13-83′) can be interpreted as follows: *the density at a point is the limit of mass divided by area, as the diameter approaches zero;* this assertion is often used as the very definition of density.

THEOREM 11. *Let f be continuous in the standard region R, let $f(x, y) \geq 0$ in R and let*

$$\iint_R f(x, y)\, dA = 0$$

Then $f(x, y) \equiv 0$ in R.

PROOF. Let (x_0, y_0) be a point of R, let ΔR be a region in R containing (x_0, y_0). Then by Theorems 8 and 9

$$0 \leq \iint_{\Delta R} f(x, y)\, dA, \qquad \iint_{\Delta R} f(x, y)\, dA \leq \iint_R f(x, y)\, dA$$

(see Remark 1 above). But the last integral is 0. Hence

$$\iint_{\Delta R} f(x, y)\, dA = 0$$

By (13-83′) we now conclude:

$$f(x_0, y_0) = \lim_{\text{diam of } \Delta R \to 0} \frac{\iint_{\Delta R} f(x, y)\, dA}{\Delta A} = 0$$

Hence, $f(x, y) \equiv 0$ in R.

The theorem asserts: if the total mass is 0, then the density is 0 everywhere.

Remark 3. There is one further question concerning mass and density. It is very natural to start with the definition of density mentioned above: *density is limit of mass divided by area.* If for a given mass distribution in R, one finds that, by this definition, there is a continuous density $\delta = f(x, y)$, then the total mass must be the double integral of f over R. The proof is left as an exercise (Problem 12 below).

PROBLEMS

1. Mass is distributed over region R with density $\delta = f(x, y)$ as given (in a fixed system of units). Find the total mass in each case:
   (a) $f(x, y) = 1 - x^2 - y^2$, R: $x^2 + y^2 \leq 1$
   (b) $f(x, y) = e^{-x-y}$, R: square of vertices $(\pm 1, \pm 1)$
   (c) $f(x, y) = 3x + 2y$, R: triangle of vertices $(0, 0)$, $(1, 1)$, $(1, 2)$
   (d) $f(x, y) = 3\sqrt{x^2 + y^2} - 2 - x^2 - y^2$, R: $1 \leq x^2 + y^2 \leq 2$
2. Electric charge, in coulombs, is spread over a circular plate with charge density $f(x, y) = k \sin \sqrt{x^2 + y^2}$, $0 \leq x^2 + y^2 \leq 4\pi^2$ (in coulombs per cm^2, x and y in centimeters). Find the total charge.
3. In a certain country it is found that the population density is well approximated by the formula $\delta = 200$ for $r \leq 5$, $\delta = 200e^{(5-r)/10}$, for $5 \leq r \leq 100$, where r is the distance in miles from the capitol. Find the total population.
4. *Probability density.* If the result of an experiment is a pair of real numbers (x, y), one may assign a probability density $f(x, y)$ (here called the "joint probability density of the random variables x and y") such that the double integral of f over region R gives the probability p that (x, y) falls in R. Here p is a number between 0 and 1, inclusive. One says that x and y are *independent* if $f(x, y)$ can be written as a product $g(x)h(y)$.
   (a) Let $f(x, y) = e^{-(x^2+y^2)/2}/(2\pi)$ for all (x, y). Show that x, y are independent and find the probability that (x, y) lies in the circle $x^2 + y^2 \leq a^2$.
   (b) Let $f(x, y) = 4xy$ for $0 \leq x \leq 1$, $0 \leq y \leq 1$, $f(x, y) = 0$ otherwise (so that f is discontinuous along two edges of the square). Find the probability that (x, y) lies inside the circle $(x - 1)^2 + (y - \frac{1}{2})^2 = 1$. (*Hint.* Interpret the probability as total mass.)
5. Mass is spread over the square with vertices $(0, 0)$, $(1, 0)$, $(0, 1)$, $(1, 1)$ with density $f(x, y) = 3x + 5y$. Find the mean density and find the points (ξ, η) at which the density equals its mean value.
6. Let R be the square of Problem 5. Justify the inequalities:

   (a) $\iint\limits_R \sin \dfrac{x^2 + y^2}{2}\, dx\, dy \leq \dfrac{1}{3}$ (b) $1 \leq \iint\limits_R \sqrt{1 + x^4 + y^4}\, dx\, dy \leq \dfrac{5}{3}$
7. Prove Theorem 8.
8. A mass distribution in the square R of Problem 5 is such that for each circular

region R_0 of center (x_0, y_0) and radius r_0 in R, the mass in R_0 is $kr_0^2(4x_0^2 + r_0^2)$, where k is a positive constant. Find the density f.

9. Let $f(x, y)$ be continuous in $R: a \leq x \leq b,\ c \leq y \leq d$. Let $F(x, y) = \int_a^x \int_b^y f(u, v)\, dv\, du$ in R. Prove:

(a) $$\int_{x_0}^{x_0+h} \int_{y_0}^{y_0+k} f(x, y)\, dy\, dx = F(x_0 + h, y_0 + k) - F(x_0 + h, y_0) - F(x_0, y_0 + k) + F(x_0, y_0)$$

(b) $f(x, y) = \dfrac{\partial^2 F}{\partial y\, \partial x} = \dfrac{\partial^2 F}{\partial x\, \partial y}$ inside R.

(c) $$\frac{\partial^2 F}{\partial y\, \partial x}(x_0, y_0) = \lim_{h \to 0} \frac{F(x_0 + h, y_0 + h) - F(x_0 + h, y_0) - F(x_0, y_0 + h) + F(x_0, y_0)}{h^2}$$

inside R (see Section 12-20).

10. (a) For a solid region beneath a surface $z = F(x, y)$, for (x, y) in region R_{xy}, one can introduce the concept of density δ as mass per unit volume. Show that, if the density is given as a function of x and y alone, $\delta = f(x, y)$, then it is reasonable to obtain the total mass of the solid as

$$M_0 = \iint_{R_{xy}} f(x, y)F(x, y)\, dA$$

(b) Deduce the result of part (a) as a consequence of the representation of the mass M_0 of a solid of density $f(x, y, z)$ as a triple integral:

$$M_0 = \iiint f(x, y, z)\, dV$$

(See Problem 9 following Section 13-7.)

11. The density δ of the earth's atmosphere in summer is described approximately by the formula $\delta = ke^{-ah}\,\text{gm/cm}^3$, where $k = 1.2\ 10^{-3}\ \text{gm/cm}^3$, h is the altitude above the earth (that is, above sea level) in km, and $a = \frac{1}{9}$. Find a formula for $M_0(h)$, the total mass up to altitude h, and graph $M_0(h)$. Find lim $M_0(h)$ as $h \to \infty$ and state the significance of this limit.

‡12. To establish the result referred to in Remark 3 at the end of Section 13-8, let it be assumed that we have a standard region R_0 in the plane and that, for each standard region R contained in R_0, we know the mass in R, which we denote by $\mu(R)$. For consistency, we then require that, if R is subdivided into standard regions $R_1, \ldots, R_k$, then $\mu(R) = \mu(R_1) + \cdots + \mu(R_k)$. Furthermore, we assume that at each point (x_0, y_0) of R, $\mu(\Delta R)/\Delta A$ has a limit as the diameter of ΔR approaches 0, where ΔR is a standard region contained in R_0, containing (x_0, y_0), and having area ΔA. We denote this limit by $f(x_0, y_0)$ and assume that f is continuous in R. Prove: for every standard region R contained in R_0, one has

$$\mu(R) = \iint_R f(x, y)\, dA$$

13-9 SURFACE AREA

In Section 7-8 we derived a formula for the area of a surface of revolution. We now seek a formula for the area of a general surface in space.

We consider a surface given by the equation

$$z = F(x, y), \qquad (x, y) \text{ in } R \tag{13-90}$$

where R is a standard region in the xy-plane (Figure 13-36). We assume that F has continuous first partial derivatives in an open region containing R. We might attempt to find the area of this surface by inscribing polyhedra, just as we found arc length by inscribing polygons. However, this process has been found to be unusable (in general, it does not lead to a limit). Hence, some other way must be found.

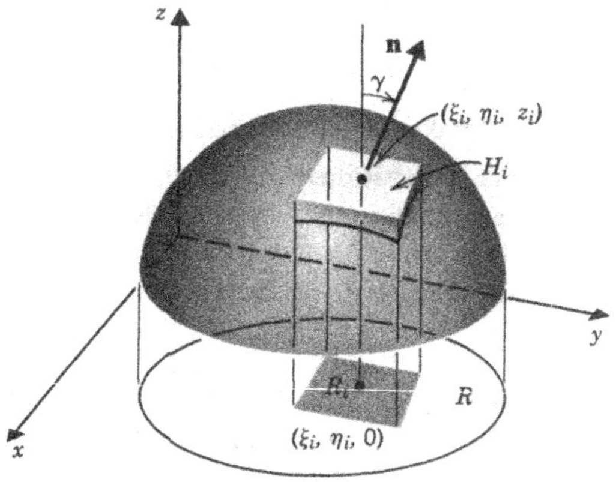

Figure 13-36 Surface area.

The method we choose is based on the approximation of the surface, near a point, by the tangent plane. We subdivide our region R by means of parallels to the axes and consider only those subregions not meeting the boundary. Each subregion R_i considered is then a rectangle of sides $\Delta_i x$, $\Delta_i y$. We choose (ξ_i, η_i) in R_i and construct the tangent plane to the surface at the corresponding point P_i: (ξ_i, η_i, z_i), $z_i = F(\xi_i, \eta_i)$. We then ask for the area of the portion of the tangent plane above R_i; that is, the area of the parallelogram H_i in the tangent plane whose projection on the xy-plane is R_i. The equation of the tangent plane (Section 12-18) is

$$z - z_i = F_x(x - \xi_i) + F_y(y - \eta_i)$$

where F_x and F_y are evaluated at (ξ_i, η_i). From the meaning of the partial derivatives, it follows that the parallelogram H_i has sides which, when properly directed, represent the vectors

$$\mathbf{u} = \Delta_i x\, \mathbf{i} + F_x\, \Delta_i x\, \mathbf{k}, \qquad \mathbf{v} = \Delta_i y\, \mathbf{j} + F_y\, \Delta_i y\, \mathbf{k}$$

Accordingly, as in Section 11-15, the area of H_i is $\|\mathbf{u} \times \mathbf{v}\|$. Now $\mathbf{u} \times \mathbf{v} = \Delta_i x\, \Delta_i y(-F_x \mathbf{i} - F_y \mathbf{j} + \mathbf{k})$. Therefore

$$\text{area of } H_i = \Delta_i x\, \Delta_i y\, \sqrt{F_x^{\,2} + F_y^{\,2} + 1}$$

and the total area of all parallelograms is

$$\sum_{i=1}^{n} \sqrt{[F_x(\xi_i, \eta_i)]^2 + [F_y(\xi_i, \eta_i)]^2 + 1}\; \Delta_i A$$

For very small subregions R_i the area of H_i should be a very good approximation to the area of the surface above R_i. Accordingly, it is reasonable to take the limit of the sums (if it exists) as the total surface area:

$$\begin{aligned} S = \text{total surface area} &= \iint_R \sqrt{[F_x(x, y)]^2 + [F_y(x, y)]^2 + 1}\, dA \\ &= \iint_R \sqrt{\left(\frac{\partial z}{\partial x}\right)^2 + \left(\frac{\partial z}{\partial y}\right)^2 + 1}\, dA \end{aligned} \tag{13-91}$$

We take this equation to be a definition of S; it can be more fully justified as the natural geometric value for surface area. For a surface of revolution, the formula (13-91) leads to the one given in Section 7-8 (Problem 2 below).

We remark that the vectors **u, v** occurring above are a basis for the base space for the tangent plane, so that $\mathbf{u} \times \mathbf{v}$ is a normal vector **n** to the surface. The cosine of the angle γ between $\mathbf{u} \times \mathbf{v}$ and **k** is

$$\cos \gamma = \mathbf{k} \cdot \frac{\mathbf{u} \times \mathbf{v}}{\|\mathbf{u} \times \mathbf{v}\|} = \frac{1}{\sqrt{F_x^{\,2} + F_y^{\,2} + 1}}$$

Therefore, we can write (13-91) as follows:

$$S = \iint_R \sec \gamma\, dA, \quad \gamma = \measuredangle(\mathbf{n}, \mathbf{k}), \quad \mathbf{n} = \text{normal at } (x, y, F(x, y)) \tag{13-91$'$}$$

Equivalently, we can write:

$$dS = \sec \gamma\, dA$$

This relation, in the form,

$$dA = \cos \gamma\, dS$$

merely states that the area of the projection on the xy-plane of a little piece of surface is the area dS of the piece of surface times the cosine of the angle between the tangent plane and the xy-plane (or between the normal **n** and **k**) (see Problem 3 below).

PROBLEMS

1. Find the surface area of each of the following surfaces:
   (a) $z = 3 + x - 2y$, $0 \le x \le 1$, $0 \le y \le x$
   (b) $z = 5 + x + y$, $0 \le y \le 1$, $y - 1 \le x \le 1 - y$
   (c) $z = (1/\sqrt{2}) \ln \cos (x - y)$, $0 \le x \le \pi/2$, $x - (\pi/4) \le y \le x$
   (d) $z = (1/\sqrt{2}) \ln \sin (x + y)$, $0 \le y \le 3\pi/4$, $(\pi/2) - y \le x \le (3\pi/4) - y$
   (e) $z = \sqrt{4 - x^2 - y^2}$, $1 \le x^2 + y^2 \le 2$
   (f) $z = \sqrt{2 - x^2 - y^2}$, $0 \le x^2 + y^2 \le 1$
   (g) $z = xy$, $x^2 + y^2 \le 1$, $x \ge 0$, $y \ge 0$
   (h) $z = x^2 - y^2$, $0 \le y \le 1/\sqrt{2}$, $y \le x \le \sqrt{1 - y^2}$
   (i) $z = 3 - x^2 - y^2$, $x^2 + y^2 \le 1$, $x \ge 0$, $y \ge 0$

(j) $z = \sqrt{x^2 + y^2}$, $0 \le x^2 + y^2 \le 1$

(k) $z = \frac{1}{4}(3x + 5\ln y)$, $0 \le x \le 1$, $1 \le y \le 2$

(l) $z = (\frac{2}{3})[(x^3/a)^{1/2} + (y^3/b)^{1/2}]$, for (x, y) in the triangle formed by the x- and y-axes and the line $(x/a) + (y/b) = \lambda$, where $a > 0$, $b > 0$, and $\lambda > 0$.

2. (a) Show that for a surface $z = G(r, \theta)$, (r, θ) in $R_{r\theta}$, given in cylindrical coordinates (r, θ, z), the surface area is given by

$$S = \iint_{R_{r\theta}} \sqrt{G_r^2 + \frac{1}{r^2} G_\theta^2 + 1}\, r\, dr\, d\theta$$

(b) Show that, in cylindrical coordinates, an equation $z = G(r)$, $a \le r \le b$, $0 \le \theta \le 2\pi$, describes a surface of revolution and, by (a), its surface area is

$$S = 2\pi \int_a^b \sqrt{G_r^2 + 1}\, r\, dr$$

as in Section 7-8 (Equation (7-87)).

(c) Find the area of the surface $z = (a^2 \cos\theta + b^2 \sin\theta)/r$, $\alpha \le \theta \le \beta$, $h \le r \le k$. (*Hint.* set $c^4 = a^4 + b^4$, $u^2 = c^4 + r^4$ in the integral with respect to r.)

(d) Find the area of the surface $z = ar + b\theta$, $\alpha \le \theta \le \beta$, $0 \le r \le c$.

‡3. Let triangle $P_1P_2P_3$ lie in plane H_0. By the (orthogonal) projection of the triangle on plane H_1, $\mathrm{Proj}_{H_1} P_1P_2P_3$, we mean the triangle $Q_1Q_2Q_3$, where Q_k is the foot of the perpendicular to H_1 through P_k $(k = 1, 2, 3)$.

(a) Show by vector geometry (Section 11-15) that the area of $\mathrm{Proj}_{H_1} P_1P_2P_3$ equals $|\cos\varphi|$ times the area of $P_1P_2P_3$, where $\varphi = \measuredangle(H_0, H_1)$—the angle between the normal vectors to H_0 and H_1.

(b) Let H_1, H_2, H_3 be the yz-, xz-, and xy-planes, respectively. Let A be the area of $P_1P_2P_3$, A_k the area of $\mathrm{Proj}_H\, P_1P_2P_3$ $(k = 1, 2, 3)$. Prove that

$$A = \sqrt{A_1^2 + A_2^2 + A_3^2}$$

(*Hint.* Show that A_1, A_2, A_3 are related to the components of the vector $\overrightarrow{P_1P_2} \times \overrightarrow{P_1P_3}$.)

4. *Surface area for surfaces in parametric form.* In Section 12-18 it is pointed out that, under proper assumptions, equations

$$(*)\quad x = f(u, v), \qquad y = g(u, v), \qquad z = h(u, v), \qquad (u, v) \text{ in } R_{uv}$$

can be considered as parametric equations of a surface in xyz-space. Near a particular point (u_0, v_0) we reason, as in the derivation of (13-91), that the mapping (*) can be approximated by linear equations $dx = f_u\, du + f_v\, dv, \ldots$. The linear equations map a rectangle of sides du, dv onto a parallelogram whose sides, properly directed, represent the vectors $\mathbf{w}_1\, du$, $\mathbf{w}_2\, dv$, where

$$\mathbf{w}_1 = f_u\mathbf{i} + g_u\mathbf{j} + h_u\mathbf{k}, \qquad \mathbf{w}_2 = f_v\mathbf{i} + g_v\mathbf{j} + h_v\mathbf{k}$$

(see Problem 8 following Section 12-18). Hence the parallelogram has area $du\, dv\, \|\mathbf{w}_1 \times \mathbf{w}_2\|$.

(a) Show that the reasoning described leads to the following formula for surface area.

$$S = \iint_{R_{uv}} \sqrt{\left(\frac{\partial(y, z)}{\partial(u, v)}\right)^2 + \left(\frac{\partial(z, x)}{\partial(u, v)}\right)^2 + \left(\frac{\partial(x, y)}{\partial(u, v)}\right)^2}\, du\, dv$$

Remark. This formula is the 2-dimensional analogue of the formula for arc length:

$$L = \int_a^b \sqrt{[x'(t)]^2 + [y'(t)]^2}\, dt$$

(b) The equations $x = a \sin \varphi \cos \theta$, $y = a \sin \varphi \sin \theta$, $z = a \cos \varphi$, $0 \le \varphi \le \pi$, $0 \le \theta \le 2\pi$, are parametric equations of a spherical surface of radius a; here the parameters φ, θ are *spherical coordinates*, as in Section 11-20. Show that in terms of the spherical coordinates, the area of a portion of the spherical surface is given by

$$S = a^2 \iint_{R_{\varphi\theta}} \sin \varphi\, d\varphi\, d\theta$$

Thus the element of area in spherical coordinates is $dS = a^2 \sin \varphi\, d\varphi\, d\theta$.

(c) The equations $x = (2 + \cos u) \cos v$, $y = (2 + \cos u) \sin v$, $z = \sin u$, $0 \le u \le 2\pi$, $0 \le v \le 2\pi$, are parametric equations for the surface of a torus. Find the element of surface area in terms of u and v.

5. To find the length of a curve $y = f(x)$, $a \le x \le b$, where f' is continuous in the interval, one can reason as in the derivation of (13-91): one subdivides the interval as for the definite integral and approximates the length of the curve by the sum of the lengths of line segments $y = m_i x + b_i$, $x_{i-1} \le x \le x_i$, where each segment is tangent to the curve at $(\xi_i, f(\xi_i))$, $x_{i-1} \le \xi_i \le x_i$. Show that this leads to the sum

$$\sum_{i=1}^{n} \sqrt{1 + [f'(\xi_i)]^2}\, \Delta_i x$$

and, hence, to the integral $\int_a^b \sqrt{1 + [f'(x)]^2}\, dx$ as in Theorem 28, Section 4-27.

13-10 OTHER APPLICATIONS OF MULTIPLE INTEGRALS

Thus far we have considered applications of multiple integrals related to volume, surface area, mass, and density. We remark that they can also be applied to area in the plane. For, if we take $f(x, y) \equiv 1$, then our double integral becomes

$$\iint_R 1\, dA = \iint_R dA = \lim \sum \Delta_i A = \text{Area of } R \qquad (13\text{-}100)$$

For a region R described in polar coordinates by the inequalities $\alpha \le \theta \le \beta$,

$0 \leq r \leq F(\theta)$, we find that, as in Section 13-4,

$$\begin{aligned} \text{Area of } R &= \iint\limits_R dA = \int_\alpha^\beta \int_0^{F(\theta)} r\,dr\,d\theta = \int_\alpha^\beta \frac{r^2}{2}\Big|_0^{F(\theta)} d\theta \\ &= \frac{1}{2}\int_\alpha^\beta [F(\theta)]^2\,d\theta \end{aligned} \tag{13-101}$$

This result was found previously in Section 7-2.

In the same way, we deduce that in space

$$\iiint\limits_R 1\,dV = \iiint\limits_R dV = \text{Volume of } R \tag{13-102}$$

Before turning to other applications, we point out an extension of the basic formula for a double integral as a limit. Let $f(x, y)$ and $g(x, y)$ be continuous in the standard region R_{xy}, let $F(u, v)$ be a function continuous on a set E in the uv-plane, and such that the composite function $F[f(x, y), g(x, y)]$ is defined in R_{xy}. Then

$$\iint\limits_{R_{xy}} F[f(x, y), g(x, y)]\,dA = \lim_{\text{mesh}\to 0} \sum_{i=1}^{n} F[f(\xi_i', \eta_i'), g(\xi_i'', \eta_i'')]\,\Delta_i A \tag{13-103}$$

where (ξ_i', η_i') and (ξ_i'', η_i'') are in R_i.

Under the hypotheses stated, the composite function $F[f(x, y), g(x, y)] = G(x, y)$ is continuous in R_{xy} and, if (ξ_i', η_i') were required to coincide with (ξ_i'', η_i''), then (13-103) would be the same as

$$\iint\limits_{R_{xy}} G(x, y)\,dA = \lim_{\text{mesh}\to 0} \sum_{i=1}^{n} G(\xi_i, \eta_i)\,\Delta_i A$$

as usual. However, by allowing (ξ_i', η_i') and (ξ_i'', η_i'') to be different (but close together, since both are in R_i and the mesh is approaching zero), it is not at once clear that (13-103) holds true. We met the same question for definite integrals—for example, in Section 4-27, in discussing arc length. In Section 12-25 (Theorem L) a proof of the counterpart of (13-103) for definite integrals is given. The proof can be repeated with minor changes for double integrals. The rule and proof can also be extended to integrals involving more than two functions f and g—for example, to $F[f(x, y), g(x, y), h(x, y)]$.

The importance of (13-103) will be seen in several applications given below.

Moments. Let mass be distributed over the standard region R, with continuous density $f(x, y)$ as in Section 13-8. We are led to definitions of the moments M_{kx} and M_{ky} by a reasoning similar to that of Section 7-10. For M_{kx} we subdivide R into regions R_i, as in Figure 13-37. By the Mean Value Theorem (Theorem 7, Section 13-8), R_i has mass

$$\iint\limits_{R_i} f(x, y)\,dA = f(\xi_i', \eta_i')\,\Delta_i A, \qquad (\xi_i', \eta_i') \text{ in } R_i$$

The kth moment M_{kx} of the mass in R_i should be equal to the mass times some average value of x^k in R_i—less than the maximum x^k and greater than

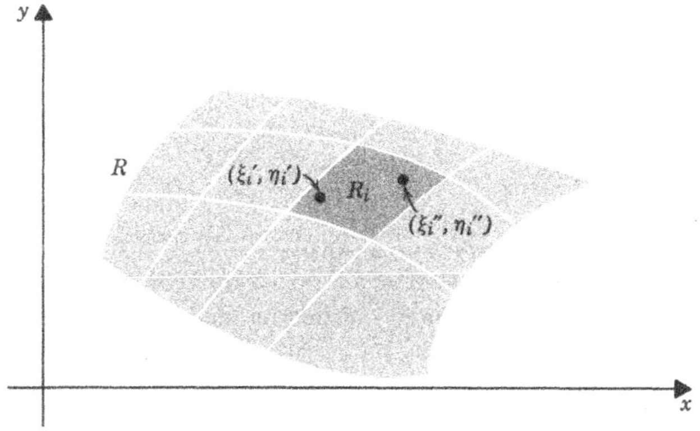

Figure 13-37 Definition of moments.

the minimum x^k in R_k. Hence, this moment should equal the mass times $\xi_i''^k$, where (ξ_i'', η_i'') is an appropriately chosen point in R_i (Figure 13-37). Therefore, the total x-moment equals

$$\sum_{i=1}^{n} \xi_i''^k f(\xi_i', \eta_i')\, \Delta_i A$$

By (13-103), with $g(x, y) \equiv x^k$ and $F(u, v) = uv$, as the mesh approaches zero, this sum has as limit the double integral of $x^k f(x, y)$ over R. Therefore, this double integral is the desired moment:

$$M_{kx} = \iint_R x^k f(x, y)\, dA \qquad (k = 0, 1, 2, \ldots) \qquad (13\text{-}104')$$

In the same way, we are led to the kth y-moment:

$$M_{ky} = \iint_R y^k f(x, y)\, dA \qquad (k = 0, 1, 2, \ldots) \qquad (13\text{-}104'')$$

As in Section 7-10, these equations are regarded as definitions of the moments.

Center of Mass. As in Section 7-10, the center of mass $(\bar{x}, \bar{y})$ is a point at which all mass can be placed and yield the same first moments M_{1x} and M_{1y}. Hence

$$M_0\bar{x} = M_{1x}, \qquad M_0\bar{y} = M_{1y}$$

or

$$\bar{x} = \frac{\iint_R x f(x, y)\, dA}{\iint_R f(x, y)\, dA}, \qquad \bar{y} = \frac{\iint_R y f(x, y)\, dA}{\iint_R f(x, y)\, dA} \qquad (13\text{-}105)$$

It is common practice to abbreviate $f(x, y)\, dA$, where f is density, by dm and to call this expression the *mass element;* also to write just one integral sign, since the context makes the meaning clear. Thus one writes

$$\bar{x} = \frac{\int_R x\, dm}{\int_R dm}, \qquad \bar{y} = \frac{\int_R y\, dm}{\int_R dm} \tag{13-105'}$$

With this notation, the formulas for center of mass in various dimensions all become the same. The more general concept of a Stieltjes integral allows one to apply these formulas to both discrete and continuous mass distributions, and to mixtures of the two types.

As the universal character of the formulas suggests, the center of mass has a location independent of the coordinate system; this is proved for various cases in Chapter 7, and the proofs generalize immediately to the case of (13-105) or (13-105′).

Moment of Inertia. The moment of inertia of our mass distribution in the plane about a line L in the plane is defined as the value of M_{2y}, when Cartesian coordinates are chosen with L as x-axis. Thus the moment of inertia about L is

$$I_L = M_{2y} = \iint_R y^2 f(x, y)\, dA = \int_R y^2\, dm \tag{13-106}$$

We can even write, more geometrically,

$$I_L = \int_R d_L{}^2\, dm \tag{13-106'}$$

where d_L is the distance from a general point to L ($d_L = |y|$ when L is the x-axis) (see Figure 13-38). The Parallel Axis Theorem of Section 7-10 continues to hold true. We can write it as follows:

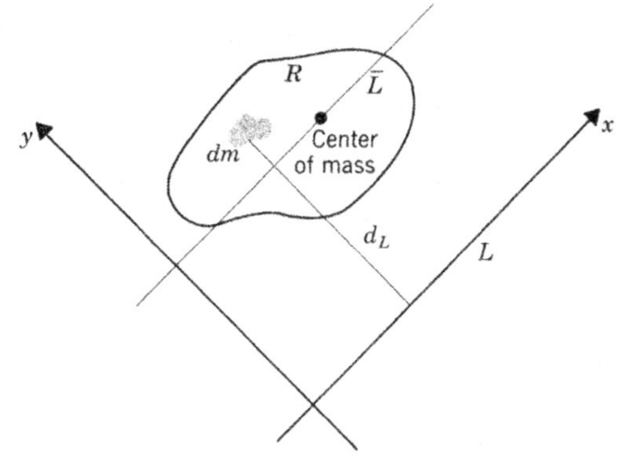

Figure 13-38 Moment of inertia.

$$I_L = I_{\bar{L}} + M_0 d^2 \tag{13-107}$$

where $\bar{L}$ is a parallel to L through the center of mass and d is the distance between L and $\bar{L}$. To prove this result, we choose L as x-axis. Then I_L

is given by (13-106) and

$$I_{\overline{L}} = \int_R (y - \overline{y})^2\,dm = \int_R y^2\,dm - 2\overline{y}\int_R y\,dm + \overline{y}^2\int_R dm$$

$$= \int_R y^2\,dm - 2\overline{y}M_0\overline{y} + M_0\overline{y}^2 = I_L - M_0\overline{y}^2$$

Since $d = |\overline{y}|$, (13-107) follows.

The *radius of gyration* of our mass distribution about L is the positive number ρ such that $M_0\rho^2 = I_L$. Thus, when L is the x-axis,

$$\rho_L{}^2 = \rho_x{}^2 = \frac{M_{2y}}{M_0} = \frac{\int_R y^2\,dm}{\int_R dm}$$

These formulas all extend to 3-dimensional space, with $dm = f(x, y, z)\,dV$. The distance from (x, y, z) to the z-axis is $\sqrt{x^2 + y^2}$ (Section 11-20). Hence, from (13-106′), we obtain the *moment of inertia about the z-axis L as*

$$I_L = \int_R (x^2 + y^2)\,dm = \int_R (x^2 + y^2)f(x, y, z)\,dV \tag{13-108}$$

Mass Distributions on Surfaces. For a surface as in (13-90) we can consider a continuous distribution of mass on the surface. This would lead us to a density in mass per unit of surface area. Let $\delta = f(x, y, z)$ be the density function; this is a function of three variables, but it is defined only on the surface; that is, for $z = F(x, y)$. We assume that $f(x, y, F(x, y))$ is continuous in R. If we subdivide R as above, the mass of the portion of surface above R_i should be the area $\Delta_i S$ of this portion times the mean density. From (13-91) and the Mean Value Theorem,

$$\Delta_i S = \iint_{R_i} \sqrt{F_x{}^2 + \cdots}\;dA = \sqrt{[F_x(\xi_i'', \eta_i'')]^2 + \cdots}\;\Delta_i A$$

Hence, the total mass is

$$\sum_{i=1}^{n} f(\xi_i', \eta_i', F(\xi_i', \eta_i'))\sqrt{[F_x(\xi_i'', \eta_i'')]^2 + [F_y(\xi_i'', \eta_i'')]^2 + 1}\;\Delta_i A$$

By (13-103) this sum has as limit a double integral, and we can write:

$$\text{total mass} = \iint_R f(x, y, F(x, y))\sqrt{[F_x(x, y)]^2 + [F_y(x, y)]^2 + 1}\;dA \tag{13-109}$$

The expression $f(x, y, F(x, y))\sqrt{F_x{}^2 + F_y{}^2 + 1}\,dA$ can also be interpreted as dm, as above. One can then obtain expressions for moments of a mass distribution on a surface $\mathcal{S}$:

$$M_{kx} = \int_{\mathcal{S}} x^k\,dm = \iint_R x^k f(x, y, F(x, y))\sqrt{F_x{}^2 + F_y{}^2 + 1}\;dA$$

$$M_{ky} = \int_{\mathcal{S}} y^k\,dm, \qquad M_{kz} = \int_{\mathcal{S}} z^k\,dm$$

The center of mass $(\bar{x}, \bar{y}, \bar{z})$ is then given by

$$\bar{x} = \frac{M_{1x}}{M_0} = \frac{\int\limits_{\mathcal{S}} x\, dm}{\int\limits_{\mathcal{S}} dm}, \qquad \bar{y} = \frac{\int\limits_{\mathcal{S}} y\, dm}{\int\limits_{\mathcal{S}} dm}, \qquad \bar{z} = \frac{\int\limits_{\mathcal{S}} z\, dm}{\int\limits_{\mathcal{S}} dm}$$

The integral on the right of (13-109) is also written as

$$\iint\limits_{\mathcal{S}} f(x, y, z)\, dS$$

Regardless of the interpretation of f as density, such an integral is called a *surface integral*. When $\mathcal{S}$ is not of the type $z = F(x, y)$ considered here, one can usually decompose it into several pieces of one of the types $z = F(x, y)$, $x = F(y, z)$, or $y = F(x, z)$ and then can proceed as above to replace the surface integral by a double integral.

PROBLEMS

1. For each of the following regions in the xy-plane (with r, θ as polar coordinates), represent the area as a double integral and evaluate:

(a) $1 \le y \le 2,\ -y^2 \le x \le e^y$ (b) $1 \le x + y \le 2,\ x \ge 0,\ y \ge 0$

(c) $0 \le r \le \sin^2 4\theta$ (d) $1 \le r \le 2,\ r \le e^\theta \le e^r$

2. Mass is spread over region R with density $f(x, y)$. Find the mass and center of mass for each of the following cases:

(a) $f(x, y) = |x| + |y|$, R: the square of vertices $(\pm 1, \pm 1)$

(b) $f(x, y) = x^2$, R: $x^2 + y^2 \le 1$

(c) $f(x, y) = (x^2 + y^2)^2$, R: $x \ge 0,\ y \ge 0,\ x^2 + y^2 \le 1$

(d) $f(x, y) = x/\sqrt{x^2 + y^2}$, $f(0, 0) = 0$, R: $x \ge 0,\ y \ge 0,\ 1 \le x^2 + y^2 \le 2$

3. (a) . . . (d) For each of the mass distributions of Problem 2, find the moment of inertia about the x-axis.

4. (a) In Section 7-10 the following formulas are given:

$$M_{kx} = \int_a^b x^k f(x)\, \delta(x)\, dx, \qquad M_{ky} = \frac{1}{k+1} \int_a^b [f(x)]^{k+1} \delta(x)\, dx$$

for mass distributed over the region $a \le x \le b$, $0 \le y \le f(x)$ with density $\delta(x)$. Derive these formulas with the aid of double integrals.

(b) In Problem 3 following Section 7-12 the following formulas are given:

$$M_{kx} = \frac{\delta}{k+2} \int_\alpha^\beta [f(\theta)]^{k+2} \cos^k \theta\, d\theta, \qquad M_{ky} = \frac{\delta}{k+2} \int_\alpha^\beta [f(\theta)]^{k+2} \sin^k \theta\, d\theta$$

for mass distributed over the region $\alpha \le \theta \le \beta$, $0 \le r \le f(\theta)$ with constant density δ. Derive these formulas with the aid of double integrals and generalize to the case when δ depends on θ.

5. Find the center of mass of the mass distribution on a surface as specified:
   (a) A hemisphere of radius a with constant density.
   (b) A hemisphere of radius a with density proportional to the distance from the diametral plane.
   (c) The hyperbolic paraboloid $z = xy$, $0 \le x \le 1$, $0 \le y \le 1$ with density $\delta = (1 + x^2 + y^2)^{-1/2}$.

6. *Double integrals of vector functions.* Let $\mathbf{v} = f(x, y)\mathbf{i} + g(x, y)\mathbf{j} = \mathbf{F}(x, y)$ be a vector function (or vector field) defined in the standard region R. Let f and g be continuous, so that the vector function $\mathbf{F}$ is also continuous. We define the double integral of $\mathbf{F}$ by the equation

$$\iint_R \mathbf{F}(x, y)\, dA = \lim_{\text{mesh}\to 0} \sum_{i=1}^{n} \mathbf{F}(\xi_i, \eta_i)\, \Delta_i A$$

As for functions of one variable (Section 4-31), we conclude that the integral exists and equals

$$\left(\iint_R f(x, y)\, dA\right)\mathbf{i} + \left(\iint_R g(x, y)\, dA\right)\mathbf{j}$$

The basic properties of the vector double integral are now deducible from those of ordinary double integrals.

   (a) Evaluate $\iint_R (y\mathbf{i} - x\mathbf{j})\, dx\, dy$, $R\colon 0 \le x \le 1$, $0 \le y \le 1$.
   (b) Show that if a mass distribution in R has continuous density $\delta = f(x, y)$, then the center of mass P_c is given by the equations

$$\overrightarrow{OP_c} = \frac{1}{M_0} \iint_R f(x, y)\, \mathbf{r}\, dA, \qquad \mathbf{r} = \overrightarrow{OP} = x\mathbf{i} + y\mathbf{j}$$

7. *Pressure and force.* Let pressure p act normally on a flat surface, which is a region R in the xy-plane, so that $p = p(x, y)$.
   (a) Show that the total force normal to the surface is $\iint_R p(x, y)\, dA$.
   (b) How should the formula be modified for pressure on a curved surface such as the surface of a sphere?

8. *Ellipse of inertia.* Let mass be distributed in standard region R with continuous density $f(x, y)$. Let L be a line through the origin O and let $\mathbf{u} = \cos\alpha\,\mathbf{i} + \sin\alpha\,\mathbf{j}$ be a unit vector along L.
   (a) Show that $I_L = A\cos^2\alpha + B\cos\alpha\sin\alpha + C\sin^2\alpha$, where

$$A = \int_R y^2\, dm = M_{2y}, \quad B = -2I_{xy} = -2\int_R xy\, dm, \quad C = \int_R x^2\, dm = M_{2x}$$

   I_{xy} is called a *product of inertia.*
   (b) Let $B = 0$. Show that, if $C < A$, then I_L has its smallest value, C, for $\alpha = \pi/2$ (L: the y-axis). Show that, if $C = A$, then I_L has the same value for all α.
   (c) Show that if the coordinate axes are rotated through angle β to obtain new

coordinates x', y', as in Section 6-5, then

$$I_{x'y'} = \int_R x'y'\,dm = \sin\beta\cos\beta\,(A - C) + I_{xy}(\cos^2\beta - \sin^2\beta)$$

and, hence, $I_{x'y'} = 0$ provided that β is chosen so that

$$\tan 2\beta = \frac{B}{A - C}$$

Remark. The curve $Ax^2 + Bxy + Cy^2 = 1$ is an ellipse, called the *ellipse of inertia.* The result in (c) shows that by rotating the axes as in Section 6-5 (see Problem 6 following that section) to eliminate the xy-term, one is also choosing axes for which the corresponding product of inertia is 0. As in (b) the directions of the major and minor axes of the ellipse give the directions of the lines through O providing the largest and smallest moment of inertia. These directions are called the *principal axes of inertia.* Each line L meets the ellipse at a point $x = r\cos\alpha$, $y = r\sin\alpha$ for which $1 = Ax^2 + Bxy + Cy^2 = r^2(A\cos^2\alpha + \cdots) = r^2 I_L$. Hence, I_L is $1/r^2$ or $1/(x^2 + y^2)$ (in the chosen system of units).

9. In each of the following exercises, first show how a triple integral can be used to find the quantity requested and then find the value.

(a) The mass of a cube $0 \le x \le a$, $0 \le y \le a$, $0 \le z \le a$ whose density is $\delta = k(xz + yz)$.

(b) The volume of the region $1 \le x^2 + y^2 + z \le 2$, $1 \le x + y \le 2$, $1 \le x \le 2$.

(c) The center of mass of the cube of part (a).

(d) The mean temperature in a spherical region $x^2 + y^2 + z^2 \le a^2$ if the temperature is $T = k(x^2 + y^2 + z^2)^{1/2}e^{-(x^2+y^2+z^2)/a^2}$

‡(e) The total gravitational potential at a point P, b units from the center of a solid sphere of radius a ($a < b$), having density proportional to the square of the distance from the center. (*Hint.* From physics, the potential at P of a particle of mass m at Q is $-km/\|\overrightarrow{PQ}\|$, where k is the constant of gravitation.)

‡(f) The total volume of an incompressible fluid crossing the sphere $x^2 + y^2 + z^2 = a^2$ between times t_1 and t_2, given that the velocity field is $c(3xt\mathbf{i} + 2yt\mathbf{j} - 5zt\mathbf{k})$.

10. The triple integral over a region R in R^3 of a vector function $\mathbf{F}(x, y, z) = f(x, y, z)\mathbf{i} + g(x, y, z)\mathbf{j} + h(x, y, z)\mathbf{k}$, where f, g, h are continuous in the standard region R, is defined as follows:

$$\iiint_R \mathbf{F}(x, y, z)\,dV = \iiint_R f(x, y, z)\,dV\,\mathbf{i} + \iiint_R g(x, y, z)\,dV\,\mathbf{j} + \iiint_R h(x, y, z)\,dV\,\mathbf{k}$$

(a) Prove: if $\mathbf{a}$ is a constant vector function, then

$$\iiint_R (\mathbf{a}\cdot\mathbf{F})\,dV = \mathbf{a}\cdot\iiint_R \mathbf{F}\,dV$$

and

$$\iiint_R (\mathbf{a}\times\mathbf{F})\,dV = \mathbf{a}\times\iiint_R \mathbf{F}\,dV$$

‡(b) Justify the formula

$$cm \iiint_R \delta(x, y, z) \frac{\mathbf{r}}{r^3}\, dV \qquad (\mathbf{r} = x\mathbf{i} + y\mathbf{j} + z\mathbf{k},\ r = \|\mathbf{r}\|)$$

for the gravitational force exerted by a solid of density $\delta(x, y, z)$ on a particle of mass m at the origin (not in R), where c is the constant of gravitation. Find this force for a homogeneous sphere of radius a, whose center is at a distance b from the origin.

(c) The *linear momentum* of a particle of mass m and velocity $\mathbf{v}$ is $m\mathbf{v}$. Justify the formula $\iiint_R \delta(x, y, z)\mathbf{v}(x, y, z)\, dV$ for the total linear momentum of a solid of density δ and velocity $\mathbf{v}$ at (x, y, z). Evaluate this expression for a homogeneous solid sphere of radius a and velocity $\mathbf{v} = \boldsymbol{\omega} \times \mathbf{r}$, where $\boldsymbol{\omega}$ is constant and $\mathbf{r} = x\mathbf{i} + y\mathbf{j} + z\mathbf{k}$; here the sphere is rotating about a line L through the origin, $\boldsymbol{\omega}$ is a vector along L and $\|\boldsymbol{\omega}\|$ is the angular velocity.

(d) The angular momentum about O of a particle P of mass m and velocity $\mathbf{v}$ is $\mathbf{r} \times m\mathbf{v}$, where $\overrightarrow{OP} = \mathbf{r}$. Justify the formula $\iiint_R \delta(x, y, z)\mathbf{r} \times \mathbf{v}\, dV$ for the total angular momentum about O of a solid in motion and evaluate for the rotating solid sphere of part (c).

‡11. We use index notation for coordinates and vectors in space. Let a solid body fill the standard region R in space, with continuous density f. Let L: $\overrightarrow{OP} = t\mathbf{u}$ be a line through the origin O, where $\mathbf{u}$ is a unit vector.

(a) Show that $I_L = \sum_{i,j=1}^{3} \gamma_{ij} u_i u_j$, where $\gamma_{ii} = \int_R (x_1^2 + x_2^2 + x_3^2 - x_i^2)\, dm$ and, for $i \neq j$, $\gamma_{ij} = -\int_R x_i x_j\, dm$. The matrix $\Gamma = (\gamma_{ij})$ is called the *inertia matrix;* one also calls Γ the *inertia tensor,* since the particular matrix depends on our choice of coordinate axes in space and this dependence follows the rules of tensor analysis [see Part (c)].

(b) Regard $\mathbf{u}$ as a column vector, whose transpose $\mathbf{u}'$ is a row vector. Show that $I_L = \mathbf{u}'\Gamma\mathbf{u}$.

(c) Let new Cartesian coordinates (y_1, y_2, y_3) be introduced in space by the equation $\mathbf{x} = A\mathbf{y}$, where A is a 3 by 3 orthogonal matrix (Section 11-21). Show that the inertia matrix Γ_x in the x-coordinates and the matrix Γ_y in the y-coordinates are related by the equation $\Gamma_y = A'\Gamma_x A$.

13-11 LINE INTEGRALS

The concept of line integral is introduced in Chapter 7 for certain special cases; in particular, the line integral

$$\frac{1}{2} \oint_C -y\, dx + x\, dy$$

is discussed and is shown to represent the area enclosed by the simple closed curve C. We now take up line integrals in general. In the following section we shall see that they are closely related to double integrals.

Let C be a path in the xy-plane given by equations

$$x = f(t), \qquad y = g(t), \qquad a \le t \le b \tag{13-110}$$

where f and g have continuous first derivatives. Thus C is a smooth path. We allow C to cross itself, as in Figure 13-39. We can assign a direction to C:

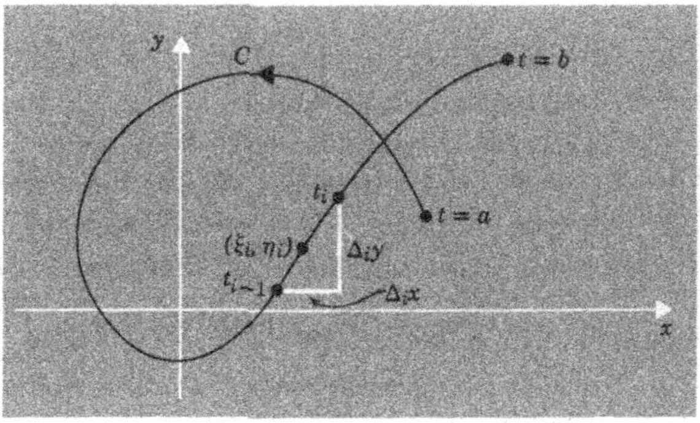

Figure 13-39 Definition of line integral.

either that of increasing t or that of decreasing t. Unless otherwise indicated, we shall choose the direction of increasing t. We also define arc length s along C, for example, by the equation

$$s = \int_a^t \sqrt{[f'(u)]^2 + [g'(u)]^2}\, du$$

We recall that on a given path we can replace the given parameter t by another one by an equation $t = \varphi(\tau)$, $\alpha \le \tau \le \beta$, where φ' is continuous and has no zeros. The path with the parameter τ is said to be *equivalent* to the old one (see Section 4-29). When $ds/dt > 0$ for $a \le t \le b$, t is related to s by such an equation $t = \varphi(s)$ and, hence, one can use s as parameter to obtain an equivalent path.

Now let $P(x, y)$ be a function defined at all points of C; generally, P will be defined in an open region containing C, but for the definition to follow $P(x, y)$ need only be defined on a set E including all points of C. We also assume that $P(x, y)$ is continuous on E.

We now subdivide the interval $[a, b]$ as for the definite integral:

$$a = t_0 < t_1 < \cdots < t_n = b$$

We let $x_i = f(t_i)$, $y_i = g(t_i)$ and

$$\Delta_i t = t_i - t_{i-1},\ \Delta_i x = x_i - x_{i-1},\ \Delta_i y = y_i - y_{i-1},\ i = 1, \ldots, n$$

as in Figure 13-39. We choose ζ_i in the interval $[t_{i-1}, t_i]$ and write

$$\xi_i = f(\zeta_i), \qquad \eta_i = g(\zeta_i)$$

so that (ξ_i, η_i) is a point on the path between t_{i-1} and t_i (Figure 13-39). We form the sum

$$\sum_{i=1}^{n} P(\xi_i, \eta_i)\, \Delta_i x$$

and seek the limit of these sums as the mesh of the subdivision of $[a, b]$ approaches 0. If that limit exists, it is called the *line integral* of P *with respect to* x *along* C and we write

$$\int_C P(x, y)\, dx = \lim_{\text{mesh}\to 0} \sum_{i=1}^{n} P(\xi_i, \eta_i)\, \Delta_i x \tag{13-111}$$

It should be noted that the parameter along C enters here only in helping us to subdivide the path and in giving a meaning to "mesh $\to$ 0." It follows that a change of parameter on C has no effect on the line integral. This statement will be illustrated and proved in detail below.

If $Q(x, y)$ is continuous on a set E containing C, the line integral $\int Q\, dy$ along C is defined similarly:

$$\int_C Q(x, y)\, dy = \lim_{\text{mesh}\to 0} \sum_{i=1}^{n} Q(\xi_i, \eta_i)\, \Delta_i y$$

Most commonly, one is considering a sum of line integrals $\int P\, dx$ and $\int Q\, dy$; one writes such a sum as

$$\int_C P\, dx + Q\, dy$$

THEOREM 12. *Let C be a smooth path in the xy-plane and let $P(x, y)$ and $Q(x, y)$ be continuous on a set including C. Then the line integral*

$$\int_C P\, dx + Q\, dy$$

exists. If C has parametric equations $x = f(t)$, $y = g(t)$, $a \le t \le b$, then

$$\int_C P\, dx + Q\, dy = \int_a^b \left[P(x, y)\frac{dx}{dt} + Q(x, y)\frac{dy}{dt} \right] dt \tag{13-112}$$

where in the integral on the right $x = f(t)$ and $y = g(t)$; that is,

$$\int_C P(x, y)\, dx + Q(x, y)\, dy \\ = \int_a^b \{P[f(t), g(t)]f'(t) + Q[f(t), g(t)]g'(t)\}\, dt \tag{13-112$'$}$$

PROOF. We consider a subdivision of $[a, b]$ as above. Let $F(t) = P[f(t), g(t)]$. Then the sum $\Sigma P(\xi_i, \eta_i)\, \Delta_i x$ can be written

$$\sum_{i=1}^{n} F(\zeta_i)[f(t_i) - f(t_{i-1})] = \sum_{i=1}^{n} F(\zeta_i) f'(\zeta_i')\, \Delta_i t$$

Here we have applied the ordinary Mean Value Theorem to $f(t_i) - f(t_{i-1})$. As the mesh approaches 0, the last sum has as limit the integral

$$\int_a^b F(t)f'(t)\,dt = \int_a^b P[f(t), g(t)]f'(t)\,dt$$

(see Theorem L in Section 12-25). Hence, $\int_C P\,dx$ exists and has as value the integral just obtained. Similarly,

$$\int_C Q\,dy = \int_a^b Q[f(t), g(t)]g'(t)\,dt$$

If we add these results we obtain (13-112′) and, hence, (13-112); at the same time we have proved existence of the line integral $\int P\,dx + Q\,dy$ along C.

EXAMPLE 1. Let C be the path $x = \cos t$, $y = \sin t$, $0 \leq t \leq \pi/2$. Let $P(x, y) = -y^2e^x$, let $Q(x, y) = xye^x$. Then

$$\int_C P\,dx + Q\,dy = \int_0^{\pi/2} (\sin^3 t\, e^{\cos t} + \sin t \cos^2 t\, e^{\cos t})\,dt$$

$$= \int_0^{\pi/2} e^{\cos t} \sin t\,dt = -e^{\cos t}\Big|_0^{\pi/2} = e - 1$$

Line Integral as Work. We write

$$\mathbf{F} = P\mathbf{i} + Q\mathbf{j}$$

and interpret $\mathbf{F}$ as a force vector, acting on a particle moving in the given direction along C. At each point (x, y) on C the vector $\mathbf{P}(x, y)\mathbf{i} + Q(x, y)\mathbf{j}$ is specified, so that our particle is subject to a force varying with position (Figure

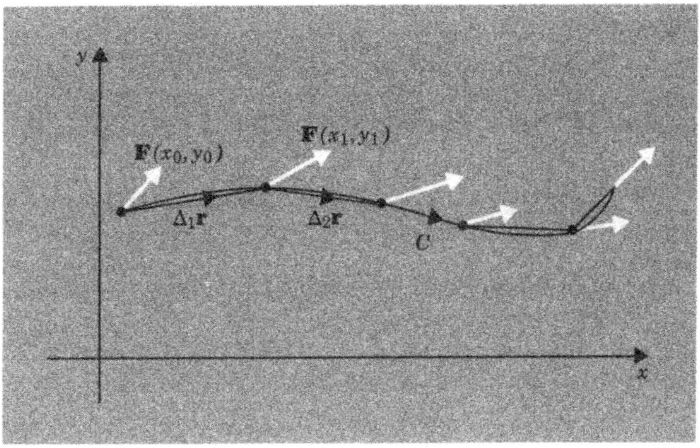

Figure 13-40 Work.

13-40). We also write, for a subdivision, as above,

$$\Delta_i\mathbf{r} = \Delta_i x\mathbf{i} + \Delta_i y\mathbf{j}, \qquad i = 1, \ldots, n$$

so that $\Delta_i\mathbf{r}$ is the displacement vector from one subdivision point to the next.

We can write

$$\int_C P\,dx + Q\,dy = \lim \sum_{i=1}^{n} P(\xi_i, \eta_i)\,\Delta_i x + Q(\xi_i, \eta_i)\,\Delta_i y$$

$$= \lim \sum_{i=1}^{n} \mathbf{F}(\xi_i, \eta_i) \cdot \Delta_i \mathbf{r}$$

If we take (ξ_i, η_i) at (x_{i-1}, y_{i-1}), as we are permitted to do, then the force vectors in the sum are as in Figure 13-40. Now the inner product of force and displacement is the work done by the force (assumed constant) during the displacement (Section 1-10). Also **F** varies continuously along C. Therefore, for a fine enough subdivision, **F** will be nearly constant along each subinterval and the path in the subinterval is approximately the line segment joining the points. Hence, the sum $\Sigma\mathbf{F}(\xi_i, \eta_i) \cdot \Delta_i\mathbf{r}$ should be a good approximation to the total work done by **F** when the particle traces C completely. Therefore, the limit of these sums, which is the line integral, should be the total work done by **F**, and we are led to the general definition:

$$\textit{work done by } \mathbf{F} = \int_C P\,dx + Q\,dy = \int_C (P\mathbf{i} + Q\mathbf{j}) \cdot (dx\,\mathbf{i} + dy\,\mathbf{j})$$
$$= \int_C \mathbf{F} \cdot d\mathbf{r} \qquad (13\text{-}113)$$

Here we have also used vector notation to abbreviate. The formula (13-113) can also be written

$$\int_C P\,dx + Q\,dy = \int_a^b \left(\mathbf{F} \cdot \frac{d\mathbf{r}}{dt}\right) dt = \int_a^b (\mathbf{F} \cdot \mathbf{v})\,dt$$

where $\mathbf{v} = d\mathbf{r}/dt$, the velocity vector of the particle. If we replace **F** by $m\mathbf{a}$ or $m\,d\mathbf{v}/dt$ and let $v = \|\mathbf{v}\|$, we reason, as in Section 7-13, that

$$\text{work done by } \mathbf{F} = \int_a^b \frac{d}{dt}\left(\frac{1}{2}mv^2\right) dt = \frac{1}{2}mv^2 \Big|_{t=a}^{t=b}$$

That is, *work equals gain in kinetic energy.*

THEOREM 13. *Under the hypotheses of Theorem* 12, *let a new parameter be introduced by the equation*

$$t = \varphi(\tau), \qquad \alpha \le \tau \le \beta$$

where φ' *is continuous and* $\varphi'(\tau) > 0$ *in* $[\alpha, \beta]$, *so that the equations*

$$x = f[\varphi(\tau)], \qquad y = g[\varphi(\tau)], \qquad \alpha \le \tau \le \beta$$

define a path C_1 *equivalent to* C. *Then*

$$\int_C P\,dx + Q\,dy = \int_{C_1} P\,dx + Q\,dy$$

PROOF. By Theorem 12,

$$\int_{C_1} P\,dx = \int_\alpha^\beta P[f\{\varphi(\tau)\}, g\{\varphi(\tau)\}] \frac{d}{d\tau} f\{\varphi(\tau)\}\,d\tau$$

$$= \int_\alpha^\beta P[f\{\varphi(\tau)\}, g\{\varphi(\tau)\}] f'\{\varphi(\tau)\}\varphi'(\tau)\,d\tau$$

If we set $t = \varphi(\tau)$ in the last integral then, as in Section 4-21, it becomes

$$\int_a^b P[f(t), g(t)]f'(t)\,dt = \int_C P(x, y)\,dx$$

Therefore

$$\int_{C_1} P\,dx = \int_C P\,dx$$

A similar argument applies to the term in dy and the theorem follows.

EXAMPLE 2. Let C be the path $x = t$, $y = 1 - t$, $0 \leq t \leq 1$. Let $P(x, y) = x^2y$, $Q(x, y) = x^2 - y$. Then

$$\int_C P\,dx + Q\,dy = \int_0^1 [t^2(1 - t)(1) + \{t^2 - (1 - t)\}(-1)]\,dt$$

$$= \int_0^1 (1 - t - t^3)\,dt = \frac{1}{4}$$

EXAMPLE 3. Let C_1 be the path $x = 2 \sin \tau$, $y = 1 - 2 \sin \tau$, $0 \leq \tau \leq \pi/6$. Then, with P and Q as in Example 2,

$$\int_{C_1} P\,dx + Q\,dy$$

$$= \int_0^{\pi/6} [4 (\sin^2 \tau)(1 - 2 \sin \tau)2 \cos \tau + (4 \sin^2 \tau - 1 + 2 \sin \tau)(-2 \cos \tau)]\,d\tau$$

$$= \int_0^{\pi/6} (1 - 2 \sin \tau - 8 \sin^3 \tau)2 \cos \tau\,d\tau = \frac{1}{4}.$$

We note that in both examples the path follows the line segment from (0, 0) to (1, 1), as in Figure 13-41. The paths are equivalent; the relationship between the parameters is $t = 2 \sin \tau$, $0 \leq \tau \leq \pi/6$. Since P and Q are the same in the two examples, the line integrals should be equal, as they are found to be.

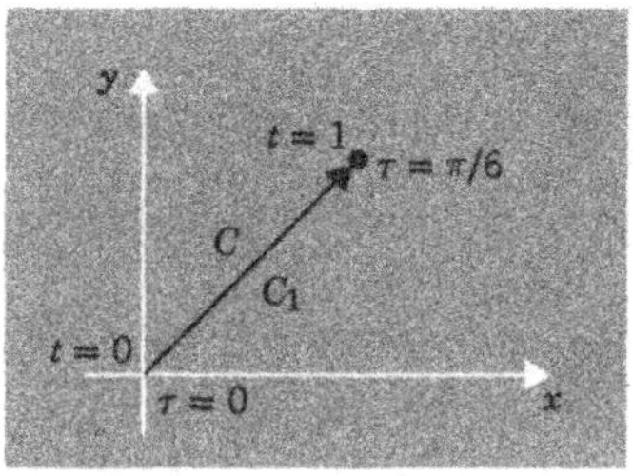

Figure 13-41 Examples 2 and 3.

As long as $ds/dt = \|\mathbf{v}\| \neq 0$ along our path, we can use s itself as parameter, with say $s = 0$ for $t = a$ and $s = L$ for $t = b$. If this is done,

$$\begin{aligned}
\int_C P\,dx + Q\,dy &= \int_0^L \left[P(x, y)\frac{dx}{ds} + Q(x, y)\frac{dy}{ds}\right] ds \\
&= \int_0^L [P(x, y)\cos\alpha + Q(x, y)\sin\alpha]\,ds \\
&= \int_0^L (P\mathbf{i} + Q\mathbf{j}) \cdot (\cos\alpha\,\mathbf{i} + \sin\alpha\,\mathbf{j})\,ds = \int_0^L \mathbf{F} \cdot \mathbf{T}\,ds \\
&= \int_0^L F_T\,ds
\end{aligned} \tag{13-114}$$

where

$$\mathbf{T} = \cos\alpha\,\mathbf{i} + \sin\alpha\,\mathbf{j} = \frac{dx}{ds}\mathbf{i} + \frac{dy}{ds}\mathbf{j} = \frac{1}{\|\mathbf{v}\|}\mathbf{v}$$

is the unit tangent vector in the chosen direction (see Section 6-6), and F_T is the tangential component of the vector function $\mathbf{F} = P\mathbf{i} + Q\mathbf{j}$. These formulas agree with the expression for work in Section 7-13. We note that $|F_T| \leq \|\mathbf{F}\|$ and, hence, we deduce the inequalities

$$\left|\int_C P\,dx + Q\,dy\right| \leq \int_0^L \sqrt{P^2 + Q^2}\,ds \leq KL \tag{13-115}$$

if $P^2 + Q^2 \leq K^2$ on C, $K \geq 0$, and L is the length of C.

Thus far we have restricted our paths to be smooth (f' and g' continuous). All results extend to the case of a piecewise smooth path: that is, a path formed of a finite number of smooth pieces, such as a path following the edges of a square. The line integral over such a path can be obtained by adding the integrals for the separate pieces or, if convenient, by one integration with respect to a suitable parameter; in the latter case, the functions dx/dt and dy/dt are usually only piecewise continuous, but this causes no difficulty. One may also use different parameters for different portions of the path.

An integral such as $\int_0^L F_T\,ds$ can be regarded as a kind of line integral. In general, one defines

$$\int_C S(x, y)\,ds = \lim \sum_{i=1}^{n} S(\xi_i, \eta_i)\,\Delta_i s$$

as above, with s as the parameter instead of t. However, as in Theorem 12,

$$\int_C S(x, y)\,ds = \int_0^L S[x(s), y(s)]\,ds$$

One can also change to an equivalent parametrization, with parameter t, $a \leq t \leq b$. Then

$$\int_C S(x, y)\, ds = \int_a^b S[f(t), g(t)] \frac{ds}{dt}\, dt$$

where $ds/dt = [\{f'(t)\}^2 + \{g'(t)\}^2]^{1/2}$ as above.

For a simple closed path C, one usually writes a line integral as

$$\oint_C P\, dx + Q\, dy \qquad \text{or} \qquad \oint_C P\, dx + Q\, dy$$

The arrows indicate the direction chosen (see Figure 13-42).

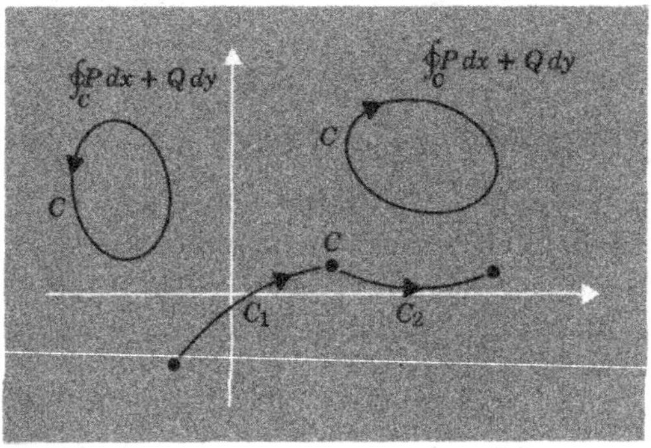

Figure 13-42 Paths for line integrals.

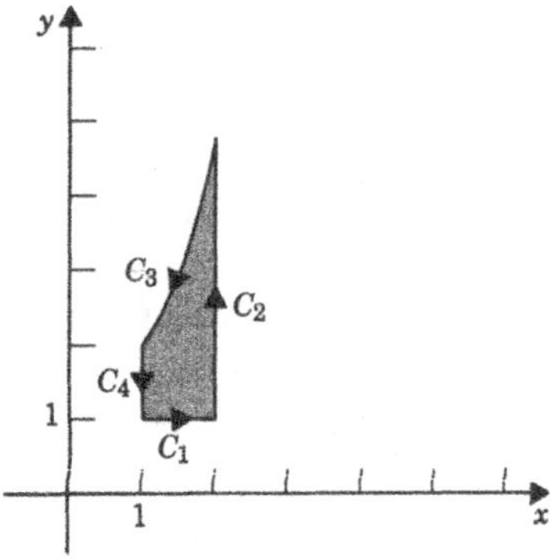

Figure 13-43 Example 4.

Here one can use a parametrization starting at any convenient point (x_0, y_0) on C: say $x = f(t), y = g(t), a \leq t \leq b$, with $f(a) = f(b) = x_0, g(a) = g(b) = y_0$.

Line integrals have several properties similar to those of definite integrals:

I. *If c_1 and c_2 are constants, then*

$$c_1 \int_C P_1\, dx + Q_1\, dy + c_2 \int_C P_2\, dx + Q_2\, dy$$
$$= \int_C (c_1P_1 + c_2P_2)\, dx + (c_1Q_1 + c_2Q_2)\, dy$$

II. *If C' is obtained from C by reversing direction, then*

$$\int_C P\, dx + Q\, dy = -\int_{C'} P\, dx + Q\, dy$$

III. *If C is formed of two parts C_1, C_2,* as in Figure 13-42, *then*

$$\int_{C_1} P\, dx + Q\, dy + \int_{C_2} P\, dx + Q\, dy = \int_C P\, dx + Q\, dy$$

These properties follow at once from (13-112), for example.

Because the value of a line integral is unaffected by change of parameter, as in Theorem 13, we tend to regard equivalent paths as the same path. For example, we think of a path following a semicircle from one end point to the other as a definite path, even though we have not specified the parameter; we can think of arc length as the parameter, or any $t = \varphi(s)$ where $\varphi' > 0$. Very often one is working with a simple path C from point A: (x_0, y_0) to point B: (x_1, y_1). One then writes the line integral as

$$\int_{\substack{A\\C}}^{B} P\,dx + Q\,dy \qquad \text{or} \qquad \int_{\substack{(x_0,y_0)\\C}}^{(x_1,y_1)} P\,dx + Q\,dy$$

and gives an equation for C. Thus for Example 2, we could write

$$\int_{\substack{(0,1)\\C}}^{(1,0)} x^2y\,dx + (x^2 - y)\,dy, \qquad \text{along } y = 1 - x.$$

We conclude the section with an example illustrating some of the points just made.

EXAMPLE 4. Evaluate

$$\int_C y^2\,dx - xy\,dy$$

where C is the boundary of the region $1 \leq x \leq 2$, $1 \leq y \leq 1 + x^2$, as in Figure 13-43. We can write the integral as the sum of the integrals along $C_1, \ldots, C_4$, as in the figure. (Each of these paths is smooth. The original path is only piecewise smooth.) We take x as parameter along C_1; that is, we represent the path as $x = t$, $y = 1$, $1 \leq t \leq 2$. Since $dy = 0$ along C_1,

$$\int_{C_1} y^2\,dx - xy\,dy = \int_{C_1} y^2\,dx = \int_1^2 dx = 1$$

(We do not bother to replace x by t.) Similarly, since $x = 2$ along C_2

$$\int_{\substack{(2,1)\\C_2}}^{(2,5)} y^2\,dx - xy\,dy = \int_1^5 (-xy)\,dy = \int_1^5 (-2y)\,dy = -24$$

Along C_3, with direction reversed, we use x as parameter,

$$\int_{\substack{(2,5)\\C_3}}^{(1,2)} y^2\,dx - xy\,dy = -\int_{\substack{(1,2)\\C_3'}}^{(2,5)} y^2\,dx - xy\,dy$$

$$= -\int_1^2 [(1 + x^2)^2 - x(1 + x^2)2x]\,dx$$

$$= -\int_1^2 (1 - x^4)\,dx = \frac{26}{5}$$

Finally

$$\int_{\substack{(1,2)\\C_4}}^{(1,1)} y^2\,dx - xy\,dy = -\int_{\substack{(1,1)\\C_4'}}^{(1,2)} (-xy)\,dy = \int_1^2 y\,dy = \frac{3}{2}$$

Therefore, the given line integral equals -16.3.

PROBLEMS

1. Evaluate the line integrals:

(a) $\int_C 2y\,dx - 3x\,dy$, $C: x = 1 - t,\ y = 2 + 3t,\ 0 \le t \le 1$.

(b) $\int_C 5y\,dx + 3x\,dy$, $C: x = 3 + 2t,\ y = 5 - t,\ 0 \le t \le 2$.

(c) $\int_{C\,(-1,1)}^{(1,1)} xy\,dx - y^2\,dy$, along the parabola $y = x^2$.

(d) $\int_{C\,(3,-1)}^{(4,-2)} \frac{y}{x}\,dx - \frac{x}{y}\,dy$ along the line $y = 2 - x$.

(e) $\int_C \frac{x\,dx + y\,dy}{x^2 + y^2}$, $C: x = \cos t,\ y = \sin t,\ 0 \le t \le 2\pi$.

(f) $\int_C xy\,dx - x^3\,dy$, $C: x = \cos t,\ y = \sin t,\ 0 \le t \le 2\pi$.

(g) $\int_C y^2\,dx + xy\,dy$, C: the triangular path from $(1, 0)$ to $(1, 1)$ to $(0, 0)$ to $(1, 0)$.

(h) $\oint_C y\,dx + 2x\,dy$, C: the boundary of the quarter-circle $x^2 + y^2 \le 1$, $-y \le x \le y,\ y \ge 0$.

(i) $\int_C 5y\,dx - 2x\,dy$, $C: x = \cos 5t,\ y = \sin 5t,\ 0 \le t \le 5\pi$.

(j) $\int_C 2y\,dx + 3x\,dy$, $C: x = \sin t + \sin 3t,\ y = \cos t - \cos 3t,\ 0 \le t \le 2\pi$.

2. Find the work done by the force $\mathbf{F}$ given on a particle moving on the path C given:

(a) $\mathbf{F} = 2\mathbf{i} - 5\mathbf{j}$, C: polygonal path from $(0, 0)$ to $(1, 1)$ to $(1, 2)$ to $(2, 2)$.

(b) $\mathbf{F} = 3\mathbf{i} + 2\mathbf{j}$, C: polygonal path from $(1, 1)$ to $(1, -1)$ to $(-1, -1)$ to $(-1, 1)$.

(c) $\mathbf{F} = x\mathbf{i}$, $C: y = x^2$ from $(0, 0)$ to $(1, 1)$.

(d) $\mathbf{F} = y\mathbf{i}$, C as in (c).

(e) $\mathbf{F} = x\mathbf{i} + y\mathbf{j}$, C: circle $x^2 + y^2 = 1$ traced once in the counterclockwise direction.

(f) $\mathbf{F} = y\mathbf{i} - x\mathbf{j}$, C as in (e).

3. Evaluate the line integrals:

(a) $\int_C xy\,ds$, $C: x = t,\ y = t,\ 0 \le t \le 1$.

(b) $\int_C (x - y)\,ds$, C as in Problem 2(a).

(c) $\int_C x^2\,ds$, C: $x = \cos 2t$, $y = \sin 2t$, $0 \le t \le 2\pi$.

(d) $\int_C \sqrt{y}\,ds$, C: $y = x^2$ from $(0, 0)$ to $(1, 1)$.

4. Verify that the paths C, C_1 are equivalent and that the line integrals along C and along C_1 are equal for each of the following cases:

(a) $\int 2y\,dx + 3x\,dy$, C: $x = t^2$, $y = 1 - t^4$, $1 \le t \le 2$, C_1: $x = 2^\tau$, $y = 1 - 4^\tau$, $0 \le \tau \le 2$.

(b) $\int 2\,dx + 3\,dy$, C: $x = t$, $y = 2/(3 - t)$, $0 \le t \le 1$, C_1: $x = (3\tau - 2)/\tau$, $y = \tau$, $2/3 \le \tau \le 1$.

5. Prove: (a) $\int_{C\,(x_0,y_0)}^{(x_1,y_1)} y\,dx + x\,dy = x_1y_1 - x_0y_0$, C: $x = f(t)$, $y = g(t)$, $a \le t \le b$.

(b) $\int_{C\,(x_0,y_0)}^{(x_1,y_1)} e^x \sin y\,dx + e^x \cos y\,dy = e^{x_1} \sin y_1 - e^{x_0} \sin y_0$, C: $x = f(t)$, $y = g(t)$, $a \le t \le b$.

Remark. In these two cases the value of the line integral depends only on the initial point (x_0, y_0) and terminal point (x_1, y_1); it does not depend on the path followed. When this happens, one says that the line integral is *independent of path* (see Section 13-14 below).

6. Prove: (a) On the path C: $y = f(x)$, $a \le x \le b$, in the direction of increasing x,

$$\int_C P(x, y)\,dx + Q(x, y)\,dy = \int_a^b [P(x, f(x)) + Q(x, f(x))f'(x)]\,dx$$

(b) On the path C: $x = g(y)$, $c \le y \le d$, in the direction of increasing y,

$$\int_C P(x, y)\,dx + Q(x, y)\,dy = \int_c^d [P(g(y), y)g'(y) + Q(g(y), y)]\,dy$$

7. Let C be a simple path from (x_0, y_0) to (x_1, y_1), where the points are distinct. Let C_1 be a path which starts at (x_0, y_0) and ends at (x_1, y_1) and follows C, but doubles back and forth a number of times: for example, as in Figure 13-44, from B_0 to B_3, then back to B_2, then ahead to B_4, then back to B_3, then ahead to B_5, then back to B_4, then ahead to B_1. Show that

$$\int_{C\,B_0}^{B_1} P\,dx + Q\,dy = \int_{C_1\,B_0}^{B_1} P\,dx + Q\,dy$$

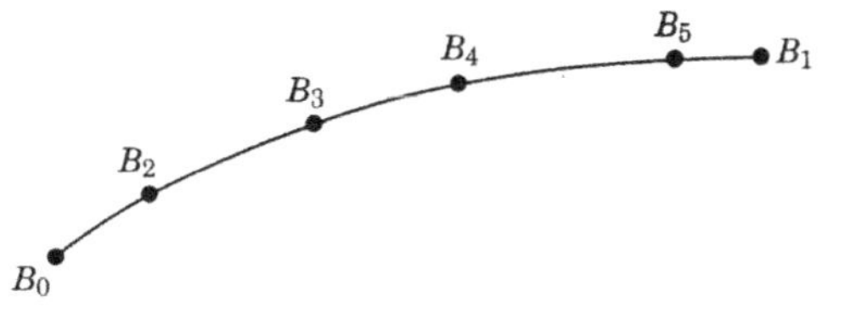

Figure 13-44 Problem 7.

13-12 GREEN'S THEOREM

The following theorem provides the fundamental relationship between line integrals and double integrals.

THEOREM 14. *(Green's theorem). Let R be a region in the xy-plane enclosed by a piecewise smooth simple closed path C. Let P(x, y) and Q(x, y) have continuous first partial derivatives in an open region containing C and R. Then*

$$\oint_C P\,dx + Q\,dy = \iint_R \left(\frac{\partial Q}{\partial x} - \frac{\partial P}{\partial y}\right) dA$$

PROOF. First let R be of the form $a \leq x \leq b$, $\varphi_1(x) \leq y \leq \varphi_2(x)$, as in Figure 13-45. Then

$$\begin{aligned}\iint_R \frac{\partial P}{\partial y}\,dA &= \int_a^b \int_{\varphi_1(x)}^{\varphi_2(x)} \frac{\partial P}{\partial y}\,dy\,dx = \int_a^b P(x, y)\Big|_{y=\varphi_1(x)}^{y=\varphi_2(x)} dx \\ &= \int_a^b [P(x, \varphi_2(x)) - P(x, \varphi_1(x))]\,dx \qquad (13\text{-}120) \\ &= \int_a^b P(x, \varphi_2(x))\,dx - \int_a^b P(x, \varphi_1(x))\,dx\end{aligned}$$

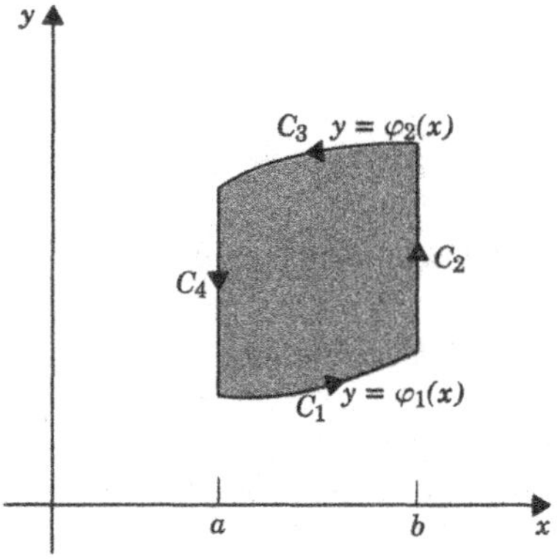

Figure 13-45 The special case for Green's theorem.

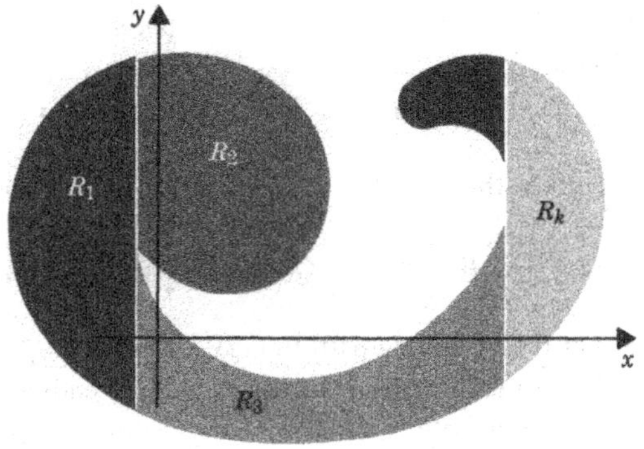

Figure 13-46 The general case for Green's theorem.

Now the line integral $\int P\,dx$ around C is the sum of the four integrals on $C_1, \ldots, C_4$. Since $dx = 0$ on C_2 and C_4, and x can be used as parameter on C_1 and C_3, we have

$$\begin{aligned}\oint_C P(x, y)\,dx &= \int_{C_1} P(x, y)\,dx + \int_{C_3} P(x, y)\,dx = \int_{C_1} P\,dx - \int_{C_3'} P\,dx \\ &= \int_a^b P(x, \varphi_1(x))\,dx - \int_a^b P(x, \varphi_2(x))\,dx\end{aligned} \qquad (13\text{-}121)$$

From (13-120) and (13-121),

$$\iint_R \left(\frac{-\partial P}{\partial y}\right) dA = \oint_C P(x, y)\, dx \tag{13-122}$$

We can now extend the rule (13-122) to any region R which can be subdivided into regions $R_1, \ldots, R_k$ of the special type of Figure 13-45, with the aid of lines $x =$ constant; see Figure 13-46. For by (13-122) for the special type of region

$$\iint_{R_i} \left(\frac{-\partial P}{\partial y}\right) dA = \oint_{C_i} P(x, y)\, dx, \qquad i = 1, \ldots, k$$

where C_i is the boundary of R_i. If we add these relations for $i = 1, \ldots, k$, the double integrals add up to the left side of (13-122) and the line integrals add up to the right side; for when we add two line integrals on paths having a portion in common, the integrals on the common part cancel by Property II of Section 13-11, and only the line integral around the outer boundary C remains. For the most general region specified in the theorem, the formula (13-122) now follows by a limit process, which we do not attempt to carry out. (The result we have proved covers all cases of practical interest.)

In the same way, we prove:

$$\iint_R \frac{\partial Q}{\partial x}\, dA = \oint_C Q(x, y)\, dy \tag{13-122$'$}$$

If we add (13-122) and (13-122′), we obtain the conclusion of Theorem 14.

EXAMPLE 1

$$\frac{1}{2}\oint_C -y\, dx + x\, dy = \frac{1}{2}\iint_R (1 + 1)\, dA = \iint_R dA = \text{area of } R$$

This is the basic rule for area developed in Section 7-3.

EXAMPLE 2. For $k = 0, 1, 2, \ldots,$

$$\frac{1}{k+2}\oint_C -x^k y\, dx + x^{k+1}\, dy = \frac{1}{k+2}\iint_R [(k+1)x^k + x^k]\, dA = \iint_R x^k\, dA$$

This is the basic rule for the moment M_{kx} given in Section 7-10.

EXAMPLE 3. Let R be the square of vertices $(\pm 1, \pm 1)$, let C be the boundary of R. Then

$$\oint_C xy\, dx - 3x^2\, dy = \iint_R (-6x - x)\, dA = -7 \iint_R x\, dA$$

Now $\iint_R x\, dA$ is equal to $A\bar{x}$, where $(\bar{x}, \bar{y})$ is the centroid of R. By symmetry, $\bar{x} = 0$. Hence, the line integral equals 0.

13-13 CURL AND DIVERGENCE, VECTOR FORM OF GREEN'S THEOREM

We consider an open region D in 3-dimensional space and consider two kinds of functions in D:

(a) the scalar functions $F(x, y, z)$;
(b) the vector functions $\mathbf{v} = f(x, y, z)\mathbf{i} + g(x, y, z)\mathbf{j} + h(x, y, z)\mathbf{k}$.

In physics such functions are often referred to as *fields: scalar fields* or *vector fields*. For the scalar field F one thinks of the values of F as numbers attached to the various points of D (like house numbers on the houses of a city); for the vector field one thinks of directed line segments attached to the various points as in Figure 13-47. Such fields arise in many applications. For example, in a fluid motion the velocity vectors of the various points form a vector field $\mathbf{v}$; the temperature at the various points forms a scalar field F. The force vector at different points due to gravitation (or to a magnet) is a vector field.

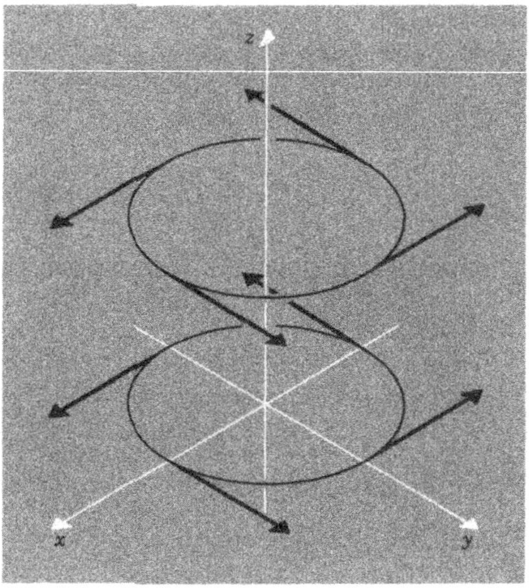

Figure 13-47 Velocity field (pure rotation).

To avoid repetition, we shall assume that all functions appearing in this section have continuous partial derivatives of the orders which appear. Our discussion emphasizes 3-dimensional space, but it will also be extended to 2-dimensional space. We can think of 2-dimensional scalar fields as functions F that happen not to depend on z, and of 2-dimensional vector fields as fields $\mathbf{v}$ which have 0 as z-component and for which the x- and y-components depend only on x and y, so that $\mathbf{v} = P(x, y)\mathbf{i} + Q(x, y)\mathbf{j}$. The 2-dimensional fields are called *planar fields*.

If we start with a scalar field $F(x, y, z)$, then we can obtain a vector field by forming ∇F, the gradient vector of F. Such a field is called a *gradient field*.

EXAMPLE 1. Let $\mathbf{v} = 2x\mathbf{i} + 2y\mathbf{j} + 2z\mathbf{k}$. Then $\mathbf{v} = \mathbf{F}(x, y, z)$ is a vector field in R^3. In this case $\mathbf{v}$ is a gradient field, for

$$\mathbf{v} = \nabla(x^2 + y^2 + z^2)$$

Now let $\mathbf{v} = f(x, y, z)\mathbf{i} + g(x, y, z)\mathbf{j} + h(x, y, z)\mathbf{k}$ be a vector field in the open region D of R^3 and let f, g, h have first partial derivatives in D. Since f, g, h are the components of $\mathbf{v}$, we can write

$$v_x = f(x, y, z), \qquad v_y = g(x, y, z), \qquad v_z = h(x, y, z)$$

(In this section and the next two sections, subscripts x, y, z will *not* be used for partial derivatives, but merely to label components.) From the first partial derivatives of f, g, and h we form a new vector field:

$$\mathbf{w} = \text{curl } \mathbf{v} = \left(\frac{\partial h}{\partial y} - \frac{\partial g}{\partial z}\right)\mathbf{i} + \left(\frac{\partial f}{\partial z} - \frac{\partial h}{\partial x}\right)\mathbf{j} + \left(\frac{\partial g}{\partial x} - \frac{\partial f}{\partial y}\right)\mathbf{k}$$

$$= \left(\frac{\partial v_z}{\partial y} - \frac{\partial v_y}{\partial z}\right)\mathbf{i} + \left(\frac{\partial v_x}{\partial z} - \frac{\partial v_z}{\partial x}\right)\mathbf{j} + \left(\frac{\partial v_y}{\partial x} - \frac{\partial v_x}{\partial y}\right)\mathbf{k}$$

We can also write symbolically

$$\mathbf{w} = \text{curl } \mathbf{v} = \nabla \times \mathbf{v} = \begin{vmatrix} \mathbf{i} & \mathbf{j} & \mathbf{k} \\ \dfrac{\partial}{\partial x} & \dfrac{\partial}{\partial y} & \dfrac{\partial}{\partial z} \\ v_x & v_y & v_z \end{vmatrix} = \begin{vmatrix} \dfrac{\partial}{\partial y} & \dfrac{\partial}{\partial z} \\ v_y & v_z \end{vmatrix} \mathbf{i} + \cdots$$

and the symbolic formulas are easier to remember.

The definition just given appears artificial. However, the *curl* of a vector field has a profound geometrical and physical meaning. We can think of vector field $\mathbf{v}$ as the field of velocity vectors of the particles of a moving fluid. Then curl $\mathbf{v}$ measures the extent to which the motion is like a rotation about a particular axis. When curl $\mathbf{v}$ is identically $\mathbf{0}$, the motion is termed *irrotational;* the same term is used for any vector field whose curl is $\mathbf{0}$. (See also the concluding paragraph of the discussion of Stokes' theorem in the plane, below.)

EXAMPLE 2. $\mathbf{v} = -y\mathbf{i} + x\mathbf{j}$. Here

$$\text{curl } \mathbf{v} = \begin{vmatrix} \mathbf{i} & \mathbf{j} & \mathbf{k} \\ \dfrac{\partial}{\partial x} & \dfrac{\partial}{\partial y} & \dfrac{\partial}{\partial z} \\ -y & x & 0 \end{vmatrix} = 2\mathbf{k}$$

In this case, as Figure 13-47 suggests, we have a pure rotation about the z-axis, at an angular velocity of 1 radian per unit of time.

THEOREM 15. *Let the vector field* $\mathbf{v}$ *be defined and be continuously differentiable in the open region D. If* $\mathbf{v}$ *is a gradient field, then* curl $\mathbf{v} = \mathbf{0}$ *in D.*

PROOF. Let $\mathbf{v} = \operatorname{grad} F = \dfrac{\partial F}{\partial x}\mathbf{i} + \dfrac{\partial F}{\partial y}\mathbf{j} + \dfrac{\partial F}{\partial z}\mathbf{k}$. Then

$$\operatorname{curl} \mathbf{v} = \begin{vmatrix} \mathbf{i} & \mathbf{j} & \mathbf{k} \\ \dfrac{\partial}{\partial x} & \dfrac{\partial}{\partial y} & \dfrac{\partial}{\partial z} \\ \dfrac{\partial F}{\partial x} & \dfrac{\partial F}{\partial y} & \dfrac{\partial F}{\partial z} \end{vmatrix}$$

$$= \left(\frac{\partial^2 F}{\partial y\,\partial z} - \frac{\partial^2 F}{\partial z\,\partial y}\right)\mathbf{i} + \left(\frac{\partial^2 F}{\partial z\,\partial x} - \frac{\partial^2 F}{\partial x\,\partial z}\right)\mathbf{j} + \left(\frac{\partial^2 F}{\partial x\,\partial y} - \frac{\partial^2 F}{\partial y\,\partial x}\right)\mathbf{k}$$

$$= \mathbf{0}$$

by the basic rule on mixed partial derivatives (Theorem 13, Section 12-20).

Under certain restrictions on the region D, the converse of this theorem is true: if $\operatorname{curl} \mathbf{v} = \mathbf{0}$ in D, then $\mathbf{v}$ is a gradient field in D. The result is valid in particular when D is all of R^3 or when D is the interior of a convex solid such as a sphere or cube, but it is not valid for the interior of a torus. Further discussion of this point and a proof of the converse in the 2-dimensional case are given in the following sections.

The condition that $\mathbf{v} = f\mathbf{i} + g\mathbf{j} + h\mathbf{k}$ be a gradient field is equivalent to the condition that $f\,dx + g\,dy + h\,dz$ be the differential of a function F. For when $f = \partial F/\partial x$, $g = \partial F/\partial y$, $h = \partial F/\partial z$, we can write

$$f\,dx + g\,dy + h\,dz = \frac{\partial F}{\partial x}\,dx + \frac{\partial F}{\partial y}\,dy + \frac{\partial F}{\partial z}\,dz = dF$$

conversely, if $dF = f\,dx + g\,dy + h\,dz$, then (by Theorem 2, Section 12-9), $f = \partial F/\partial x$, $g = \partial F/\partial y$, $h = \partial F/\partial z$. Thus we can restate Theorem 15 and its converse (with restrictions on D understood) as follows.

The expression $f(x, y, z)\,dx + g(x, y, z)\,dy + h(x, y, z)\,dz$ is the differential dF of a function F in D if, and only if,

$$\frac{\partial f}{\partial y} = \frac{\partial g}{\partial x}, \qquad \frac{\partial g}{\partial z} = \frac{\partial h}{\partial y} \qquad \text{and} \qquad \frac{\partial f}{\partial z} = \frac{\partial h}{\partial x} \tag{13-130}$$

One refers to an expression $f\,dx + g\,dy + h\,dz$ which is the differential dF of a function F as an *exact differential.*

EXAMPLE 3. If a, b are constants, the expression

$$(ay + bz)\,dx + (3x + 2z)\,dy + (2y + x)\,dz$$

is an exact differential if, and only if,

$$a = 3, \qquad 2 = 2, \qquad b = 1$$

Thus the condition is $a = 3$, $b = 1$. With these choices, we have

$$(3y + z)\,dx + (3x + 2z)\,dy + (2y + x)\,dz = dF$$

or

$$(3y\,dx + 3x\,dy) + (z\,dx + x\,dz) + (2z\,dy + 2y\,dz) = dF$$

Clearly, we can take

$$F = 3xy + xz + 2yz + \text{const}$$

In general, the function F is determined only up to addition of a constant. For the addition of a constant to F has no effect on dF and, if $dF_1 = dF_2$, then $d(F_1 - F_2) = 0$ so that $\nabla(F_1 - F_2) = \mathbf{0}$ in D and, hence, by Theorem 5 of Section 12-9, $F_1 - F_2$ is constant.

EXAMPLE 4. Show that

$$e^{x^2z}(y + 2x^2yz)\,dx + xe^{x^2z}\,dy + x^3ye^{x^2z}\,dz$$

is an exact differential and find the function F for which it is the differential. The conditions (13-130) become

$$e^{x^2z}(1 + 2x^2z) = e^{x^2z}(1 + 2x^2z), \qquad x^3e^{x^2z} = x^3e^{x^2z}$$
$$e^{x^2z}(3x^2y + 2x^4yz) = e^{x^2z}(3x^2y + 2x^4yz)$$

and, hence, are satisfied for all (x, y, z). It is hard to see what F is. However, we can reason:

$$\frac{\partial F}{\partial z} = x^3ye^{x^2z}$$

so that $F = xye^{x^2z}$ plus—not necessarily a constant—a function whose derivative with respect to z is 0, that is, a function $\varphi(x, y)$:

$$F = xye^{x^2z} + \varphi(x, y)$$

The conditions

$$\frac{\partial F}{\partial x} = e^{x^2z}(y + 2x^2yz), \qquad \frac{\partial F}{\partial y} = xe^{x^2z}$$

become

$$e^{x^2z}(y + 2x^2yz) + \frac{\partial\varphi}{\partial x} = e^{x^2z}(y + 2x^2yz), \qquad xe^{x^2z} + \frac{\partial\varphi}{\partial y} = xe^{x^2z}$$

Thus $\partial\varphi/\partial x = 0$, $\partial\varphi/\partial y = 0$ and φ is actually a constant:

$$F(x, y, z) = xye^{x^2z} + c \qquad (c = \text{const})$$

We can specialize Theorem 15 and the subsequent discussion to vector fields in R^2: that is, to the planar field $\mathbf{v} = f(x, y)\mathbf{i} + g(x, y)\mathbf{j}$. Thus $\mathbf{v}$ has no z-component and the x- and y-components of $\mathbf{v}$ depend only on x and y. We again conclude that if $\mathbf{v}$ is a gradient field:

$$\mathbf{v} = f\mathbf{i} + g\mathbf{j} = \nabla F = \frac{\partial F}{\partial x}\mathbf{i} + \frac{\partial F}{\partial y}\mathbf{j}$$

with $F = F(x, y)$, then $\operatorname{curl}\mathbf{v} = \mathbf{0}$. But now

$$\operatorname{curl}\mathbf{v} = \begin{vmatrix} \mathbf{i} & \mathbf{j} & \mathbf{k} \\ \dfrac{\partial}{\partial x} & \dfrac{\partial}{\partial y} & \dfrac{\partial}{\partial z} \\ f(x, y) & g(x, y) & 0 \end{vmatrix} = \left(\frac{\partial g}{\partial x} - \frac{\partial f}{\partial y}\right)\mathbf{k}$$

Hence, the condition curl $\mathbf{v} = \mathbf{0}$ becomes

$$\frac{\partial f}{\partial y} = \frac{\partial g}{\partial x} \tag{13-131}$$

The equation (13-131) is a sufficient condition that $f(x, y)\,dx + g(x, y)\,dy$ be an exact differential:

$$f(x, y)\,dx + g(x, y)\,dy = dF, \qquad F = F(x, y) \tag{13-132}$$

As before, (13-131) implies (13-132) only in an appropriate open region D in the xy-plane (z can be completely ignored). The condition on D is that it be *simply connected* (Section 7-4; see Section 13-14).

EXAMPLE 5. Show that $(2x + 3y)\,dx + (3x + 4y)\,dy$ is an exact differential dF and find F. We verify that condition (13-131) is satisfied:

$$\frac{\partial}{\partial y}(2x + 3y) = \frac{\partial}{\partial x}(3x + 4y) = 3$$

We then reason as in Example 4:

$$\frac{\partial F}{\partial x} = 2x + 3y, \qquad F = x^2 + 3xy + \varphi(y)$$

$$\frac{\partial F}{\partial y} = 3x + 4y, \qquad \text{so that } 3x + \varphi'(y) = 3x + 4y$$

$$\varphi'(y) = 4y, \qquad \varphi(y) = 2y^2 + c, \qquad F(x, y) = x^2 + 3xy + 2y^2 + c$$

Stokes' Theorem in the Plane. Green's theorem can now be formulated in vector form. We let $\mathbf{u} = P(x, y)\mathbf{i} + Q(x, y)\mathbf{j}$ as in Theorem 14. Thus $\mathbf{u}$ is a vector field in D. The line integral $\oint P\,dx + Q\,dy$ which appears in Green's theorem can be written as $\oint u_T\,ds$ as in (13-114). In the double integral the integrand is the z-component of curl $\mathbf{u}$. Hence, Green's theorem takes the form:

$$\oint_C u_T\,ds = \iint_R \operatorname{curl}_z \mathbf{u}\,dA \tag{13-133}$$

This equation is a 2-dimensional form of Stokes' theorem. We extend this to 3-dimensional space in Section 13-15.

The line integral $\oint u_T\,ds$ on the left of (13-133) is known as the *circulation of the vector field* $\mathbf{u}$ on C; it measures the extent to which corresponding fluid motion is like a rotation around C. By the Mean Value Theorem (Theorem 7, Section 13-8), we can write (13-133) as follows

$$\oint u_T\,ds = \operatorname{curl}_z \mathbf{u}\Big|_{(\xi,\eta)} A$$

where A is the area of R. If we divide by A and pass to the limit as in Theorem 10 (Section 13-8), letting R shrink to a point (x_0, y_0), we conclude that

$$\operatorname{curl}_z \mathbf{u}\Big|_{(x_0, y_0)} = \lim_{\text{diam of } R \to 0} \frac{1}{A} \oint_C u_T\,ds$$

Thus the z-component of curl $\mathbf{u}$ is a *limit of circulation per unit area*. This result shows that the curl, indeed, does have a basic geometric and physical meaning.

Divergence of a Vector Field. Let $\mathbf{v} = f(x, y, z)\mathbf{i} + g(x, y, z)\mathbf{j} + h(x, y, z)\mathbf{k}$ as above. From $\mathbf{v}$ we construct a scalar function (or scalar field), the *divergence of* $\mathbf{v}$, by the equation:

$$G(x, y, z) = \operatorname{div} \mathbf{v} = \frac{\partial f}{\partial x} + \frac{\partial g}{\partial y} + \frac{\partial h}{\partial z} \tag{13-134}$$

Again the definition appears to be artificial. However, the divergence has also a basic physical and geometrical significance. As we shall show, for the case of a velocity field $\mathbf{v}$, div $\mathbf{v}$ measures the extent to which the fluid is expanding. In particular, div $\mathbf{v} = 0$ for the motion of an incompressible fluid. When div $\mathbf{v} = 0$, the vector field $\mathbf{v}$ is called *solenoidal*.

EXAMPLE 6. $\mathbf{v} = -y\mathbf{i} + x\mathbf{j}$. This is the field of Example 2 considered above (Figure 13-47) and is the velocity field of a purely rotational motion. We find that

$$\operatorname{div} \mathbf{v} = \frac{\partial}{\partial x}(-y) + \frac{\partial}{\partial y}(x) = 0$$

Since the motion involves no expansion or contraction, the result is expected.

We can use a symbolic notation for div $\mathbf{v}$:

$$\operatorname{div} \mathbf{v} = \nabla \cdot \mathbf{v} = \left(\frac{\partial}{\partial x}\mathbf{i} + \frac{\partial}{\partial y}\mathbf{j} + \frac{\partial}{\partial z}\mathbf{k}\right) \cdot (v_x\mathbf{i} + v_y\mathbf{j} + v_z\mathbf{k})$$

and this helps in remembering the definition.

THEOREM 16. *Let* $\mathbf{v} = \operatorname{curl} \mathbf{w}$ *in D, where the components of* $\mathbf{w}$ *have continuous partial derivatives through the second order in D. Then* div $\mathbf{v} = 0$ *in D.*

PROOF. If $\mathbf{v} = \operatorname{curl} \mathbf{w}$, then

$$\mathbf{v} = \left(\frac{\partial w_z}{\partial y} - \frac{\partial w_y}{\partial z}\right)\mathbf{i} + \left(\frac{\partial w_x}{\partial z} - \frac{\partial w_z}{\partial x}\right)\mathbf{j} + \left(\frac{\partial w_y}{\partial x} - \frac{\partial w_x}{\partial y}\right)\mathbf{k}$$

and

$$\begin{aligned}\operatorname{div} \mathbf{v} &= \frac{\partial}{\partial x}\left(\frac{\partial w_z}{\partial y} - \frac{\partial w_y}{\partial z}\right) + \frac{\partial}{\partial y}\left(\frac{\partial w_x}{\partial z} - \frac{\partial w_z}{\partial x}\right) + \frac{\partial}{\partial z}\left(\frac{\partial w_y}{\partial x} - \frac{\partial w_x}{\partial y}\right) \\ &= \frac{\partial^2 w_z}{\partial x\,\partial y} - \frac{\partial^2 w_y}{\partial x\,\partial z} + \frac{\partial^2 w_x}{\partial y\,\partial z} - \frac{\partial^2 w_z}{\partial y\,\partial x} + \frac{\partial^2 w_y}{\partial z\,\partial x} - \frac{\partial^2 w_x}{\partial z\,\partial y} = 0\end{aligned}$$

by the rule on mixed partial derivatives.

Again, there is a converse: if div $\mathbf{v} = 0$ in D, then $\mathbf{v} = \operatorname{curl} \mathbf{w}$ for some $\mathbf{w}$; again restrictions must be imposed on D (not the same as those for the converse of Theorem 15). The converse is valid, in particular, for all of R^3 and for the interior of a convex solid. For more information, see *Advanced Calculus* by W. Kaplan (Addison-Wesley, Reading, Mass., 1952).

The divergence can be defined for vector fields in R^n for every n:

$$\operatorname{div} \mathbf{v} = \frac{\partial v_1}{\partial x_1} + \frac{\partial v_2}{\partial x_2} + \cdots + \frac{\partial v_n}{\partial x_n}$$

The concept has important applications in the kinetic theory of gases, with n of the order of 10^{23}!

Divergence of a Gradient. If $\mathbf{v} = \nabla F$, where F has derivatives through the second order in D, then

$$\operatorname{div} \mathbf{v} = \operatorname{div} \operatorname{grad} F = \frac{\partial}{\partial x}\left(\frac{\partial F}{\partial x}\right) + \frac{\partial}{\partial y}\left(\frac{\partial F}{\partial y}\right) + \frac{\partial}{\partial z}\left(\frac{\partial F}{\partial z}\right)$$

$$= \frac{\partial^2 F}{\partial x^2} + \frac{\partial^2 F}{\partial y^2} + \frac{\partial^2 F}{\partial z^2} = \nabla^2 F$$

Hence, the divergence of a gradient is the same as the *Laplacian* (Section 12-19). Symbolically

$$\nabla \cdot \nabla = \nabla^2$$

The Divergence Theorem. We again consider Green's theorem but now let $\mathbf{v} = Q\mathbf{i} - P\mathbf{j} = -\mathbf{u}^{\dashv}$, where $\mathbf{u} = P\mathbf{i} + Q\mathbf{j}$. Then

$$\operatorname{div} \mathbf{v} = \frac{\partial Q}{\partial x} - \frac{\partial P}{\partial y}$$

and this is the integrand in the double integral. For the line integral we introduce the vector

$$\mathbf{n} = -\mathbf{T}^{\dashv} = \frac{dy}{ds}\mathbf{i} - \frac{dx}{ds}\mathbf{j}$$

so that $\mathbf{n}$ is a unit normal vector along the path, pointing to the exterior of R, as in Figure 13-48. Then

$$v_n = \mathbf{v} \cdot \mathbf{n} = Q\frac{dy}{ds} + P\frac{dx}{ds} = \mathbf{u} \cdot \mathbf{T} = u_T$$

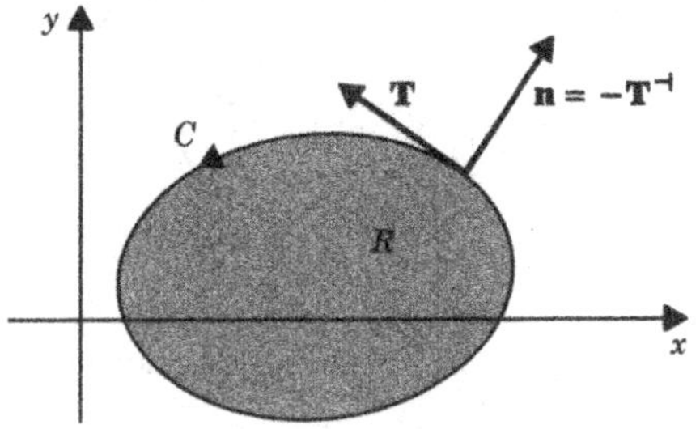

Figure 13-48 Divergence Theorem.

Hence, Green's theorem takes the form:

$$\oint_C v_n \, ds = \iint_R \operatorname{div} \mathbf{v} \, dA \qquad (13\text{-}135)$$

This formula is valid for each planar vector field satisfying the required continuity conditions. Equation (13-135) is known as the *divergence theorem in the plane.*

The line integral $\oint v_n\, ds$ on the left of (13-135) is known as the *flux* of **v** across C. If **v** is the field of velocity vectors of a 2-dimensional fluid motion, the flux measures how rapidly the fluid is leaving R, or entering R. An argument like that for the curl shows that

$$\operatorname{div} \mathbf{v}\Big|_{(x_0,y_0)} = \lim \frac{\int_C v_n\, ds}{A}$$

Thus the divergence is a limit of flux per unit area, again a basic geometric and physical idea (see Problem 3 below).

The Gradient, Curl, Divergence and Laplacian as Linear Mappings. Each of the four operations listed can be applied to the elements of an appropriate vector space to yield a linear mapping of that vector space into another vector space. For example, we let U be the vector space of all scalar functions in D having continuous derivatives of all orders and let V be the vector space of all vector fields in D whose components have continuous partial derivatives of all orders. Then

∇ or grad maps U into V
curl maps V into V
div maps V into U
∇^2 maps U into U

Each of these mappings is linear (Problem 9 below). We remark that $\nabla^2 f =$ div grad f, so that ∇^2 is the product of the two linear mappings div and grad.

As for all linear mappings, we now ask for the kernel and range of each of these mappings. For a region D, such as the interior of a sphere (or of a circle in two dimensions), one can make the following statements in view of our discussion or, for the last statement, by advanced theory.

The range of ∇ is the set of gradient fields, and this is the kernel of curl (hence, the same as the irrotational fields).

The range of curl is the set of solenoidal fields, and this is the kernel of div.

The kernel of ∇^2 is the set of harmonic functions in D.

The range of ∇^2 is U and the range of div is U.

All these relationships are suggested in the diagram of Figure 13-49.

PROBLEMS

1. Evaluate by Green's Theorem:

(a) $\oint_C (a_1x + b_1y)\, dx + (a_2x + b_2y)\, dy$, a_1, b_1, a_2, b_2 constants, C arbitrary.

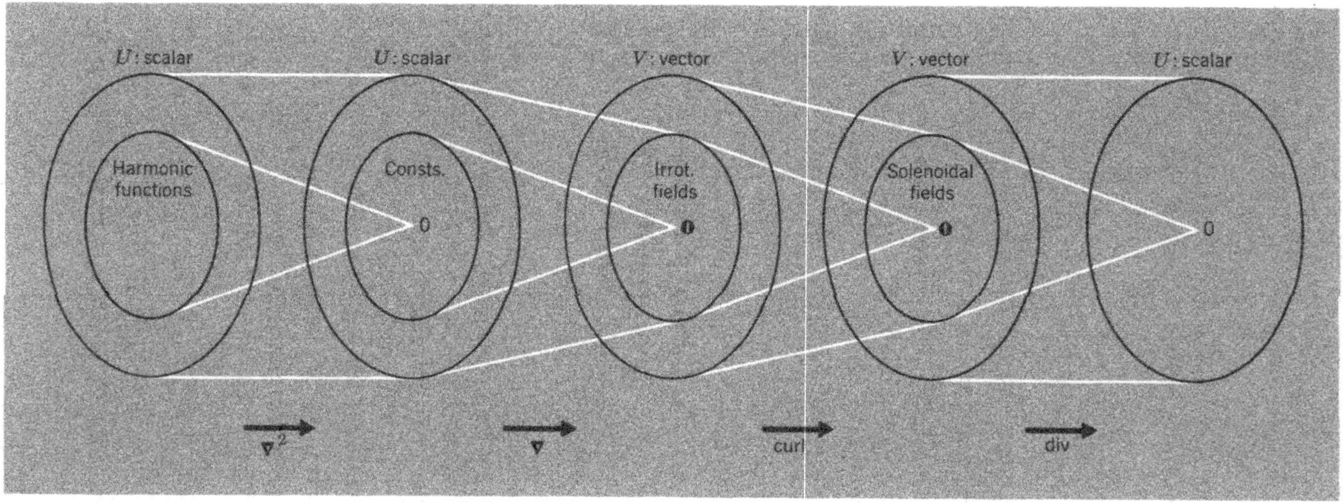

Figure 13-49 The linear mappings ∇, curl, div, and ∇^2.

(b) $\oint_C P(x)\,dx + Q(y)\,dy$, C arbitrary

(c) $\oint_C x^2y\,dx - xy^2\,dy$, C: $x^2 + y^2 = 1$

(d) $\oint_C e^x(x + \sin y)\,dx + e^x(x + \cos y)\,dy$, C: square of vertices $(\pm 1, \pm 1)$

2. Let C be a smooth simple closed curve enclosing region R; let $\mathbf{n}$ be the unit normal on C pointing outward, let $\partial/\partial n$ signify the directional derivative in the direction of $\mathbf{n}$. Let f and g satisfy appropriate differentiability conditions. Prove:

(a) $\oint_C f\frac{\partial g}{\partial n}\,ds = \iint_R (\nabla f \cdot \nabla g + f\nabla^2 g)\,dA$ [See Problem 12].

(b) $\oint_C f\frac{\partial f}{\partial n}\,ds = \iint_R (\|\nabla f\|^2 + f\nabla^2 f)\,dA$.

(c) If f is harmonic in R, then $\oint_C f\frac{\partial f}{\partial n}\,ds = \iint_R \|\nabla f\|^2\,dA$.

(d) If f is harmonic in R and $\oint_C f\frac{\partial f}{\partial n}\,ds = 0$, then f is constant in R.

(e) $\oint_C \left(f\frac{\partial g}{\partial n} - g\frac{\partial f}{\partial n}\right) ds = \iint_R (f\nabla^2 g - g\nabla^2 f)\,dA$.

(f) If f and g are harmonic in R, then $\oint_C f\frac{\partial g}{\partial n}\,ds = \oint_C g\frac{\partial f}{\partial n}\,ds$.

3. In this problem reason physically, rather than attempting a precise mathematical derivation. We consider a 2-dimensional fluid motion in the xy-plane. This can be thought of as the motion of the particles on the surface of a river or, as a motion of a fluid in 3-dimensional space in which all particles move parallel to the xy-plane and have a velocity $\mathbf{v}$ independent of the z-coordinate. Thus $\mathbf{v} = \mathbf{v}(x, y)$ or, if the velocity changes with time, $\mathbf{v} = \mathbf{v}(x, y, t)$.

(a) Assume $\mathbf{v}$ is a constant vector. Let C be a directed line segment with unit normal vector $\mathbf{n}$. Show that $\int_C \mathbf{v}\cdot\mathbf{n}\,ds$ is equal to the rate at which the fluid is crossing C (Figure 13-50), in units of area per unit time; a negative rate signifies that the fluid crosses toward the side opposite to $\mathbf{n}$. When is the rate 0?

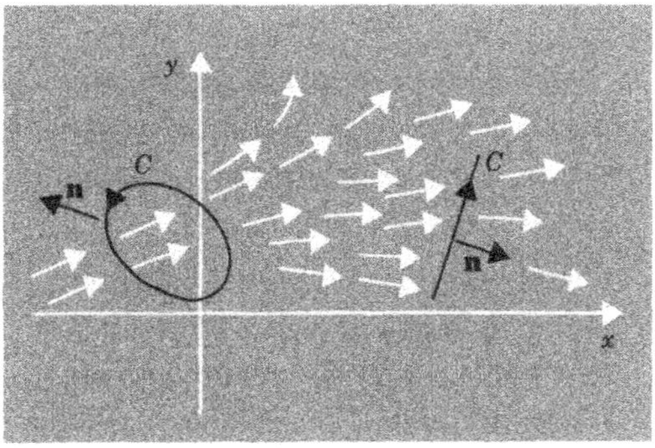

Figure 13-50 Two-dimensional fluid motion.

(b) Extend the result of (a) to a general field **v** and path C and show that, when C is closed and encloses region R, $\oint_C \mathbf{v}\cdot\mathbf{n}\,ds$ is the net rate of loss of fluid from R. Conclude from this result that the fluid is incompressible if, and only if, $\operatorname{div}\mathbf{v}\equiv 0$.

(c) Let the fluid have density $\delta = \delta(x, y, t)$. Let C enclose R of area A. Show that $\int_C \delta\mathbf{v}\cdot\mathbf{n}\,ds$ is the net rate of loss of mass from R. Let R shrink down on a point (x_0, y_0) to conclude that

$$\lim_{\text{diam of } R\to 0} \frac{1}{A}\int_C \delta\mathbf{v}\cdot\mathbf{n}\,ds$$

is the rate of decrease of density at (x_0, y_0). Conclude that

$$\frac{\partial\delta}{\partial t} + \operatorname{div}(\delta\mathbf{v}) = 0$$

This is the 2-dimensional *continuity equation of hydrodynamics*. Conclude that if δ is constant in time and position, so that the fluid is incompressible, then $\operatorname{div}\mathbf{v} = 0$.

4. By finding the circulation of **v** around the circle C: $x^2 + y^2 = a^2$, dividing by πa^2 and letting $a \to 0$, find curl **v** at $(0, 0)$. Sketch the vector field in each case.
   (a) $\mathbf{v} = -y\mathbf{i} + x\mathbf{j}$ (rotation) (b) $\mathbf{v} = x\mathbf{i} + y\mathbf{j}$ (radial motion)
   (c) $\mathbf{v} = x\mathbf{i}$ (d) $\mathbf{v} = x\mathbf{j}$

5. Find the curl of each of the following vector fields:
   (a) $y\mathbf{i} + z\mathbf{j} + x\mathbf{k}$
   (b) $(3x - 2y)\mathbf{i} + (3y - 2z)\mathbf{j} + (3z - 5x)\mathbf{k}$
   (c) $x^2yz\mathbf{i} + (x^3y - y^2z^2)\mathbf{j} + xyz^2\mathbf{k}$
   (d) $xy\cos z\,\mathbf{i} + (x^2 - y^2)\sin z\,\mathbf{j} + (x^2 - xy)\cos z\,\mathbf{k}$

6. (a) . . . (d) Find the divergence of the vector fields of Problem 5.

7. Show that each of the following vector fields is irrotational and find F such that $\mathbf{v} = \nabla F$:
   (a) $\mathbf{v} = yz\mathbf{i} + xz\mathbf{j} + xy\mathbf{k}$ (b) $\mathbf{v} = e^x\mathbf{i} + \cos y\,\mathbf{j} + z^3\mathbf{k}$
   (c) $\mathbf{v} = \dfrac{1}{x^2 + y^2}(x\mathbf{i} + y\mathbf{j})$ (d) $\mathbf{v} = (x^2 + y^2 + z^2)^{-1/2}(x\mathbf{i} + y\mathbf{j} + z\mathbf{k})$

8. Show that each of the following is an exact differential dF and find F:
   (a) $(2x + y)\,dx + (2y + x + z)\,dy + (y - 2z)\,dz$
   (b) $(6x - y + 2z)\,dx + (-2y - x - z)\,dy + (2x - y + 2z)\,dz$
   (c) $(x^2 - 2y)\,dx + (y^2 - 2x)\,dy$
   (d) $(\sin xy + xy\cos xy)\,dx + x^2\cos xy\,dy$
   (e) $(e^x + 2xy)\,dx + (x^2 + \sin y)\,dy$
   (f) $(1 + \ln x + e^y)\,dx + (xe^y + \sec^2 y)\,dy$

9. Prove that each of the following operations is a linear mapping, as described at the end of Section 13-13:
   (a) ∇ (b) curl (c) div (d) ∇^2

10. Show that $\mathbf{v}$ is solenoidal and find $\mathbf{w}$ so that curl $\mathbf{w} = \mathbf{v}$: (*Hint.* Assume $w_z = 0$.)
    (a) $\mathbf{v} = x\mathbf{i} + y\mathbf{j} - 2z\mathbf{k}$ (b) $\mathbf{v} = (x^2 + y^2)\mathbf{k}$
11. Let $u(x, y)\mathbf{i} - v(x, y)\mathbf{j}$ be an irrotational solenoidal vector field in the open region D; let u, v have continuous partial derivatives through the second order.
    (a) Show that $(\partial u/\partial x) = (\partial v/\partial y)$ and $(\partial u/\partial y) = -(\partial v/\partial x)$.
    (b) Show that u and v are harmonic in D.

    [The equations in (a) are called the *Cauchy-Riemann equations*. There is a vast literature concerning the functions satisfying them. They are discussed in books on functions of a complex variable.]
12. Prove, under appropriate hypotheses:
    (a) $\nabla(fg) = f\nabla g + g\nabla f$
    (b) $\nabla \cdot (f\mathbf{v}) = f(\nabla \cdot \mathbf{v}) + (\nabla f \cdot \mathbf{v})$
    (c) $\operatorname{div}(\mathbf{u} \times \mathbf{v}) = \mathbf{v} \cdot \operatorname{curl} \mathbf{u} - \mathbf{u} \cdot \operatorname{curl} \mathbf{v}$

13-14 EXACT DIFFERENTIALS AND INDEPENDENCE OF PATH

In Section 13-13 we defined an expression $P\,dx + Q\,dy$ to be an *exact differential* in the open region D if $P\,dx + Q\,dy = dF$ for some function F in D. We also showed that if $dF = P\,dx + Q\,dy$ and appropriate continuity conditions hold true, then

$$\frac{\partial P}{\partial y} = \frac{\partial Q}{\partial x} \tag{13-140}$$

We also stated that if (13-140) holds true and D is simply connected, then $P\,dx + Q\,dy$ is an exact differential. In this section we establish this last result, explain why simple-connectedness is needed, and show how F can be found from P and Q.

Definition. Let $P(x, y)$ and $Q(x, y)$ be defined and continuous in the open region D. We say that the line integral $\int P\,dx + Q\,dy$ is *independent of path in D if the value of the line integral depends only on the end points and not on the particular path joining these end points; that is, if*

$$\int_{A\,C}^{B} P\,dx + Q\,dy = \int_{A\,C_1}^{B} P\,dx + Q\,dy$$

for all choices of A and B and of paths C, C_1 from A to B in D.

THEOREM 17. *Let $F(x, y)$ have continuous first partial derivatives in the open region D. Let $\nabla F = P\mathbf{i} + Q\mathbf{j}$, so that $\partial F/\partial x = P$, $\partial F/\partial y = Q$ and $dF = P\,dx + Q\,dy$ and, hence, $P\,dx + Q\,dy$ is an exact differential. Then $\int P\,dx + Q\,dy$ is independent of path in D and*

$$\int_{(x_0,y_0)\,C}^{(x_1,y_1)} P\,dx + Q\,dy = F(x_1, y_1) - F(x_0, y_0) = F(x, y)\Big|_{(x_0,y_0)}^{(x_1,y_1)}$$

PROOF. We need consider only smooth paths, since the general piecewise smooth path is decomposable into several smooth paths, and the assertions to be proved follow from the special case by simple addition. Let C be a smooth path $x = f(t)$, $y = g(t)$, $a \le t \le b$. Then

$$\int_{(x_0,y_0)\atop C}^{(x_1,y_1)} P\,dx + Q\,dy = \int_a^b \left(P\frac{dx}{dt} + Q\frac{dy}{dt}\right) dt = \int_a^b \left(\frac{\partial F}{\partial x}\frac{dx}{dt} + \frac{\partial F}{\partial y}\frac{dy}{dt}\right) dt$$

$$= \int_a^b \frac{d}{dt} F[x(t), y(t)]\,dt$$

by the chain rule (Section 12-10, Theorem 6). Hence

$$\int_{(x_0,y_0)\atop C}^{(x_1,y_1)} P\,dx + Q\,dy = F[x(t), y(t)]\Big|_{t=a}^{t=b} = F(x_1, y_1) - F(x_0, y_0)$$

The value obtained does not depend on the path C. Therefore, the theorem is proved.

EXAMPLE 1

$$\int_{(1,3)\atop C}^{(2,8)} y\,dx + x\,dy = \int_{(1,3)\atop C}^{(2,8)} d(xy) = (xy)\Big|_{(1,3)}^{(2,8)} = 16 - 3 = 13$$

As the example indicates, Theorem 17 can be stated concisely as follows:

$$\int_{A\atop C}^{B} dF = F(B) - F(A) \tag{13-141}$$

The choice of path is irrelevant and one can even write the integral as $\int_A^B dF$, omitting any reference to C.

THEOREM 18 (*Converse of Theorem 17*). *Let $P(x, y)$ and $Q(x, y)$ be continuous in the open region D and let $\int P\,dx + Q\,dy$ be independent of path in D. Then there is a function $F(x, y)$ defined in D for which $dF = P\,dx + Q\,dy$, or equivalently,*

$$\frac{\partial F}{\partial x} = P, \qquad \frac{\partial F}{\partial y} = Q\text{—that is, } \nabla F = P\mathbf{i} + Q\mathbf{j}$$

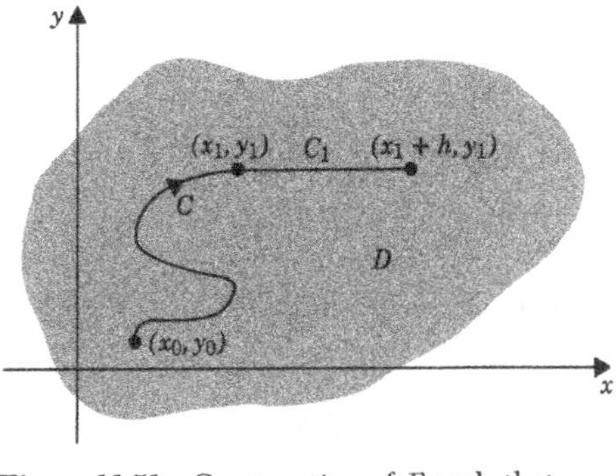

Figure 13-51 Construction of F such that $dF = P\,dx + Q\,dy$.

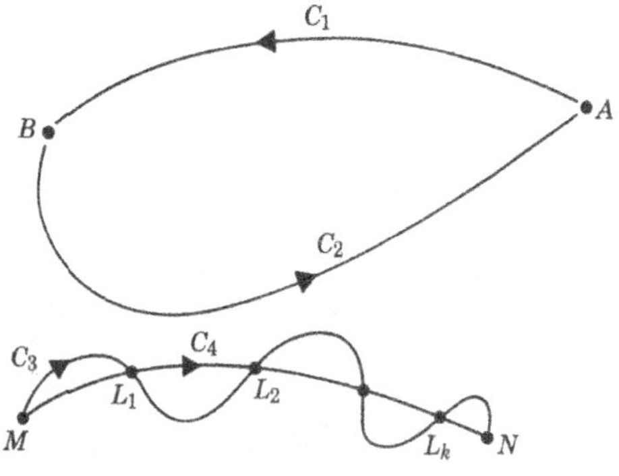

Figure 13-52 Proof of Theorem 19.

PROOF. Let (x_0, y_0) be a fixed point in D. For each point (x_1, y_1) of D, we define $F(x_1, y_1)$ by the equation

$$F(x_1, y_1) = \int_{\substack{(x_0,y_0)\\C}}^{(x_1,y_1)} P\,dx + Q\,dy \tag{13-142}$$

where C is a path from (x_0, y_0) to (x_1, y_1) in D, as in Figure 13-51. Since the line integral is independent of path, it does not matter which path C from (x_0, y_0) to (x_1, y_1) we use in (13-142).

Now let $|h|$ be so small and positive that the line segment C_1 from (x_1, y_1) to $(x_1 + h, y_1)$ is in D. Then

$$\begin{aligned} F(x_1 + h, y_1) &= \int_{\substack{(x_0,y_0)\\C}}^{(x_1,y_1)} P\,dx + Q\,dy + \int_{\substack{(x_1,y_1)\\C_1}}^{(x_1+h,y_1)} P\,dx + Q\,dy \\ &= F(x_1, y_1) + \int_{x_1}^{x_1+h} P(x, y_1)\,dx \\ &= F(x_1, y_1) + hP(\xi, y_1), \qquad (\xi \text{ between } x_1 \text{ and } x_1 + h) \end{aligned}$$

by the Mean Value Theorem for integrals. Therefore

$$\frac{F(x_1 + h, y_1) - F(x_1, y_1)}{h} = P(\xi, y_1)$$

If we let h approach 0, we conclude that

$$\frac{\partial F}{\partial x}(x_1, y_1) = P(x_1, y_1)$$

Similarly, $\partial F/\partial y = Q$ in D and the theorem is proved.

Remark. The function F is not unique, for we can clearly add a constant. Also if

$$\frac{\partial F_1}{\partial x} = \frac{\partial F}{\partial x} \quad \text{and} \quad \frac{\partial F_1}{\partial y} = \frac{\partial F}{\partial y} \quad \text{in } D$$

then $\nabla(F_1 - F) = \mathbf{0}$ in D, so that $F_1 = F +$ constant. Hence, all solutions are given by $F +$ const, where F is one solution. (The kernel of the mapping ∇ consists of all constant functions in D).

THEOREM 19. *Let $P(x, y)$ and $Q(x, y)$ be continuous in the open region D. Then the line integral $\int P\,dx + Q\,dy$ is independent of path in D if, and only if,*

$$\oint_C P\,dx + Q\,dy = 0 \tag{13-143}$$

for every piecewise smooth simple closed path C in D.

PROOF. Let us suppose that the line integral is independent of path and let C be a simple closed path as in Figure 13-52. We choose distinct points A and B on C. Then we can decompose C into a path C_1 from A to B and a path C_2 from B to A. Hence

$$\oint_C P\,dx + Q\,dy = \int_{A \atop C_1}^{B} P\,dx + Q\,dy + \int_{B \atop C_2}^{A} P\,dx + Q\,dy$$
$$= \int_{A \atop C_1}^{B} P\,dx + Q\,dy - \int_{A \atop C_2'}^{B} P\,dx + Q\,dy$$

By independence of path the last two integrals are equal and (13-143) holds true.

Conversely, let (13-143) hold true for all simple closed paths in D. If C_3 and C_4 are two simple paths from M to N in D (Figure 13-52), crossing only at points $L_1, \ldots, L_k$, then the integral along C_3 from M to L_1 equals the integral along C_4 from M to L_1, since the integral along C_4 from M to L_1 followed by that along C_3, with direction reversed, from L_1 to M is 0 by (13-143). Similarly, the integral from L_1 to L_2 along C_3 is the same as that along C_4, and so on. Adding these results, we conclude that

$$\int_{M \atop C_3}^{N} P\,dx + Q\,dy = \int_{M \atop C_4}^{N} P\,dx + Q\,dy \tag{13-144}$$

The same result holds true even if C_3 and C_4 are not simple or cross infinitely often, as is shown in advanced texts. Accepting this result, we conclude from (13-144) that the line integral $\int P\,dx + Q\,dy$ is independent of path in D.

THEOREM 20. *Let $P(x, y)$ and $Q(x, y)$ have continuous first partial derivatives in the open region D. If the line integral $\int P\,dx + Q\,dy$ is independent of path in D, then*

$$\frac{\partial P}{\partial y} = \frac{\partial Q}{\partial x} \qquad \text{in } D$$

That is, curl $(P\mathbf{i} + Q\mathbf{j}) = \mathbf{0}$ *in D.*

PROOF. By Theorem 18 there is a function F such that $P = \partial F/\partial x$, $Q = \partial F/\partial y$. Hence

$$\frac{\partial P}{\partial y} = \frac{\partial^2 F}{\partial y\,\partial x}, \qquad \frac{\partial Q}{\partial x} = \frac{\partial^2 F}{\partial x\,\partial y}$$

Since all these derivatives are continuous, we know that $\partial^2 F/\partial x \partial y = \partial^2 F/\partial y \partial x$ (Theorem 13 in Section 12-20). Therefore, $\partial P/\partial y = \partial Q/\partial x$ in D.

We now come to the main theorem of the section. The theorems thus far have made no reference to the type of region D. In the next theorem we require that D be *simply connected;* that is, that D have no "holes," as suggested in Figure 13-53*a*. A region that is not simply connected is called *multiply connected;* it is *doubly connected* if it has one hole, *triply connected* if it has two holes, and so on. For our purposes here we can define an open region D to be simply connected if every simple closed path C in D forms the boundary of a bounded closed region R contained in D. This property clearly holds true for the region in Figure 13-53*a*. It does not hold true for *b* and *c*, since in these cases for some choices of C we cannot have R lying

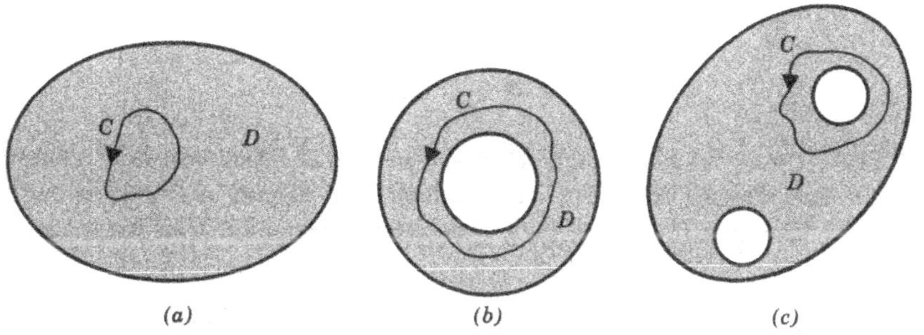

Figure 13-53 Simply connected and multiply connected open regions. (a) Simply connected; (b) doubly connected; (c) triply connected.

wholly in D. In general, when C goes around a hole, there is trouble. The reason for the importance of the property of simple connectedness is that, when it holds true, one can apply Green's theorem freely; when it fails, Green's theorem is inapplicable to certain simple closed curves. For example, we are not permitted to apply Green's theorem to the line integral

$$\oint_C \frac{\sin x}{x^2+y^2}\,dx + \frac{\cos x}{x^2+y^2}\,dy$$

where C is the circle $x^2 + y^2 = 1$. Here our coefficients P and Q are defined and differentiable in the open region D, consisting of the whole xy-plane minus one point: the origin $(0, 0)$. This exceptional point is a "hole" in D, so that D is doubly connected. Since C goes around the hole, we cannot apply Green's theorem.

THEOREM 21. *Let D be a simply connected open region in the xy-plane. Let P(x, y) and Q(x, y) have continuous first partial derivatives in D and let*

$$\frac{\partial P}{\partial y} = \frac{\partial Q}{\partial x} \qquad \text{in } D.$$

Then $P\,dx + Q\,dy$ is an exact differential; that is, there is a function F in D such that

$$\frac{\partial F}{\partial x} = P, \qquad \frac{\partial F}{\partial y} = Q$$

Remark. In vector language, the theorem states: if curl $(P\mathbf{i} + Q\mathbf{j}) = \mathbf{0}$ in the simply connected domain D, then $P\mathbf{i} + Q\mathbf{j} = \nabla F$ in D. Theorems 20 and 21 together roughly state: in simply connected domains in the xy-plane, the irrotational vector fields and the gradient fields are the same.

PROOF OF THEOREM 21. Let C be a piecewise smooth simple closed path in D, as in Figure 13-53*a*. Since D is simply connected, C is the boundary of a bounded closed region R lying in D, and Green's theorem is applicable:

$$\oint_C P\,dx + Q\,dy = \iint_R \left(\frac{\partial Q}{\partial x} - \frac{\partial P}{\partial y}\right) dA$$

Since $\partial Q/\partial x = \partial P/\partial y$ in D, the right side is 0. Hence

$$\oint_C P\,dx + Q\,dy = 0$$

for every piecewise smooth simple closed path in D. Therefore, by Theorem 19, the line integral $\int P\,dx + Q\,dy$ is independent of path in D and, accordingly, by Theorem 18, $P\,dx + Q\,dy$ is an exact differential. Thus the theorem is proved.

For Example 1, $P = y$, $Q = x$ and $\partial Q/\partial x = \partial P/\partial y$. Thus $y\,dx + x\,dy$ must be an exact differential. In this case we see at once that $y\,dx + x\,dy = d(xy)$. In other cases, we may not recognize the function F so easily. We can then use (13-142) in order to find F.

EXAMPLE 2. $2xy^2e^{x^2y}\,dx + e^{x^2y}(1 + x^2y)\,dy$. Here

$$P = 2xy^2e^{x^2y}, \qquad Q = e^{x^2y}(1 + x^2y)$$

$$\frac{\partial P}{\partial y} = e^{x^2y}(4xy + 2x^3y^2), \qquad \frac{\partial Q}{\partial x} = e^{x^2y}(2xy + 2xy(1 + x^2y))$$

and $\partial P/\partial y = \partial Q/\partial x$ for all (x, y). Our region D is the whole xy-plane; there are no holes, and D is simply connected. Therefore, $P\,dx + Q\,dy = dF$. By (13-142), we take

$$F(x_1, y_1) = \int_{(0,0)}^{(x_1,y_1)} 2xy^2e^{x^2y}\,dx + e^{x^2y}(1 + x^2y)\,dy$$

and the choice of C is at our disposal. For example, let us take C to be the

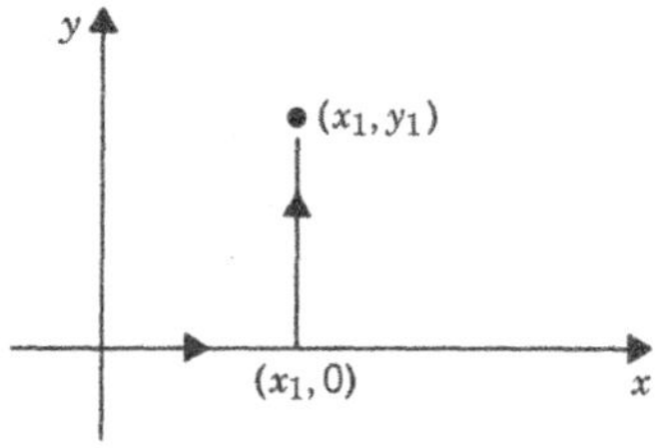

Figure 13-54 Path for Example 2.

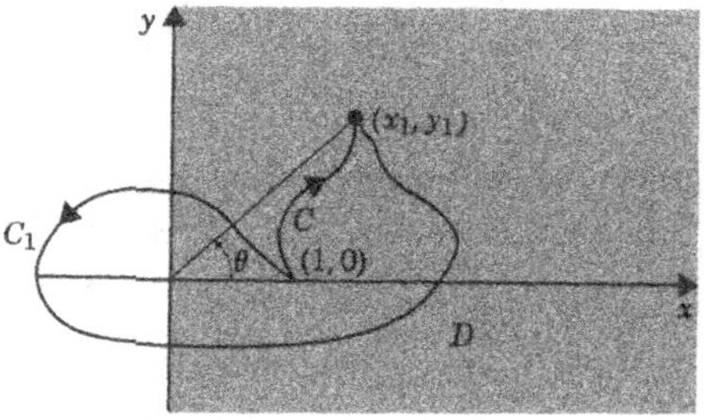

Figure 13-55 Example 3.

broken line from $(0, 0)$ to $(x_1, 0)$ to (x_1, y_1) as in Figure 13-54. Then we find that

$$F(x_1, y_1) = \int_0^{x_1} 0\,dx + \int_0^{y_1} e^{x_1^2y}(1 + x_1^2y)\,dy$$

In the second integral we treat x_1 as a constant. If $x_1 \neq 0$, we integrate by parts to find that

$$F(x_1, y_1) = \frac{e^{x_1^2y}}{x_1^2}(1 + x_1^2y)\Big|_{y=0}^{y=y_1} - \int_0^{y_1} e^{x_1^2y}\,dy = y_1e^{x_1^2y_1}$$

For $x_1 = 0$, we find that

$$F(0, y_1) = \int_0^{y_1} dy = y_1$$

Hence, $F(x_1, y_1) = y_1 e^{x_1^2 y_1}$ for all (x_1, y_1); that is, $F(x, y) = ye^{x^2y}$. The general function having the given differential $P\,dx + Q\,dy$ is $ye^{x^2y} + c$.

Check: $\dfrac{\partial F}{\partial x} = 2xy^2e^{x^2y} = P, \qquad \dfrac{\partial F}{\partial y} = e^{x^2y}(1 + x^2y) = Q$

EXAMPLE 3 $\qquad P = \dfrac{-y}{x^2 + y^2}, \qquad Q = \dfrac{x}{x^2 + y^2}$

Here

$$\frac{\partial P}{\partial y} = \frac{y^2 - x^2}{(x^2 + y^2)^2}, \qquad \frac{\partial Q}{\partial x} = \frac{y^2 - x^2}{(x^2 + y^2)^2}$$

and $\partial P/\partial y = \partial Q/\partial x$. But, we cannot choose D as the entire xy-plane, since P and Q are discontinuous at the origin $(0, 0)$. We are permitted to use any simply connected open region D which does not contain the origin. For example, we can let D be the "right half-plane" $x > 0$, as in Figure 13-55. Then we can take

$$F(x_1, y_1) = \int_{(1,0)}^{(x_1,y_1)} \frac{-y\,dx + x\,dy}{x^2 + y^2}$$

We can evaluate F in detail, as in Example 2. However, we recognize the line integral as that for the angle function of Sections 5-4 and 6-6. The value $F(x_1, y_1)$ is simply the change in the polar angle θ as we go from $(1, 0)$ to (x_1, y_1), and θ varies continuously on the path. Since we are restricted to the right half-plane, we can take $\theta = 0$ at $(1, 0)$ and then, by continuity, we have $-\pi/2 < \theta < \pi/2$ in D. Therefore

$$F(x_1, y_1) = \text{polar angle } \theta \text{ at } (x_1, y_1), \text{ where } -\frac{\pi}{2} < \theta < \frac{\pi}{2}$$

We can also write

$$F(x_1, y_1) = \text{Tan}^{-1} \frac{y_1}{x_1}$$

since the principal value of the inverse tangent gives just the value we want.

If we allowed D to be the whole xy-plane minus the origin, then the value of F would depend on the choice of C. For the path C_1 of Figure 13-55, the value would be 2π more than along C. We note also that, for a simple closed path C_2 going around the "hole" at $(0, 0)$,

$$\oint_{C_2} \frac{-y\,dx + x\,dy}{x^2 + y^2} = 2\pi$$

since the polar angle increases by 2π on this path.

In other examples, it may happen that D has several holes, but the integral $\int P\,dx + Q\,dy$ around each hole is 0. When this happens, $P\,dx + Q\,dy = dF$ in all of D, even though D is not simply connected [Problems 3(e) and 4 below].

PROBLEMS

1. Evaluate:

(a) $\displaystyle\int_{C\,(1,1)}^{(4,8)} y\,dx + x\,dy$, C: $x = t^2$, $y = t^3$, $1 \le t \le 2$.

(b) $\displaystyle\int_{C\,(1,1)}^{(e,e^e)} y\,dx + x\,dy$, C: $x = e^{t^3}$, $y = e^{e^t}$, $0 \le t \le 1$.

(c) $\displaystyle\oint y\,dx + x\,dy$, C: $x = 100\cos t - \cos 10t$, $y = 100\sin t - \sin 10t$.

(d) $\displaystyle\int_{C\,(1,0)}^{(0,17)} y\,dx + x\,dy$ on any path C you choose.

(e) $\displaystyle\int_{C\,(1,1)}^{(3,2)} 2xy\,dx + x^2\,dy$, (f) $\displaystyle\int_{C\,(0,0)}^{(3,5)} e^{xy}(xy+1)\,dx + e^{xy}x^2\,dy$.

2. Test for exactness and, when the expression is exact, find the functions having the given differential:

(a) $(x^2 + 3x^2y^5)\,dx + (5x^3y^4 + e^y)\,dy$

(b) $\cos(x+2y)\,dx + 2\cos(x+2y)\,dy$

(c) $y\,dx - x\,dy$

(d) $e^x(1+\sin^2 y)\,dx + 2e^x \sin y \cos y\,dy$

(e) $(x^2y^3 - y)\,dx + (3xy^4 + x^2)\,dy$

(f) $(x^3 + xy^2)\,dx + (x^2y + y^3)\,dy$

(g) $e^x \sin y\,dx + xe^x \cos y\,dy$

(h) $y\sin(xy)\,dx + x\sin(xy)\,dy$

(i) $\dfrac{y\,dx - x\,dy}{y^2}$

(j) $\dfrac{-y\,dx + (x-1)\,dy}{(x-1)^2 + y^2}$

(k) $\dfrac{x\,dx + y\,dy}{(x^2+y^2)^2}$

(l) $e^{F(x,y)}(F_x\,dx + F_y\,dy)$

3. (a) Let R be a region bounded by two disjoint piecewise smooth simple closed curves C_1, C_2 as in Figure 13-56. Let P and Q have continuous first partial derivatives in an open region containing R. Show that

$$\oint_{C_1} P\,dx + Q\,dy - \oint_{C_2} P\,dx + Q\,dy = \iint_R \left(\frac{\partial Q}{\partial x} - \frac{\partial P}{\partial y}\right) dA$$

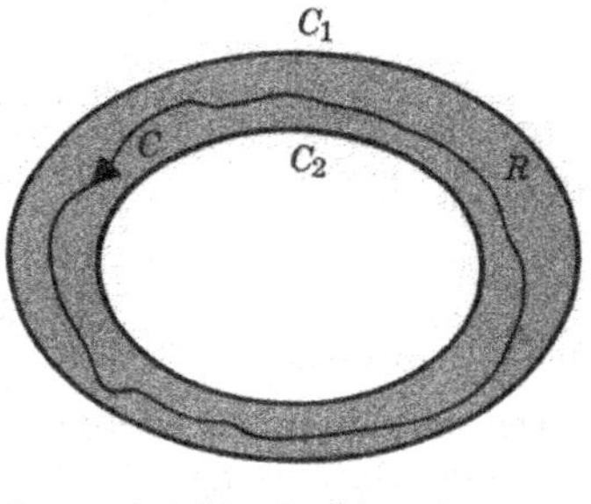

Figure 13-56 Problem 3.

(*Hint.* Split R into two regions for which Green's theorem is applicable and add the results.)

(b) Extend the result of part (a) to a region bounded by several curves $C_1, \ldots, C_n$.

(c) Show that if R is as in Figure 13-56 and $\partial Q/\partial x = \partial P/\partial y$ in R, then

$$\oint_{C_1} P\,dx + Q\,dy = \oint_{C_2} P\,dx + Q\,dy = \oint_{C} P\,dx + Q\,dy$$

where C is a piecewise smooth simple closed path separating C_1 and C_2.

(d) Show that if D is multiply connected, $\partial Q/\partial x = \partial P/\partial y$ in D and C_1, C_2 are piecewise smooth simple closed paths in D going around the same holes of D, then

$$\oint_{C_1} P\,dx + Q\,dy = \oint_{C_2} P\,dx + Q\,dy$$

(e) Let D be a triply connected open region. Let $\partial Q/\partial x = \partial P/\partial y$ in D. Let

$$\oint_{C_1} P\,dx + Q\,dy = k_1, \qquad \oint_{C_2} P\,dx + Q\,dy = k_2$$

where C_1 encloses just one of the holes of D, C_2 encloses only the other. Show that for each two points (x_0, y_0), (x_1, y_1) of D the line integral $\int P\,dx + Q\,dy$ from (x_0, y_0) to (x_1, y_1) along a path in D, although generally dependent on path, has for two different paths values differing by $n_1k_1 + n_2k_2$, where n_1, n_2 are integers (possibly 0 or negative). Show that if k_1 and k_2 happen to be 0, the line integral is independent of path in D.

4. For $P(x, y) = x/(x^2 + y^2)$, $Q(x, y) = y/(x^2 + y^2)$, the natural region D is the xy-plane minus the origin. Prove: $\partial P/\partial y = \partial Q/\partial x$ in D and, even though D has a hole, $P\,dx + Q\,dy = dF$, so that $\int P\,dx + Q\,dy$ is independent of path in D.

13-15 THE DIVERGENCE THEOREM AND STOKES' THEOREM IN SPACE

The divergence theorem in the plane (Equation (13-135) asserts that, for a planar vector field $\mathbf{v}$,

$$\iint_R \operatorname{div} \mathbf{v}\,dA = \oint_C v_n\,ds$$

It is natural to surmise that there is an analogous formula for a vector field $\mathbf{v}$ in space:

$$\iiint_R \operatorname{div} \mathbf{v}\,dV = \iint_{\mathcal{S}} v_n\,dS \qquad (13\text{-}150)$$

where R is a solid region bounded by the surface $\mathcal{S}$, $\mathbf{n}$ is an outer unit normal vector to $\mathcal{S}$, as in Figure 13-57, $v_n = \mathbf{v} \cdot \mathbf{n}$, and dS is the element of surface

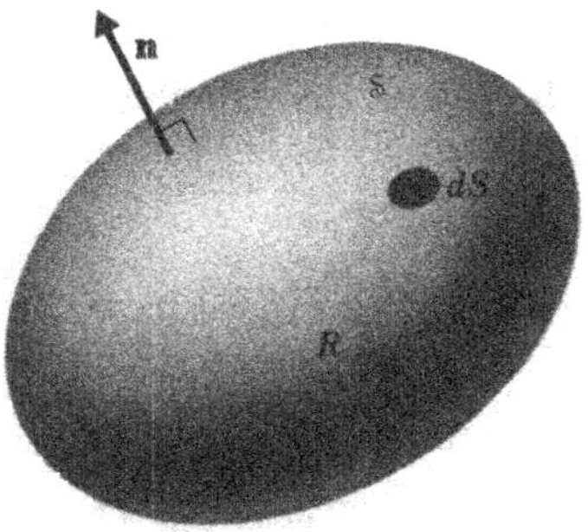

Figure 13-57 Divergence Theorem in space.

area on $\mathcal{S}$ (Section 13-9); the right side of (13-150) is a *surface integral* (Section 13-10). Formula (13-150) is called the *divergence theorem* (or *Gauss's theorem*) in space. It can be proved under quite general conditions [see texts on advanced calculus or the book, *Foundations of Potential Theory*, by O. D. Kellogg (Springer, Berlin, 1929)]. We note that for $\mathbf{v} = \mathbf{r} = x\mathbf{i} + y\mathbf{j} + z\mathbf{k}$ we have

$$\operatorname{div} \mathbf{v} = \frac{\partial}{\partial x}(v_x) + \frac{\partial}{\partial y}(v_y) + \frac{\partial}{\partial z}(v_z) = 1 + 1 + 1 = 3$$

Hence, after division by 3, formula (13-150) becomes

$$\iiint_R dV = \frac{1}{3} \iint_{\mathcal{S}} r_n \, dS \tag{13-151}$$

This is a useful expression for volume (see Section 11-15). For a rectangular solid region R: $a_1 \le x \le a_2$, $b_1 \le y \le b_2$, $c_1 \le z \le c_2$ (Figure 13-58) we can

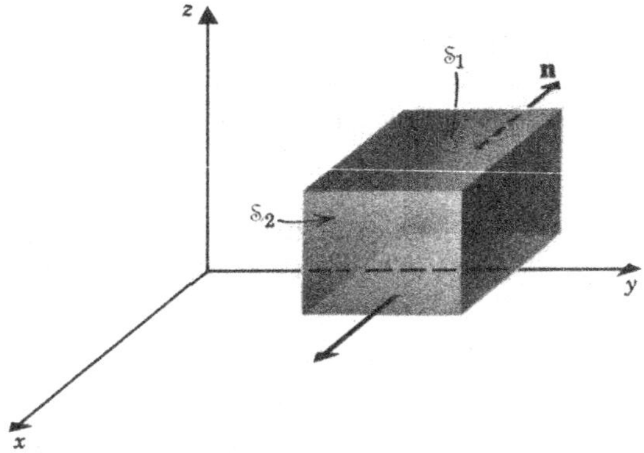

Figure 13-58 Divergence Theorem for rectangular solid.

verify (13-150) directly. On the boundary surface $\mathcal{S}_1$: $x = a_1$ the normal $\mathbf{n}$ is $-\mathbf{i}$ and $v_n = \mathbf{v} \cdot \mathbf{n} = -v_x(a_1, y, z)$. Similarly on the boundary surface $\mathcal{S}_2$: $x = a_2$, v_n is $v_x(a_2, y, z)$. On both $\mathcal{S}_1$ and $\mathcal{S}_2$, $dS = dy\,dz$, and thus

$$\iint_{\mathcal{S}_1} v_n \, dS + \iint_{\mathcal{S}_2} v_n \, dS = \int_{c_1}^{c_2} \int_{b_1}^{b_2} [v_x(a_2, y, z) - v_x(a_1, y, z)] \, dy \, dz$$

$$= \int_{c_1}^{c_2} \int_{b_1}^{b_2} \int_{a_1}^{a_2} \frac{\partial v_x}{\partial x} \, dx \, dy \, dz = \iiint_R \frac{\partial v_x}{\partial x} \, dV$$

Similar results apply to the surfaces $\mathcal{S}_3$: $y = b_1$ and $\mathcal{S}_4$: $y = b_2$ and to the surfaces $\mathcal{S}_5$: $z = c_1$ and $\mathcal{S}_6$: $z = c_2$. Adding the integrals for all six surfaces gives

$$\iint_{\mathcal{S}} v_n \, dS = \iiint_R \left(\frac{\partial v_x}{\partial x} + \frac{\partial v_y}{\partial y} + \frac{\partial v_z}{\partial z} \right) dV = \iiint_R \operatorname{div} \mathbf{v} \, dV$$

as expected. Other special cases are left to the exercises.

Physical Interpretation of Divergence Theorem. The surface integral on the right of (13-150) is called the *flux* of the vector field **v** across the surface $\mathcal{S}$ in the direction of **n**. When **v** is the velocity field of a fluid motion in space, for each small portion dS of $\mathcal{S}$ the product $v_n\, dS$ is the volume of a cylindrical solid of base dS and altitude v_n; since v_n is distance per unit time $v_n\, dS$ is *volume* per unit time. Thus for a fluid, the flux of **v** across $\mathcal{S}$ is the rate at which fluid is *leaving* R, measured in volume per unit time. If we apply (13-150) to a small region ΔR containing a point (x_0, y_0, z_0) and also apply property 8 of Section 13-6, we conclude that at (x_0, y_0, z_0)

$$\operatorname{div} \mathbf{v} = \lim_{\text{diam of } \Delta R \to 0} \frac{\displaystyle\iint_{\Delta \mathcal{S}} v_n\, dS}{\Delta V} \tag{13-152}$$

where $\Delta\mathcal{S}$ is the boundary surface of ΔR. Thus as in two dimensions (Section 13-13) the divergence of a vector field can be interpreted as *limit of flux per unit volume*. For the case of an *incompressible* fluid, the flux across $\Delta\mathcal{S}$ is 0, since the net rate of loss of fluid from ΔR is 0; hence, for an incompressible fluid of velocity field **v**, one has

$$\operatorname{div} \mathbf{v} \equiv 0$$

For a general fluid one can also let $\mathbf{v} = \delta\mathbf{u}$, where **u** is the velocity and $\delta(x, y, z, t)$ is the density. The flux of **v** across $\mathcal{S}$ now equals the rate of loss of *mass* from R. The right side of (13-152) gives the limit of the rate of loss of mass per unit volume; but this is simply $-\partial\delta/\partial t$, the rate of decrease of the density δ. Therefore by (13-152)

$$\operatorname{div}(\delta\mathbf{u}) + \frac{\partial\delta}{\partial t} = 0 \tag{13-153}$$

This is the *continuity equation of fluid mechanics.*

EXAMPLE 1. Let $\mathbf{v} = 3x\mathbf{i} - 2y\mathbf{j} + 5z\mathbf{k}$. Evaluate $\iint_{\mathcal{S}} v_n dS$ on the boundary $\mathcal{S}$ of a solid torus R in space, where **n** is the outer normal.

Solution. By the divergence theorem

$$\iint_{\mathcal{S}} v_n\, dS = \iiint_R \operatorname{div} \mathbf{v}\, dV = \iiint_R (3 - 2 + 5)\, dV = 6V$$

where V is the volume of the torus.

EXAMPLE 2. Let $\mathbf{v} = x^2\mathbf{i} + xy\mathbf{j} + xz\mathbf{k}$. Evaluate $\iint_{\mathcal{S}} v_n\, dS$ over the surface of the cube $0 \le x \le 1, 0 \le y \le 1, 0 \le z \le 1$, where **n** is the outer normal.

Solution. By the divergence theorem

$$\iint_{\mathcal{S}} v_n\, dS = \iiint_R (2x + x + x)\, dV = 4\int_R x\, dm = 4M_0\bar{x}$$

where M_0 is the mass of the cube when we take a constant density of 1, and $(\bar{x}, \bar{y}, \bar{z})$ is the corresponding center of mass. Clearly $M_0 = 1$ and $\bar{x} = \frac{1}{2}$, so that the integral equals 2.

Line Integrals in Space and Stokes' Theorem. The concept of line integral (Section 13-11) extends easily to paths in space. If C is a piecewise smooth path: $\mathbf{r} = \mathbf{F}(t) = f(t)\mathbf{i} + g(t)\mathbf{j} + h(t)\mathbf{k}$, $a \leq t \leq b$, and a continuous vector field $\mathbf{v} = X\mathbf{i} + Y\mathbf{j} + Z\mathbf{k}$ is defined along C, then the line integral is given formally by the equation

$$\int_C X\,dx + Y\,dy + Z\,dz = \lim \sum (X\,\Delta x + Y\,\Delta y + Z\,\Delta z)$$

where we have omitted the usual subscripts and other details, to be filled in as in Section 13-11. We can also define the line integral directly in terms of arc length and the unit tangent vector $\mathbf{T}$ along C:

$$\int_C X\,dx + Y\,dy + Z\,dz = \int_C \mathbf{v}\cdot\mathbf{T}\,ds = \int_C v_T\,ds = \int_0^L v_T\,ds$$

Here $\mathbf{T} = x'(s)\mathbf{i} + y'(s)\mathbf{j} + z'(s)\mathbf{k}$. In terms of the given parameter on C,

$$\int_C X\,dx + Y\,dy + Z\,dz = \int_a^b \left[X(f(t), g(t), h(t))\frac{dx}{dt} + Y(\cdots)\frac{dy}{dt} + Z(\cdots)\frac{dz}{dt}\right] dt$$

The discussion of these relations is the same as in Section 13-11. As in that section, the line integral can be interpreted as the *work* done by the force $X\mathbf{i} + Y\mathbf{j} + Z\mathbf{k}$ along C.

In Section 13-13 we saw that Green's theorem could be written in the form:

$$\oint_C v_T\,ds = \iint_R \operatorname{curl}_z \mathbf{v}\,dA$$

We can interpret the double integral on the right as a surface integral over a surface $\mathcal{S}$ that happens to be a region R in the xy-plane; we can also replace $\operatorname{curl}_z \mathbf{v}$ by $\operatorname{curl}_n \mathbf{v}$, where $\operatorname{curl}_n \mathbf{v}$ denotes the component of curl $\mathbf{v}$ in the direction of the normal $\mathbf{n}$ (here chosen as $\mathbf{k}$) to the surface $\mathcal{S}$. With these replacements the equation becomes

$$\int_C v_T\,ds = \iint_{\mathcal{S}} \operatorname{curl}_n \mathbf{v}\,dS \tag{13-154}$$

(We have also dropped the circular arrow on the line integral, as explained below.) Equation (13-154) is *Stokes' theorem in space*. It is valid for a differentiable vector field $\mathbf{v}$, an arbitrary piecewise smooth surface $\mathcal{S}$ in space, having normal vector $\mathbf{n}$, whose edge is formed of the piecewise smooth simple closed path C, properly directed; see Figure 13-59. For a full discussion one is referred

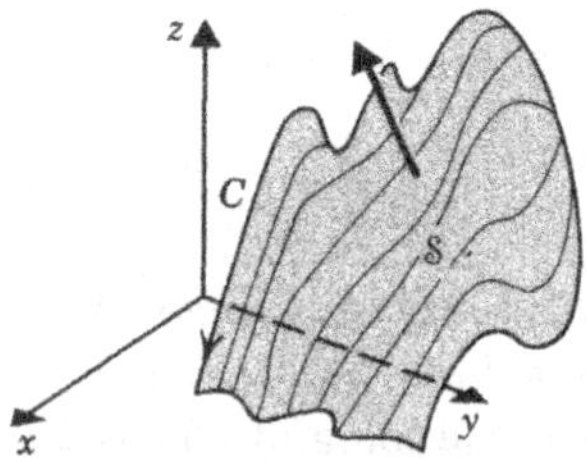

Figure 13-59 Stokes' theorem in space.

to texts on advanced calculus. We remark here that choice of the normal $\mathbf{n}$ on $\mathcal{S}$ and of the direction on C requires considerable care. In fact, for some surfaces (said to be *nonorientable*), there is no proper way of defining $\mathbf{n}$ on $\mathcal{S}$, so that the theorem is inapplicable! The most famous example of this phenomenon is the "one-sided" surface called the *Möbius strip* (Figure 13-60); this can be created physically by taking a long rectangular strip of paper, giving it a half-twist and then gluing the ends together.

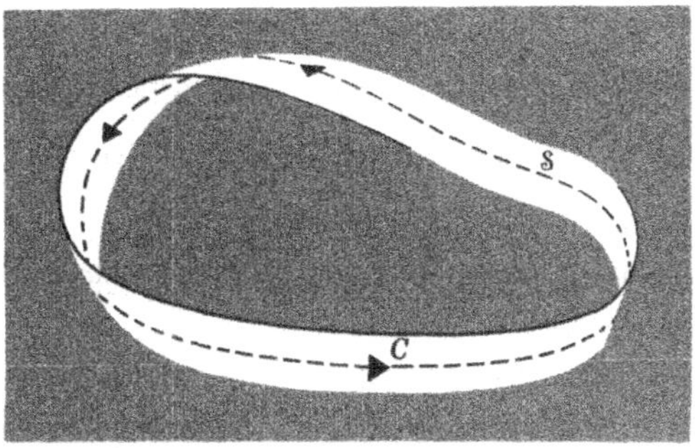

Figure 13-60 Möbius strip, nonorientable.

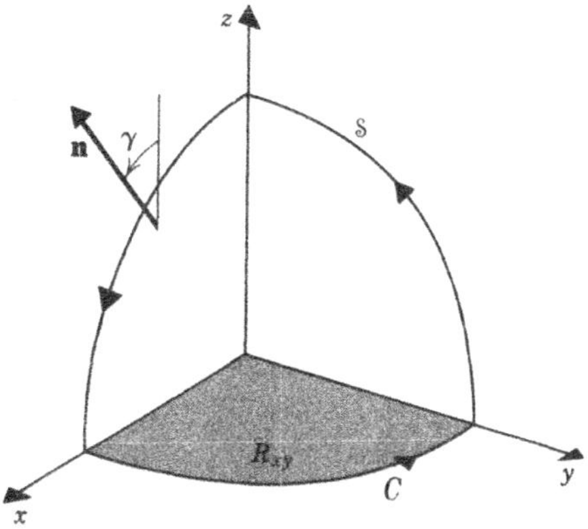

Figure 13-61 Example 3.

EXAMPLE 3. Let $\mathcal{S}$ be the portion of the sphere $x^2 + y^2 + z^2 = 1$ in the first octant, as in Figure 13-61, with normal $\mathbf{n} = x\mathbf{i} + y\mathbf{j} + z\mathbf{k}$. Let C be the edge of $\mathcal{S}$, formed of three circular arcs and directed as in the figure. (This is the proper direction on C for Stokes' theorem.) Let $\mathbf{v} = x^2\mathbf{i} + 2xy\mathbf{j} + xz\mathbf{k}$. Then Stokes' theorem asserts that

$$\int_C x^2\,dx + 2xy\,dy + xz\,dz = \iint_{\mathcal{S}} \operatorname{curl}_n (x^2\mathbf{i} + 2xy\mathbf{j} + xz\mathbf{k})\,dS$$

We evaluate each side separately to verify the equality. The line integral can be written as the sum of three integrals on the paths: $x = \cos t$, $y = \sin t$, $z = 0, 0 \leq t \leq \pi/2$; $x = 0, y = \cos t, z = \sin t, 0 \leq t \leq \pi/2$; $x = \sin t, y = 0$, $z = \cos t$, $0 \leq t \leq \pi/2$. Hence

$$\int_C v_T\,ds = \int_0^{\pi/2} [\cos^2 t\,(-\sin t) + 2\cos^2 t \sin t]\,dt$$
$$+ \int_0^{\pi/2} 0\,dt + \int_0^{\pi/2} [\sin^2 t \cos t + \sin t \cos t\,(-\sin t)]\,dt = \frac{1}{3}$$

Now $\operatorname{curl} \mathbf{v} = -z\mathbf{j} + 2y\mathbf{k}$, so that $\operatorname{curl}_n \mathbf{v} = (\operatorname{curl} \mathbf{v}) \cdot \mathbf{n} = yz$. Also $dS = \sec\gamma\,dx\,dy$, as in Section 13-9, and $\sec\gamma = 1/z$, since $\cos\gamma = \mathbf{n} \cdot \mathbf{k} = z$. (Note that $\|\mathbf{n}\| = 1$.) Therefore, the surface integral is

$$\iint_{\mathcal{S}} yz\,dS = \iint_{R_{xy}} y\,dx\,dy$$

where R_{xy} is the region $x^2 + y^2 \leq 1$, $x \geq 0$, $y \geq 0$. We find easily that the double integral equals $\frac{1}{3}$, so that the theorem is verified.

PROBLEMS

1. Evaluate $\iint_{\mathcal{S}} v_n\, dS$, where $\mathcal{S}$ is the boundary surface of the solid region R given, $\mathbf{v}$ is the vector field given, $\mathbf{n}$ is the unit outer normal to the surface:
   (a) $R: x^2 + y^2 + z^2 \leq 1$; $\mathbf{v} = xy^2\mathbf{i} + yz^2\mathbf{j} + xy\mathbf{k}$.
   (b) R: the tetrahedron of vertices $(0, 0, 0)$, $(1, 0, 0)$, $(0, 1, 0)$, $(0, 0, 1)$; $\mathbf{v} = x^2\mathbf{i} + xz\mathbf{j} + xy\mathbf{k}$.
   (c) R: the rectangular solid $0 \leq x \leq 1$, $0 \leq y \leq 2$, $0 \leq z \leq 3$; $\mathbf{v} = e^x \cos y\,\mathbf{i} + e^x \sin y\,\mathbf{j} + ze^x\,\mathbf{k}$.
   (d) R: the solid ellipsoid $2x^2 + 3y^2 + z^2 \leq 1$; $\mathbf{v} = x^2\mathbf{i} + xy\mathbf{j} + z^2\mathbf{k}$.
2. Let $\mathbf{v}$ be a differentiable vector field in space such that for every spherical surface $\mathcal{S}$ in space $\iint_{\mathcal{S}} v_n\, dS = 0$, where $\mathbf{n}$ is the outer unit normal vector on $\mathcal{S}$ (pointing away from the center of the sphere). Prove that $\operatorname{div} \mathbf{v} \equiv 0$.
3. Verify the correctness of the divergence theorem for a region R and vector field $\mathbf{v}$ as described (assume differentiability as needed):
   (a) R: the tetrahedron of vertices (a, b, c), $(a + h, b, c)$, $(a, b + k, c)$, $(a, b, c + l)$; $\mathbf{v} = Z(x, y, z)\mathbf{k}$. (*Hint.* The oblique face of the tetrahedron has equation

$$\frac{x-a}{h} + \frac{y-b}{k} + \frac{z-c}{l} = 1$$

   Evaluate the triple integral of $\operatorname{div} \mathbf{v}$ in the order $dz\, dy\, dx$.)
   (b) R: an arbitrary tetrahedron, $\mathbf{v}$: an arbitrary vector field. [*Hint.* For a tetrahedron as in part (a) the theorem is valid for $\mathbf{v} = X(x, y, z)\mathbf{i}$ and for $\mathbf{v} = Y(x, y, z)\mathbf{j}$ by the same reasoning as in part (a). Hence, conclude that the theorem holds for a general $\mathbf{v}$ for a tetrahedron as in part (a). For an arbitrary tetrahedron, decompose it into tetrahedra as in part (a), apply the divergence theorem to each and add the results.]
   (c) R: a region $0 \leq z \leq \psi(x, y)$, $\varphi_1(x) \leq y \leq \varphi_2(x)$, $a \leq x \leq b$; $\mathbf{v} = Z(x, y, z)\mathbf{k}$
4. Let $\mathbf{v}$ be a twice differentiable vector field in space. Let C be a circle in space. Let $\mathcal{S}_1$ and $\mathcal{S}_2$ be portions of two spherical surfaces, each having C as edge (Figure 13-62). Let a unit normal vector $\mathbf{n}$ be defined on $\mathcal{S}_1$, pointing away from the center, and on $\mathcal{S}_2$, pointing toward the center.
   Show by the divergence theorem that

$$\iint_{\mathcal{S}_1} (\operatorname{curl} \mathbf{v}) \cdot \mathbf{n}\, dS = \iint_{\mathcal{S}_2} (\operatorname{curl} \mathbf{v}) \cdot \mathbf{n}\, dS$$

   How is this result suggested by Stokes' theorem?

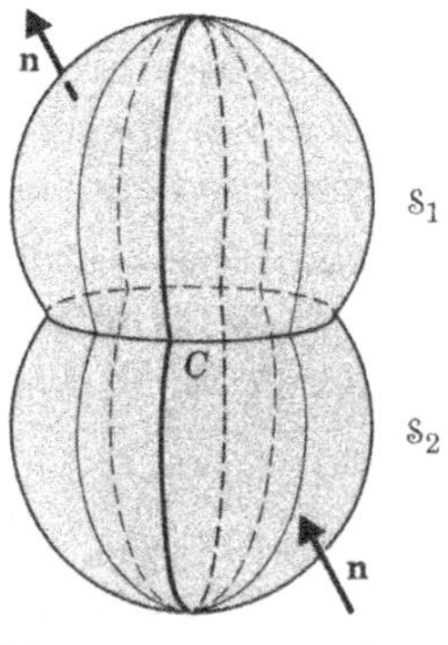

Figure 13-62 Problem 4.

5. The definition of independence of path for line integrals in space is the same as for line integrals in the plane (Section 13-14). Throughout this problem D is an open region in space, F is a function continuous in D, $\mathbf{v} = X\mathbf{i} + Y\mathbf{j} + Z\mathbf{k}$ is a vector field continuous in D. All paths are assumed piecewise smooth.

(a) Prove: if $\nabla F = \mathbf{v}$, then $\int X\,dx + Y\,dy + Z\,dz$ is independent of path in D (cf. Theorem 17 in Section 13-14).

(b) Prove: if $\int X\,dx + Y\,dy + Z\,dz$ is independent of path in D, then there is a function $G(x, y, z)$ in D such that $\nabla G = \mathbf{v}$ (cf. Theorem 18 in Section 13-14).

(c) Prove: $\int X\,dx + Y\,dy + Z\,dz$ is independent of path in D, if and only if $\int_C X\,dx + Y\,dy + Z\,dz = 0$ for every closed path in D (cf. Theorem 19 in Section 13-14).

(d) Prove: if X, Y, Z have continuous first partial derivatives in D and $\int X\,dx + Y\,dy + Z\,dz$ is independent of path in D, then $\operatorname{curl} \mathbf{v} \equiv \mathbf{0}$ in D; that is, $\partial X/\partial y = \partial Y/\partial x$, $\partial X/\partial z = \partial Z/\partial x$, $\partial Y/\partial z = \partial Z/\partial y$ (cf. Theorem 20 in Section 13-14).

(e) Let us call D simply connected if every piecewise smooth simple closed path in D forms the edge of a smooth surface $\mathcal{S}$ in D to which Stokes' theorem is applicable. (This definition differs from the standard one, but for normal applications the difference is of no significance). Prove: if D is simply connected, $\mathbf{v}$ is differentiable and $\operatorname{curl} \mathbf{v} \equiv \mathbf{0}$ in D, then $\mathbf{v} = \operatorname{grad} F$ for some F in D (cf. Theorem 21 in Section 13-14).

14
ORDINARY DIFFERENTIAL EQUATIONS

Differential equations have appeared at several points in earlier chapters, especially in Chapter 7. We study them afresh in this chapter, without reference to the earlier work, except for proofs of a few theorems.

14-1 BASIC CONCEPTS

By differentiating the equation $y = x^2 + \sin x$, we obtain the equation

$$y' = 2x + \cos x$$

for the derivative of the given function. From the equation $x^2 + y^2 = 1$ we obtain, by implicit differentiation, $2x + 2yy' = 0$ or

$$y' = -\frac{x}{y}$$

In both examples, differentiation has led us to a *relation* between the derivative y', y itself, and x, where y is a function of x. Such a relation is said to be an *ordinary differential equation of first order*. We can write the equation in the form $G(x, y, y') = 0$. In this chapter we consider only equations which are explicitly solvable for y'; in that case the equation has the form:

$$y' = F(x, y) \tag{14-10}$$

Our examples suggest that each differential equation (14-10) arises by differentiating an equation defining y as a function of x. However, differential equations arise in many other ways, especially from problems in the sciences. One can easily write down many first-order ordinary differential equations

$$\begin{array}{ll} \text{(i) } y' = y + x^2 & \text{(ii) } y' = y^2 \\ \text{(iii) } y' = x^2 - y^2 & \text{(iv) } y' = e^y + \tan x \end{array} \tag{14-11}$$

It is not at once evident how each of these equations follows from an equation relating x and y. In fact, our fundamental problem is to find, for each equation of form (14-10), all functions $y = f(x)$ such that the relation (14-10) is satisfied:

that is, to find all functions $f(x)$ such that

$$f'(x) \equiv F[x, f(x)]$$

for all x in the interval in which f is defined. Such a function f is said to be a *solution* of the differential equation (14-10). For example, the function $y = -x^2 - 2x - 2$ is a solution of part (i) of Equation (14-11), since

$$y = -x^2 - 2x - 2, \qquad y' = -2x - 2 = y + x^2$$

for all x. However, this is not the only solution of the differential equation. For example, $y = e^x - x^2 - 2x - 2$ is also a solution, since if $y = e^x - x^2 - 2x - 2$, then $y' = e^x - 2x - 2 = y + x^2$, for all x. We shall in due course find all solutions of this differential equation: $y' = y + x^2$. The collection of all solutions of a differential equation is called the *general solution* of the equation.

The differential equation (iii) in (14-11) happens to be a difficult one, and it is not at all easy to "guess" solutions. (Try it!) This leads us to ask whether we are sure that every equation of form (14-10) must have solutions—say, at least one. This question is answered by the famous Existence Theorem, stated in detail at the end of this section. In essence, the Existence Theorem states that under "reasonable" hypotheses on F, every differential equation $y' = F(x, y)$ has solutions; in fact, for every point (x_0, y_0) in the xy-plane (with the exception of certain points where F misbehaves) *there is one and only one solution $y = f(x)$ of (14-10) whose graph passes through (x_0, y_0)*.

There is one case in which the problem at hand reduces to a familiar one: namely, the equation

$$y' = F(x) \tag{14-12}$$

where F is continuous for $a < x < b$. Here we know all the solutions: they are given by

$$y = \int F(x)\,dx + C$$

or by

$$y = G(x) + C \tag{14-13}$$

where $G(x)$ is one indefinite integral of F over the interval $a < x < b$. For example, we can take

$$G(x) = \int_{x_0}^{x} F(u)\,du, \qquad a < x < b$$

where x_0 is a point of the interval. (That $G' = F$ follows from the Fundamental Theorem of Calculus, Section 4-17.) The graphs of the functions (14-13) have the appearance of Figure 14-1. Changing C merely raises or lowers the graph. Hence, it is clear that there is precisely one solution through each point of the open region $a < x < b$, $-\infty < y < \infty$.

EXAMPLE 1. Find the solution of $y' = xe^x$ such that $y = 2$ when $x = 3$.

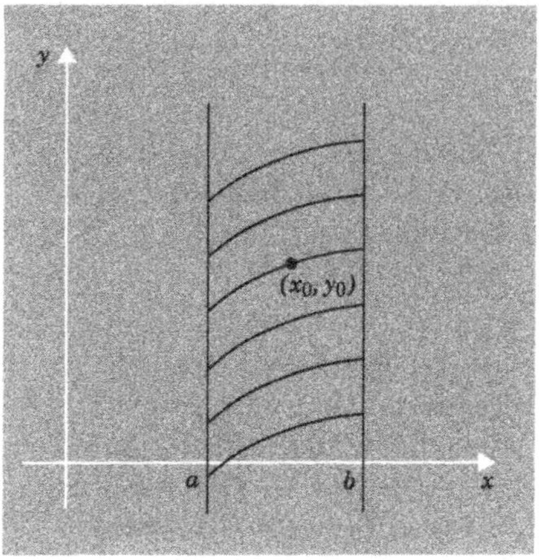

Figure 14-1 Solutions of $y' = F(x)$.

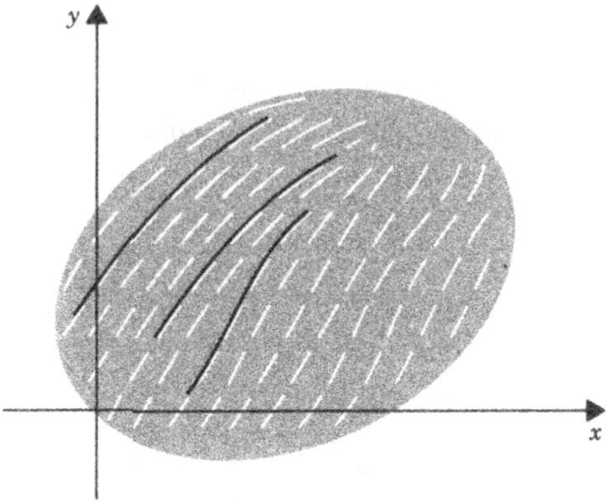

Figure 14-2 Field of line elements determined by equation $y' = F(x, y)$.

Solution. The general solution is

$$y = \int xe^x\, dx + C = xe^x - e^x + C$$

Setting $y = 2$, $x = 3$, we find that $2 = 3e^3 - e^3 + C$, so that $C = 2(1 - e^3)$ and the solution sought is

$$y = xe^x - e^x + 2(1 - e^3)$$

For the general equation $y' = F(x, y)$ we have no such simple formula giving all solutions. However, we can give a graphical meaning to the equation and, hence, see why we expect solutions. Let F be given in a certain open region D in the xy-plane. Then at each point (x, y) of D, Equation (14-10) gives a corresponding value of y'; that is, we know the slope of the tangent to a solution, even though we do not know the solution. Let us draw a little piece of this tangent line near (x, y) and let us repeat the process at many points of D. The result is a pattern as in Figure 14-2. We call this pattern a *field of line elements*. As long as F behaves reasonably, the slopes of these line elements vary gradually as one varies the point (x, y). Hence, the line elements themselves suggest smooth curves having the line elements as tangent lines; each such curve is (approximately) a solution of the differential equation (14-10). A few curves are shown in Figure 14-2. It is plausible that there should be one such curve through each point of D. That is precisely what the Existence Theorem tells us.

The graphical process corresponds to a well-known physical experiment: If one places iron filings on a horizontal glass plate and then holds a magnet below the plate, each filing aligns itself in accordance with the magnetic force field. Hence the filings form a field of line elements and one can physically see the curves they form, called the "lines of force." In this experiment the laws of magnetism lead to a differential equation whose solutions we are observing (Figure 14-3).

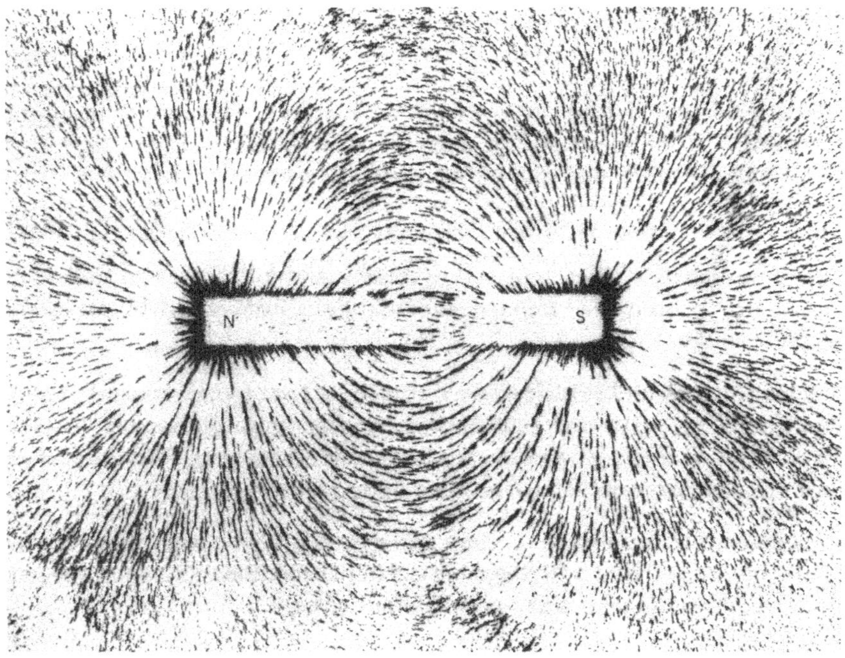

Figure 14-3 Field of line elements created by a magnet.

Thus far we have considered only differential equations of first order. An ordinary differential equation of second order is an equation that can be written in the form $G(x, y, y', y'') = 0$. Here we consider only equations explicitly solvable for y''; that is, we consider equations of the form

$$y'' = F(x, y, y') \tag{14-14}$$

More generally, an ordinary differential equation of nth order is an equation of form

$$y^{(n)} = F(x, y, y', \ldots, y^{(n-1)}) \tag{14-15}$$

where n is a positive integer. The word "ordinary" refers to the fact that no partial derivatives appear; when they do appear, one has a *partial* differential equation. We shall generally omit the word "ordinary," since with rare exceptions we consider only ordinary differential equations in this chapter. The equations

$$y'' = e^x, \qquad y'' = xy' - y^2, \qquad y''' = x \cos y - yy'', \qquad y^{(iv)} = (y'')^2$$

are differential equations of orders 2, 2, 3, and 4, respectively.

By a solution of (14-15) we mean a function $y = f(x)$, having derivatives through order n on some interval, such that

$$f^{(n)}(x) \equiv F(x, f(x), \ldots, f^{(n-1)}(x))$$

on the interval. We again ask whether an equation of form (14-15) has solutions. A study of the special case

$$y^{(n)} = F(x)$$

as above for $n = 1$ leads us to expect an Existence Theorem of the form: provided F is well-behaved, for a given set of values $y_0, y_0', \ldots, y_0^{(n-1)}$ at a chosen x_0, there is one, and only one, solution of (14-15) such that

$$f(x_0) = y_0, \qquad f'(x_0) = y_0', \qquad \ldots, \qquad f^{(n-1)}(x_0) = y_0^{(n-1)} \tag{14-16}$$

In applications x is often replaced by time t, and for this reason the values $y_0, \ldots, y_0^{(n-1)}$ are called *initial values* and the equations (14-16) are called *initial conditions.*

EXAMPLE 2. A rocket is climbing vertically, subject to a downward gravitational force (approximately constant), a constant upward force and an additional upward thrust proportional to time t after launching. Find the altitude and velocity t seconds after launching.

Solution. Let the altitude at time t be y, so that the velocity is dy/dt, and $y = 0$, $dy/dt = 0$ for $t = 0$ (these are the initial conditions). Newton's Second Law (force equals mass times acceleration) gives

$$m\frac{d^2y}{dt^2} = a + bt - mg$$

where m, a, b, and g are positive constants. Integrating twice we find that

$$m\frac{dy}{dt} = \frac{bt^2}{2} + (a - mg)t + c_1$$

$$my = \frac{bt^3}{6} + (a - mg)\frac{t^2}{2} + c_1t + c_2,$$

where c_1 and c_2 are constants. From the given initial conditions we conclude that $c_1 = 0$ and $c_2 = 0$, so that

$$y = \frac{1}{6m}[bt^3 + 3(a - mg)t^2]$$

$$\frac{dy}{dt} = \frac{1}{2m}[bt^2 + 2(a - mg)t]$$

From the form of the solution we see that for an effective launching we must have $a \geq mg$. (Why?)

This example illustrates how Newton's Second Law leads to differential equations. A great many other differential equations arise from physical problems in this way. Other laws of physics, such as those governing electric currents and heat conduction, also lead to differential equations. One reason why we should expect physical laws to correspond to differential equations is that most fundamental laws of physics do not state exactly how a system will behave but state *restrictions* on its behavior; each law specifies not one function but a set of functions. The solutions of a differential equation also form a set of functions, and it turns out that, in many cases, this set describes all the ways in which a particular physical system can behave. The actual behavior is determined by how the system is *started;* that is, by giving the proper *initial values* which specify a particular solution of the differential equation.

The concept of differential equation must be generalized further in order to meet the needs of physical problems: namely, to include *simultaneous differential equations.* These are differential equations relating several functions of one variable and their derivatives. We give an example:

$$\frac{d^2x}{dt^2} = 3\frac{dx}{dt} - 5x\frac{dy}{dt} + y \sin t, \qquad \frac{d^2y}{dt^2} = 3\frac{dx}{dt} - y \tag{14-17}$$

Here x and y are unspecified functions of t. A solution of this pair of simultaneous differential equations is a pair of functions: $x = f(t)$, $y = g(t)$, satisfying the equations identically over an interval. If we write $dx/dt = z$, $dy/dt = w$, then the two given equations can be replaced by four equations:

$$\begin{aligned} \frac{dx}{dt} &= z \\ \frac{dy}{dt} &= w \\ \frac{dz}{dt} &= 3z - 5xw + y \sin t \\ \frac{dw}{dt} &= 3z - y \end{aligned} \tag{14-17$'$}$$

For most purposes a form such as (14-17′) is preferable to a form such as (14-17) (which involves derivatives of higher than first order). This process leads us to the *standard form* for simultaneous differential equations:

$$\frac{dx_i}{dt} = F_i(x_1, \ldots, x_n, t), \qquad (i = 1, \ldots, n) \tag{14-18}$$

We refer to (14-18) as a *system of first order differential equations, in standard form.* A solution of (14-18) is a set of n functions

$$x_i = f_i(t) \qquad (i = 1, \ldots, n)$$

satisfying (14-18) over an interval, so that

$$f_i'(t) \equiv F_i(f_1(t), \ldots, f_n(t), t) \qquad (i = 1, \ldots, n)$$

over the interval.

We remark that (14-18) can be simplified by vector notation:

$$\frac{d\mathbf{x}}{dt} = \mathbf{F}(\mathbf{x}, t) \tag{14-18$'$}$$

Here **x**, **F** are functions (values in V_n). Vectors and matrices aid greatly in the study of systems. The nth order equation (14-15) can also be written in the form (14-18) [and, hence, as a vector equation (14-18′)]. For let

$y_1 = y, y_2 = y', \ldots, y_n = y^{(n-1)}$. Then (14-15) is equivalent to the system

$$\frac{dy_1}{dx} = y_2, \quad \frac{dy_2}{dx} = y_3, \ldots, \quad \frac{dy_{n-1}}{dx} = y_n, \quad \frac{dy_n}{dx} = F(x, y_1, \ldots, y_n) \qquad (14\text{-}15')$$

This reasoning shows that (14-18) or (14-18′) can be taken as the basic form for the study of ordinary differential equations.

When the vectors in (14-18′) are 3-dimensional, we can interpret the equation physically in terms of a fluid motion in space. For such a motion there is a velocity $\mathbf{v} = d\mathbf{x}/dt$ at each point, in general, changing with time; the differential equation describes this changing pattern. The solutions of the differential equation are simply the paths of the fluid particles. When $\mathbf{F}$ depends only on $\mathbf{x}$, not on t, the equation becomes

$$\frac{d\mathbf{x}}{dt} = \mathbf{F}(\mathbf{x}) \qquad (14\text{-}18'')$$

We call such an equation (or the corresponding system (14-18)) *autonomous*. The corresponding fluid motion is steady; the velocity pattern is unchanging in time and provides a field of line elements tangent to the solution paths. An example is provided by a smoothly flowing river.

Guided by our experience with the first order equation (14-10), we would expect the first order vector equation (14-18′) to have a unique solution with given initial value $\mathbf{x}_0$ of $\mathbf{x}$ for a chosen initial value t_0 of t; equivalently, we expect (14-18) to have a unique solution with given initial values of $x_1, \ldots, x_n$ for $t = t_0$. For the special system (14-15′), we expect a unique solution with given initial values of $y_1, \ldots, y_n$ for $x = x_0$. Accordingly, by the way $y_1, \ldots, y_n$, were defined, (14-15) should have a unique solution with given initial values of $y, y', \ldots, y^{(n-1)}$ at $x = x_0$. This last statement is just the Existence Theorem for (14-15), as formulated above. Accordingly, we see that there is really only one basic Existence Theorem needed and that it can be stated for (14-18) [or, in vector language, for (14-18′)]. We now state the theorem formally for a system of first order equations.

EXISTENCE THEOREM. *Let the functions $F_i(x_1, \ldots, x_n, t)$ $(i = 1, \ldots, n)$ be defined in an open region D in $(n+1)$-dimensional space R^{n+1} (in which $x_1, \ldots, x_n, t$ are the coordinates). Let these functions be continuous and have continuous first partial derivatives with respect to $x_1, \ldots, x_n$ in D. Then for each point $(x_1^0, \ldots, x_n^0, t_0)$ in D there is a set of functions*

$$x_i = f_i(t), \qquad i = 1, \ldots, n, \qquad |t - t_0| < h$$

which form a solution of (14-18) on the interval $|t - t_0| < h$ and satisfy the initial conditions

$$f_i(t_0) = x_i^0, \qquad (i = 1, \ldots, n)$$

This solution is unique; that is, if $x_i = g_i(t)$, $|t - t_0| < k$, is also a solution satisfying the same initial conditions:

$$g_i(t_0) = x_i^0, \qquad (i = 1, \ldots, n)$$

then $f_i(t) \equiv g_i(t)$ *for* $i = 1, \ldots, n$ *over the common part of the two t-intervals.*

For a proof of the Existence Theorem, see Chapter 12 of *Ordinary Differential Equations* by W. Kaplan (Addison-Wesley, Reading, Mass., 1958).* For various special cases, proofs are given here as these cases are discussed.

PROBLEMS

1. Let $y = f(x)$ be a differentiable function defined by the equation given. Find a differential equation of first order satisfied by this function.

   (a) $y = e^x$, (b) $y = \ln x$ (c) $y = \sin x$

   (d) $y = \operatorname{Tan}^{-1} x$ (e) $2x^2 + 3y^2 = 1$ (f) $x^4 + y^4 = 1$

   (g) $x^3 + x^2y + y^3 = 0$ (h) $xe^y + y\cos x + x^2 + y^2 = 0$

2. Find a differential equation of second order satisfied by the function $y = f(x)$ given:

   (a) $y = 5x$ (b) $y = x^3$ (c) $y = e^x$ (d) $y = \cos x$

3. Each of the following equations contains an arbitrary constant c. Show that for each c, subject to the restrictions given, the equation defines a function $y = f(x)$ and that, for each point (x_0, y_0) in the open region D specified, there is exactly one such function whose graph passes through the point; graph several such functions. Find a first-order differential equation satisfied by these functions.

   (a) $y = x \ln x + c$, D: $x > 0$ (b) $y = \sin x + c$, D: all (x, y)

   (c) $y = \tan(x + c)$, D: all (x, y) (d) $y = \ln(x + c)$, D: all (x, y)

   (e) $x^2 + y^2 = c$, D: $y > 0$; $c > 0$ (f) $xy = c$, D: $x > 0$

   Remark. Each of the equations (a), . . . , (f) is said to be a *primitive* of the corresponding differential equation.

4. Find the solution satisfying the given initial conditions:

   (a) $y' = x^5$, $y = 1$ for $x = 0$ (b) $y' = \sin 5x$, $y = 1$ for $x = \pi$

   (c) $y' = x^2e^{x^3}$, $y = 1$ for $x = 0$ (d) $y' = \tan x$, $y = 1$ for $x = \pi/4$

   (e) $y'' = x^3$, $y = 1$ and $y' = -1$ for $x = 0$

   (f) $y'' = e^{2x}$, $y = 3$ and $y' = 0$ for $x = 0$

   (g) $y''' = 0$, $y = 1$, $y' = 2$ and $y'' = 3$ for $x = 0$

   (h) $y^{(iv)} = x^7$, $y = 0$, $y' = 0$, $y'' = 0$, $y''' = 0$ for $x = 0$

5. A particle of mass m moves on a line, the x-axis, subject to a force F depending on time t, as specified, with the initial conditions given. Find the position at time t_1. Throughout, k, ω, b are positive constants and $v = dx/dt$.

   (a) $F = kt^3$, $x = x_0$, $v = v_0$ for $t = 0$

   (b) $F = k\sin\omega t$, $x = 0$, $v = 0$ for $t = 0$

   (c) $F = ke^{-bt}$, $x = 0$, $v = 0$ for $t = 0$

   (d) $F = k\sec^2 bt$, $x = x_0$, $v = 0$ for $t = 0$

*This book will be referred to as ODE.

6. Verify that the given pair of functions is a solution of the system of first-order equations:

(a) $(dx/dt) = y$, $(dy/dt) = -4x$; $x = \cos 2t$, $y = -2 \sin 2t$

(b) $(dx/dt) = 2x - y$, $(dy/dt) = y$; $x = e^t$, $y = e^t$

7. Write as a system of first-order differential equations in standard form:

(a) $y'' = e^{xy'} - y$ (b) $y'' = x^2 + y'^2$

(c) $y''' = xy^2 - y'y''$ (d) $\dfrac{d^4x}{dt^4} = x^2 - t^2\dfrac{d^2x}{dt^2}$

14-2 GRAPHICAL METHOD AND METHOD OF STEP-BY-STEP INTEGRATION

In Section 14-1 we pointed out that a differential equation $y' = F(x, y)$ determines a field of line elements, whose graph suggests the solutions sought. One can develop this idea systematically by means of *isoclines* to obtain a very useful method for obtaining the solutions graphically.

An *isocline* of the differential equation $y' = F(x, y)$ is a curve along which F has a constant value, say m. Thus it is a level curve of F. The isoclines are (with rare exceptions) not themselves solutions of the differential equation, but as we shall see, they help us to find the solutions.

EXAMPLE 1. $y' = x^2 - y^2$. Here the isoclines are the curves $x^2 - y^2 = m$, for different values of m. Thus they form a family of hyperbolas; for $m = 0$, the isocline is the graph of $x^2 - y^2 = 0$, formed of the two lines $x = \pm y$ (degenerate hyperbola); these lines form the asymptotes of all the hyperbolas. We now graph a number of isoclines as in Figure 14-4. We select an isocline, say the one for $m = 1$, and draw many parallel short-line segments of slope

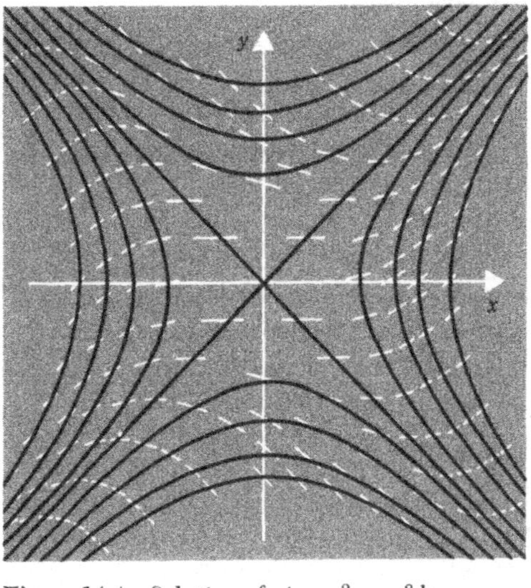

Figure 14-4 Solution of $y' = x^2 - y^2$ by means of isoclines.

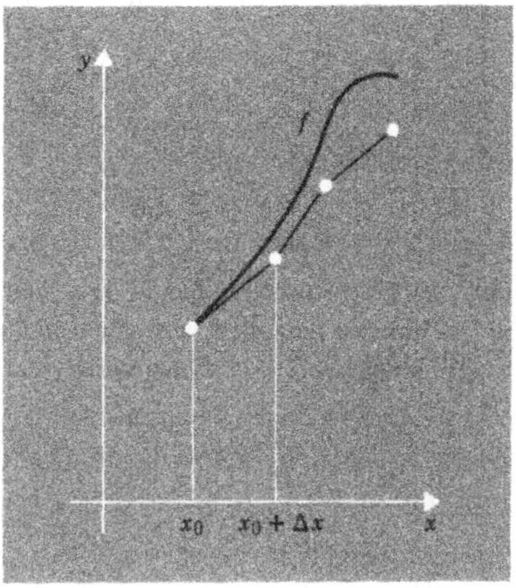

Figure 14-5 Step-by-step integration.

m, all centered on the isocline. We do the same for the isocline $m = \frac{1}{2}$, then for $m = 0$, $m = -\frac{1}{2}$, and so on. If we use some care in drawing the segments, they will fit together to form broken lines suggesting smooth curves, which are the solutions. Thus we can graphically approximate the solutions in a region. The accuracy can be improved by choosing more closely spaced values of m. Unfortunately, the method has no practical extension to equations of higher than first order.

The graphical method has a numerical counterpart that permits us to compute solutions to as much accuracy as is desired.

For the general equation $y' = F(x, y)$ we start with an initial point (x_0, y_0) and try to find the solution $y = f(x)$ through the initial point. We begin by attempting to determine $f(x + \Delta x)$. Since y' is known at x_0, we can use the differential $dy = y' \Delta x$ to approximate the change Δy in y. Since at x_0 we have $y' = F(x_0, y_0)$, we thus compute Δy approximately by the formula:

$$\Delta y \sim F(x_0, y_0)\,\Delta x$$

If we write $x_1 = x_0 + \Delta x$, $y_1 = y_0 + \Delta y$, then we have now found a point (x_1, y_1) that is close to our solution curve; the error is caused by our following the *tangent line* instead of the curve itself (Figure 14-5); we can reduce the error by decreasing $|\Delta x|$.

Having found the point (x_1, y_1), we can use it as a new initial point, again change x by an amount Δx (not necessarily the same as before), and find Δy by the formula $\Delta y \sim F(x_1, y_1)\,\Delta x$; we thus obtain a new point (x_2, y_2), where $x_2 = x_1 + \Delta x$, $y_2 = y_1 + \Delta y$. The process can be repeated indefinitely. We obtain a sequence of points (x_n, y_n), where

$$y_n = y_{n-1} + F(x_{n-1}, y_{n-1})(x_n - x_{n-1}) \qquad (14\text{-}20)$$

and the values $x_0, x_1, x_2, \ldots$ are chosen as desired (as an increasing or decreasing sequence). By choosing them close together we in general improve the accuracy; however, in so doing, we also require a correspondingly large number of steps to cover a given interval. The whole process is called *step-by-step numerical integration of the differential equation*. The method produces the sequence of points $(x_0, y_0), (x_1, y_1), \ldots$; by joining successive points we obtain a broken line that approximates our solution. For a proof that, under appropriate hypotheses on F, the error in the approximation over a given interval can be made as small as desired, one is referred to *Introduction to Numerical Analysis* by F. B. Hildebrand (McGraw-Hill, N.Y., 1956), especially Chapter 6.

EXAMPLE 2. $y' = 2x - 3y$, $y = 1$ for $x = 0$. Here we take $\Delta x = 0.1$ and seek the solution over the interval $[0, 1]$. The results are given in Table 14-1. The last column shows the exact solution; by methods developed later in this chapter, this is found to be

$$y = (11e^{-3x} - 2 + 6x)/9$$

The agreement is fairly good; the maximum error is 0.074. In the computation decimals had to be rounded off. This process itself creates errors that may be serious, especially through accumulation.

Table 14-1

x	y	$y' = 2x - 3y$	$\Delta y = y' \Delta x$, $\Delta x = 0.1$	exact y
0.0	1.000	−3.00	−0.300	1.000
0.1	0.700	−1.90	−0.190	0.743
0.2	0.510	−1.13	−0.113	0.578
0.3	0.397	−0.59	−0.059	0.471
0.4	0.338	−0.21	−0.021	0.411
0.5	0.317	0.05	0.005	0.382
0.6	0.322	0.23	0.023	0.378
0.7	0.345	0.37	0.037	0.393
0.8	0.382	0.45	0.045	0.421
0.9	0.427	0.52	0.052	0.459
1.0	0.479			0.505

In the example we can also take Δx negative (for example, $\Delta x = -0.1$) to obtain an approximate solution in the interval $[-1, 0]$, for example.

The method of step-by-step integration can be extended to equations of higher order and to systems. For example, for the second-order equation

$$y'' = y - x$$

with initial values $y = 1$, $y' = 0$ for $x = 0$, we introduce the corresponding system

$$\frac{dy}{dx} = z, \qquad \frac{dz}{dx} = y - x$$

and then approximate Δy, Δz for given Δx, as before. Table 14-2 shows a

Table 14-2

x	y	z	Δy	Δz
0.0	1.000	0.000	0.000	0.100
0.1	1.000	0.100	0.010	0.090
0.2	1.010	0.190	0.019	0.081
0.3	1.029	0.271		

typical calculation, with $\Delta x = 0.1$. In each case $\Delta y = z\,\Delta x = 0.1z$, $\Delta z = (y - x)\,\Delta x = 0.1(y - x)$.

For a general system of first-order equations

$$\frac{dx_i}{dt} = F_i(x_1, \ldots, x_n, t), \qquad i = 1, \ldots, n$$

the corresponding approximation equations are

$$\Delta x_i \sim F_i(x_1, \ldots, x_n, t)\,\Delta t$$

We return to the discussion of numerical methods in Section 14-23. These are of major importance because of the availability of rapid computers.

Remark. For the equation $y' = F(x)$, the method of step-by-step integration reduces to a familiar procedure for evaluating a definite integral. Let $y = y_0$ for $x = x_0$. Then the successive values of y are given by

$$\begin{aligned} y_1 &= y_0 + F(x_0)(x_1 - x_0) \\ y_2 &= y_1 + F(x_1)(x_2 - x_1) \\ &\vdots \\ y_n &= y_{n-1} + F(x_{n-1})(x_n - x_{n-1}) \end{aligned} \tag{14-21}$$

If we write the first equation as $y_1 - y_0 = \ldots$, the second as $y_2 - y_1 = \ldots$ and so on, and add the equations, we obtain

$$y_n = y_0 + \sum_{i=1}^{n} F(x_{i-1})(x_i - x_{i-1})$$

If we take $x_0 = a$ and $x_n = b$, then $x_0, \ldots, x_n$ provide a subdivision of the interval $[a, b]$ and the sum is an approximation to the definite integral of F from a to b. If we let the mesh of the subdivision approach 0, then the sum approaches the integral as limit and hence $y_n \to y$, where

$$y = y_0 + \int_a^b F(x)\, dx \tag{14-22}$$

If $f(x)$ is the solution of the differential equation $y' = F(x)$ such that $f(x_0) = y_0$, then $f'(x) = F(x)$, so that

$$f(b) = f(x_0) + \int_{x_0}^b f'(x)\, dx = y_0 + \int_a^b F(x)\, dx$$

Hence, (14-22) does give the exact value $y = f(b)$.

A similar method for indefinite integrals is discussed in Section 4-14.

PROBLEMS

1. Solve graphically with the aid of isoclines:
   (a) $y' = 1 + y$ (solutions $y = ce^x - 1$)
   (b) $y' = x - y$ (solutions $y = x - 1 + ce^{-x}$)
   (c) $y' = y/x$, $x > 0$ (d) $y' = x/(y - 1)$, $y > 1$
   (e) $y' = x^2 + y^2$ (f) $y' = x^2 + y^2 - 1$
   (g) $y' = xy$ (h) $y' = y + y^2$
2. Solve by step-by-step numerical integration as directed and compare the results with the exact solution given:
   (a) $y' = y^2$, starting at $x = 2$, $y = -1$ and continuing to $x = 5$ with $\Delta x = 0.5$; exact solution is $y = 1/(1 - x)$.

(b) $y' = -y$, starting at $x = 0$, $y = 1$ and continuing to $x = 3$ with $\Delta x = 0.5$; exact solution is $y = e^{-x}$.

(c) $y' = -xy$, starting at $x = 0$, $y = 1$ and continuing to $x = 1$ with $\Delta x = 0.2$; exact solution is $y = e^{-x^2/2}$.

(d) $y' = x^2 - 3y$, starting at $x = 0$, $y = 0$ and continuing to $x = 5$ with $\Delta x = 1$; exact solution is $y = (9x^2 - 6x + 2 - 2e^{-3x})/27$.

3. For the differential equation $y' = -2y + 2$ with initial condition $y = 0$ for $x = 0$, the exact solution is $y = 1 - e^{-2x}$. For each of the following choices of Δx, find a solution by step-by-step integration over the interval $[0, 2]$, graph it as a broken line and compare with the graph of the exact solution.

(a) $\Delta x = 1$ (b) $\Delta x = 0.5$ (c) $\Delta x = 0.2$ (d) $\Delta x = 0.1$

4. For each of the following systems of first-order equations, find an approximate solution by step-by-step integration and compare with the exact solution given:

(a) $dx/dt = x + y$, $dy/dt = 4x + y$, starting at $t = 0$, $x = 1$, $y = -2$, using $\Delta t = 0.5$ and continuing to $t = 2$; exact solution is $x = e^{-t}$, $y = -2e^{-t}$.

(b) $dx/dt = y + t$, $dy/dt = xy - t^3 + 1$, starting at $t = 0$, $x = 0$, $y = 0$ using $\Delta t = 0.5$ and continuing to $t = 3$; exact solution is $x = t^2$, $y = t$.

5. Solve by step-by-step integration as directed and compare with the exact solution given:

(a) $y'' = -y$, starting at $x = 0$, $y = 0$, $y' = 1$, using $\Delta x = 0.5$ and continuing to $x = 3$; exact solution is $y = \sin x$.

(b) $y'' = y^2 - \cos x - \cos^2 x$, starting at $x = 0$, $y = 1$, $y' = 0$ with $\Delta x = 0.5$ and continuing to $x = 3$; exact solution is $y = \cos x$.

(c) $y''' = yy' + y'' - y^2$, starting at $x = 0$, $y = 1$, $y' = 1$, $y'' = 1$ using $\Delta x = 0.5$ and continuing to $x = 2$; exact solution is $y = e^x$.

(d) $y^{(iv)} = xy' - 4y + 24$, starting at $x = 0$, $y = 0$, $y' = 0$, $y'' = 0$, $y''' = 0$, using $\Delta x = 0.2$ and continuing to $x = 1$; exact solution is $y = x^4$.

14-3 EXACT FIRST-ORDER EQUATIONS

First-order differential equations often appear in the form:

$$P(x, y)\,dx + Q(x, y)\,dy = 0 \tag{14-30}$$

We say that this equation is in *differential form*. After division by $Q(x, y)\,dx$, the equation can be written,

$$\frac{dy}{dx} = -\frac{P(x, y)}{Q(x, y)} \tag{14-30$'$}$$

hence, as an equation of form $y' = F(x, y)$. One can also divide (14-30) by $P(x, y)\,dy$ to obtain an equation:

$$\frac{dx}{dy} = -\frac{Q(x, y)}{P(x, y)} = G(x, y) \tag{14-30$''$}$$

This we can regard as a first-order equation for x as a function of y. Actually, it is simplest to work with the form (14-30), which treats x and y on the same basis. It will be seen that for many such equations the solutions are given by an equation of form:

$$H(x, y) = c \tag{14-31}$$

For each c, this is an *implicit equation* (Sections 3-8 and 12-14), which defines certain functions $y = f(x)$ and also certain functions $x = g(y)$; under reasonable hypotheses, these functions are solutions of the corresponding differential equation (14-30′) or (14-30″). More generally, we regard as a solution of (14-30) every function $y = f(x)$ or $x = g(y)$ which satisfies (14-30) identically over some interval. Through (14-31) these functions are given geometrically as the *level curves* of the function H (Section 12-2).

We note that (14-30′) becomes meaningless at points (x, y) at which $Q(x, y) = 0$, whereas (14-30″) becomes meaningless at points (x, y) at which $P(x, y) = 0$. At points (x, y) at which *both equations* hold true:

$$P(x, y) = 0 \qquad \text{and} \qquad Q(x, y) = 0$$

neither (14-30′) nor (14-30″) has meaning. Such points are called *singular points* of the equation (14-30); the same term is used for points at which one or both of P and Q is discontinuous. In solving equations of form (14-30), one often has reason to multiply or divide the equation by a function $v(x, y)$ (the *integrating factor*). This operation can introduce new singular points [where $v(x, y) = 0$] or remove apparent singular points by cancellation. The operation can also introduce extraneous solutions of (14-30). We shall point out these complications, where significant, in connection with examples.

EXAMPLE 1. $y\,dx + x\,dy = 0$. The equation is the same as $d(xy) = 0$. Hence, the solutions are given by

$$xy = c$$

a family of hyperbolas (Figure 14-6); for $c = 0$, the hyperbola is degenerate, reducing to two straight lines: $x = 0$ and $y = 0$. Each of these lines provides

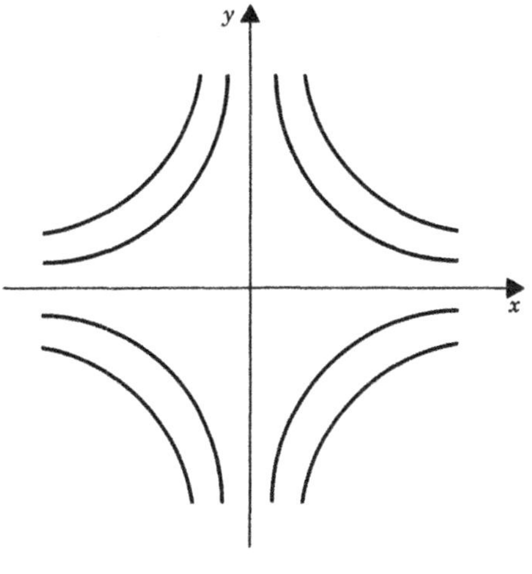

Figure 14-6 Solutions of the equation $y\,dx + x\,dy = 0$.

a solution of the equation in differential form. For example, if $x = 0$ (constant function of y), then $dx = (dx/dy)\,dy = 0\,dy = 0$ and $y\,dx + x\,dy = y \cdot 0 + 0\,dy = 0$ for all y and dy. We note that the equation has a singular point at the origin: $(0, 0)$; here both of the equations

$$\frac{dy}{dx} = -\frac{y}{x}, \qquad \frac{dx}{dy} = -\frac{x}{y}$$

are meaningless. At all other points, at least one of the two forms is usable and the solutions are given, respectively, by $y = c/x$ or $x = c/y$ through such points. Note that we have two solutions passing through the singular point, but only one solution through each other point.

EXAMPLE 2. $y\,dx - x\,dy = 0$. Here we divide by y^2 to obtain

$$\frac{y\,dx - x\,dy}{y^2} = 0$$

This is the same as $d(x/y) = 0$ and, hence, the solutions are given by

$$\frac{x}{y} = c$$

a family of straight lines through the origin (Figure 14-7). In dividing by y, we have introduced singular points along the line $y = 0$, the x-axis. However, $y = 0$ is a solution of the original equation. Hence, we must also include this as a solution, and we now have *all* lines through the origin. The origin itself is a singular point of the original equation.

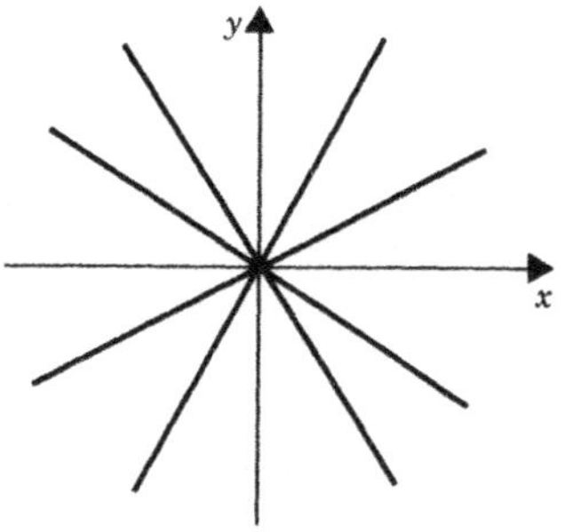

Figure 14-7 Solutions of the equation $y\,dx - x\,dy = 0$.

We note that we could also have divided the equation by $-x^2$:

$$\frac{-y\,dx + x\,dy}{x^2} = 0 \qquad \text{or} \qquad d\left(\frac{y}{x}\right) = 0$$

and hence

$$\frac{y}{x} = c$$

Thus we obtain the solutions: $y = cx$ and $x = 0$.

We also could have divided by xy:

$$\frac{dx}{x} - \frac{dy}{y} = 0, \qquad d(\ln x - \ln y) = 0, \qquad \ln x - \ln y = c$$

$$\ln\frac{x}{y} = c, \qquad \frac{x}{y} = e^c, \qquad x = Cy$$

The steps are valid only for $x > 0$ and $y > 0$; for $x < 0$ one must replace $\ln x$ by $\ln(-x)$, for $y < 0$ one must replace $\ln y$ by $\ln(-y)$; however, the one equation $x = Cy$ covers all these cases. For $x = 0$ or $y = 0$, one has the special straight-line solutions noted above.

The examples raise a number of questions:

1. For a given equation $P(x, y)\,dx + Q(x, y)\,dy = 0$, how do we recognize when the equation can be written in the form $dH(x, y) = 0$ and, if it can be so written, how do we find H?
2. Having found H, do we know that the equation $H(x, y) = c$ implicitly defines solutions of the differential equation and that it gives all solutions?
3. Can one always find an integrating factor $v(x, y)$ such that, after multiplication by v, the given differential equation can be written in the form $dH(x, y) = 0$?

The questions suggest an important definition: The differential equation $P(x, y)\,dx + Q(x, y)\,dy = 0$ is said to be *exact*, in the open region D, if there is a function H, differentiable in D, such that $dH = P\,dx + Q\,dy$ in D.

We now answer the first question:

THEOREM 1. *Let the function $P(x, y)$ and $Q(x, y)$ have continuous first partial derivatives in the open region D. If the differential equation $P\,dx + Q\,dy = 0$ is exact in D, then*

$$\frac{\partial P}{\partial y} = \frac{\partial Q}{\partial x} \quad \textit{in } D \tag{14-32}$$

Conversely, if (14-32) holds true in D and D is simply connected, then the equation $P\,dx + Q\,dy = 0$ is exact in D.

Remark. By virtue of this theorem, condition (14-32) is called the Test for Exactness.

PROOF OF THEOREM 1. The theorem is an immediate consequence of results of Section 13-14. As shown in that Section, when (14-32) holds true and D is simply connected, one can always find H by the equation

$$H(x, y) = \int_{(x_0, y_0)}^{(x, y)} P\,dx + Q\,dy \tag{14-33}$$

where (x_0, y_0) is a fixed point of D and the line integral is taken on any convenient path from (x_0, y_0) to (x, y) in D.

If the Test for Exactness (14-32) is satisfied, but D is not simply connected, one can apply Theorem 1 to each simply connected portion D_1 of D to obtain a function H_1 in D_1 so that $dH_1 = P\,dx + Q\,dy$ in D_1. For the given differential equation in D, this means that we may need several different representations for the solutions in different parts of D. However, it often happens that one function H can be found for all of D even though D is not simply connected (see Problem 4 following Section 13-14).

EXAMPLE 3. $(x^2 + 3x^2y)\,dx + (\cos y + x^3)\,dy = 0$. Here the Test for Exactness is satisfied:

$$\frac{\partial P}{\partial y} = 3x^2 = \frac{\partial Q}{\partial x}$$

for all (x, y). Hence, the equation is exact in the whole xy-plane (a simply connected open region). We can easily guess the function H. We can also use (14-33) and choose a path from $(0, 0)$ to (x, y), as in Figure 14-8:

$$H(x, y) = \int_{(0,0)}^{(x,0)} P\,dx + Q\,dy + \int_{(x,0)}^{(x,y)} P\,dx + Q\,dy$$

$$= \int_0^x t^2\,dt + \int_0^y (\cos u + x^3)\,du = \frac{x^3}{3} + \sin y + x^3y$$

We readily verify that $dH = P\,dx + Q\,dy$.

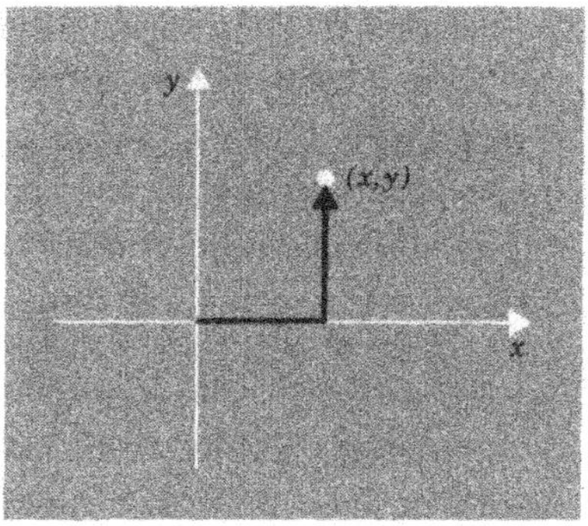

Figure 14-8 Path for Example 3.

We have now answered Question 1. For Question 2, we state here simply that the answer is affirmative: for an exact equation the solutions, other than those passing through singular points, are the functions defined implicitly by the equation $H(x, y) = c$. The proof is left as an exercise (Problem 7 below). For Question 3 the complete answer is more complicated. Here we give a partial answer: if P and Q have continuous first partial derivatives in D and (x_0, y_0) is a point of D which is not a singular point of the differential equation, then in a sufficiently small neighborhood of (x_0, y_0) there is an integrating factor $v(x, y)$. For a proof one is referred to Chapter 12 of ODE. Finding integrating factors is a very difficult problem. For many special types of equations, rules for integrating factors have been found. Illustrations are given in the next example and in the exercises.

EXAMPLE 4. $(4xy^3 + 3y)\,dx + (5x^2y^2 + 3x)\,dy = 0$

Solution. We write the equation as follows:

$$4xy^3\,dx + 5x^2y^2\,dy + 3(y\,dx + x\,dy) = 0$$

The quantity in parentheses is $d(xy)$. If we multiply the equation by a function of xy, say $g(xy)$, the last part of the equation will become $3g(xy)\,d(xy) = 3g(u)\,du$, where $u = xy$. Hence, this last part will remain exact; it is the differential of $3G(xy)$, where $G'(u) = g(u)$. Because of the form of the first part of the equation, we try $g(xy) = (xy)^m$ for some constant m, to obtain

$$4x^{m+1}y^{m+3}\,dx + 5x^{m+2}y^{m+2}\,dy + 3(xy)^m\,d(xy) = 0$$

To make the first part of the new equation exact, we apply the Test for Exactness:

$$\frac{\partial}{\partial y}(4x^{m+1}y^{m+3}) \equiv \frac{\partial}{\partial x}(5x^{m+2}y^{m+2})$$

or

$$4(m+3)x^{m+1}y^{m+2} \equiv 5(m+2)x^{m+1}y^{m+2}$$

This equation is satisfied for all (x, y) if $4m + 12 = 5m + 10$ or $m = 2$. Our equation becomes

$$4x^3y^5\,dx + 5x^4y^4\,dy + 3(xy)^2\,d(xy) = 0$$

We recognize the first two terms as $d(x^4y^5)$ and the whole equation can be written as $d[x^4y^5 + (xy)^3] = 0$. The solutions are given by

$$x^4y^5 + x^3y^3 = c$$

PROBLEMS

1. Graph the given level curves, find the corresponding differential equation and note singular points:

   (a) $x^2 + y^2 = c$ (b) $x^2 - y^2 = c$ (c) $x \sin y = c$ (d) $y^3 - x^2 = c$

2. Show that the equation is exact, find the general solution and find the solution satisfying the given initial conditions:

   (a) $2xy\,dx + x^2\,dy = 0$; $x = 1$, $y = 2$ (b) $e^y\,dx + xe^y\,dy = 0$; $x = 1$, $y = 0$

   (c) $(2x + 3y)\,dx + (3x + 4y)\,dy = 0$; $x = 2$, $y = -1$

   (d) $(4x - 5y)\,dx + (2y - 5x)\,dy = 0$; $x = 3$, $y = 4$

   (e) $2xe^{x^2}\sin y\,dx + (e^{x^2}\cos y - \sin y)\,dy = 0$; $x = 0$, $y = -\pi/4$

   (f) $\left(\dfrac{1}{y} - \dfrac{y}{x^2 + y^2}\right)dx + \left(\dfrac{x}{x^2 + y^2} - \dfrac{x}{y^2}\right)dy = 0$; $x = 1$, $y = 1$

3. Find the general solution with the aid of an integrating factor:

   (a) $(3xy^2 - 4y)\,dx + (2x^2y - 2x)\,dy = 0$

   (b) $(xy + y)\,dx + (x + xy^2)\,dy = 0$

   (c) $(x^2 + y^2 + x)\,dx + y\,dy = 0$

(d) $(3x^2y - y^3)\,dx + (3xy^2 - 5x^3)\,dy = 0$. (*Hint.* Try $v = y^n$.)

(e) $(3y^2 + 5x)\,dx + 4xy\,dy = 0$. [*Hint.* Try $v = x^n$ for $x > 0$ and $v = (-x)^n$ for $x < 0$.]

(f) $(2x + x^2 + \sin 3y)\,dx + 3\cos 3y\,dy = 0$. [*Hint.* Try $v = g(x)$.]

4. Let the differential equation $P\,dx + Q\,dy = 0$ be given, where $P(x, y)$ and $Q(x, y)$ have continuous first partial derivatives in the simply connected open region D.

(a) Prove: $v(x, y)$ is an integrating factor, if and only if

$$v(x, y)\left(\frac{\partial P}{\partial y} - \frac{\partial Q}{\partial x}\right) + P\frac{\partial v}{\partial y} - Q\frac{\partial v}{\partial x} = 0 \qquad \text{in } D$$

‡(b) Prove: $\partial P/\partial y - \partial Q/\partial x = r(x)Q(x, y)$ in D, where r is continuous, implies that the equation has an integrating factor which depends on x alone.

(c) Prove: if $\alpha(x)$, $\beta(x)$, $\gamma(x)$ and $\gamma'(x)$ are continuous for $a < x < b$ and $\gamma(x) \neq 0$ in (a, b), then the equation (linear equation in y)

$$[\alpha(x)y + \beta(x)]\,dx + \gamma(x)\,dy = 0$$

has an integrating factor depending on x alone.

‡5. Let D be the xy-plane minus the origin, so that D is not simply connected. Show that the differential equation

$$\frac{-y\,dx + x\,dy}{x^2 + y^2} = 0$$

satisfies the Test for Exactness in D, but there is no function H such that $dH = (-y\,dx + x\,dy)/(x^2 + y^2)$ in all of D (see Section 13-14).

‡6. Let $\psi(z)$ be a function such that ψ' is continuous and has no zeros for $-\infty < z < \infty$. Show that the functions $H(x, y)$ and $H_1(x, y) = \psi[H(x, y)]$ have the same level curves. Compare the differential equations for the two sets of level curves. [*Remark.* This result shows that the representation of solutions of a differential equation in the form $H(x, y) = c$ is not unique. For the same reason, the integrating factor is not unique.]

‡7. Let $dH(x, y) = P(x, y)\,dx + Q(x, y)\,dy$, so that the equation $P\,dx + Q\,dy = 0$ is exact. Let P and Q be continuous in the open region D.

(a) Prove: if $y = f(x)$ is a solution of the differential equation, then the curve $y = f(x)$ is a level curve of H. {*Hint.* Show that $H[x, f(x)]$ is constant.}

(b) Prove: if $Q(x_0, y_0) \neq 0$, then there is a unique level curve of H through (x_0, y_0), of form $y = f(x)$, and that f is a solution of the differential equation. [*Hint.* Apply the Implicit Function Theorem of Section 12-15.]

14-4 EQUATIONS WITH VARIABLES SEPARABLE AND EQUATIONS OF FORM $y' = g(y/x)$

A commonly occurring exact equation is one of the form:

$$P(x)\,dx + Q(y)\,dy = 0 \tag{14-40}$$

An equation that can be written in this form is said to have *variables separable.*

We note that an equation of form

$$\frac{dy}{dx} = F(x)G(y) \tag{14-41}$$

has variables separable. In order to "separate the variables"—that is, to write the equation in the form (14-40)—one must divide by $G(y)$; the points where $G(y) = 0$ must then be studied separately.

If $P(x)$ is continuous for $a < x < b$ and $Q(y)$ is continuous for $c < y < d$, then (14-40) is exact in the open region D: $a < x < b$, $c < y < d$. For we can take

$$H(x, y) = \int P(x)\, dx + \int Q(y)\, dy$$

using any convenient choices of the indefinite integrals. Thus the solutions of (14-40) (except perhaps at singular points) are obtained by integrating:

$$\int P(x)\, dx + \int Q(y)\, dy = c \tag{14-42}$$

EXAMPLE 1. $y\, dx + x^2 dy = 0$.

Solution. We separate variables by dividing by x^2y, so that $1/(x^2y)$ is the integrating factor:

$$\frac{dx}{x^2} + \frac{dy}{y} = 0, \qquad \int\frac{dx}{x^2} + \int\frac{dy}{y} = c, \qquad -\frac{1}{x} + \ln y = c$$

For $y < 0$ we must replace $\ln y$ by $\ln(-y)$. We can avoid this complication by writing

$$\int\frac{dy}{y} = \ln\left|\frac{y}{c}\right|$$

Then our solutions are given by

$$-\frac{1}{x} + \ln\left|\frac{y}{c}\right| = 0, \qquad \left|\frac{y}{c}\right| = e^{1/x}, \qquad y = \pm ce^{1/x}$$

and the one equation $y = Ce^{1/x}$ describes all these solutions. Since we divided by x^2y, the points where $x = 0$ or $y = 0$ must still be examined. We see that both $x = 0$ and $y = 0$ are solutions. The origin is a singular point. The solutions are graphed in Figure 14-9. The strange configuration at the origin should be noted (Problem 6 below).

EXAMPLE 2. $y' = x(y^2 - 1)$. We separate variables:

$$\frac{dy}{dx} = x(y^2 - 1), \qquad \frac{dy}{y^2 - 1} = x\, dx \qquad (y \neq \pm 1)$$

$$\int\frac{dy}{y^2 - 1} = \int x\, dx, \qquad \frac{1}{2}\ln\frac{y - 1}{y + 1} = \frac{x^2}{2} + c$$

$$\frac{y - 1}{y + 1} = e^{2c + x^2} = Ce^{x^2} \qquad (C = e^{2c})$$

$$y = \frac{1 + Ce^{x^2}}{1 - Ce^{x^2}} \tag{14-43}$$

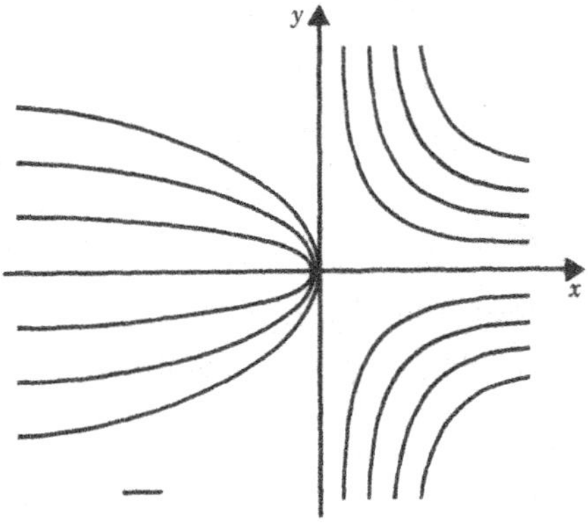

Figure 14-9 Solutions of $y\,dx + x^2\,dy = 0$.

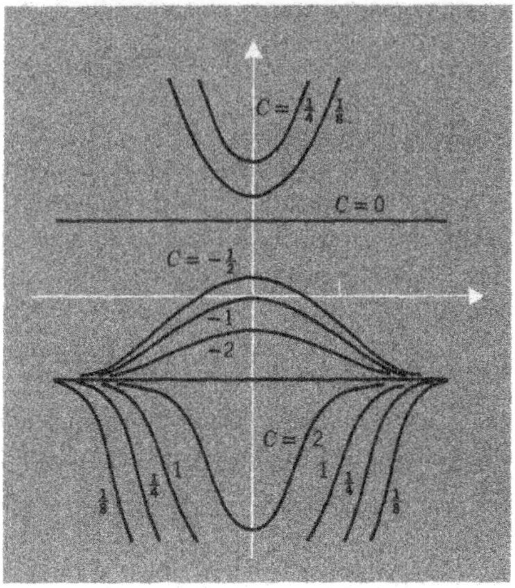

Figure 14-10 Solutions of $y' = x(y^2 - 1)$.

We divided by $y^2 - 1 = (y + 1)(y - 1)$ and note that $y = 1$ and $y = -1$ are solutions of the given differential equation. In solving above, we introduced a new arbitrary constant $C = e^{2c}$. From its definition, C should be positive. However, we verify that the functions (14-43) satisfy the differential equation for every choice of C. The reason for the paradox is that in the integration of $1/(y^2 - 1)$ we should have used the logarithm of the absolute value, as in Example 1. If we take this into account, we find that C can have all values except 0. The value $C = 0$ happens to correspond to one of the two horizontal solutions: $y = 1$. The other line: $y = -1$ can be considered to be the limiting case $C \to \infty$.

We can avoid changing arbitrary constants by writing

$$\int \frac{dy}{y^2 - 1} = \frac{1}{2} \ln \frac{y - 1}{y + 1} - \frac{1}{2} \ln C = \frac{1}{2} \ln \frac{y - 1}{C(y + 1)} = \int x\,dx = \frac{x^2}{2}$$

If we solve for y, we obtain (14-43) again.

The solutions are graphed in Figure 14-10. We note that for $0 < C < 1$ each function (14-43) is actually formed of three separate solutions, for $C = 1$ it is formed of two solutions; for $C < 0$ or $C > 1$ each function (14-43) is a solution for $-\infty < x < \infty$.

Equations of Form $y' = g(y/x)$. Differential equations of the form named (often termed "homogeneous first-order equations") are reducible to equations with variables separable. Examples are the following:

$$y' = \frac{x - y}{x + y} = \frac{1 - (y/x)}{1 + (y/x)}, \qquad y' = \frac{e^{y/x}x^2}{y^2} = e^{y/x}\left(\frac{y}{x}\right)^{-2}$$

The general method is to introduce a new dependent variable $v = y/x$ and obtain a differential equation for v in terms of x. We write formally:

$$xv = y, \qquad xv' + v = y' = g\left(\frac{y}{x}\right) = g(v)$$

$$x\frac{dv}{dx} + v - g(v) = 0$$

$$\frac{dv}{v - g(v)} + \frac{dx}{x} = 0$$

Thus the variables are separated. If we obtain the solutions in terms of x and v, then we can replace v by y/x to obtain the solutions in terms of x and y. Under appropriate continuity assumptions (and a consideration of zeros in denominators), the method provides the desired solutions.

EXAMPLE 3. $y' = \dfrac{x - y}{x + y} = \dfrac{1 - v}{1 + v}$. With $v = y/x$,

$$xv' + v = y' = \frac{1 - v}{1 + v}, \qquad xv' = \frac{1 - v}{1 + v} - v = \frac{1 - 2v - v^2}{1 + v}$$

$$\int \frac{1 + v}{v^2 + 2v - 1}\,dv + \int \frac{dx}{x} = 0, \qquad \frac{1}{2}\ln\frac{v^2 + 2v - 1}{c} + \ln x = 0$$

$$v^2 + 2v - 1 = \frac{c}{x^2}, \qquad y^2 + 2xy - x^2 = c$$

The solutions are graphed in Figure 14-11. We note that the origin is a singular point for the corresponding equation in differential form: $(x - y)\,dx - (x + y)\,dy = 0$.

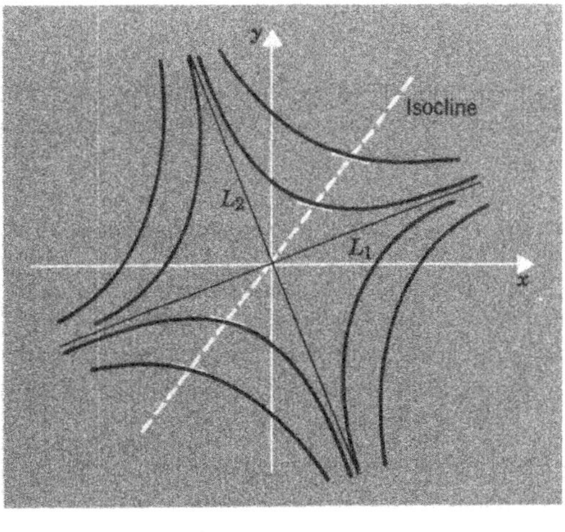

Figure 14-11 Solutions of $y' = (x - y)/(x + y)$.

Remark. For equations of the form $y' = g(y/x)$, the corresponding field of line elements is such that along each line through the origin all line elements are parallel, as indicated in Figure 14-11. (Thus the isoclines are the lines through the origin; see Section 14-2.) If for a particular m, $g(m) = m$, then the line through the origin of slope m is a solution; this is illustrated by the lines L_1 and L_2 in Figure 14-11. From the property of the line elements just noted, it also follows that, if $y = f(x)$ is a solution, then $ky = f(kx)$ also defines a solution (k = const $\neq 0$); that is, the solutions are similar curves (see Problem 10 below).

PROBLEMS

1. Find the general solution:

(a) $e^x\,dx + 2y\,dy = 0$ (b) $\cos 2x\,dx + ye^{y^2}\,dy = 0$

(c) $e^{x+2y}x\,dx + e^{y-x}\,dy = 0$

(d) $(x^2y + x^2 - xy - x)\,dx + (xy + 2y)\,dy = 0$

(e) $y' = 1 + y$ (f) $y' = 1 + y^2$

(g) $y' = x^2y^2$ (h) $y' = e^y \cos x$

2. Find the general solution:

(a) $y' = \dfrac{2x - 3y}{3x - 5y}$ (b) $y' = \dfrac{4x - y}{x - 2y}$ (c) $y' = e^{-y/x} + \dfrac{y}{x}$

(d) $y' = \dfrac{x^2 + y^2}{xy}$ (e) $y' = \dfrac{y^2 + xy - x^2}{x^2}$ (f) $y' = \dfrac{2y^2 - x^2}{xy}$

3. *Growth.* The rate of growth of a population in a fixed environment is governed mainly by the size of the population, hence, is well described by a differential equation $dx/dt = f(x)$, where t is time and x is a real number measuring the size of the population. Discuss the growth pattern for each of the following choices of f; throughout k and b are positive constants.

(a) $f(x) = kx$ (b) $f(x) = kx^2$

(c) $f(x) = kx(b - x)$ (d) $f(x) = kx^2(b - x)$

4. *Air resistance.* A body falling against air resistance is subject to a force depending mainly on velocity, so that Newton's Second Law leads to a differential equation $dv/dt = f(v)$, where v is the velocity (measured positively downward) and t is the time. Discuss the type of variation of velocity with time for each of the following choices of f; throughout, g, a, m, and b are positive constants.

(a) $f(v) = (g/m) - bv$ (b) $f(v) = (g/m) - bv^2$ (c) $f(v) = (g/m) - be^{av}$

5. The Existence Theorem (Section 14-1) does not apply to the equation $y' = y^{1/3}$ at points $(x_0, 0)$, since the function $y^{1/3}$ does not have a continuous derivative for $y = 0$. Find the solutions, graph, and show that the uniqueness of solutions is violated along the x-axis.

6. The equation of Example 1 in Section 14-4 has a singular point at $(0, 0)$, and the Existence Theorem is not applicable to the corresponding equation $y' = -y/x^2$ at this point. Show that the equation has infinitely many solutions through the origin for $x \leq 0$ but just one solution through the origin for $x \geq 0$.

7. Let a family of curves in the plane be given, forming the general solution of a differential equation $y' = F(x, y)$ in an open region D. A second family of curves in D is said to form the set of *orthogonal trajectories* of the first family if at each point (x, y) the curve of the first family through the point is orthogonal to the curve of the second family through the point. By analytic geometry, perpendicular lines have slopes m_1 and $m_2 = -1/m_1$. Hence, a curve $y = f(x)$ of the second family satisfies the equation $y' = -1/F(x, y)$ at each point on its graph and, therefore, the orthogonal trajectories form the general solution of the differential equation $y' = -1/F(x, y)$ in D (at least, where $F \neq 0$). One can also write the given equation in the form $P(x, y)\, dx + Q(x, y)\, dy = 0$, and then the equation of the orthogonal trajectories is $-Q(x, y)\, dx + P(x, y)\, dy = 0$; thus $\mathbf{u} = P\mathbf{i} + Q\mathbf{j}$ is replaced by $\mathbf{u}^{\dashv} = -Q\mathbf{i} + P\mathbf{j}$. (What is the geometric meaning of $\mathbf{u}$?) We observe that the first family is also the family of orthogonal trajectories of the second family. Important examples of orthogonal trajectories are the sets of lines parallel to the coordinate axes, and the curves $r =$ const and $\theta =$ const in polar coordinates (in fact, orthogonal trajectories are often the basis for curvilinear coordinates in the plane).

For each of the differential equations below, state the differential equation for the orthogonal trajectories and find its general solution. Also graph the solutions of the given equation and the orthogonal trajectories.

(a) $y' = e^x$ (b) $y' = x^2 + 1$ (c) $y\, dx + x\, dy = 0$ (d) $y\, dx - x\, dy = 0$
(e) $x^2\, dy + y\, dx = 0$ (f) $(x + y)\, dx + (x - y)\, dy = 0$

‡8. For an equation $y' = F(x)G(y)$ in which G has zeros at y_1 and y_2, the lines $y = y_1$ and $y = y_2$ are solutions, and it appears that solutions starting between these lines are "trapped" between them, as in Figure 14-10. To establish this, assume that $F(x)$ is continuous for all x, that $G'(y)$ is continuous for all y and that $G(y_1) = G(y_2) = 0$, $G(y) > 0$ for $y_1 < y < y_2$. Let $y_1 < y_0 < y_2$ and show that a solution $y = f(x)$ with initial value (x_0, y_0) is defined for $-\infty < x < \infty$ and satisfies $y_1 < f(x) < y_2$ for all x. {*Hint.* Show that f is defined implicitly by the equation $g(y) = \varphi(x)$, where

$$g(y) = \int_{y_0}^{y} \frac{1}{G(u)}\, du, \qquad \varphi(x) = \int_{x_0}^{x} F(t)\, dt$$

Show that there are positive constants K_1, K_2 such that $G(y) \leq K_1(y - y_1)$ and $G(y) \leq K_2(y_2 - y)$ for $y_1 \leq y \leq y_2$ and, hence, conclude that $g(y)$ is monotone strictly increasing for $-\infty < y < \infty$ with limit $+\infty$ as $y \to y_2-$, limit $-\infty$ as $y \to y_1+$. Now use the fact that $f(x) = g^{-1}[\varphi(x)]$.}

‡9. Let $P(x, y)$ and $Q(x, y)$ have continuous first derivatives for all (x, y) and let P and Q be homogeneous of degree n, so that $P(tx, ty) = t^n P(x, y)$, $Q(tx, ty) = t^n Q(x, y)$. Show that the equation $P\, dx + Q\, dy = 0$ can be written in the form $y' = g(y/x)$ and that $(xP + yQ)^{-1}$ is an integrating factor for the equation.

‡10. Prove: if $y = f(x)$ is a solution of the differential equation $y' = g(y/x)$, then so also is $y = h(x) = (1/k)f(kx)$, where k is a nonzero constant. Interpret the result as showing that the solutions of $y' = g(y/x)$ are a family of similar curves.

14-5 THE LINEAR EQUATION OF FIRST ORDER

A first-order equation of form

$$s(x)y' + p(x)y = q(x)$$

is said to be linear. Whenever $s(x) \neq 0$ we can divide by $s(x)$ to obtain an equation of form

$$y' + p(x)y = q(x) \tag{14-50}$$

and we shall only consider the linear equation in the form (14-50). Equations of this form are exceptionally important for applications.

The solutions of Equation (14-50) can be found with the aid of an integrating factor, as in Section 7-16. We here illustrate the application of another method: *variation of parameters.* We first consider the equation

$$y' + p(x)y = 0 \tag{14-51}$$

called the *associated homogeneous equation* of Equation (14-50). [The word "homogeneous" here has no relation to the equations of form $y' = g(y/x)$ considered in the preceding section.] The solutions of (14-51) are found easily by separating variables:

$$\int \frac{dy}{y} + \int p(x)\,dx = 0, \qquad \ln\frac{y}{c} + \int p(x)\,dx = 0$$

$$y = ce^{-\int p(x)\,dx} \tag{14-52}$$

We assume a fixed choice of the indefinite integral of p. Now to obtain the solutions of equation (14-50), we replace the arbitrary constant in (14-52) by a function of x and seek solutions in the form:

$$y = v(x)e^{-\int p(x)\,dx} \tag{14-53}$$

Substitution in (14-50) gives

$$v'e^{-\int p(x)\,dx} - ve^{-\int p\,dx}p + pve^{-\int p\,dx} = q(x)$$

or

$$v'(x) = q(x)e^{\int p(x)\,dx}$$

Hence, (14-53) is a solution of (14-50) precisely when

$$v(x) = \int q(x)e^{\int p(x)\,dx}\,dx + c$$

That is, all solutions of (14-50) are given by

$$y = e^{-\int p(x)\,dx}\int q(x)e^{\int p(x)\,dx}\,dx + ce^{-\int p(x)\,dx} \tag{14-54}$$

Our procedure has been purely formal. However, all steps are valid if $p(x)$ and $q(x)$ are continuous over an interval $a < x < b$ and hence, (14-54) gives the general solution of (14-50) over this interval.

We remark also that (14-54) provides a unique solution through each point (x_0, y_0) with $a < x_0 < b$. For since the coefficient of c has no zeros, evaluation of both sides of (14-54) at (x_0, y_0) leads, to an equation of form $y_0 = h + kc$, where $k \neq 0$, and this equation has a unique solution for c. Therefore, we have proved the following *existence theorem* (Section 14-1).

THEOREM 2. *Let $p(x)$ and $q(x)$ be continuous for $a < x < b$. Then for each point (x_0, y_0) with $a < x_0 < b$ there exists a unique solution $y(x)$, $a < x < b$, of the linear differential equation*

$$y' + p(x)y = q(x)$$

such that $y(x_0) = y_0$.

EXAMPLE. $y' + y \tan x = \sin x \cos x$.

Solution. We imitate the method used in deriving (14-54). The associated homogeneous equation is

$$y' + y \tan x = 0$$

We solve by separating variables:

$$\int \frac{dy}{y} + \int \tan x \, dx = 0, \qquad \ln \frac{y}{c} - \ln \cos x = 0, \qquad y = c \cos x$$

We then set $y = v \cos x$ in the given equation:

$$v' \cos x - v \sin x + v \cos x \tan x = \sin x \cos x$$

$$v' \cos x = \sin x \cos x, \qquad v' = \sin x, \qquad v = -\cos x + c$$

$$y = v \cos x = (c - \cos x) \cos x = c \cos x - \cos^2 x$$

The steps are all valid if we restrict ourselves to an interval in which $\tan x$ is continuous—for example, to the interval $-\pi/2 < x < \pi/2$. (Note that in such an interval $\cos x$ has no zeros and hence the cancellation of that factor as above is permitted.) The solutions in this interval are graphed in Figure 14-12.

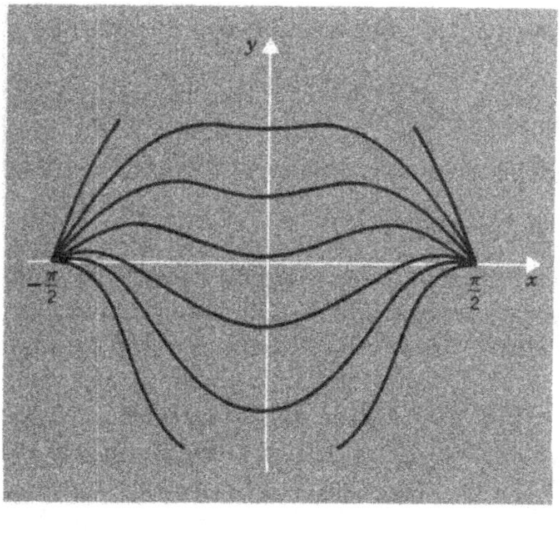

Figure 14-12 Solutions of $y' + y \tan x = \sin x \cos x$.

Remark. For the linear equation (14-50) it is often convenient to consider the equation in an interval $a \leq x < b$, $a < x \leq b$, or $a \leq x \leq b$. All of the discussion applies to these cases. For the interval $a \leq x < b$, for example, the solution $f(x)$ has a derivative to the right at a and, if we denote this derivative by $f'(a)$, then

$$f'(a) + p(a)f(a) = q(a)$$

Discussion of the Linear Equation. The theory of the linear equation is exceptionally simple. The equation can be considered in a given interval of x; all solutions exist over that interval and form a very simple pattern, as in Figure 14-12. There is one general formula (14-54) for the solutions.

One can consider the equation from the point of view of linear algebra. Let p and q be continuous on a fixed interval. Then we consider the vector space V of all functions having continuous first derivatives on the interval and the vector space W of all functions that are continuous on the interval. For each function f in V, we form $g = T[f]$, where

$$g(x) = f'(x) + p(x)f(x)$$

That is, $T[f] = g = f' + pf$. Then T is a *linear mapping* of V into W. For if $T[f_1] = g_1$ and $T[f_2] = g_2$, then

$$g_1 = f_1' + pf_1, \qquad g_2 = f_2' + pf_2$$

$$c_1g_1 + c_2g_2 = (c_1f_1 + c_2f_2)' + p(c_1f_1 + c_2f_2)$$

and hence,

$$T[c_1f_1 + c_2f_2] = c_1T[f_1] + c_2T[f_2]$$

The *kernel* of the linear mapping T is given by the set of f such that $T[f] = 0$; that is, by all $y = f(x)$ such that

$$y' + p(x)y = 0$$

But this equation is just the homogeneous equation (14-51) associated with (14-50). Therefore, *the kernel of T consists of all solutions of the homogeneous equation*. Hence, the kernel consists of all functions (14-52) for arbitrary c. If we write

$$y_1(x) = e^{-\int p(x)\,dx}$$

(for one fixed choice of the indefinite integral), then (14-52) is of the form

$$y = cy_1(x)$$

where y_1 is not identically 0 and c may be any constant. Accordingly, *the kernel of T is a one-dimensional vector space*. A basis for this vector space is any nonzero member of the kernel; that is, $y_1(x)$ itself, or $cy_1(x)$ for some nonzero scalar c.

To find the general solution of (14-50) we must find all functions $y = f(x)$ such that $T[f] = q$, for a given q in W; that is, we seek the T-pre-image of q. By linear algebra (Section 9-13), this T-pre-image (if not empty) forms a linear variety consisting of all functions $y^* + u$, where y^* is one function such that $T[y^*] = g$ and u is an arbitrary function in the kernel of T. Thus the T-pre-image of q consists of all functions

$$y = y^*(x) + cy_1(x) \qquad (14\text{-}54')$$

where $y^*(x)$ is one solution of the given differential equation $y' + py = q$ and $y_1(x)$ is one nonzero solution of the associated homogeneous differential equation. Now (14-54) also gives the general solution of (14-50) and (14-54) also consists of two terms. The second term we recognize as $cy_1(x)$. For $c = 0$, the function $y(x)$ defined by (14-54) reduces to the first term. Hence, the first term must be a solution $y^*(x)$ of (14-50). Thus (14-54) is the same as (14-54′).

In particular, we have shown that, for every q in W, the T-pre-image of q is not empty. Therefore, *T maps V onto all of W.*

Applications of the Linear Equation. Because of the simplicity of the theory of the linear equation, the equation is exceptionally useful in describing natural phenomena. In fact, it is common practice to construct devices that behave in accordance with a linear differential equation of first order. Similar statements apply to linear equations of higher order, discussed in the later sections of this chapter.

In many applications, the independent variable is time t, and one writes the equation as follows:

$$\frac{dx}{dt} + p(t)x = q(t) \tag{14-55}$$

For a *control system*, such as a thermostat, $p(t)$ is usually given for $0 \leq t < \infty$ and q is allowed to be any one of a class of continuous functions on $[0, \infty)$ [often a subspace of $\mathcal{C}[0, \infty)$]. One considers q as the *input*, the solution x as the *output*. The formula (14-54′) becomes

$$x = x^*(t) + cx_1(t) \tag{14-56}$$

Here $x_1(t)$ is a nontrivial solution of the homogeneous equation:

$$x'(t) + p(t)x(t) = 0 \tag{14-57}$$

For most applications, $x_1(t) \to 0$ as $t \to \infty$, and one then calls $x_1(t)$ a *transient*. For t sufficiently large, one is then concerned with the behavior of $x^*(t)$, called the *steady state*. For example, the equation

$$x'(t) + x = \sin t - \cos t$$

has as solutions

$$x = \sin t + ce^{-t}$$

The term ce^{-t} does approach 0 as $t \to \infty$, for every c. Thus for t very large x approximates a pure sine function. We note that in this example the input can be written as $\sqrt{2}\sin(t - (\pi/4))$; hence, the steady-state output resembles the input, with the same *period* (or *frequency*), but with a change in *amplitude* and *phase*. [In general, $x = A\sin(\omega t + \alpha)$ has amplitude A, frequency ω, period $2\pi/\omega$, and phase α.] Much of the theory of control systems concerns the way sinusoidal inputs are related to their outputs, as in this example. Particular examples are illustrated in the exercises.

PROBLEMS

1. Find the general solution:

(a) $y' + 3y = e^x$ (b) $y' + 2y = e^{-x}$

(c) $y' + 5y = xe^x$ (d) $y' - y = xe^x$

(e) $xy' + y = \sin x$ (f) $xy' + y = 6x^2$

(g) $xy' + 2y = \ln x$ (h) $xy' - 3y = x^5$

(i) $y' + y \cot x = e^{2x}$ (j) $y' + y \sec x = \tan x$

(k) $e^x y' - y = 1$ (l) $xy' - xy = -x \ln x - xe^x + 1$

2. Let a be a nonzero constant. For each of the following choices of $q(t)$ verify that the linear equation $x'(t) + ax(t) = q(t)$ has a particular solution $x^*(t)$ as given, so that the general solution is $x = x^*(t) + ce^{-at}$. Throughout k, b, and ω are nonzero constants, $P(t)$ is a polynomial of degree $n > 0$.

	$q(t)$	$x^*(t)$
(a)	k	k/a
(b)	$e^{bt}, \quad b \neq -a$	$e^{bt}/(a + b)$
(c)	e^{-at}	te^{-at}
(d)	$\sin \omega t$	$(a \sin \omega t - \omega \cos \omega t)/(a^2 + \omega^2)$
(e)	$\cos \omega t$	$(a \cos \omega t + \omega \sin \omega t)/(a^2 + \omega^2)$
(f)	$P(t), \quad \deg P = n$	$\dfrac{P(t)}{a} - \dfrac{P'(t)}{a^2} + \cdots + (-1)^n \dfrac{P^{(n)}(t)}{a^{n+1}}$
(g)	$e^{bt}P(t), \quad b \neq -a, \ \deg P = n$	$e^{bt}\left[\dfrac{P(t)}{a+b} - \dfrac{P'(t)}{(a+b)^2} + \cdots + (-1)^n \dfrac{P^{(n)}(t)}{(a+b)^{n+1}}\right]$
(h)	$e^{-at}P(t)$	$e^{-at}Q(t)$, where $Q'(t) = P(t)$

3. (a) Let $q(t)$ be continuous for $t \geq 0$ and let a be a constant. Show that the solution of the equation $x' + ax = q$ with initial value x_0 for $t = 0$ is

$$x = \int_0^t e^{a(u-t)}\, q(u)\, du + x_0 e^{-at}$$

(b) Show that the solution of part (a) can be written:

$$x = \int_0^t e^{-av}\, q(t - v)\, dv + x_0 e^{-at}$$

4. A control mechanism is governed by the differential equation $x' + 3x = q(t)$. Show that if $q(t)$ is sinusoidal: $q(t) = A \sin(\omega t + \alpha)$, then the steady-state output $x^*(t)$ is also sinusoidal: $x = B \sin(\omega t + \beta)$ and find its amplitude B and phase β. [*Hint.* Use the results of Problem 2.]

5. The motion of a descending parachutist is governed by the differential equation $m(dv/dt) + kv = mg$, where m is the mass of man and parachute, v is the velocity (measured positively downward) and k and g are positive constants. For $t = 0$, the velocity is v_0. Describe the motion in each of the three cases:

(a) $0 \le v_0 < mg/k$ (b) $v_0 = mg/k$ (c) $v_0 > mg/k$

6. Newton's law of cooling states that the rate of change of temperature x of a body is proportional to the difference between x and the temperature z of the surrounding medium. Show that, if z remains close to a value z_0 but fluctuates sinusoidally at high frequency about this value then, as t increases, x essentially ignores the fluctuations in z and approaches z_0 (see Problems 2 and 4).

7. For a particle moving along a line (the x-axis) against friction proportional to velocity, and subject to a spring force $-k^2x$ and an external force $F(t)$, Newton's Second Law gives the differential equation

$$m\frac{d^2x}{dt^2} + h\frac{dx}{dt} + k^2x = F(t)$$

h, k, and m are positive constants. If h is exceptionally large, it is reasonable to ignore the first term, so that one has a first-order differential equation for x. Discuss the motion for the case $F(t) = ae^{-bt}$ where a and b are positive constants, with m replaced by 0 in the differential equation.

8. In the theory of electricity, it is shown that a circuit containing an inductance, L, resistance, capacitance, and driving electromotive force, the charge stored q satisfies the equation

$$L\frac{d^2q}{dt^2} + R\frac{dq}{dt} + \frac{1}{C}q = E(t)$$

where L, R, and C are positive constants and $E(t)$ is the variable emf (Figure 14-13). The current i is given by $i = dq/dt$. When there is no inductance, the term in L is omitted, and one has a first-order differential equation for q; when there is no capacitor, the term in $1/C$ is omitted, and one has a first order differential equation for i. Discuss the behavior of charge and/or current for $t \ge 0$ as requested, for the following cases:

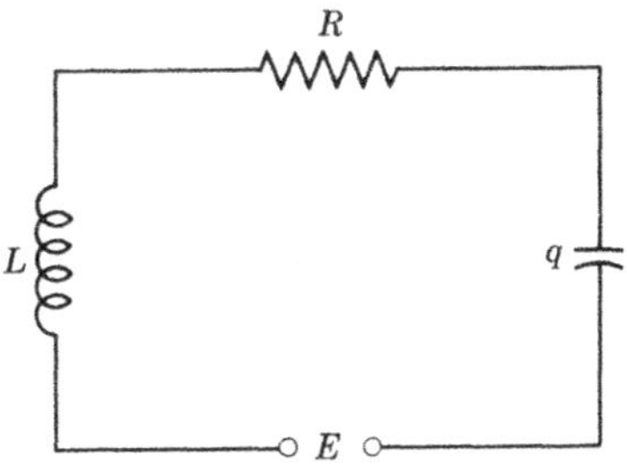

Figure 14-13 Simple electric circuit.

(a) q and i; $L = 0$, $E = E_0 = \text{const}$, $q = 0$ for $t = 0$;
(b) q and i; $L = 0$, $E = E_0 \sin \omega t$ where $E_0 = \text{const}$, $q = 0$ for $t = 0$;
(c) i; no capacitor, $E = E_0 = \text{const}$, $i = 0$ for $t = 0$;
(d) i; no capacitor, $E = E_0 \sin \omega t$ where $E_0 = \text{const}$, $i = 0$ for $t = 0$.

14-6 LINEAR DIFFERENTIAL EQUATIONS OF ORDER n

In the preceding section we have discussed first-order linear equations. We now consider the *linear differential equation of order n*: that is, the equation of form:

$$A_n(x)y^{(n)} + \cdots + A_1(x)y' + A_0(x)y = Q(x) \qquad (14\text{-}60)$$

We shall consider this equation only in an interval in which the coefficient $A_n(x)$ is not 0. Hence, we can divide by $A_n(x)$ and obtain an equation of form:

$$y^{(n)} + a_{n-1}(x)y^{(n-1)} + \cdots + a_1(x)y' + a_0(x)y = q(x) \qquad (14\text{-}61)$$

We shall assume that the coefficients $a_{n-1}(x), \ldots, a_0(x)$ and the right-hand member $q(x)$ are continuous on an interval J. When $a_{n-1}, \ldots, a_0$ are constants [or $A_n, \ldots, A_0$ are constants, with $A_n \neq 0$, in (14-60)] we say that the equation has *constant coefficients.*

EXAMPLES (i) $y'' - y = x^2$

(ii) $y''' + xy'' - y = 0$

(iii) $4y^{(iv)} - 2y''' + y'' - 2y' + y = \ln x$

The first and third examples have constant coefficients. The interval J would be taken to be the infinite interval $(-\infty, \infty)$ for Examples (i) and (ii), and the interval $(0, \infty)$ for Example (iii).

In Section 14-1 we stated a general Existence Theorem for differential equations. For the linear equation (14-61) we can state a stronger result:

THEOREM 3. *Let $a_0(x), \ldots, a_{n-1}(x)$, $q(x)$ be defined and continuous on the interval J. Then there is a solution $y(x)$ of Equation (14-61) on the interval J with prescribed initial values*

$$y(x_0) = y_0, \quad \ldots, \quad y^{(n-1)}(x_0) = y_0^{(n-1)} \qquad (14\text{-}62)$$

at a point x_0 of J. Furthermore, the solution is unique: that is, if $y^*(x)$ is a solution of (14-61) in an interval J^* including x_0 and $y^*(x_0) = y_0, \ldots, y^{*(-1)}(x_0) = y_0^{(n-1)}$, then $y(x) = y^*(x)$ over the common part of the two intervals.

For $n = 1$, the theorem follows from Theorem 2 in Section 14-5. For $n > 1$ we shall prove the theorem for special cases below. For the general case, see Chapter 12 of ODE.

Theorem 3 differs from the Existence Theorem of Section 14-1 in that the solution is guaranteed to exist *over the whole interval J.* For nonlinear equations, this property generally fails (Problem 5 below). Because of Theorem 3, for a linear differential equation we shall use the word "solution" only for a function $y(x)$ satisfying the equation over the whole interval J.

We note that every solution $y(x)$ of (14-61) has continuous derivatives through order n on the interval J (Problem 6 below).

If in (14-61) the function $q(x)$ is identically 0 on J, then Equation (14-61) is said to be *homogeneous.* This is illustrated by Example (ii) above. For a given (nonhomogeneous) equation (14-61) we obtain the *associated homogeneous equation* by replacing $q(x)$ by 0:

$$y^{(n)} + a_{n-1}(x)y^{(n-1)} + \cdots + a_1(x)y' + a_0(x)y = 0 \qquad (14\text{-}63)$$

For $n = 1$ we saw in Section 14-5 that the solutions of the homogeneous equation form a one-dimensional vector space, and that the solutions of the nonhomogeneous equation form a linear variety whose base space consists of the solutions of the associated homogeneous equation. The following discussion shows that these results generalize to the equation (14-61) of order n.

THEOREM 4. *Let $a_0(x), \ldots, a_{n-1}(x)$ be continuous on the interval J. Then the solutions of the homogeneous equation (14-63) on J form a vector space $\mathcal{H}$. Furthermore, if f is in $\mathcal{H}$ and f satisfies the initial conditions*

$$f(x_0) = 0, \quad \ldots, \quad f^{(n-1)}(x_0) = 0 \tag{14-64}$$

at some point x_0 of J, then f is the zero function on J.

PROOF. We let $\mathcal{C}$ denote the vector space of all functions continuous on J; we let $\mathcal{C}^{(n)}$ be the vector space of all functions having continuous derivative through order n on J. For f in $\mathcal{C}^{(n)}$, we define $L[f] = g$ in $\mathcal{C}$ by the equation:

$$g(x) = f^{(n)}(x) + a_{n-1}(x)f^{(n-1)}(x) + \cdots + a_0(x)f(x) = L[f]$$

Then L is a linear mapping; the proof is just like that for the mapping T in Section 14-5. The functions of $\mathcal{H}$ are precisely the functions f in $\mathcal{C}^{(n)}$ such that $L[f] = 0$; that is, *$\mathcal{H}$ is the kernel of L*. Therefore $\mathcal{H}$ is a vector space, a subspace of $\mathcal{C}^{(n)}$.

By Theorem 3, there is just one function in $\mathcal{H}$ satisfying the initial conditions (14-64). But the zero function is in $\mathcal{H}$ and clearly satisfies (14-64)! Therefore, if f is in $\mathcal{H}$ and satisfies (14-64) then $f(x) \equiv 0$ on J.

The Associated Vector-Functions. To each function $y(x)$ in $\mathcal{C}^{(n)}$, we associate the continuous vector-function $\mathbf{v}(x) = (y(x), y'(x), \ldots, y^{(n-1)}(x))$ on J. For each fixed x_0, $\mathbf{v}(x_0)$ is a vector of V_n and, if $y(x)$ is a solution of (14-63), the components of $\mathbf{v}(x_0)$ are just the corresponding initial values $y(x_0)$, $y'(x_0), \ldots, y^{(n-1)}(x_0)$. The whole function $\mathbf{v}(x)$ can be considered as a member of the vector space of all continuous vector-functions on J, with values in V_n (see Section 9-16).

LEMMA 1. *Let $y_1(x), \ldots, y_k(x)$ be functions in $\mathcal{C}^{(n)}$. Then $y_1(x), \ldots, y_k(x)$ are linearly dependent over J if, and only if, the associated vector-functions $\mathbf{v}_1(x), \ldots, \mathbf{v}_k(x)$ are linearly dependent over J.*

PROOF. If $y_1(x), \ldots, y_k(x)$ are linearly dependent over J, then there are scalars $c_1, \ldots, c_k$, not all 0, so that $c_1y_1(x) + \cdots + c_ky_k(x) \equiv 0$ in J. We differentiate repeatedly and conclude that

$$c_1y_1^{(i)}(x) + \cdots + c_ky_k^{(i)}(x) \equiv 0 \quad \text{in } J, \qquad i = 0, \ldots, n-1 \tag{14-65}$$

But this implies that

$$c_1\mathbf{v}_1(x) + \cdots + c_k\mathbf{v}_k(x) \equiv \mathbf{0} \quad \text{in } J \tag{14-65'}$$

so that $\mathbf{v}_1(x), \ldots, \mathbf{v}_k(x)$ are linearly dependent vector-functions. Conversely, if $\mathbf{v}_1(x), \ldots, \mathbf{v}_k(x)$ are linearly dependent vector-functions, then (14-65′) holds true with $c_1, \ldots, c_k$ not all 0. Hence, (14-65) holds true; for $i = 0$; but this is just the relation $c_1y_1(x) + \cdots + c_ky_k(x) \equiv 0$. Therefore, $y_1(x), \ldots, y_k(x)$ are linearly dependent.

LEMMA 2. *Let $y_1(x), \dots, y_k(x)$ be solutions of (14-63) over J. If the associated vector-functions $\mathbf{v}_1(x), \dots, \mathbf{v}_k(x)$ are linearly dependent vectors in V_n for one value x_0, then $\mathbf{v}_1(x), \dots, \mathbf{v}_k(x)$ are linearly dependent vector-functions over J.*

PROOF. Let us suppose that $c_1, \dots, c_k$ are not all 0 and that

$$c_1\mathbf{v}_1(x_0) + \cdots + c_k\mathbf{v}_k(x_0) = \mathbf{0}$$

Then, by Theorem 4, the function $y(x) = c_1y_1(x) + \cdots + c_ky_k(x)$ is a solution of (14-63) and by the preceding equation

$$y^{(i)}(x_0) = c_1y_1^{(i)}(x_0) + \cdots + c_ky_k^{(i)}(x_0) = 0 \qquad \text{for } i = 0, \dots, n-1$$

Therefore, by Theorem 4, $y(x) \equiv 0$ on J. Accordingly,

$$c_1y_1(x) + \cdots + c_ky_k(x) \equiv 0$$

so that $y_1(x), \dots, y_k(x)$ are linearly dependent over J and, by Lemma 1, so are $\mathbf{v}_1(x), \dots, \mathbf{v}_k(x)$.

Remark. The conclusion of Lemma 2 is false if we omit the hypothesis that $y_1(x), \dots, y_k(x)$ are solutions of a homogeneous linear differential equation; see remark following Example 2 below.

THEOREM 5. *The vector space $\mathcal{H}$ of Theorem 4 has dimension n. Therefore the homogeneous linear equation (14-63) has n linearly independent solutions on J and, if $y_1(x), \dots, y_n(x)$ are n linearly independent solutions on J, the general solution is given by*

$$y = c_1y_1(x) + \cdots + c_ny_n(x)$$

PROOF. Let us fix x_0 in J. We choose n linearly independent vectors $\mathbf{u}_1, \dots, \mathbf{u}_n$ in V_n [for example, the basis vectors $(1, 0, \dots, 0), \dots, (0, \dots, 0, 1)$]. For each $i = 1, \dots, n$, we choose $y_i(x)$, a solution of (14-63) with initial values at x_0 such that for the associated $\mathbf{v}_i(x)$ one has $\mathbf{v}_i(x_0) = \mathbf{u}_i$. Hence, $\mathbf{v}_1(x), \dots, \mathbf{v}_n(x)$ are linearly independent in V_n for $x = x_0$; therefore, by Lemma 2, $\mathbf{v}_1(x), \dots, \mathbf{v}_n(x)$ are linearly independent vector-functions over J and, hence, by Lemma 1, $y_1(x), \dots, y_n(x)$ are linearly independent over J. Accordingly, $\dim \mathcal{H} \geq n$.

On the other hand, we cannot choose more than n linearly independent solutions $y_1(x), \dots, y_k(x)$ of (14-63). For if we could, the associated vector-functions $\mathbf{v}_1(x), \dots, \mathbf{v}_k(x)$ would be linearly independent and, hence, by Lemma 2, $\mathbf{v}_1(x_0), \dots, \mathbf{v}_k(x_0)$ would be linearly independent vectors in V_n. But there can be no more than n linearly independent vectors in V_n, since $\dim V_n = n$. Hence, k cannot exceed n and $\dim \mathcal{H} = n$.

The Wronskian. If $y_1(x), \dots, y_k(x)$ are solutions of (14-63), one can form the n by k matrix $Y(x)$ whose column vectors are $\mathbf{v}_1(x), \dots, \mathbf{v}_k(x)$:

$$Y(x) = (y_j^{(i-1)}(x)) = \begin{pmatrix} y_1(x) & \cdots & y_k(x) \\ \vdots & & \\ y_1^{(n-1)}(x) & \cdots & y_k^{(n-1)}(x) \end{pmatrix}$$

From Lemmas 1 and 2, we see that $y_1(x), \ldots, y_k(x)$ *are linearly independent over* J, *if, and only if,* $Y(x)$ *has rank* k *for every* x *in* J; *if the rank is* k *for one* x, *then it is* k *for every* x *in* J; *if the rank is less than* k *for one* x, *then it is less than* k *for all* x *in* J.

For $k = n$, $Y(x)$ is a square matrix and its determinant $W(x)$ is called the *Wronskian* of $y_1(x), \ldots, y_n(x)$:

$$W(x) = \begin{vmatrix} y_1(x) & \cdots & y_n(x) \\ \vdots & & \\ y_1^{(n-1)}(x) & \cdots & y_n^{(n-1)}(x) \end{vmatrix} \tag{14-66}$$

Hence, either $W(x) \equiv 0$ on J or $W(x)$ has no zeros on J; in the former case, the solutions are linearly dependent; in the latter case, they are linearly independent. When they are linearly independent, they serve as a basis for $\mathcal{H}$ as in Theorem 5.

Let $y_1(x), \ldots, y_n(x)$ be solutions of (14-63) on J and let J_1 be an interval contained in J. Then $y_1(x), \ldots, y_n(x)$ are solutions of (14-63) on J_1. If $y_1(x), \ldots, y_n(x)$ are linearly independent on J, then they are clearly linearly independent on subinterval J_1. For if $W(x)$ has no zeros on J, it has none on J_1.

Conversely, if $y_1(x), \ldots, y_n(x)$ are linearly independent on J_1, then they must be linearly independent on J; for $c_1y_1(x) + \cdots + c_ny_n(x) \equiv 0$ on J would imply $c_1y_1(x) + \cdots + c_ny_n(x) \equiv 0$ on J_1.

EXAMPLE 1. $y'' - y = 0$. We take J to be $(-\infty, \infty)$. The functions $y_1(x) = e^x$ and $y_2(x) = e^{-x}$ are solutions and have the Wronskian

$$W(x) = \begin{vmatrix} e^x & e^{-x} \\ e^x & -e^{-x} \end{vmatrix} = -2$$

Since $W(x)$ has no zeros, the solutions are linearly independent and the general solution is

$$y = c_1e^x + c_2e^{-x}$$

EXAMPLE 2. $(x - 1)y'' - xy' + y = 0$. We divide by $(x - 1)$ to obtain an equation of form (14-63):

$$y'' - \frac{x}{x-1}y' + \frac{y}{x-1} = 0$$

Because of the discontinuity at $x = 1$, we see that J can be taken as $(1, \infty)$ or $(-\infty, 1)$ but *not* as $(-\infty, \infty)$. We find (by design of the equation!) that $y_1 = x$, $y_2 = e^x$ are solutions. Their Wronskian is

$$W(x) = \begin{vmatrix} x & e^x \\ 1 & e^x \end{vmatrix} = e^x(x - 1)$$

On either $(1, \infty)$ or $(-\infty, 1)$, $W(x)$ has no zeros. Hence, on either interval the general solution is $y = c_1x + c_2e^x$.

Remark. Example 2 illustrates two functions $y_1(x)$, $y_2(x)$ on $(-\infty, \infty)$ which are such that the associated vector-functions $\mathbf{v}_1(x) = (x, 1)$, $\mathbf{v}_2(x) = (e^x, e^x)$ are linearly dependent in V_2 for one x, namely, $x = 1$: $\mathbf{v}_1(1) = (1, 1)$, $\mathbf{v}_2(1) = (e, e)$. But $\mathbf{v}_1(x)$, $\mathbf{v}_2(x)$ are not linearly dependent vector-functions over $(-\infty, \infty)$; for if they were, they would be linearly dependent in V_2 for each fixed x, and this condition fails for $x = 0$: $\mathbf{v}_1(0) = (0, 1)$, $\mathbf{v}_2(0) = (1, 1)$. Thus the conclusion of Lemma 2 fails. However, $y_1(x)$ and $y_2(x)$ are not solutions of a differential equation of form (14-63) on the whole interval $(-\infty, \infty)$. The difficulty all comes from the discontinuity at $x = 1$, where the coefficient of y'', in the equation as first given, is zero.

THEOREM 6. *Under the hypotheses of Theorem* 3, *let* $\mathcal{S}$ *be the set of all solutions of* (14-61) *on* J, *let* $\mathcal{H}$ *be the set of all solutions of the associated homogeneous equation* (14-63) *on* J. *Then* $\mathcal{S}$ *is nonempty and, if* y_p *is one function in* $\mathcal{S}$,

$$\mathcal{S} = \{y_p + \mathcal{H}\}$$

Thus $\mathcal{S}$ *is a linear variety in* $\mathcal{C}^{(n)}$, *with base space* $\mathcal{H}$.

PROOF. That $\mathcal{S}$ is nonempty follows from Theorem 3. That $\mathcal{S}$ is a linear variety as described follows by linear algebra (Theorem 20, Section 9-13). In particular, if L is as in the proof of Theorem 4, then $\mathcal{S}$ is the L-pre-image of $q(x)$, and L has kernel $\mathcal{H}$ and, hence, $\mathcal{S} = \{y_p + \mathcal{H}\}$.

EXAMPLE 3. $y'' - y = -2 \sin x$. The associated homogeneous equation is $y'' - y = 0$, as in Example 1. Its general solution is $y = c_1e^x + c_2e^{-x}$. A particular solution of the given nonhomogeneous equation is $y_p = \sin x$ (by design!). Hence, the general solution is

$$y = \sin x + c_1e^x + c_2e^{-x}$$

Discussion. A basis for $\mathcal{H}$—that is, a set of linearly independent solutions $y_1(x), \ldots, y_n(x)$ for the associated homogeneous equation—is called a *fundamental set* of solutions of the homogeneous equation. The corresponding general solution $y = c_1y_1(x) + \cdots + c_ny_n(x)$ is called the *complementary function* and abbreviated as $y_c(x)$. Thus the general solution of the nonhomogeneous equation (14-61) is $y_p(x) + y_c(x)$: that is, a *particular solution plus the complementary function.* The essential difficulty lies in finding one fundamental set; there are actually infinitely many. For equations with constant coefficients, it is fairly simple to find a fundamental set; see Section 14-9 below. When the coefficients $a_j(x)$ in (14-61) are not constant, it can be exceedingly difficult to determine a fundamental set of solutions in exact form, and we often must resort to numerical methods or power series (Sections 14-22 and 14-23 below). Once the complementary function is known, there is a standard procedure for finding a particular solution $y_p(x)$ of the nonhomogeneous equation; see Section 14-7 below.

PROBLEMS

1. For each of the following differential equations, verify that the functions given are solutions, determine whether they are linearly independent and whether they form a basis for the set $\mathcal{H}$ of solutions:

   (a) $y''' + y' = 0$; $1, \cos x, \sin x$ on $(-\infty, \infty)$.

   (b) $y''' - 2y'' - 5y' + 6y = 0$; e^x, e^{-2x}, e^{3x} on $(-\infty, \infty)$.

   (c) $y^{(4)} + y''' - 3y'' - y' + 2y = 0$; e^x, xe^x, e^{-2x}, e^{-x} on $(-\infty, \infty)$.

   (d) $xy''' - y'' = 0$; $1, x, x^3$ on $(0, \infty)$.

   (e) $x^2y'' + 3xy' + y = 0$; $x^{-1}, x^{-1}\ln x$ on $(0, \infty)$.

   (f) $x^2y'' + 4xy' + 2y = 0$; x^{-1}, x^{-2} on $(-\infty, 0)$.

   (g) $y''' - 4y' = 0$; e^{2x}, e^{-2x} on $(-\infty, \infty)$.

   (h) $y''' + y' = 0$; $\cos x, \sin x$ on $(-\infty, \infty)$.

   (i) $y'' - y = 0$; $e^x, e^{-x}, \cosh x, \sinh x$ on $(-\infty, \infty)$.

   (j) $x^2y'' - 2xy' + 2y = 0$; $2x^2 - x$, $x^2 + 2x$, x^2 on $(-\infty, \infty)$.

2. (a) Prove: if $y_1(x)$ and $y_2(x)$ are solutions of the equation $y'' + a_1(x)y' + a_0(x)y = 0$ on J, then their Wronskian $W(x) = y_1(x)y_2'(x) - y_2(x)y_1'(x)$ satisfies the equation $W'(x) + a_1(x)W(x) = 0$, so that

   (*) $$W(x) = W(x_0)e^{-\int_{x_0}^{x} a_1(t)\,dt}$$

   Conclude: if W has one zero in J, then $W(x) \equiv 0$.

   ‡(b) Prove: if $W(x)$ is the Wronskian of solutions $y_1(x), \ldots, y_n(x)$ of (14-63) on J, then $W'(x) + a_{n-1}(x)W(x) = 0$, so that

   (**) $$W(x) = W(x_0)e^{-\int_{x_0}^{x} a_{n-1}(t)\,dt}$$

   Conclude: if W has one zero in J, then $W(x) \equiv 0$. (*Hint.* See Theorem 20 in Section 10-17).

3. (a) . . . (f). Verify the rules (*) or (**) of Problem 2 for the equations and solutions of Problem 1(a) . . . (f).

4. Find the general solution:

   (a) $y''' + y' = x$. [*Hint.* See Problem 1(*a*); seek a particular solution of form $ax^2 + bx + c$.]

   (b) $y''' + y' = \cos x$. [*Hint.* Try $y_p = ax\cos x + bx\sin x$.]

   (c) $xy''' - y'' = x$. [*Hint.* Set $u = y''$.]

   (d) $xy''' - y'' = x^2e^x$.

5. Find all solutions and discuss the intervals in which they are defined:

   (a) $y' = y^2$ (b) $y' = e^y$

6. Prove: if $a_0(x), \ldots, a_{n-1}(x)$, and $q(x)$ are continuous on the interval J, then each solution of (14-61) on J has a continuous nth derivative.

7. Let $$y_1(x) = \begin{cases} 0 & \text{if } x < 0 \\ x^2 & \text{if } x \geq 0 \end{cases}, \qquad y_2(x) = \begin{cases} x^2 & \text{if } x < 0 \\ 0 & \text{if } x \geq 0 \end{cases}$$

   Show that y_1' and y_2' are continuous on $(-\infty, \infty)$ and that $\mathbf{v}_1(x) = (y_1(x), y_1'(x))$,

$\mathbf{v}_2(x) = (y_2(x),\ y_2'(x))$ are linearly dependent for each fixed x, but are linearly independent vector-functions on $(-\infty, \infty)$.

8. Given functions $y_1(x), \ldots, y_n(x)$ having nth order derivatives on an interval, one can define their Wronskian by (14-66). If $W(x)$ has no zeros, they form a linearly independent set. (a) Show that the functions $y = x$ and $y = x^2$ are linearly independent on $(-\infty, \infty)$, but their Wronskian has a zero at $x = 0$. (b) Show that these two functions can not be solutions of a differential equation $y'' + p(x)y' + q(x)y = 0$ on $(-\infty, \infty)$, where p and q are continuous.

14-7 VARIATION OF PARAMETERS

We now discuss a method for determining a particular solution y_p of a nonhomogeneous linear differential equation

$$y^{(n)} + a_{n-1}(x)y^{(n-1)} + \cdots + a_0(x)y = q(x) \tag{14-70}$$

assuming that we already know a basis for the space $\mathcal{H}$ of solutions of the associated homogeneous equation:

$$y^{(n)} + a_{n-1}(x)y^{(n-1)} + \cdots + a_0(x)y = 0 \tag{14-71}$$

Thus we assume that we know the complementary function:

$$y_c(x) = c_1y_1(x) + \cdots + c_ny_n(x) \tag{14-72}$$

We assume that $a_{n-1}(x), \ldots, a_0(x)$, $q(x)$ are continuous on the interval J, so that all theorems of the preceding section apply.

The method of variation of parameters for finding y_p is based on the assumption that we can find functions $u_1(x), \ldots, u_n(x)$ such that

$$y_p = u_1(x)y_1(x) + \cdots + u_n(x)y_n(x) \tag{14-73}$$

is a solution of (14-70). Since we wish to determine n functions, we expect to impose n conditions on the $u_j(x)$. Of course, one of these conditions will be that y_p satisfies (14-70). We are free to choose the other conditions so as to simplify our calculations. Since part of our difficulty comes from solving differential equations of order greater than one, we shall choose the remaining conditions so that we need never consider relations involving any $u_j^{(k)}$ for $k > 1$. From Equation (14-73) we obtain

$$y_p' = u_1y_1' + \cdots + u_ny_n' + u_1'y_1 + \cdots + u_n'y_n$$

To be sure that no u_j'' appears in later calculations, we require that

$$u_1'y_1 + \cdots + u_n'y_n \equiv 0 \qquad \text{on } J$$

We then have

$$y_p' = u_1y_1' + \cdots + u_ny_n'$$

and therefore

$$y_p'' = u_1y_1'' + \cdots + u_ny_n'' + u_1'y_1' + \cdots + u_n'y_n'$$

As above, we require that

$$u_1'y_1' + \cdots + u_n'y_n' \equiv 0 \qquad \text{on } J$$

and obtain

$$y_p'' = u_1y_1'' + \cdots + u_ny_n''$$

Continuing in this way, we require that

$$u_1'y_1^{(h)} + \cdots + u_n'y_n^{(h)} \equiv 0 \qquad \text{on } J \qquad \text{for } h = 0, 1, \ldots, n-2$$

and obtain

$$y_p^{(k)} = u_1y_1^{(k)} + \cdots + u_ny_n^{(k)} \qquad \text{for } k = 0, 1, \ldots, n-1$$

On substituting the last expressions in (14-70), we obtain

$$u_1[y_1^{(n)} + \cdots + a_0(x)y_1] + \cdots + u_n[y_n^{(n)} + \cdots + a_0(x)y_n] + u_1'y_1^{(n-1)} + \cdots + u_n'y_n^{(n-1)} = q(x)$$

But $y_1(x), \ldots, y_n(x)$ are solutions of the homogeneous equation (14-71). Therefore, the last equation reduces to

$$u_1'y_1^{(n-1)} + \cdots + u_n'y_n^{(n-1)} = q(x)$$

Collecting the conditions we have imposed on $u_1, \ldots, u_n$, we see that we seek functions $u_1, \ldots, u_n$ such that on J

$$\left.\begin{aligned} &u_1'y_1 + \cdots + u_n'y_n \equiv 0 \\ &\vdots \\ &u_1'y_1^{(n-2)} + \cdots + u_n'y_n^{(n-2)} \equiv 0 \\ &u_1'y_1^{(n-1)} + \cdots + u_n'y_n^{(n-1)} \equiv q(x) \end{aligned}\right\} \tag{14-74}$$

Since $y_1(x), \ldots, y_n(x)$ are a basis for $\mathcal{H}$, their Wronskian

$$W(x) = \begin{vmatrix} y_1(x) & \cdots & y_n(x) \\ \vdots & & \\ y_1^{(n-1)}(x) & \cdots & y_n^{(n-1)}(x) \end{vmatrix}$$

has no zeros on J. But we can regard (14-74) as n linear equations in the n unknowns $u_1', \ldots, u_n'$, and $W(x)$ is the determinant of coefficients. Therefore, we can solve Equations (14-74) uniquely for $u_1', \ldots, u_n'$ and obtain

$$u_j' = \frac{W_j(x)}{W(x)}, \qquad j = 1, \ldots, n \tag{14-75}$$

where $W_j(x)$ differs from $W(x)$ in having $0, \ldots, 0, q(x)$ as its jth column. The functions $W(x)$ and $W_j(x)$ are continuous, so that we can integrate Equations (14-75) to obtain $u_1(x), \ldots, u_n(x)$. By the way they were determined, the function (14-73) satisfies (14-70) on J. Since we are seeking only one solution, we choose just one indefinite integral in going from each u_j' to u_j.

Remark. If the differential equation is written in the form

$$a_n(x)y^{(n)} + \cdots + a_0(x)y = q(x)$$

then one must replace $q(x)$ by $q(x)/a_n(x)$ in (14-74).

EXAMPLE 1. Find a particular solution of the equation $y^{(4)} - y = q(x)$, where $q(x)$ is continuous on $[0, 1]$.

Solution. We can easily verify that $\sin x$, $\cos x$, $\sinh x$, $\cosh x$ are solutions of the associated homogeneous equation: $y^{(4)} - y = 0$. Their Wronskian is

$$W(x) = \begin{vmatrix} \sin x & \cos x & \sinh x & \cosh x \\ \cos x & -\sin x & \cosh x & \sinh x \\ -\sin x & -\cos x & \sinh x & \cosh x \\ -\cos x & \sin x & \cosh x & \sinh x \end{vmatrix} = 4$$

Therefore, $y_c = c_1 \sin x + c_2 \cos x + c_3 \sinh x + c_4 \cosh x$, and a particular solution will be given by

$$y_p = u_1 \sin x + u_2 \sin x + u_3 \sinh x + u_4 \cosh x$$

We use (14-75) to find the u_j':

$$u_1' = -\frac{1}{4}q(x)\begin{vmatrix} \cos x & \sinh x & \cosh x \\ -\sin x & \cosh x & \sinh x \\ -\cos x & \sinh x & \cosh x \end{vmatrix} = -\frac{1}{2}q(x)\cos x$$

$$u_2' = \frac{1}{4}q(x)\begin{vmatrix} \sin x & \sinh x & \cosh x \\ \cos x & \cosh x & \sinh x \\ -\sin x & \sinh x & \cosh x \end{vmatrix} = \frac{1}{2}q(x)\sin x$$

$$u_3' = \frac{1}{2}q(x)\cosh x, \qquad u_4' = -\frac{1}{2}q(x)\sinh x$$

Hence

$$\begin{aligned} y_p &= \sin x \int_0^x \frac{-q(t)\cos t}{2}\,dt + \cos x \int_0^x \frac{q(t)\sin t}{2}\,dt \\ &\quad + \sinh x \int_0^x \frac{q(t)\cosh t}{2}\,dt + \cosh x \int_0^x \frac{-q(t)\sinh t}{2}\,dt \\ &= \frac{1}{2}\int_0^x q(t)[\sin (t - x) - \sinh (t - x)]\,dt \end{aligned}$$

is a particular solution of $y^{(4)} - y = q(x)$.

14-8 COMPLEX-VALUED SOLUTIONS OF LINEAR DIFFERENTIAL EQUATIONS

Here we consider complex-valued functions of a real variable: that is, functions of the form $f(x) = u(x) + iv(x)$, where $u(x)$ and $v(x)$ are real-valued.

The calculus can be extended to such functions by the formulas

$$f'(x) = u'(x) + iv'(x) \qquad \text{and} \qquad \int f(x)\,dx = \int u(x)\,dx + i\int v(x)\,dx$$

The familiar rules continue to hold true. If a and b are real, one defines

$$e^{(a+ib)x} = e^{ax}e^{ibx} = e^{ax}(\cos bx + i\sin bx)$$

and then verifies that

$$[e^{(a+ib)x}]' = (a + ib)e^{(a+ib)x}$$

For proofs of these statements, one is referred to Section 5-8; see also Section 8-23.

Now let $L[y] = y^{(n)} + a_{n-1}(x)y^{(n-1)} + \cdots + a_0(x)y$ as above, so that our linear differential equation is the equation:

$$L[y] = q(x) \tag{14-80}$$

If we now allow $a_0(x), \ldots, a_{n-1}(x)$ and $q(x)$ to be complex functions of x, then (14-80) becomes a complex linear differential equation for the complex function $y(x)$.

EXAMPLE 1. The equation

$$y'' - 3iy' - 2y = 0 \tag{14-81}$$

is a homogeneous complex linear equation of second order.

The theorems of Section 14-6 all extend to the complex equation (14-80), with essentially the same proofs. In all the vector space considerations, one must use *complex* vector spaces; that is, vector spaces in which the scalars are complex. For example, the solutions of (14-81) form a 2-dimensional complex vector space. The functions $y_1(x) = e^{ix}$ and $y_2(x) = e^{2ix}$ are solutions and are linearly independent; that is, there are no complex scalars other than $c_1 = 0$, $c_2 = 0$ for which $c_1y_1(x) + c_2y_2(x) \equiv 0$. The Wronskian is

$$\begin{vmatrix} e^{ix} & e^{2ix} \\ ie^{ix} & 2ie^{2ix} \end{vmatrix} = ie^{3ix} = i(\cos 3x + i\sin 3x)$$

and this has no zeros on $(-\infty, \infty)$. Accordingly, the general solution of (14-81) is $y = c_1e^{ix} + c_2e^{2ix}$, where c_1 and c_2 are arbitrary complex constants.

In most applications one is concerned with equations in which $a_0(x), \ldots, a_{n-1}(x)$ and $q(x)$ are real-valued. But often in such a case it is convenient to consider them as complex functions (which happen to have 0 imaginary part). We can then discuss the general complex solution of such a real equation. From the complex solutions one can then recover the real solutions—as those complex solutions that have 0 imaginary part.

EXAMPLE 2. $y'' + y = 0$. The functions e^{ix} and e^{-ix} are solutions and form a basis for the complex solutions, as one verifies. The general complex solution is

$$y = c_1e^{ix} + c_2e^{-ix} = (\alpha_1 + i\beta_1)e^{ix} + (\alpha_2 + i\beta_2)e^{-ix}$$

where $\alpha_1, \beta_1, \alpha_2, \beta_2$ are arbitrary real constants. We seek the solutions with 0 imaginary part:

$$y = (\alpha_1 + i\beta_1)(\cos x + i \sin x) + (\alpha_2 + i\beta_2)(\cos x - i \sin x)$$
$$= (\alpha_1 + \alpha_2) \cos x + (\beta_2 - \beta_1) \sin x + i[(\alpha_1 - \alpha_2) \sin x + (\beta_1 + \beta_2) \cos x]$$

The imaginary part is 0 when $\alpha_2 = \alpha_1$, $\beta_2 = -\beta_1$ so that the real solutions are given by

$$y = 2\alpha_1 \cos x - 2\beta_1 \sin x = C_1 \cos x + C_2 \sin x$$

where C_1, C_2 are arbitrary real constants.

We could also have reasoned that another basis for the complex solutions is given by

$$\cos x = \frac{e^{ix} + e^{-ix}}{2}, \qquad \sin x = \frac{e^{ix} - e^{-ix}}{2i}$$

For these linear combinations of e^{ix}, e^{-ix} span the same vector space as e^{ix}, e^{-ix}. The general complex solution is then $c_1 \cos x + c_2 \sin x$; this is real precisely when c_1 and c_2 are real.

Finally we can reason in general: if $u(x) + iv(x)$ is a solution of a real equation (14-80), then

$$L[u + iv] = L[u] + iL[v] = q(x)$$

Hence, $L[u] = q(x)$, $L[v] = 0$. Thus the complex solution provides a real solution of the given equation and a real solution of the associated homogeneous equation. If $q(x)$ happens to be 0, u and v are both solutions of the homogeneous equation. This is illustrated by Example 2 with $u(x) + iv(x) = e^{ix} = \cos x + i \sin x$. Both $\cos x$ and $\sin x$ are real solutions of $y'' + y = 0$.

PROBLEMS

1. Find the general (real) solution with the aid of the given complementary function:
   (a) $y'' - 4y = e^x$, $y_c = c_1 e^{2x} + c_2 e^{-2x}$
   (b) $y'' + y = \cos x$, $y_c = c_1 \cos x + c_2 \sin x$
   (c) $y'' + 3y' + 2y = e^{5x}$, $y_c = c_1 e^{-x} + c_2 e^{-2x}$
   (d) $y'' + 2y' + 2y = e^x$, $y_c = c_1 e^{-x} \cos x + c_2 e^{-x} \sin x$
   (e) $y''' + y' = x^2$, $y_c = c_1 + c_2 \cos x + c_3 \sin x$
   (f) $y''' + y' = e^x$, $y_c = c_1 + c_2 \cos x + c_3 \sin x$
   (g) $x^2 y'' - 3xy' + 3y = x$, $y_c = c_1 x + c_2 x^3$, $x > 0$
   (h) $x^2 y'' - 3xy' + 3y = 1/x$, $y_c = c_1 x + c_2 x^3$, $x > 0$
2. Show that the functions given form a basis for the set $\mathcal{H}$ of all complex solutions:
   (a) $y'' + 4y = 0$; $y_1 = e^{2ix}$, $y_2 = e^{-2ix}$
   (b) $y'' + 4y = 0$; $y_1 = \cos 2x$, $y_2 = \sin 2x$
   (c) $y'' + 2y' + 2y = 0$; $y_1 = e^{(-1+i)x}$, $y_2 = e^{(-1-i)x}$
   (d) $y'' + 2y' + 2y = 0$; $y_1 = e^{-x} \cos x$, $y_2 = e^{-x} \sin x$

(e) $y' + iy = 0$; $y_1 = e^{-ix}$

(f) $(1 - ix)y'' - xy' + y = 0$; $y_1 = x$, $y_2 = e^{ix}$ [on $(-\infty, \infty)$]

3. Prove: if $a_0(x), \dots, a_{n-1}(x)$ are real-valued and $y = u(x) + iv(x)$ is a solution of $L[y] = 0$, then $u(x) - iv(x)$ is also a solution.

14-9 HOMOGENEOUS LINEAR DIFFERENTIAL EQUATIONS WITH CONSTANT COEFFICIENTS

We consider the differential equation

$$a_n y^{(n)} + \cdots + a_0 y = 0 \tag{14-90}$$

in which $a_0, \dots, a_n$ are constants and $a_n \neq 0$. We can divide by a_n to make the coefficient of $y^{(n)}$ equal to 1, but that is not essential for the following discussion. The theorems of Section 14-6 apply to (14-90) for the interval J: $(-\infty, \infty)$ and assure us that the solutions of (14-90) form an n-dimensional vector space $\mathcal{H}$. Our problem is to find a basis for $\mathcal{H}$.

For most of this section we can consider (14-90) either as a complex linear differential equation or as a real linear differential equation, and we need not say which case is intended. For certain special points there is a difference between the two cases; we shall note the difference where it arises.

We introduce the symbol D for the derivative d/dx and regard D as a linear transformation on the vector space $\mathcal{C}^{(\infty)}$ of functions on $(-\infty, \infty)$ having derivatives of all orders. As in Section 9-20, we can form powers of D such as D^2 (the second derivative) and D^3 (the third derivative). We can also form polynomials in D such as $3D^2 - 5D + 1$; here 1 stands for the identity transformation I. Each of these polynomials is also a linear transformation on the vector space $\mathcal{C}^{(\infty)}$. For example

$$(3D^2 - 5D + 1)y = 3y'' - 5y' + y$$

for each function $y(x)$ in $\mathcal{C}^{(\infty)}$. If $P(\lambda)$ is a polynomial in λ, then we denote by $P(D)$ the linear transformation obtained by replacing λ by D. For example, if $P(\lambda) = 3\lambda^2 - 5\lambda + 1$, then $P(D)$ is the linear transformation $3D^2 - 5D + 1$. It is shown in Section 9-20 that the polynomials in D can be added and multiplied as in ordinary algebra. For example, $D^2 - 4 = (D + 2)(D - 2)$ (where, for example, 4 stands for $4I$).

Our differential equation (14-90) can now be written as

$$(a_n D^n + \cdots + a_1 D + a_0)y = 0 \tag{14-90$'$}$$

or as

$$P(D)y = 0 \tag{14-90$''$}$$

where

$$P(\lambda) = a_n \lambda^n + \cdots + a_0 \tag{14-91}$$

We call $P(\lambda)$ the *characteristic polynomial* associated with the differential equation (14-90). The equation $P(\lambda) = 0$:

$$a_n \lambda^n + \cdots + a_0 = 0 \tag{14-92}$$

is called the *characteristic equation* associated with (14-90); its roots, the zeros of $P(\lambda)$, are called the *characteristic roots* of Equation (14-90).

Now by algebra, Equation (14-92) has n complex roots $\lambda_1, \ldots, \lambda_n$, some of which may coincide, and there is a corresponding factorization of $P(\lambda)$:

$$P(\lambda) = a_n(\lambda - \lambda_1)(\lambda - \lambda_2) \cdots (\lambda - \lambda_n)$$

Hence, also

$$P(D) = a_n(D - \lambda_1)(D - \lambda_2) \cdots (D - \lambda_n)$$

Therefore, Equation (14-90) can be written:

$$a_n(D - \lambda_1) \cdots (D - \lambda_n)y = 0 \qquad (14\text{-}90''')$$

We note the following useful identity, called the *exponential shift:*

$$P(D)(e^{\alpha x}y) = e^{\alpha x}P(D + \alpha)y \qquad (14\text{-}93)$$

Here α is a constant. For example,

$$(D^2 - 1)(e^{3x}y) = e^{3x}[(D + 3)^2 - 1]y = e^{3x}(D^2 + 6D + 8)y$$

We verify the correctness of this relation by writing

$$\begin{aligned}(D^2 - 1)(e^{3x}y) &= D^2(e^{3x}y) - e^{3x}y = D[D(e^{3x}y)] - e^{3x}y \\ &= D(e^{3x}\,Dy + 3e^{3x}y) - e^{3x}y \\ &= e^{3x}D^2y + 6e^{3x}\,Dy + 9e^{3x}y - e^{3x}y \\ &= e^{3x}(D^2y + 6Dy + 8y)\end{aligned}$$

Another form of the exponential shift is as follows:

$$e^{\beta x}P(D)y = P(D - \beta)(e^{\beta x}y) \qquad (14\text{-}93')$$

The proofs of these identities are left to the exercises (Problem 13 below).

Now let λ_1 be a characteristic root of multiplicity k. Then $P(D) = Q(D)(D - \lambda_1)^k$, where $Q(\lambda)$ is a polynomial of degree $n - k$. By (14-93),

$$P(D)(e^{\lambda_1 x}u) = e^{\lambda_1 x}P(D + \lambda_1)u = e^{\lambda_1 x}Q(D + \lambda_1)D^k u$$

Hence, if u is chosen so that $D^k u = 0$, then $y = e^{\lambda_1 x}u$ is a solution of $P(D)y = 0$. But $D^k u = 0$ for every polynomial of degree at most $k - 1$:

$$u = b_0 + b_1x + \cdots + b_{k-1}x^{k-1}$$

Hence

$$y = e^{\lambda_1 x}(b_0 + b_1x + \cdots + b_{k-1}x^{k-1})$$

is a solution of $P(D)y = 0$ for every choice of the constants $b_0, \ldots, b_{k-1}$. In particular, the functions

$$e^{\lambda_1 x}, \qquad xe^{\lambda_1 x}, \qquad \ldots, \qquad x^{k-1}e^{\lambda_1 x} \qquad (14\text{-}94)$$

are solutions of $P(D)y = 0$. We proceed in this way for each characteristic root. Each root of multiplicity k leads to k solutions as in (14-94), and in all we obtain n different solutions $y_1(x), \ldots, y_n(x)$ of the differential equation. We show in the next section that these n solutions are linearly independent on $(-\infty, \infty)$, so that they form a basis for $\mathcal{H}$.

EXAMPLE 1. $D^2(D-1)^2(D+3)^3y = 0$. Here we have given the differential equation in "factored form." If we multiply out, we obtain the equation

$$(D^7 + 7D^6 + 10D^5 - 18D^4 - 27D^3 + 27D^2)y = 0$$

or

$$y^{(7)} + 7y^{(6)} + 10y^{(5)} - 18y^{(4)} - 27y''' + 27y'' = 0$$

The characteristic equation is $\lambda^2(\lambda-1)^2(\lambda+3)^3 = 0$, the characteristic roots are $0, 0, 1, 1, -3, -3, -3$. Hence, the general solution is

$$y = c_1e^{0x} + c_2xe^{0x} + c_3e^x + c_4xe^x + c_5e^{-3x} + c_6xe^{-3x} + c_7x^2e^{-3x}$$

or, more simply,

$$y = c_1 + c_2x + e^x(c_3 + c_4x) + e^{-3x}(c_5 + c_6x + c_7x^2)$$

EXAMPLE 2. $y'' + 4y = 0$ or $(D^2+4)y = 0$. The characteristic equation is $\lambda^2 + 4 = 0$, the characteristic roots are $2i, -2i$. Hence, it is natural to first form the general complex solution

$$y = c_1e^{2ix} + c_2e^{-2ix} \qquad (c_1, c_2 \text{ arbitrary complex constants})$$

However, since the differential equation has real coefficients, we are also interested in the real solutions. As pointed out in Section 14-8, when $u(x) + iv(x)$ is a complex solution of a homogeneous equation with real coefficients, both $u(x)$ and $v(x)$ are real solutions. Here $e^{2ix} = \cos 2x + i \sin 2x$ is a complex solution; hence, both $\cos 2x$ and $\sin 2x$ are real solutions. They are linearly independent since their Wronskian is

$$\begin{vmatrix} \cos 2x & \sin 2x \\ -2\sin 2x & 2\cos 2x \end{vmatrix} = 2$$

Hence, the general real solution is

$$y = c_1 \cos 2x + c_2 \sin 2x \qquad (c_1, c_2 \text{ arbitrary real constants})$$

The same conclusion can be deduced in other ways, as shown in Section 14-8.

EXAMPLE 3. $y'' - 2y' + 2y = 0$. The characteristic equation is $\lambda^2 - 2\lambda + 2 = 0$, the characteristic roots are $1 \pm i$, the general complex solution is

$$y = c_1e^{(1+i)x} + c_2e^{(1-i)x} \qquad (c_1, c_2 \text{ complex})$$

Since

$$e^{(1+i)x} = e^xe^{ix} = e^x(\cos x + i \sin x) = e^x \cos x + ie^x \sin x$$

$e^x \cos x$ and $e^x \sin x$ are real solutions; we verify that they are linearly independent and, hence, that

$$y = c_1e^x \cos x + c_2e^x \sin x \qquad (c_1, c_2 \text{ real})$$

is the general real solution.

EXAMPLE 4. $(D^2-1)(D^2+4D+5)^2y = 0$. The characteristic roots are $1, -1, -2 \pm i, -2 \pm i$. We seek only the general real solution. Since $e^{(-2+i)x}$ and $xe^{(-2+i)x}$ are complex solutions, we obtain the corresponding real solutions

$e^{-2x}\cos x$, $e^{-2x}\sin x$, $xe^{-2x}\cos x$, $xe^{-2x}\sin x$. These together with e^x and e^{-x} are a basis for the real solutions. The general real solution is

$$y = c_1e^x + c_2e^{-x} + e^{-2x}[(c_3 + c_4x)\cos x + (c_5 + c_6x)\sin x]$$

We summarize the procedures for finding real solutions of a real homogeneous equation with real coefficients: To each real characteristic root λ of multiplicity k, one assigns the k functions $e^{\lambda x}, xe^{\lambda x}, \ldots, x^{k-1}e^{\lambda x}$. The complex roots will always come in conjugate pairs of equal multiplicities. If $\alpha \pm i\beta$ is a pair of complex roots of multiplicity k, then one assigns the $2k$ functions

$$e^{\alpha x}\cos\beta x, \quad e^{\alpha x}\sin\beta x, \quad xe^{\alpha x}\cos\beta x, \quad xe^{\alpha x}\sin\beta x, \quad \ldots, \quad x^{k-1}e^{\alpha x}\cos\beta x, \quad x^{k-1}e^{\alpha x}\sin\beta x$$

In all, n functions $y_1(x), \ldots, y_n(x)$ are assigned, and the general real solution is

$$y = c_1y_1(x) + \cdots + c_ny_n(x)$$

‡14-10 LINEAR INDEPENDENCE OF SOLUTIONS OF THE HOMOGENEOUS LINEAR EQUATION WITH CONSTANT COEFFICIENTS

We wish to prove that the methods of the preceding sections always provide us with a basis for the set $\mathcal{H}$ of solutions of the homogeneous linear equation with constant coefficients (14-90).

THEOREM 7. *For each set of distinct constants $\lambda_1, \ldots, \lambda_m$ and positive integers $k_1, \ldots, k_m$, with $k_1 + \cdots + k_m = n$, the n functions*

$$e^{\lambda_1 x}, xe^{\lambda_1 x}, \ldots, x^{k_1-1}e^{\lambda_1 x}, e^{\lambda_2 x}, xe^{\lambda_2 x}, \ldots, e^{\lambda_m x}, \ldots, x^{k_m-1}e^{\lambda_m x} \qquad (14\text{-}100)$$

are linearly independent functions in $\mathcal{C}^{(\infty)}(-\infty, \infty)$.

PROOF. Let $c_1, \ldots, c_n$ be constants such that

$$c_1e^{\lambda_1 x} + \cdots + c_nx^{k_m-1}e^{\lambda_m x} \equiv 0 \qquad \text{on } (-\infty, \infty)$$

We can write this equation as

$$e^{\lambda_1 x}p_1(x) + \cdots + e^{\lambda_m x}p_m(x) \equiv 0 \qquad (14\text{-}101)$$

where $p_1(x), \ldots, p_m(x)$ are polynomials. We assert that (14-101) implies $p_1(x) \equiv 0, \ldots, p_m(x) \equiv 0$ and, hence, $c_1 = 0, \ldots, c_n = 0$, so that the n functions of the theorem are linearly independent.

We prove our assertion by induction. For $m = 1$, (14-101) becomes $e^{\lambda_1 x}p_1(x) \equiv 0$, which implies $p_1(x) \equiv 0$, as asserted. Let us assume that the assertion has been proved for $m = N$ and then consider the case $m = N + 1$. We are given that

$$e^{\lambda_1 x}p_1(x) + \cdots + e^{\lambda_N x}p_N(x) + e^{\lambda_{N+1} x}p_{N+1}(x) \equiv 0 \qquad (14\text{-}101')$$

We multiply by $e^{-\lambda_{N+1} x}$ to obtain the equation

$$e^{\mu_1 x}p_1(x) + \cdots + e^{\mu_N x}p_N(x) + p_{N+1}(x) \equiv 0 \qquad (14\text{-}102)$$

where $\mu_1 = \lambda_1 - \lambda_{N+1}, \ldots, \mu_N = \lambda_N - \lambda_{N+1}$. We differentiate (14-102) k_{N+1} times. We note that

$$D[e^{\mu_j x}p_j(x)] = e^{\mu_j x}(D + \mu_j)p_j(x) = e^{\mu_j x}[p_j'(x) + \mu_j p_j(x)]$$

by the exponential shift (14-93). Since $\mu_j = \lambda_j - \lambda_{N+1} \neq 0$ for $j = 1, \ldots, N$, our result is of form $e^{\mu_j x}q_j(x)$ where $q_j(x)$ is of the same degree as $p_j(x)$. Therefore, differentiating k_{N+1} times has no effect on the degrees of the polynomials in (14-102), except for the last term—that term is a polynomial of degree $k_{N+1} - 1$ and its k_{N+1}-th derivative is zero. Therefore, after differentiating k_{N+1} times, we obtain an equation of form:

$$e^{\mu_1 x}q_1(x) + \cdots + e^{\mu_N x}q_N(x) \equiv 0$$

By the induction hypothesis, it follows that $q_1(x) \equiv 0, \ldots, q_N(x) \equiv 0$. Since the q's have the same degrees as the corresponding p's, we conclude that $p_1(x) \equiv 0, \ldots, p_N(x) \equiv 0$ and, hence, by (14-101′) also $p_{N+1}(x) \equiv 0$. Therefore, the induction is complete and the theorem is proved.

Remark. The proof given is valid in both real and complex cases. For the complex case it is to be noted that

$$e^{(\alpha+i\beta)x} = e^{\alpha x}(\cos \beta x + i \sin \beta x)$$

has no zeros on $(-\infty, \infty)$. For at a zero one would have both $\cos \beta x = 0$ and $\sin \beta x = 0$, which is impossible since $\cos^2 \beta x + \sin^2 \beta x \equiv 1$.

THEOREM 8. *In Theorem 7, let $2s \leq n$ and let $\lambda_1 = \alpha_1 + \beta_1 i$, $\lambda_2 = \alpha_1 - \beta_1 i, \ldots, \lambda_{2s-1} = \alpha_s + \beta_s i$, $\lambda_{2s} = \alpha_s - \beta_s i$ and let $\lambda_{2s+1}, \ldots, \lambda_m$ be real. Then the n real functions*

$$\begin{array}{l} e^{\alpha_1 x}\cos \beta_1 x,\ e^{\alpha_1 x}\sin \beta_1 x, \ldots, x^{k_1-1}e^{\alpha_1 x}\cos \beta_1 x,\ x^{k_1-1}e^{\alpha_1 x}\sin \beta_1 x \\ \vdots \\ e^{\alpha_s x}\cos \beta_s x,\ e^{\alpha_s x}\sin \beta_s x, \ldots, x^{k_s-1}e^{\alpha_s x}\cos \beta_s x,\ x^{k_s-1}e^{\alpha_s x}\sin \beta_s x \\ e^{\lambda_{2s+1}x}, \quad \ldots, \quad x^{k_m-1}e^{\lambda_m x} \end{array} \tag{14-103}$$

are linearly independent on $(-\infty, \infty)$.

PROOF. We first consider the case of complex coefficients. We need only prove that the n functions (14-103) span the same subspace of the complex vector space $\mathfrak{C}^{(\infty),c}$ as that spanned by the functions (14-100). Now

$$x^r e^{(\alpha+\beta i)x} = x^r e^{\alpha x}\cos \beta x + i x^r e^{\alpha x}\sin \beta x$$
$$x^r e^{(\alpha-\beta i)x} = x^r e^{\alpha x}\cos \beta x - i x^r e^{\alpha x}\sin \beta x$$

and

$$x^r e^{\alpha x}\cos \beta x = x^r e^{\alpha x}\frac{e^{\beta i x} + e^{-\beta i x}}{2} = \frac{1}{2}x^r e^{(\alpha+i\beta)x} + \frac{1}{2}x^r e^{(\alpha-i\beta)x}$$

$$x^r e^{\alpha x}\sin \beta x = x^r e^{\alpha x}\frac{e^{\beta i x} + e^{-\beta i x}}{2i} = \frac{1}{2i}x^r e^{(\alpha+i\beta)x} - \frac{1}{2i}x^r e^{(\alpha-i\beta)x}$$

Thus we see that the functions of each set are expressible as linear combination of the functions of the other set. Therefore, both sets span the same subspace and the n functions (14-103) are linearly independent with respect to complex coefficients. But this implies that they are linearly independent with respect to real coefficients (Problem 4 below). Therefore, the theorem is proved.

Remark. Questions related to this section are discussed in Section 5-11.

14-11 NONHOMOGENEOUS LINEAR DIFFERENTIAL EQUATIONS WITH CONSTANT COEFFICIENTS

We consider the equation

$$a_n y^{(n)} + \cdots + a_0 y = q(x) \tag{14-110}$$

where $a_0, \ldots, a_n$ are constants, $a_n \neq 0$ and $q(x)$ is continuous on an interval J. The general solution on J is then given by

$$y = c_1 y_1(x) + \cdots + c_n y_n(x) + y_p(x) = y_c(x) + y_p(x)$$

where $y_c(x)$ is the general solution of the associated homogeneous equation and $y_p(x)$ is one particular solution of (14-110). Since the equation has constant coefficients, we can determine $y_c(x)$, as in Section 14-9; the general solution on $(-\infty, \infty)$ as found in that section also provides us with the general solution on the interval J. Having found $y_c(x)$, we can then find $y_p(x)$ by variation of parameters as in Section 14-7.

EXAMPLE. $(D^2 - 1)y = e^{3x}$. For the associated homogeneous equation we readily obtain the general solution $y = c_1 e^x + c_2 e^{-x}$. To obtain the particular solution we set

$$y_p = v_1 e^x + v_2 e^{-x}$$

Then the method of Section 14-7 leads us to the equations

$$v_1' e^x + v_2' e^{-x} = 0$$
$$v_1' e^x - v_2' e^{-x} = e^{3x}$$

We solve to find $v_1' = e^{2x}/2$, $v_2' = -e^{4x}/2$ and, hence, can take $v_1(x) = e^{2x}/4$, $v_2(x) = -e^{4x}/8$, so that

$$y_p(x) = \frac{e^{3x}}{4} - \frac{e^{3x}}{8} = \frac{e^{3x}}{8}$$

The general solution [on $(-\infty, \infty)$] is

$$y = c_1 e^x + c_2 e^{-x} + (e^{3x}/8)$$

Because of the importance of equations of form (14-110) in applications, a number of other methods for finding particular solutions have been developed. Some of them are illustrated below. We note that we can write (14-110), as before, in the form $L[y] = q$, where L is a *linear* mapping. Hence, if $L[f_1(x)] = q_1(x)$ and $L[f_2(x)] = q_2(x)$, then

$$L[c_1 f_1 + c_2 f_2] = c_1 q_1 + c_2 q_2$$

Accordingly, to find a particular solution y_p of $L[y] = q = c_1 q_1 + c_2 q_2$, we can find particular solutions f_1, f_2 of $L[f_1] = q_1$ and $L[f_2] = q_2$ and then take $y_p(x) = c_1 f_1(x) + c_2 f_2(x)$. This process is sometimes called the *principle of superposition*.

Method of Undetermined Coefficients. To find a particular solution y_p of $P(D)y = q(x)$ with $q(x) = e^{\mu x}r(x)$, where $r(x)$ is a polynomial of degree m, one reasons that $q(x)$ itself is a solution of an equation $Q(D)y = 0$, where $Q(D) = (D - \mu)^{m+1}$. Hence, if $P(D)y_p = q(x)$, then

$$Q(D)P(D)y_p = Q(D)q(x) = 0$$

Therefore, y_p is a solution of the homogeneous linear equation $Q(D)P(D)y = 0$. The general solution of this equation is formed of two parts: the first part is the general solution of $P(D)y = 0$; the second part is formed of extra terms because of the factor $Q(D)$. In seeking y_p, we can ignore the first part, since it is simply $y_c(x)$ for $P(D)y = 0$; the second part provides an expression for y_p with certain coefficients to be determined.

For example, for the equation

$$(D - 1)(D + 2)^2 y = e^x(x^2 - 1)$$

we take $Q(D) = (D - 1)^3$, since $e^x(x^2 - 1)$ is a solution of $(D - 1)^3 y = 0$. The general solution of

$$(D - 1)^4(D + 2)^2 y = 0$$

is $y = [c_1e^x + c_2e^{-2x} + c_3xe^{-2x}] + [k_1xe^x + k_2x^2e^x + k_3x^3e^x]$. The first part is $y_c(x)$ and is to be ignored. We substitute $y = k_1xe^x + k_2x^2e^x + k_3x^3e^x$ in the given equation to determine k_1, k_2, and k_3. After substitution we obtain

$$27k_3x^2e^x + (36k_3 + 18k_2)xe^x + (6k_3 + 12k_2 + 9k_1)e^x = e^x(x^2 - 1)$$

We compare coefficients of x^2e^x, xe^x, and e^x (why?) and find that $27k_3 = 1$, $36k_3 + 18k_2 = 0$, $6k_3 + 12k_2 + 9k_1 = -1$. Therefore, $k_1 = -1/27$, $k_2 = -2/27$, $k_3 = 1/27$, and

$$y_p = (e^x/27)(-x - 2x^2 + x^3)$$

Application of Exponential Shift. We illustrate procedures that simplify the method of undetermined coefficients. For the equation $(D - 1)(D + 2)^2 y = e^x(x^2 - 1)$ considered above, we multiply by e^{-x} and use the exponential shift:

$$e^{-x}(D - 1)(D + 2)^2 y = x^2 - 1, \qquad D(D + 3)^2(e^{-x}y) = x^2 - 1$$

We set $u = e^{-x}y$ and seek u so that $D(D + 3)^2u = x^2 - 1$ or $(D + 3)^2(Du) = x^2 - 1$. We set $v = Du$ and seek v so that $(D + 3)^2 v = x^2 - 1$ or $v'' + 6v' + 9v = x^2 - 1$. We can find a solution that is a polynomial of degree 2: $v = k_1 + k_2x + k_3x^2$ (undetermined coefficients). We find $k_1 = -1/27$, $k_2 = -4/27$, $k_3 = 1/9$. Having found $v = Du$, we obtain u by integration and then $y = e^x u$.

Method of Factorization. To find a particular solution of the equation $P(D)y = q(x)$, we can factor $P(D)$ into factors of first degree in D and make corresponding substitutions to obtain a set of first-order linear differential equations, which can be solved in turn as in Section 14-5. For example, for the equation $(D-1)(D-2)y = e^x$, we set $(D-2)y = u$ and $(D-1)u = e^x$ (or $u' - u = e^x$). The method of Section 14-5 gives $u = e^x(x+c)$. We ignore the c and take $u = xe^x$. Then $(D-2)y = xe^x$ or $y' - 2y = xe^x$ is solved similarly to give $y = -xe^x - e^x$; here the second term is included in $y_c(x)$ and can be ignored. Hence, we obtain $y_p(x) = -xe^x$.

PROBLEMS

1. Find the general real solution:

   (a) $(D^2 - 9) = 0$ (b) $(D^2 - 4D - 5)y = 0$
   (c) $(D^2 - 4D + 4)y = 0$ (d) $(D^2 + 16)y = 0$
   (e) $D^2(D^2 + 4)y = 0$ (f) $(D-1)^2(D^2 + 2D + 5)y = 0$
   (g) $(D^3 - D^2 - D + 1)y = 0$ (h) $(D^3 - 1)^2 y = 0$
   (i) $(D^4 - 8)y = 0$ (j) $y^{(4)} - 5y'' + 4y = 0$
   (k) $y^{(4)} - 6y'' + 4y = 0$ (l) $y^{(6)} - y^{(2)} = 0$

2. Find the general real solution:

   (a) $(D^2 - 9)y = e^{2x}$ (b) $(D^2 - 9)y = x - 2$
   (c) $(D^2 - 9)y = e^{3x}$ (d) $(D^2 - 9)y = x^2 e^{3x}$
   (e) $y''' - y'' - y' + y = x + 1$ (f) $(D^3 + D^2 - D - 1)y = x + e^x$
   (g) $y'' - 2y' + y = e^{3x}$ (h) $y'' + 9y = \cos 2x$
   (i) $y'' + 9y = \sin 2x$ (j) $y'' + 9y = \cos 3x$
   (k) $y'' + 9y = \sin 3x$ (l) $y'' + 9y = x \sin 3x$

3. Find the solution satisfying the given initial conditions:

   (a) $y''' - y'' - y' + y = 0$; $y(0) = 1$, $y'(0) = y''(0) = 0$.
   (b) $y'' - 2y' + y = 0$; $y(0) = 0$, $y'(0) = 1$.
   (c) $(D^2 - 9)y = e^{2x}$; $y = 1$ and $y' = 0$ for $x = 0$.
   (d) $(D^2 + 9)y = \sin 2x$; $y = 0$ and $y' = 0$ for $x = 0$.
   (e) $y''' + 7y'' + y = 0$; $y = 0$, $y' = 0$, $y'' = 0$ for $x = 0$. [*Hint.* Find an easy way!]

4. Prove: if $f_1, \ldots, f_m$ are real-valued functions on the interval J that are linearly independent for complex scalars, then they are linearly independent for real scalars.

5. Prove: if $u(x) + iv(x)$ is a complex solution of a linear differential equation with real coefficients $L[y] = q(x)$, where $q(x)$ is complex-valued, then $L[u] = \mathrm{Re}(q(x))$ and $L[v] = \mathrm{Im}(q(x))$. Hence, to find a particular solution of $L[y] = e^{\alpha x}\cos\beta x$ or $L[y] = e^{\alpha x}\sin\beta x$, one can first find a particular solution y_p of $L[y] = e^{(\alpha + i\beta)x}$ and then take real and imaginary parts, respectively.

6. Let a linear differential equation with constant coefficients $P(D)y = q$ be given.

   (a) Show that if $q = e^{\alpha x}$ and $P(\alpha) \neq 0$, then a particular solution is $e^{\alpha x}/P(\alpha)$.
   (b) Show that if $q = e^{\alpha x}\cos\beta x$, $P(D)$ has real coefficients and $P(\alpha + \beta i) \neq 0$, then a particular solution is $\mathrm{Re}[e^{(\alpha+\beta i)x}/P(\alpha + \beta i)]$ (see Problem 5).

(c) Show that if $q = e^{\alpha x} \sin \beta x$, $P(D)$ has real coefficients and $P(\alpha + \beta i) \neq 0$, then a particular solution is $\operatorname{Im}[e^{(\alpha+\beta i)x}/P(\alpha + \beta i)]$.

7. Apply the results of Problem 6 to obtain a particular solution:

(a) $(D^2 + 8)y = e^{3x}$ (b) $(D^2 + 6D + 9)y = e^{3x}$

(c) $(D^2 - 4)y = \cos 3x$ (d) $(D^2 - 4)y = \sin 3x$

(e) $(D^2 + 3D + 1)y = e^x \cos 2x$ (f) $(D^2 + 3D + 1)y = e^x \sin 2x$

8. (a) . . . (l) Apply the method of undetermined coefficients to find y_p for parts (a), . . . , (l) of Problem 2. The exponential shift will be found helpful in some cases. For parts (g), . . . , (i), one should first solve with the appropriate function $e^{i\omega x}$ on the right, as in Problem 5.

9. (a) . . . (f) Apply the method of undetermined coefficients to find $y_p(x)$ for parts (a), . . . , (f) of Problem 7. Here one can use superposition and the results of Problem 5.

10. Apply the method of factorization to find y_p for parts (a), . . . , (j) of Problem 2.

11. Use the results of previous problems to find a particular solution:

(a) $(D^2 - 9)y = 3e^{2x} + 5e^{3x}$ (b) $(D^2 + 9)y = 5 \cos 2x + \sin 3x$

‡12. (a) Apply the method of factorization and induction to prove that the general solution of the equation $P(D)y = 0$ is given by the expression $c_1 e^{\lambda_1 x} + \cdots$ in accordance with the rules of Section 14-9. First verify this for $n = 1$, then assume that it has been proved for equations of order n and write the equation of order $n + 1$ as $(D - \lambda_1) \cdots (D - \lambda_n)(D - \lambda_{n+1})y = 0$. Now set $(D - \lambda_{n+1})y = u$, so that $(D - \lambda_1) \cdots (D - \lambda_n)u = 0$ and we know the expression for u. Find y from the equation $y' - \lambda_{n+1}y = u$ and show that it has the form asserted.

(b) Show also by induction, as in part (a), that the equation $P(D)y = 0$ has a unique solution satisfying initial conditions $y(x_0) = y_0, \ldots, y^{(n-1)}(x_0) = y_0^{(n-1)}$. Note that from the initial values $y(x_0), \ldots, y^{(n)}(x_0)$ one obtains the initial values $u(x_0), \ldots, u^{(n-1)}(x_0)$, where $(D - \lambda_{n+1})y = u$.

Remark. The results of Problem 12 provide a new proof of the existence theorem, Theorem 3 of Section 14-6, for homogeneous linear differential equations with constant coefficients. The methods described also extend to the nonhomogeneous equation.

13. Prove by induction that $D^n(e^{\alpha x}y) = e^{\alpha x}(D + \alpha)^n y$ and, hence, prove (14-93). Prove (14-93′) from (14-93).

14-12 APPLICATIONS OF LINEAR DIFFERENTIAL EQUATIONS

In Section 14-5 we illustrated how linear differential equations of first order are used to represent the behavior of physical systems. We now extend the discussion to linear equations of higher order. We shall emphasize the case of equations with constant coefficients; most of the ideas carry over to equations with variable coefficients.

There are many physical systems which are described by a homogeneous linear differential equation of order n with constant real coefficients:

$$a_n \frac{d^n x}{dt^n} + \cdots + a_1 \frac{dx}{dt} + a_0 x = 0 \tag{14-120}$$

Here x is a function of time t, most commonly considered in the interval $(-\infty, \infty)$ or in the interval $[0, \infty)$ ($t = 0$ being the initial time for a process continuing indefinitely). The value of x may be a coordinate of a particle or of a reference point on a rigid body; it may be a voltage or current or some other electrical quantity; it may be a temperature; it may denote a key variable in a biological or economic process.

The equation (14-120) has one very simple solution: $x \equiv 0$ for all t. We call this the *zero solution*. In applications in mechanics we call $x = 0$ the *equilibrium position* of the mechanical system. (In general, an equilibrium position of a mechanical system is one at which all forces are zero, so that, if undisturbed, the system can remain at rest in that position forever.) An example is provided by the motion of a pendulum (Figure 14-14). Here θ is

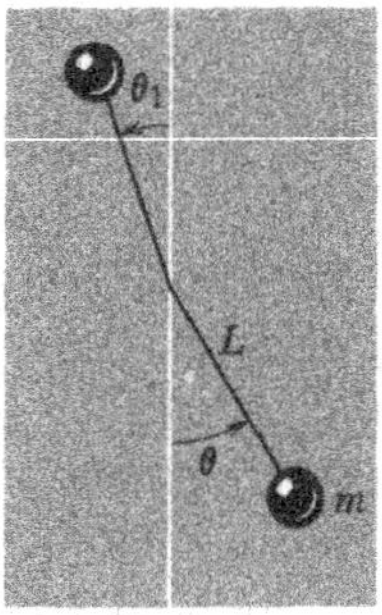

Figure 14-14 Simple pendulum.

an angular coordinate measuring the displacement of the pendulum from the vertical position. For small oscillations ($|\theta|$ remaining close to 0), the motion is governed by the equation

$$mL \frac{d^2\theta}{dt^2} + mg\theta = 0 \tag{14-121}$$

(Problem 3 below). Here the zero solution $\theta \equiv 0$ corresponds to the pendulum remaining at rest in the vertical position.

The zero solution $x \equiv 0$ of (14-120) is said to be *stable** if each solution of (14-120) approaches 0 at $t \to \infty$—in other words, if every solution of (14-120) is a transient. When the zero solution is stable, we also say that $x = 0$ is a *stable equilibrium point*.

Now for Equation (14-121) the solutions are given by

$$\theta = c_1 \cos \beta t + c_2 \sin \beta t, \qquad \beta = \sqrt{\frac{g}{L}} \tag{14-122}$$

Here the solutions are generally not transients and the equilibrium is not stable.

*The concept of stability is treated here in a simplified form. For a full treatment, see advanced texts on differential equations.

However, the solutions (14-122) have the property that they are *bounded:* $|\theta(t)| \leq k =$ constant on the solution. When the solutions have this property, we say that the equilibrium position is *neutrally stable.* If one takes friction into account, then Equation (14-121) is replaced by

$$mL\frac{d^2\theta}{dt^2} + mh\frac{d\theta}{dt} + mg\theta = 0 \tag{14-121$'$}$$

where h is a positive constant. If h is very small, so that $h^2 < 4gL$, the solutions of (14-121′) are given by

$$\theta = e^{\alpha t}(c_1 \cos \beta t + c_2 \sin \beta t) \tag{14-122$'$}$$

where

$$\alpha = -\frac{h}{2L}, \qquad \beta = \frac{\sqrt{4gL - h^2}}{2L}$$

Now $|\cos \beta t| \leq 1$ and $|\sin \beta t| \leq 1$, so that

$$|\theta(t)| \leq e^{\alpha t}(|c_1| + |c_2|)$$

and $e^{\alpha t} \to 0$ as $t \to \infty$, since $\alpha = -h/(2L) < 0$. Therefore, for the more realistic equation (14-121′), the zero solution $\theta \equiv 0$ is stable, If one experiments with a pendulum (a weight suspended by a string suffices), one observes that each motion does die out as time increases. We can try to reduce friction to a minimum and thereby slow down the rate at which the oscillations die out. However, we find that we cannot eliminate friction completely; if we could, we could achieve the solution (14-122) and have a perpetual motion machine!

One can also consider the motion of a pendulum near the position at which the bob is directly above the point of support ($\theta = \pi$ in Figure 14-14). We can introduce a new angular coordinate θ_1 relative to this position (say $\theta_1 = \theta - \pi$) and obtain a differential equation like (14-121) for the motion near $\theta_1 = 0$:

$$mL\frac{d^2\theta_1}{dt^2} - mg\theta_1 = 0 \tag{14-121$''$}$$

(Problem 3 below). Here the solutions are

$$\theta_1 = c_1 e^{\beta t} + c_2 e^{-\beta t}, \qquad \text{where } \beta = \sqrt{\frac{g}{L}} \tag{14-122$''$}$$

Thus, except for the solutions having $c_1 = 0$, on each solution θ_1 deviates more and more from 0 for large t (theoretically approaching ∞ as $t \to \infty$). The equilibrium is clearly not stable and not even neutrally stable. The introduction of friction does not affect the form of the solutions [Problem 3(c) below] and, hence, does not provide stability. If one experiments (using a rod instead of a string), one is quickly convinced that the upper equilibrium position of the bob is unstable. The solutions with $c_1 = 0$ correspond to motions with exactly the right initial speed, so that one approaches the vertical position gradually—the slightest error will force us to overshoot or to undershoot, so that this motion cannot be physically realized.

We now ask the general question: When is the zero solution of (14-120) stable? The answer is very simple: *The zero solution of* (14-120) *is stable precisely when every characteristic root is either a negative real number or a complex number with negative real part.*

For $\lambda < 0$ implies that $e^{\lambda t} \to 0$ as $t \to \infty$ and also that, for every positive integer k, $t^k e^{\lambda t} \to 0$ as $t \to \infty$; for $\lambda = \alpha + i\beta$, $\alpha < 0$ implies that $e^{\alpha t}\cos\beta t \to 0$ and $e^{\alpha t}\sin\beta t \to 0$ and, in general, $t^k e^{\alpha t}\cos\beta t \to 0$ and $t^k e^{\alpha t}\sin\beta t \to 0$ as $t \to \infty$. On the other hand, if $\lambda \geq 0$, then neither $e^{\lambda t}$ nor $t^k e^{\lambda t}$ has limit 0 as $t \to \infty$. Similarly, for $\lambda = \alpha + i\beta$ with $\alpha \geq 0$, none of $e^{\alpha t}\cos\beta t$, $e^{\alpha t}\sin\beta t$, $t^k e^{\alpha t}\cos\beta t$, $t^k e^{\alpha t}\sin\beta t$ has limit 0 as $t \to \infty$. The proofs are left to Problem 4 below.

EXAMPLE 1. $3x'' + 2x' + 9x = 0$. The characteristic roots are $(-1 \pm \sqrt{-26})/3 = \alpha \pm \beta i$ with $\alpha = -\frac{1}{3}$. Hence, the zero solution is stable.

The example suggests that for every differential equation of second order $ax'' + bx' + cx = 0$ with $a > 0$, the zero solution is stable precisely when $b > 0$ and $c > 0$. This conjecture is easily verified to be true [Problem 2(a) below].

EXAMPLE 2. $x''' + 3x'' + 4x' + 32x = 0$. The characteristic equation is

$$\lambda^3 + 3\lambda^2 + 4\lambda + 32 = 0 \qquad \text{or} \qquad (\lambda + 4)(\lambda^2 - \lambda + 8) = 0$$

The roots are -4 and $(1 \pm \sqrt{-31})/2 = \alpha \pm \beta i$ with $\alpha = \frac{1}{2}$. Hence, the zero solution is not stable. This example illustrates that, for an equation of third order, positiveness of the coefficients does not insure stability [Problem 2(b) below].

In general, testing for stability of the zero solution of (14-120) leads us to a problem of algebra: to determine whether all roots of an algebraic equation have negative real parts. This problem has been studied intensively, and a variety of criteria have been established (See for example Chapter XV of *Theory of Matrices*, Vol. II, by Gantmacher, Chelsea Pub. Co., N.Y., 1959).

Input and Output. The nonhomogeneous equation

$$a_n \frac{d^n x}{dt^n} + \cdots + a_1 \frac{dx}{dt} + a_0 x = q(t) \tag{14-123}$$

to which Equation (14-120) is associated, also has many physical applications. A great many of these concern control systems and can be discussed in terms of input and output as in Section 14-5. The right-hand member $q(t)$ is the *input;* the solution $x(t)$ is the *output.* When the zero solution of the associated homogeneous equation is stable, one calls the control system described by (14-123) stable. Thus for a stable system the outputs are of the form $x_p(t) + x_c(t)$, where $x_c(t)$ is a transient. The effect of the transient term becomes negligible for t sufficiently large, so that in effect one has just one output (called *steady state*) for each input.

Boundary Value Problems. The solutions of a linear differential equation of order n depend on n arbitrary constants. Up to this point we have used the n constants only to meet prescribed initial conditions. However, in a number of important applications of equations of order greater than 1, the initial conditions are replaced by a set of n conditions on the values of the solution and certain of its derivatives at *two* points, the end points of a closed interval J. The problem of finding a solution of the equation in J, satisfying the given conditions at the end points, is called a *boundary value problem* for the interval J; the n conditions are called *boundary conditions*. A boundary value problem may have no solution, a unique solution, or many solutions.

EXAMPLE 3. It is shown in physics that a vibrating string, such as a violin string, can emit pure tones of frequency ν, where ν is such that the boundary value problem

$$\begin{aligned} &a^2y'' + \nu^2 y = 0, \qquad 0 \le x \le L \\ &y(0) = 0, \qquad y(L) = 0 \end{aligned} \tag{14-124}$$

has a solution $y(x)$ not identically zero. Each solution of (14-124) gives the shape of the string when emitting the tone of frequency ν. In (14-124) a is a constant, essentially determining the "speed of sound" along the string; L is the length of the string. Also ν is a positive constant, but its value is not known; we must find all possible values of ν for which the boundary value problem (14-124) has a solution other than $y(x) \equiv 0$.

Now the general solution of the equation $a^2y'' + \nu^2y = 0$ is

$$y = c_1 \cos \frac{\nu}{a} x + c_2 \sin \frac{\nu}{a} x$$

The boundary conditions give

$$0 = c_1, \qquad 0 = c_1 \cos \frac{\nu L}{a} + c_2 \sin \frac{\nu L}{a}$$

Hence, $c_1 = 0$ and c_2 can be an arbitrary nonzero constant, provided that $\sin(\nu L/a) = 0$. This last equation tells us that $\nu L/a$ must be a multiple of π:

$$\frac{\nu L}{a} = n\pi, \qquad \nu = \frac{n\pi a}{L}, \qquad n = 1, 2, \ldots$$

Therefore, the possible frequencies of pure tones are multiples of a *fundamental frequency* $\pi a/L$. When $n = 2$, one obtains the first overtone (the octave in music); when $n = 3$, the second overtone (octave plus a fifth), and so on. For each choice of n, the solutions give the shape of the string:

$$y = c_2 \sin \frac{n\pi}{L} x, \qquad n = 1, 2, \ldots \qquad (c_2 \neq 0)$$

These shapes are shown for $n = 1$, 2, and 3 in Figure 14-15. They can easily be seen in a vibrating violin string.

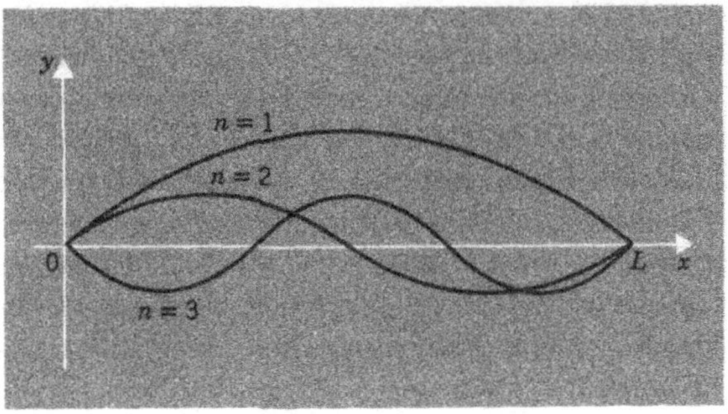

Figure 14-15 Shapes of vibrating string.

14-13 VIBRATIONS OF A MASS-SPRING SYSTEM

Let a particle of mass m move on the x-axis, subject to a spring force in accordance with Hooke's law: force is proportional to displacement (Figure 14-16). The displacement is x, measured from the position of the particle when the spring is neither stretched nor compressed. By Hooke's law, the spring force is $-k^2x$, where k is a positive constant. We also allow for friction

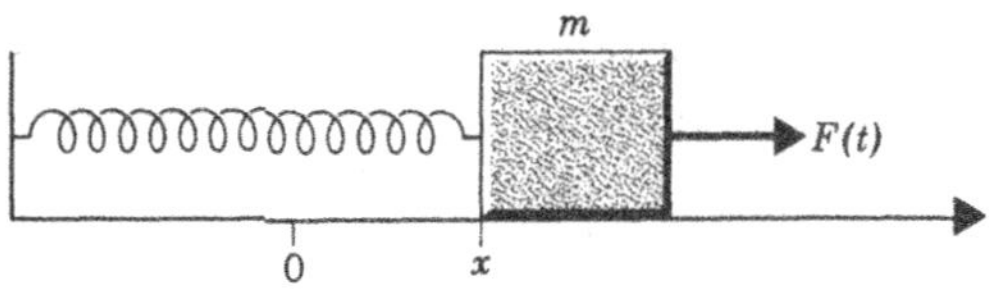

Figure 14-16 Mass-spring system.

proportional to velocity, hence, a force of form $-hx'(t)$, where h is a positive constant or 0. Finally we allow for an external driving force $F(t)$. By Newton's Second Law, the motion of the particle is governed by the equation

$$m\frac{d^2x}{dt^2} = -k^2x - h\frac{dx}{dt} + F(t)$$

or

$$m\frac{d^2x}{dt^2} + h\frac{dx}{dt} + k^2x = F(t) \tag{14-130}$$

This is a linear differential equation of second order with constant coefficients. Therefore, all solutions can be found by the methods of Sections 14-9 and 14-11.

When $F(t) = 0$, the motion is unforced and the differential equation is the homogeneous equation:

$$m\frac{d^2x}{dt^2} + h\frac{dx}{dt} + k^2x = 0 \tag{14-131}$$

Let first $h = 0$ (no friction), so that we have the equation:

$$m\frac{d^2x}{dt^2} + k^2x = 0 \tag{14-132}$$

The characteristic equation is $m\lambda^2 + k^2 = 0$, the characteristic roots are $\pm\beta i$, where $\beta = k/\sqrt{m}$ and the solutions are

$$x = c_1 \cos \beta t + c_2 \sin \beta t \tag{14-133}$$

Here we can write $c_1 = A \sin \gamma$, $c_2 = A \cos \gamma$ [so that A, γ are polar coordinates of the point (c_2, c_1)]. Thus

$$x = A(\sin \gamma \cos \beta t + \cos \gamma \sin \beta t) = A \sin(\beta t + \gamma) \tag{14-133'}$$

This equation shows that the motion follows a sine curve, changed in scale and position, as in Figure 14-17. We say that the particle moves in *simple harmonic motion*. The time for a complete cycle (from one maximum to the next) is $2\pi/\beta$; this is called the *period* of the oscillation. The number β is called the *frequency* (measured in radians per unit time). The maximum value of x is A, which we call the *amplitude* of the oscillation. As for the pendulum (Section 14-12), the equilibrium position $x = 0$ is only neutrally stable.

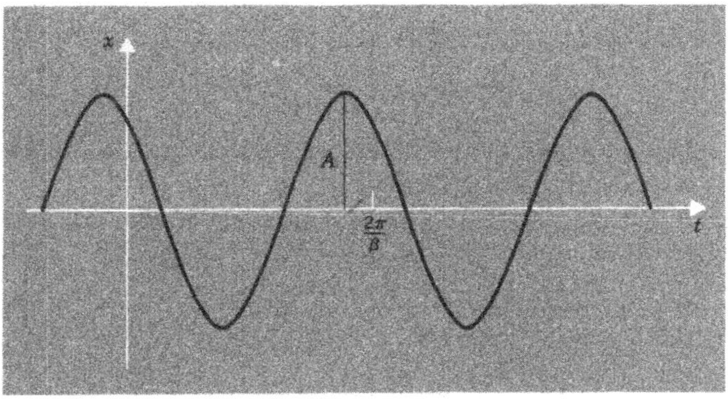

Figure 14-17 Simple harmonic motion.

For $h > 0$, the characteristic equation becomes $m\lambda^2 + h\lambda + k^2 = 0$, and the roots are

$$\frac{-h \pm \sqrt{h^2 - 4k^2 m}}{2m}$$

Here we have three cases:

I. $h^2 - 4k^2m > 0$. The roots λ_1, λ_2 are real and distinct. We note that, since m, h, and k^2 are positive, no characteristic root can be positive. Hence, $\lambda_1 < 0$, $\lambda_2 < 0$ and the zero solution is stable. The solutions are given by $x = c_1 e^{\lambda_1 t} + c_2 e^{\lambda_2 t}$ and x approaches 0 as $t \to \infty$. A typical solution is graphed in Figure 14-18. We refer to the motion as *exponential decay*. The friction force (proportional to h) is so large that vibrations have disappeared. In general, friction has a tendency, at least, to slow down or to dampen vibrations. Here we say the motion is *overdamped*.

II. $h^2 - 4k^2m = 0$. The roots λ_1, λ_2 are real, equal and negative and the zero solution is stable. The solutions have form $x = c_1 e^{\lambda_1 t} + c_2 t e^{\lambda_1 t}$ and, since $\lambda_1 < 0$, they again approach 0 as $t \to \infty$. The motion is much like that of

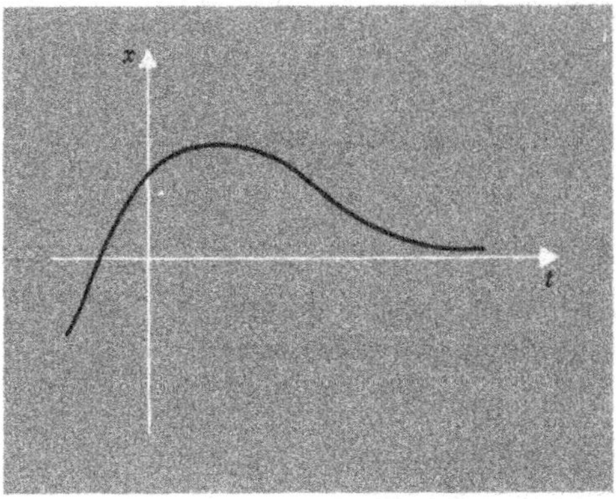

Figure 14-18 Exponential decay.

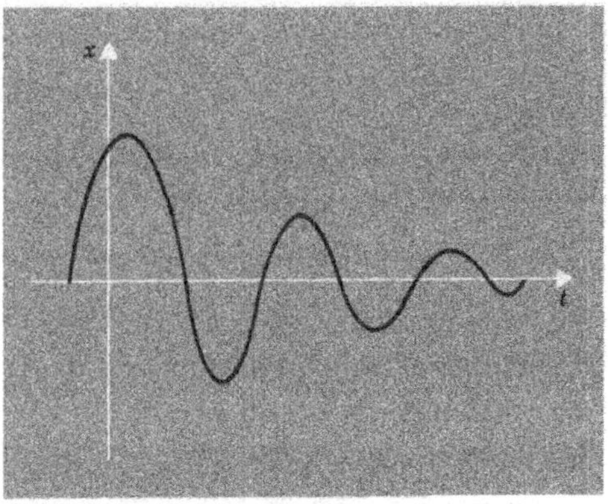

Figure 14-19 Damped vibrations.

Figure 14-18. We here say that the vibrations are *critically damped*. This is a borderline case, as the next paragraph shows.

III. $h^2 - 4k^2m < 0$. Now the characteristic roots are complex: $\alpha \pm \beta i$, where

$$\alpha = -\frac{h}{2m} < 0, \qquad \beta = \frac{\sqrt{4k^2m - h^2}}{2m} > 0$$

The solutions have form $x = c_1 e^{\alpha t} \cos \beta t + c_2 e^{\alpha t} \sin \beta t = e^{\alpha t} A \sin (\beta t + \gamma)$, just as for simple harmonic motion. The motion is like that of Figure 14-17 except that, because $\alpha < 0$, the factor $e^{\alpha t}$ forces the vibrations to gradually die down, approaching 0 as $t \to \infty$, as in Figure 14-19, and the zero solution is stable. We call the motion a *damped vibration*.

Effect of Driving Force. When we have an external driving force, the motion of the particle is described by adding to the complementary function a particular solution. We consider only the case of a sinusoidal driving force of frequency $\omega > 0$; that is, we consider the equation:

$$m\frac{d^2x}{dt^2} + h\frac{dx}{dt} + k^2x = B \sin \omega t \qquad (14\text{-}134)$$

We consider two examples and leave further analysis to problems.

EXAMPLE 1 $$\frac{d^2x}{dt^2} + 3\frac{dx}{dt} + 2x = 2 \sin 2t$$

By the methods of Sections 14-9 and 14-11 the general solution is found to be

$$x = c_1 e^{-t} + c_2 e^{-2t} + \frac{1}{10}(-\sin 2t - 3 \cos 2t)$$

As t increases, the first two terms approach 0 (hence, they are called transients) and, hence, for large positive t only the last two terms are important. They represent a simple harmonic motion of the same frequency as the driving force.

EXAMPLE 2 $$\frac{d^2x}{dt^2} + 4x = 24 \sin 2t$$

We find the general solution to be $x = c_1 \cos 2t + c_2 \sin 2t - 6t \cos 2t$. The first two terms represent a simple harmonic motion, not damped. However, the last term represents an oscillation steadily growing in size and, hence, dominating the motion for large positive t; this term is graphed in Figure 14-20. We have here an example of *resonance* or *sympathetic vibrations*. The

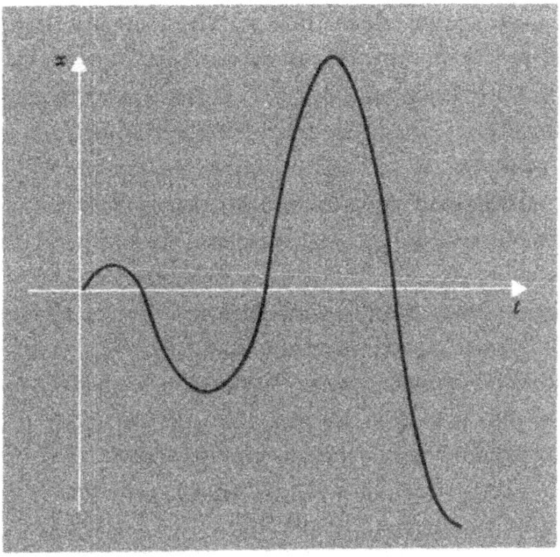

Figure 14-20 Resonance.

driving force has the same frequency, 2, as the "natural" vibrations $x = a \cos 2t + b \sin 2t$. The forcing term and the natural vibrations reinforce each other and lead to the ever-growing oscillations. This is illustrated by the disaster created by soldiers marching in step across a bridge.

PROBLEMS

1. Test for stability of the zero solution:
   (a) $x'' - 3x' + 4x = 0$ (b) $x'' + 4x' + 5x = 0$
   (c) $x'' + 2x' + x = 0$ (d) $x'' + 16x = 0$
   (e) $x''' + 2x'' + 5x' + 4x = 0$ (f) $2x''' + x'' + x' + 2x = 0$
2. (a) Prove: if $a > 0$, then the zero solution of $ax'' + bx' + cx = 0$ is stable if, and only if, $b > 0$ and $c > 0$.
   ‡(b) Prove: if $a > 0$, then the zero solution of $ax''' + bx'' + cx' + dx = 0$ is stable if, and only if, $b > 0$, $c > 0$, $d > 0$ and $bc - ad > 0$. [*Hint*. By part (a), one has stability precisely when the characteristic polynomial can be factored as $(\lambda - \lambda_1)(a\lambda^2 + B\lambda + C)$ where $\lambda_1 < 0$, $B > 0$ and $C > 0$.]
3. (a) Apply Newton's Second Law to derive the differential equation (14-121) for the motion of the pendulum. (Replace $\sin\theta$ by θ for small $|\theta|$.)

(b) Show with the aid of Figure 14-14 that, if $\theta_1 = \theta - \pi$, so that $\theta_1 = 0$ when the bob of the pendulum is above the support, then the differential equation for the motion is (14-121″), for θ_1 close to 0.

(c) For the case of part (b), show that, if a frictional force $mh(d\theta_1/dt)$ is introduced, then the equilibrium position remains unstable.

4. (a) Prove: if $\alpha < 0$ and k is a nonnegative integer, then $t^k e^{\alpha t} \to 0$ as $t \to \infty$.

(b) Prove: if α and k are as in part (a), then $t^k e^{\alpha t} \cos \beta t \to 0$ and $t^k e^{\alpha t} \sin \beta t \to 0$ as $t \to \infty$.

(c) Prove: if $\alpha \geq 0$ and k is a nonnegative integer, then $t^k e^{\alpha t}$ does not have limit 0 as $t \to \infty$, and hence the functions of part (b) do not have limit 0 as $t \to \infty$.

5. The equation (*) $a_n x^{(n)} + \cdots + a_0 x = k$, where $a_n, \ldots, k$ are constants and $a_0 \neq 0$, has the particular solution $x = k/a_0$. Since x remains constant, this is again called an *equilibrium position.*

(a) Show that $u = x - (k/a_0)$ satisfies the homogeneous equation $a_n u^{(n)} + \cdots + a_0 u = 0$ and that the constant solution $x \equiv k/a_0$ becomes the zero solution $u \equiv 0$.

(b) We call the constant solution $x = k/a_0$ of (*) stable if $x(t) \to k/a_0$ as $t \to \infty$ for every solution of (*). Show that the constant solution of (*) is stable precisely when the associated homogeneous equation has a stable zero solution.

6. Find all solutions of each of the following boundary value problems.

(a) $y'' = 0$ on $[0, 1]$, $y(0) = 3$, $y(1) = 5$ (Interpret geometrically.)

(b) $y'' = 0$, $y(0) + y'(0) = 1$, $y(1) - y'(1) = 2$

(c) $y'' + y = 0$, $y(0) = 1$, $y(2\pi) = 1$

(d) $y'' + 4y = 0$, $y'(0) = 1$, $y'(\pi) = -1$

7. In each of the following equations, determine whether the motion is simple harmonic, overdamped, critically damped, or a damped vibration, and plot a typical nonzero solution.

(a) $(d^2x/dt^2) + 2(dx/dt) + 2x = 0$

(b) $(d^2x/dt^2) + 9x = 0$

(c) $(d^2x/dt^2) + 4(dx/dt) + 3x = 0$

(d) $(d^2x/dt^2) + 8(dx/dt) + 16x = 0$

(e) $(d^2x/dt^2) + 16x = 0$

(f) $(d^2x/dt^2) + 4(dx/dt) + 5x = 0$

8. Example 1 in Section 14-13 suggests that, in general, Equation (14-134) has a solution of the form $a \cos \omega t + b \sin \omega t$, where a and b are constants.

(a) By substituting this expression in Equation (14-134), show that such a solution does exist with

$$a = \frac{-Bh\omega}{(k^2 - m\omega^2)^2 + h^2\omega^2}, \qquad b = \frac{B(k^2 - m\omega^2)}{(k^2 - m\omega^2)^2 + h^2\omega^2}$$

provided that the denominators are not 0. Show also that the latter case arises only when $h = 0$, $\omega = (k^2/m)^{1/2}$.

(b) Show that, in the exceptional case of part (a), the driving function and natural vibrations both have frequency ω. Show, by substituting in the equation, that the equation has a particular solution of form $x = at \cos \omega t$, with $a = -B/(2m\omega)$. This is the case of resonance.

9. Determine the solution of the equation of Example 1 in Section 14-13 such that $x = x_0$ and $dx/dt = v_0$ when $t = 0$.

10. Show that in Case I of Section 14-13 (overdamped motion) each nonzero solution has at most one critical point (where $dx/dt = 0$).

11. *Torsional vibrations.* A body of mass m when suspended by a vertical wire can move by rotating about the wire as axis, so that the wire is twisted or untwisted. It is shown in mechanics that the motion is governed by the equation

$$I\frac{d^2\theta}{dt^2} + C\theta = 0$$

where I and C are positive constants and θ is the angle through which the wire has been twisted. Describe the motion for the case in which $\theta = \pi/2$ and $d\theta/dt = 0$ for $t = 0$.

12. A 100-lb mass stretches a spring 5 ft under its own weight. With what frequency will the spring oscillate when supporting the weight?

13. A weight of 100 gm stretches a spring 4 cm. When the weight is in equilibrium, it is acted upon by a force that impacts a velocity 8 cm/sec downward. Show that the weight travels a distance of 2 cm during the time interval 0 to $\pi/8$ sec before it starts to return. (Here use $g = 980$ cm/sec^2.)

14. When a certain weight supported by a vertical spring is set into motion, the period is 1.5 secs. When an additional 8 lb is added, the period becomes 2.5 secs. How much weight was originally on the spring?

15. Show that for Equation (14-131), in Case III (damped vibrations), the frequency β is less than the frequency of the simple harmonic motion allowed when h is replaced by 0. Explain physically why you would expect this result.

16. Show that the local maxima of $e^{\alpha t} \cos \beta t$ do not occur at points on the curve $x = e^{\alpha t}$.

17. The motion of a certain spring is described by

$$x''(t) + 25x = 4 \cos 3t, \qquad x(0) = 0 \text{ and } x'(0) = 0$$

Show that $x = \frac{1}{2} \sin 4t \sin t$ and graph $x(t)$. We say that the motion is an *amplitude modulation* of the motion $x = \sin 4t$. In acoustics such fluctuations are called *beats*. One can experience beats by setting into vibration two tuning forks of nearly equal frequency.

18. A forced vibration is described by

$$x''(t) + 4x = a \cos^3 bt \qquad (a > 0, b > 0)$$

Show that there are two values of b for which we obtain resonance.

19. We consider a simple LRC electric circuit as in Problem 8 following Section 14-5.
   (a) Let $L = 0.25$ H, $R = 0$, $C = 10^{-4}$ F, $E = 40$ V; for $t = 0$ let $q = 0$ and $i = 0$. Determine the maximum charge on the capacitor.
   (b) Let $L = 0.5$ H, $R = 5\ \Omega$, $C = 0.08$ F, $E = 0$. For $t = 0$, $q = 0$ and $i = 10$ A. Determine the maximum charge on the capacitor.

20. When the input $q(t)$ in (14-123) is a periodic function of t, then one can expand

$q(t)$ in a Fourier series, as in Section 8-24. One obtains a Fourier series for the output by finding a sinusoidal output for each term of the Fourier series and adding the results. It can be shown that this series converges to a solution of (14-123). Obtain an output as a Fourier series for each of the following equations.

(a) $x''(t) + x'(t) + 2x(t) = \sin t - 3\cos 4t$

(b) $x''(t) + 2x'(t) + x(t) = \sum_{n=1}^{\infty} (2n-1)^{-1} \sin (2n-1)t$

(c) $x''(t) + x'(t) + x(t) = \sum_{n=1}^{\infty} n^{-3} \sin nt$

21. Show that the output for the equation

$$x'' + 4x = \sum_{n=1}^{\infty} \frac{\cos nt}{n^2}$$

exhibits resonance.

14-14 SIMULTANEOUS LINEAR DIFFERENTIAL EQUATIONS

In Section 14-1 the concept of simultaneous differential equations is introduced and the standard form for such equations is defined:

$$\frac{dx_i}{dt} = F_i(x_1, \ldots, x_n, t) \qquad (i = 1, \ldots, n) \tag{14-140}$$

For simultaneous equations not in standard form, introduction of appropriate new variables will generally lead to equations in standard form, as illustrated in Section 14-1.

We are here interested in linear differential equations, for which the standard form is as follows:

$$\frac{dx_i}{dt} = \sum_{j=1}^{n} a_{ij}x_j + q_i(t) \qquad (i = 1, \ldots, n) \tag{14-141}$$

Here the coefficients a_{ij} may be functions of t, but we shall emphasize the case of constant coefficients a_{ij}. The following are examples of systems of linear differential equations in standard form.

EXAMPLE 1

$$\frac{dx}{dt} = 3x + 8y + t + 10$$

$$\frac{dy}{dt} = x + y + 2t + 2$$

EXAMPLE 2

$$\frac{dx_1}{dt} = -5x_1 + 5x_2 + 7x_3$$

$$\frac{dx_2}{dt} = -2x_1 + 2x_2 + 3x_3$$

$$\frac{dx_3}{dt} = -4x_1 + 5x_2 + 4x_3$$

The treatment of systems of linear differential equations is greatly simplified by matrix notation. For (14-141) we introduce the vector space V_n and write $\mathbf{x}$ for $(x_1, \ldots, x_n)$, A for the matrix (a_{ij}), $\mathbf{q}(t)$ for $(q_1(t), \ldots, q_n(t))$; all vectors will be considered as *column* vectors. Now (14-141) formally becomes

$$\frac{d\mathbf{x}}{dt} = A\mathbf{x} + \mathbf{q}(t) \tag{14-142}$$

Here the matrix A may depend on t: $A = A(t)$, but as stated above, we shall emphasize the case when A is constant. Example 2 when written in form (14-142) becomes:

$$\frac{d\mathbf{x}}{dt} = \begin{pmatrix} -5 & 5 & 7 \\ -2 & 2 & 3 \\ -4 & 5 & 4 \end{pmatrix} \mathbf{x}, \qquad \mathbf{x} = \operatorname{col}(x_1, x_2, x_3)$$

By definition, a solution of (14-140) is an n-tuple of functions $x_1(t), \ldots, x_n(t)$ which satisfy Equations (14-140) identically over some interval J. Thus, in vector notation, a solution of (14-142) is a vector function $\mathbf{x} = \mathbf{f}(t)$ which satisfies (14-142) identically:

$$\mathbf{f}'(t) \equiv A\mathbf{f}(t) + \mathbf{q}(t)$$

over an interval of t. For example

$$\mathbf{f}(t) = \begin{pmatrix} 2e^t \\ e^t \\ e^t \end{pmatrix}$$

is a solution of Example 2; that is, the given equations are satisfied for

$$x_1 = 2e^t, \qquad x_2 = e^t, \qquad x_3 = e^t$$

The general Existence Theorem of Section 14-1 assures us that, under appropriate hypotheses, there is a unique solution of (14-140) with given initial values at a chosen t_0. For linear differential equations we have a stronger theorem (Theorem 9 below). We now proceed to develop the theory in strict analogy with Section 14-6; the reason for the analogy is that, as pointed out in Section 14-1, the single equation of order n with dependent variable x can be written as a system of first-order equations in which the n dependent variables are $x, x', \ldots, x^{(n-1)}$. Our first theorem is a parallel to Theorem 3 of Section 14-6:

THEOREM 9. (*Existence Theorem for systems of linear differential equations*). *Let* $A(t) = (a_{ij}(t))$ *be a continuous* n *by* n *matrix function of* t *in the interval* J; *let* $\mathbf{q}(t)$ *be a continuous vector function in* J; *let* t_0 *be a number in* J. *Then for each choice of the vector* $\mathbf{x}_0$ *in* V_n, *there is a solution* $\mathbf{x} = \mathbf{f}(t)$ *of* (14-142) *in the interval* J *such that* $\mathbf{f}(t_0) = \mathbf{x}_0$. *The solution is unique: that is, if* $\mathbf{g}(t)$ *is a second solution of* (14-142) *in an interval* J_1 *containing* t_0 *and* $\mathbf{g}(t_0) = \mathbf{x}_0$, *then* $\mathbf{f}(t) \equiv \mathbf{g}(t)$ *wherever* J *and* J_1 *overlap.*

For a proof see Chapter 12 of ODE. For special cases proofs will appear in the following sections. We stress that, as in Section 14-6, the existence theorem for linear differential equations assures existence of each solution over the whole interval J. For this reason, we shall assume each solution to be defined over the whole interval.

If in (14-142) $\mathbf{q}(t) \equiv \mathbf{0}$, then we say that (14-142) is *homogeneous;* otherwise (14-142) is called *nonhomogeneous.* For each nonhomogeneous equation (14-142) there is an *associated homogeneous equation:*

$$\frac{d\mathbf{x}}{dt} = A\mathbf{x} \tag{14-143}$$

These terms also apply to the simultaneous equations (14-141); when all $q_i(t)$ are identically 0, the equations are homogeneous; the associated homogeneous system is obtained by replacing all q_i by 0:

$$\frac{dx_i}{dt} = \sum_{j=1}^{n} a_{ij}x_j \qquad (i = 1, \ldots, n) \tag{14-144}$$

Each solution of (14-143) is a vector function $\mathbf{f}$ defined on the interval J, with values in V_n. *The set of all such vector functions on J forms a vector space V* (Section 9-16). Since our vector functions form a vector space, the concepts of subspace, basis, and linear independence can be applied; in particular, if $\mathbf{x}_1(t), \ldots, \mathbf{x}_n(t)$ are in V, then these n vector functions are linearly independent if, and only if,

$$c_1\mathbf{x}_1(t) + \cdots + c_n\mathbf{x}_n(t) \equiv \mathbf{0} \qquad \text{on } J$$

implies $c_1 = 0, \ldots, c_n = 0$.

We remark that the vector functions in V which are *continuous* on J form a subspace of V; we denote this subspace by $\mathbf{C}$; similarly, the vector functions in V having a continuous derivative on J form a subspace $\mathbf{C}^{(1)}$ of V. Also, if $A(t)$ is continuous on J, then the mapping L defined by

$$L[\mathbf{x}] = \frac{d\mathbf{x}}{dt} - A\mathbf{x}$$

for $\mathbf{x} = \mathbf{f}(t)$ in $\mathbf{C}^{(1)}$, is a *linear mapping* of $\mathbf{C}^{(1)}$ into $\mathbf{C}$:

$$L[c_1\mathbf{x}_1 + c_2\mathbf{x}_2] = c_1\frac{d\mathbf{x}_1}{dt} + c_2\frac{d\mathbf{x}_2}{dt} - A\mathbf{x}_1 - A\mathbf{x}_2 = c_1L[\mathbf{x}_1] + c_2L[\mathbf{x}_2]$$

THEOREM 10. *Let $A(t)$ be a continuous n by n matrix function on the interval J. Then the solutions of the homogeneous equation $d\mathbf{x}/dt = A\mathbf{x}$ on J form a vector space $\mathcal{H}$. Furthermore, if $\mathbf{f}$ is in $\mathcal{H}$ and $\mathbf{f}$ satisfies the initial condition*

$$\mathbf{f}(t_0) = \mathbf{0} \tag{14-145}$$

at some point t_0 of J, then $\mathbf{f}$ is the zero vector function on J.

PROOF. In the notation given above, $\mathcal{H}$ is the set of vector functions $\mathbf{x} = \mathbf{f}(t)$ in $\mathbf{C}^{(1)}$ for which $L[\mathbf{x}] = \mathbf{0}$. Therefore, $\mathcal{H}$ is the kernel of the linear mapping L and $\mathcal{H}$ is a subspace of $\mathbf{C}^{(1)}$. By Theorem 9 there is just one vector function in $\mathcal{H}$, satisfying the initial condition (14-145). But the zero function is in $\mathcal{H}$ and clearly satisfies (14-145). Therefore, if $\mathbf{f}$ is in $\mathcal{H}$ and satisfies (14-145), then $\mathbf{f}(t) \equiv \mathbf{0}$ on J.

LEMMA. *Let $A(t)$ be as in Theorem 10. Let $\mathbf{x}_1(t), \ldots, \mathbf{x}_k(t)$ be k solutions of (14-143) on J and let $\mathbf{x}_1(t_0), \ldots, \mathbf{x}_k(t_0)$ be linearly dependent vectors in V_n for some t_0 in J. Then $\mathbf{x}_1(t), \ldots, \mathbf{x}_k(t)$ are linearly dependent vector functions in $\mathbf{C}$ over J.*

PROOF. By hypothesis we can choose $c_1, \ldots, c_k$ not all 0 so that

$$c_1\mathbf{x}_1(t_0) + \cdots + c_k\mathbf{x}_k(t_0) = \mathbf{0}$$

By Theorem 10 the function

$$\mathbf{x}(t) = c_1\mathbf{x}_1(t) + \cdots + c_k\mathbf{x}_k(t)$$

is a solution of (14-143), and by the preceding equation $\mathbf{x}(t_0) = \mathbf{0}$. Therefore, by Theorem 10, $\mathbf{x}(t) \equiv \mathbf{0}$ on J. Accordingly,

$$c_1\mathbf{x}_1(t) + \cdots + c_k\mathbf{x}_k(t) \equiv \mathbf{0} \qquad \text{on } J$$

so that $\mathbf{x}_1(t), \ldots, \mathbf{x}_k(t)$ are linearly dependent vector functions in $\mathbf{C}$.

Remark. The conclusion of the lemma is false if we omit the hypothesis that $\mathbf{x}_1(t), \ldots, \mathbf{x}_k(t)$ are solutions of an equation of form (14-143); see Problem 4 below.

THEOREM 11. *The vector space $\mathcal{H}$ of Theorem 10 has dimension n. Therefore, the general solution of Equation (14-143) is given by*

$$\mathbf{x} = c_1\mathbf{x}_1(t) + \cdots + c_n\mathbf{x}_n(t)$$

where $\mathbf{x}_1(t), \ldots, \mathbf{x}_n(t)$ are linearly independent solutions.

PROOF. We choose t_0 in J and n linearly independent vectors in V_n—say, the vectors $\mathbf{e}_1 = (1, 0, \ldots, 0), \ldots, \mathbf{e}_n = (0, \ldots, 0, 1)$. By Theorem 9, for $i = 1, \ldots, n$, we can choose a solution $\mathbf{x}_i(t)$ of (14-143) such that $\mathbf{x}_i(t_0) = \mathbf{e}_i$. By the Lemma, $\mathbf{x}_1(t), \ldots, \mathbf{x}_n(t)$ are linearly independent vector functions on J. Hence, $\mathcal{H}$ has dimension at least n. Also we cannot choose more than n linearly independent solutions $\mathbf{x}_1(t), \ldots, \mathbf{x}_k(t)$ of (14-143). For if we could, then by the Lemma, $\mathbf{x}_1(t), \ldots, \mathbf{x}_k(t_0)$ would be k linearly independent vectors in V_n, but that is impossible for $k > n$. Therefore, $\mathcal{H}$ has dimension n, and the general solution has the form stated.

Remarks. Let $\mathbf{x}_1(t), \ldots, \mathbf{x}_k(t)$ be solutions of (14-143) on J and write

$$\mathbf{x}_j(t) = (x_{1j}(t), x_{2j}(t), \ldots, x_{nj}(t))$$

so that $x_{ij}(t)$ is the ith component of the vector function $\mathbf{x}_j(t)$. We can then introduce the matrix function

$$X(t) = (x_{ij}(t)) = \begin{pmatrix} x_{11}(t) & x_{12}(t) & \cdots & x_{1k}(t) \\ \vdots & & & \\ x_{n1}(t) & x_{n2}(t) & \cdots & x_{nk}(t) \end{pmatrix} \qquad (14\text{-}146)$$

The k column vectors of $X(t)$ are the solutions $\mathbf{x}_1(t), \ldots, \mathbf{x}_k(t)$. From the Lemma we see that these solutions are linearly independent precisely when $X(t)$ has rank k for every t in J; if the rank is k for one t, then it is k for all t in J; if the rank is less than k for one t, then it is less than k for all t in J.

For $k = n$, $X(t)$ is a square matrix and its determinant $W(t)$ is called the *Wronskian* of $\mathbf{x}_1(t), \ldots, \mathbf{x}_n(t)$:

$$W(t) = \begin{vmatrix} x_{11}(t) & \cdots & x_{1n}(t) \\ \vdots & & \vdots \\ x_{n1}(t) & \cdots & x_{nn}(t) \end{vmatrix}$$

Hence, either $W(t) \equiv 0$ on J or $W(t)$ has no zeros on J; in the former case, the solutions are linearly dependent; in the latter case, they are linearly independent and serve as a basis for $\mathcal{H}$ as in Theorem 11.

In the following examples, certain solutions of homogeneous differential equations will be given without explanation. In Section 14-17, we explain how they are found.

EXAMPLE 3

$$\frac{dx}{dt} = 3x + 8y \qquad \text{on } (-\infty, \infty)$$
$$\frac{dy}{dt} = x + y$$

Here one solution is

$$x = x_1(t) = 4e^{5t}, \quad y = y_1(t) = e^{5t} \qquad \text{or} \qquad \begin{pmatrix} x \\ y \end{pmatrix} = \begin{pmatrix} 4e^{5t} \\ e^{5t} \end{pmatrix}$$

For

$$\frac{dx}{dt} = 20e^{5t} = 12e^{5t} + 8e^{5t} = 3x + 8y$$

$$\frac{dy}{dt} = 5e^{5t} = 4e^{5t} + e^{5t} = x + y$$

Similarly, a second solution is $x = x_2(t) = 2e^{-t}$, $y = y_2(t) = -e^{-t}$ or

$$\begin{pmatrix} x \\ y \end{pmatrix} = \begin{pmatrix} 2e^{-t} \\ -e^{-t} \end{pmatrix}$$

Now these two solutions are linearly independent, since their Wronskian is

$$W(t) = \begin{vmatrix} 4e^{5t} & 2e^{-t} \\ e^{5t} & -e^{-t} \end{vmatrix} = -6e^{4t}$$

and $W(t)$ has no zeros. Hence, the general solution is

$$\begin{pmatrix} x \\ y \end{pmatrix} = c_1 \begin{pmatrix} 4e^{5t} \\ e^{5t} \end{pmatrix} + c_2 \begin{pmatrix} 2e^{-t} \\ -e^{-t} \end{pmatrix}$$

or

$$x = 4c_1e^{5t} + 2c_2e^{-t}, \qquad y = c_1e^{5t} - c_2e^{-t}$$

EXAMPLE 4. $(d\mathbf{x}/dt) = \begin{pmatrix} -2 & -3 \\ 3 & 4 \end{pmatrix}\mathbf{x}$ on $(-\infty, \infty)$. We find linearly independent solutions:

$$\mathbf{x}_1(t) = \begin{pmatrix} e^t \\ -e^t \end{pmatrix}, \qquad \mathbf{x}_2(t) = \begin{pmatrix} -3te^t \\ (3t+1)e^t \end{pmatrix}$$

Hence, the general solution is

$$\mathbf{x} = c_1\begin{pmatrix} e^t \\ -e^t \end{pmatrix} + c_2\begin{pmatrix} -3te^t \\ (3t+1)e^t \end{pmatrix}$$

We, of course, can choose other bases for $\mathcal{H}$. For example, $\mathbf{x}_1(t)$ and $\mathbf{x}_2(t) + \mathbf{x}_1(t)$ are also linearly independent (why?) and, hence, the general solution is also given by

$$\mathbf{x} = c_1\begin{pmatrix} e^t \\ -e^t \end{pmatrix} + c_2\begin{pmatrix} (1-3t)e^t \\ 3te^t \end{pmatrix}$$

EXAMPLE 5

$$\frac{d\mathbf{x}}{dt} = \begin{pmatrix} -5 & 5 & 7 \\ -2 & 2 & 3 \\ -4 & 5 & 4 \end{pmatrix}\mathbf{x} \quad \text{on } (-\infty, \infty)$$

Here we find three linearly independent solutions

$$\mathbf{x}_1(t) = \begin{pmatrix} 2e^t \\ e^t \\ e^t \end{pmatrix}, \quad \mathbf{x}_2(t) = \begin{pmatrix} \cos t - 7\sin t \\ \cos t - 3\sin t \\ -\cos t - 3\sin t \end{pmatrix}, \quad \mathbf{x}_3(t) = \begin{pmatrix} 7\cos t + \sin t \\ 3\cos t + \sin t \\ 3\cos t - \sin t \end{pmatrix}$$

The general solution is $\mathbf{x} = c_1\mathbf{x}_1(t) + c_2\mathbf{x}_2(t) + c_3\mathbf{x}_3(t)$.

THEOREM 12. *Let $A(t)$ be a continuous n by n matrix function on the interval J. Let $\mathbf{q}(t)$ be a continuous n by 1 vector function on J. Let $\mathcal{S}$ be the set of all solutions of the nonhomogeneous equation* (14-142) *on J; let $\mathcal{H}$ be the set of all solutions of the associated homogeneous equation* (14-143) *on J. Then $\mathcal{S}$ is nonempty and, if $\mathbf{x}_p = \mathbf{x}_p(t)$ is one function in $\mathcal{S}$, then*

$$\mathcal{S} = \{\mathbf{x}_p + \mathcal{H}\}$$

Thus $\mathcal{S}$ is a linear variety in $\mathbf{C}^{(1)}$ with base space $\mathcal{H}$.

PROOF. That $\mathcal{S}$ is nonempty follows from Theorem 9. That $\mathcal{S}$ is a linear variety as described follows by linear algebra (Theorem 20 in Section 9-13). For $\mathcal{S}$ is the L-pre-image of $\mathbf{q}$ and L has kernel $\mathcal{H}$, where $L[\mathbf{x}] = \mathbf{x}' - A\mathbf{x}$.

EXAMPLE 6 $\quad \dfrac{d\mathbf{x}}{dt} = \begin{pmatrix} 3 & 8 \\ 1 & 1 \end{pmatrix}\mathbf{x} + \begin{pmatrix} t+10 \\ 2t+2 \end{pmatrix}$ on $(-\infty, \infty)$

The associated homogeneous equation is that of Example 3. Its general solution is

$$\mathbf{x} = c_1\begin{pmatrix} 4e^{5t} \\ e^{5t} \end{pmatrix} + c_2\begin{pmatrix} 2e^{-t} \\ -e^{-t} \end{pmatrix}$$

We seek a particular solution $\mathbf{x}_p$ of form $\begin{pmatrix} at + b \\ ct + d \end{pmatrix}$, where a, b, c, d are constants (undetermined coefficients). If we substitute in the differential equation, we find that

$$\begin{pmatrix} a \\ c \end{pmatrix} = \begin{pmatrix} 3 & 8 \\ 1 & 1 \end{pmatrix} \begin{pmatrix} at + b \\ ct + d \end{pmatrix} + \begin{pmatrix} t + 10 \\ 2t + 2 \end{pmatrix}$$

and hence

$$\begin{aligned} a &= 3(at + b) + 8(ct + d) + t + 10 \\ c &= at + b + ct + d + 2t + 2 \end{aligned}$$

For these equations to hold true for all t we must have

$$\begin{aligned} a &= 3b + 8d + 10, \qquad & 0 &= 3a + 8c + 1 \\ c &= b + d + 2, \qquad & 0 &= a + c + 2 \end{aligned}$$

We find a unique solution: $a = -3, b = 1, c = 1, d = -2$. Hence, $\mathbf{x}_p(t) = \begin{pmatrix} -3t + 1 \\ t - 2 \end{pmatrix}$ and the general solution is

$$\mathbf{x} = \begin{pmatrix} -3t + 1 \\ t - 2 \end{pmatrix} + c_1 \begin{pmatrix} 4e^{5t} \\ e^{5t} \end{pmatrix} + c_2 \begin{pmatrix} 2e^{-t} \\ -e^{-t} \end{pmatrix}$$

14-15 SOLUTIONS SATISFYING INITIAL CONDITIONS, VARIATION OF PARAMETERS

We begin by some remarks on the *calculus of matrix functions* of a real variable. An example of such a function is

$$X(t) = \begin{pmatrix} t^2 & e^{3t} \\ t^3 & \cos 5t \end{pmatrix}, \qquad -\infty < t < \infty$$

In general, we consider a function $Y(t) = (y_{ij}(t))$, defined on an interval J. The derivative dY/dt of such a function is defined simply as the matrix of derivatives of the entries:

$$\frac{dY}{dt} = \left(\frac{dy_{ij}}{dt}\right)$$

provided that all $y_{ij}(t)$ have derivatives at the value or values of t considered. For the matrix $X(t)$ given above,

$$\frac{dX}{dt} = \begin{pmatrix} 2t & 3e^{3t} \\ 3t^2 & -5 \sin 5t \end{pmatrix}, \qquad -\infty < t < \infty$$

From the definition it follows at once that the usual rules on derivative of a sum and a (scalar) constant times a function are valid. There is also a rule for products:

$$\frac{d}{dt}[Y(t)Z(t)] = Y(t)\frac{dZ}{dt} + \frac{dY}{dt}Z$$

Here the sizes of Y and Z must match, and one must keep the order given. This rule follows at once from the definition of product of two matrices (Problem 10 below). There is also a chain rule: if $Y = (y_{ij})$, where $y_{ij} = y_{ij}(u)$ and $u = f(t)$, then (assuming all functions differentiable, as usual)

$$\frac{dY}{dt} = \frac{dY}{du}\frac{du}{dt} = f'(t)\left(\frac{dy_{ij}}{du}\right)$$

This follows at once from the usual chain rule.

One can also integrate matrix functions entry-wise, and the usual rules hold true. In the following discussion, we shall meet matrix functions inside the integral sign, but the integration is applied only to vector functions. For the latter, one integrates component-wise, as in Section 4-31.

We now consider the homogeneous equation

$$\frac{d\mathbf{x}}{dt} = A\mathbf{x} \tag{14-150}$$

as above, with general solution $\mathbf{x} = c_1\mathbf{x}_1(t) + \cdots + c_n\mathbf{x}_n(t)$. We form the corresponding matrix $X(t) = (x_{ij}(t))$, as in (14-146), and note that the general solution of (14-150) is

$$\mathbf{x} = c_1\mathbf{x}_1(t) + \cdots + c_n\mathbf{x}_n(t) = \begin{pmatrix} x_{11}(t) & \cdots & x_{1n}(t) \\ \vdots & & \vdots \\ x_{n1}(t) & \cdots & x_{nn}(t) \end{pmatrix}\begin{pmatrix} c_1 \\ \vdots \\ c_n \end{pmatrix}$$

or

$$\mathbf{x} = X(t)\mathbf{c} \tag{14-151}$$

where $\mathbf{c} = (c_1, \ldots, c_n)$ is an arbitrary constant vector. As above, we can differentiate $X(t)$ by differentiating each element of the matrix. But the derivative of the jth column of X is A times that column, since the column is $\mathbf{x}_j(t)$, a solution of the homogeneous differential equation (14-150). It follows that

$$\frac{dX}{dt} = AX \tag{14-152}$$

We can verify this relation also by writing

$$\frac{dx_{ij}}{dt} = \sum_{k=1}^{n} a_{ik}x_{kj} \qquad (i = 1, \ldots, n, j = 1, \ldots, n)$$

since $\mathbf{x}_j(t) = (x_{1j}(t), \ldots, x_{nj}(t))$ is a solution for each j. Thus $X(t)$ is a solution of the *matrix differential equation* (14-152).

Conversely, if $X(t)$ is an n by n matrix solution of (14-152), then the columns of $X(t)$ provide n solutions of (14-150). If $X(t)$ is nonsingular for one t, then, by the Lemma of Section 14-14, it is nonsingular for all t in J and the columns of $X(t)$ provide n linearly independent solutions of (14-150); the general solution is given by (14-151). A nonsingular matrix solution of (14-152) is called a *fundamental matrix solution.*

From (14-151) we can find the solution of (14-150) with given initial value

$\mathbf{x}_0$ at t_0. For we write

$$\mathbf{x}_0 = X(t_0)\mathbf{c}$$

and must solve for $\mathbf{c}$. Since $X(t_0)$ is nonsingular, we can multiply by $X^{-1}(t_0)$:

$$X^{-1}(t_0)\mathbf{x}_0 = \mathbf{c}$$

Hence, we have found $\mathbf{c}$ and

$$\mathbf{x}(t) = X(t)X^{-1}(t_0)\mathbf{x}_0 \tag{14-153}$$

is the solution of (14-150) *with given initial value* $\mathbf{x}_0$ *at* t_0.

We remark that we can also choose $X(t)$ as a solution of (14-152) to have a prescribed initial value $X(t_0)$. For we need only choose the column vectors of $X(t)$ [solutions of (14-150)] to have as initial values the corresponding columns of $X(t_0)$. In particular, we can choose $X(t)$ so that $X(t_0)$ is the identity matrix I. With this choice, (14-153) simplifies to

$$\mathbf{x}(t) = X(t)\mathbf{x}_0 \qquad (X(t_0) = I) \tag{14-153'}$$

EXAMPLE 1. For Example 3 of the preceding section, the matrix differential equation is

$$\frac{dX}{dt} = \begin{pmatrix} 3 & 8 \\ 1 & 1 \end{pmatrix} X$$

and a fundamental matrix solution is

$$X = \begin{pmatrix} 4e^{5t} & 2e^{-t} \\ e^{5t} & -e^{-t} \end{pmatrix}$$

The two columns of X are the two linearly independent solutions previously found. To check this, we compute

$$\frac{dX}{dt} = \begin{pmatrix} 20e^{5t} & -2e^{-t} \\ 5e^{5t} & e^{-t} \end{pmatrix}$$

and

$$AX = \begin{pmatrix} 3 & 8 \\ 1 & 1 \end{pmatrix}\begin{pmatrix} 4e^{5t} & 2e^{-t} \\ e^{5t} & -e^{-t} \end{pmatrix} = \begin{pmatrix} 20e^{5t} & -2e^{-t} \\ 5e^{5t} & e^{-t} \end{pmatrix}$$

We see that they are equal. Also $\det X(t) = -6e^{4t}$, so that $X(t)$ is nonsingular for all t.

Let us impose initial conditions at $t = 0$. We find easily that (see Section 10-12, Example 1)

$$X(0) = \begin{pmatrix} 4 & 2 \\ 1 & -1 \end{pmatrix}, \qquad X^{-1}(0) = \frac{1}{6}\begin{pmatrix} 1 & 2 \\ 1 & -4 \end{pmatrix}$$

and, hence, the solution (14-153) becomes

$$\mathbf{x}(t) = \begin{pmatrix} 4e^{5t} & 2e^{-t} \\ e^{5t} & -e^{-t} \end{pmatrix}\frac{1}{6}\begin{pmatrix} 1 & 2 \\ 1 & -4 \end{pmatrix}\mathbf{x}(0) = \frac{1}{6}\begin{pmatrix} 4e^{5t} + 2e^{-t} & 8e^{5t} - 8e^{-t} \\ e^{5t} - e^{-t} & 2e^{5t} + 4e^{-t} \end{pmatrix}\mathbf{x}(0)$$

This is now in the form (14-153′) and, hence, the matrix function

$$Y(t) = \frac{1}{6}\begin{pmatrix} 4e^{5t} + 2e^{-t} & 8e^{5t} - 8e^{-t} \\ e^{5t} - e^{-t} & 2e^{5t} + 4e^{-t} \end{pmatrix}$$

must be the fundamental matrix solution with value $I = \begin{pmatrix} 1 & 0 \\ 0 & 1 \end{pmatrix}$ for $t = 0$. We verify that this is the case. Also

$$\mathbf{x} = \frac{1}{6}\begin{pmatrix} 4e^{5t} + 2e^{-t} \\ e^{5t} - e^{-t} \end{pmatrix}$$

is a solution of the given differential equation with initial value $\begin{pmatrix} 1 \\ 0 \end{pmatrix}$ for $t = 0$. Similarly, the second column of the matrix $Y(t)$ is a solution of the differential equation with initial value $\begin{pmatrix} 0 \\ 1 \end{pmatrix}$ for $t = 0$.

Variation of Parameters. The procedure to be described here follows that of Section 14-7. We seek a particular solution $\mathbf{x}_p$ of the nonhomogeneous equation

$$\frac{d\mathbf{x}}{dt} = A\mathbf{x} + \mathbf{q}(t) \tag{14-154}$$

We assume we have found the general solution of the associated homogeneous equation (14-150). We write this general solution in the form

$$\mathbf{x} = X(t)\mathbf{c}$$

as in (14-151), where X satisfies (14-152). We now seek our particular solution in the form

$$\mathbf{x} = X(t)\mathbf{v}(t) \tag{14-155}$$

where $\mathbf{v}(t)$ is a vector function to be found. Now by the product rule for matrix functions,

$$\frac{d\mathbf{x}}{dt} = \frac{dX}{dt}\mathbf{v}(t) + X\frac{d\mathbf{v}}{dt} = AX\mathbf{v}(t) + X\frac{d\mathbf{v}}{dt}$$

Thus, if we substitute (14-155) in (14-154), we obtain

$$AX\mathbf{v}(t) + X\frac{d\mathbf{v}}{dt} = AX\mathbf{v}(t) + \mathbf{q}(t)$$

or

$$X\frac{d\mathbf{v}}{dt} = \mathbf{q}(t) \tag{14-156}$$

This is the condition to be satisfied by $\mathbf{v}(t)$ if (14-155) is to be a solution of the nonhomogeneous equation. Since $X(t)$ is nonsingular, we obtain

$$\frac{d\mathbf{v}}{dt} = X^{-1}(t)\mathbf{q}(t)$$

$$\mathbf{v} = \int X^{-1}(t)\mathbf{q}(t)\,dt \tag{14-157}$$

Here $X^{-1}(t)\mathbf{q}(t)$ is a vector function and the integration is as in Section 4-31. We seek only one choice of the indefinite integral. We can also write

$$\mathbf{v} = \int_{t_0}^{t} X^{-1}(u)\mathbf{q}(u)\,du \tag{14-157'}$$

where t_0 is a point of J [we note that $X^{-1}(t)$ is continuous in J—see Problem 8 below].

From (14-155) our solution is given as

$$\mathbf{x}_p(t) = X(t)\int X^{-1}(t)\,\mathbf{q}(t)\,dt \tag{14-158}$$

This is the *variation of parameters formula* for a particular solution.

The general solution of (14-154) is now given by

$$\mathbf{x} = X(t)\mathbf{c} + \mathbf{x}_p(t) \tag{14-159}$$

with $\mathbf{x}_p(t)$ as in (14-158). If we use (14-157') and also choose $X(t)$ so that $X(t_0) = I$, then (14-159) becomes

$$\mathbf{x} = X(t)\mathbf{c} + X(t)\int_{t_0}^{t} X^{-1}(u)\mathbf{q}(u)\,d\mathbf{u}$$

If we put $t = t_0$ on the right, we obtain $\mathbf{c}$. Hence, $\mathbf{x}(t_0) = \mathbf{c}$. Therefore

$$\mathbf{x} = X(t)\mathbf{x}(t_0) + X(t)\int_{t_0}^{t} X^{-1}(u)\mathbf{q}(u)\,d\mathbf{u} \tag{14-159'}$$

is the solution with initial value $\mathbf{x}(t_0)$, provided that $X(t_0) = I$

EXAMPLE 2 $\quad \dfrac{d\mathbf{x}}{dt} = \begin{pmatrix} 3 & 8 \\ 1 & 1 \end{pmatrix}\mathbf{x} + \begin{pmatrix} t+10 \\ 2t+2 \end{pmatrix} \quad$ on $(-\infty, \infty)$

This is the same as Example 6 of the preceding section. By Example 1, we can take

$$X(t) = \begin{pmatrix} 4e^{5t} & 2e^{-t} \\ e^{5t} & -e^{-t} \end{pmatrix}$$

and we find that

$$X^{-1}(t) = \frac{1}{6}\begin{pmatrix} e^{-5t} & 2e^{-5t} \\ e^{t} & -4e^{t} \end{pmatrix}$$

Hence by (14-158)

$$\begin{aligned} \mathbf{x}_p(t) &= \begin{pmatrix} 4e^{5t} & 2e^{-t} \\ e^{5t} & -e^{-t} \end{pmatrix}\int \frac{1}{6}\begin{pmatrix} e^{-5t} & 2e^{-5t} \\ e^{t} & -4e^{t} \end{pmatrix}\begin{pmatrix} t+10 \\ 2t+2 \end{pmatrix} dt \\ &= \frac{1}{6}\begin{pmatrix} 4e^{5t} & 2e^{-t} \\ e^{5t} & -e^{-t} \end{pmatrix}\int \begin{pmatrix} e^{-5t}(5t+14) \\ e^{t}(-7t+2) \end{pmatrix} dt \\ &= \frac{1}{6}\begin{pmatrix} 4e^{5t} & 2e^{-t} \\ e^{5t} & -e^{-t} \end{pmatrix}\begin{pmatrix} \int e^{-5t}(5t+14)\,dt \\ \int e^{t}(-7t+2)\,dt \end{pmatrix} \end{aligned}$$

$$= \frac{1}{6}\begin{pmatrix} 4e^{5t} & 2e^{-t} \\ e^{5t} & -e^{-t} \end{pmatrix}\begin{pmatrix} e^{-5t}(-t-3) \\ e^{t}(-7t+9) \end{pmatrix}$$

$$= \begin{pmatrix} -3t+1 \\ t-2 \end{pmatrix}$$

This agrees with the previous result.

14-16 COMPLEX-VALUED SOLUTIONS OF SYSTEMS OF LINEAR DIFFERENTIAL EQUATIONS

As in Section 14-8, the theory of simultaneous linear differential equations extends immediately to the case of complex valued functions. The components $x_1(t), \ldots, x_n(t)$ of a solution $\mathbf{x}(t)$ are now complex functions of t; the values of $\mathbf{x}$ are in the complex n-dimensional vector space V_n^c. All the theorems of Section 14-14 continue to hold true, with all vector spaces interpreted as complex; in particular, linear dependence is defined with reference to complex scalars.

For an equation

$$\frac{d\mathbf{x}}{dt} = A\mathbf{x} + \mathbf{q}(t) \tag{14-160}$$

in which $A = (a_{ij})$ and $\mathbf{q} = (q_1, \ldots, q_n)$ are such that all a_{ij} and all q_i are real valued, we say that the equation has *real form*. For such an equation we can seek the complex solutions and the real solutions. As in Section 14-8, the general real solution is obtainable from the general complex solution by simply choosing those complex solutions that are real; or one can simply take the real part of the general complex solution. When the a_{ij} are real, but $\mathbf{q}(t)$ is complex, one can write

$$\mathbf{q}(t) = \mathbf{q}_1(t) + i\mathbf{q}_2(t)$$

where $\mathbf{q}_1(t)$ and $\mathbf{q}_2(t)$ have real components. Then each solution $\mathbf{x}(t)$ can be written as $\mathbf{x}_1(t) + i\mathbf{x}_2(t)$, where

$$\frac{d\mathbf{x}_1}{dt} = A\mathbf{x}_1 + \mathbf{q}_1(t), \qquad \frac{d\mathbf{x}_2}{dt} = A\mathbf{x}_2 + \mathbf{q}_2(t) \tag{14-161}$$

Thus by taking real and imaginary parts of $\mathbf{x}(t)$ we obtain solutions of the two equations (14-161) in real form.

PROBLEMS

1. For each of the given differential equations, verify that the given functions are solutions on $(-\infty, \infty)$ and are linearly independent, find the general solution

and the fundamental matrix solution $X(t)$ such that $X(0) = I$:

(a) $\dfrac{d\mathbf{x}}{dt} = \begin{pmatrix} 7 & -2 \\ 15 & -4 \end{pmatrix}\mathbf{x}$, $\mathbf{x} = \begin{pmatrix} 2e^{2t} \\ 5e^{2t} \end{pmatrix}$, $\mathbf{x} = \begin{pmatrix} e^{t} \\ 3e^{t} \end{pmatrix}$

(b) $\dfrac{d\mathbf{x}}{dt} = \begin{pmatrix} -2 & 2 \\ -15 & 9 \end{pmatrix}\mathbf{x}$, $\mathbf{x} = \begin{pmatrix} 2e^{3t} \\ 5e^{3t} \end{pmatrix}$, $\mathbf{x} = \begin{pmatrix} e^{4t} \\ 3e^{4t} \end{pmatrix}$

(c) $\dfrac{d\mathbf{x}}{dt} = \begin{pmatrix} 7 & 3 \\ -10 & -4 \end{pmatrix}\mathbf{x}$, $\mathbf{x} = \begin{pmatrix} 3e^{2t} \\ -5e^{2t} \end{pmatrix}$, $\mathbf{x} = \begin{pmatrix} -e^{t} \\ 2e^{t} \end{pmatrix}$

(d) $\dfrac{d\mathbf{x}}{dt} = \begin{pmatrix} 11 & 6 \\ -20 & -11 \end{pmatrix}\mathbf{x}$, $\mathbf{x} = \begin{pmatrix} 3e^{t} \\ -5e^{t} \end{pmatrix}$, $\mathbf{x} = \begin{pmatrix} -e^{-t} \\ 2e^{-t} \end{pmatrix}$

(e) $\dfrac{d\mathbf{x}}{dt} = \begin{pmatrix} 9 & -11 & -5 \\ 4 & -4 & -3 \\ 2 & -4 & 1 \end{pmatrix}\mathbf{x}$, $\mathbf{x} = \begin{pmatrix} 3e^{2t} \\ e^{2t} \\ 2e^{2t} \end{pmatrix}$, $\mathbf{x} = \begin{pmatrix} e^{3t} \\ e^{3t} \\ -e^{3t} \end{pmatrix}$, $\mathbf{x} = \begin{pmatrix} 2e^{t} \\ e^{t} \\ e^{t} \end{pmatrix}$

(f) $\dfrac{d\mathbf{x}}{dt} = \begin{pmatrix} 2 & 1 & 0 \\ 0 & -3 & 1 \\ -1 & -13 & 4 \end{pmatrix}\mathbf{x}$, $\mathbf{x} = \begin{pmatrix} e^{t} \\ -e^{t} \\ -4e^{t} \end{pmatrix}$,

$$\mathbf{x} = \begin{pmatrix} te^{t} \\ (1-t)e^{t} \\ (3-4t)e^{t} \end{pmatrix}, \quad \mathbf{x} = \begin{pmatrix} t^2e^{t} \\ (2t-t^2)e^{t} \\ (2+6t-4t^2)e^{t} \end{pmatrix}$$

2. Let $A(t)$ be an n by n continuous matrix function on the interval J. Let X be a fundamental matrix solution for the equation $d\mathbf{x}/dt = A\mathbf{x}$.

   (a) Show that, if C is a constant n by n matrix, then $X(t)C$ is a fundamental matrix solution if, and only if, C is nonsingular.

   (b) Show that, if $Y(t)$ is also a fundamental matrix solution, then C can be chosen so that $Y(t) = X(t)C$; namely, $C = X^{-1}(t_0)Y(t_0)$, where t_0 is a number in J. [*Hint.* The kth column of $X(t)C$ is $X(t)\mathbf{c}_k$, where $\mathbf{c}_k$ is the kth column vector of C.]

   Remark. These results show how different forms of the general solution of $d\mathbf{x}/dt = A\mathbf{x}$ are related. To verify that $\mathbf{x} = X(t)\mathbf{c}$ and $\mathbf{x} = Y(t)\mathbf{c}$ represent the same general solution, one can compute $C = X^{-1}(t_0)Y(t_0)$ and verify that $Y(t) = X(t)C$.

3. Use the results of Problem 2 to show that each of the following gives the general solution of the differential equation referred to.

   (a) $\mathbf{x} = \begin{pmatrix} 6e^{2t} + e^{t} & 10e^{2t} + 2e^{t} \\ 15e^{2t} + 3e^{t} & 25e^{2t} + 6e^{t} \end{pmatrix}\mathbf{c}$, equation of Problem 1(a).

   (b) $\mathbf{x} = \begin{pmatrix} 6e^{3t} + e^{4t} & 10e^{3t} + 2e^{4t} \\ 15e^{2t} + 3e^{4t} & 25e^{3t} + 6e^{4t} \end{pmatrix}\mathbf{c}$, equation of Problem 1(b).

4. (a) Show that the vector functions of $\mathbf{C}(-\infty, \infty)$

$$\mathbf{f}_1(t) = \begin{pmatrix} 1 \\ 1 \end{pmatrix}, \quad \mathbf{f}_2(t) = \begin{pmatrix} 1 \\ t \end{pmatrix}, \quad \mathbf{f}_3(t) = \begin{pmatrix} 1 \\ t^2 \end{pmatrix}$$

   are linearly independent, but for each t the vectors $\mathbf{f}_1(t)$, $\mathbf{f}_2(t)$, $\mathbf{f}_3(t)$ are linearly dependent vectors in V_2.

(b) Show that the vector functions of $\mathbf{C}(-\infty, \infty)$

$$\mathbf{f}_1(t) = \begin{pmatrix} \sin t \\ \sin t \\ \cos t \end{pmatrix}, \qquad \mathbf{f}_2(t) = \begin{pmatrix} \cos t \\ \sin t \\ \sin t \end{pmatrix}, \qquad \mathbf{f}_3(t) = \begin{pmatrix} \sin t \\ \cos t \\ \sin t \end{pmatrix}$$

are linearly independent, but there are values of t for which the three vectors $\mathbf{f}_1(t)$, $\mathbf{f}_2(t)$, $\mathbf{f}_3(t)$ are linearly dependent vectors in V_3.

5. Find a particular solution with the aid of the information given in Problem 1.

(a) $\dfrac{d\mathbf{x}}{dt} = \begin{pmatrix} 7 & -2 \\ 15 & -4 \end{pmatrix}\mathbf{x} + \begin{pmatrix} e^{3t} \\ 2e^{3t} \end{pmatrix}$ (b) $\dfrac{d\mathbf{x}}{dt} = \begin{pmatrix} 7 & 3 \\ -10 & -4 \end{pmatrix}\mathbf{x} + \begin{pmatrix} 1 \\ t \end{pmatrix}$

(c) $\dfrac{d\mathbf{x}}{dt} = \begin{pmatrix} 7 & -2 \\ 15 & -4 \end{pmatrix}\mathbf{x} + \begin{pmatrix} \cos 3t \\ 2\cos 3t \end{pmatrix}$ (*Hint.* First solve with $\cos 3t$ replaced by e^{3it}.)

(d) $\dfrac{d\mathbf{x}}{dt} = \begin{pmatrix} 7 & 3 \\ -10 & -4 \end{pmatrix}\mathbf{x} + \begin{pmatrix} \sin 5t \\ \cos 5t \end{pmatrix}$

(e) $\dfrac{d\mathbf{x}}{dt} = \begin{pmatrix} 9 & -11 & -5 \\ 4 & -4 & -3 \\ 2 & -4 & 1 \end{pmatrix}\mathbf{x} + \begin{pmatrix} 3 \\ 1 \\ 5 \end{pmatrix}$

(f) $\dfrac{d\mathbf{x}}{dt} = \begin{pmatrix} 9 & -11 & -5 \\ 4 & -4 & -3 \\ 2 & -4 & 1 \end{pmatrix}\mathbf{x} + \begin{pmatrix} t-1 \\ t \\ 2t+1 \end{pmatrix}$

6. Prove: if $\mathbf{f}_1(t), \ldots, \mathbf{f}_k(t)$ are defined on J with values in V_n, then the k functions are linearly dependent with respect to complex coefficients if, and only if, they are linearly dependent with respect to real coefficients.

7. Let $B(t) = (b_{ij}(t))$ be an n by n continuous matrix function on J.

(a) Show that the matrix functions $Y(t)$ which are solutions of the equation

$$\frac{dY}{dt} = YB$$

are the transposes X' of the solutions of

$$\frac{dX}{dt} = AX, \qquad \text{with } A = B'$$

(b) Let $dY/dt = YB$, as in part (a). Write out the system of differential equations satisfied by the *row vectors* of Y.

8. Prove: if $X(t)$ is a fundamental matrix solution of $d\mathbf{x}/dt = A\mathbf{x}$, where $A(t)$ is continuous on J, then $X^{-1}(t)$ is continuous on J (see Theorem 18, Section 10-13).

9. Prove: if $A(t)$, $B(t)$ and dB/dt are continuous n by n matrix functions on J with $\det B \neq 0$ on J, and $d\mathbf{x}/dt = A\mathbf{x}$, then $\mathbf{y} = B(t)\mathbf{x}$ is a solution of

$$\frac{d\mathbf{y}}{dt} = C\mathbf{y}$$

with $C(t) = B(t)A(t)B^{-1}(t) + (dB/dt)B^{-1}(t)$.

10. Let $X = X(t)$ and $Y = Y(t)$ be differentiable matrix functions on the interval J, where X is m by n and Y is n by p. Let $f(t)$ be a differentiable scalar function

on J, let c be a constant scalar. Prove the rules:

(a) $\frac{d}{dt}(X+Y) = \frac{dX}{dt} + \frac{dY}{dt}$ (b) $\frac{d}{dt}(cX) = c\frac{dX}{dt}$

(c) If $X(t) \equiv$ const, then its derivative is O.

(d) If the derivative of $X(t)$ is O, then $X(t) \equiv$ const.

(e) $\frac{d}{dt}(fX) = f\frac{dX}{dt} + \frac{df}{dt}X$ (f) $\frac{d}{dt}(XY) = X\frac{dY}{dt} + \frac{dX}{dt}Y$

(g) If $X(t)$ is nonsingular for all t, then $\frac{d}{dt}X^{-1} = -X^{-1}\frac{dX}{dt}X^{-1}$

(h) If $t(s)$ is a differentiable function for $a \leq s \leq b$ and has values in J, then

$$\frac{d}{ds}X(t(s)) = \frac{dt}{ds}\frac{dX}{dt}$$

14-17 HOMOGENEOUS LINEAR SYSTEMS WITH CONSTANT COEFFICIENTS

We consider the system

$$\frac{dx_i}{dt} = \sum_{j=1}^{n} a_{ij}x_j, \qquad i = 1, \ldots, n \tag{14-170}$$

where the a_{ij} are constants. It is simplest to allow complex values to start with. When the a_{ij} are real, one obtains real solutions as in Section 14-16. All solutions are defined on $(-\infty, \infty)$.

We write the equations in matrix form:

$$\frac{d\mathbf{x}}{dt} = A\mathbf{x}, \qquad A = (a_{ij}) \tag{14-170$'$}$$

From the examples and exercises of the preceding sections one would expect solutions of the form

$$\mathbf{x} = e^{\lambda t}\mathbf{b} \tag{14-171}$$

where λ is a constant scalar, $\mathbf{b}$ is a constant nonzero vector in V_n^c. Substituting in (14-170′), we obtain $\lambda e^{\lambda t}\mathbf{b} = Ae^{\lambda t}\mathbf{b}$ or

$$A\mathbf{b} = \lambda\mathbf{b} \tag{14-172}$$

Accordingly, λ must be an *eigenvalue* of A and $\mathbf{b}$ an *eigenvector* associated with λ (Section 10-18). Therefore, λ must be a root of the characteristic equation:

$$\det(A - \lambda I) = 0 \tag{14-173}$$

This is an algebraic equation of degree n. In expanded form it is the equation

$$\begin{vmatrix} a_{11}-\lambda & a_{12} & \cdots & a_{nn} \\ a_{21} & a_{22}-\lambda & \cdots & a_{2n} \\ \vdots & & & \\ a_{n1} & a_{n2} & \cdots & a_{nn}-\lambda \end{vmatrix} = 0 \qquad (14\text{-}173')$$

Each solution of (14-173′) is an eigenvalue λ, and we can determine at least one corresponding eigenvector $\mathbf{b} = (b_1, \ldots, b_n)$ by solving the corresponding equation

$$(A - \lambda I)\mathbf{b} = \mathbf{0} \qquad (14\text{-}174)$$

or, in expanded form,

$$\begin{aligned} (a_{11}-\lambda)b_1 + a_{12}b_2 + \cdots + a_{1n}b_n &= 0 \\ a_{21}b_1 + (a_{22}-\lambda)b_2 + \cdots + a_{2n}b_n &= 0 \\ &\vdots \\ a_{n1}b_1 + a_{n2}b_2 + \cdots + (a_{nn}-\lambda)b_n &= 0 \end{aligned} \qquad (14\text{-}174')$$

The eigenvectors associated with a particular eigenvalue λ, plus $\mathbf{0}$, fill a k-dimensional subspace of $V_n{}^c$.

Let us first assume that the eigenvalues $\lambda_1, \ldots, \lambda_n$ are distinct. Then by Theorem 31 in Section 9-23, we can find associated eigenvectors $\mathbf{b}_1, \ldots, \mathbf{b}_n$ which are necessarily *linearly independent.* The corresponding solutions (14-171)

$$\mathbf{x} = e^{\lambda_1 t}\mathbf{b}_1, \qquad \ldots, \qquad \mathbf{x} = e^{\lambda_n t}\mathbf{b}_n$$

form n linearly independent solutions of (14-170′). For their Wronskian is

$$W(t) = \begin{vmatrix} e^{\lambda_1 t}b_{11} & \cdots & e^{\lambda_n t}b_{1n} \\ \vdots & & \\ e^{\lambda_1 t}b_{n1} & \cdots & e^{\lambda_n t}b_{nn} \end{vmatrix}$$

Here we have written $\mathbf{b}_k = (b_{1k}, \ldots, b_{nk})$. We see that

$$W(t) = e^{(\lambda_1+\ldots+\lambda_n)t} \begin{vmatrix} b_{11} & \cdots & b_{1n} \\ \vdots & & \\ b_{n1} & \cdots & b_{nn} \end{vmatrix}$$

and, since $\mathbf{b}_1, \ldots, \mathbf{b}_n$ are linearly independent, $W(t)$ has no zeros. Therefore, the general solution is

$$\mathbf{x} = c_1 e^{\lambda_1 t}\mathbf{b}_1 + c_2 e^{\lambda_2 t}\mathbf{b}_2 + \cdots + c_n e^{\lambda_n t}\mathbf{b}_n = X(t)\mathbf{c}$$

where $X(t)$ is the matrix whose columns are $e^{\lambda_1 t}\mathbf{b}_1, \ldots, e^{\lambda_n t}\mathbf{b}_n$.

EXAMPLE 1 $$\frac{d\mathbf{x}}{dt} = \begin{pmatrix} 3 & 8 \\ 1 & 1 \end{pmatrix}\mathbf{x}$$

This is the same as Example 3 of Section 14-14. The characteristic equation is

$$\begin{vmatrix} 3-\lambda & 8 \\ 1 & 1-\lambda \end{vmatrix} = 0 \quad \text{or} \quad \lambda^2 - 4\lambda - 5 = 0$$

The roots are $\lambda_1 = 5$, $\lambda_2 = -1$. For $\lambda_1 = 5$, $\mathbf{b}_1 = (b_{11}, b_{21})$ is found by solving the simultaneous linear equations (we write x for b_{11}, y for b_{21}):

$$\begin{cases} (3-5)x + 8y = 0 \\ 1x + (1-5)y = 0 \end{cases} \quad \text{or} \quad \begin{cases} -2x + 8y = 0 \\ x - 4y = 0 \end{cases}$$

We see that the solutions form the one-dimensional vector space $c(4, 1)$. Hence, an associated eigenvector (**0** not allowed!) is $(4, 1) = \mathbf{b}_1$. Thus we have found one solution:

$$e^{5t}\begin{pmatrix} 4 \\ 1 \end{pmatrix} = \begin{pmatrix} 4e^{5t} \\ e^{5t} \end{pmatrix}$$

Similarly, for $\lambda_2 = -1$ we obtain the equations:

$$\begin{cases} (3+1)x + 8y = 0, \\ 1x + (1+1)y = 0, \end{cases} \quad \text{or} \quad \begin{cases} 4x + 8y = 0 \\ x + 2y = 0 \end{cases}$$

The solutions are $c(2, -1)$ and our second solution of the differential equation is $e^{-t}(2, -1)$. The general solution is

$$\mathbf{x} = \begin{pmatrix} 4e^{5t} & 2e^{-t} \\ e^{5t} & -e^{-t} \end{pmatrix}\mathbf{c}$$

as in Section 14-14.

EXAMPLE 2

$$\frac{d\mathbf{x}}{dt} = \begin{pmatrix} -5 & 5 & 7 \\ -2 & 2 & 3 \\ -4 & 5 & 4 \end{pmatrix}\mathbf{x}$$

This is Example 5 of Section 14-14. The characteristic equation is

$$\begin{vmatrix} -5-\lambda & 5 & 7 \\ -2 & 2-\lambda & 3 \\ -4 & 5 & 4-\lambda \end{vmatrix} = 0$$

After expansion this becomes $\lambda^3 - \lambda^2 + \lambda - 1 = 0$ or $(\lambda - 1)(\lambda^2 + 1) = 0$. Hence, the eigenvalues are $\lambda_1 = 1$, $\lambda_2 = i$, $\lambda_3 = -i$. For $\lambda_1 = 1$ we find an eigenvector $\mathbf{b}_1 = (x, y, z)$ by solving the equations

$$\begin{aligned} -6x + 5y + 7z &= 0 \\ -2x + y + 3z &= 0 \\ -4x + 5y + 3z &= 0 \end{aligned}$$

The elimination of z from the first and second equations, or from the second and third, gives $x - 2y = 0$. Hence, we can choose $y = 1$, $x = 2$, and find $z = 1$; the solutions form the one-dimensional vector space $c(2, 1, 1)$, and our solution is

$$e^t \begin{pmatrix} 2 \\ 1 \\ 1 \end{pmatrix} \quad \text{or} \quad \begin{pmatrix} 2e^t \\ e^t \\ e^t \end{pmatrix}$$

For $\lambda_2 = i$, we have the following equations for $\mathbf{b}_2 = (x, y, z)$:

$$\begin{aligned} (-5 - i)x + 5y + 7z &= 0 \\ -2x + (2 - i)y + 3z &= 0 \\ -4x + 5y + (4 - i)z &= 0 \end{aligned}$$

The elimination of z from the first two equations gives

$$(1 + 3i)x - (1 + 7i)y = 0$$

Hence, we can take $x = 1 + 7i$, $y = 1 + 3i$, and find $z = -1 + 3i$. Therefore, we have $\mathbf{b}_2 = (1 + 7i, 1 + 3i, -1 + 3i)$, and our solution is

$$\begin{pmatrix} (1 + 7i)e^{it} \\ (1 + 3i)e^{it} \\ (-1 + 3i)e^{it} \end{pmatrix}$$

If we replace e^{it} by $\cos t + i \sin t$ and expand, the solution becomes

$$\begin{pmatrix} \cos t - 7 \sin t \\ \cos t - 3 \sin t \\ -\cos t - 3 \sin t \end{pmatrix} + i \begin{pmatrix} 7 \cos t + \sin t \\ 3 \cos t + \sin t \\ 3 \cos t - \sin t \end{pmatrix}$$

Now we verify that the real and imaginary parts of this solution are solutions and that the three solutions

$$\begin{pmatrix} \cos t - 7 \sin t \\ \cos t - 3 \sin t \\ -\cos t - 3 \sin t \end{pmatrix}, \quad \begin{pmatrix} 7 \cos t + \sin t \\ 3 \cos t + \sin t \\ 3 \cos t - \sin t \end{pmatrix}, \quad \begin{pmatrix} 2e^t \\ e^t \\ e^t \end{pmatrix}$$

are linearly independent real solutions of the differential equation (Problem 7 below). Hence, we do not need to concern ourselves with $\lambda_3 = -i$ and have the general real solution

$$\mathbf{x} = \begin{pmatrix} 2e^t & \cos t - 7 \sin t & 7 \cos t + \sin t \\ e^t & \cos t - 3 \sin t & 3 \cos t + \sin t \\ e^t & -\cos t - 3 \sin t & 3 \cos t - \sin t \end{pmatrix} \mathbf{c}$$

The examples show that, for the case of distinct eigenvalues, our problem is solved in a straightforward manner by linear algebra. For equations with real coefficients the complex roots come in conjugate pairs of form $\alpha \pm \beta i$. For each such pair of roots one finds the solution corresponding to $\alpha + \beta i$ and takes real and imaginary parts as above to obtain the two corresponding solutions, as needed.

When some of the eigenvalues are repeated, the method generally does not yield enough linearly independent solutions. By further application of linear algebra, one can show that the missing solutions can be chosen in the form $e^{\lambda t}(p_1(t), \ldots, p_n(t))$, where λ is an eigenvalue of A and $p_1(t), \ldots, p_n(t)$ are polynomials of degree less than the multiplicity of λ. In detail, one has the following theorem.

THEOREM 13. *Let A be a constant n by n matrix. Then the differential equation*

$$\frac{d\mathbf{x}}{dt} = A\mathbf{x}$$

has a set of n linearly independent solutions $\mathbf{x}_1(t), \ldots, \mathbf{x}_n(t)$, *each having the form:*

$$\mathbf{x} = e^{\lambda t}(p_1(t), \ldots, p_n(t)) \qquad (14\text{-}175)$$

Here λ *is an eigenvalue of A and each of* $p_1(t), \ldots, p_n(t)$ *is a polynomial in t of degree less than the multiplicity of* λ. *For each eigenvalue* λ *of multiplicity m, there are m solutions of form* (14-175) *in the set* $\mathbf{x}_1(t), \ldots, \mathbf{x}_n(t)$.

For a proof of this theorem see Chapter 6 of ODE. Examples are given in Section 14-19 below. It should be noted that it can happen that all polynomials $p_j(t)$ have degree 0, even though λ is a multiple root; see Example 5 in Section 14-19.

14-18 NONHOMOGENEOUS LINEAR SYSTEMS WITH CONSTANT COEFFICIENTS: STABILITY

For the nonhomogeneous equation with constant coefficients

$$\frac{d\mathbf{x}}{dt} = A\mathbf{x} + \mathbf{q}(t) \qquad (14\text{-}180)$$

the method of variation of parameters is always available and can be applied as in Section 14-15; see Example 2 of that section. We give one more example.

EXAMPLE 1 $$\frac{d\mathbf{x}}{dt} = \begin{pmatrix} 1 & -5 \\ 2 & -1 \end{pmatrix}\mathbf{x} + \begin{pmatrix} 6 \sin 3t \\ 0 \end{pmatrix}$$

For the related homogeneous equation the characteristic equation is

$$\begin{vmatrix} 1-\lambda & -5 \\ 2 & -1-\lambda \end{vmatrix} = 0 \qquad \text{or} \qquad \lambda^2 + 9 = 0$$

The eigenvalues are $\pm 3i$. For $\lambda = 3i$ an eigenvector $\mathbf{b} = (b_1, b_2)$ must satisfy

$$(1 - 3i)b_1 - 5b_2 = 0, \qquad 2b_1 + (-1 - 3i)b_2 = 0$$

We find a solution to be $\mathbf{b} = (5, 1 - 3i)$. Hence, $(5e^{3it}, (1 - 3i)e^{3it})$ or

$$(5 \cos 3t + 5i \sin 3t, \cos 3t + 3 \sin 3t + i(-3 \cos 3t + \sin 3t))$$

is a corresponding solution of the homogeneous equation. By taking real and imaginary parts as above, we obtain two linearly independent solutions and obtain the fundamental matrix solution

$$X(t) = \begin{pmatrix} 5\cos 3t & 5\sin 3t \\ \cos 3t + 3\sin 3t & -3\cos 3t + \sin 3t \end{pmatrix}$$

Our particular solution is now given by (14-158):

$$\mathbf{x}_p = X(t)\int X^{-1}(t)\mathbf{q}(t)\,dt$$

We find easily that

$$X^{-1}(t) = \frac{1}{15}\begin{pmatrix} 3\cos 3t - \sin 3t & 5\sin 3t \\ \cos 3t + 3\sin 3t & -5\cos 3t \end{pmatrix}$$

Therefore (we omit arbitrary constants)

$$\begin{aligned}\int X^{-1}(t)\mathbf{q}(t)\,dt &= \frac{1}{15}\int \begin{pmatrix} 3\cos 3t - \sin 3t & 5\sin 3t \\ \cos 3t + 3\sin 3t & -5\cos 3t \end{pmatrix}\begin{pmatrix} 6\sin 3t \\ 0 \end{pmatrix} dt \\ &= \frac{1}{15}\int \begin{pmatrix} 18\cos 3t\sin 3t - 6\sin^2 3t \\ 6\cos 3t\sin 3t + 18\sin^2 3t \end{pmatrix} dt \\ &= \frac{1}{15}\begin{pmatrix} 3\sin^2 3t - 3t + \sin 3t\cos 3t \\ \sin^2 3t + 9t - 3\sin 3t\cos 3t \end{pmatrix}\end{aligned}$$

and $\mathbf{x}_p$ is $X(t)$ times this or

$$\mathbf{x}_p = \frac{1}{15}\begin{pmatrix} 5\sin^3 3t + 5\sin 3t\cos^2 3t + 15t(3\sin 3t - \cos 3t) \\ 10\sin^3 3t + 10\sin 3t\cos^2 3t - 30t\cos 3t \end{pmatrix}$$

Stability. For Equation (14-180) we have a discussion analogous to that of Section 14-12. The related homogeneous equation

$$\frac{d\mathbf{x}}{dt} = A\mathbf{x}$$

has always the *zero solution* $\mathbf{x} = \mathbf{0}$. This solution is said to be *stable* if all other solutions approach $\mathbf{0}$ as $t \to \infty$. In that case, the solutions of the homogeneous equation are called *transients,* and each solution of the nonhomogeneous equation is of the form "particular solution plus transient." Hence, for large positive t the particular solution dominates and can be regarded as the steady state. Now Theorem 13 gives us the general form of a basis for the solutions of the homogeneous equation. From that information we conclude, as in Section 14-12, that *the zero solution is stable precisely when all eigenvalues of A have negative real parts.*

EXAMPLE 2. Discuss the behavior for large positive t of the solutions of the equation

$$\frac{d\mathbf{x}}{dt} = \begin{pmatrix} -1 & 1 & 1 \\ -1 & -2 & 2 \\ -3 & 1 & -3 \end{pmatrix}\mathbf{x} + \begin{pmatrix} \sin t \\ 0 \\ 0 \end{pmatrix}$$

Solution. For the related homogeneous equation, the characteristic equation is

$$\begin{vmatrix} -1-\lambda & 1 & 1 \\ -1 & -2-\lambda & 2 \\ -3 & 1 & -3-\lambda \end{vmatrix} = 0 \quad \text{or} \quad \lambda^3 + 6\lambda^2 + 13\lambda + 20 = 0$$

By the result of Problem 2(b) following Section 14-13, the roots of the characteristic equation all have negative real part. Therefore, the solutions of the homogeneous equation are transients and, for large positive t, we need only consider the particular solution.

To find a particular solution, we can here use *undetermined coefficients* as follows. We try to find a solution of the form:

$$\mathbf{x} = (a_1 \cos t + b_1 \sin t,\ a_2 \cos t + b_2 \sin t,\ a_3 \cos t + b_3 \sin t)$$

Substitution in the differential equation and linear independence of $\cos t$, $\sin t$ lead to the simultaneous equations

$$a_1 - a_2 - a_3 + b_1 = 0, \quad a_1 + 2a_2 - 2a_3 + b_2 = 0,$$
$$3a_1 - a_2 + 3a_3 + b_3 = 0, \quad a_1 - b_1 + b_2 + b_3 = -1,$$
$$a_2 - b_1 - 2b_2 + 2b_3 = 0, \quad a_3 - 3b_1 + b_2 - 3b_3 = 0$$

Solving, we find $a_1 = 17/170$, $a_2 = 47/170, \ldots$ and, hence,

$$\mathbf{x} = \frac{1}{170}(17 \cos t + 51 \sin t,\ 47 \cos t - 69 \sin t,\ 21 \cos t - 67 \sin t)$$

Remarks. Example 2 is typical of applications of simultaneous linear differential equations. Most commonly one is concerned only with the stability of the zero solution of the homogeneous equation and the related question of transients. For this purpose one need only find the characteristic equation and answer the question: *Do the roots all have negative real parts?* As mentioned in Section 14-12 (see especially the book of Gantmacher cited there), there are many tests for answering the question posed. Once one knows that the solutions of the homogeneous equation are transients, one can study the nature of a particular solution of the nonhomogeneous equation. The method of undetermined coefficients can be extended considerably and suffices for the most common applications.

When some simple characteristic roots have zero real part, but no characteristic root has positive real part, the solutions of the homogeneous equation are formed of terms in $\sin \omega t$ and $\cos \omega t$ for various ω, and transients. In that case the solutions do not generally approach 0 as $t \to \infty$, but they, at least, are bounded, so that one has "neutral stability" (see Section 14-12); this is illustrated by Example 1 above. For such equations a sinusoidal input of appropriate frequency may lead to *resonance;* this is also illustrated in Example 1 (see also Problem 4 below).

PROBLEMS

1. Find the general solution:

(a) $\dfrac{d\mathbf{x}}{dt} = \begin{pmatrix} 1 & 1 \\ 4 & -2 \end{pmatrix}\mathbf{x}$ (b) $\dfrac{d\mathbf{x}}{dt} = \begin{pmatrix} 5 & 3 \\ -1 & 1 \end{pmatrix}\mathbf{x}$

(c) $\dfrac{d\mathbf{x}}{dt} = \begin{pmatrix} 3 & 3 \\ -3 & 1 \end{pmatrix}\mathbf{x}$ (d) $\dfrac{d\mathbf{x}}{dt} = \begin{pmatrix} 4 & -1 \\ 2 & 2 \end{pmatrix}\mathbf{x}$

(e) $\dfrac{d\mathbf{x}}{dt} = \begin{pmatrix} 1 & -1 & 4 \\ 3 & 2 & -1 \\ 2 & 1 & -1 \end{pmatrix}\mathbf{x}$ (f) $\dfrac{d\mathbf{x}}{dt} = \begin{pmatrix} 2 & -1 & -1 \\ 1 & 2 & 1 \\ 0 & 1 & 2 \end{pmatrix}\mathbf{x}$

2. For each of the following choices of A determine whether the zero solution of the equation $d\mathbf{x}/dt = A\mathbf{x}$ is stable. (It is not necessary to find the general solution.)

(a) $\begin{pmatrix} 3 & 4 \\ -1 & 5 \end{pmatrix}$ (b) $\begin{pmatrix} -1 & 3 \\ 2 & -7 \end{pmatrix}$

(c) $\begin{pmatrix} -2 & 1 & 1 \\ 1 & -3 & 1 \\ 1 & 1 & -4 \end{pmatrix}$ (d) $\begin{pmatrix} -1 & 0 & 2 \\ 0 & 3 & 2 \\ 2 & 0 & -1 \end{pmatrix}$

3. Let the following differential equation be given:

$$\frac{d\mathbf{x}}{dt} = \begin{pmatrix} -1 & 2 & 1 \\ 0 & -3 & -1 \\ 1 & 4 & -2 \end{pmatrix}\mathbf{x} + \mathbf{q}(t)$$

(a) Show that the zero solution of the associated homogeneous equation is stable.

(b) Find a particular solution for $\mathbf{q}(t) = (2, 1, -8)$. [*Hint.* Try $\mathbf{x}(t) = \mathbf{c}$, a constant vector.]

(c) Find a particular solution for $\mathbf{q}(t) = e^t(1, 0, 1)$. [*Hint.* Try $\mathbf{x}(t) = e^t\mathbf{c}$.]

(d) Find a particular solution for $\mathbf{q}(t) = (0, 3\cos 2t - \sin 2t, -2\cos 2t + \sin 2t)$.

4. Let the following differential equation be given

$$\frac{d\mathbf{x}}{dt} = \begin{pmatrix} 5 & -5 & -7 \\ 2 & -2 & -3 \\ 4 & -5 & -4 \end{pmatrix}\mathbf{x} + \mathbf{q}(t)$$

(a) Verify that for the associated homogeneous equation the eigenvalues are -1 and $\pm i$, so that the zero solution is neutrally stable.

(b) Verify that for $\mathbf{q}(t) = (\sin t, 0, 0)$ resonance occurs.

(c) Verify that for $\mathbf{q}(t) = (2\sin t, \sin t, \sin t)$, a particular solution is given by $\mathbf{x} = (\frac{1}{2})(2\sin t - 2\cos t,\ \sin t - \cos t,\ \sin t - \cos t)$, so that there is no resonance.

Remark. Parts (b) and (c) show that, even when $\pm i\omega$ are eigenvalues, a sinusoidal forcing function of frequency ω may, but need not necessarily lead to resonance.

5. *Coupled springs.* Let a mass m_1 be suspended by a spring. Let a second mass m_2 be suspended from the first by a second spring (Figure 14-21). Then an equilibrium will be reached under gravity. Let x_1, x_2 measure the downward displacements

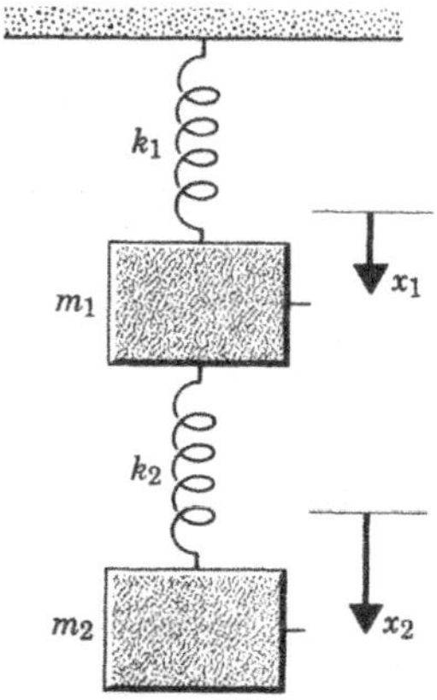

Figure 14-21 Coupled springs.

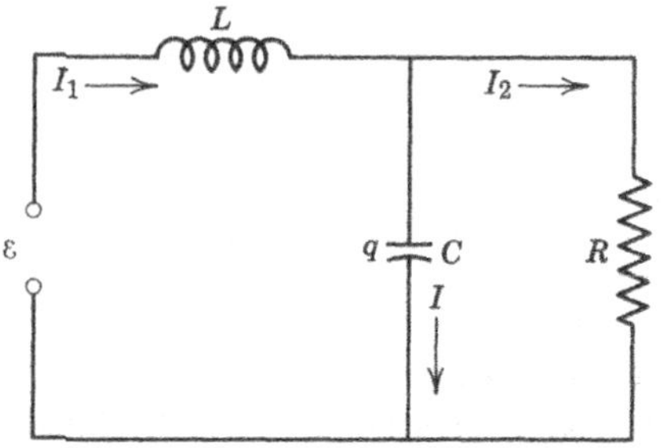

Figure 14-22 An electrical network.

of the two masses from these positions, as in the figure. If we ignore friction, then by mechanics the vertical motion of the system is governed by the differential equations

$$m_1 \frac{d^2x_1}{dt^2} = -k_1x_1 + k_2(x_2 - x_1), \qquad m_2 \frac{d^2x_2}{dt^2} = -k_2(x_2 - x_1)$$

As in Section 14-1, these equations are equivalent to a system:

$$x_1' = x_3, \qquad x_2' = x_4, \qquad m_1x_3' = -k_1x_1 + k_2(x_2 - x_1), \qquad m_2x_4' = -k_2(x_2 - x_1)$$

Take $m_1 = 3$, $m_2 = 2$, $k_1 = k_2 = 1$ (in appropriate units) and find the general solution. Show also that there are possible motions of the system in which the two masses oscillate synchronously at the same frequency ω and find two frequencies ω for which this is possible. (These oscillations are called *normal modes*. The results of the exercise can be verified by experiment.)

6. *Electrical networks*. The case of a single circuit is discussed in Problem 8 following Section 14-5. Figure 14-22 shows a simple network. From physics it follows that the currents I, I_1, I_2, and charge stored q are related by the equations

$$L\frac{dI_1}{dt} + \frac{q}{C} = \mathcal{E}, \qquad RI_2 - \frac{q}{C} = 0, \qquad \frac{dq}{dt} = I, \qquad I = I_1 - I_2$$

Here the inductance L, resistance R, and capacitance C are positive constants and $\mathcal{E}$ is a given function of t. Show that the vector $\mathbf{x} = (q, I_1)$ satisfies the differential equation $d\mathbf{x}/dt = A\mathbf{x} + \mathbf{f}(t)$, where

$$A = \begin{pmatrix} -1/(RC) & 1 \\ -1/(LC) & 0 \end{pmatrix}, \qquad \mathbf{f}(t) = \begin{pmatrix} 0 \\ \mathcal{E}/L \end{pmatrix}$$

Show also that if $\mathcal{E}$ is constant and $L < 4R^2C$, then q and I_1 follow damped vibrations.

7. Verify the steps needing justification in the solution of Example 2 in Section 14-17.

14-19 METHOD OF ELIMINATION

We illustrate the method:

EXAMPLE 1

$$\frac{dx}{dt} = 3x + 8y + t + 10$$

$$\frac{dy}{dt} = x + y + 2t + 2$$

This is the same as Example 6 in Section 14-14. We now write $D = d/dt$ and use operator notation as in Section 14-9. The equations can be written

$$\begin{aligned}(D - 3)x - 8y &= t + 10\\ -x + (D - 1)y &= 2t + 2\end{aligned} \tag{14-190}$$

We now eliminate y by multiplying the first equation by $D - 1$, the second by 8, and adding:

$$(D - 1)(D - 3)x - 8x = (D - 1)(t + 10) + 8(2t + 2) = 15t + 7$$

or

$$(D^2 - 4D - 5)x = 15t + 7$$

Hence, by undetermined coefficients (Section 14-11), we find that $x = c_1e^{5t} + c_2e^{-t} + 1 - 3t$. From the first equation of (14-190) we find that

$$8y = (D - 3)x - t - 10 = (D - 3)(c_1e^{5t} + c_2e^{-t} + 1 - 3t) - t - 10$$

and, hence, we find that $y = \frac{1}{4}c_1e^{5t} - \frac{1}{2}c_2e^{-t} + t - 2$. We can write our general solution in vector form:

$$\begin{pmatrix} x \\ y \end{pmatrix} = \frac{1}{4}c_1\begin{pmatrix} 4e^{5t} \\ e^{5t} \end{pmatrix} + \frac{1}{2}c_2\begin{pmatrix} 2e^{-t} \\ -e^{-t} \end{pmatrix} + \begin{pmatrix} 1 - 3t \\ t - 2 \end{pmatrix}$$

This is clearly equivalent to the solution found in Section 14-14.

The method can be applied to any set of simultaneous linear differential equations with constant coefficients of form:

$$\begin{aligned}&\varphi_{11}(D)x_1 + \cdots + \varphi_{1n}(D)x_n = q_1(t)\\ &\vdots\\ &\varphi_{n1}(D)x_1 + \cdots + \varphi_{nn}(D)x_n = q_n(t)\end{aligned} \tag{14-191}$$

Here each $\varphi_{ij}(D)$ stands for a polynomial in D; for example, $5D^3 - 2D^2 + D - 1$; the functions $q_j(t)$ are assumed to have derivatives of an appropriate order on a given interval.

From (14-191) we obtain an equivalent set of equations (one with the same solutions) by multiplying any equation by a nonzero constant or by subtracting from (or adding to) one equation $f(D)$ times another equation of the set, where $f(D)$ is a polynomial in D. Thus we can replace the second equation in (14-191) by

$$\begin{aligned}(\varphi_{21}(D) - f(D)\varphi_{11}(D))x_1 + \cdots + (\varphi_{2n}(D) - f(D)\varphi_{1n}(D))x_n&\\ = q_2(t) - f(D)q_1(t)&\end{aligned} \tag{14-192}$$

If this new equation holds true and the first, third, ..., nth given equation (14-191) holds true, then the second given equation must also hold true. For $f(D)$ times the first equation is then valid:

$$f(D)\varphi_{11}(D)x_1 + \cdots + f(D)\varphi_{1n}(D)x_n = f(D)q_1(t)$$

If we add this equation to (14-192), we recover the original second equation. Conversely, if (14-191) all hold true, then (14-192) holds true, by similar reasoning.

By applying the two processes described, we can always successively eliminate variables and, hence, get the general solution, as in Example 1 above [for details, see Section 7-10 of *Theory of Matrices* by S. Perlis (Addison-Wesley, Reading, Mass., 1952)].

EXAMPLE 2

$$\frac{dx}{dt} = -2x - 3y$$

$$\frac{dy}{dt} = 3x + 4y$$

This is the same as Example 4 of Section 14-14. The characteristic equation is

$$\begin{vmatrix} -2-\lambda & -3 \\ 3 & 4-\lambda \end{vmatrix} = 0 \qquad \text{or} \qquad \lambda^2 - 2\lambda + 1 = 0$$

Hence, there is a double root: $\lambda_1 = \lambda_2 = 1$. The method of Section 14-17 breaks down. We use elimination:

$$(1) \qquad (D + 2)x + 3y = 0$$

$$(2) \qquad -3x + (D - 4)y = 0$$

We multiply (1) by $(D - 4)$ and subtract from 3 times (2) to obtain a new equation:

$$(2') \qquad (D^2 - 2D + 1)x = 0$$

We find $x = c_1e^t + c_2te^t$ and substitute in (1) to obtain y:

$$3y = -(D + 2)x = -(D + 2)(c_1e^t + c_2te^t) = -3c_1e^t - c_2e^t(3t + 1)$$

In vector form, the general solution is

$$\begin{pmatrix} x \\ y \end{pmatrix} = \begin{pmatrix} e^t & te^t \\ -e^t & -\frac{1}{3}(3t + 1)e^t \end{pmatrix} \mathbf{c}$$

This is equivalent to the solution found in Section 14-14.

EXAMPLE 3

$$(1) \qquad \frac{dx_1}{dt} = 2x_1 + x_2$$

$$(2) \qquad \frac{dx_2}{dt} = -3x_2 + x_3$$

$$(3) \qquad \frac{dx_3}{dt} = -x_1 - 13x_2 + 4x_3$$

This is the same as Problem 1(f) following Section 14-16. The characteristic equation is

$$\begin{vmatrix} 2-\lambda & 1 & 0 \\ 0 & -3-\lambda & 1 \\ -1 & -13 & 4-\lambda \end{vmatrix} = 0$$

After expansion, this becomes $(\lambda - 1)^3 = 0$. Hence, there is a triple root: $\lambda_1 = \lambda_2 = \lambda_3 = 1$. We use elimination. We write the equations as

$$\begin{aligned} &(1) & (D-2)x_1 - x_2 &= 0 \\ &(2) & (D+3)x_2 - x_3 &= 0 \\ &(3) & x_1 + 13x_2 + (D-4)x_3 &= 0 \end{aligned}$$

We eliminate x_3 from the third equation by adding to it $(D - 4)$ times (2), to obtain

$$(3') \qquad x_1 + (D^2 - D + 1)x_2 = 0$$

We eliminate x_1 from (1) by subtracting from it $(D - 2)$ times (3′), to obtain

$$(1') \qquad (D^3 - 3D^2 + 3D - 1)x_2 = 0$$

We find that

$$\begin{aligned} x_2 &= c_1e^t + c_2te^t + c_3t^2e^t \\ x_1 &= -(D^2 - D + 1)x_2 = -c_1e^t - c_2e^t(t+1) - c_3e^t(t^2 + 2t + 2) \\ x_3 &= (D+3)x_2 = 4c_1e^t + c_2(4t+1)e^t + c_3(4t^2 + 2t)e^t \end{aligned}$$

Hence, the general solution is

$$\mathbf{x} = e^t \begin{pmatrix} -1 & -t-1 & -t^2 - 2t - 2 \\ 1 & t & t^2 \\ 4 & 4t+1 & 4t^2 + 2t \end{pmatrix} \mathbf{c}$$

This can be verified to be equivalent to the previous solution (see Problem 2 following Section 14-16).

EXAMPLE 4

$$\frac{d\mathbf{x}}{dt} = \begin{pmatrix} 1 & 1 & 0 \\ -1 & 3 & 0 \\ 1 & -1 & 2 \end{pmatrix} \mathbf{x}$$

Here we find the eigenvalues to be 2, 2, 2. The method of elimination [Problem 2(e) below] gives the general solution:

$$x_1 = e^{2t}(c_1 + c_2t), \qquad x_2 = e^{2t}[c_1 + c_2(1+t)], \qquad x_3 = e^{2t}(-c_2t + c_3)$$

We note that the eigenvalue has multiplicity 3, but the polynomials that appear have degree at most 1.

EXAMPLE 5

$$\frac{d\mathbf{x}}{dt} = \begin{pmatrix} 2 & 0 & 0 & 0 \\ 2 & 0 & -2 & 0 \\ -1 & 1 & 3 & 0 \\ -1 & 1 & 1 & 2 \end{pmatrix} \mathbf{x}$$

Here we find the eigenvalues to be 2, 2, 2, 1. By elimination [Problem 2(f) below] we find the general solution to be

$$x_1 = e^{2t}(3c_1 + c_2 + c_3), \qquad x_2 = e^{2t}(3c_1 + c_2 - c_4)$$
$$x_3 = c_3 e^{2t} + c_4 e^t, \qquad x_4 = e^{2t}(2c_1 + c_2 + c_3) - c_4 e^t$$

Here, despite the multiple eigenvalue, the general solution is of form $c_1 e^{\lambda_1 t}\mathbf{u}_1 + c_2 e^{\lambda_2 t}\mathbf{u}_2 + \cdots$, where $\mathbf{u}_1, \mathbf{u}_2, \ldots$ are constant vectors, just as in the case of distinct eigenvalues.

‡14-20 APPLICATION OF EXPONENTIAL FUNCTION OF A MATRIX

We here consider infinite series whose terms are n by n matrices. The sum of such a series is obtained by adding element-wise; for example,

$$\begin{pmatrix} 1 & 0 \\ 0 & 1 \end{pmatrix} + \begin{pmatrix} \frac{1}{3} & 0 \\ 0 & \frac{1}{2} \end{pmatrix} + \cdots + \begin{pmatrix} \frac{1}{3^k} & 0 \\ 0 & \frac{1}{2^k} \end{pmatrix} + \cdots$$
$$= \begin{pmatrix} \sum_{k=0}^{\infty} 3^{-k} & 0 \\ 0 & \sum_{k=0}^{\infty} 2^{-k} \end{pmatrix} = \begin{pmatrix} \frac{3}{2} & 0 \\ 0 & 2 \end{pmatrix}$$

Here the individual series are geometric series. In general, for $A_k = (a_{ij}^{(k)})$, we write

$$\sum_{k=p}^{\infty} A_k = \left(\sum_{k=p}^{\infty} a_{ij}^{(k)} \right) \tag{14-200}$$

provided that all series on the right converge.

If B is a square n by n matrix, we can define e^B by such a series:

$$e^B = I + B + \frac{1}{2!}B^2 + \cdots + \frac{1}{k!}B^k + \cdots \tag{14-201}$$

We shall see that this series converges for every B and, hence, defines e^B as an n by n matrix. For example, for $n = 2$,

$$e^I = I + I + \frac{1}{2!}I^2 + \cdots = \begin{pmatrix} 1 & 0 \\ 0 & 1 \end{pmatrix} + \begin{pmatrix} 1 & 0 \\ 0 & 1 \end{pmatrix} + \begin{pmatrix} \frac{1}{2} & 0 \\ 0 & \frac{1}{2} \end{pmatrix} + \cdots$$
$$= \begin{pmatrix} \sum_{k=0}^{\infty} \frac{1}{k!} & 0 \\ 0 & \sum_{k=0}^{\infty} \frac{1}{k!} \end{pmatrix} = \begin{pmatrix} e & 0 \\ 0 & e \end{pmatrix}$$

To verify convergence of (14-201) in general, choose a positive constant c such that $|b_{ij}| \leq c$ for all i and j. Then each element of B^2 is of the form

$$b_{i1}b_{1j} + \cdots + b_{in}b_{nj}$$

and is, hence, in absolute value at most nc^2. By induction we show in this way that each element of B^k is in absolute value at most $n^{k-1}c^k$. Hence [see (14-200)] each element of e^B is defined by a series that is dominated by the series of positive terms

$$1 + c + \frac{1}{2!}nc^2 + \cdots + \frac{1}{k!}n^{k-1}c^k + \cdots$$

that converges [to $n^{-1}(e^{nc} - 1) + 1$]. Therefore, e^B is well defined for every B.

We are here concerned especially with e^{tA}, where t is real and A is an n by n constant square matrix:

$$e^{tA} = I + tA + \frac{1}{2!}t^2A^2 + \cdots + \frac{1}{k!}t^kA^k + \cdots \tag{14-202}$$

We note that

$$\begin{aligned} Ae^{tA} &= A(I + tA + \cdots) = A + tA^2 + \cdots \\ e^{tA}A &= (I + tA + \cdots)A = A + tA^2 + \cdots \end{aligned}$$

and, hence, that

$$Ae^{tA} = e^{tA}A \tag{14-203}$$

The multiplication of the series term-by-term by A, on the left or right, is justified by familiar reasoning on limits of sums and products.

Now from the definition (14-202) it follows that each entry of e^{tA} is itself a power series in t, converging for all real t by the reasoning given above. But such a power series in t may be differentiated term-by-term, hence, by replacing t^k by kt^{k-1} for all k; this is proved for series with real coefficients in Section 8-18. For series with complex terms, the assertion is then proved by first taking real and imaginary parts. We conclude that e^{tA}, as a function of t, has a derivative:

$$\begin{aligned} \frac{d}{dt}e^{tA} &= A + \frac{1}{2!}2tA^2 + \cdots + \frac{1}{k!}kt^{k-1}A^k + \cdots \\ &= A + tA^2 + \cdots + \frac{1}{k!}t^kA^{k+1} + \cdots \\ &= A\left(I + tA + \cdots + \frac{1}{k!}t^kA^k + \cdots\right) = Ae^{tA} \end{aligned}$$

Therefore, we have the important rule for *the derivative of the exponential function* e^{tA}:

$$\frac{d}{dt}e^{tA} = Ae^{tA} \tag{14-204}$$

This is the matrix analogue of the familiar rule $(e^{at})' = ae^{at}$ of calculus.

From (14-204) we can now prove the rule:

$$e^{sA}e^{tA} = e^{(s+t)A} \tag{14-205}$$

For we can write $s + t = h$. Then

$$e^{sA}e^{tA} = e^{(h-t)A}e^{tA}$$

and, for fixed h, by the chain rule [see Problem 10(h) following Section 14-16],

$$\begin{aligned}\frac{d}{dt}(e^{(h-t)A}e^{tA}) &= -Ae^{(h-t)A}e^{tA} + e^{(h-t)A}Ae^{tA}\\ &= -Ae^{(h-t)A}e^{tA} + Ae^{(h-t)A}e^{tA} = O\end{aligned}$$

Therefore, $e^{(h-t)A}e^{tA}$, as a matrix function of t, is constant. For $t = 0$, its value is e^{hA}. Hence

$$e^{hA} = e^{(h-t)A}e^{tA} \qquad \text{or} \qquad e^{(s+t)A} = e^{sA}e^{tA}$$

as asserted. From (14-205) we conclude that e^{sA} and e^{tA} *commute:*

$$e^{sA}e^{tA} = e^{tA}e^{sA}$$

For both sides are equal to $e^{(s+t)A}$. If we take $t = -s$ in (14-205), we conclude that

$$e^{sA}e^{-sA} = e^{0A} = e^{O} = I \tag{14-206}$$

Hence, in general, e^{-tA} *is the inverse of* e^{tA}.

We now apply our results to differential equations:

THEOREM 14. *Let A be a constant n by n matrix. Then the fundamental matrix solution $X(t)$ for the differential equation*

$$\frac{d\mathbf{x}}{dt} = A\mathbf{x}$$

such that $X(0) = I$ is

$$X(t) = e^{tA}$$

Accordingly, the solution with initial value $\mathbf{x}(0)$ at $t = 0$ is

$$\mathbf{x} = e^{tA}\mathbf{x}(0)$$

PROOF. By (14-204) above, if $X = e^{tA}$, then

$$\frac{dX}{dt} = \frac{d}{dt}(e^{tA}) = Ae^{tA} = AX$$

Also $X(0) = e^{0A} = e^{O} = I + O + \cdots + O + \cdots = I$. Therefore, the theorem is proved.

It thus appears that we have one general formula for the general solution of a homogeneous linear system of differential equations with constant coefficients. However, the infinite series expression (14-202) for e^{tA} is awkward to work with and, for practical purposes, this formula is not adequate. Considerable effort has been expended on finding other ways of computing e^{tA} and one can, in fact, always find e^{tA} by matrix operations not involving infinite series. When A has distinct eigenvalues, our fundamental matrix solution (see Sections 14-15 and 14-17) is

$$Y(t) = \begin{pmatrix} b_{11}e^{\lambda_1 t} & \cdots & b_{1n}e^{\lambda_n t} \\ \vdots & & \vdots \\ b_{n1}e^{\lambda_1 t} & \cdots & b_{nn}e^{\lambda_n t} \end{pmatrix}$$

If we multiply on the right by $Y^{-1}(0)$, we obtain the fundamental matrix solution whose value at $t = 0$ is I (see Problem 2 following Section 14-16). Hence, this must be the same as $X(t) = e^{tA}$:

$$e^{tA} = \begin{pmatrix} b_{11}e^{\lambda_1 t} & \cdots & b_{1n}e^{\lambda_n t} \\ \vdots & & \\ b_{n1}e^{\lambda_1 t} & \cdots & b_{nn}e^{\lambda_n t} \end{pmatrix} \begin{pmatrix} b_{11} & \cdots & b_{1n} \\ \vdots & & \\ b_{n1} & \cdots & b_{nn} \end{pmatrix}^{-1} \tag{14-207}$$

where $\lambda_1, \ldots, \lambda_n$ are the distinct eigenvalues of A and $\mathbf{b}_1, \ldots, \mathbf{b}_n$ are the associated eigenvectors. Where A has multiple eigenvalues, this formula has to be modified.

Remark. Theorem 14 shows that the equation $d\mathbf{x}/dt = A\mathbf{x}$ has a solution with given initial value at $t = 0$. The proof given does not rely on the existence theorem (Theorem 9, Section 14-14). We can also show that this solution is unique, without using the existence theorem. For if $\mathbf{x}(t)$ is a solution, then,

$$\frac{d\mathbf{x}}{dt} = A\mathbf{x}$$

$$e^{-tA}\frac{d\mathbf{x}}{dt} - e^{-tA}A\mathbf{x} = \mathbf{0}$$

$$\frac{d}{dt}(e^{-tA}\mathbf{x}) = \mathbf{0}$$

$$e^{-tA}\mathbf{x} = \mathbf{c} = \mathbf{x}(0)$$

$$\mathbf{x} = e^{tA}\mathbf{x}(0)$$

Here we used (14-203), (14-204), and (14-206).

Our proof shows in effect that e^{-tA} is an integrating factor for the differential equation.

PROBLEMS

1. (a) . . . (f) Use the method of elimination to find the general solution of each of the equations of Problem 5 following Section 14-16.
2. Verify that there is a multiple eigenvalue and find the general solution:

(a) $\dfrac{d\mathbf{x}}{dt} = \begin{pmatrix} 2 & 0 \\ 0 & 2 \end{pmatrix}\mathbf{x}$ (b) $\dfrac{d\mathbf{x}}{dt} = \begin{pmatrix} -7 & 4 \\ -25 & 13 \end{pmatrix}\mathbf{x}$

(c) $\dfrac{d\mathbf{x}}{dt} = \begin{pmatrix} -1 & 7 & 1 \\ -2 & 6 & 1 \\ -1 & 3 & 2 \end{pmatrix}\mathbf{x}$ (d) $\dfrac{d\mathbf{x}}{dt} = \begin{pmatrix} 13 & -20 & -8 \\ 6 & -9 & -4 \\ 6 & -10 & -3 \end{pmatrix}\mathbf{x}$

(e) The equation of Example 4 in Section 14-19.
(f) The equation of Example 5 in Section 14-19.

3. Find the general solution by elimination.

(a) $\dfrac{d\mathbf{x}}{dt} = \begin{pmatrix} 1 & 2 \\ 2 & 4 \end{pmatrix}\mathbf{x} + \begin{pmatrix} e^t - e^{2t} \\ e^t + 2e^{2t} \end{pmatrix}$

(b) $\dfrac{d\mathbf{x}}{dt} = \begin{pmatrix} 2 & 6 \\ 2 & 3 \end{pmatrix}\mathbf{x} + \begin{pmatrix} \cos 3t + \sin 3t \\ \sin 3t \end{pmatrix}$

(c) $\dfrac{d\mathbf{x}}{dt} = \begin{pmatrix} 3 & 1 & 0 \\ 8 & 1 & 0 \\ -1 & -1 & 4 \end{pmatrix}\mathbf{x} + \begin{pmatrix} e^t \\ 2e^t \\ 5e^t \end{pmatrix}$

(d) $\dfrac{d\mathbf{x}}{dt} = \begin{pmatrix} 2 & 4 & 1 \\ 2 & 0 & 7 \\ 8 & 9 & 8 \end{pmatrix}\mathbf{x} + \begin{pmatrix} t^3 - t \\ t^2 + 1 \\ t^3 + 3t - 5 \end{pmatrix}$

‡4. Prove: if A is a diagonal matrix with diagonal elements $\lambda_1, \dots, \lambda_n$, then e^{tA} is a diagonal matrix with diagonal elements $e^{\lambda_1 t}, \dots, e^{\lambda_n t}$.

‡5. Let A be a diagonal matrix with distinct eigenvalues $\lambda_1, \dots, \lambda_n$. Let $g(\lambda) = \det(\lambda I - A)$. Let

$$L_k(A) = \frac{1}{g'(\lambda_k)}(A - \lambda_1 I)\cdots(A - \lambda_{k-1}I)(A - \lambda_{k+1}I)\cdots(A - \lambda_n I)$$

Show that $L_k(A)$ has 0 at all entries except the kk entry, where a 1 occurs. Hence, show that

$$e^{tA} = e^{\lambda_1 t}L_1(A) + \cdots + e^{\lambda_n t}L_n(A)$$

(*Hint.* See Problem 4).

‡6. Let B be an n by n matrix with distinct eigenvalues. Then (see Section 10-21) for some matrix U, $B = UAU^{-1}$, where A is as in Problem 4. Show that

$$L_k(B) = UL_k(A)U^{-1}, \qquad e^{tB} = Ue^{tA}U^{-1}$$

and, hence, that

$$e^{tB} = \sum_{i=1}^{n} e^{\lambda_i t}L_i(B)$$

‡7. Using Problem 6, determine e^{tB} when $B = \begin{pmatrix} 1 & 2 & -5 \\ 0 & -1 & 4 \\ 0 & 0 & 2 \end{pmatrix}$

‡8. Suppose that C is an n by n matrix with λ as its only eigenvalue. Then, $(C - \lambda I)^n = O$ (see Section 10-20) and, therefore, $(C - \lambda I)^k = O$ for $k \geq n$. Show that

$$e^{tC} = e^{\lambda t}\sum_{k=0}^{n-1} \frac{t^k}{k!}(C - \lambda I)^k$$

‡9. Use Problem 8 to determine e^{tC} when

$$C = \begin{pmatrix} 2 & 1 & 3 \\ 0 & 2 & -5 \\ 0 & 0 & 2 \end{pmatrix}$$

‡10. Let A be a real constant n by n matrix with distinct eigenvalues $\alpha_1 \pm i\beta_1, \ldots, \alpha_s \pm i\beta_s, \lambda_{2s+1}, \ldots, \lambda_n$, where $\lambda_{2s+1}, \ldots, \lambda_n$ are real. Let $\mathbf{u}_1, \ldots, \mathbf{u}_s$ be eigenvectors of A associated with $\alpha_1 + i\beta_1, \ldots, \alpha_s + i\beta_s$, let $\mathbf{u}_{2s+1}, \ldots, \mathbf{u}_n$ be eigenvectors of A associated with $\lambda_{2s+1}, \ldots, \lambda_n$. Let $\mathbf{u}_{2s+1}, \ldots, \mathbf{u}_n$ have real components; let $\mathbf{u}_1 = \mathbf{u}_1' + i\mathbf{u}_1'', \ldots, \mathbf{u}_s = \mathbf{u}_s' + i\mathbf{u}_s''$, where $\mathbf{u}_1', \ldots, \mathbf{u}_s''$ have real components. Show that the n functions

$$(*) \qquad \begin{aligned} &e^{\alpha_k t}(\cos \beta_k t\, \mathbf{u}_k' - \sin \beta_k t\, \mathbf{u}_k'') = \operatorname{Re} e^{(\alpha_k + i\beta_k)t}\, \mathbf{u}_k, \qquad k = 1, \ldots, s \\ &e^{\alpha_k t}(\sin \beta_k t\, \mathbf{u}_k' + \cos \beta_k t\, \mathbf{u}_k'') = \operatorname{Im} e^{(\alpha_k + i\beta_k)t}\, \mathbf{u}_k, \qquad k = 1, \ldots, s \\ &e^{\lambda_{2s+1} t}\mathbf{u}_{2s+1}, \quad \ldots, \quad e^{\lambda_n t}\mathbf{u}_n \end{aligned}$$

are real solutions of $d\mathbf{x}/dt = A\mathbf{x}$, linearly independent with respect to real coefficients on $(-\infty, \infty)$. [*Hint.* Show first that $\mathbf{u}_k^* = \mathbf{u}_k' - i\mathbf{u}_k''$ is an eigenvector of A associated with $\alpha_k - i\beta_k$. Then show that the n functions (*) are expressible as linear combinations of the n functions

$$(**) \qquad \begin{aligned} &e^{(\alpha_k + i\beta_k)t}\mathbf{u}_k, \qquad e^{(\alpha_k - i\beta_k)t}\mathbf{u}_k^* \qquad (k = 1, \ldots, s) \\ &e^{\lambda_{2s+1} t}\mathbf{u}_{2s+1}, \quad \ldots, \quad e^{\lambda_n t}\mathbf{u}_n \end{aligned}$$

and also that the functions (**) are expressible as linear combinations of the function (*), so that both sets span $\mathcal{H}$. Hence, the functions are linearly independent with respect to complex coefficients. Now show that this implies that they are linearly independent with respect to real coefficients.]

11. Prove: for all square matrices A, e^A is nonsingular. [*Hint*: e^{tA} is a fundamental matrix solution of $d\mathbf{x}/dt = A\mathbf{x}$.]

14-21 AUTONOMOUS LINEAR SYSTEMS OF ORDER TWO

In this section we consider the linear system

$$\begin{aligned} \frac{dx_1}{dt} &= a_{11}x_1 + a_{12}x_2 \\ \frac{dx_2}{dt} &= a_{21}x_1 + a_{22}x_2 \end{aligned} \qquad \text{or} \qquad \frac{d\mathbf{x}}{dt} = A\mathbf{x} \tag{14-210}$$

Here $A = (a_{ij})$ and the a_{ij} are real constants. Since t does not appear on the right-hand side, the system is autonomous (Section 14-1). Each solution provides a path

$$x_1 = x_1(t), \qquad x_2 = x_2(t), \qquad -\infty < t < \infty \tag{14-211}$$

in the x_1x_2-plane. Thus, t is interpreted as a *parameter* along the path. Physically, t is usually *time*, and we shall often refer to it thus.

If in (14-211) we replace t by $t + h$, where h is a constant, then we obtain another path passing through the same points in the x_1x_2-plane, with a delay in time of h, or an anticipation in time of h, according as h is negative or positive. The delayed or anticipated path is also a solution of (14-210). For by (14-210) $\mathbf{x}'(t) = A\mathbf{x}(t)$ for all t, and hence, $\mathbf{x}'(t+h) = A\mathbf{x}(t+h)$: thus, by the chain rule,

$$\frac{d}{dt}\mathbf{x}(t+h) = \mathbf{x}'(t+h) = A\mathbf{x}(t+h)$$

For a given point $(x_1^{(0)}, x_2^{(0)})$ in the x_1x_2-plane we can always find a solution (14-211) passing through this point. For we need only choose the solution (14-211) with initial values $x_1(0) = x_1^{(0)}$, $x_2(0) = x_2^{(0)}$. If

$$x_1 = x_1^*(t), \qquad x_2 = x_2^*(t)$$

is another solution passing through $(x_1^{(0)}, x_2^{(0)})$, say for $t = t_0$, then we must have

$$x_1^*(t) = x_1(t - t_0), \qquad x_2^*(t) = x_2(t - t_0)$$

For both $(x_1^*(t), x_2^*(t))$ and $(x_1(t - t_0), x_2(t - t_0))$ are solutions of (14-210), and they agree for $t = t_0$; hence, by the uniqueness of solutions, they agree for all t. Therefore, apart from delay or anticipation in time, there is just one solution path through each point $(x_1^{(0)}, x_2^{(0)})$ in the plane.

Remark. One can eliminate t from (14-210) to obtain the single first order equation

$$\frac{dx_2}{dx_1} = \frac{a_{21}x_1 + a_{22}x_2}{a_{11}x_1 + a_{12}x_2} \tag{14-212}$$

The solutions (14-211) are parametric representations of the solutions of (14-212). We note that (14-212) is of the type $y' = g(y/x)$ (homogeneous first-order equation) studied in Section 14-4. Accordingly, as pointed out in that section, apart from rays starting at (0, 0), the solutions are a family of similar curves (see also Problem 4 below).

We now proceed to study the configurations formed by the solution paths. We first consider an example.

EXAMPLE 1
$$\frac{d\mathbf{x}}{dt} = \begin{pmatrix} 1 & -2 \\ 3 & -4 \end{pmatrix}\mathbf{x}$$

The characteristic equation is

$$\begin{vmatrix} 1-\lambda & -2 \\ 3 & -4-\lambda \end{vmatrix} = 0 \qquad \text{or} \qquad \lambda^2 + 3\lambda + 2 = 0$$

The eigenvalues are -1 and -2. As in Section 14-17, we find linearly independent solutions $\mathbf{x} = (e^{-t}, e^{-t})$ and $\mathbf{x} = (2e^{-2t}, 3e^{-2t})$. Hence, the general solution is

$$\mathbf{x} = \begin{pmatrix} e^{-t} & 2e^{-2t} \\ e^{-t} & 3e^{-2t} \end{pmatrix}\begin{pmatrix} c_1 \\ c_2 \end{pmatrix}$$

or

$$x_1 = c_1e^{-t} + 2c_2e^{-2t}, \qquad x_2 = c_1e^{-t} + 3c_2e^{-2t}$$

For $c_1 = 1$, $c_2 = 1$, we obtain the particular solution

$$x_1 = e^{-t} + 2e^{-2t}, \qquad x_2 = e^{-t} + 3e^{-2t}$$

We note that $x_1 \to 0$ and $x_2 \to 0$ as $t \to \infty$, and that $x_1 \to \infty$ and $x_2 \to \infty$ as $t \to -\infty$. Other properties of the graph are found with the aid of derivatives, as in Section 6-4. This solution and the ones for other choices of c_1, c_2 are graphed in Figure 14-23. The arrows point in the direction of increasing t.

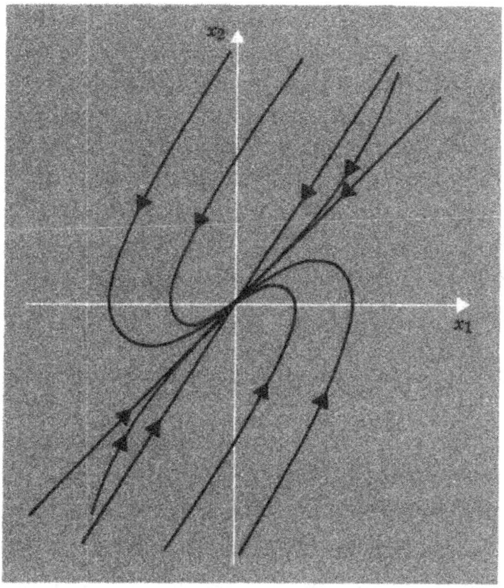

Figure 14-23 Node.

We note that four of the solutions are rays. Also for $c_1 = 0$, $c_2 = 0$ we obtain the *zero solution* $x_1 = 0$, $x_2 = 0$, which remains at the origin for all t; hence, we refer to the origin as an *equilibrium point*. Apart from the rays and the zero solutions, the solutions form a family of similar curves; having drawn one such curve, one can draw others by replacing each point (x_1, x_2) on the first curve by (kx_1, kx_2) for constant k (positive or negative). It is clear that all solutions approach the origin as $t \to \infty$, so that the equilibrium point is *stable*. Since the eigenvalues λ_1, λ_2 are both negative, this is to be expected (Section 14-18).

For the general system (14-210), we can classify the possible configurations according to the nature of the eigenvalues. The main cases are the following:

I, *Node.* (a) $\lambda_2 < \lambda_1 < 0$ or (b) $0 < \lambda_1 < \lambda_2$

II, *Saddle Point.* $\lambda_1 < 0$, $\lambda_2 > 0$

III, *Focus.* $\lambda = \alpha \pm i\beta$, and (a) $\alpha < 0$ or (b) $\alpha > 0$

IV, *Center.* $\lambda = \pm\beta i$, $\beta > 0$

This list omits certain borderline cases such as $\lambda_1 < 0$ and $\lambda_2 = 0$. We illustrate some of the missing cases in the exercises below. For applications, the cases listed are those of major interest.

In the discussion to follow, various properties of the solutions will be stated without proof. For further explanation one is referred to Chapter 11 of ODE.

CASE I: NODE. (a) *Eigenvalues unequal and negative.* Here our solutions are given by

$$\begin{aligned} x_1 &= c_1 h_1 e^{\lambda_1 t} + c_2 k_1 e^{\lambda_2 t} \\ x_2 &= c_1 h_2 e^{\lambda_1 t} + c_2 k_2 e^{\lambda_2 t} \end{aligned} \qquad \lambda_2 < \lambda_1 < 0 \tag{14-213}$$

Here (h_1, h_2) is an eigenvector associated with λ_1 and (k_1, k_2) is an eigenvector associated with λ_2. This case is illustrated in Example 1 above (Figure 14-23), and the chief properties observed hold true in general. For example, $x_1 \to 0$ and $x_2 \to 0$ as $t \to \infty$, since the eigenvalues are negative, and the origin is a stable equilibrium point. There are four solutions that are rays: those for $c_1 = \pm 1$, $c_2 = 0$ and those for $c_1 = 0$, $c_2 = \pm 1$. For the first two, $x_1 = \pm h_1 e^{\lambda_1 t}$, $x_2 = \pm h_2 e^{\lambda_1 t}$, and the rays pass through $(\pm h_1, \pm h_2)$; for the second two, the rays pass through $(\pm k_1, \pm k_2)$. Thus the two eigenvectors serve to locate the four ray solutions. For a solution other than a ray or the zero solution, we observe that

$$\frac{x_2}{x_1} = \frac{c_1 h_2 e^{\lambda_1 t} + c_2 k_2 e^{\lambda_2 t}}{c_1 h_1 e^{\lambda_1 t} + c_2 k_1 e^{\lambda_2 t}} = \frac{c_1 h_2 + c_2 k_2 e^{(\lambda_2 - \lambda_1)t}}{c_1 h_1 + c_2 k_1 e^{(\lambda_2 - \lambda_1)t}}$$

and, since $\lambda_2 < \lambda_1$, $x_2/x_1 \to h_2/h_1$ as $t \to \infty$. This implies that the solution approaches the origin in a direction tangent to the ray solution $x_1 = c_1 h_1 e^{\lambda_1 t}$, $x_2 = c_1 h_2 e^{\lambda_1 t}$; this is illustrated in Figure 14-23.

CASE I: NODE. (b) *Eigenvalues unequal and positive.* Here $0 < \lambda_1 < \lambda_2$; the discussion is the same as that for Case Ia except that the direction of time is reversed and $t \to \infty$ is to be replaced by $t \to -\infty$ throughout. The configuration is as in Figure 14-23, except for the reversal of direction of the paths. The origin becomes an *unstable* equilibrium point.

CASE II: SADDLE POINT. *Eigenvalues of opposite sign.* The solutions are again given by (14-213), except that now $\lambda_1 < 0 < \lambda_2$. We again have ray solutions $\mathbf{x} = \pm e^{\lambda_1 t}(h_1, h_2)$ and $\mathbf{x} = \pm e^{\lambda_2 t}(k_1, k_2)$. Each of the first two approaches the origin as $t \to \infty$, since $\lambda_1 < 0$; each of the second two approaches the origin as $t \to -\infty$, since $\lambda_2 > 0$. There is again a zero solution, so that the origin is an equilibrium point. For each solution not a ray or the zero solution, as $t \to \pm\infty$ one or both of x_1, x_2 must have an infinite limit; for as $t \to \infty$, $e^{\lambda_1 t} \to 0$ and $e^{\lambda_2 t} \to \infty$, and there is a similar reasoning for $t \to -\infty$. A typical example is graphed in Figure 14-24. The configuration is called a *saddle point.* In this case, the origin is clearly an *unstable* equilibrium point.

CASE III: FOCUS. (a) $\lambda = \alpha \pm i\beta$, $\alpha < 0$, $\beta > 0$. If $(h_1 + ih_2, k_1 + ik_2)$ is an eigenvector corresponding to $\lambda = \alpha + i\beta$, then a complex solution is the vector function $((h_1 + ih_2)e^{(\alpha + i\beta)t}, (k_1 + ik_2)e^{(\alpha + i\beta)t})$ and, hence, linearly inde-

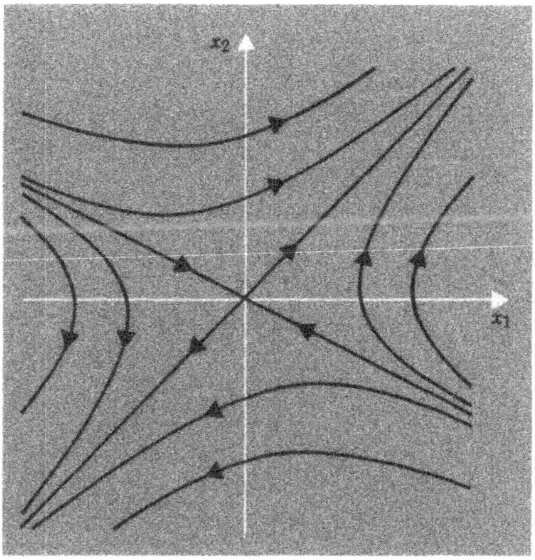

Figure 14-24 Saddle point.

pendent real solutions are given by

$$\begin{pmatrix} e^{\alpha t}(h_1 \cos \beta t - h_2 \sin \beta t) \\ e^{\alpha t}(k_1 \cos \beta t - k_2 \sin \beta t) \end{pmatrix}, \quad \begin{pmatrix} e^{\alpha t}(h_2 \cos \beta_t + h_1 \sin \beta t) \\ e^{\alpha t}(k_2 \cos \beta t + k_1 \sin \beta t) \end{pmatrix}$$

and the general real solution is given by

$$\begin{aligned} x_1 &= c_1 e^{\alpha t}(h_1 \cos \beta t - h_2 \sin \beta t) + c_2 e^{\alpha t}(h_2 \cos \beta t + h_1 \sin \beta t) \\ x_2 &= c_1 e^{\alpha t}(k_1 \cos \beta t - k_2 \sin \beta t) + c_2 e^{\alpha t}(k_2 \cos \beta t + k_1 \sin \beta t) \end{aligned} \tag{14-214}$$

The following example illustrates the nature of the solutions.

EXAMPLE 2
$$\frac{d\mathbf{x}}{dt} = \begin{pmatrix} 0 & 1 \\ -2 & -2 \end{pmatrix} \mathbf{x}$$

The eigenvalues are found to be $-1 \pm i$. An eigenvector corresponding to $-1 + i$ is found to be $(-1, 1 - i)$. Hence, $h_1 = -1, h_2 = 0, k_1 = 1, k_2 = -1$, and the general solution (14-214) becomes

$$\begin{aligned} x_1 &= -c_1 e^{-t} \cos t - c_2 e^{-t} \sin t, \\ x_2 &= c_1 e^{-t}(\cos t + \sin t) + c_2 e^{-t}(-\cos t + \sin t) \end{aligned}$$

The solution with $c_1 = 1$, $c_2 = 0$ is

$$x_1 = e^{-t} \cos t, \; x_2 = e^{-t}(\cos t + \sin t) \tag{14-215}$$

If the factor e^{-t} were not present, we would have the path $x_1 = -\cos t$, $x_2 = \cos t + \sin t$, which follows an ellipse. For

$$x_1 + x_2 = \sin t, \; (x_1 + x_2)^2 = \sin^2 t = 1 - \cos^2 t = 1 - x_1^{\,2}$$

or

$$2x_1^{\,2} + 2x_1x_2 + x_2^{\,2} = 1$$

By the rule of Section 6-5 ($B^2 - 4\,AC < 0$), this second-degree curve is an ellipse. We verify also from the parametric equations that, as t increases, the

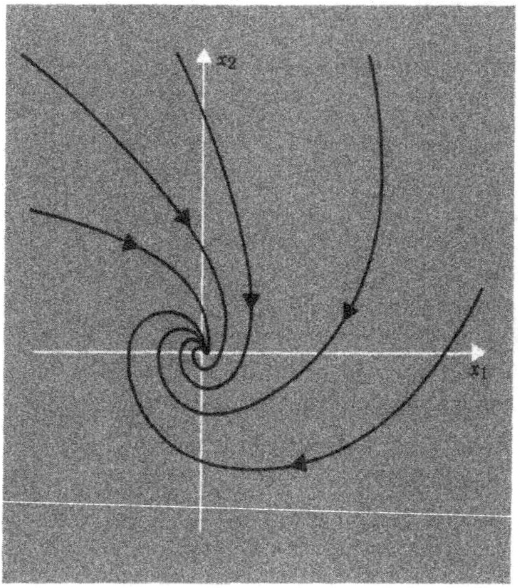

Figure 14-25 Focus (stable).

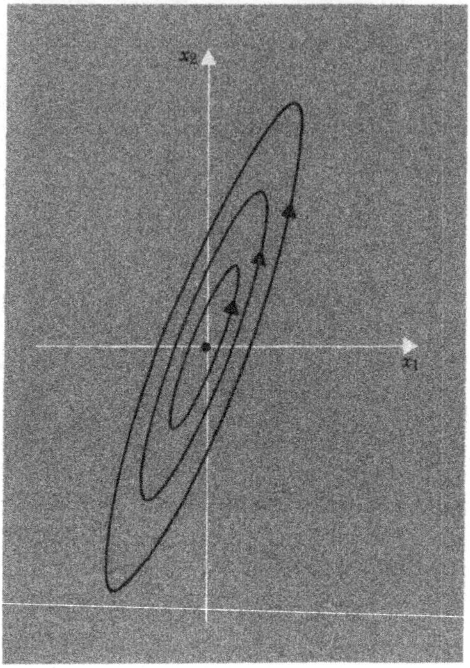

Figure 14-26 Center.

ellipse is traced in the clockwise direction. The actual path (14-215) differs from the elliptical path because of the factor e^{-t}. As t increases, e^{-t} decreases, approaching 0. Therefore, the path must spiral in towards the origin, approaching it as $t \to \infty$. By the similarity of the paths, all other paths are spirals also, and we obtain the configuration of Figure 14-25, called a *focus*. The origin is clearly a stable equilibrium point.

It can be shown that the example is typical of Case III*a*.

CASE III: FOCUS. (b) $\lambda = \alpha \pm i\beta$, $\alpha > 0$, $\beta > 0$. Here the solutions are the same as the preceding, except that the factor $e^{\alpha t}$ increases as t increases. Hence, one obtains the same configuration with arrows reversed. The origin is now *unstable*.

CASE IV: CENTER. $\lambda = \pm\beta i$, $\beta > 0$. This case is similar to Case III, except that the exponential factor $e^{\alpha t}$ is missing, and the paths are similar ellipses. We give an illustration:

EXAMPLE 3
$$\frac{d\mathbf{x}}{dt} = \begin{pmatrix} 2 & -1 \\ 5 & -2 \end{pmatrix} \mathbf{x}$$

The eigenvalues are $\pm i$. For $\lambda = i$, an eigenvector is $(1, 2 - i)$. Hence, as for Case III, we find the general solution:

$$x_1 = c_1 \cos t + c_2 \sin t, \quad x_2 = c_1(2 \cos t + \sin t) + c_2(2 \sin t - \cos t)$$

For $c_1 = 1$, $c_2 = 0$, the solution is $x_1 = \cos t$, $x_2 = 2 \cos t + \sin t$. Eliminating

t, we obtain

$$5x_1^2 - 4x_1x_2 + x_2^2 = 1$$

Since $16 - 20 = -4 < 0$, the path follows an ellipse. The solutions are graphed in Figure 14-26.

In the case of a center, each solution is bounded, and hence we would call the origin a *neutrally stable* equilibrium point.

Application to Nonlinear Equations. Let an autonomous equation

$$\frac{d\mathbf{y}}{dt} = \mathbf{F}(\mathbf{y}) \tag{14-216}$$

be given, where $\mathbf{y}$ is now a vector of V_n and $\mathbf{F}$ is defined and has a continuous Jacobian matrix $\mathbf{F}_\mathbf{y}$ in an open region D of V_n. Let $\mathbf{F}(\mathbf{y}^0) = \mathbf{0}$. Then $\mathbf{y} \equiv \text{const} = \mathbf{y}^0$ is one solution of the differential equation (14-216); that is, $\mathbf{y}^0$ is an equilibrium point. In many problems, one wishes to study the behavior of the solutions near such an equilibrium point, as we have done for the linear system (14-210). Since $\mathbf{y}$ is to remain close to $\mathbf{y}^0$, it is reasonable to write, as in Section 12-12,

$$\Delta\mathbf{F} = \mathbf{F}(\mathbf{y}) - \mathbf{F}(\mathbf{y}^0) \sim d\mathbf{F} = A(\mathbf{y} - \mathbf{y}^0)$$

where A is the Jacobian matrix $\mathbf{F}_\mathbf{y}$ evaluated at $\mathbf{y}^0$. Since we are assuming $\mathbf{F}(\mathbf{y}^0) = \mathbf{0}$, our relation becomes

$$\mathbf{F}(\mathbf{y}) \sim A(\mathbf{y} - \mathbf{y}^0)$$

and the differential equation (14-216) is being approximated, for $\mathbf{y}$ near $\mathbf{y}^0$, by the equation

$$\frac{d\mathbf{y}}{dt} = A(\mathbf{y} - \mathbf{y}^0) \tag{14-217}$$

We finally set $\mathbf{x} = \mathbf{y} - \mathbf{y}^0$ (that is, we take a new origin at the equilibrium point $\mathbf{y}^0$) and (14-217) becomes the *homogeneous linear equation:*

$$\frac{d\mathbf{x}}{dt} = A\mathbf{x} \tag{14-218}$$

Here A is a constant matrix, so that the theory of Section 14-17 applies.

It has been shown that, except in certain critical cases, the approximating equation (14-218) does give us the main features of the solutions of the original nonlinear differential equation (14-216) near the equilibrium point. In particular, if all eigenvalues of A have negative real part, so that the zero solution of (14-218) is stable, then the equilibrium position $\mathbf{y}^0$ is also stable. (See, for example, Chapter 13 of the book *Theory of Ordinary Linear Differential Equations* by E. A. Coddington and N. Levinson (McGraw-Hill, N.Y., 1955). If $n = 2$, then the approximating linear equation (14-218) has form (14-210) and one can classify cases as above. It can be shown that, if the linear equation has a node, then so does the nonlinear equation; the same holds true for a focus or a saddle point, but fails for a center (see Chapter 11 of ODE).

Application to Nonlinear Vibrations. In various physical problems, one is led to a nonlinear differential equation

$$\frac{d^2x}{dt} + f\left(x, \frac{dx}{dt}\right)\frac{dx}{dt} + g(x) = 0 \tag{14-219}$$

resembling the equation for vibrations considered in Section 14-13. If one writes $x_1 = x$, $x_2 = dx/dt$, then the equation is replaced by a system of two equations

$$\frac{dx_1}{dt} = x_2, \qquad \frac{dx_2}{dt} = -x_2 f(x_1, x \text{ [illegible] } _1)$$

which are a special case of (14-216). Hence, the solutions of (14-219) can be studied in the x_1x_2-plane, usually termed the *phase plane.*

PROBLEMS

1. (a) Verify the graph in Figure 14-23 of the solutions of Example 1.
   (b) Verify that the system $dx_1/dt = 2x_1 + 3x_2$, $dx_2/dt = 2x_1 + 2x_2$ is of type II, find the general solution and verify the graph of the solutions, given in Figure 14-24.
   (c) Verify the graph in Figure 14-25 of the solutions of Example 2.
   (d) Verify the graph in Figure 14-26 of the solutions of Example 3.
2. For each of the following choices of the matrix A, find the general solution of the differential equation $d\mathbf{x}/dt = A\mathbf{x}$, and state whether one has a node, a focus, a saddle point, or a center, and state whether the origin is a stable equilibrium point.

   (a) $\begin{pmatrix} -3 & 4 \\ -2 & 3 \end{pmatrix}$ (b) $\begin{pmatrix} -5 & 4 \\ -2 & 1 \end{pmatrix}$ (c) $\begin{pmatrix} -2 & -2 \\ 4 & 2 \end{pmatrix}$ (d) $\begin{pmatrix} 4 & -2 \\ 1 & 1 \end{pmatrix}$

   (e) $\begin{pmatrix} -1 & 1 \\ -5 & 3 \end{pmatrix}$ (f) $\begin{pmatrix} 3 & 3 \\ -6 & 3 \end{pmatrix}$ (g) $\begin{pmatrix} -6 & 4 \\ -2 & 0 \end{pmatrix}$ (h) $\begin{pmatrix} -11 & 16 \\ -8 & 13 \end{pmatrix}$
3. For each of the following choices of the matrix A, the differential equation $d\mathbf{x}/dt = A\mathbf{x}$ illustrates a case not included in the cases I to IV discussed in the text. Find the general solution and graph the solutions:

   (a) $\begin{pmatrix} -1 & 1 \\ -4 & 3 \end{pmatrix}$ (b) $\begin{pmatrix} -1 & 2 \\ -1 & 2 \end{pmatrix}$ (c) $\begin{pmatrix} -3 & 0 \\ 0 & -3 \end{pmatrix}$

   (d) $\begin{pmatrix} -2 & 2 \\ -1 & 1 \end{pmatrix}$ (e) $\begin{pmatrix} 0 & 0 \\ 0 & 0 \end{pmatrix}$
4. To show the similarity of the graphs of the solutions of equation (14-210), prove: if $x_1 = f_1(t)$, $x_2 = f_2(t)$ is a solution, then so is $x_1 = kf_1(t)$, $x_2 = kf_2(t)$ for each constant k. Interpret the result in terms of vector spaces.
5. For each of the following nonlinear systems, show that the origin is an equilibrium point, find the approximating linear system at the origin, determine the type of configuration for the linear system and, hence, determine whether the origin is a stable equilibrium point for the given nonlinear system.

(a) $\dfrac{dx_1}{dt} = -3 \sin x_1 + 4(e^{x_2} - 1), \qquad \dfrac{dx_2}{dt} = -2 \sin x_1 + 3(e^{x_2} - 1)$

[*Hint*. See Problem 2(a).]

(b) $\dfrac{dx_1}{dt} = -5x_1 + 4x_2 + x_1x_2, \qquad \dfrac{dx_2}{dt} = -2x_1 + x_2 + x_1^2 - x_2^2$

[*Hint*. See Problem 2(b).]

(c) $\dfrac{dx_1}{dt} = \dfrac{x_1 - x_2}{1 + x_1 + x_2}, \qquad \dfrac{dx_2}{dt} = \dfrac{3x_1 + x_2}{1 + x_1 + x_2}$

(d) $\dfrac{dx_1}{dt} = e^{x_1x_2} \sin (x_1 + x_2^2), \qquad \dfrac{dx_2}{dt} = \cos (x_1 - x_2) - 1 + 2x_2$

14-22 POWER SERIES SOLUTIONS

We recall that the Taylor series (Section 8-20) of a function $f(x)$ with center x_0 is the series

$$f(x_0) + f'(x_0)(x - x_0) + f''(x_0)\frac{(x - x)^2}{2!} + \cdots + f^{(n)}(x_0)\frac{(x - x_0)^n}{n!} + \cdots \qquad (14\text{-}220)$$

Under appropriate conditions, this series converges to $f(x)$ in an interval with center x_0.

Now if $y = f(x)$ is a solution of a differential equation

$$y' = F(x, y) \qquad (14\text{-}221)$$

and F is appropriately differentiable, then we can obtain the value $f(x_0) = y_0$ from a given initial condition and can obtain all higher derivatives of f at x_0 from the differential equation itself. First the equation gives us

$$f'(x_0) = F(x_0, y_0) = y_0'$$

Next we *differentiate* the differential equation to obtain an expression for y''. By the chain rule, $y'' = F_x + F_y y'$, so that

$$f''(x_0) = F_x(x_0, y_0) + F_y(x_0, y_0)y_0' = y_0''$$

By repeated differentiation, we can find $y^{(n)}$ and, hence, obtain $f^{(n)}(x_0)$ in terms of the values previously found. Therefore, *we can in principle construct the Taylor series* (14-220) and, hence, if we know that the series converges to $f(x)$, we can obtain the desired solution of (14-221) as the sum of a power series. A similar reasoning applies to equations of higher order and to simultaneous equations. In all cases, the initial values get us started, and the differential equation gives us information by which we can successively obtain as many derivatives of the solution as desired. We proceed to illustrate the formal process by examples.

EXAMPLE 1. $y' = x + y + y^2$, $y = 0$ for $x = 0$.

Here $y' = 0 + 0 + 0 = 0$ for $x = 0$. In general,

$$\begin{aligned} y'' &= 1 + y' + 2yy' \\ y''' &= y'' + 2y'^2 + 2yy'' \\ y^{(4)} &= y''' + 6y'y'' + 2yy''' \\ y^{(5)} &= y^{(4)} + 8y'y''' + 6y''^2 + 2yy^{(4)} \\ y^{(6)} &= y^{(5)} + 10y'y^{(4)} + 20y''y''' + 2yy^{(5)}, \quad \ldots \end{aligned}$$

Hence, with $x = 0$, $y = 0$, $y' = 0$, we find that $y'' = 1$, $y''' = 1$, $y^{(4)} = 1$, $y^{(5)} = 7$, $y^{(6)} = 27$. Hence, our solution is

$$f(x) = 0 + 0x + \frac{x^2}{2!} + \frac{x^3}{3!} + \frac{x^4}{4!} + \frac{7x^5}{5!} + \frac{27x^6}{6!} + \cdots$$

We note that it is difficult to find a general formula for the nth term of this series.

EXAMPLE 2. $y'' = xy$, with $y = 1$, $y' = 2$ for $x = 3$.
Here $y'' = 3$ for $x = 3$ and

$$y''' = xy' + y, \quad y^{(4)} = xy'' + 2y', \quad y^{(5)} = xy''' + 3y'', \quad \ldots$$

and clearly, in general,

$$y^{(n)} = xy^{(n-2)} + (n-2)y^{(n-3)}.$$

Accordingly at $x = 3$, we have $y = 1$, $y' = 2$, $y'' = 3$ and $y''' = 7$, $y^{(4)} = 13$, $y^{(5)} = 30$, $y^{(6)} = 67$, and our solution is

$$f(x) = 1 + 2(x-3) + \frac{3(x-3)^2}{2!} + \frac{7(x-3)^3}{3!} + \frac{13(x-3)^4}{4!} + \frac{30(x-3)^5}{5!} + \frac{67(x-3)^6}{6!} + \cdots$$

EXAMPLE 3. $(dx/dt) = xy$, $(dy/dt) = x - y$; $x = 1$, $y = 0$ for $t = 0$.
Here $x'(0) = 0$, $y'(0) = 1$. In general (with $' = d/dt$)

$$\begin{aligned} &x' = xy, \quad y' = x - y, \quad x'' = xy' + x'y, \quad y'' = x' - y', \\ &x''' = xy'' + 2x'y' + x''y, \quad y''' = x'' - y'', \quad \ldots \end{aligned}$$

and, hence, for $t = 0$: $x = 1$, $y = 0$, $x' = 0$, $y' = 1$, $x'' = 1$, $y'' = -1$, $x''' = -1$, $y''' = 2$, and our solution is

$$x = 1 + 0t + \frac{t^2}{2!} - \frac{t^3}{3!} + \cdots,$$

$$y = 0 + t - \frac{t^2}{2!} + 2\frac{t^3}{3!} + \cdots$$

From these examples it is clear that formally the power series for the solution is uniquely determined and can be found explicitly for as many terms as are desired. It can be shown that, for each of these cases, the series converges in some interval about the chosen initial value x_0 and does represent the unique solution with the given initial values. In general, for a first-order equation (14-221), if we know that $F(x, y)$ is analytic in a neighborhood of (x_0, y_0)—that is, representable by its Taylor series (Section 12-21) in such a neighborhood, then the formal power series obtained for the solution does converge and represents the solution in some interval $|x - x_0| < h$. There are similar statements for equations of higher order and simultaneous equations. For proofs of these assertions, see Chapter 12 of ODE.

Case of Linear Equations. For linear differential equations, the formal process of finding power series solutions can be simplified, and in many cases one can find formulas for the nth term of the series for the general solution. Also one can be much more specific about the interval in which the series solution is valid. For information on this point, see Section 12-14 of ODE.

EXAMPLE 4. $y'' + 3xy' + y = 0$. We seek a solution in the form of a power series with center $x = 0$, but leave the initial values of y and y' unspecified. We write

$$y = c_0 + c_1 x + \cdots + c_n x^n + \cdots$$

so that

$$y' = c_1 + 2c_2 x + \cdots + nc_n x^{n-1} + \cdots$$
$$y'' = 2c_2 + 6c_3 x + \cdots + n(n-1)c_n x^{n-2} + \cdots$$

We substitute in the differential equation:

$$2c_2 + 6c_3 x + \cdots + n(n-1)c_n x^{n-2} + \cdots + 3x(c_1 + 2c_2 x + \cdots + nc_n x^{n-1} + \cdots) + c_0 + c_1 x + \cdots + c_n x^n + \cdots = 0$$

We then reason that the left side can be arranged as a power series in x, and the sum of such a series can reduce to 0 over an interval only if the coefficient of x^n is 0 for each n (Section 8-17). The coefficient of x^0 is $2c_2 + c_0$ and that of x^1 is $6c_3 + 4c_1, \ldots$ so that we can conclude:

$$2c_2 + c_0 = 0, \qquad 6c_3 + 4c_1 = 0, \qquad 12c_4 + 7c_2 = 0$$

and in general

$$n(n-1)c_n + (3n-5)c_{n-2} = 0$$

Hence

$$c_2 = -\frac{1}{2}c_0, \qquad c_3 = -\frac{4}{6}c_1, \qquad c_4 = -\frac{7}{12}c_2$$

and, in general, there is a *recursion formula:*

$$c_n = -\frac{3n-5}{n(n-1)}c_{n-2}$$

Accordingly, if n is even,

$$c_n = \frac{-(3n-5)}{n(n-1)} \cdot \frac{-(3n-11)}{(n-2)(n-3)} \cdots \frac{-7}{4 \cdot 3} \cdot \frac{-1}{2 \cdot 1} c_0$$

and if n is odd,

$$c_n = \frac{-(3n-5)}{n(n-1)} \cdot \frac{-(3n-11)}{(n-2)(n-3)} \cdots \frac{-4}{3 \cdot 2} \cdot c_1$$

Here c_0 and c_1 can be chosen arbitrarily; they are just the initial values of y and y' for $x = 0$. Our general solution can now be written

$$y = c_0\left(1 + \sum_{\substack{n=2 \\ n \text{ even}}}^{\infty} \frac{(-1)(-7)\cdots(-3n+5)}{n!} x^n\right) + c_1\left(x + \sum_{\substack{n=3 \\ n \text{ odd}}}^{\infty} \frac{(-4)(-10)\cdots(-3n+5)}{n!} x^n\right)$$

or, with $n = 2k$ in the first series and $n = 2k - 1$ in the second,

$$y = c_0\left(1 + \sum_{k=1}^{\infty} \frac{(-1)^k 1 \cdot 7 \cdots (6k-5)}{(2k)!} x^{2k}\right) + c_1\left(x + \sum_{k=2}^{\infty} \frac{(-1)^k 4 \cdot 10 \cdots (6k-8)}{(2k-1)!} x^{2k-1}\right)$$

The test ratio for the first series is $(6k-5)x^2/[2k(2k-1)]$; this has limit 0 for all x, as $k \to \infty$. Therefore, the series converges for all x. The same situation prevails for the second series. Accordingly, all the formal steps are valid and the two parentheses represent solutions $y_1(x)$ and $y_2(x)$ for all x. The two functions are linearly independent, since neither is a constant times the other. Therefore, we, in fact, have found the general solution of the differential equation.

In the above example, we note that the equation can be written

$$y'' = -3xy' - y = F(x, y, y')$$

and the right-hand side is well-defined and continuous for all x, y, y'. In a number of important linear differential equations, solving for the highest derivative leads to discontinuities at precisely the x-values of greatest interest. The following example illustrates this difficulty. We show how, despite the difficulty, solutions can be found with the aid of power series.

EXAMPLE 5. $x^2y'' + xy' + (x^2 - k^2)y = 0$, where k is a nonnegative constant. This equation is known as the *Bessel equation*. If we solve for y'', we obtain

$$y'' = \frac{-xy' - (x^2 - k^2)y}{x^2} = F(x, y, y')$$

and F is discontinuous for $x = 0$, so that the existence theorem is not applicable with 0 as initial value for x; one says that the equation has a *singular point* at $x = 0$. It has been found that a solution can be found of the form

$$y_1(x) = x^k(1 + c_1x^2 + c_2x^4 + \cdots + c_nx^{2n} + \cdots) \qquad (14\text{-}222)$$

where the series converges for all x. If k is a nonnegative integer, this is actually a power series for $y_1(x)$. However, k may be a fraction such as $\frac{1}{2}$, and then our solution is not itself a power series in x. To verify the correctness of our assertion, we simply substitute (14-222) in the differential equation and try to determine the c's:

$$y_1(x) = x^k + c_1x^{k+2} + c_2x^{k+4} + \cdots + c_nx^{k+2n} + \cdots,$$

$$y_1'(x) = kx^{k-1} + c_1(k+2)x^{k+1} + c_2(k+4)x^{k+3} + \cdots + c_n(k+2n)x^{k+2n-1} + \cdots$$

$$y_1''(x) = k(k-1)x^{k-2} + c_1(k+2)(k+1)x^k + c_2(k+4)(k+3)x^{k+2} + \cdots + c_n(k+2n)(k+2n-1)x^{k+2n-2} + \cdots$$

Hence

$$\begin{aligned} x^2y_1'' + xy_1' + (x^2 - k^2)y_1 &= k(k-1)x^k + c_1(k+2)(k+1)x^{k+2} \\ &+ c_2(k+4)(k+3)x^{k+4} + \cdots + c_n(k+2n)(k+2n-1)x^{k+2n} + \cdots \\ &+ kx^k + c_1(k+2)x^{k+2} + c_2(k+4)x^{k+4} + \cdots + c_n(k+2n)x^{k+2n} \\ &+ \cdots + x^{k+2} + c_1x^{k+4} + c_2x^{k+6} + \cdots + c_nx^{k+2n+2} + \cdots \\ &- k^2x^k - k^2c_1x^{k+2} - k^2c_2x^{k+4} - \cdots - k^2c_nx^{k+2n} - \cdots = 0 \end{aligned}$$

The lowest power of x occurring is x^k. Its coefficient is $k(k-1) + k - k^2 = 0$. Hence, this term drops out. For x^{k+2} the coefficient is

$$c_1(k+2)(k+1) + c_1(k+2) + 1 - k^2c_1 = c_1(4k+4) + 1$$

and on equating this to zero gives

$$c_1 = \frac{-1}{4(k+1)}$$

The coefficient of x^{k+2n} is

$$c_n(k+2n)(k+2n-1) + c_n(k+2n) + c_{n-1} - k^2c_n$$

and equating this to zero gives

$$c_n = -\frac{c_{n-1}}{4n(k+n)}$$

Therefore, with c_1 as above,

$$\begin{aligned} c_n &= \frac{-1}{4n(k+n)} \cdot \frac{-1}{4(n-1)(k+n-1)} \cdots \frac{-1}{4 \cdot 2(k+2)} \cdot \frac{-1}{4(k+1)} \\ &= \frac{(-1)^n}{4^n n!(k+1)(k+2)\cdots(k+n)} \end{aligned}$$

and

$$y_1(x) = x^k\left(1 + \sum_{n=1}^{\infty} \frac{(-1)^n x^{2n}}{4^n n!(k+1)(k+2)\cdots(k+n)}\right)$$

By the ratio test the series converges for all x and, hence, the solution is valid whenever x^k is twice differentiable (for all x if k is a nonnegative integer; for $x > 0$, at least, otherwise).

If k is not a positive integer or 0, a second solution $y_2(x)$ is obtained from $y_1(x)$ by replacing k by $-k$; the two functions $y_1(x)$, $y_2(x)$ are then linearly independent (say for $x > 0$). If k is a positive integer or 0, a second linearly independent solution can be found in a quite different form. The function $y_1(x)$ is, apart from a constant factor, *a Bessel function of the first kind.*

For more information on linear equations with singular points, see Chapter 9 of ODE.

PROBLEMS

1. Find the terms through the term in $(x - x_0)^5$ of the solution, in the form of a power series with given initial value at x_0:
   (a) $y' = y + y^2 - e^{2x}$, $y = 1$ for $x = x_0 = 0$.
   (b) $y' = 1 + \sec^2 x + (y^2/4)$, $y = 0$ for $x = 0$.
   (c) $y' = y + e^{-y} - \ln x$, $y = 0$ for $x = 1$.
   (d) $y' = -y + xy^3 + y^3$, $y = \frac{1}{2}$ for $x = 1$.
   (e) $y'' = e^{-x}yy'$, $y = 1$ and $y'' = 1$ for $x = 0$.
   (f) $y'' = x^{-3} + y^3$, $y = 1$ and $y' = -1$ for $x = 1$.
2. Find the term through $(t - t_0)^3$ of the power series solution with given initial value at t_0:
   (a) $x'(t) = x^2 + y^2 - 1 + \cos t$, $y'(t) = -x + y - \cos t$, $x = 0$ and $y = 1$ for $t = t_0 = 0$.
   (b) $x' = x - t + e^{-t}y$, $y' = e^x + y - e^t$, $x = 1$ and $y = e$ for $t = 1$.
   (c) $x' = tz^2 + e^x - t$, $y' = t^3z + 2e^{2x}$, $z' = y - z^2 - t^3$, $x = 0$, $y = 1$, and $z = 1$ for $t = 1$.
   (d) $x' = 3t^2 + x^2 - y^2$, $y' = 3x^2z^{-2}$, $z' = 2yz^{-1}$, $x = 1$, $y = 1$ and $z = 1$ for $t = 1$.
3. Find the *curvature* at the specified point of the solution of the given differential equation passing through that point:
   (a) $y' = x^2 + 2y^3$ at $(1, 2)$ (b) $y' = e^{xy} + x^2$ at $(2, 5)$
4. Find the general solution in terms of power series with center at $x = 0$ and verify that the series found converge for all x:
   (a) $y'' + xy' - y = 0$ (b) $y'' + x^2y' + xy = 0$
   (c) $x^2y'' - 4xy' + 6y = 0$ (d) $xy'' - 3y' = 0$
5. Let a, b, and c be constants, where c is not a negative integer or 0. Find a solution of the differential equation:

$$x(1 - x)y'' + [c - (a + b + 1)x]y' - aby = 0$$

   of form $y = 1 + \sum_{n=1}^{\infty} c_n x^n$ and show that the series converges for $|x| < 1$.

Remark. The differential equation has a singular point for $x = 0$, also for $x = 1$. The differential equation is called the *hypergeometric equation*, and the series solution is called the *hypergeometric series*. For $a = 1$ and $b = c$ it reduces to the geometric series $\sum_{n=0}^{\infty} x^n$.

6. Let the differential equation $x^2y'' + x^3y' - 6y = 0$ be given (with singular point at $x = 0$).

(**a**) Find a solution of the form $x^3\left(1 + \sum_{k=1}^{\infty} c_k x^{2k}\right)$ and show that the series converges for all x.

(b) Find a solution of the form $x^{-2}\left(1 + \sum_{k=1}^{\infty} c_k x^{2k}\right)$ and show that the series converges for all x.

(c) Show that in general a solution of the form $x^m\left(1 + \sum_{n=1}^{\infty} c_n x^n\right)$ can be found only for $m = -2$ and $m = 3$.

7. (a) To expand $y = \tan x$ in a power series with center $x = 0$ one can reason that $y' = \sec^2 x = 1 + y^2$ and $y = 0$ for $x = 0$, and then proceed as for Example 1. Show in this way that $\tan x = x + (x^3/3) + (2x^5/15) + \cdots$

(**b**) Proceed as in part (a) for $\sec x$ with the aid of the differential equation $y' = y^2 \sin x$.

(c) Find the series for $\sec x$ with the aid of the second-order differential equation $y'' = 2y^3 - y$.

8. Let A_0 and A_1 be n by n matrices. Show that there is a (formal) power series solution of the differential equation:

$$\frac{d\mathbf{x}}{dt} = (A_0 + tA_1)\mathbf{x}$$

with $\mathbf{x}(0) = \mathbf{c}_0$ of form $\sum_{m=0}^{\infty} t^m \mathbf{c}_m$, where $\mathbf{c}_{m+1} = \{1/(m+1)\}[A_0\mathbf{c}_m + A_1\mathbf{c}_{m-1}]$, $m = 1, 2, \ldots$ $\mathbf{c}_1 = A_0\mathbf{c}_0$, and find the series explicitly for the case $A_1 = O$.

14-23 NUMERICAL SOLUTION OF DIFFERENTIAL EQUATIONS

In Section 14-2 we introduce a general numerical procedure for obtaining solutions. This procedure, called "*step-by-step integration*" (and also known as *Euler's method*) gives the solution of

$$y' = F(x, y) \tag{14-230}$$

with given initial point (x_0, y_0) as a set of tabulated values $y_0, y_1, y_2, \ldots$ at the points $x_0, x_1, x_2, \ldots$. The x-values are chosen as desired, as an increasing or decreasing sequence, and then the y-values are calculated successively by the formula

$$y_{n+1} = y_n + F(x_n, y_n)(x_{n+1} - x_n) \tag{14-231}$$

For many applications this method is sufficiently accurate. In any case, the accuracy can be made as good as desired by taking the x-values very close together; however, to cover a given interval of x, one must then use many steps and, hence, must pay for the accuracy through an increased amount of calculation.

There are a number of other methods for obtaining the solution numerically. For a given amount of calculation (measured, for example, by time required on a digital computer), they in general provide much better accuracy than the step-by-step method and, hence, are preferred for most applications. We here briefly describe two such methods and at the end of the section give references for fuller discussion of these and other methods.

We restrict attention to the equation (14-230) and for simplicity assume that the x-values form an increasing sequence and are equally spaced: $x_1 = x_0 + h$, $x_2 = x_0 + 2h, \ldots, x_n = x_0 + nh, \ldots,$ with $h > 0$.

Runge's method. This method uses the formulas:

$$y_{n+1} = y_n + \frac{h}{6}(m_0 + 4m_1 + m_3)$$

where

$$\begin{aligned} m_0 &= F(x_n, y_n), \qquad m_1 = F\left(x_n + \frac{1}{2}h, y_n + \frac{1}{2}m_0 h\right) \\ m_2 &= F(x_n + h, y_n + m_0 h), \qquad m_3 = F(x_n + h, y_n + m_2 h) \end{aligned} \tag{14-232}$$

One can think of the formulas as giving $y_{n+1} = y_n + mh$, where m is an appropriate weighted average of the slopes $y' = F(x, y)$ at three appropriately chosen points. For the equation $y' = F(x)$, the formulas are equivalent to Simpson's rule (Section 7-22).

EXAMPLE 1. $y' = y + 1 = F(x, y)$ with $y(0) = 0$. We choose $h = 0.1$ and $x_0 = 0$, $x_1 = 0.1$, $x_2 = 0.2, \ldots$. Then for $n = 0$, formulas (14-232) give

$$m_0 = F(0, 0) = 1, \qquad m_1 = F(0.05, 0.05) = 1.05$$
$$m_2 = F(0.1, 0.1) = 1.1, \qquad m_3 = F(0.1, 0.11) = 1.11$$
$$y_1 = 0 + \frac{0.1}{6}(1 + 4.2 + 1.11) = 0.1052$$

Similarly, for $n = 1$,

$$m_0 = F(0.1, 0.1052) = 1.1052, \qquad m_1 = F(0.15, 0.1605) = 1.1605$$
$$m_2 = F(0.2, 0.2157) = 1.2157, \qquad m_3 = F(0.2, 0.2268) = 1.2268$$
$$y_2 = 0.1052 + \frac{0.1}{6}(1.1052 + 4.6420 + 1.2268) = 0.2214$$

Now the exact solution is verified to be $y = e^x - 1$ and from tables (Appendix III of Volume I), $y_1 = 0.1052$, $y_2 = 0.2214$. Hence, we have perfect agreement to 4 decimal places.

Adams' Method. We here describe a special case of the general Adams method. For equation (14-230) we assume the values y_0, y_1, y_2, y_3 have been found (for example, by Runge's method). The method to be described then successively gives us $y_4, y_5, \ldots$. In the formulas we use the special notations $\nabla, \nabla^2, \nabla^3$ for *first, second,* and *third differences:* given a sequence $z_1, z_2, \ldots,$ then

$$\begin{aligned}\nabla z_n &= z_n - z_{n-1},\\ \nabla^2 z_n &= \nabla(\nabla z_n) = \nabla z_n - \nabla z_{n-1} = z_n - 2z_{n-1} + z_{n-2}\\ \nabla^3 z_n &= \nabla^2(\nabla z_n) = \nabla^2 z_n - \nabla^2 z_{n-1}\end{aligned}$$

We note that $\nabla^2 z_n$ depends in part on z_{n-2}, and $\nabla^3 z_n$ in part on z_{n-3}. If now we have found $y_0, y_1, y_2, \ldots, y_n$ $(n \geq 3)$, then we can calculate y'_{n-3}, y'_{n-2}, y'_{n-1}, and y'_n and, hence, can find $\nabla y'_n, \nabla^2 y'_n, \nabla^3 y'_n$. The next value y_{n+1} is given by the formula

$$y_{n+1} = y_n + h\left(y'_n + \frac{1}{2}\nabla y'_n + \frac{5}{12}\nabla^2 y'_n + \frac{3}{8}\nabla^3 y'_n\right) \qquad (14\text{-}233)$$

EXAMPLE 2. $y' = y + 1$, with initial point $(0, 0)$. This is the same problem as Example 1. We assume that we have computed $y_1 = 0.1052$, $y_2 = 0.2214$, $y_3 = 0.3499$ as in Table 14-3. The table also shows $y' = y + 1$ at the corre-

Table 14-3

x	y	y'	$\nabla y'$	$\nabla^2 y'$	$\nabla^3 y'$
0.0	0.0000	1.0000			
0.1	0.1052	1.1052	0.1052		
0.2	0.2214	1.2214	0.1162	0.0110	
0.3	0.3499	1.3499	0.1285	0.0123	0.0013
0.4	0.4919	1.4919	0.1420	0.0135	0.0012
0.5	0.6488	1.6488	0.1569	0.0149	0.0014
0.6	0.8222	1.8222	0.1734	0.0165	0.0016
0.7	1.0132				

sponding points and the first, second, and third difference of y'_n. We note that the column of first differences is obtained by subtracting each value of y' from the one below it and by entering the result opposite the *lower* of the two y' entries; a similar rule applies to each of the last two columns.

We now find that

$$\begin{aligned}y_4 &= y_3 + 0.1\left(y'_3 + \frac{1}{2}\nabla y'_3 + \frac{5}{12}\nabla^2 y'_3 + \frac{3}{8}\nabla^3 y'_3\right)\\ &= 0.3499 + 0.1\left[1.3499 + \frac{1}{2}(0.1285) + \frac{5}{12}(0.0123) + \frac{3}{8}(0.0013)\right]\\ &= 0.4919\end{aligned}$$

From this value we can find y'_4 and the successive differences as shown in the table. Then

$$y_5 = y_4 + 0.1\left(y'_4 + \frac{1}{2}\nabla y'_4 + \frac{5}{12}\nabla^2 y'_4 + \frac{3}{8}\nabla^3 y'_4\right)$$

$$= 0.4919 + 0.1\left[1.4919 + \frac{1}{2}(0.1420) + \frac{5}{12}(0.0135) + \frac{3}{8}(0.0012)\right]$$

$$= 0.6488$$

We note that the exact values are $y_4 = e^{0.4} - 1 = 0.49182$ and $y_5 = e^{0.5} - 1 = 0.64872$. Hence, the error is in the fourth decimal place.

There are general procedures for estimating errors in the various numerical methods. For information on this question and on the whole topic, one is referred to the following books.

F. B. Hildebrand, *Introduction to Numerical Analysis* (McGraw-Hill, N.Y., 1956).

Peter Henrici, *Discrete Variable Methods in Ordinary Differential Equations* (Wiley, N.Y., 1962).

W. E. Milne, *Numerical Solution of Differential Equations* (Wiley, N.Y., 1953).

In each numerical method, we end up with an expression for y_{n+1} in terms of h and certain preceding data. If $F(x, y)$ is analytic [represented by a power series in x and y converging in a neighborhood of (x_n, y_n)], it can be shown that we are in fact finding y_{n+1} as the sum of a power series in h, whose coefficients depend on the value and successive partial derivatives of F at (x_n, y_n). The power series method of Section 14-24 yields a similar expression for the true value of y_{n+1}. In each numerical method, the resulting series is found to agree with the true series up to terms of a certain degree in h. For step-by-step integration, the series agrees only up to the terms of first degree; in Runge's method, they agree up to terms of third degree; in Adams' method, they agree up to terms of fourth degree. These statements indicate that, for sufficiently small h, Runge's method should be much more accurate than step-by-step integration, and Adams' method should be more accurate than Runge's method. The texts cited above explain these relationships in detail.

PROBLEMS

1. Let $f(x)$ be the solution of $y' = x + 2 - y$ through the initial point $(0, 0)$.
   (a) Verify that $f(x) = x + 1 - e^{-x}$ and that $f(0) = 0$, $f(0.1) = 0.195163$, $f(0.2) = 0.381269$, $f(0.3) = 0.559182$, $f(0.4) = 0.729680$, $f(0.5) = 0.893469$.
   (b) Find the solutions numerically for $x = 0.1, 0.2, 0.3, 0.4, 0.5$ by step-by-step integration with $h = 0.1$.
   (c) Find the solution numerically for $x = 0.1$ and 0.2 by Runge's method and compare with the exact values.
   (d) Find the solution numerically for $x = 0.4$ and 0.5 by Adams' method, using the exact values for $x = 0, 0.1, 0.2, 0.3$, and compare the results with the exact values.
   (e) Find $f(x)$ as a power series $\sum_{n=0}^{\infty} c_n x^n$ and calculate $f(0), f(0.1), f(0.2), f(0.3)$ using terms through degree 5 and compare with the exact values.

2. (a), (b), (c), (d), (e) Proceed as in Problem 1 with the differential equation $y' = 2 + (x - y)^2$. For step (a) use $f(x) = x + \tan x$, $f(0) = 0$, $f(0.1) = 0.20033$, $f(0.2) = 0.40271$, $f(0.3) = 0.60934$, $f(0.4) = 0.82279$, $f(0.5) = 1.04630$.

3. The solution of $y' = 2x^2 + (y/x)$ through $(1, 1)$ is $y = x^3$. Study the effect of taking smaller steps in numerical solutions as follows:

   (a) Find y for $x = 5$ in one step of step-by-step integration ($h = 4$).

   (b) Find y for $x = 5$ by two equal steps of step-by-step integration ($h = 2$).

   (c) Find y for $x = 5$ by four equal steps of step-by-step integration ($h = 1$).

4. Carry out steps (a) and (b) of Problem 3, using Runge's method instead of step-by-step integration.

ANSWERS TO SELECTED PROBLEMS

Section 9-1, page 645

1. **(a)** (7, 3), **(c)** (9, 6), **(e)** (26, 9), **(g)** (0, 0).

2. **(c)** (6, 3, 6), (6, 4, 2), (−1, −2, 5).

3. **(a)** (10, −1, 14, 16), **(c)** (0, 0, 0, 0).

4. **(a)** Not a vector space (no zero element), **(c)** vector space,
(e) vector space, **(g)** not a vector space (no zero element).

5. **(a)** $(4 + 2i, -2 + 6i)$, **(c)** i.

Section 9-2, page 650

4. **(a)** Subspace, line through O and (1, 1, 1).
(c) Not a subspace (no zero element), line through (2, 0, 0) parallel to the line in part (a).
(e) Not a subspace (not closed under addition), line segment joining $(-1, -\frac{1}{2}, -3)$ to $(1, \frac{1}{2}, 3)$.

8. and **10.** **(a)** Subspace, **(c)** subspace, **(e)** and **(g)** not a subspace.

Section 9-3, page 654

1. **(a)** Intersection is subspace of all functions with 1 and 0 as zeros.
(c) Intersection is subspace of all constant polynomials.
(e) Intersection is empty, hence not a subspace.

Section 9-4, page 659

1. **(a)** {(2, 2), (3, 2), (3, 3), (4, 3)},
(c) Line segment joining (1, 1) to (2, 1),
(e) Line: $y = 2$, **(g)** Line: $x = 2 + 3t$, $y = 1 + 5t$,
(i) Circle: $(x - 2)^2 + (y - 3)^2 = 1$,
(k) Square: $0 \le x \le 1$, $0 \le y \le 1$, **(m)** The xy-plane.

2. **(a)** Plane parallel to and one unit above xy-plane.
(c) Line: $x = 2 + t$, $y = 2 + t$, $z = 2$.
(e) Plane parallel to given plane.
(g) Plane: $\overrightarrow{OP} = t(1, 1, 2) + s(1, 0, 0)$, **(i)** 3-dimensional space.

4. **(a)** $\{\sin x + \{C\}\}$ **(c)** $\{6x^2 + \{C_1 + C_2x\}\}$.
(e) $\{\frac{1}{6}x^3 + \{C_1 + C_2x + C_3x^2\}\}$.
(g) $\{1 + W\}$, where W is space of functions with 1 as a zero.

Section 9-5, page 664

1. (a) Leader: $(1, 2)$, base space: $t(3, -5)$; or $\{(1, 2) + \{t(3, -5)\}\}$ (line).
   (c) Leader: $(2, 0)$, base space: $t(2, 3)$; or $\{(2, 0) + \{t(2, 3)\}\}$ (line).
   (e) Line: $\{(2, 1, 3) + \{t(-1, 1, 1)\}\}$.
   (g) Plane: $\{(1, 1, 1) + \{t(1, -1, 0) + s(1, 0, -1)\}\}$.
   (i) Plane: $\{(2, 3, 4) + \{t(5, -2, 1) + s(2, 4, -1)\}\}$.
   (k) Line: $\{(1, 2, 5, 2) + \{t(1, -1, 1, 2)\}\}$.
   (m) Plane: $\{(0, 0) + \{s(1, 0) + t(0, 1)\}\}$, ($xy$-plane).
   (o) Plane: $\{(2, 5, 5) + \{s(2, 0, 0) + t(0, 1, 0)\}\}$.
   (q) Plane: $\{(1, 1, 0, 0) + \{s(3, 0, 1, 0) + t(0, 5, 0, 1)\}\}$.
   (s) $\{(2, 0, 0, 0) + \{s(1, -2, 0, 0) + t(1, 0, -2, 0) + r(1, 0, 0, -2)\}\}$.
2. (a) Linear variety $= \{1 + \{x^2 g(x)\}\}$, $g(x)$ any polynomial, (c) no variety.
3. (a) Linear variety $\{x^3 - x + \{C\}\}$.
   (c) Linear variety, actually a subspace $= \{W + U\}$, where W is all scalar multiples of f, where $f(1) = f(2) = f(3) = 1$ and $f = 0$ elsewhere, U is space of all functions that have 1, 2, 3 as zeros.
   (e) A subspace so a linear variety.
4. (a) Linear variety $= \{[-1 + 2^{-n}] + W\}$, $W =$ space of null sequences.
   (c) Subspace, so a linear variety.

Section 9-7, page 670

2. (a), (c) and (e) are linearly independent, (g) is linearly dependent.
3. (a), (e) Linearly independent, (c) Linearly dependent.
4. (a), (c), (e) are bases,
   (g) all vectors are linearly combinations of $\ln x$, hence, $\ln x$ is the basis vector.
5. (a), (e) are bases, (c) $\{(1, 2, 3), (3, 2, 1), (2, 3, 1)\}$.
9. (a) Linearly dependent, (c) and (e) are linearly independent.

Section 9-10, page 679

1. The sets (a) and (g) are bases, sets (c) and (e) are not.
2. (a) X is a linearly independent set, $\{(1, 1, 1, 0), (1, 2, 3, 4), (0, 0, 1, 0), (0, 0, 0, 1)\}$, for example, is a basis for V_4.
   (c) X is a linearly independent set, $\{x + x^3, x^2 + x^6, 1 + x - x^3, 1, x^2, x^4, x^5\}$, for example, is a basis for $\mathcal{P}_6$.
3. (a) 2, (c) 2.
4. (a) $\{x - 1, x(x - 1), x^2(x - 1), x^3(x - 1), x^4(x - 1)\}$ is a basis for V, $\dim V = 4$, this set with 1 is a basis for $\mathcal{P}_5$.
   (c) $\{1, x^2 - 6x, 2x^3 - 9x^2, x^4 - 4x^3, 4x^5 - 15x^4\}$ is a basis for W, $\dim W = 4$. This set together with x is a basis for $\mathcal{P}_5$.

Section 9-11, page 686

2. The mappings (a), (c), (e) are linear, while the mappings (g), (i) are not linear.
5. There are linear mappings satisfying (a), (c), (e).

Section 9-13, page 690

1. (a) Range $= V_3$, Kernel $= V_0$.
(c) Range $= V_1$, Kernel: all continuous functions with 0 as a zero.
(e) Range $= V_2$, Kernel: all continuous functions with 0 and 1 as zeros.
(g) Range $= \mathcal{C}[0, 1]$, Kernel $= V_0$.

2. (a) Constant functions, (c) constant functions.
(e) $\mathcal{P}_2$, (g) $\mathcal{P}$, (i) all polynomials with 5 as constant term.

3. (a) All polynomials with constant term 0. (c) same as (a).
(e) All polynomials with 2, 3, and 4 as zeros.

6. (a) Range $= V_2$, Kernel $=$ Span $((-1, -1, 1))$, pre-image of $(1, 0)$ is $\{(1, 0, 0) + t(-1, -1, 1)\}$.
(c) Range $= V_2$, Kernel $= V_0$, Pre-image of $(1, 0)$ is $(\frac{1}{2}, -\frac{1}{2})$.

Section 9-15, page 695

1. (a) Kernel $= V_0$, Range $= V_3$, nullity $= 0$, rank $= 3$, pre-image of $(1, 0, 0)$ is $(1, 1, 1)$; (c) Kernel $=$ Span $((1, -1, -1))$, Range $=$ Span $((1, 0, 2), (1, 1, 1))$, nullity $= 1$, rank $= 2$, pre-image of $(1, 0, 0)$ is empty.

2. (a) Range $=$ Span $((1, 2, 3), (2, 1, 0))$, Kernel $=$ Span $((0, 1, 0, 0), (1, 0, -2, -3))$, rank $=$ nullity $= 2$, pre-image of $(1, 0, 0)$ is empty, pre-image of $(1, 1, 1)$ is $\frac{1}{2}(1, 0, 0, -1)$.
(c) Range $= V_3$, Kernel $=$ Span $((-21,70,11,7))$, null $= 1$, rank $= 3$, pre-image of $(1, 0, 0) = \{(0, 1, 0, 0) + \text{Kernel}\}$, pre-image of $(1, 1, 1) = \{\frac{1}{14}(7, 7, -3, 0) + \text{Kernel}\}$.

Section 9-18, page 701

1. (a) rank $S = 3$, Kernel $S = V_0$; rank $T = 2$, Kernel $T =$ Span $(\mathbf{e}_1 - \mathbf{e}_2 - \mathbf{e}_3)$.
(c) $(S + T)(\mathbf{e}_1) = 5\mathbf{e}_1 - 3\mathbf{e}_2 + \mathbf{e}_3$, $(S + T)(\mathbf{e}_2) = \mathbf{e}_1 - \mathbf{e}_3$, $(S + T)(\mathbf{e}_3) = 4\mathbf{e}_1 - \mathbf{e}_2 - 3\mathbf{e}_3$, Range $(S + T) = V_3$, Kernel $(S + T) = V_0$, rank $S + T = 3$, null $(S + T) = 0$. $(S + M)(\mathbf{e}_1) = \mathbf{e}_1 + \mathbf{e}_3$. $(S + M)(\mathbf{e}_2) = \mathbf{e}_2 - 2\mathbf{e}_3$, $(S + M)(\mathbf{e}_3) = \mathbf{e}_1 - \mathbf{e}_2 + \mathbf{e}_3$, Range $(S + M) = V_3$, Kernel $(S + M) = V_0$, rank $(S + M) = 3$, null $S + M = 0$.
(f) $A(\mathbf{e}_1) = \mathbf{e}_1$, $A(\mathbf{e}_2) = \mathbf{e}_2$, $A(\mathbf{e}_3) = \mathbf{e}_3$; $B(\mathbf{e}_1) = \mathbf{e}_1$, $B(\mathbf{e}_2) = \mathbf{e}_2$, $B(\mathbf{e}_3) = -\mathbf{e}_3$.

2. (a)

	$\mathbf{e}_1$	$\mathbf{e}_2$	$\mathbf{e}_3$	rank
TS	$3\mathbf{e}_1 - 4\mathbf{e}_3$	$-14\mathbf{e}_1 - 2\mathbf{e}_2 + 25\mathbf{e}_3$	$6\mathbf{e}_1 - 8\mathbf{e}_3$	2
ST	$5\mathbf{e}_1 - 7\mathbf{e}_2 + 11\mathbf{e}_3$	$6\mathbf{e}_1 - 8\mathbf{e}_2 + 15\mathbf{e}_3$	$-\mathbf{e}_1 + \mathbf{e}_2 - 4\mathbf{e}_3$	2
TT	$17\mathbf{e}_1 - 4\mathbf{e}_2 - 10\mathbf{e}_3$	$17\mathbf{e}_1 + 2\mathbf{e}_2 - 29\mathbf{e}_3$	$-6\mathbf{e}_2 + 19\mathbf{e}_3$	2
TTT	$34\mathbf{e}_1 - 26\mathbf{e}_2 + 37\mathbf{e}_3$	$-17\mathbf{e}_1 - 38\mathbf{e}_2 + 143\mathbf{e}_3$	$51\mathbf{e}_1 + 12\mathbf{e}_2 - 106\mathbf{e}_3$	2

(c) Fixed vectors for S, T, and M is only the zero vector; for N the fixed vectors are $\{t(3, 1, -5)\}$.

3. (a) $S^{-1}(\mathbf{e}_1) = 4\mathbf{e}_1 + \mathbf{e}_2 - 5\mathbf{e}_3$, $S^{-1}(\mathbf{e}_2) = -5\mathbf{e}_1 + \mathbf{e}_2 + 5\mathbf{e}_3$, $S^{-1}(\mathbf{e}_3) = -\mathbf{e}_1 + \mathbf{e}_3$.
(c) M has no inverse since its rank is $2 < 3$.

Section 9-22, page 713

1. (a) $T^{-1}(x, y) = (x, \frac{1}{2}y)$, (c) $T^{-1}(x, y, z) = (x - y + z, y - z, z)$,
(e) $T^{-1}(x, y, z) = (-x + y + z, x, x - z)$.

Section 9-23, page 717

1. (a) $\lambda = 2$ with the nonzero scalar multiples of $\mathbf{e}_1$ as associated eigenvectors, and $\lambda = 3$ with nonzero scalar multiples of $\mathbf{e}_2$ as associated eigenvectors.
(c) $\lambda = 4$, eigenvectors: multiples of $\mathbf{e}_1$; $\lambda = 3$, eigenvectors: multiples of $\mathbf{e}_2$.
(e) $\lambda = 0$, eigenvectors: multiples of $(1, -3)$; $\lambda = 5$, eigenvectors: multiples of $(1, 2)$, (g) eigenvalues k, eigenvectors $\mathbf{e}_k$, $k = 1, 2, 3$.

Section 10-2, page 722

1. (a) For A, $m = 2 = n$; for B, $m = 2$, $n = 3$.
(c) $a_{11} = 2$, $a_{22} = 3$, $a_{21} = 1$, $b_{23} = 0$, $b_{13} = 5$.
(e) Row vectors for A are $(2, 1)$ and $(1, 3)$; Row vectors for B are $(1, 3, 5)$, $(1, -1, 0)$.
(g) Column vectors for A are col $(2, 1)$, col $(1, 3)$; column vectors for B are col $(1, 1)$, col $(3, -1)$, col $(5, 0)$.

2. (a) $x_1 + x_2 = y_1$, $-x_1 + 2x_2 = y_2$, $\begin{pmatrix} 1 & 1 \\ -1 & 2 \end{pmatrix}$.

(c) $x_1 + x_2 = y_1$, $2x_1 - 3x_2 = y_2$, $-x_1 + 5x_2 = y_3$, $\begin{pmatrix} 1 & 1 \\ 2 & -3 \\ -1 & 5 \end{pmatrix}$.

3. (a) $(2, 1)$, $(1, 3)$, $(3, 4)$, $(2, 6)$, $(-1, -8)$.
(c) $(3, 5, 7)$, $(1, 2, 3)$, $(4, 7, 10)$, $(2, 4, 6)$, $(0, -1, -2)$.

Section 10-5, page 728

1. (a) col $(5, 1)$, (c) col $(3, 2)$, (e) col $(2, 8, 1)$,
(g) col $(3, 4, 4)$, (i) col $(0, 3, 10)$, (k) col $(6, 5, 14, 9)$.

3. (a) col $(9, 15)$, (c) col $(81, 135)$, (e) col $(0, 0)$, (g) col $(0, 0)$.

4. (a) col $(2 + 2i, 2 - 2i)$, (c) Range $A = \mathcal{V}_2^c$, (e) col $(0, 0)$.

5. (a) Range $= \mathcal{V}_2$, Kernel $= \mathcal{V}_0$,
(c) Range $= \mathcal{V}_3$, Kernel $=$ Span $((1, 7, 4, -5))$.

6. (g) $\{(1, 6, 4, -4) + t(1, 7, 4, -5)\}$.

7. (a) Span $((9, -7, 3, -1))$ (c) Span $((5, -22, 44, 2))$.

8. (a) Has $(1, 2, 3)$ in its range.

9. (a) $\{(1, -1) + \{t(2, -3)\}\}$, (c) $\{(\frac{1}{3})(1, 1, 0) + \{t(8, -13, 6)\}\}$.

11. (a) $\begin{pmatrix} 1 & -1 & 2 \\ 1 & -1 & 3 \end{pmatrix}$, (c) $\begin{pmatrix} 1 & -1 & 0 \\ 1 & 0 & -1 \\ 0 & 1 & -1 \end{pmatrix}$, (e) impossible.

12. (a) $\begin{pmatrix} 1 & -1 & 0 \\ 2 & -2 & 3 \end{pmatrix}$.

Section 10-6, page 739

1. (a) $x = 2t$, $y = t$, (line), (c) $(\frac{1}{2}, -\frac{1}{2})$ (point),
(e) no solution, (g) $(2, 0)$ (point),
(i) $\{(2, 0, 0, 1) + \{t(1, 0, 1, 0) + s(0, 1, 1, 0)\}\}$ (plane),
(k) $x_1 = 2 + s_1$, $x_2 = 3 + s_2$, $x_3 = s_3$, $x_4 = s_4$, $x_5 = -3s_1 + 2s_2 - 3s_3 + s_4$.

2. (a) Yes, pre-image of b is of dimension 0,
(c) impossible: $r + k \neq n$, (e) impossible $r + k \neq n$,
(g) impossible: $r_1 < r$, (i) impossible: $r_1 > \min(m, n)$,
(k) impossible: $r_1 > m$,
(m) yes, kernel of dimension 3, pre-image of **b** is empty,
(o) impossible: $r + k \neq n$, (q) impossible, $r + k \neq n$,
(s) yes, kernel of dimension 6, pre-image of **b** is empty.

3. (a) $\begin{pmatrix} 1 & 1 \\ 1 & -1 \end{pmatrix}\begin{pmatrix} x_1 \\ x_2 \end{pmatrix} = \begin{pmatrix} 2 \\ 3 \end{pmatrix}$, solution: $(\frac{5}{2}, -\frac{1}{2})$,

(c) $\begin{pmatrix} 1 & 1 & 0 \\ 1 & 0 & 1 \\ 2 & 1 & 1 \end{pmatrix}\begin{pmatrix} x_1 \\ x_2 \\ x_3 \end{pmatrix} = \begin{pmatrix} 2 \\ 3 \\ 5 \end{pmatrix}$, solution: $(0, 2, 3) + t(1, -1, -1)$.

4. (a) $\begin{pmatrix} 1 & 1 & 1 \\ 2 & 2 & 2 \\ -1 & -1 & -1 \\ 3 & 3 & 3 \end{pmatrix}\begin{pmatrix} x_1 \\ x_2 \\ x_3 \end{pmatrix} = \begin{pmatrix} 1 \\ 0 \\ 0 \\ 0 \end{pmatrix}$, (c) $\begin{pmatrix} 1 & 1 \\ 2 & 2 \\ 3 & 3 \\ 4 & 4 \\ 5 & 5 \end{pmatrix}\begin{pmatrix} x_1 \\ x_2 \end{pmatrix} = \begin{pmatrix} 1 \\ 0 \\ 0 \\ 0 \\ 0 \end{pmatrix}$.

5. (a) $x_1 = (3 + i)t$, $x_2 = (i - 2)t$, (c) no solution.

Section 10-9, page 746

1. (a) $\begin{pmatrix} 0 & 4 \\ 1 & 5 \end{pmatrix}$, (c) $\begin{pmatrix} 3 & 12 \\ 0 & 9 \end{pmatrix}$, (e) $\begin{pmatrix} 2 \\ 1 \\ 6 \end{pmatrix}$,

(g) $\begin{pmatrix} 0 & 11 \\ 2 & 11 \end{pmatrix}$, (i) $\begin{pmatrix} 3 & 8 \\ 3 & 6 \end{pmatrix}$, (k) $\begin{pmatrix} 5 & 10 & 2 \\ 0 & 3 & 0 \end{pmatrix}$,

(m) $\begin{pmatrix} 25 \\ 2 \end{pmatrix}$, (o) $\begin{pmatrix} 1 & 1 & 8 \\ 2 & 3 & 5 \\ 1 & 5 & 8 \end{pmatrix}$, (q) $\begin{pmatrix} 8 & 28 & 28 \\ 6 & 9 & 22 \\ 12 & 16 & 24 \end{pmatrix}$,

(s) $\begin{pmatrix} 1 & 3 & 6 \\ 0 & 1 & 3 \\ 0 & 0 & 1 \end{pmatrix}$.

3. (a) $\begin{pmatrix} 1 + 23x + x^3 + 3x^4 & 15x + 5x^3 \\ -36 + x + 11x^2 + 9x^3 & -25 + 3x - 3x^2 + 2x^4 \\ 64 + 5x^2 + 19x^3 & 48 + 16x^2 + 3x^3 + x^5 \end{pmatrix}$.

(c) $\begin{pmatrix} 2\cos x + 7 + 2x\sin x + 4x^2 + 2x & 14x + (2x^2+4x+2)\sin x + \sin^2 x \\ (1+x)\cos x - x^2 \sin x \cos x + x + x^2 & (1+x)\sin x \\ -\sin x \cos x \ln x + 2e^x - \ln x + x^2 e^x & 4xe^x - 2x\ln x + xe^x \sin x \\ -\cos x - x\sin x\cos x + 2 - 3x + x^2 & 4x - 4x^2 + (x-1)\sin x \end{pmatrix}$

4. (a) $\begin{pmatrix} 1 & 0 \\ 4 & 3 \end{pmatrix}$, (c) $\begin{pmatrix} 5 & 0 \\ 6 & 1 \\ 2 & 0 \end{pmatrix}$, (e) $\begin{pmatrix} -1 & 9 \\ 0 & 6 \end{pmatrix}$, (g) $(1, 13)$.

Section 10-11, page 753

1. (a) $\begin{pmatrix} 2 & 4 & 0 & 0 \\ 3 & 5 & 0 & 0 \\ 0 & 0 & 1 & 1 \\ 0 & 0 & 0 & 1 \end{pmatrix}$, (c) $\begin{pmatrix} 8 & 4 & 1 \\ 11 & 3 & 2 \\ 5 & 2 & -1 \end{pmatrix}$,

3. (a) $\begin{pmatrix} 11 & 2 \\ 4 & 5 \end{pmatrix}$ (c) $3I$, (e) $\begin{pmatrix} 34 & 9 \\ 18 & 7 \end{pmatrix}$, (g) $\begin{pmatrix} 10 & -7 \\ 35 & 10 \end{pmatrix}$.

4. (a) $J = I$, $N = 0$, (c) $J = A$, $N = 0$, (e) $J = \begin{pmatrix} 1 & 0 \\ 0 & 0 \end{pmatrix}$, $N = \begin{pmatrix} 0 & 0 \\ 1 & 0 \end{pmatrix}$.

12. (a) $(x-1)(x-2)$, (c) $(x-1)^2$, (e) $(x-1)^2$, (g) $(x-1)^2$, (i) $(x-1)(x-2)(x-3)$.

14. (a) x^3, (c) $(x-2)^3$, (e) $(x+1)^2(x-1)(x-2)$, (g) $(x+1)^2(x-1)$.

Section 10-12, page 759

1. (a) $\begin{pmatrix} 1 & 0 \\ 0 & \frac{1}{2} \end{pmatrix}$, (c) $\begin{pmatrix} -1 & -1 & 1 \\ 4 & 5 & -\frac{7}{2} \\ -2 & -3 & 2 \end{pmatrix}$, (e) $\begin{pmatrix} 1 & -2 & 5 \\ 0 & 1 & -4 \\ 0 & 0 & 1 \end{pmatrix}$,

(g) $\begin{pmatrix} 1 & 0 & 0 \\ -\alpha & 1 & 0 \\ \alpha\gamma - \beta & -\gamma & 1 \end{pmatrix}$, (i) $\frac{1}{26}\begin{pmatrix} 7 - 9i & 1 - 5i \\ -5 - i & 3 + 11i \end{pmatrix}$.

3. (a) $\begin{pmatrix} \frac{1}{2} & 0 \\ 0 & 1 \end{pmatrix}$, (c) $\begin{pmatrix} 1 & 0 \\ -c & 1 \end{pmatrix} = C_{-c}$,

(e) direct sum of B_{-c} and C_{-c}, (g) direct sum B_{-c} and C_{-d},

(i) direct sum of D^{-1} and D^{-1}, where $D^{-1} = \frac{1}{6}\begin{pmatrix} 6 & 0 & 0 \\ 0 & 3 & 0 \\ 0 & 0 & 2 \end{pmatrix}$.

4. (a) Not nonsingular, (c) $\frac{1}{8}\begin{pmatrix} 4 & 0 & 0 \\ -6 & 4 & 0 \\ 13 & -10 & 4 \end{pmatrix}$,

(e) direct sum of X and $(\frac{1}{2})I$, where $X = \begin{pmatrix} -1 & -2 & 11 \\ 0 & -1 & 4 \\ 0 & 0 & 1 \end{pmatrix}$,

(g) direct sum of X and X.

7. (a) $\begin{pmatrix} 0 & 1 & 0 & 0 & 0 \\ 1 & 0 & 0 & 0 & 0 \\ -1 & 1 & 0 & 0 & 1 \\ -3 & -2 & 1 & 0 & 0 \\ 0 & -1 & 0 & 1 & 0 \end{pmatrix}$, (c) $\begin{pmatrix} 1 & -2 & 1 & 0 & 0 \\ 0 & 1 & -2 & 1 & 0 \\ 0 & 0 & 1 & -2 & 1 \\ 0 & 0 & 0 & 1 & -2 \\ 0 & 0 & 0 & 0 & 1 \end{pmatrix}$.

Section 10-15, page 777

1. (a) Minus, (c) minus, (e) plus, (g) minus.

2. (a) -29, (c) 21, (e) -2, (g) 30, (i) 108,

(k) $-x^4 - x^3 - 4x^2$, (m) x^5, (o) $\dfrac{x(1-y)(2-y)\cdots(5-y) - y(1-x)\cdots(5-x}{x-y}$

7. (a) $x(x+1)\cdots(x+n-1)$.

8. (a) $x_2 - x_1$, (c) $(x_4 - x_1)(x_4 - x_2)(x_4 - x_3)(x_3 - x_1)(x_3 - x_2)(x_2 - x_1)$.

10. (a) Singular, (c) nonsingular, (e) singular.

11. (a) Linearly independent, (c) linearly independent, (e) linearly dependent.

12. (a) 2, (c) 3, (e) 4.

13. (a) $\frac{1}{10}\begin{pmatrix} 4 & -1 \\ -2 & 3 \end{pmatrix}$, (c) $\begin{pmatrix} -3 & 2 & 2 \\ 9 & -5 & -6 \\ -7 & 4 & 5 \end{pmatrix}$.

14. (a) 0, (c) 2, (e) no solution.

Section 10-16, page 787

1. $x_1 = 1 - 4x_4$, $x_2 = -2 + 7x_4$, $x_3 = 26x_4$.

3. $x_1 = 4$, $x_2 = -\frac{15}{4}$, $x_3 = -\frac{7}{4}$, $x_4 = \frac{3}{2}$.

5. No solution.

8. (a) $x = -13.67$, $y = 12$, (c) $x = 7.67$, $y = -4$.

9. See Problem 7, following Section 10-12.

10. (a) $\frac{1}{2}\begin{pmatrix} 2 & -3 & 8 \\ 4 & 12 & -12 \\ 4 & 11 & -10 \end{pmatrix}$.

Section 10-17, page 790

2. (a) 1, (c) 1, (e) 1.

4. (a) Never singular, (c) -1, $6/5$.
(e) $(\pi/4) + n(\pi/2)$, $(n = 0, \pm 1, \pm 2, \ldots)$,
(g) ± 1.

Section 10-18, page 794

1. (a) $\lambda = 2$, $t\mathbf{e}_1 (t \neq 0)$; $\lambda = 3$, $t\mathbf{e}_2 (t \neq 0)$.
(c) $\lambda = 1$, $t(1, -1)(t \neq 0)$; $\lambda = 3$, $t\mathbf{e}_2 (t \neq 0)$.
(e) No real eigenvalues: $\lambda = (3 + \sqrt{19}\,i)/2$, $t(-5, -\frac{1}{2} - (\sqrt{19}/2)i)(t \neq 0)$;
$\lambda = (3 - \sqrt{19}\,i)/2$, $t(-5, -\frac{1}{2} + (\sqrt{19}/2)i)(t \neq 0)$.

(g) $\lambda = 3$, $t(1, -1)$, $(t \neq 0)$.
(i) $\lambda = 1$, $t\mathbf{e}_1(t \neq 0)$; $\lambda = 2$, $t(2, 1, 0)(t \neq 0)$; $\lambda = 3$, $t(3, 4, 1)(t \neq 0)$.

2. (a) $\lambda = 1$, $t(i, 1)(t \neq 0)$.

3. (a) $\lambda = 1$, $K_1 = \text{Span}(\mathbf{e}_1)$, $W_1 = \text{Span}(\mathbf{e}_1, \mathbf{e}_2)$;
$\lambda = 2$, $K_2 = W_2 = \text{Span}((11, 4, 1))$.
(c) $\lambda = 0$, $K_0 = \text{Span}(\mathbf{e}_1, \mathbf{e}_2)$, $W_0 = V_3$.
(e) $\lambda = 0$, $K_0 = \text{Span}(\mathbf{e}_1)$, $W_0 = V_3$.

Section 10-21, page 801

1. (a) $\begin{pmatrix} 0 & 1 & 0 & 0 & 0 & 0 \\ 0 & 0 & 2 & 0 & 0 & 0 \\ 0 & 0 & 0 & 3 & 0 & 0 \\ 0 & 0 & 0 & 0 & 4 & 0 \\ 0 & 0 & 0 & 0 & 0 & 5 \\ 0 & 0 & 0 & 0 & 0 & 0 \end{pmatrix}$, (c) $\begin{pmatrix} 0 & 1 & 0 & 0 & 0 & 0 \\ 0 & 0 & 1 & 0 & 0 & 0 \\ 0 & 0 & 0 & 1 & 0 & 0 \\ 0 & 0 & 0 & 0 & 1 & 0 \\ 0 & 0 & 0 & 0 & 0 & 1 \\ 0 & 0 & 0 & 0 & 0 & 0 \end{pmatrix}$.

2. (a) $\begin{pmatrix} 2 & 1 & -1 & 0 \\ 3 & 2 & 2 & 2 \\ 6 & 0 & -3 & -1 \end{pmatrix}$, (c) $\frac{1}{2}\begin{pmatrix} 2 & 7 & 7 & 2 \\ 8 & 1 & 1 & 8 \\ 4 & -7 & -9 & 2 \end{pmatrix}$.

9. (a) $\begin{pmatrix} 1 & 0 & 0 \\ 0 & 2 & 0 \\ 0 & 0 & -1 \end{pmatrix}$, (c) $\begin{pmatrix} 1 & 1 & 0 \\ 0 & 1 & 0 \\ 0 & 0 & -3 \end{pmatrix}$, (e) $\begin{pmatrix} 1 & 1 & 0 & 0 \\ 0 & 1 & 0 & 0 \\ 0 & 0 & 1 & 0 \\ 0 & 0 & 0 & -1 \end{pmatrix}$.

Section 11-3, page 809

1. (a) 25, (c) $\sqrt{29}$, (e) -12, (g) 5, (i) $\text{Cos}^{-1} 5/\sqrt{58}$.

4. (e) $\alpha = \gamma = \pi/4$, $\beta = \pi/2$.

Section 11-5, page 817

1. (a) Coincident.

2. (a) $\overrightarrow{OP} = (3, 5, 1) + t(-1, -5, 6)$, Q on line,
(c) $\overrightarrow{OP} = (4, -1, 5) + t(2, 3, -8)$, Q not on line.

3. (a) Skew, (c) meet at $(1, 1, 1)$.

4. (a) Collinear, (c) collinear.

5. (a) Not orthogonal, (c) not orthogonal, (e) orthogonal.

6. (a) $\text{Cos}^{-1} 6/\sqrt{161}$, (c) $\text{Cos}^{-1} 5/\sqrt{259}$.

8. (a) $(\frac{3}{2}, \frac{7}{2}, 4)$, (c) $(\frac{5}{4}, \frac{11}{4}, 3)$ and $(\frac{1}{2}, \frac{1}{2}, 0)$.

Section 11-9, page 829

1. (a) $12\mathbf{i} + 4\mathbf{k}$, (c) 4, (e) 4, (g) $6\mathbf{i} - 18\mathbf{j} + 8\mathbf{k}$,
(i) $-4\mathbf{i} - 4\mathbf{j} + 12\mathbf{k}$.

2. (a) $(1, 0, 1) + t(-11, 1, -6)$.

3. (a) Parallel, noncoincident, (c) skew.

8. (a) Negative, (c) positive, (e) linearly dependent.

Section 11-10, page 837

2. **(a)** Same plane.

3. **(a)** $(2, 3, 1)$, **(c)** $(1, -1, 0)$, **(e)** $(1, 0, 0)$.

4. **(a)** $2(x - 1) - (y - 3) + 5(z - 2) = 0$, **(c)** $-x + y + z + 2 = 0$.

5. **(a)** $\overrightarrow{OP} = s(1, 1, 1) + t(3, 1, 4)$, **(c)** $\overrightarrow{OP} = (1, 2, 1) + s(2, 0, 1) + t(2, -1, 4)$,
(e) $\overrightarrow{OP} = (2, 1, 2) + s(5, 6, 7) + t(1, 0, 0)$.

6. **(a)** $2(x - 1) + y - 4z = 0$, **(c)** $(x - 2) + 4(y - 1) - (z - 2) = 0$.

7. **(a)** $\overrightarrow{OP} = 5\mathbf{i} + s(-3\mathbf{i} + \mathbf{j}) + t(\mathbf{i} + \mathbf{k})$,
(c) $\overrightarrow{OP} = \mathbf{i} + 3\mathbf{j} + \mathbf{k} + s(7\mathbf{i} - 2\mathbf{j}) + t(\mathbf{i} + 2\mathbf{k})$.

8. **(a)** Different planes.

9. **(a)** $2\mathbf{i} + \mathbf{j} - 4\mathbf{k}$, **(c)** $\mathbf{i} + 4\mathbf{j} - \mathbf{k}$.

Section 11-13, page 843

1. **(a)** $6(x - 6) - 6(y - 4) + 8(z + 2) = 0$,
(c) $(x - 1) - 2(y - 2) + 3(z - 3) = 0$, **(e)** $y - 2 = 0$,
(g) $(x - 7) - 2(y + 1) - (z - 2) = 0$,
(i) $-2(x - 1) + 6(y - 2) + (z - 1) = 0$.

2. **(a)** $\overrightarrow{OP} = (-2, 1, 3) + t(2, 3, 1)$, **(c)** $\overrightarrow{OP} = (7, 0, 0) + t(5, 2, -3)$.

3. **(a)** $(-\frac{1}{2}, -\frac{1}{2}, 2)$.

4. **(a)** $\overrightarrow{OP} = (0, \frac{5}{3}, \frac{2}{3}) + t(3, -5, 1)$, **(c)** parallel lines.

5. **(a)** $y = 2, 3(z - 2) - 5(x - 1) = 0$.

11. Line: $(\frac{5}{3}, \frac{1}{3}, -\frac{2}{3}, 0, -\frac{1}{3}) + t(1, 0, -6, 2, 3)$.

Section 11-16, page 854

1. **(a)** $\mathbf{i}' = (\frac{1}{3})(2, 2, 1)$, $\mathbf{j}' = (1/\sqrt{2})(-1, 1, 0)$, $\mathbf{k}' = (1/3\sqrt{2})(-1, -1, 4)$.

5. **(a)** $\sqrt{3} + \sqrt{11}$, **(c)** $\sqrt{11} + \sqrt{17}$.

6. **(a)** $2\sqrt{2}$, **(c)** 0.

7. **(a)** $(3 + \sqrt{3})/2$, **(c)** $2 + 2\sqrt{3}$.

9. **(a)** $4/3$.

Section 11-18, page 860

1. **(a)** $(1, 2, 0), (2, 3, 1), (1, 2, 3)$;
(c) $x_1 = -1 + (\frac{1}{3})(y_1 + y_2 + y_3)$, $x_2 = 1 + (\frac{1}{3})(-y_1 - y_2 + 2y_3)$,
$x_3 = 1 + (\frac{1}{3})(y_1 - 2y_2 + y_3)$;
(d) x_1-axis goes into line: $(1, 2, 0) + t(1, 1, 1)$ which is parallel to diagonal in first octant, x_2-axis goes into line: $(1, 2, 0) + t(-1, 0, 1)$ which is parallel to bisector of second quadrant of x_1x_3-plane, x_3-axis goes into line: $(1, 2, 0) + t(1, -1, 0)$ which is parallel to bisector of second quadrant in x_1x_2-plane.
(e) $y_1 = 1, y_3 = 0, y_1 - 2y_2 + 4y_3 = 0$.
(f) Region: $3 \leq y_1 + y_2 + y_3 \leq 3 + 3h$, $-3 \leq -y_1 - y_2 + 2y_3 \leq -3 + 3h$, $-3 \leq y_1 - 2y_2 + y_3 \leq -3 + 3h$. A parallelepiped with edge vectors $h(1, 1, 1)$, $h(-1, 0, 1)$, $h(1, -1, 0)$, and volume $= 3h^3 = \det(A)h^3$.

3. **(b)** Plane: $2y_1 - y_2 - y_3 = 0$, **(c)** Line: $t(2, 3, 1)$, **(d)** point: $(8, 13, 3)$.

6. (b) $y = (1, 1, 1) + t(2, -1, 2)$,
   (c) parallelogram: $y = (1, 1, 1) + hs(1, 2, 1) + ht(2, -1, 2)$, $0 \le s \le 1$, $0 \le t \le 1$, area $= 5\sqrt{2}h^2$.

Section 11-19, page 865

1. (a) Sphere of radius $\frac{7}{2}$, center at $(-1, -2, \frac{9}{2})$. (c) Point: $(1, 2, 2)$.
   (e) Cylinder of radius 2 with axis along z-axis.
   (g) Elliptic cylinder with axis along z-axis.
   (i) Parabolic cylinder with axis along z-axis.
   (k) Hyperbolic cylinder with xz and yz-planes as asymptotes.
   (m) Circular cone with axis along the z-axis.
   (o) Circular cone with axis along x-axis.
   (q) Elliptical cone with axis along z-axis.
2. (a) $x^2 + y^2 = 4z^2$, (c) $16x^4 + 16y^4 = z^4$,
   (e) $2y^2 + z^2 = 4x^2$, (g) $x + y = z$.
3. (a) $x^4 + y^4 = 3$ in $z = 1$.

Section 11-20, page 868

1. (a) Cylinder, (c) plane, (e) cylinder of radius 2,
   (g) solid rod, (i) region above xy-plane beneath nappe of a cone,
   (k) circular plate.
2. (a) Sphere, (c) nappe of a cone,
   (e) Surface of revolution of a circle rotated about a tangent to the circle.
   (g) solid nappe of a cone, (i) spiral surface.
3. (a) $r^2 = 5$, (c) $r^2 + 2z^2 = 4$, (e) $r^2(1 + \sin^2\theta) + 3z^2 = 6$,
   (g) $r^2\cos^2\theta - z^2 = 4$. [For spherical coordinate equations replace r^2 by $\rho^2\sin^2\varphi$ and z^2 by $\rho^2\cos^2\varphi$.]
4. (a) $(r_1^2 + r_2^2 - 2r_1r_2\cos(\theta_1 - \theta_2) + (z_1 - z_2)^2)^{1/2}$.

Section 11-22, page 873

3. (a) $\frac{1}{3}\begin{pmatrix} 1 & 2 & 2 \\ 2 & 1 & -2 \\ 2 & -2 & 1 \end{pmatrix}$
4. (a) $(-1, -2, -3), (-1, -2, -2), (1, -3, -3), (2, 0, -2), (-2, -4, -6)$;
   (b) $x_1' + x_2' + x_3' = 0$, $(x_1' + 1)^2 + (x_2' + 2)^2 = (x_3' + 3)^2$, $x_1'^2 + x_2'^2 + x_3'^2 = 26$.
5. (a) Ellipsoid, (c) hyperbolic paraboloid, (e) hyperbolic paraboloid.

Section 12-2, page 881

2. (a) Open region, (c) open set, (e) open region, (g) open region.
4. (a) 1, (c) $a^3 + 2b^3$, (c) $-x^3 - 2y^3$.
7. (a) 0, (c) 12, (e) $2f(v)$.
8. (a) 4, (c) 0.
9. (a) Open set, (c) open set.

Section 12-6, page 887

1. **(a)** $-\frac{1}{3}$, **(c)** 0, **(e)** undefined, **(g)** $(a - b)/(b - c)$, **(i)** $f(x, y, z)$.

3. **(a)** $(1, 0)$, $(-1, 0)$, $(0, 1)$, $(-e/\sqrt{2}, e/\sqrt{2})$.
(c) Circle center at origin, radius 1.

4. $\mathbf{i}$, $\mathbf{j}$, $2\mathbf{i} - \mathbf{j}$, $(x + y)(x\mathbf{i} + y\mathbf{j})$.

6. $3\mathbf{i} + 2\mathbf{j}$, $3\mathbf{j} - 2\mathbf{i}$, $13\mathbf{j}$, $(3x - 2y)\mathbf{i} + (3y + 2x)\mathbf{j}$.

8. $\mathbf{i}, \mathbf{j} + (\pi/2)\mathbf{k}$, $-\mathbf{i} + \pi\mathbf{k}$.

10. $(0, 0, 1, z)$.

12. **(a)** $2\mathbf{i} + 5\mathbf{j} + \mathbf{k}$, $3\mathbf{i} + 6\mathbf{j} + 4\mathbf{k}$, $3\mathbf{i} + 4\mathbf{j} + 3\mathbf{k}$, $12\mathbf{i} + \mathbf{j} + 24\mathbf{k}$.
(b) Linear variety of dimension 2 with $2\mathbf{i} + 5\mathbf{j} + \mathbf{k}$ as leader and Span $(\mathbf{i} + \mathbf{j} + 3\mathbf{k}, \mathbf{i} - \mathbf{j} + 2\mathbf{k})$ as base space.

15. **(a)** $x^2 - 3xy + y^2$, **(c)** $(x^2 + y^2)^2 - x + y$.

16. **(a)** $[2\mathbf{u} \cdot \mathbf{u} - 3\mathbf{u} \cdot \mathbf{i}]\mathbf{i} + [\mathbf{u} \cdot (3\mathbf{u} + 2\mathbf{i})]\mathbf{j}$.

17. **(a)** Origin, **(c)** parabola: $y = 1 - x^2$, **(e)** Span $(3, -1)$.

Section 12-7, page 895

1. **(a)** Continuous everywhere. **(c)** Continuous except on line: $x = y$.
(e) Continuous for $2x + 3y > 0$. **(g)** Continuous for $0 \leq |x| \leq |y|$, $y \neq 0$.
(i) Continuous for $z \neq 0$. **(k)** Continuous for $t > 0$.
(m) Continuous everywhere. **(o)** Continuous everywhere.
(q) Continuous except for $\mathbf{x} = \mathbf{0}$.

2. **(a)** 4, **(c)** 3.

5. **(a)** In each region of the plane after removing the lines $x = 4y$, $x = -y$.
(b) Yes.

Section 12-8, page 900

1. **(a)** 2, 3, **(c)** $2x$, $-2y$, **(e)** $4xy$, $2x^2$,
(g) $ye^{xy} \cos(x - 2y) - e^{xy} \sin(x - 2y)$, $xe^{xy} \cos(x - 2y) + 2e^{xy} \sin(x - 2y)$,
(i) $2x(1 - y)(x^2 - 2xy - y^2)^{-1}$, $(-2x - 2y)(x^2 - 2xy - y^2)^{-1}$,
(k) $-x^{-2}\sqrt[x]{y} \ln y$, $(xy)^{-1}\sqrt[x]{y}$,
(m) $-3x^2(ye^z + 1)^{-1}$, $-e^z(ye^z + 1)^{-1}$, **(o)** $-zye^y$, $-zxe^y(1 + y)$.

2. **(a)** -1, **(c)** -1, **(e)** 1.

3. **(a)** $z - 3 = 4(x - 2) - 2(y - 1)$, **(c)** $z = (x - 2) + (y + 1)$.

4. **(a)** $4\mathbf{i} + 2\mathbf{j}$, **(c)** $\mathbf{i} - \mathbf{j}$, **(e)** $\cos(x + 2y)\mathbf{i} + 2\cos(x + 2y)\mathbf{j}$.

5. **(a)** Local min at origin, **(c)** local max at origin,
(e) No local max or min, **(g)** local min along line $x + y = 0$.

6. **(a)** $3x^2y^2z\mathbf{i} + 2x^3yz\mathbf{j} + x^3y^2\mathbf{k}$,
(c) $\cos(u + 2v - 3w)(\mathbf{i} + 2\mathbf{j} - 3\mathbf{k})$,
(e) $2(x_1\mathbf{e}_1 + x_2\mathbf{e}_2 + \cdots + x_n\mathbf{e}_n)$.

Section 12-9, page 907

1. **(a)** $2x\,dx + 2dy$, **(c)** $(x + 1)ye^x\,dx + xe^x\,dy$,
(e) $(2y)(x + y)^{-2}\,dx - 2x(x + y)^{-2}\,dy$,

(g) $2x\,dx + 2y\,dy + 2z\,dz$,
(i) $3v^2w^3\,du + 6uvw^3\,dv + 9uv^2w^2\,dw$,
(k) $\cos(uvw)[vw\,du + uw\,dv + uv\,dw]$.

2. (a) 1.2, (c) 4.6, (e) 2.8.

3. (a) $\Delta x + 2\,\Delta y + \Delta x\,\Delta y$,
(c) $(\frac{1}{2})\,\Delta x - (\frac{3}{4})\,\Delta y - \Delta x\,\Delta y(4 + 2\,\Delta y)^{-1} + 3(\Delta y)^2(8 + 4\,\Delta y)^{-1}$.

4. (a) 4.01, (c) .06, (e) 80.94.

Section 12-10, page 912

1. (a) $6x(dx/dt) - 2y(dy/dt)$,
(c) $yu^2(dx/dt) + xu^2(dy/dt) + 2xyu(du/dt)$,
(e) $\cos(xu + ty)[u(dx/dt) + x(du/dt) + y + t(dy/dt)]$.

2. (a) $2tf_x + (3t^2 + 1)f_y$,
(c) $-(x + 2t)(2x + t)^{-1}f_x + (y - 2t)(2y - t)^{-1}f_y$.

3. (a) $z_x = 6xyf_u + (3x^2 - 6xy)f_v$, $z_y = (3x^2 - 3y^2)f_u - 3x^2f_v$,
(c) $z_x = y^2f_u + 2xyf_v + 2xy^2f_w$, $z_y = 2xyf_u + x^2f_v + 2x^2yf_w$.

7. (a) $z_x = 11$, $z_y = 20$, 5.3.

9. (a) max $= \sqrt{2}$, min $= -\sqrt{2}$.

10. (a) $z_x = (-yz)(xy + 3z^2)^{-1}$, $z_y = -xz(xy + 3z^2)^{-1}$.

Section 12-11, page 917

1. (a) $26\sqrt{2}$, (c) $\frac{3}{5}$, (e) $24/\sqrt{13}$.

2. (a) $-32/\sqrt{13}$.

4. (a) max $= 2$, min $= 1$, (c) max $= \frac{3}{2}$, min $= \frac{1}{2}$.

6. Makes angle $\text{Tan}^{-1}(\frac{182}{43})$ with the outer wall.

Section 12-13, page 925

1. (a) $\begin{pmatrix} 1 & -2 \\ 2 & 3 \end{pmatrix}$ (c) $\begin{pmatrix} x_2x_3 & x_1x_3 & x_1x_2 \\ 2x_1x_2 & x_1^2 + 2x_2x_3 & x_2^2 \\ 2x_1 + 2x_3 & 0 & 2x_1 + 2x_3 \end{pmatrix}$

(e) $\begin{pmatrix} 4x^3 + y^3 & 3xy^2 \\ 2xy^2 & 2x^2y - 12y^3 \end{pmatrix}$ (g) $\begin{pmatrix} 2x & z & y - 2z \\ y - z & x & -x - 4z \end{pmatrix}$

2. (a) $du_1 = -2dx_2 + 2dx_3$, $du_2 = 2dx_3$, $du_3 = (-\pi/2)\,dx_1 - 2dx_2 + \pi dx_3$,
$(-0.86, 0.2, -3.05)$.

3. (a) $\begin{pmatrix} -\sin(x + 2y) & -2\sin(x + 2y) \\ \cos(x + 2y) & 2\cos(x + 2y) \end{pmatrix}$, rank = 1, range: circle radius 1, center at origin.

(c) $\begin{pmatrix} 1 & 0 & 1 \\ 1 & -1 & 0 \\ 2x - y + z & -x - z & x - y \end{pmatrix}$, rank 2, range: surface $uv = w$.

4. (a) $\begin{pmatrix} 3x^2 - 3y^2 & -6xy \\ 6xy & 3x^2 - 3y^2 \end{pmatrix}$ Jacobian has rank 2 everywhere except at the origin.

(c) $(2x \quad 2y)$ rank 1 except at origin. (e) rank 1 except at 5/2.

5. (a) $\begin{pmatrix} -1 & -1 & 0 \\ 11 & 2 & 18 \end{pmatrix}$, (c) $\begin{pmatrix} 27 & 66 \\ 11 & 12 \end{pmatrix}$.

6. (a) $\begin{pmatrix} -22 & 45 \\ 106 & -49 \end{pmatrix}$, (c) $\begin{pmatrix} -2 & 2 & 0 \\ 2 & -3 & -5 \\ 0 & 4 & 20 \end{pmatrix}$.

Section 12-15, page 938

1. (a) $z_x = [-y - 2xe^z][x^2e^z + 2yz]^{-1}$, $\quad z_y = [-x - z^2][x^2e^z + 2yz]^{-1}$.

(c) $[2x - yzt][xyt - 2z]^{-1}$.

(e) $x_t = [-4xyt^3 + 4y^3t + 2yt^2 - 2y^2t + 2xy - 4xt + 2y - x + 1]Q^{-1}$
$y_t = [-2y^2t^3 + 2y^2t - 4x^2t + xt^2 + 4xy - 2ty + 3x - t]Q^{-1}$
where $Q = 2yt^4 - 2yt^2 + 4xy - 4tx - 2yt + 3t^2 - 2x + t - 1$.

(g) $x_u = [\cos u][x(\sin u + \sin y) + y(\cos v + \cos y)]Q^{-1}$
$x_v = [\sin v][y(\sin u + \sin y) + x(\cos v + \cos y)]Q^{-1}$
where $Q = \cos(x - y) + (\sin u)(\sin x - \sin y) + (\cos v)(\cos x + \cos y)$
$+ \cos^2 v - \sin^2 u$.

(i) $z_x =$

$$\begin{vmatrix} -y + xz & -tz - 2x & -y^2 \\ x - yz^2 & -2x - t - 2xu & x^2 \\ 2t + z & y - z + 2x & -1 \end{vmatrix} \begin{vmatrix} -y + xz & xt + 2z & -y^2 \\ x - yz^2 & -2ytz - y & x^2 \\ 2t + z & x + t & -1 \end{vmatrix}^{-1}$$

2. (a) $-\begin{pmatrix} 2 & 3 \\ 3 & 5 \end{pmatrix}^{-1} \begin{pmatrix} 1 & 0 \\ 2 & 5 \end{pmatrix}$

(c) $-\begin{pmatrix} 6 & 2 & 3 \\ 1 & 0 & 1 \\ 5 & 1 & 3 \end{pmatrix}^{-1} \begin{pmatrix} 4 & 5 \\ 1 & 2 \\ 0 & 2 \end{pmatrix}$

3. (a) $x_1 = 3.5$, $x_2 = -2.5$

5. (b-i) $2x - 3y$

Section 12-16, page 944

1. (a) $9, \frac{1}{9}$; (c) $9, \frac{1}{9}$; (e) $-15e^xy^2$, $(-\frac{1}{15})e^{-x}y^{-2}$;

(g) J, $1/J$, where $J = -6x^5z + 6x^4y^2 + 9x^4z^2 - 15x^3yz^2 + 12x^4yz - 3x^2yz^3 - 3xy^2z^3 - 3x^5y$.

2. (a) $u = 2 + x + 3(y - 1)$, $v = 1 + 3x - 6(y - 1)$;
$x = \frac{2}{5}(u - 2) + \frac{1}{5}(v - 1)$, $y = 1 + \frac{1}{5}(u - 2) - \frac{1}{15}(v - 1)$.

(c) $u = 2 + 4(x - 1) + (y - 1) + (z - 1)$, $v = (x - 1) - (y - 1)$,
$w = 1 + 2(x - 1) + (z - 1)$; $x = \frac{1}{3}(u + v - w + 2)$,
$y = \frac{1}{3}(u - 2v - w + 2)$, $z = \frac{1}{3}(-2u - 2v + 5w + 2)$.

Section 12-18, page 952

1. (a) Tangent: $\mathbf{r} = (\frac{1}{4})(2\sqrt{2}\,\mathbf{i} + 2\sqrt{2}\,\mathbf{j} + \pi\mathbf{k}) + (\tau/2)(-\sqrt{2}\,\mathbf{i} + \sqrt{2}\,\mathbf{j} + 2\mathbf{k})$,
normal: $2\sqrt{2}x - 2\sqrt{2}y - 4z = -\pi$.

(c) Tangent: $\mathbf{r} = (\sqrt{2}/2)(\mathbf{i} + \mathbf{j} + \mathbf{k}) + \tau(-\mathbf{i} + \mathbf{j} - \mathbf{k})$
normal: $2x - 2y + 2z = \sqrt{2}$.

2. (a) Tangent: $2x + 4y - z = 5$, normal: $x = 1 + 2t$, $y = 2 + 4t$, $z = 5 - t$.

(c) Tangent: $2x + 4y + e^5z = 11$, normal: $x = 1 + 2t$, $y = 2 + 4t$, $z = e^{-5} + e^5t$.

(e) Tangent: $3x + y + z = 11$, normal: $x = 3 - 3t$, $y = 1 - t$, $z = 1 - t$.

(g) Tangent: $5x - 3y - 4z = 0$, normal: $x = 5 + 5t$, $y = 3 - 3t$, $z = 4 - 4t$.

3. (a) $x = 1$, $y = 1 - t$, $z = 1 + t$, (c) $x = 1 + 3t$, $y = 1 - t$, $z = 1 - 2t$.

4. (b) $\mathbf{v} = -(\sqrt{2}/2)\mathbf{i} + (\sqrt{2}/2)\mathbf{j} + \mathbf{k}$, $\mathbf{a} = (\sqrt{2}/2)(-\mathbf{i} - \mathbf{j})$,
$2\sqrt{2}x - 2\sqrt{2}y + 4z = \pi$.

9. (b) Tangent: $x + y = \sqrt{2}$, normal: $x = \sqrt{2}/2 + t$, $y = \sqrt{2}/2 + t$, $z = 1$.

11. (b) Tangent: $x + y + \sqrt{2}z = 2\sqrt{2} + 2$,
normal: $x = (2\sqrt{2} + 1)/2 + t$, $y = (2\sqrt{2} + 1)/2 + t$, $z = \sqrt{2}/2 + \sqrt{2}t$.

Section 12-20, page 959

1. (a) $f_{xx} = 12xy^2$, $f_{xy} = 12x^2y - 6y$, $f_{yy} = 4x^3 - 6x$;
(c) $f_{xy} = 2xze^{yz}$, $f_{yz} = (1 + yz)x^2e^{yz}$, $f_{xz} = 2xye^{yz}$.

Section 12-21, page 965

1. (a) $1 - \frac{1}{2}(x + y)^2 + R_3$, (c) $2x + y - (1/2!)(2x + y)^2 + R_2$,
(e) $1 + xy + (1/2!)x^2y^2 + R_4$,
(g) $\sin 1 + (\cos 1)(x + y) - \frac{1}{2}(\sin 1)(x + y)^2 + R_2$,
(h) $\cos 1 - (\sin 1)(2x + y) - \frac{1}{2}(\cos 1)(2x + y)^2 + \frac{1}{6}(\sin 1)(2x + y)^3 + R_3$.

2. (a) $x \cos \mu^2xy - \mu^2x^2y \sin \mu^2xy - \mu^2xy^2 \sin \mu^2xy$,
(c) $x + (1/2!)[2e^{\mu y}xy + \mu xy^2e^{\mu y}]$,
(e) $1 + (1/3!)[8\mu^3x^3y^3 \sin \mu^2xy - 12\mu x^2y^2 \cos \mu^2xy]$.

Section 12-24, page 977

1. (a) Not closed, boundary: the rational numbers 0, 1, 1/2, 1/3, ...
(c) Not closed, boundary: all real numbers in the interval $|x| \leq 1$.
(e) Not closed, boundary, the entire plane.
(g) Not closed, boundary: line segments: $x = 0$, $-1 \leq y \leq 2$ and $x = 9$, $-82 \leq y \leq 162$, and the curves: $y = 2(1 + x^2)$ and $y = -1 - x^2$ for $0 \leq x \leq 9$.

3. (a) (0, 0) saddle point. (c) (−1, 0) local min.
(e) (0, 0) local min. (g) (−3/2, −1/2) local min.
(i) (0, 0) local max.
(k) (1/3, 1/3) local min, (0, 0), (1, 0), and (0, 1) saddle points.
(m) Points on $x = y + 1$ are local min for surface.
(o) (0, 0) saddle point, (0, ±1) local min.

4. (a) Max at (3/5, 4/5), min at (−3/5, −4/5).
(c) Max at (±1, 0), min at (0, 0).
(e) Max at $(1/2, 1/\sqrt{2})$ and $(-1/2, -1/\sqrt{2})$, min at $(-1/2, 1/\sqrt{2})$ and $(1/2, -1/\sqrt{2})$.
(g) Max at (0, 0), min at (−3/2, −1/2).

7. (a) $\alpha_1 = A/C$, $\alpha_2 = B/C$, where $A = (\Sigma y_i)(\Sigma x_i) - n\Sigma x_iy_i$
$B = (\Sigma x_iy_i)(\Sigma x_i) - (\Sigma y_i)(\Sigma x_i^2)$, $C = (\Sigma x_i)^2 - n\Sigma x_i^2$

(c) $\alpha_1 = A/C$, $\alpha_2 = B/C$, where
$A = (\Sigma y_i \sin x_i)(\Sigma \cos x_i \sin x_i) - (\Sigma y_i \cos x_i)(\Sigma \sin^2 x_i)$
$B = (\Sigma y_i \cos x_i)(\Sigma \cos x_i \sin x_i) - (\Sigma y_i \sin x_i)(\Sigma \cos^2 x_i)$
$C = (\Sigma \cos x_i \sin x_i)^2 - (\Sigma \sin^2 x_i)(\Sigma \cos^2 x_i)$

9. **(a)** Local max: $(3/5, 4/5)$, local min: $(-3/5, -4/5)$, max $= 5$, min $= -5$.
(c) Local max: $(\pm 2^{-1/4}, \pm 2^{-1/4})$, local min: $(\pm 1, 0)$ and $(0, \pm 1)$, max $= 2^{1/2}$, min $= 1$.
(e) Local min: $(2, 2)$, min $= 2$, no local or absolute max.
(g) Local max: $(3^{-1/2}, 3^{-1/2}, 3^{-1/2})$, local min $(-3^{-1/2}, -3^{-1/2}, -3^{-1/2})$, max $= 3^{1/2}$, min $= -3^{1/2}$.
(i) Local max: $(3^{-1/2}, 3^{-1/2}, 3^{-1/2})$, $(-3^{-1/2}, -3^{-1/2}, -3^{-1/2})$, local min at points of the circle: $x + y + z = 0$, $x^2 + y^2 + z^2 = 1$, max $= 1$, min $= -\frac{1}{2}$.

10. $(1, 3, 4)$.

11. **(a)** Local max: $(\pm 5^{-1/2}, \mp 5^{-1/2})$, local min: $(\pm 7^{-1/2}, \pm 7^{-1/2})$, max $= \frac{2}{5}$, min $= \frac{2}{7}$.
(c) Local min: $(\pm 1, \pm 1)$, min $= 2$, there is no local max.

Section 12-25, page 986

1. **(a)** Subsequences with n even or with n odd.
(c) Sequences with subscript of the form $30k$.

5. **(a)** $\delta = \epsilon/3$

9. **(a)** $(1 - \cos y)/y$, 0 for $y = 0$,
(c) $(1/2)\sqrt{1 + y} - (1/8)\sqrt{1 + 4y} + (1/2)y \ln[2(1 + \sqrt{1 + y})/(1 + \sqrt{1 + 4y})]$.

Section 13-1, page 995

1. **(a)** $e - 1$ **(c)** $3/4$, **(e)** $\cos 1 + \sin 1 - \sin 2$,
(g) $4\sqrt{2}/21$, **(i)** $\pi/6$, **(k)** $\frac{1}{2}$, **(m)** $2\sqrt{6}\pi/9$, **(o)** 8.

2. **(a)** $\int_1^2 \int_{1-x}^{1+y} f\,dy\,dx = \int_{-1}^3 \int_{\psi(y)}^2 f\,dx\,dy$, where $\psi(y) = 1 - y$ for $-1 \le y \le 0$, $= 1$ for $0 \le y \le 2$, $= y - 1$ for $2 \le y \le 3$.

(c) $\int_{-5/3}^3 \int_{\varphi(x)}^{(x+3)/2} f\,dy\,dx = \int_{-1/2}^3 \int_{\psi(y)}^y f\,dx\,dy$, where
$\varphi(x) = -1 - x$ for $-\frac{5}{3} \le x \le -\frac{1}{2}$, $= x$ for $-\frac{1}{2} \le x \le 3$ and
$\psi(y) = -1 - y$ for $-\frac{1}{2} \le y \le \frac{2}{3}$, $= 2y - 3$ for $\frac{2}{3} \le y \le 3$.

(e) $$\int_0^1 \int_{-\sqrt{x-x^2}}^{\sqrt{x-x^2}} f\,dy\,dx = \int_{-1/2}^{1/2} \int_{\frac{1-\sqrt{1-4y^2}}{2}}^{\frac{1+\sqrt{1-4y^2}}{2}} f\,dx\,dy$$

(g) $$\int_0^1 \int_{1-x}^{\sqrt{1-x^2}} f\,dy\,dx = \int_0^1 \int_{1-y}^{\sqrt{1-y^2}} f\,dx\,dy,$$

(i) $$\int_0^2 \int_{|x-1|}^1 f\,dy\,dx = \int_0^1 \int_{1-y}^{1+y} f\,dx\,dy,$$

3. **(a)** $\int_0^{1/2} \int_{1/2}^{1-y} f\,dx\,dy$ **(c)** $\int_0^1 \int_0^{\sqrt{1-y^2}} f\,dx\,dy$

(e) $\int_{-1}^{0}\int_{0}^{x+1} f\,dy\,dx$ (g) $\int_{0}^{2}\int_{|y-1|}^{1} f\,dx\,dy$

4. (a) $c^2(\sqrt{2} - 1)/2$

5. (a) 4.56 using midpoints, 4.1 if use lower left-hand points.

7. (a) 1.468, (c) 2.763.

Section 13-3, page 1008

5. (a) 7, (c) −3, (e) 1.

Section 13-5, page 1016

1. (a) $16\pi/3$, (c) 1/16, (e) 13/10, (g) $\pi^4 3^{-1} 2^{-6}$.
2. (a) $a^2 h\pi$, (c) 2π, (e) $\pi h a^2/3$.

Section 13-7, page 1029

1. (a) $(8/35)[3^{5/2} - 2^{7/2} + 1]$, (c) 1/56, (e) $\pi/8$.

2. (a) $\int_{4}^{5}\int_{2}^{3}\int_{0}^{1} e^{xy}\,dz\,dy\,dx$,

   (c) $\int_{0}^{1}\int_{x}^{1}\int_{\sqrt{x^2-y^2}}^{1} \sin(xy)\,dz\,dy\,dx$.

3. (a) 1/8, (c) $(e^{5\pi} + 1)/130$.

4. (a) $2\pi/3$, (c) −1/24.

Section 13-8, page 1036

1. (a) $\pi/2$, (c) 2.
2. $-4\pi^2 k$ coulombs.
4. (a) $1 - e^{-a^2/2}$.
5. Mean density is 4, density equals mean for all points in square on $3x + 5y = 4$.

Section 13-9, page 1039

1. (a) $\sqrt{6}/2$, (c) $(\pi/2)\ln(1 + 2^{1/2})$, (e) $4\pi\sqrt{3}$.
   (g) $(2^{3/2} - 1)\pi/6$, (i) $\pi[5^{3/2} - 1]/24$,
   (k) $\frac{5}{4}\{\ln[(2^{3/2} + 2)/(5^{1/2} + 1)] + 5^{1/2} - 2^{1/2}\}$.

2. (c) $$\tfrac{1}{2}(\beta - \alpha)\left\{\sqrt{a^4 + b^4 + k^2} - \sqrt{a^4 + b^4 + h^2} - \sqrt{a^4 + b^4}\ln\left[\frac{k}{h}\,\frac{\sqrt{a^4 + b^4} + \sqrt{a^4 + b^4 + k^2}}{\sqrt{a^4 + b^4} + \sqrt{a^4 + b^4 + h^2}}\right]\right\}.$$

Section 13-10, page 1046

1. (a) $e^2 - e + 7/3$, (c) $3\pi/8$.
2. (a) 4, center of mass is (0.0).
   (c) $\pi/12$, center of mass is $(12/7\pi, 12/7\pi)$.
3. (a) 5/3, (c) $\pi/32$.

5. (a) $(0, 0, a/2)$, (c) $(1/2, 1/2, 1/4)$.

6. (a) $\frac{1}{2}(\mathbf{i} - \mathbf{j})$.

9. (a) $ka^5/2$, (c) $(\frac{7}{12}a, \frac{7}{12}a, \frac{2}{3}a)$, (e) $-kM_0/b$, M_0 = mass of sphere.

Section 13-11, page 1058

1. (a) $-23/2$, (c) 0, (e) 0, (g) $-1/6$, (i) -87.5π.

2. (a) -6, (c) $1/2$, (e) 0.

3. (a) $\sqrt{2}/3$, (c) 2π.

Section 13-13, page 1069

1. (a) $(a_2 - b_1)A$, (c) $-\pi/2$.

4. (a) $2\mathbf{k}$, (c) $\mathbf{0}$.

5. (a) $-\mathbf{i} - \mathbf{j} - \mathbf{k}$, (c) $(xz^2 + 2y^2z)\mathbf{i} + (x^2y - yz^2)\mathbf{j} + (3x^2y - x^2z)\mathbf{k}$.

6. (a) 0, (c) $x^3 + 4xyz - 2yz^2$.

7. (a) xyz, (c) $(1/2)\ln(x^2 + y^2)$.

8. (a) $x^2 + y^2 - z^2 + xy + zy$, (c) $\frac{1}{3}(x^3 + y^3) - 2xy$,
(e) $e^x + x^2y - \cos y$.

10. (a) $yz\mathbf{i} - xz\mathbf{j}$.

Section 13-14, page 1080

1. (a) 31, (c) 0, (e) 17.

2. (a) $x^3/3 + x^3y^5 + e^y + c$; (c), (e), (g) not exact.
(i) $x/y + c$ for $y > 0$ or for $y < 0$.
(k) $-[2(x^2 + y^2)]^{-1} + c$ for $(x, y) \neq (0, 0)$.

Section 13-15, page 1086

1. (a) $8\pi/15$, (c) $6(e - 1)(1 + \sin 2)$.

Section 14-1, page 1095

1. (a) $y' = y$, (c) $y' = \cos x$, (e) $2x + 3yy' = 0$.
(g) $(x^2 + 3y^2)y' + (3x^2 + 2xy) = 0$.

2. (a) $y'' = 0$, (c) $y'' = y$.

3. (a) $y' = 1 + \ln x$, (c) $y' = 1 + y^2$, (e) $yy' + x = 0$.

4. (a) $1 + (x^6/6)$, (c) $(1/3)(e^{x^3} + 2)$,
(e) $1 - x + (x^5/20)$, (g) $1 + 2x + (3/2)x^2$.

5. (a) $(k/20m)t_1^5 + v_0t_1 + x_0$, (c) $(k/mb^2)(bt_1 + e^{-bt_1} - 1)$.

7. (a) $w' = e^{xw} - y$, $y' = w$,
(c) $z' = xy^2 - wz$, $w' = z$, $y' = w$.

Section 14-3, page 1105

2. (a) $x^2y = 2$, (c) $x + 2y = 0$. (e) $e^{x^2}\sin y + \cos y = 0$.

3. (a) $x^3y^2 - 2x^2y = c$, (c) $(x^2 + y^2)^{1/2}e^x = c$,
(e) $2x|x|^{3/2} + 2|x|^{3/2}y^2 = c$.

Section 14-4, page 1110

1. (a) $y^2 = c - e^x$, (c) $(2x - 1)e^{2x} - 4e^{-y} = c$.
(e) $y = ce^x - 1$, (g) $y(c + x^3/3) + 1 = 0$.

2. (a) $5y^2 - 6xy + 2x^2 = c$, (c) $e^{y/x} - \ln x = c$,
(e) $x^2y + x^3 = c(y - x)$.

7. (a) $y' = -e^{-x}$, $y = e^{-x} + c$,
(c) $-x\,dx + y\,dy = 0$, $y^2 = x^2 + c$,
(e) $x^2\,dx - y\,dy = 0$, $c = 2x^3 - 3y^2$.

Section 14-5, page 1116

1. (a) $4y = e^{-3x}[c + e^{4x}]$, (c) $y = ce^{-2x} + \frac{1}{36}e^x(6x - 1)$,
(e) $xy = c - \cos x$, (g) $4x^2y = x^2(2\ln x - 1) + c$,
(i) $5y\sin x = 2e^{2x}\sin x - e^{2x}\cos x - c$,
(k) $y = -1 + ce^{-e^{-x}}$.

5. (a) The motion is downward with decreasing acceleration and with velocity increasing and approaching mg/k as t approaches ∞.
(c) The motion is as in (a), except that the velocity is decreasing.

8. (a) Current decreases monotonically from E_0/R to 0 as $t \to \infty$.
(c) Current increases with time and approaches E_0/R as $t \to \infty$.

Section 14-6, page 1123

1. (a), (c), (e), (g) linearly independent, (a), (c), (e) form a basis,
(i) linearly dependent.

4. (a) $x^2/2 + c_1 + c_2\cos x + c_3\sin x$,
(c) $(1/6)x^3\ln x + c_1 + c_2x + c_3x^3$.

5. (a) $y = 0$ and $y = -(x + c)^{-1}$, the latter is defined for $x > -c$ and for $x < -c$.

Section 14-8, page 1128

1. (a) $-\frac{1}{3}e^x + c_1e^{2x} + c_2e^{-2x}$,
(c) $\frac{1}{42}e^{5x} + c_1e^{-x} + c_2e^{-2x}$,
(e) $\frac{1}{3}x^3 - 2x + c_1 + c_2\cos x + c_3\sin x$,
(g) $-\frac{1}{2}x\ln x + c_1x + c_2x^3$.

Section 14-11, page 1136

1. (a) $c_1e^{3x} + c_2e^{-3x}$, (c) $c_1e^{2x} + c_2xe^{2x}$,
(e) $c_1 + c_2x + c_3\cos 2x + c_4\sin 2x$,
(g) $c_1e^x + c_2xe^x + c_3e^{-x}$,
(i) $c_1e^{\sqrt[4]{8}x} + c_2e^{-\sqrt[4]{8}x} + c_3\cos\sqrt[4]{8}x + c_4\sin\sqrt[4]{8}x$,
(k) $c_1e^{ax} + c_2e^{-ax} + c_3e^{bx} + c_4e^{-bx}$, where $a = \sqrt{3 + \sqrt{5}}$, $b = \sqrt{3 - \sqrt{5}}$.

2. (a) $-\frac{1}{5}e^{2x} + c_1e^{3x} + c_2e^{-3x}$.
(c) $\frac{1}{6}xe^{3x} + c_1e^{3x} + c_2e^{-3x}$.

(e) $2 + x + c_1e^x + c_2xe^x + c_3e^{-x}$.
(g) $\frac{1}{4}e^{3x} + c_1e^x + c_2xe^x$.
(i) $\frac{1}{5}\sin 2x + c_1\cos 3x + c_2\sin 3x$.
(k) $-\frac{1}{6}x\cos 3x + c_1\cos 3x + c_2\sin 3x$.

3. (a) $\frac{1}{4}[3e^x - 2xe^x + e^{-x}]$.
(c) $\frac{1}{15}[-3e^{2x} + 10e^{3x} + 8e^{-3x}]$, (e) 0.

7. (a) $\frac{1}{17}e^{3x}$, (c) $-\frac{1}{13}\cos 3x$,
(e) $\frac{1}{101}(e^x\cos 2x + 10e^x\sin 2x)$.

11. (a) $-\frac{3}{5}e^{2x} + \frac{5}{6}xe^{3x}$.

Section 14-13, page 1145

1. (a) not stable, (c) stable, (e) stable.

6. (a) $3 + 2x$, (c) $\sin x + \cos t$.

7. (a) damped, (c) overdamped, (e) simple harmonic.

11. $(\pi/2)\cos(C/I)^{1/2}t$.

14. 4.5 lb.

20. (a) $(1/2)[\sin t - \cos t] + (1/106)[21\cos 4t - 6\sin 4t]$.

(c) $\sum_{n=1}^{\infty}(n^7 - n^5 + n^3)^{-1}[(1 - n^2)\sin nt - n\cos nt]$.

Section 14-16, page 1159

1. (a) $\begin{pmatrix} 2e^{2t} & e^t \\ 5e^{2t} & 3e^t \end{pmatrix}\begin{pmatrix} c_1 \\ c_2 \end{pmatrix}, \begin{pmatrix} 6e^{2t} - 5e^t & 2(e^t - e^{2t}) \\ 15(e^{2t} - e^t) & -5e^{2t} + 6e^t \end{pmatrix}$

(c) $\begin{pmatrix} 3e^{2t} & -e^t \\ -5e^{2t} & 2e^t \end{pmatrix}\begin{pmatrix} c_1 \\ c_2 \end{pmatrix}, \begin{pmatrix} 6e^{2t} - 5e^t & 3(e^{2t} - e^t) \\ 10(e^t - e^{2t}) & 6e^t - 5e^{2t} \end{pmatrix}$

(e) $\begin{pmatrix} 3e^{2t} & e^{3t} & 2e^t \\ e^{2t} & e^{3t} & e^t \\ 2e^{2t} & -e^{3t} & e^t \end{pmatrix}\begin{pmatrix} c_1 \\ c_2 \\ c_3 \end{pmatrix}, \begin{pmatrix} 3e^{2t} & e^{3t} & 2e^t \\ e^{2t} & e^{3t} & e^t \\ 2e^{2t} & -e^{3t} & e^t \end{pmatrix}\begin{pmatrix} 2 & -3 & -1 \\ 1 & -1 & -1 \\ -3 & 5 & 2 \end{pmatrix}$.

5. (a) $\frac{1}{2}e^{3t}(3, 7)$.
(c) $(1/130)(-27\cos 3t + 21\sin 3t, -61\cos 3t + 33\sin 3t)$.
(e) $(-8, -4, -5)$.

Section 14-18, page 1169

1. (a) $\begin{pmatrix} -e^{-3t} & e^{2t} \\ 4e^{-3t} & e^{2t} \end{pmatrix}\begin{pmatrix} c_1 \\ c_2 \end{pmatrix}$,

(c) $e^{2t}\begin{pmatrix} 3\cos bt & 3\sin bt \\ -\cos bt - b\sin bt & b\cos bt - \sin bt \end{pmatrix}\begin{pmatrix} c_1 \\ c_2 \end{pmatrix}$,

(e) $\begin{pmatrix} -e^t & e^{-2t} & e^{3t} \\ 4e^t & -e^{-2t} & 2e^{3t} \\ e^t & -e^{-2t} & e^{3t} \end{pmatrix}\begin{pmatrix} c_1 \\ c_2 \\ c_3 \end{pmatrix}$.

2. (a) not stable, (c) stable.

3. (b) $(1/9)(-20, -7, 12)$, (d) $(\sin 2t, \cos 2t, \sin 2t)$.

5. Frequencies are 1 and $1/\sqrt{6}$, synchronous oscillations (normal modes) illustrated by $x_1 = \sin t$, $x_2 = -\sin t$, and $x_1 = 2\sin t/\sqrt{6}$, $x_2 = 3\sin t/\sqrt{6}$.

Section 14-20, page 1177

2. (a) $e^{2t}\begin{pmatrix} 1 & 0 \\ 0 & 1 \end{pmatrix}\begin{matrix} c_1 \\ c_2 \end{matrix}$, (c) $\begin{pmatrix} 3e^{2t} & e^{2t}(3t+1) & 2e^{3t} \\ e^{2t} & e^{2t}(t+1) & e^{3t} \\ 2e^{2t} & e^{2t}(2t-1) & e^{3t} \end{pmatrix}\begin{pmatrix} c_1 \\ c_2 \\ c_3 \end{pmatrix}$.

3. (a) $\begin{pmatrix} 2 & e^{5t} \\ -1 & 2e^{5t} \end{pmatrix}\begin{pmatrix} c_1 \\ c_2 \end{pmatrix} + \begin{pmatrix} \frac{1}{4}e^t - e^{2t} \\ -\frac{1}{2}e^t \end{pmatrix}$,

(c) $\begin{pmatrix} 5e^{-t} & e^{5t} & 0 \\ -20e^{-t} & 2e^{5t} & 0 \\ -3e^{-t} & -3e^{5t} & e^{4t} \end{pmatrix}\begin{pmatrix} c_1 \\ c_2 \\ c_3 \end{pmatrix} - \frac{1}{4}e^t\begin{pmatrix} 1 \\ 2 \\ 23 \end{pmatrix}$.

Section 14-21, page 1186

2. (a) $c_1e^{-t}(2, 1) + c_2e^t(1, 1)$, saddle point, unstable.
(c) $c_1(-\cos 2t, \cos 2t - \sin 2t) + c_2(-\sin 2t, \cos 2t + \sin 2t)$, center, neutrally stable.
(e) $e^t[c_1(\cos t, 2\cos t - \sin t) + c_2(\sin t, 2\sin t + \cos t)]$, focus, unstable.
(g) $c_1e^{-2t}(1, 1) + c_2e^{-4t}(2, 1)$, stable node.

3. (a) $e^t[c_1(1, 2) + c_2(t, 2t + 1)]$, (c) $c_1(e^{-3t}, 0) + c_2(0, e^{-3t})$, (e) (c_1, c_2).

5. (a) Same as 2(a), saddle point, unstable, (c) focus, unstable.

Section 14-22, page 1192

1. (a) $1 + x + x/2! + x^3/3! + x^4/4! + x^5/5! + \cdots = e^x$.
(c) $(x-1) - (x-1)^2/2 + (x-1)^3/3 - (x-1)^4/4 + (x-1)^5/5 + \cdots = \ln x$.
(e) Same as (a).

2. (a) $x = t - t^3/6 + \cdots = \sin t$, $y = 1 - t^2/2 + \cdots = \cos t$.
(c) $x = (t-1) - (t-1)^2/2 + (t-1)^3/3 - \cdots = \ln t$, $y = t^3$, $z = t^{-1}$.

3. (a) $411/(290)^{3/2}$.

4. (a) $y(x) = c_0[1 + \sum_{k=1}^{\infty} (-1)^k x^{2k}(-1)1 \cdots (2k-3)/(2k)!] + c_1x$.

(c) $y = c_2x^2 + c_3x^3$.

5. $c_n = a(a+1)\cdots(a+n-1)b(b+1)\cdots(b+n-1)Q^{-1}$, where $Q = n!c(c+1)\cdots(c+n-1)$.

6. (a) $c_k = (-1)^k 3\cdot 4\cdots(2k+1)/[2\cdot 4\cdots(2k)\cdot 7\cdot 9\cdots(2k+5)]$.

Section 14-23, page 1196

1. (b) .20, .39, .571, .744, .910, (c) 0.195167, 0.381277,
(d) .729677, .893462, (e) 0, 0, .195163, .381270, .559183.

3. (a) 13, (c) 77.8.

4. (a) 118.7.

INDEX

Made in the USA
Monee, IL
06 November 2026

67383147R00332